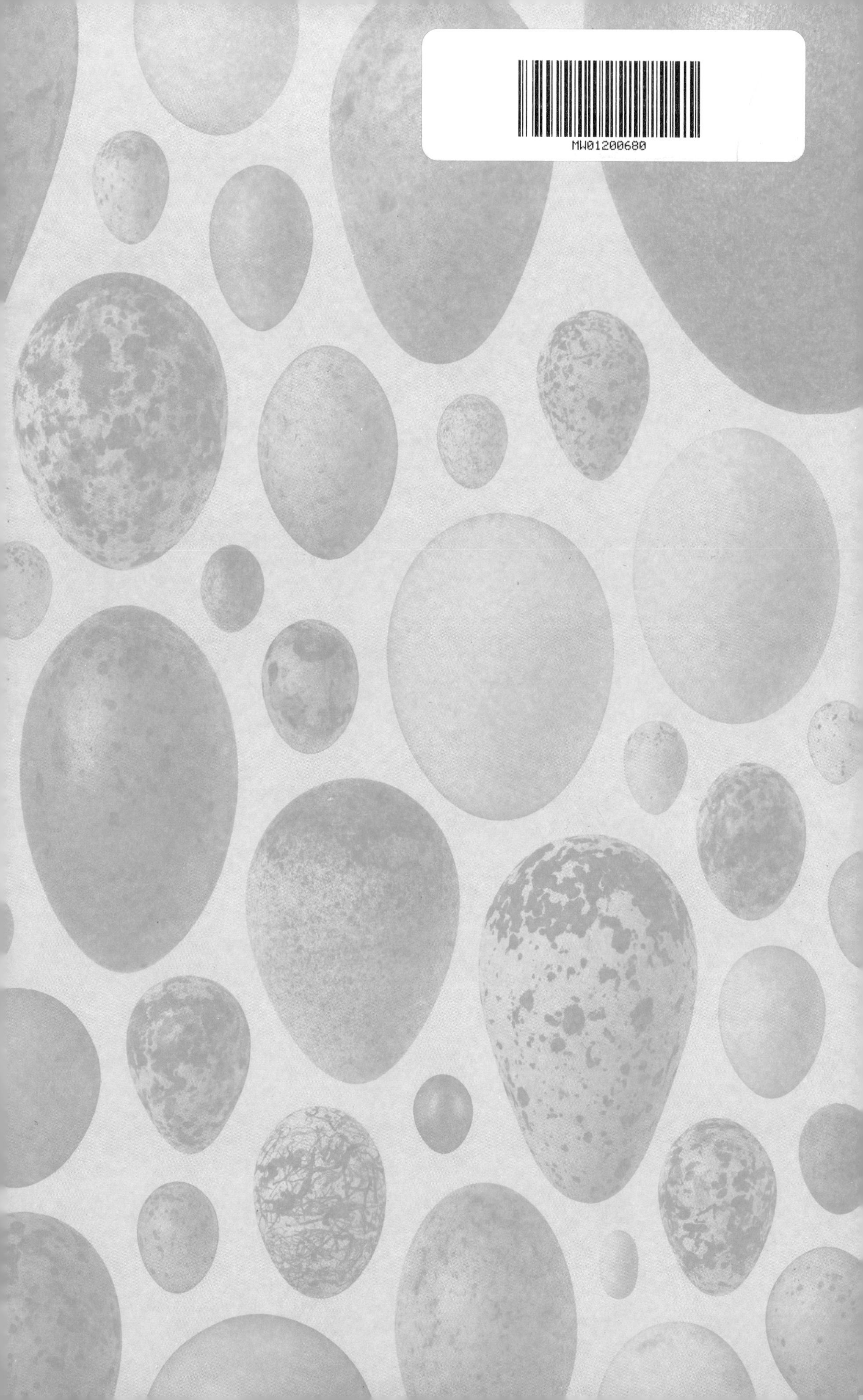
MW01200680

THE BOOK OF EGGS

MARK E. HAUBER

THE BOOK OF EGGS

A LIFESIZE GUIDE TO THE EGGS OF SIX HUNDRED OF THE WORLD'S BIRD SPECIES

EDITORS
JOHN BATES & BARBARA BECKER

PHOTOGRAPHY
JOHN WEINSTEIN

THE UNIVERSITY OF CHICAGO PRESS
Chicago and London

MARK E. HAUBER is professor in the animal behavior and conservation program, and editor of the journals *Ethology* and *The Auk: Ornithological Advances*, based at the Department of Psychology at Hunter College and the Graduate Centet of the City University of New York.

JOHN BATES is associate curator of birds at the Field Museum in Chicago, and BARBARA BECKER, a former member of the museum's team, is now an independent consultant.

The University of Chicago Press, Chicago 60637
The University of Chicago Press, Ltd., London

Printed in China

23 22 21 20 19 18 17 16 15 14   1 2 3 4 5

ISBN-13: 978-0-226-05778-1 (cloth)
ISBN-13: 978-0-226-05781-1 (e-book)
DOI: 10.7208/chicago/9780226057811.001.0001

Library of Congress Cataloging-in-Publication Data

Hauber, Mark E., 1972– author.
The book of eggs : a life-size guide to the eggs of six hundred of the world's bird species / Mark E. Hauber.
pages : illustrations, maps ; cm
Includes index.
ISBN-13: 978-0-226-05778-1 (cloth : alk. paper)
ISBN-13: 978-0-226-05781-1 (e-book) 1. Birds—Eggs. 2. Birds—Nests. 3. Birds—Breeding. 4. Birds—Eggs—Classification. 5. Birds—Eggs—North America—Pictorial works. I. Title.
QL675.H38 2014
598.14'68—dc23

2013042253

∞ This paper meets the requirements of ANSI/NISO Z39.48-1992 (Permanence of Paper).

JACKET AND LITHOCASE IMAGES John Weinstein

Color origination by Ivy Press Reprographics

This book was conceived, designed, and produced by
Ivy Press
210 High Street, Lewes
East Sussex BN7 2NS
United Kingdom
www.ivypress.co.uk

Creative Director PETER BRIDGEWATER
Publisher JASON HOOK
Art Director MICHAEL WHITEHEAD
Editorial Director CAROLINE EARLE
Senior Editor STEPHANIE EVANS
Commissioning Editor KATE SHANAHAN
Designer GLYN BRIDGEWATER
Commissioned Photographs JOHN WEINSTEIN
Illustrators IVAN HISSEY, ADAM HOOK, CORAL MULA
Artwork RICHARD PETERS

Typeset in Fournier and News Gothic

CONTENTS

Foreword by John Bates 6

Introduction 8

Egg anatomy & physiology *14*

Egg size & shape *16*

Egg coloration & patterning *18*

Nests & eggs *20*

Breeding strategies: clutch size *22*

Breeding strategies: nest parasitism *24*

Science & egg collections *28*

The eggs *32*

WATER BIRDS *34*

LARGE NON-PASSERINE LAND BIRDS *198*

SMALL NON-PASSERINE LAND BIRDS *310*

PASSERINES *378*

Appendices *642*

Glossary 644

Resources & useful information 646

The classification of birds 648

Index by common name 650

Index by scientific name 653

Acknowledgments 656

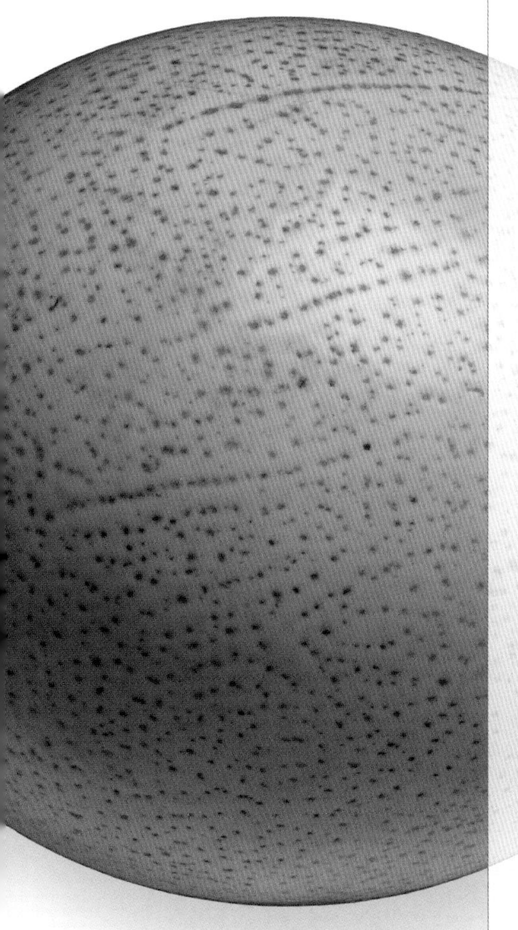

RIGHT **Male ostrich** tending eggs in the Serengeti, Tanzania. The beige eggshells protect the embryos from sun damage and the thickness of the shell protects the embryos from the weight of the parent.

FOREWORD

When one thinks of the beauty, function, and fragility that have evolved in the natural world, bird eggs have to hold a place at the top of the list. Birds are all around us, every single individual hatched from an egg laid by a parent bird who then warmed them so the chick could successfully develop. Eggs are a key component of bird reproduction, which has allowed birds to flourish and diversify around the planet.

The eggs of one species, the domestic chicken, also typify nature that we, almost unconsciously, come into contact with on a daily basis: we scramble them, poach them and color them for holidays, without much thought to the fact that this is essentially modified predation. Beyond a source of nutrition, bird eggs and their components have provided building materials for cosmetics and other chemical industries. Microbiologists working to develop new vaccines grow many bacteria and viruses by piercing through chicken eggshells to inoculate the otherwise sterile interiors. Clearly, bird eggs deserve all the attention that this book can offer!

ABOVE **The egg of the domestic chicken** is not only a major source of human nutrition but also a resource for vaccine researchers.

Throughout this book, you will be introduced to basic breeding behaviors of birds and the myriad of ways in which evolution has changed, augmented, and enhanced the basic process of laying and caring for eggs. Eggs may have rigid structural limits, but simply flipping through these pages reveals that there has been a wide range of exceedingly successful experimentation in bird egg evolution in the form of size, shape, and coloration.

EGGS IN MUSEUMS

As we celebrate the eggs of birds and their diversity, from albatrosses and anhingas to wood ducks and Zebra Finches, we also celebrate the irreplaceable value of collections of eggs like those from which these magnificently detailed photographs were taken. The collections represented here are housed at the Field Museum of Natural History in Chicago, and the Western Foundation of Vertebrate Zoology in Camarillo, California. The Field Museum collection contains 23,000 sets of eggs from over 1,600 species, the result of the diligent efforts of the collectors who, largely between 1890 and 1930, found each nest, carefully prepared each egg, and recorded valuable natural history data.

This collection made it possible to create this book, but it also reminds us of the scientific responsibility we have to understand bird biology. Throughout the text you will learn that we still do not know basic aspects about the breeding biology of many species of birds, even common ones. We hope this recognition leads people of all kinds into the field for more study, because ultimately it will be our scientific understanding that will allow us to effectively monitor and conserve birds living in a world that is increasingly influenced by human activities.

JOHN BATES

BELOW **Australasian Gannets** breeding at a colony of 1,500 pairs at Cape Kidnappers, New Zealand; many seabird species form colonies to lay eggs in the safety of dense neighborhoods.

LEFT & BELOW **Tinamou eggs** are bright, shiny, and vulnerable; what makes these eggs produce a ceramic sheen remains a mystery for scientists.

INTRODUCTION

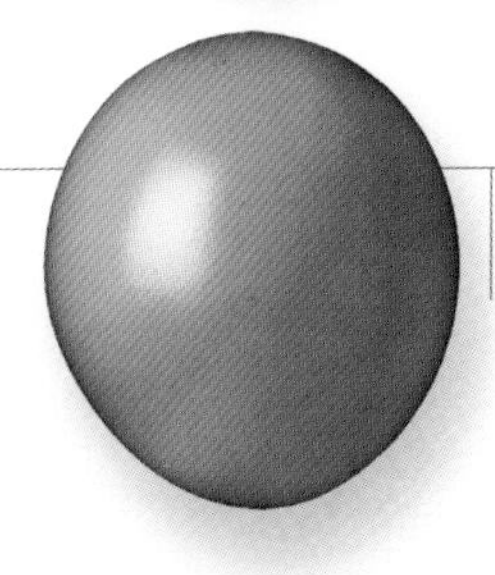

The mystery and beauty of the vast diversity of avian eggs easily captures the human imagination. One explanation for this may be that people and birds share each other's sensory worlds: we both communicate mainly through voices, colors, shapes, and other visible and audible displays to both our own kind and to other species. Bird songs remind us of the beauty in the world; watching bright and colorful birds at home and away both calms and alerts our senses; we read about and view with awe the long transoceanic flights of albatrosses, and the mating dances of cranes and birds-of-paradise. This book brings you the eggs of birds in a way that includes but also goes far beyond robin egg's blue, to capture the diversity that exists across the avian world.

PACKAGING LIFE

Avian eggs are a true biological masterpiece, and at the same time, a puzzle. Why is it that all birds, large and small, put their entire future (embryo, hormones, antibiotics, vitamins, and lipids) into a fragile package?

Humans and the majority of other mammals retain their fertilized eggs inside their bodies, where the developing young receive the nourishment and protection they need to grow directly from the mother. In contrast, a female bird packages everything that is needed to form a chick into its egg—and then ejects the egg from her body. Outside, the parent or parents

BELOW **The Emu's** eggshell is deep green and textured like an avocado; it takes nearly two months for this egg to hatch.

LEFT **Once the Maleo's** egg is laid, it is never touched again by the parents: it lies buried in warm sand until the chick hatches.

must provide the warmth, shelter, and protection, typically in a nest, that the eggs need so that the embryos can fully develop and hatch into viable chicks. The stories of the eggs in this book highlight the vast array of strategies and choices birds make to find appropriate mates and nesting sites, to warm and protect the eggs, and to assure adequate food, water, and shelter for the chicks.

The diversity of birds that successfully reproduce via the egg is astonishing. Birds live on every continent and successfully breed in every terrestrial habitat. In the frigid Antarctic where winter temperatures are below minus 40 degrees and wind speeds may reach 200 miles (over 320 km) per hour, the male Emperor Penguin stands in place, carrying its single egg on top of its feet for two months to warm it before the chick hatches. In Chile, Gray Gulls breed in the world's driest deserts, where few predators can venture; the eggs and chicks are safe, but the parents must commute daily to the ocean to obtain food and water for themselves and their offspring.

We still have much to learn about the biology of birds' eggs, but there is no doubt that reproduction through eggs has been a very successful system for birds for millions of years. This book is a journey through the strategies that different bird species, and the tactics that different individuals, have evolved and adapted to successfully reproduce via the fragile egg.

RIGHT **A fossilized clutch of dinosaur eggs** (hadrosaur), found in China, was laid in the Jurassic era (some 150 million years ago), around the time when the first bird-like dinosaurs appeared.

WHICH CAME FIRST: THE CHICKEN OR THE EGG?

This age-old question is an easy one to answer: the egg. Why? An egg is simply a reproductive cell, and animals (including the dinosaurs, the direct ancestors of modern birds) were laying eggs for millions of years before the first chicken evolved. Of course, few non-avian eggs (then or now) looked like modern bird eggs: with a soft membrane and no hard casing in some cases, they are transparent, gelatinous, and quick to dry out. Most must remain in or near water in order for the young to develop and hatch.

Birds' eggs are "amniotic," meaning they contain features (including a hard shell and porous membranes) that allow them to be laid and develop on dry land. In the evolution of biodiversity on Earth, the appearance of the amniotic egg made possible a major shift: animals could leave water in order to reproduce. This trait helped amniotes—reptiles, dinosaurs, birds, and mammals—became a dominant form of animal life on Earth.

The story of when and how birds' eggs evolved also tells us something about birds' relationship with dinosaurs. The first amniotic eggs were laid by "basal amniotes," small lizard-like animals that appeared in the Carboniferous period about 325 million years ago. Relatively quickly, the basal amniotes diverged into two groups: the synapsids, which eventually evolved into mammals; and the sauropsids, which gave rise to the turtles, lizards, snakes, crocodilians, pterosaurs, dinosaurs, and birds.

UNDERSTANDING THE FOSSIL RECORD

Most scientists now accept that birds are a specialized subgroup of one branch of dinosaurs, the therapods. While dinosaurs flourished into many different niches of life throughout the Cretaceous, some therapods evolved features allowing them to fly. *Archaeopteryx*, dating to about 150 million years ago, combined reptile features (teeth, clawed fingers, and a long tail) with wings and flight feathers resembling modern birds'. While this well-known fossil is no longer considered to be a direct ancestor of modern birds, many other bird fossils show that the earliest forms were small, perhaps arboreal, and able to glide. Modern birds likely evolved from there.

By 100 million years ago, the two major groups of modern birds had split from each other. Based on differences in their skulls, the Paleognathes ("old jaws") include the flightless ostriches, rheas, cassowaries, emus, kiwis, and the extinct moas and elephantbirds, as well as the flighted tinamous; the Neognathes ("new jaws") include the rest of the birds we know today (see page 649). By the end of the Eocene, about 34 million years ago, all of the modern bird orders (and many of the families) roamed and flew the Earth as members of distinct and recognizable lineages.

We do not know what color dinosaur eggs were, but from fossilized remains scientists have established that the surface was rarely smooth, and instead textured with holes and bumps—much like eggs of modern cassowaries and emus. Recent research has begun to reconstruct the structure, chemical make-up, and color of fossil feathers; in future

BELOW **Modern crocodiles** and alligators are the closest living relatives of birds. New fossils and modern research techniques offer fresh insights into whether the first bird eggs were smooth and white—like these reptiles'—or diversely shaped and colored, as are many bird eggs today.

the same could be done for egg shells. Many modern avian species lay smooth and white eggs. But in reconstructing the evolutionary history of bird egg coloration, we must not overlook the huge variety of egg colors and surface texture among some of the earliest lineages such as emus (dark green), rheas (light blue), and tinamous (polished blues, browns, and greens); even the extinct moas had at least one species, the Upland Moa, which laid a dark blue-green egg.

Egg color is mostly produced by pigments laid down on an initially white egg while it is still in the female's body. Since pigmentation appears to be a secondary step in the formation of hard-shelled eggs, some scientists have argued that the first bird eggs would have been white, as crocodile eggs are today, and only later did pigmentation appear among birds. But this line of thought assumes that the eggs of the dinosaurs were also white, which is unproven. In addition, many seabirds, including penguins and gannets, have a distinct blue hue to the eggshell matrix between the inner and outer edge of the shells; this suggests that eggs may also become pigmented during the 24 hours prior to laying, while the calcite (calcium carbonate) crystals are in the process of generating the hard egg shell.

While any one female bird typically lays consistently colored and patterned eggs, when resources are limited, the same female may lay not only smaller but paler, less pigmented eggs. The implication is that we can only speculate about the color and pattern of the eggs not only of the first birds, but also of the egg laid by a bird in the hand just yesterday.

HOW THIS BOOK IS ARRANGED

The species whose eggs are represented in this book are organized into four chapters, representing a reasonable although not entirely phylogenetic presentation. Within the sections, the arrangement is taxonomic, by order and then by family. The four chapters are:

Water Birds including ducks, geese, loons, herons, gulls, shorebirds, and relatives

Large Non-passerine Land Birds including ostriches, emus, tinamous, chickens and other fowl, and birds of prey

Small Non-passerine Land Birds including doves, cuckoos, swifts, hummingbirds, woodpeckers, and parrots

Passerines "perching birds," including New and Old World flycatchers, jays, swallows and swifts, warblers, sparrows, blackbirds, and finches.

The classifications of birds are always in a state of flux as we gain a better understanding of the avian tree of life through both genomic and paleontological research. For those with an interest, a tree of modern relationships is presented on pages 648–9.

Each of the 600 featured birds is shown with a life-size color photograph of its egg. Additional photographs provide close-up details of shell texture and patterning and a view of the typical clutch size (when more than one egg). The actual sizes depicted are based on the specimen (with average dimensions cited in the caption, where the egg's typical appearance is also described). In some cases, eggs become discolored during incubation; others fade over time (many of these museum specimens are over a century old). As a result, the description may not match the image precisely. In a few cases where eggs are rare or unknown, there continues to be disagreement in the literature as to what is "typical" of a species; these cases are pointed out. The text for each entry explores the egg in relation to the breeding strategy of that species, although for some very little is known about their eggs, nests, or breeding behaviors; much research remains to be done.

FACING PAGE AND ABOVE
The engravings here represent the four chapters into which the 600 species of birds in this book are organized—the water birds and the land or terrestrial large and small birds.

Accompanying the text and color photographs is a map showing the breeding range of each bird, a summary of its breeding habitat, nest type and placement, and its current conservation status. Small engravings of the bird offer a quick visual aid; the range of measurements is provided for an appreciation of the average adult size. The collection catalog numbers from the Field Museum (FMNH), or the Western Foundation for Vertebrate Zoology (WFVZ) can be used to find out via the museums' online egg databases when and where these eggs were collected. See Resources & Useful Information, page 647, for the web site addresses.

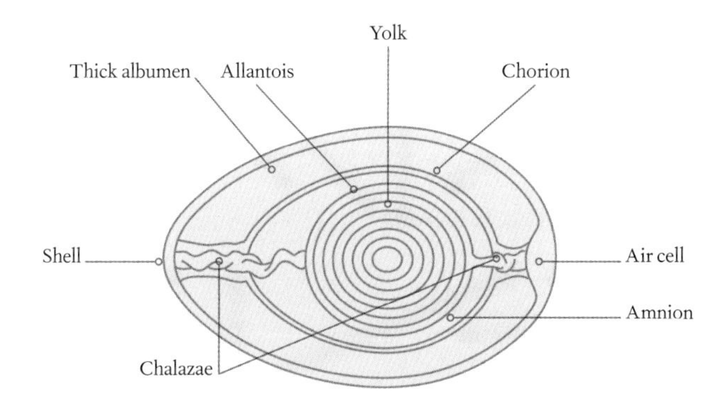

RIGHT **The freshly laid egg** contains many compartments, separated by the different layers of the calcified eggshell and a series of internal membranes.

EGG ANATOMY & PHYSIOLOGY

BELOW **Egg formation** requires a full day; in most birds only the left ovary is functional, limiting the rate of egg laying to one per day.

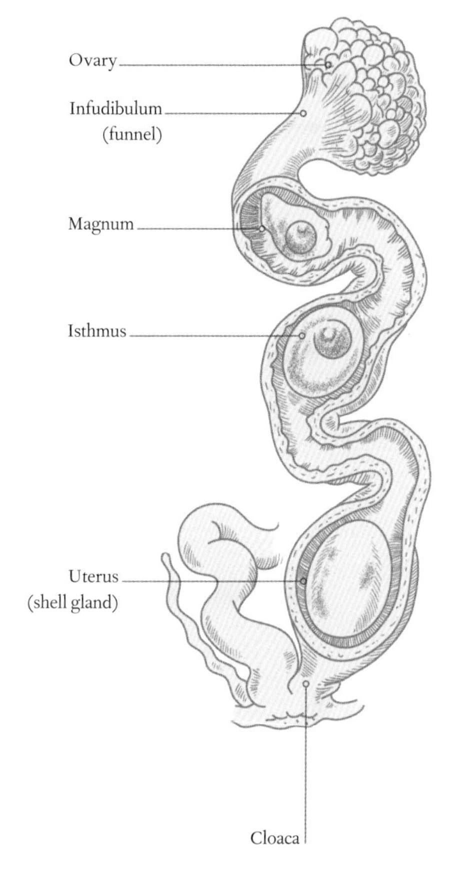

The avian egg is the equivalent of a furnished apartment at a vacation resort: it contains all the ingredients and structures required for safely housing the developing embryo, but it needs regular parental servicing and support to function fully. The internal architecture of the egg includes both the genetic machinery and the biochemical structures required to build a viable hatchling.

The already fertilized embryo is enclosed in the amnion, nourished by the yolk and the allantois. The yolk contains high amounts of fat, cholesterol, protein, vitamins, and minerals needed by the developing embryo, while the allantois aids respiration by storing nitrogeneous waste.

These compartments are enclosed within the chorion, and surrounded by the albumen (commonly known as the "egg white"), which supplies hydration for the embryo, and acts as a shock-absorber against sudden movements by the egg. Twisted threads called chalazae (the textured strings sometimes seen in raw or soft-cooked eggs) attach the compartment to the shell for added stability.

Membranes on the inner and outer surface of the shell act as physical and biological barriers to desiccation and bacterial infestation. In addition, both the albumen and the eggshell cuticle contain enzymes and other proteins that have active antimicrobial properties. These enzymes are activated by heat, and so sitting on the eggs at night, even before full incubation begins, may serve to protect the eggs from infections. The whole of this package is contained in the hardened eggshell, which is made up mostly of calcium carbonate. However, this eggshell is semi-permeable, with microscopic pores that penetrate the shell, providing channels for gas exchange necessary for respiration by the developing embryo.

ABOVE **Common Tern** hatchling with sibling breaking out of its shell; females lay only one egg a day so different eggs in the same nest may hatch at different times.

Because of the hard shell, the egg must be fertilized while it is still inside the female's body, before the shell has formed. The chicken's ability to lay eggs whether they are fertilized or not accounts in part for its eggs being such a food staple for humans.

While the egg itself provides much that the embryo needs, the parents must still provide critical services. Typically, one or both parents provide the external heat necessary to jump-start embryonic metabolism, and maintain the necessary microclimate, including high levels of humidity, to keep the eggs from desiccating. They select nest sites and build or usurp nests for the eggs to shield them from predators, sun, dryness, and other threats. They also rotate the eggs in the nest to assure even heating or cooling, and to prevent embryonic malformation.

EGGS & ENVIRONMENTAL TOXINS

The avian egg is a compact and adaptable product of evolutionary engineering. Yet human activities are capable of compromising and even breaking down this highly functional reproductive system. Toxic chemicals such as DDT, introduced into the environment during the 1960s, interfered with some birds' ability to produce the calcium needed to harden the eggshells. The result was thin-shelled eggs unable to bear the weight of the incubating adults; the eggs were crushed, ending in reproductive failure. It took 30 years for populations of Peregrine Falcons, Osprey, and Brown Pelicans to recover their numbers after the banning of DDT.

BELOW **Embryonic development** of the chicken is rapid; most organs are formed by day 10, and grow and mature to hatch less than a week later.

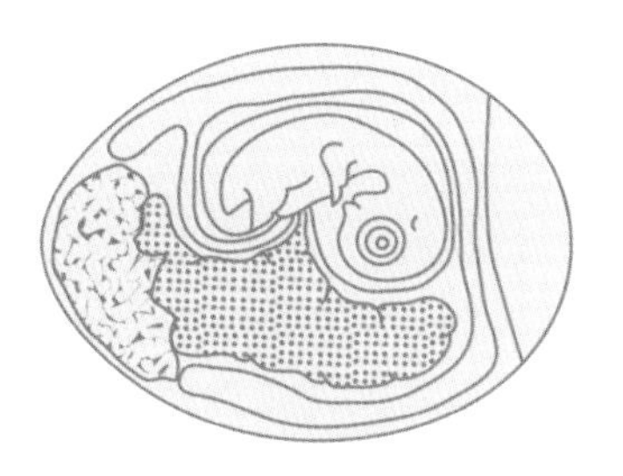

10 days

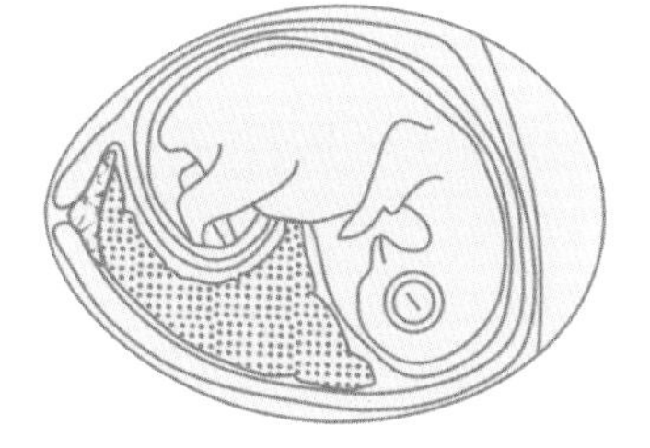

15 days

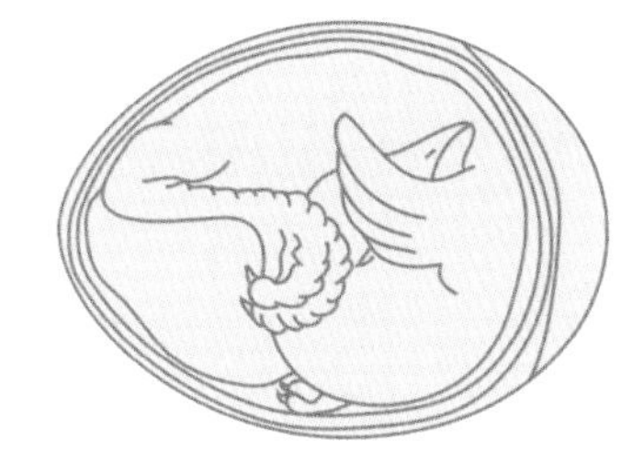

20 days

Costa's Hummingbird

Blue-winged Warbler

Pied Kingfisher

EGG SIZE & SHAPE

ABOVE **The term "ovoid," or "egg-shaped,"** applies to all bird eggs; different species—and different eggs of the same species—may be spherical, oval, elliptical, or conical.

Bird eggs come in extreme sizes and shapes. The ostrich's egg is the largest and heaviest of any living bird; weighing in at over 4 lb (2 kg), it represents over 30 chicken-egg equivalents. At the other end of the scale, the smallest eggs are laid by hummingbirds; they are 1/5,000th the size of an ostrich egg and may weigh less than a paper clip. Some species of the extinct elephantbirds laid eggs that were well over twice the size of an ostrich egg.

SIZE MATTERS

In general, larger birds lay larger eggs; they also have thicker shells, which makes mechanical sense, with each egg having to withstand some of the weight of the incubating adult sitting on top of it. But when considered in relation to adult body weight, egg sizes tell a different story. In this context, ostrich eggs are rather small for the adult's weight and hummingbird eggs are rather large. Some of largest eggs of any bird relative to the female's body weight are laid by the several kiwi species of New Zealand.

Thus, relative egg size, compared to adult body weight, is far from constant and varies extensively with ecology and evolutionary history. For example, birds that lay larger clutches also tend to lay smaller eggs, and species whose chicks hatch fully feathered and ready to follow the parents, tend to produce larger eggs than species whose chicks hatch blind and naked and require prolonged parental care. An egg's size often varies with its internal content, including lipids in the yolk, and the concentrations of hormones, vitamins, and maternal antibodies in

Great Elephantbird

the yolk and the albumin. Eggs with smaller yolks typically hatch earlier, with the young soon having to obtain additional nutrition for growth on their own, or by begging for provisions actively from the parents.

BELOW **The eggs of some seabirds** that nest on steep cliffs and ledges are often conical; if bumped, they roll in a small circle and so avoid falling off.

THE ADVANTAGES OF BEING "EGG-SHAPED"

Most eggs have a clearly identifiable blunt end (or pole), typically formed where the shell is physically closer to the cloaca in the oviduct; the sharp end forms nearer the ovary. The shell is typically thinner near the blunt pole and thicker near the equator and sharp pole. But as the embryo grows, the calcium from these thicker shell regions is recruited as building material for the developing bones in the skeleton, so that at the time of hatching, most regions of the shell are equally thin. Chicken embryos typically face the blunt pole with their beaks, and start breaking the shell in this region in preparation for hatching.

Common Murre

"Egg-shaped," or ovoid, has many advantages: despite the fragile shell, an ovoid can withstand surprising compression (for example, from the incubating adult's weight) before it breaks. An ovoid is also easier for the female to push out of her body. The symmetrical shape and smooth texture of the shell is achieved by muscles rotating the entire egg in the oviduct during shell formation; when this process is interrupted, for example due to trauma or aggression by predators or competitors, an asymmetrical and rough-textured egg may be formed, causing difficulties, or even death, to the female during laying.

Ostrich

RIGHT **Shorebirds, such as the Killdeer**, nest on open ground, and lay heavily pigmented and camouflaged eggs to reduce the likelihood of predators spotting their clutch.

EGG COLORATION & PATTERNING

The photographs in this book reveal the astonishing colors and patterns of avian eggs. Eggs are made primarily of calcium carbonate, which is white to the human eye. While many bird eggs are also white, all the additional variation in eggshells is the result of the interaction of physical and chemical properties of the shell and just two major pigments: biliverdins, which are responsible for the blue-green hues of eggs, and protoporphyrins, which make the rusty colors, from yellow to red to brown. Spotted, lined, blotched, or scrawled eggs have higher concentrations of protoporphyrins. When the two pigments combine in different proportions they can create hues from violet to green.

PIGMENT POWER

How is it possible that just two pigments, interacting with the crystalline structure of the eggshell, can generate the diversity of shell coloration and patterning seen in nature? The surprising answer is that we simply do not know yet. All the studies that have attempted to extract pigment-like compounds from avian eggshells have produced chemicals consistent with the structure of biliverdin and protoporphyrin, or chemicals whose structure could not be identified even with the latest analytical instruments. Biologists and chemists now must combine forces to solve this conundrum. Despite this outstanding question, some reasons for egg colors are relatively straightforward to explain: for example, the typically whitish eggs of the Horned Grebe quickly become stained red from the wet plant matter used to cover the eggs when the incubating parent is off the nest.

The shell gland compartment of the oviduct has to take resources away from the laying female and divert them into producing colorful pigment molecules, which then combine to generate the background color and the spotting, streaking, and blotching on the eggshell. Scientists therefore argue that white eggs are "cheaper" than colorful ones and more likely to be produced by birds whose eggs are hidden in a deep nest (or nest hole) or under the cryptic plumage of a dedicated incubating parent. This is the case for the white eggs of woodpeckers, hummingbirds, ducks, and owls. Sometimes, patterning on eggs can be critical to identifying individual eggs laid in a colony, such as the fantastic variation among individual eggs of Common Murres. In this species, variation has evolved to allow parents to pick out their own egg from a thousand on a crowded cliff face.

A critical piece in the conversation about egg colors is that we have only recently begun to learn about what the birds themselves see. All birds have four photo receptor proteins, compared to three in humans, which provides them with instantly more accurate and detailed color perception, relative to humans, including seeing color in the ultraviolet (UV) region, not visible to people. Researchers are now surveying and analyzing eggshells with physical instruments, such as UV-filter camera-lenses and reflectance spectrometers to reveal unexpected variation between and within eggshells, that birds can see, but, until recently, scientists had not.

BELOW **Egg appearance** is generated by pigments deposited within and atop the eggshell (as in the left and central columns), but also by powdery materials which wear off during incubation (shown here in the right column).

NESTS & EGGS

Birds have evolved a fantastic package—the egg—to nourish and support the development of their embryos after they are ejected from the mother's body when the egg is laid. As part of the same set of successful reproductive strategies, birds have also developed an extraordinary array of structures to shelter, protect, and help warm the eggs and chicks: the nest.

ABOVE **Hummingbird nests** are tightly woven baskets, deep and sturdy, but their walls are also flexible, expanding to make room for the growing nestlings.

The simplest definition of the nest is any structure or space that surrounds and houses the egg(s). Constructed nests are a form of tool use by animals: birds take materials found in the environment (including human castoffs), manipulate and modify them, and generate a novel use. This basic definition of tool use covers most bird nests, from the loose stick nest of a Rock Dove breeding on a window ledge, to the hundred-unit nesting aggregations of colonially breeding weavers.

Nests have evolved to help protect the eggs and the nestlings from competitors, predators, and also parasites. Enclosed nests, whether constructed domes or a hole in a tree, keep the eggs and chicks out of the sight of predators and away from sun and rain. Building a nest in a hidden place, among dense tufts of grasses or foliage, ensures discovery by competitors, and many nest predators, is minimized. A tight or difficult-to-access nest can also provide protection from parasites: Orchard Orioles, for example, protect their clutch by sitting tightly on the nest, so that the contents are not accessible to parasitic Bronzed Cowbirds attempting to lay their eggs into the nest cup.

Finally, nesting sites and nests themselves can serve as an aphrodisiac; in many bird species, males attract females to their territory by building a nest or defending a nest site. Females choose a male based on his ability to build a good nest in a safe location; these traits of the nest architecture and location work together to assure greater reproductive success for both the talented male and the choosy female.

ABOVE **Bald Eagles** are long-term residents; instead of building nests anew, they renovate and refurbish the nest used in the previous year.

HATCHING EGGS WITHOUT NESTS

Not all birds use a nest; some simply deposit the eggs on bare ground, a cliff ledge, in leaf litter, at the bottom of a tree cavity, or in a shallow ground scraping. The heat for embryogenesis is still provided by the incubating adult. Many seabirds, including Gannets and Emperor Penguins, cannot afford to lose feathers from the chest to develop a barren skin, called the brood patch, because they need to keep diving deep into the cold seawater for fish throughout the nesting period; in these cases the heat is transferred to the eggs in other ways, particularly through the warm blood carried in the veins of the webbed feet.

ABOVE **Southern Masked Weavers** fulfill the definition of tool use by animals: their nest is a physical object manipulated to achieve a specific function.

Most unusually, brush turkeys and other birds in the family Megapodidae lay their eggs into heaping mounds of rotting vegetation or warm sandy soils near sunbathed beaches or volcanic slopes. There, the biochemical, solar, or geothermal energy provides the heat for the eggs to maintain embryonic development.

RIGHT **Members of a Waved or Galapagos Albatross** pair take several days for each turn to incubate their single egg until it finally hatches, two months later.

BREEDING STATEGIES: CLUTCH SIZE

Birds lay eggs in sets, or clutches. Clutch size is the number of eggs laid in one nesting attempt; it varies across species, from a single egg to as many as 20. In part, this is determined by evolutionary history; all albatrosses lay just one egg, hummingbirds consistently lay two, while partridges can lay 10–15 or more. But this is not always the case. Clutch size also can vary within species, within populations, and even for an individual from year to year, based on such variables as habitat, latitude, altitude, nest type, size of a nesting colony, food availability, and body size and health of the mother. Such trade-offs and the choices involved are a critical part of a bird's breeding strategies at all levels.

CLUTCHES LARGE OR SMALL

Broadly speaking, natural selection leads birds to lay as many eggs as they can successfully rear. For example, birds that breed in northern latitudes tend to have larger clutches while related birds from the tropics lay fewer eggs; although temperate regions have a shorter breeding season, there is plenty of food during that time to feed more chicks. Birds that lay multiple clutches each year may have fewer eggs in their final clutches; this may be due to reduced resources in the health of the mother but also could be related to the availability of food late in the breeding season. Likewise, birds with precocial young (those that leave the nest soon after hatching and can feed themselves thereafter) lay more eggs per clutch than do birds with altricial young (those that require prolonged brooding and care). Parents with altricial young may find more success with a smaller brood.

Most small birds lay one egg each day until their clutch is complete; others, typically large birds, lay only one egg or lay consecutive eggs two to three days apart. Female kiwis, whose eggs weigh some 25 percent of their body weight, may go weeks between laying the first and second egg.

SURVIVAL OF THE FITTEST

Parent birds can control to some extent the rate and timing of egg development to influence the success of their breeding attempt. Some start to incubate each egg on the day when it is laid, resulting in asynchronous hatching; the chicks hatch over a series of days, and the first-hatched chicks are larger and more dominant. In Cattle Egrets, asynchronous hatching, and the resulting size and dominance hierarchy of the chicks, is further enhanced by the mother depositing different amounts of testosterone into the egg yolk: the first two eggs receive nearly twice as much as the last, third-laid egg. The result is that in seasons with poor food supplies, the hungry, aggressive, and large first chicks attack, peck, and drive off the nest the small, timid last chick, assuring that the resulting brood of just two chicks receives sufficient food from the parents to fledge successfully.

In contrast, synchronous hatching takes place when the parents wait for all the eggs to be laid and start incubation on the last day, so that each egg hatches within hours of the others. This might assure equal chances of survival to more of the young. Mallard embryos take additional steps to assure simultaneous hatching; they listen to maternal calls and vocalize to one another, communicating their developmental state and, thus, the estimated time of hatching.

LEFT **Great Tits**, despite their small size, lay large clutches of eggs. They are an example of a bird of northern latitudes where larger clutch sizes are more common.

BREEDING STATEGIES: NEST PARASITISM

Given what we know about the demands placed on bird parents, it is, perhaps, no surprise, that some birds have evolved, independently of each other, to forgo most of the costs and hassles involved with parental care. Brood parasitism is a breeding strategy where eggs are laid into another bird's nest; the foreign eggs are then incubated by the foster-parent(s), who also feed and protect the parasitic young until they become fully independent. Brood parasitism is not unique to birds; some catfish are parasitic and exploit mouth-brooding cichlids to raise their young; many termites, ants, wasps, and bees are also parasitic, usurping the parental care of other individuals. The effect of brood parasitism, as in all types of parasitism, is the loss of reproductive success for the host: foster-parents raising parasites misdirect expensive parental care to a genetically foreign young instead of their own progeny.

ABOVE **Female Lesser Cuckoos** lay eggs that are slightly larger but mimic superbly the host Thick-billed Warbler's eggs in color and maculation.

Avian brood parasitism comes in two main flavors: intraspecific brood parasites are those which lay their eggs into nests of other birds of their own species (conspecifics); intraspecific brood parasites may also have their own nest and look after their own eggs and young. Many gamebirds, ducks, and rails have some intraspecific brood parasites, but increasingly more songbirds are also recognized as such. Researchers have successfully induced intraspecific parasitism by inducing nest loss, as if the eggs in the nest were taken by predators: Zebra Finches, for example, frequently engaged in parasitism of other finches in their colonies, and in captivity, when their nest was destroyed during the laying period, these females sought out and laid their remaining eggs in active nests of other females. The benefit of parasitism to individuals within the same species is to increase the number of their offspring produced, without the expense of full-term parental care. At the same time, the chances for the parasitic egg and young to survive are high: they are incubated and raised among others of their own species, so they are likely to receive just the right amount of incubation and nutritionally suitable food items during parental feeding visits.

The second flavor of nest parasitism is interspecific brood parasitism: this is the case when parasites lay their eggs in other species' nests. Some interspecific parasites may be facultative—that is, they act as parasites only when the opportunity presents itself. For facultative brood parasites, some individuals may always be parasitic, while others may be parasitic only occasionally, laying some eggs in a nest they've built themselves. For other species, called obligate brood parasites, all individuals are parasitic and never build a nest, so must by necessity lay eggs in other species' nests.

GIVING UP ON PARENTING

Obligate brood parasitism has evolved several times, independently, among birds: for example, there is a single duck species, the Black-headed Duck, which is an obligate brood parasite; in addition, all honeyguides, species in at least two separate lineages of Old World and New World cuckoos,

LEFT **In this nest**, the Song Sparrow's egg (at the bottom) is outnumbered by eggs laid by two to five different cowbirds, in a situation called multiple parasitism.

RIGHT **The brood parasitic Common Cuckoo** hatchling evicts host eggs from the nest to eliminate competition with nest mates for parental provisions.

and of Old World finches (indigobirds, whydahs, and the Cuckoo-Finch) and New World blackbirds (cowbirds), are obligately brood parasitic. These birds never build a nest or incubate their own eggs.

Most brood parasitic young represent more than a fair challenge for the foster-parents: nestling cowbirds and whydahs beg more intensively than their nest mates, and receive a more than equitable share of the parental food deliveries. This then typically causes the smaller and younger host nestlings to starve and perish in the parasitized brood. Nestlings of the Common Cuckoo and Greater Honeyguide go a step further. They eliminate all host eggs and nestlings soon after they themselves hatch; they toss host eggs and nestlings beyond the rim of the nest, or slaughter the host chicks with their sharply hooked beaks. The end result is that these chicks grow up alone in the nest, monopolizing all parental food deliveries.

THE HOST FIGHTS BACK

Surprisingly, some potential hosts of brood parasitic birds do not face the cost of looking after foreign young; this happens when parasites lay their eggs into nests where the parents feed the chicks with unsuitable foods. For example House Finches provision their own and the Brown-headed Cowbird hatchling with seeds, instead of insects: because cowbirds can't digest seeds, the cowbird chick starves with a crop full of food.

When the parasite chick survives, however, potential hosts have evolved several traits to reduce or eliminate parasitism. Many species aggressively defend their nests against intruding parasites, and loudly mob, even physically attack, the parasite; many of the same, and also other, hosts species have also evolved the ability to assess, discriminate, and reject foreign eggs and chicks from the nest; and so when parasites lay an egg in their nest (or scientists take one of the bird's own eggs and change its color with a highlighter pen), the odd-looking eggs are grabbed and tossed, or pecked, pierced, and carried, out of the nest. Other brood parasites listen to the begging calls produced by their own chicks, and refuse to feed nestlings that sound different from their own.

TRICKERY & MIMICRY

Clearly, both parasitism and antiparasitic strategies represent strong enough evolutionary forces to drive the counter-evolution of behavioral and morphological traits of both hosts and parasites. The result is an arms-race process: hosts evolve to combat parasites, parasites evolve to evade the hosts, the hosts become better tuned to detect parasitism, and the parasites overcome the increasingly sophisticated host defenses. The best example of such a co-evolutionary process is the mimetic eggs of different host races of the Common Cuckoo, each specializing on one of the cuckoo's many host species: in this system, a female cuckoo lays a specific type of egg which best matches the colors and patterns acceptable to the respective host. Often, the mimicry is so close that neither the host, nor the researcher, can tell the difference until the cuckoo chick hatches in the nest, and starts tossing the other eggs and chicks from the brood.

LEFT **This European Robin is tricked** into feeding the oversized Common Cuckoo chick when it begs incessantly and display its large, bright gape.

SCIENCE & EGG COLLECTIONS

The scientific role of egg collections has come a long way since the first days when collectors prided themselves simply on assembling collections based on diversity and rarity. The Victorian age term for this was "a cabinet of curiosities." Most egg collecting in the United States took place between 1800 and 1930, after which time it went out of fashion as concern grew about potentially hurting populations through such collecting. Today, egg collecting requires special permits from national and state governments.

EACH EGG IS UNIQUE

Why, then, are egg collections housed at various public and private institutions so valuable for research and education? And why are eggs and data about them still being collected by researchers? One answer is that such collections and associated data can answer many questions about bird biology that cannot otherwise be answered. Through the years, collectors became astute natural historians recording important details about timing, habitat, nest structure and placement for each set of eggs collected. Thus, the historical egg sets and their associated data in collections provide detailed, tangible documentation about the historical nesting biology of each species represented. Such data are extremely useful in answering questions far into the future. Does climate change affect the nesting biology of birds? Clutch size, laying date, coloration patterns, nest placement, and eggshell composition are examples of things that could change through time in response to environmental changes. Comparing data from historical collections of eggs could help answer such a question for those species for which there are collections. For that to happen, comparative modern data are needed. Eggs collected today document the current situation and if that data is preserved in collections it can be accessed and studied in the future as well.

COLLECTING FOR TOMORROW'S SCIENTISTS

The truth is that we cannot anticipate all the things that such collections might be used for in the future. The egg collectors who gathered Peregrine Falcon eggs in the 1890s had no idea that 70 years later the eggs they collected would be used to document the thinning of Peregrine Falcon eggs that happened with the advent of DDT; the difference in the thickness of shells between these historic and 1960s peregrine eggs was striking. Thus, those eggs were used as part of the argument to ban the use of this pesticide, which subsequently led to the recovery of the species.

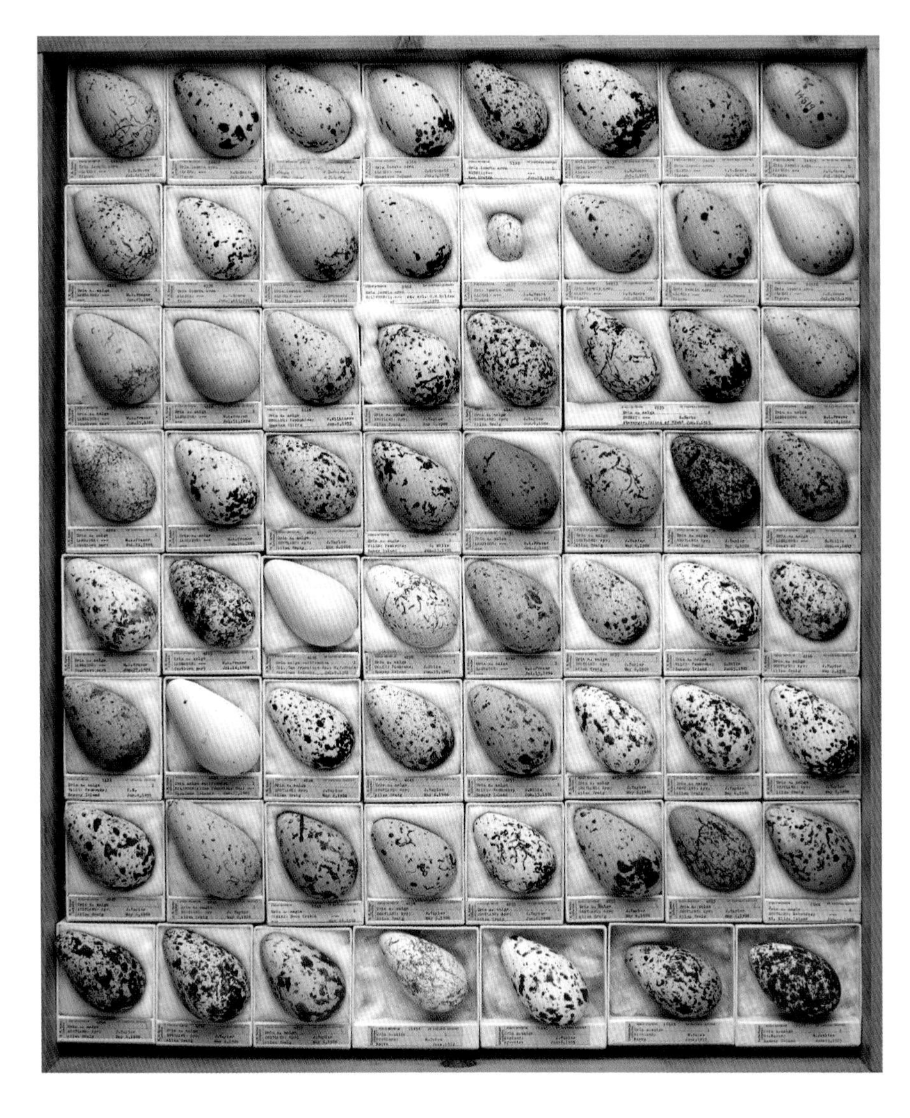

LEFT **This tray of eggs** of Common and Thick-billed Murres from the Field Museum's collection shows the individual variation in egg coloration that allows pairs in these colony-nesting birds to relocate their own eggs. The small egg is a "runt"—rarely do such eggs hatch.

What other things can be learned from eggs? Systematic analysis of the ultraviolet component of eggshell coloration only began about 15 years ago. Until then the color and maculation of the eggshell was described, quantified, and preserved for natural history records and scientific analyses only as seen by the human eye. Today, we know that differences in the ultraviolet hues of eggshells—invisible to human eyes—are used by birds for some critical purposes. For example, some hosts of brood parasites use ultraviolet hues to reject foreign eggs from their nests. Museum-based analyses of full-spectrum reflectance between different eggs of cuckoos and their hosts in Africa, Australia, and Europe have revealed the existence of host-specialization that had remained cryptic when only examined by the human eye.

ABOVE **Some remains of the bones and eggs** of extinct elephantbirds are only a few hundred years old.

PRESERVING EGGS FOR POSTERITY

When eggs are first collected, their content is emptied through a small hole drilled into the shell. And while this content—a mix of yolk, albumin, and embryonic material—is typically discarded, or used for the analyses of protein, hormone, and vitamin concentrations in the egg of that species, the shell is then dried and stored in a dark, dry, and

RIGHT **Moa exhibit in 2013** at the Auckland War Memorial Museum, New Zealand, showing the variation in size of the skeletons, skulls, and eggs of New Zealand's extinct flightless birds.

cool place. This procedure, however, leaves intact some of the egg membranes inside the shell, which dry out and remain attached to the inside of the shell. Modern genetic analyses, capable of isolating and amplifying single copies of DNA, can tap into the preserved eggshell and its internal membranes as a source of genetic material, even from an egg collection 200 years old; these resulting genetic data then can be used for studies of systematics and taxonomy, population and conservation genetics, and parentage analysis.

OOLOGY: THE SCIENCE OF EGGS

Serendipitous uses for scientific exploration of egg collections also involve the material which museum curators and other collectors might consider discarding: broken fragments recovered at the bottom of holding containers and full egg specimens donated but without sufficiently detailed data to be included in the permanent collection. Such fragments can be used to study past exposure of birds (through their eggs) to environmental toxins and heavy metals. In a recent and unique example of the value of museum collections, broken-off pieces found at the bottom of the holding case of the colorful, deep green eggs of the now extinct Upland Moa of New Zealand, were the only ones allowed by the understanding curator of the Otago Museum to be taken for destructive sampling (dissolving in sulphuric acid); these samples served then to identify that the same two chemical pigments—biliverdin and protoporphyrin—that produce the myriad of colors in living birds' eggshells are the same ones used by extinct birds to color their eggs.

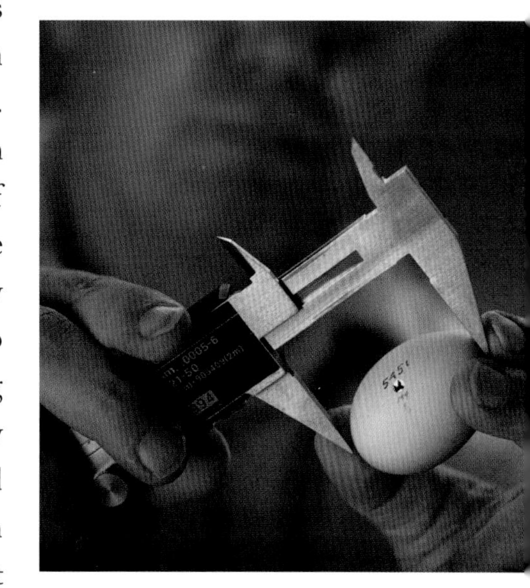

ABOVE **Eggshells prepared, preserved, and protected** in museum collections provide an invaluable source of scientific data and visual inspiration for researchers and the public alike.

EGGS FOR THE FUTURE

There are over 10,000 species of birds and for some species the eggs have still never been described much less collected and deposited in a collection where they can be compared and contrasted with those of other species. Eggs are such a critical component of bird biology that assembling and using egg collections to monitor and understand how birds are adapting to changing environments is something that needs to continue. The long-term scientific value is and will continue to be substantial.

THE EGGS

WATER BIRDS

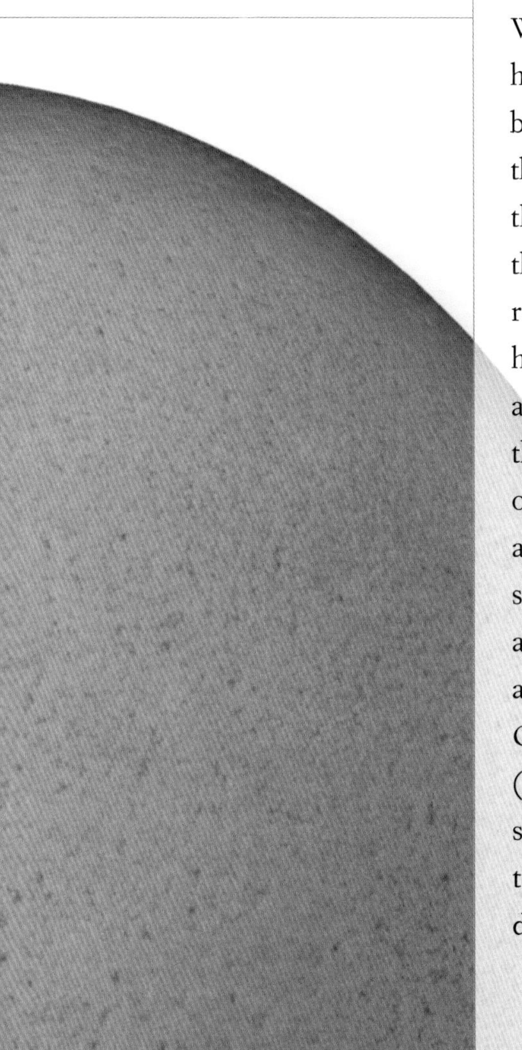
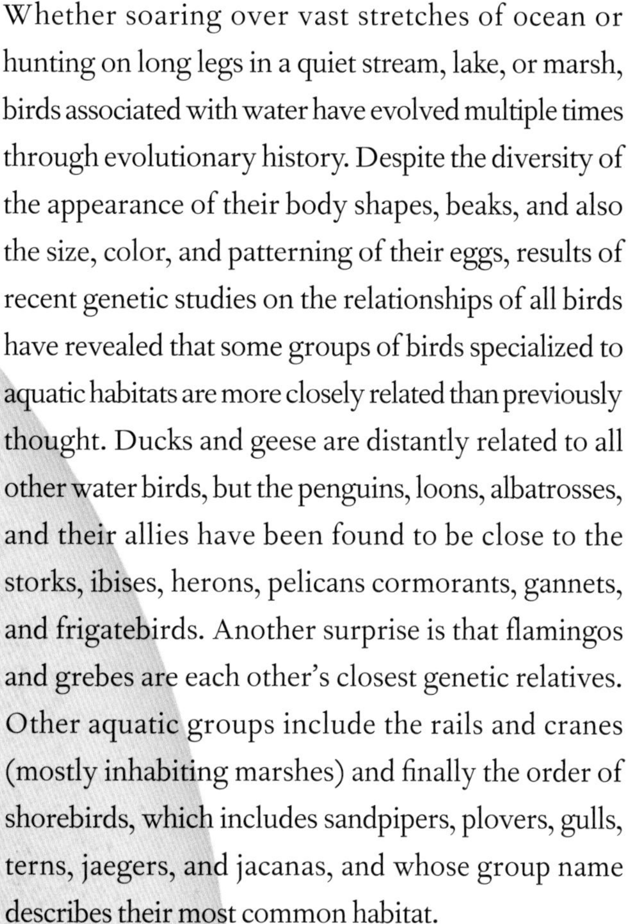

Whether soaring over vast stretches of ocean or hunting on long legs in a quiet stream, lake, or marsh, birds associated with water have evolved multiple times through evolutionary history. Despite the diversity of the appearance of their body shapes, beaks, and also the size, color, and patterning of their eggs, results of recent genetic studies on the relationships of all birds have revealed that some groups of birds specialized to aquatic habitats are more closely related than previously thought. Ducks and geese are distantly related to all other water birds, but the penguins, loons, albatrosses, and their allies have been found to be close to the storks, ibises, herons, pelicans cormorants, gannets, and frigatebirds. Another surprise is that flamingos and grebes are each other's closest genetic relatives. Other aquatic groups include the rails and cranes (mostly inhabiting marshes) and finally the order of shorebirds, which includes sandpipers, plovers, gulls, terns, jaegers, and jacanas, and whose group name describes their most common habitat.

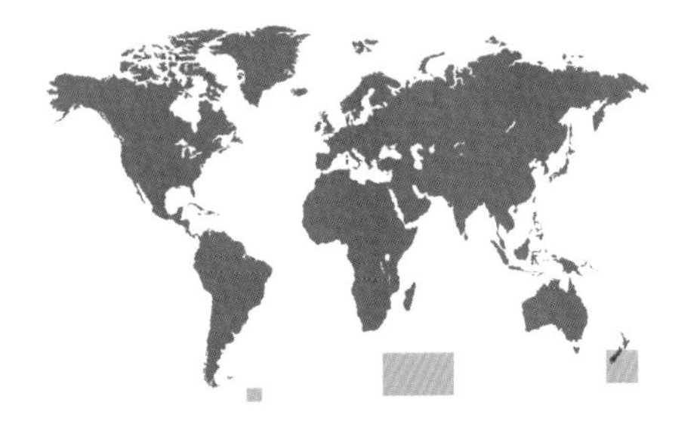

ORDER	Procellariiformes
FAMILY	Diomedeidae
BREEDING RANGE	Southern Ocean, subantarctic islands
BREEDING HABITAT	Grassy patches on sloping island ridges, facing the sea
NEST TYPE AND PLACEMENT	Large ground nest, built up from mud and vegetation
CONSERVATION STATUS	Vulnerable
COLLECTION NO.	FMNH 2860

ADULT BIRD SIZE
42–53 in (107–135 cm)

INCUBATION
77 days

CLUTCH SIZE
1 egg

DIOMEDEA EXULANS

WANDERING ALBATROSS

PROCELLARIIFORMES

The Wandering Albatross, also known as the Snowy or White-winged Albatross, mates for life. Pairs attempt to breed every other year, nesting on remote southern islands that are free of most terrestrial predators. Breeding is a colonial affair, with these majestic birds settling in loose colonies where their neighbors are visible but separated by adequate pecking distance.

Laying just one egg, the parents invest heavily into raising the single chick. The hatchling has to be guarded for many weeks by at least one of the parents, while the other forages for food both for itself and to feed its progeny. This can be a successful strategy in the long run, provided that the adults survive. However, longline commercial fishing methods often snag and drown adult birds, and, because they lay only one egg, their populations do not recover quickly from declines.

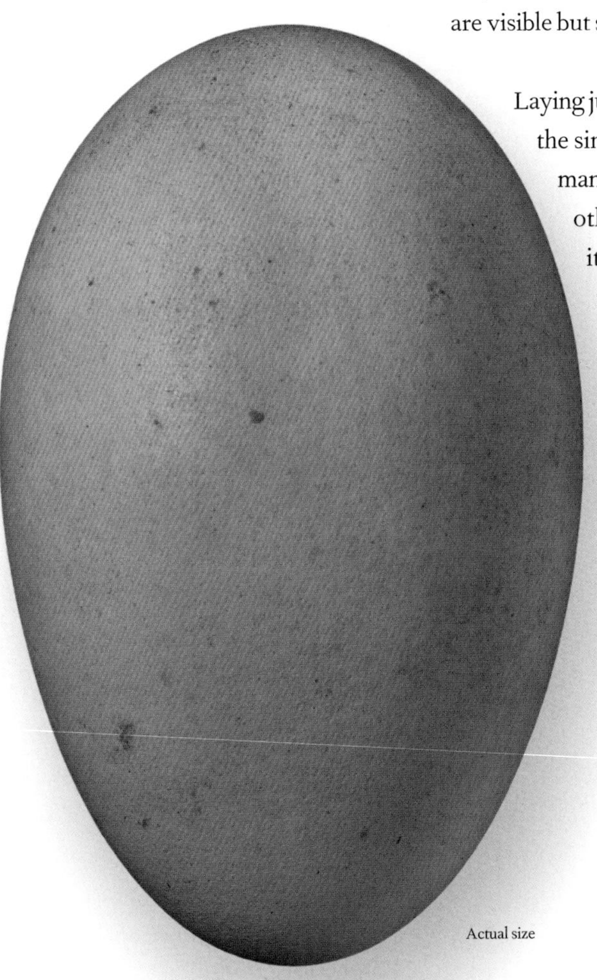

The egg of the Wandering Albatross is elongated yet blunt on both ends; it is 4 x 2 in (100 x 50 mm) in size, and sparsely and finely spotted. It becomes soiled during the lengthy 11-week incubation period, during which parents take turns of several days sitting on the nest.

Actual size

ORDER	Procellariiformes
FAMILY	Diomedeidae
BREEDING RANGE	Tropical North Pacific
BREEDING HABITAT	Ocean islands, sandy flats and plateaus facing the open sea
NEST TYPE AND PLACEMENT	Simple scraping in the sand
CONSERVATION STATUS	Vulnerable
COLLECTION NO.	FMNH 4884

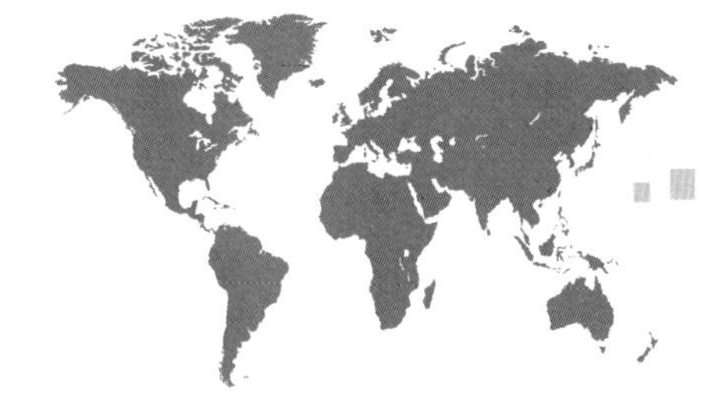

ADULT BIRD SIZE
25–29 in (64–74 cm)

INCUBATION
65 days

CLUTCH SIZE
1 egg

DIOMEDEA NIGRIPES

BLACK-FOOTED ALBATROSS

PROCELLARIIFORMES

In contrast to most albatross species, which live in the temperate and subantarctic southern regions, the Black-footed Albatross occurs and breeds in the tropics just north of the equator. Its largest colonies, on the Laysan Islands off Hawaii, are currently protected by a no-longline-fishing zone 50 miles (80 km) offshore, so that breeding birds can safely feed and return to change guard with their mates incubating the egg or protecting the hatchling. One parent, whose mate had probably died, spent seven weeks without food or water before it was forced to abandon the unhatched egg and feed itself.

Adults begin breeding when they are seven years old (a long time for birds), forming a pair bond and mating every two years. Although largely isolated from predators on their remote breeding islands, failed nesting attempts due to adverse weather, accidents, or shortage of food can significantly reduce the reproductive rates of these birds, making them more vulnerable to habitat deterioration and adult mortality.

The egg of the Black-footed Albatross is dull white, speckled with rusty spots; it measures 4¼ x 2¾ in (108 x 70 mm) in size. Incubation lasts over two months; the parents take turns of two to three weeks each to attend the egg and feed the young chick.

Actual size

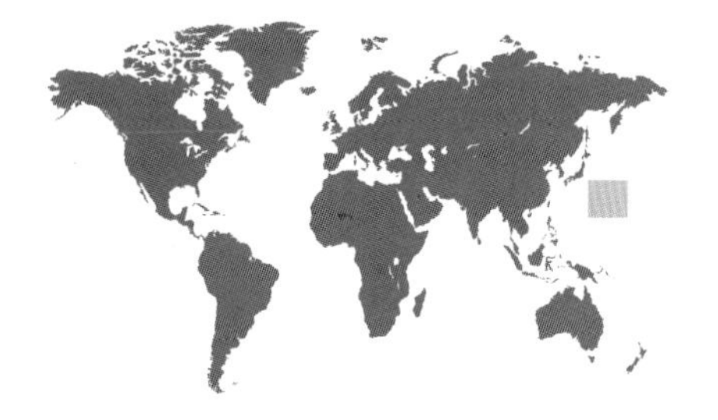

ORDER	Procellariiformes
FAMILY	Diomedeidae
BREEDING RANGE	Tropical and temperate North Pacific
BREEDING HABITAT	Remote oceanic islands, large open flats
NEST TYPE AND PLACEMENT	Shallow mound of dry vegetation, placed in short grassy patches
CONSERVATION STATUS	Vulnerable
COLLECTION NO.	FMNH 4887

ADULT BIRD SIZE
33–37 in (84–94 cm)

INCUBATION
65 days

CLUTCH SIZE
1 egg

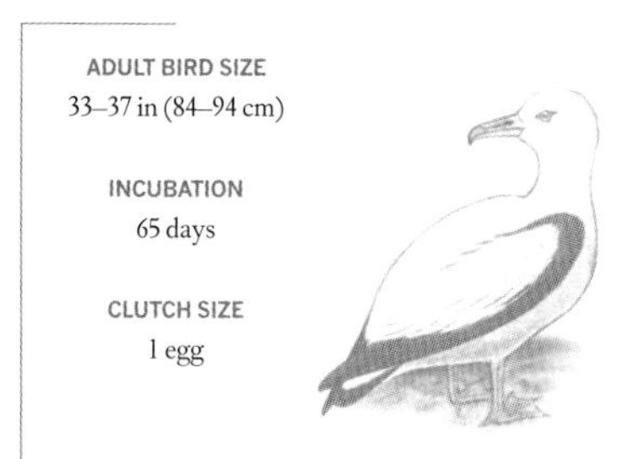

PHOEBASTRIA ALBATRUS

SHORT-TAILED ALBATROSS

PROCELLARIIFORMES

The egg of the Short-tailed Albatross is dirty white in background, covered in a dense aggregation of red spots at the blunt end, and measures 4½ x 2⅞ in (116 x 74 mm) in size. Adults alternate with one another during the two-month-long incubation period.

Whether caught in fishing nets, hunted for its feathers, or killed by cats introduced to their remote island breeding colonies, the Short-tailed Albatross is at risk both at sea and while breeding. In addition, because adults typically do not begin to breed until ten years of age and lay just one egg per nesting attempt, populations are slow to recover from adult mortality or nest failure.

Nesting on remote oceanic islands is typically a safe and successful strategy for most seabirds—unless the island happens to be an active volcano! The larger of the two main colonies of this albatross is on the active volcanic island of Torishima, Japan. The good news is that populations there are growing (from as few as ten individuals to over 2,400 in the last 70 years). A pair recently appeared and bred on Midway Island off Hawaii.

Actual size

ORDER	Procellariiformes
FAMILY	Diomedeidae
BREEDING RANGE	Circumpolar Southern Ocean
BREEDING HABITAT	Subantarctic islands, on vegetated cliff edges
NEST TYPE AND PLACEMENT	Ground nest of a built-up mound of peat and soil
CONSERVATION STATUS	Near threatened
COLLECTION NO.	FMNH 15004

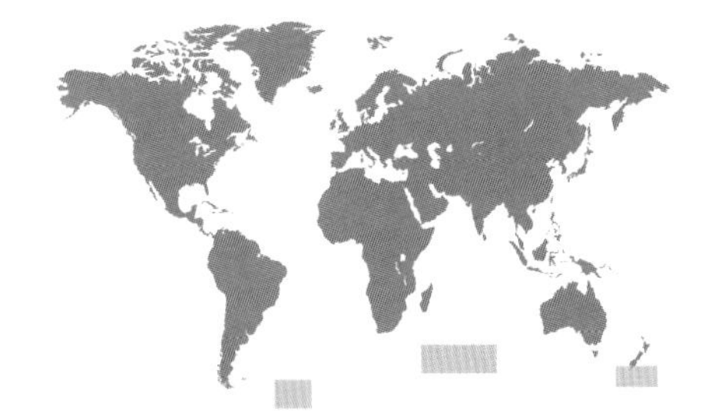

ADULT BIRD SIZE
31–39 in (79–99 cm)

INCUBATION
65–72 days

CLUTCH SIZE
1 egg

PHOEBETRIA PALPEBRATA

LIGHT-MANTLED ALBATROSS

PROCELLARIIFORMES

Feeding and ranging strictly over the Southern Ocean, this species, which also goes by the names Gray-mantled or Light-mantled Sooty Albatross, is as close to an Antarctic breeding albatross as they come. Like other albatrosses, these birds mate for life, working together to nest, incubate, guard, and feed the young. Nesting alone or in small colonies, parents alternate in week-long bouts of guarding the egg or ranging away to find food. The chick eventually grows larger than the parents before it is abandoned to depart and feed itself from the ocean.

If the single egg in the nest is damaged or lost, the parents abandon their annual breeding attempt, and only return in the following year to try again. This leads to a slow rate of population growth, an important conservation concern if ecological and anthropogenic losses of eggs, chicks, and adults increase.

The egg of the Light-mantled Albatross is 4¼ x 2⅔ in (107 x 67 mm) in size, elongated, and white in color, with light brown-reddish blotching. After the long incubation period, the chick takes several days to hatch.

Actual size

ORDER	Procellariiformes
FAMILY	Procellariidae
BREEDING RANGE	North Atlantic and North Pacific
BREEDING HABITAT	Remote oceanic islands, under vegetation or in burrows
NEST TYPE AND PLACEMENT	Ground nests, covered by vegetation or tree roots, or inside crevices and burrows
CONSERVATION STATUS	Least concern
COLLECTION NO.	FMNH 4942

ADULT BIRD SIZE
10–11 in (25–28 cm)

INCUBATION
42–46 days

CLUTCH SIZE
1 egg

BULWERIA BULWERII

BULWER'S PETREL

PROCELLARIIFORMES

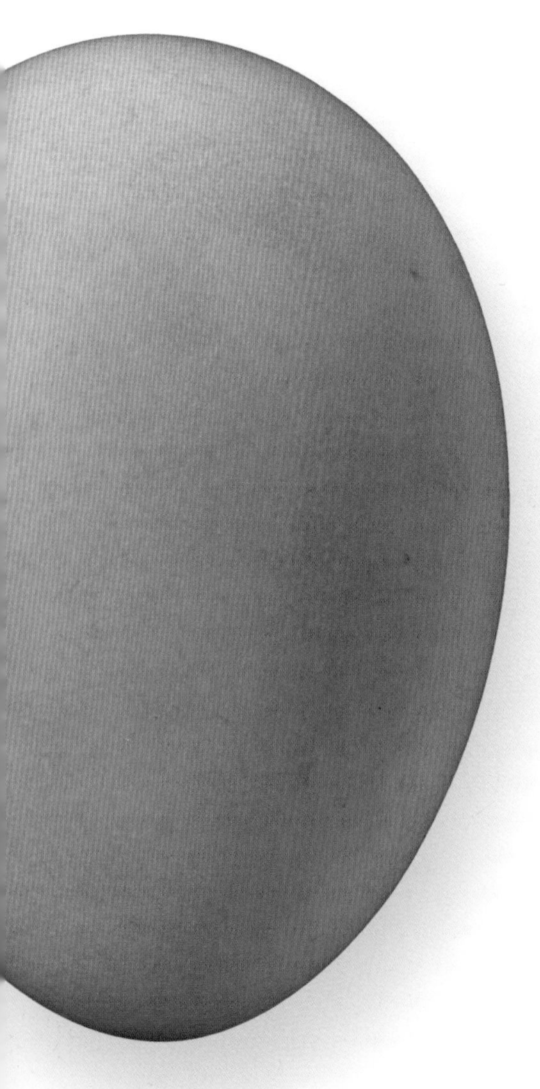

Petrels are strictly nocturnal when it comes to breeding: they arrive after sunset at their isolated island nesting colonies and depart before sunrise. They use a keen sense of smell to locate nests and mates, as well as to find food. In the vast open seas they can sniff out the pungent, oil substances released by potential prey. The Bulwer's Petrel uses its long and thin wings to fly buoyantly just above the ocean, searching for plankton and other small prey to pick up from the water surface.

Adult birds are strictly philopatric to their nesting site, which means that they return year after year to the exact same site, and in the process, encounter their old mates, too! This species does not excavate its own burrows and instead uses naturally available rock crevices and subterraneous holes for nesting. Both parents take turn in incubating the eggs and provisioning the chick.

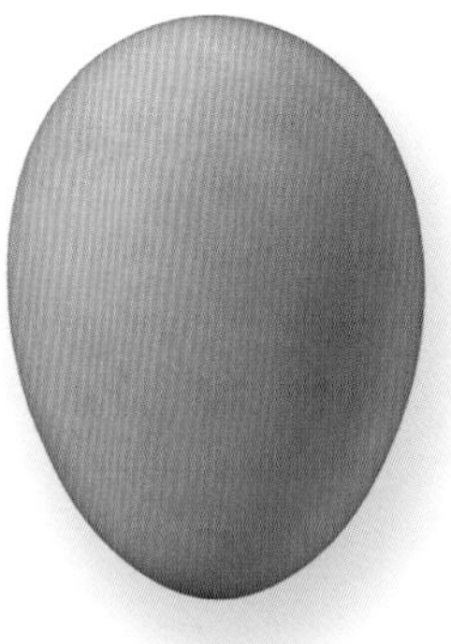

Actual size

The egg of the Bulwer's Petrel is beige-white in background color, immaculate, roundish, and measures 1$\frac{2}{3}$ x 1$\frac{3}{16}$ in (42 x 30 mm) in dimensions. A few young, less experienced adults may lay two eggs, but the first is typically moved out of the nest chamber so that only one egg hatches.

ORDER	Procellariiformes
FAMILY	Procellariidae
BREEDING RANGE	Circumpolar southern seas and oceans
BREEDING HABITAT	Remote subantarctic islands, cliffs and plateaus near the sea
NEST TYPE AND PLACEMENT	Colonial breeding, the ground nest is an aggregation of pebbles under a rock overhang or inside crevices
CONSERVATION STATUS	Least concern
COLLECTION NO.	FMNH 15009

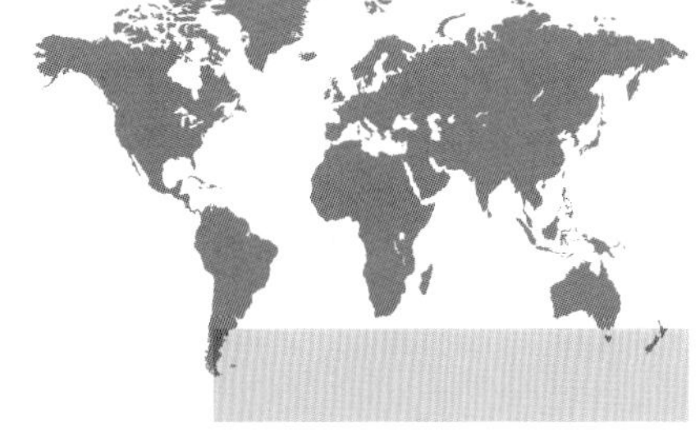
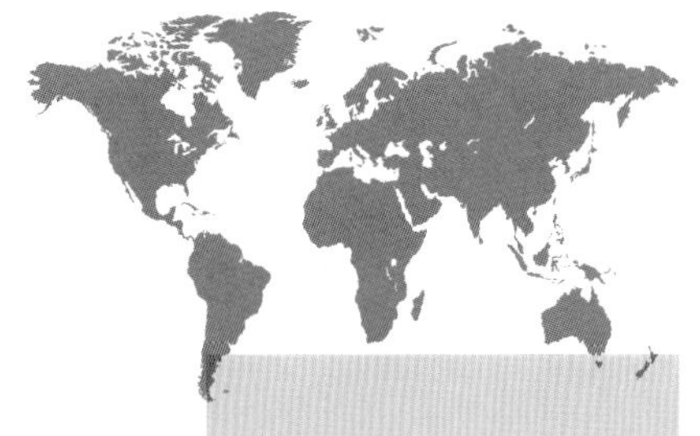

ADULT BIRD SIZE
15–16 in (38–41 cm)

INCUBATION
45 days

CLUTCH SIZE
1 egg

DAPTION CAPENSE

CAPE PETREL

PROCELLARIIFORMES

The Cape Petrel, also called the Cape Pigeon or the Pintado Petrel, is one of the most abundant seabirds that can be seen following fishing boats and cruise ships in the southern Pacific seas. Its notable black-and-white checkered plumage makes it appealing and easy to recognize from long distances and the tendency to aggregate closely with others of its own species (conspecifics) and other petrels (heterospecifics) near potential food sources makes the Cape Petrel an even more memorable sight at sea.

Despite its relatively small size, the Cape Petrel is in evolutionary terms most closely related to the giant petrels. Similar to fulmars, adults defend their nests by vomiting foul-smelling stomach oils on approaching predators. Their main enemies are typically other seabirds, including jaegers (skuas); the remote oceanic islands where the petrels nest support no mammalian predators.

The egg of the Cape Petrel is clear white in color, and measures 2⅛ x 1½ in (53 x 38 mm) in size. Although parents take turns incubating the eggs, the male takes the first shift and his subsequent incubation bouts typically last a day longer than the female's, perhaps to balance out the female's efforts in laying the egg in the first place.

Actual size

ORDER	Procellariiformes
FAMILY	Procellariidae
BREEDING RANGE	Subarctic regions of the North Atlantic and Pacific Oceans
BREEDING HABITAT	Open ledges, cliffs, and more recently building rooftops
NEST TYPE AND PLACEMENT	On ground, a scrape in ground, or mound of vegetation
CONSERVATION STATUS	Least concern
COLLECTION NO.	FMNH 770

ADULT BIRD SIZE
18–19 in (46–48 cm)

INCUBATION
50–54 days

CLUTCH SIZE
1 egg

FULMARUS GLACIALIS

NORTHERN FULMAR

PROCELLARIIFORMES

The Northern Fulmar, also known as the Arctic Fulmar, is one of the most numerous northern hemisphere petrel species and, unlike most seabirds, its populations have been increasing for nearly two centuries. Today these birds still nest on typical remote and inaccessible cliffs, ledges, and rocky plateaus in loose colonies facing the sea. However, near human settlements, they have started to exploit the relative safety of building roofs for nesting sites.

Northern Fulmars are conspicuous and yet somewhat confusing in their appearance, as they come in several color morphs, from clear white to dark gray. The genetic basis of these color morphs is related to the different forms of the gene involved in the production of melanin pigments.

Actual size

The egg of the Northern Fulmar is white in color, immaculate, and measures 2⅞ x 2 in (74 x 51 mm) in size. Both parents take turns to incubate the egg, and one always stays with the hatchling for the first two weeks of its life to provide it with safety.

ORDER	Procellariiformes
FAMILY	Procellariidae
BREEDING RANGE	Southern Ocean, from Antarctica to isolated southern islands
BREEDING HABITAT	In loose colonies, near cliffs, plateaus, and shores of islands
NEST TYPE AND PLACEMENT	Ground nests of piled-up dried seaweed and grasses
CONSERVATION STATUS	Least concern
COLLECTION NO.	FMNH 4910

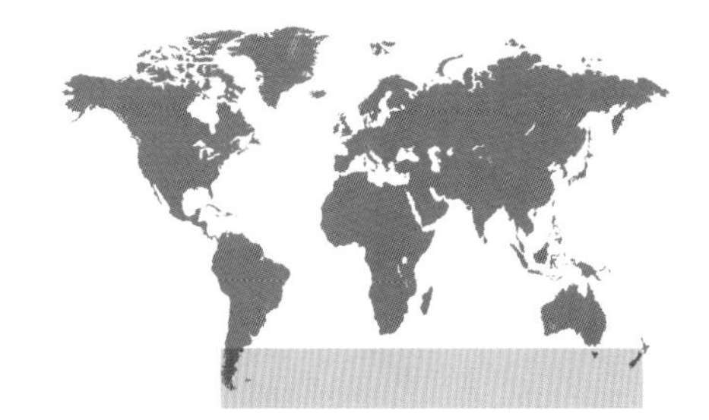

ADULT BIRD SIZE
34–39 in (86–99 cm)

INCUBATION
55–66 days

CLUTCH SIZE
1 egg

MACRONECTES GIGANTEUS

ANTARCTIC GIANT PETREL

PROCELLARIIFORMES

Matching the size of albatrosses, these are the largest members of the petrel family, as the name suggests. They also share other life history traits with albatrosses: long lifespan; delayed maturation of up to seven years before the onset of pair bonding and breeding; and extended parental care, which includes a four-month period between hatching and fledging. One of the parents always attends the egg, not only to keep it warm but also to protect it from avian nest predators, including other petrels and jaegers (skuas).

Although in general appearance the Antarctic Giant Petrel is variable from light to dark gray, its foraging strategies can result in a bright red head, with feathers covered in dried blood after feeding on dead seals and whales washed ashore. Introduced mammals, including small rodents, are not recognized by the parents as a potential danger to the chick, and can devastate a once-thriving colony.

The egg of the Antarctic Giant Petrel is large, 4$\frac{1}{16}$ x 2$\frac{3}{4}$ in (103 x 70 mm) in dimensions, and immaculate white in color. Sitting on a bed of drying and rotting seaweed, the eggshell becomes increasingly soiled during the approximate two-month incubation period.

Actual size

ORDER	Procellariiformes
FAMILY	Procellariidae
BREEDING RANGE	Temperate Southern Ocean, temperate northeast Atlantic
BREEDING HABITAT	Isolated oceanic islands
NEST TYPE AND PLACEMENT	Cavities in grassy fields or among rocks
CONSERVATION STATUS	Least concern
COLLECTION NO.	FMNH 15006

ADULT BIRD SIZE
10–12 in (25–30 cm)

INCUBATION
52–58 days

CLUTCH SIZE
1 egg

PUFFINUS ASSIMILIS

LITTLE SHEARWATER

PROCELLARIIFORMES

The egg of the Little Shearwater is clear white in color and 2 x 1⅜ in (51 x 35 mm) in size. The egg is vulnerable to predation by introduced mammalian predators, including rats that travel on ships and swim ashore when these vessels visit remote oceanic islands.

The smallest of the shearwaters, this gregarious species spends most of its life silently at sea; it can only be heard when landing and making its way to its cavity nest in large breeding colonies on remote oceanic islands. Its flocks can be seen in tropical and subtropical waters around the globe during the non-breeding season.

The parents incubate and attend the chick, alternating shifts with one another. Despite the small size of the species, the incubation period is more similar to those of larger seabirds, lasting nearly two months; it then takes two more months to raise the chick before it finally fledges. Once their reproductive season ends, these birds' nesting cavities are typically taken over by other shearwaters or petrels for another round of nesting and breeding.

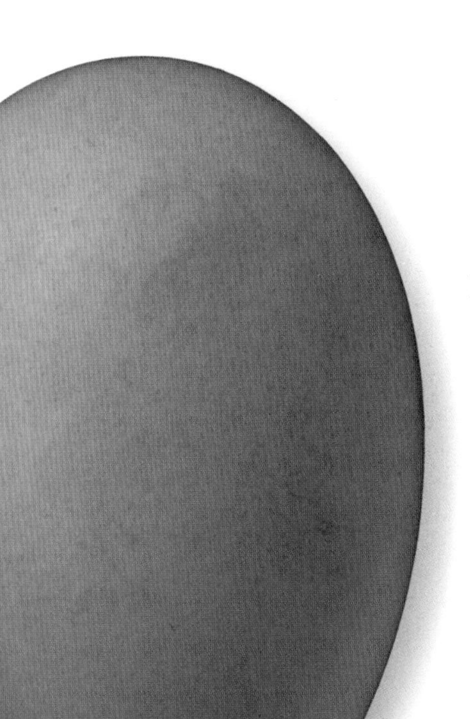

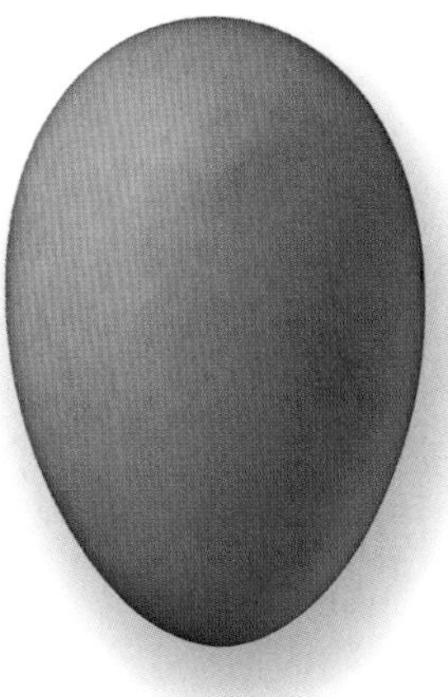

Actual size

ORDER	Procellariiformes
FAMILY	Hydrobatidae
BREEDING RANGE	North Atlantic and western Mediterranean
BREEDING HABITAT	Remote islands off the continental shores
NEST TYPE AND PLACEMENT	Ground burrows and rock crevices near or facing the open sea
CONSERVATION STATUS	Least concern
COLLECTION NO.	FMNH 7977

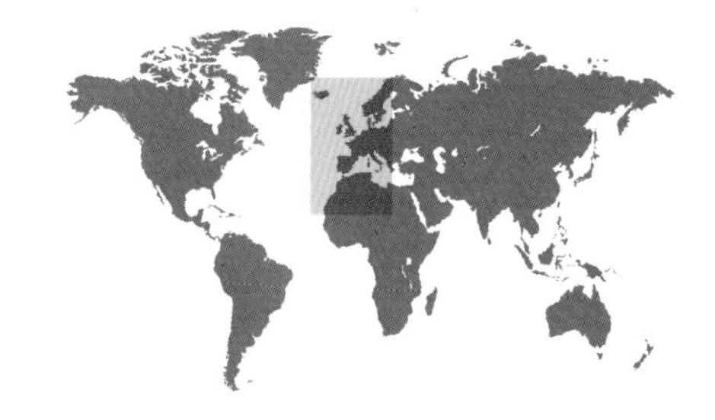

ADULT BIRD SIZE
6–6⅓ in (15–16 cm)

INCUBATION
38–50 days

CLUTCH SIZE
1 egg

HYDROBATES PELAGICUS

EUROPEAN STORM-PETREL

PROCELLARIIFORMES

Strictly nocturnal during the breeding season, this gregarious species breeds off the shores of Europe and migrates to South Africa during the winter. At sea, storm-petrels resemble swallows or bats in both size and shape, hovering over and repeatedly touching down onto the surface of the sea to pick up small crustaceans and fish as their prey.

Like many other petrel relatives, the European Storm-Petrel lays just a single egg that requires two months of incubation. The chick is fed digested prey in the form of an energy-rich oil regurgitated by the parents and must wait several days between feeding bouts.

Actual size

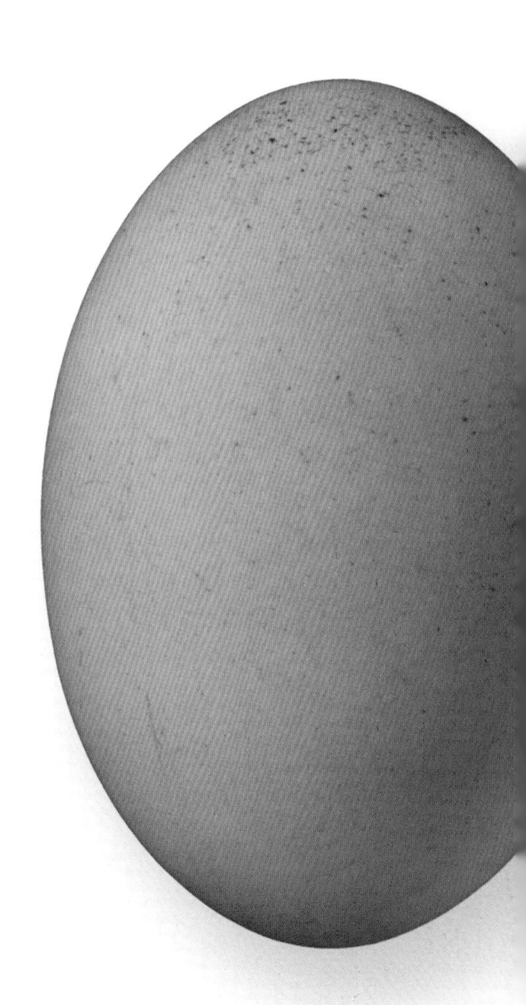

The egg of the European Storm-Petrel is white, and 1⅛ x ⅞ in (28 x 21 mm) in size. The parents take turns to incubate the egg for stretches of 4–7 days, an impressive task for this small bird.

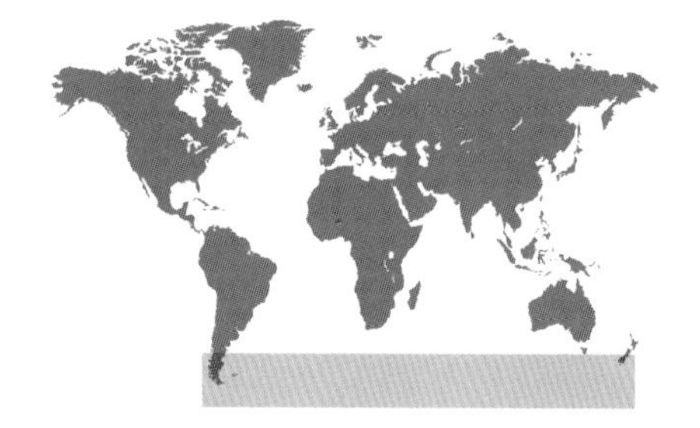

ORDER	Procellariiformes
FAMILY	Hydrobatidae
BREEDING RANGE	Antarctic coastline, subantarctic islands
BREEDING HABITAT	Oceanic shores, rocky or grassy plateaus
NEST TYPE AND PLACEMENT	Rock crevices and ground burrows, near the sea
CONSERVATION STATUS	Least concern
COLLECTION NO.	FMNH 21223

ADULT BIRD SIZE
6⅓–7⅓ in (16–19 cm)

INCUBATION
40–50 days

CLUTCH SIZE
1 egg

OCEANITES OCEANICUS

WILSON'S STORM-PETREL

PROCELLARIIFORMES

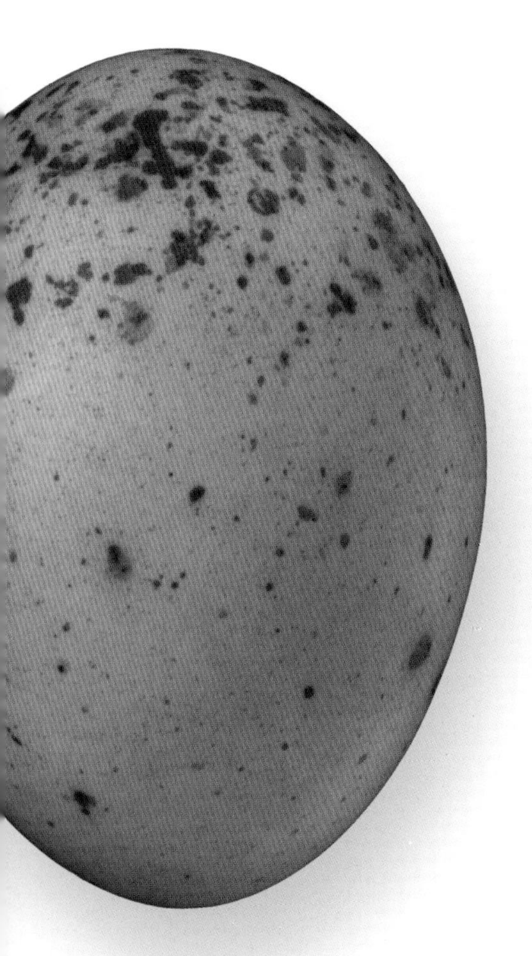

Unlike other storm-petrels, this species flies with less flutter and in more direct lines. Strictly pelagic during the non-breeding season, it can be seen in the North Pacific and Atlantic but it breeds predominantly on the shores of Antarctica and its islands.

The thin legs of the Wilson's Storm-Petrel make walking on land difficult and so the species breeds near cliffs facing the ocean, often in large colonies. Late spring snowstorms may interfere with its ability to leave or access the nest, causing breeding failure. The parents do not come to land for nesting during the daytime or even on clear-lit nights, in order to avoid attracting the attention of larger birds intent on preying on the adults or their eggs and chicks. In the darkness, individuals use a keen sense of smell to relocate nesting burrows, mates, and chicks.

Actual size

The egg of the Wilson's Storm-Petrel is white with reddish spots near the blunt end, and is 1⅓ x 1 in (33 x 24 mm) in size. It can be left unattended and cooled for up to two days without this interfering with the successful development of the embryo and the hatching of the chick.

ORDER	Procellariiformes
FAMILY	Hydrobatidae
BREEDING RANGE	Northern Pacific Ocean
BREEDING HABITAT	Rocky cliffs and grassy plateaus of offshore islands
NEST TYPE AND PLACEMENT	Ground cavities, rock or tree-root crevices
CONSERVATION STATUS	Least concern
COLLECTION NO.	FMNH 4961

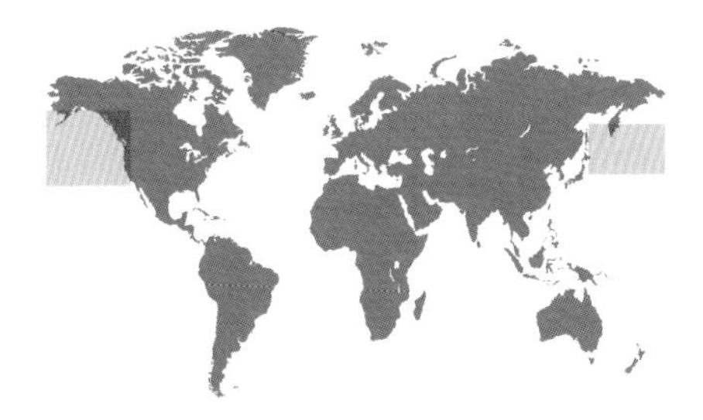

ADULT BIRD SIZE
8–9 in (20–23 cm)

INCUBATION
43–57 days

CLUTCH SIZE
1 egg

OCEANODROMA FURCATA

FORK-TAILED STORM-PETREL

PROCELLARIIFORMES

Like many other petrel-like birds, this species lays just a single egg in the nest. Unlike with other relatives, however, a lost egg can be replaced in 80 percent of the nests, although these second (or even third) eggs are smaller than the initial egg. Chick viability depends on the size of the egg, and so chicks hatched from these replacement eggs survive less well.

Egg neglect, or prolonged lack of incubation of the egg, is prevalent in this species, especially during stormy weather in the cold, northern climates where it breeds. However, eggs have been known to hatch successfully even after going an incredibly long time without a parent to keep them warm. One egg hatched despite experiencing seven consecutive days and a cumulative total of four weeks without incubation.

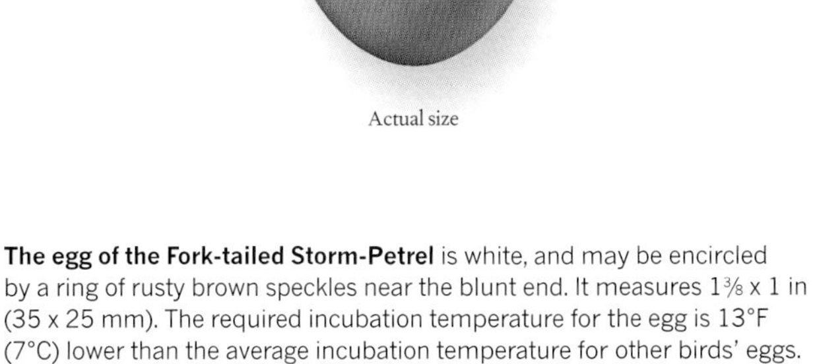

Actual size

The egg of the Fork-tailed Storm-Petrel is white, and may be encircled by a ring of rusty brown speckles near the blunt end. It measures 1⅜ x 1 in (35 x 25 mm). The required incubation temperature for the egg is 13°F (7°C) lower than the average incubation temperature for other birds' eggs.

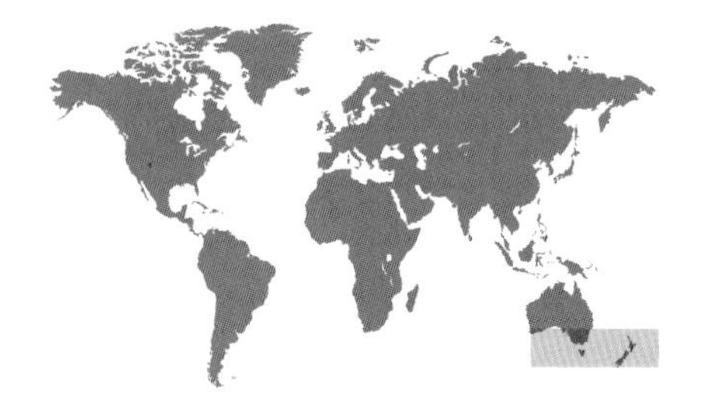

ORDER	Sphenisciformes
FAMILY	Spheniscidae
BREEDING RANGE	South Australia, New Zealand and their outlying islands
BREEDING HABITAT	Sandy and rocky seashore, but occasionally forest floor several hundred yards from the coast, and also urban parks
NEST TYPE AND PLACEMENT	Ground cavities, rock crevices, under dense bushes
CONSERVATION STATUS	Least concern
COLLECTION NO.	FMNH 2862

ADULT BIRD SIZE
16–17 in (40–43 cm)

INCUBATION
40–42 days

CLUTCH SIZE
2 eggs

EUDYPTULA MINOR

LITTLE PENGUIN

SPHENISCIFORMES

Clutch

The egg of the Little Penguin measures $2\frac{3}{16}$ x $1\frac{2}{3}$ in (56 x 42 mm) in size and is mostly white-beige, but quickly becomes soiled during incubation. The two eggs are laid within 2–7 days of each other, and the chicks hatch apart, establishing a natural size (and dominance) hierarchy in the nest.

The smallest of the penguins, these birds are widespread in Australia and New Zealand. Locally referred to as Fairy Penguins or Little Blue Penguins, in places they have become commensals of humans, breeding in urban parks in coastal towns. Luckily, these penguins are nocturnal when coming and going from their nests, so human traffic patterns interfere with them less than if they moved about in the day.

Due to their small size, they are vulnerable to introduced predators, including foxes, dogs, and cats which are evolutionarily recent arrivals to their breeding grounds, brought by European settlers. But for unknown reasons, some colonies have entirely disappeared and other populations of this species have severely declined on a number of islands that are free of mammalian predators. Unlike many seabirds, Little Penguins breed twice or more per season, which helps offset population losses.

Actual size

ORDER	Sphenisciformes
FAMILY	Spheniscidae
BREEDING RANGE	Remote subantarctic islands in the Southern Ocean
BREEDING HABITAT	Muddy or rocky flats near the beach
NEST TYPE AND PLACEMENT	No nest; the egg is carried inside a brood pouch between the feet
CONSERVATION STATUS	Least concern
COLLECTION NO.	WFVZ 148139

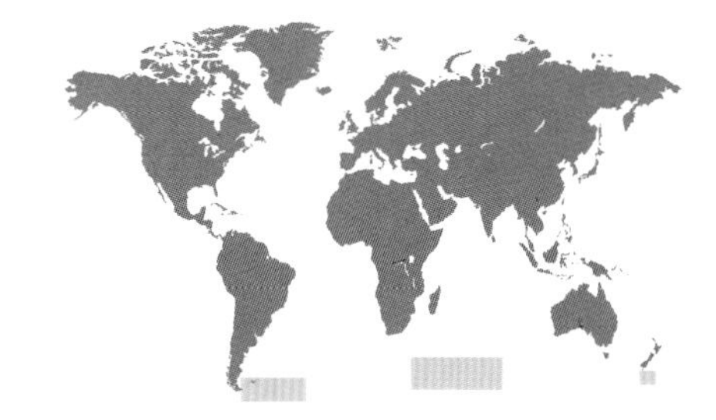

ADULT BIRD SIZE
35–37 in (90–94 cm)

INCUBATION
55 days

CLUTCH SIZE
1 egg

APTENODYTES PATAGONICUS

KING PENGUIN

SPHENISCIFORMES

This is the second largest penguin species, exceeded only by the Emperor Penguins of Antarctica. Both sexes display colorful facial and neck feathers providing cues for recognizing individuals in the large colonies, which can number in the hundreds of thousands to millions of birds. Individuals with brighter facial feathering have better mating success in this species.

The breeding cycle of this penguin is highly variable. A typical spring-summer-fall cycle starts when eggs are laid in the early spring, and ends in fall with the chick reaching 90 percent of the adult weight, at which point the parents leave it. But if egg laying is delayed to late spring, the cycle can take up to 14 months: one of the longest reproductive cycles of any bird. Once the egg hatches, and the chick is a month old, the parents leave them in crèches (avian kindergartens) where a select few adults look after the young, while most parents head out to sea to fish for their own young's next meal.

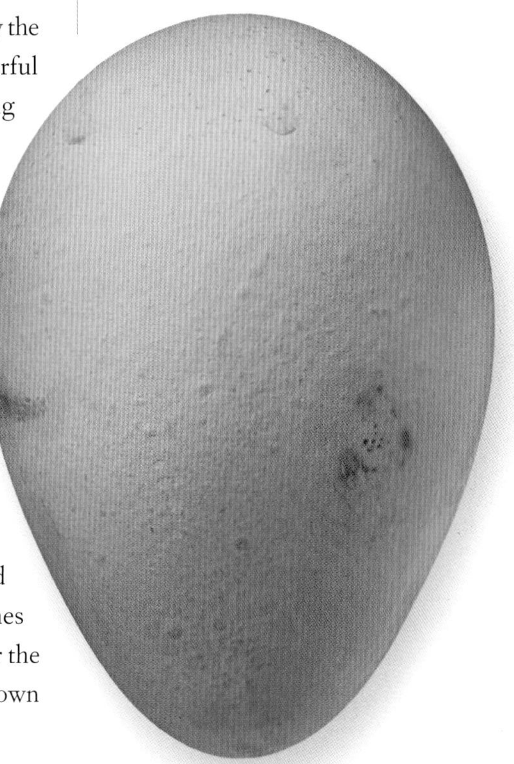

Actual size

The egg of the King Penguin is clear white or pale green, pear-shaped, initially soft but quickly hardening, and measures 3⅞ x 2¾ in (100 x 70 mm) in size. The parents keep the egg inside their brood pouch for 6–18 days before transferring it to their mate, and heading out to sea to feed.

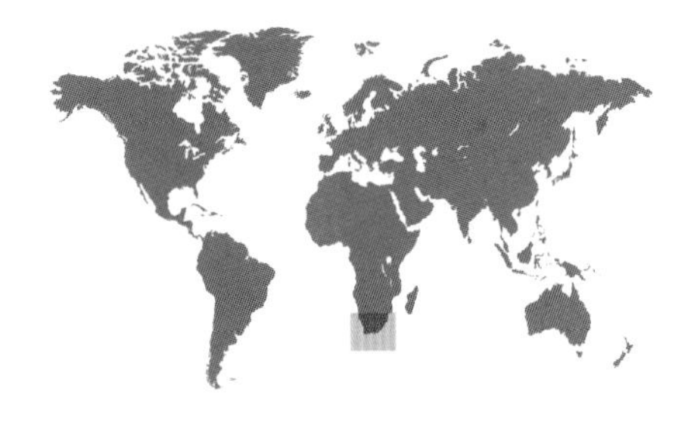

ORDER	Sphenisciformes
FAMILY	Spheniscidae
BREEDING RANGE	Southern African shores and islands
BREEDING HABITAT	Grassy, guano-covered, and rocky landings near the sea
NEST TYPE AND PLACEMENT	Ground cavities, crevices, under thick vegetation or human-made structures
CONSERVATION STATUS	Endangered
COLLECTION NO.	FMNH 20535

ADULT BIRD SIZE
24–28 in (60–70 cm)

INCUBATION
40 days

CLUTCH SIZE
2 eggs

SPHENISCUS DEMERSUS

JACKASS PENGUIN

SPHENISCIFORMES

Clutch

Jackass (or African or Black-footed) Penguins are the only penguin species native to the southern tip of Africa. They breed further north than any other penguin except the Galapagos Penguin. Typically nesting on remote beaches on offshore islands, some of their colonies today are formed in close proximity to human settlements, including the infamous prison island, Robben Island, where Nelson Mandela was held in captivity for decades.

Typically burrowing into thick layers of guano sediments, much of which was removed in the nineteenth century for use as fertilizer, today some of these penguins have to make do with burrowing into loose, sandy soils; as a result, overheating, flooding, and the collapse of burrows have begun to contribute to other habitat-loss-related population declines of this uniquely African penguin species.

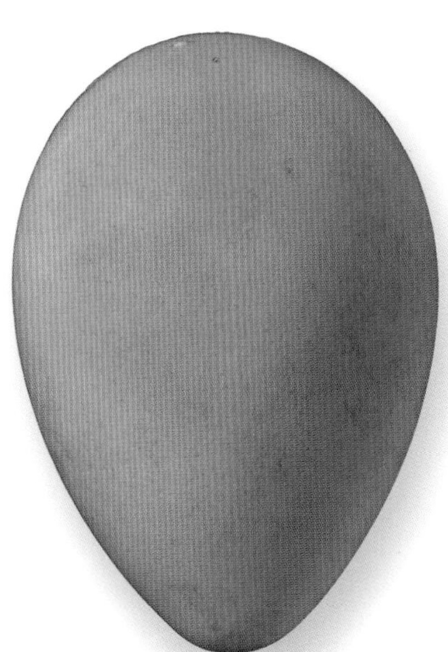

Actual size

The egg of the Jackass Penguin is white and measures 2¾ x 2³⁄₁₆ in (72 x 56 mm) in size. Penguin eggs have long been considered a delicacy; today, even with protections in place, many eggs are lost to people who collect them for food.

ORDER	Gaviiformes
FAMILY	Gaviidae
BREEDING RANGE	Arctic and subarctic Pacific coasts, Norwegian coastline
BREEDING HABITAT	Wetlands, grassy lake shores
NEST TYPE AND PLACEMENT	Large ground nest of piled-up mud and vegetation, within a few steps of the open lake surface
CONSERVATION STATUS	Near threatened
COLLECTION NO.	FMNH 4081

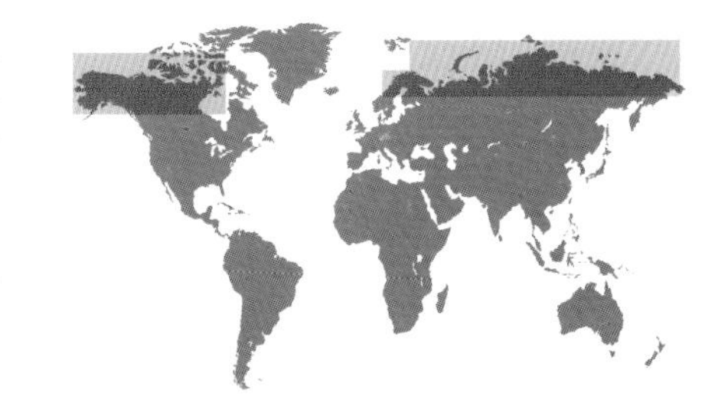

ADULT BIRD SIZE
31–35 in (80–90 cm)

INCUBATION
27–29 days

CLUTCH SIZE
2 eggs

GAVIA ADAMSII

YELLOW-BILLED LOON

GAVIIFORMES

Yellow-billed Loons are the largest and heaviest of all the loons. Specialist fish-eaters that often dive to great depths (for which reason in Britain they are known as White-billed Divers), loons also consume small insects and crustaceans, especially when feeding their young. Parents lead their free-swimming hatchlings away from the nest but the chicks are still dependent on them for several weeks.

Clutch

Loons are perhaps best known for their wailing, dueting calls, heard throughout the long Arctic days, made by both sexes during the mating season. These duets are used both to communicate between the pair members, and to keep intruders away from the lakes on which that pair has set up its breeding territory. High-quality lakes, best suited for successful breeding, are relatively deep, maintain a stable water level, and are densely populated with fish. Loons nesting near or on the shores of these lakes produce the most young each year.

The egg of the Yellow-billed Loon is strongly oval, light purple-brown in background color, and dotted with darker blotches. It measures 3½ x 2⅛ in (89 x 55 mm) in size, and blends in well with the background soil and vegetation when exposed during the brief breaks taken by the parents between incubation bouts.

Actual size

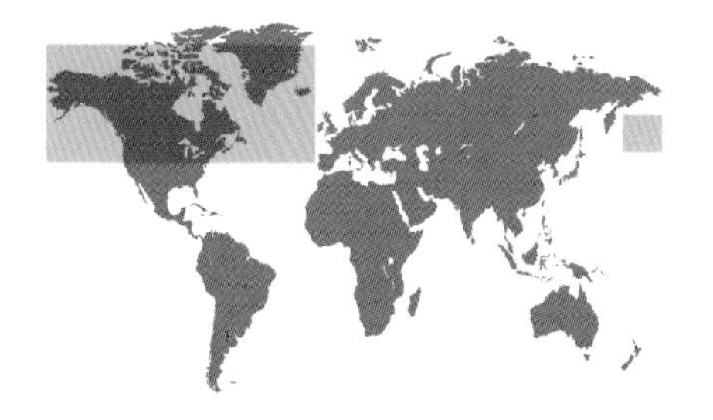

ORDER	Gaviiformes
FAMILY	Gaviidae
BREEDING RANGE	North America landmasses and islands of the North Atlantic
BREEDING HABITAT	On shores of lakes, preferring to nest on islands
NEST TYPE AND PLACEMENT	Bulky ground nest made up of twigs, grasses, reeds, and dried water plants
CONSERVATION STATUS	Least concern
COLLECTION NO.	FMNH 4074

ADULT BIRD SIZE
26–36 in (66–91 cm)

INCUBATION
28–30 days

CLUTCH SIZE
2 eggs

GAVIA IMMER

COMMON LOON

GAVIIFORMES

Clutch

Agile in the water, loons are awkward and slow-moving on land. The Common Loon, or Great Northern Loon or Diver, prefers to nest on islands in lakes to keep both itself and its eggs safe from ground predators. Nevertheless, many nests are destroyed when Arctic Foxes or raccoons swim out to the islands, or avian predators including gulls and jaegers (skuas) attack from the air.

The parents are highly territorial and keep others of their own species off the waters of their breeding lakes. Once the chicks hatch, they are mobile and able to swim within a day, but they rely on their parents to feed them small prey for several weeks before they develop into efficient fish hunters. The two chicks hatch 24 hours apart, and quickly establish an age- and size-based hierarchy. In years when the parents cannot bring enough food to their chicks, only the alpha young survives to independence.

Actual size

The egg of the Common Loon is elongated, green-brown in color, blotched throughout, and measures 3⅜ x 2⅛ in (87 x 55 mm) in size. The parents take turns to incubate the eggs, which hatch within 24 hours of each other, and the chicks leave together with the parents.

ORDER	Gaviiformes
FAMILY	Gaviidae
BREEDING RANGE	Eurasian Arctic, with small population in western Alaska
BREEDING HABITAT	Isolated deep freshwater lakes in tundra and taiga regions
NEST TYPE AND PLACEMENT	Ground nest of heaped plant matter, leaves, and sticks on lake shores
CONSERVATION STATUS	Least concern
COLLECTION NO.	FMNH 4085

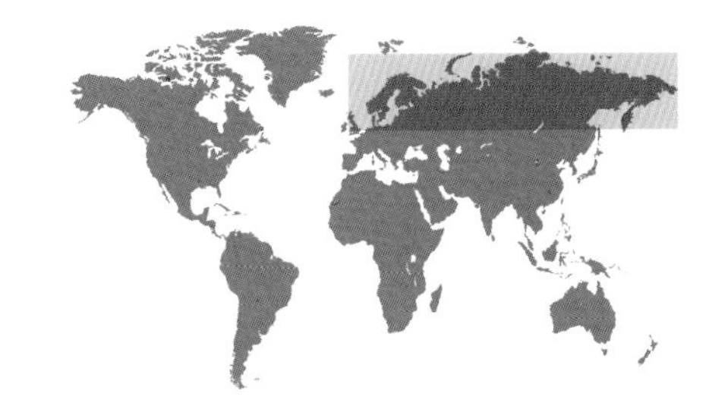

ADULT BIRD SIZE
23–29 in (58–74 cm)

INCUBATION
27–29 days

CLUTCH SIZE
2 eggs

GAVIA ARCTICA

ARCTIC LOON

GAVIIFORMES

Clutch

Originally described by Carl Linnaeus, the species was reclassified and split into two (together with the Pacific Loon, page 54) based on genetic and morphological traits. The 'new' Arctic Loon, otherwise known as the Black-throated Loon or Black-throated Diver, nonetheless still has an extensive distribution and many breeding populations. While some populations are decreasing, its conservation status remains Least Concern according to current international guidelines.

A strongly migratory species, it often overwinters in large groups, but it breeds as isolated pairs in the far north. The adults molt into their distinctive nuptial plumage just before departing the wintering grounds, and arrive in full color at the nesting sites. Both males and females defend their territory, which is always centered on a deep freshwater lake; the parents also join efforts to incubate the eggs and provision for weeks the mobile but still-needy young.

The egg of the Arctic Loon is brown-green in color, and speckled with darker blotches throughout. It is 3 x 1 ⅞ in (76 x 47 mm) in dimensions. Its camouflaging coloration helps to keep it safe from aerial predators including gulls and eagles.

Actual size

ORDER	Gaviiformes
FAMILY	Gaviidae
BREEDING RANGE	North America: Alaska and Canada, eastern Siberia
BREEDING HABITAT	On shores and islands of deep lakes in the tundra
NEST TYPE AND PLACEMENT	Ground nest of piled-up vegetation, within a few paces of the water
CONSERVATION STATUS	Least concern
COLLECTION NO.	FMNH 4090

ADULT BIRD SIZE
23–29 in (58–74 cm)

INCUBATION
23–25 days

CLUTCH SIZE
1–2 eggs

GAVIA PACIFICA

PACIFIC LOON

GAVIIFORMES

Clutch

Loons are extreme specialists and are well-adapted fish hunters in many aquatic habitats in the far northern hemisphere, including both lakes and oceans. Pacific Loons or Divers are known to forage cooperatively by swimming in groups under schools of fish and pushing them up toward the surface. The Pacific Loon is a strictly marine species, moving inland only for three brief summer months to lay eggs and raise its young.

Nesting close to the water is essential for breeding loons: because their feet are positioned so far back, they walk awkwardly and cannot take flight from land. Instead, they require open water to accelerate before taking to the air and flying off. Loon nests serve as home not only to the eggs but also to the young chicks; the parents often return to the nest with the chicks, after a full day on the lake.

Actual size

The egg of the Pacific Loon is light buff or green in color, maculated with variably sized brown speckles across the surface, and 3 x 1⅞ in (76 x 47 mm) in size. The eggs are laid a few days apart, but typically hatch within a day of one another.

ORDER	Gruiformes
FAMILY	Rallidae
BREEDING RANGE	South and Central America
BREEDING HABITAT	Moist lowland forests, swamps, and mangroves
NEST TYPE AND PLACEMENT	3–10 ft (1–3 m) above ground, on flat branches or in thickets, lined with twigs and leaves
CONSERVATION STATUS	Least concern
COLLECTION NO.	FMNH 2382

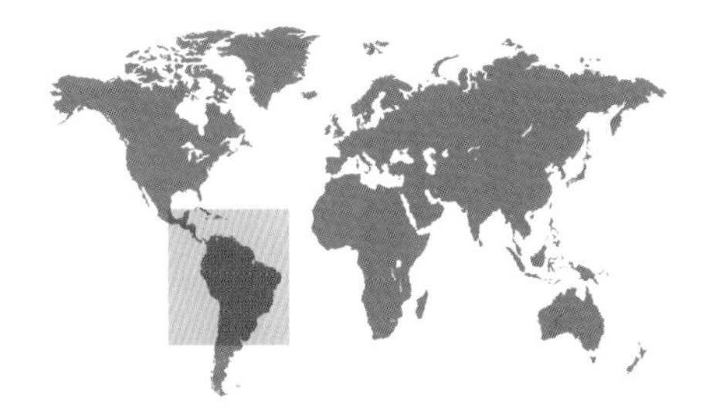

ADULT BIRD SIZE
15–16 in (37–41 cm)

INCUBATION
20 days

CLUTCH SIZE
3–7, typically 5 eggs

ARAMIDES CAJANEUS

GRAY-NECKED WOOD-RAIL

GRUIFORMES

These birds maintain long-term pair bonds, often inhabiting the same patch of forest or swamp for several years. Unlike other rails and coots, this species, true to its name, is arboreal. It can be seen resting on a tree branch to perch during daylight hours whereas it sets out to feed at night. When it comes to feeding, mated birds are selfish, keeping their partner at a distance with threat displays while they consume invertebrate or small vertebrate prey.

As devoted parents, individuals rely on cooperating with one another and take long (six- to eight-hour) turns at sitting on the nest and incubating the eggs. Once hatched, the chicks remain in the nest for a couple of days, brooded and protected by the parents, before the whole family departs following the parents' duetting calls.

Clutch

The egg of the Gray-necked Wood-Rail is oval in shape, whitish in background color, heavily blotched around the blunt pole with brown speckles, and is 2 ½ x 1⅜ in (52 x 36 mm) in size. To help the female to gather enough energy to lay the eggs, the male can be seen passing food items to the female just before starting to nest.

Actual size

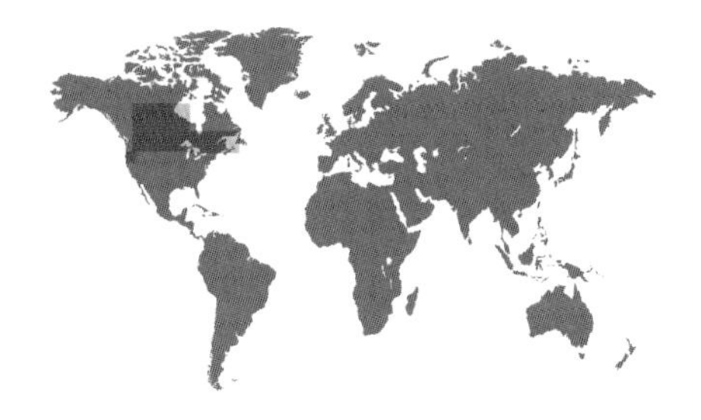

ORDER	Gruiformes
FAMILY	Rallidae
BREEDING RANGE	North America
BREEDING HABITAT	Wet meadows, sedge marshes
NEST TYPE AND PLACEMENT	Ground nest woven from grasses and leaves, under a canopy of dry vegetation
CONSERVATION STATUS	Least concern
COLLECTION NO.	FMNH 5621

ADULT BIRD SIZE
6–7 in (15–18 cm)

INCUBATION
16–18 days

CLUTCH SIZE
5–10 eggs

COTURNICOPS NOVEBORACENSIS

YELLOW RAIL

GRUIFORMES

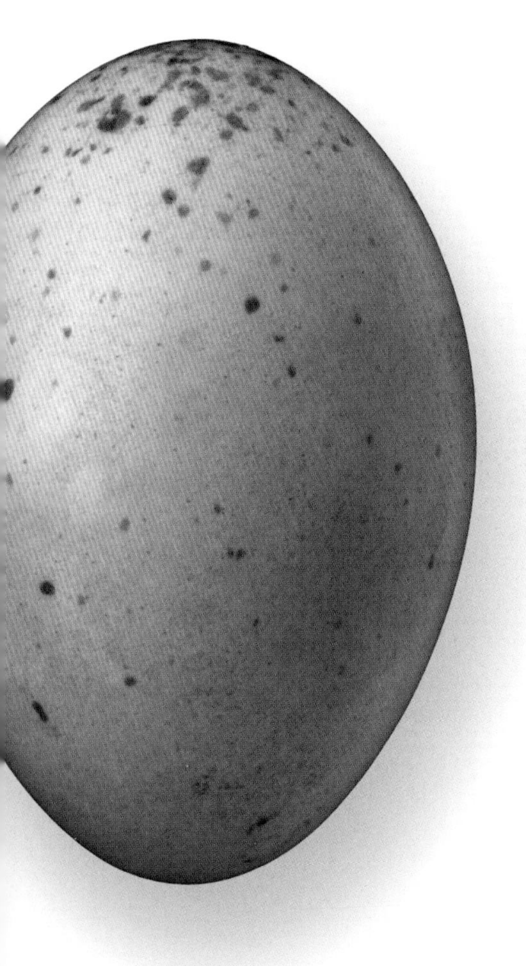

Clutch

A master of disguise, this small North American rail is mostly active and vocal at night. Rather than taking flight from predators, it relies on its camouflaged plumage to stay safe. It finds mates on its northern breeding grounds after spending the winter in southern swamps and flooded fields along the Atlantic and Gulf coasts of the United States.

The male does not contribute to parental duties, although he remains in a territory which overlaps the female's nesting site. In turn, the female keeps her nest hidden from predators by building it under a cover of vegetation; if the cover is disturbed, she replaces it by pulling more dry leaf matter over the nest. After the chicks hatch, some females crush the eggshells and push them to the bottom of the nest to hide the shell fragments from view; others remove the eggshells from the nest and deposit them along walkways leading away from the nest.

Actual size

The egg of the Yellow Rail is oval or elongate, creamy in background color and heavily speckled with larger reddish spots forming a ring near one end, and smaller black spots scattered over the rest of the shell. It measures 1⅛ x ⅞ in (29 x 21 mm) in size. If the first clutch is destroyed by predators, the female will lay a set of replacement eggs.

ORDER	Gruiformes
FAMILY	Rallidae
BREEDING RANGE	North and Central America, the Caribbean
BREEDING HABITAT	Swamps and marshes, reservoirs and lakes, city parks
NEST TYPE AND PLACEMENT	Floating nest of dried vegetation matter, anchored by reed or cattail stems.
CONSERVATION STATUS	Least concern
COLLECTION NO.	FMNH 5661

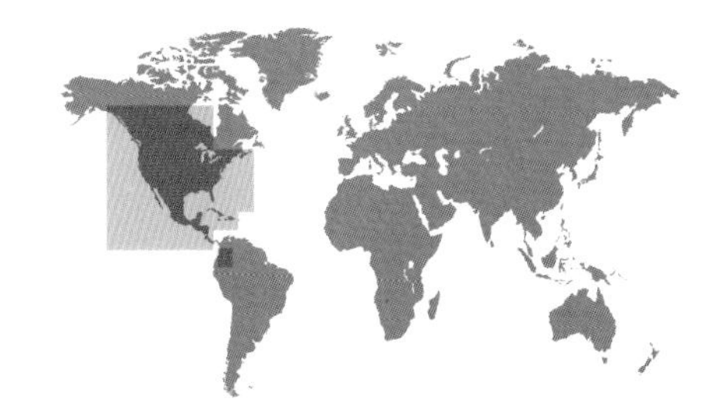

ADULT BIRD SIZE
16 in (40 cm)

INCUBATION
23–25 days

CLUTCH SIZE
8–12 eggs

FULICA AMERICANA

AMERICAN COOT

GRUIFORMES

Parenting for American Coots is an intensive affair: females lay many eggs and the chicks cannot feed themselves. Instead, they beg loudly and display the bright red skin patch on their head in return for small aquatic insects caught by their parents. Coot parents eventually discriminate against and even attack the less healthy and vigorous chicks, leaving only the two or three largest and strongest ones to survive.

Clutch

To avoid the manifold costs of parental care, some coots lay their eggs into other coots' nests, so emancipating themselves from the time and expenditure of energy required to be a parent. In response, incubating coots seek the eggs that look different from their own and push these to the edge of the nest, where they receive less heat and do not hatch as readily.

The egg of the American Coot is buff or gray in background color, and heavily speckled with large black blotches; it measures 2 x 1⅛ in (50 x 30 mm) in size. Individual females lay consistently patterned eggs, which allows them to recognize their own and to reject foreign eggs in the nest.

Actual size

ORDER	Gruiformes
FAMILY	Rallidae
BREEDING RANGE	Sporadic small populations across North America and the Caribbean, and the Pacific region of South America
BREEDING HABITAT	Shallow freshwater or salt marshes
NEST TYPE AND PLACEMENT	Ground nest, in dense swamp vegetation or flooded grass patches
CONSERVATION STATUS	Near threatened
COLLECTION NO.	FMNH 2993

ADULT BIRD SIZE
4–6 in (10–15 cm)

INCUBATION
16–20 days

CLUTCH SIZE
6–8 eggs

LATERALLUS JAMAICENSIS

BLACK RAIL

GRUIFORMES

Clutch

The smallest of North American rails, just the size of a sparrow, this species is both secretive and rare. Birds are active and vocal only at night, and their populations are declining, most likely due to habitat destruction. Relatively little is known about their migration patterns, breeding biology, or social behaviors. Territorial during the breeding season, some males may associate with two or more females for mating, a breeding system called "polygyny."

Black Rails build bowl-shaped nests of loosely woven vegetation, typically well hidden under a mat of dense swamp plants or a roof of woven leafy materials. The precocial, downy young can walk away from the nest and swim within a day of hatching.

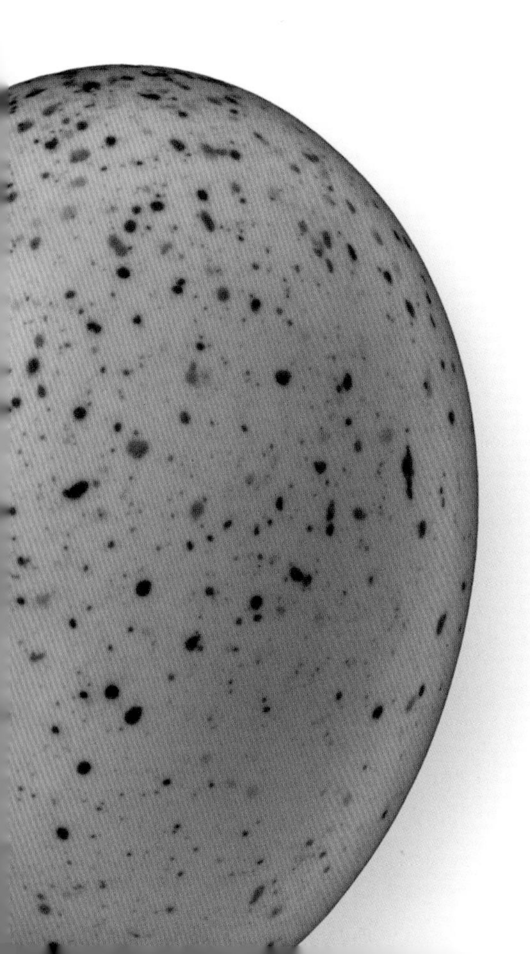

Actual size

The egg of the Black Rail is roundish, creamy white in color, and speckled with fine reddish-black spots across its surface. It measures ⅞ x ⅔ in (23 x 17 mm) in size. Both parents take turns to incubate the eggs, in shifts that last approximately an hour, but the intensity and exact duration of parental care still remain unknown for this poorly studied species.

ORDER	Gruiformes
FAMILY	Rallidae
BREEDING RANGE	Tropical and subtropical South America
BREEDING HABITAT	Grassy fields, swamps, abandoned agricultural areas
NEST TYPE AND PLACEMENT	Ground nest, under tufts and canopy of vegetation in grassy fields
CONSERVATION STATUS	Least concern
COLLECTION NO.	FMNH 3486

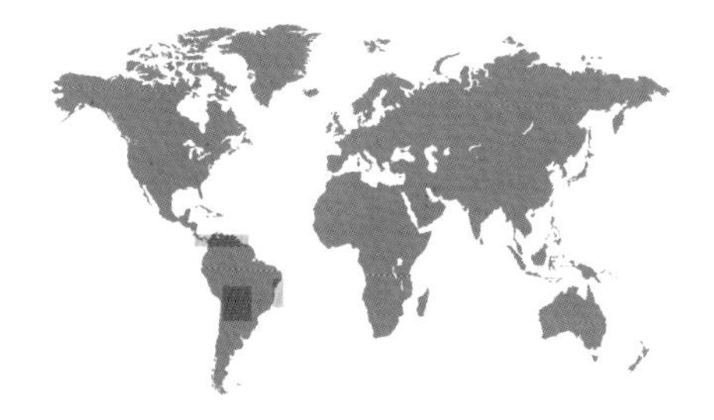

ADULT BIRD SIZE
7–8 in (18–20 cm)

INCUBATION
24 days

CLUTCH SIZE
3–7 eggs

NEOCREX ERYTHROPS

PAINT-BILLED CRAKE

GRUIFORMES

This small, handsomely feathered rail gets its name from its bright red beak. Yet Paint-billed Crakes are active mostly at night and hard to observe, so that records of their behavior and breeding are sparse. For instance, the first scientific observation of its young, and hence the first documented record of breeding activity in all of Central America, was only made in 1999 in Costa Rica.

Paint-billed Crakes are territorial, so playing recordings of male calls can bring the territory-owner male out of the dense vegetation where it usually hides. So little is known about this species that it is unclear what the male's, and his mate's, other roles are in breeding. During incubation, the attending parent sits tightly on the nest and relies on visual camouflage to avoid detection by predators.

Clutch

Actual size

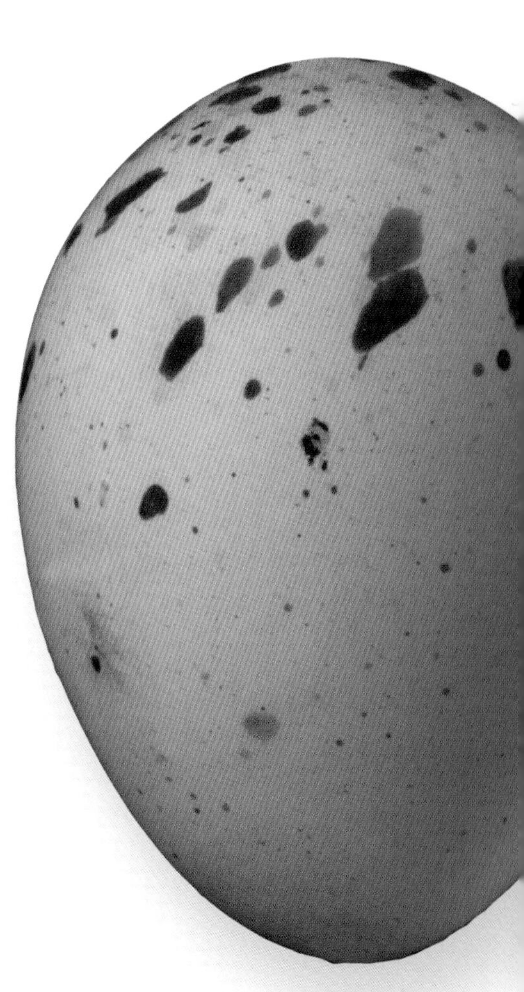

The egg of the Paint-billed Crake is creamy buff with large reddish speckles near the blunt end of the eggshell; it is small and measures 1⅛ x ⅞ in (28 x 21 mm) in dimension. The incubating parent is so dedicated to covering the eggs that agricultural workers often flush the bird only at the last moment when they are about to step on the nest.

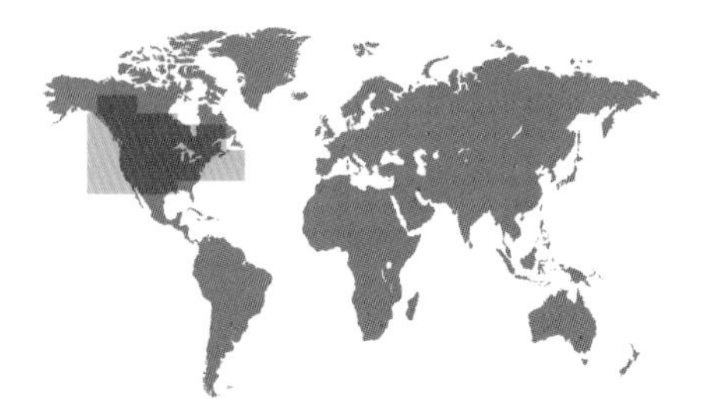

ORDER	Gruiformes
FAMILY	Rallidae
BREEDING RANGE	Temperate North America
BREEDING HABITAT	Marshes, swamps, and other wetlands
NEST TYPE AND PLACEMENT	Cup of reeds, grasses, and sedges, placed near water
CONSERVATION STATUS	Least concern
COLLECTION NO.	FMNH 5619

ADULT BIRD SIZE
8–10 in (20–25 cm)

INCUBATION
19 days

CLUTCH SIZE
8–13 eggs

PORZANA CAROLINA

SORA

GRUIFORMES

Clutch

Heard more often than seen, as it slips through dense marsh vegetation, this species is the most common and widespread rail in North America. In recent decades, Sora populations have declined due to loss of large and undisturbed wetland habitats.

The female Sora constructs the nest by pulling plants over it to form a protective canopy. Often there is a covered "walkway" of vegetation leading to the nest. Incubation begins early, usually by the time the third egg is laid. Precocial, the many downy young can walk and swim within a day of hatching. But because the eggs hatch asynchronously over a period of days, an instant size, age, and dominance hierarchy is generated among the hatchlings, which in turn influences survival to independence by favoring the older chicks.

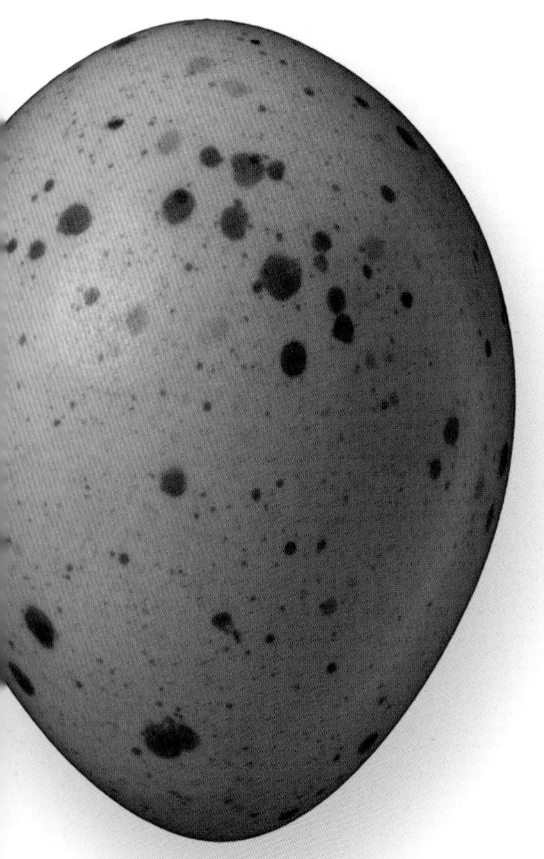

Actual size

The egg of the Sora is smooth and glossy, brownish to olive-buff colored, and lightly spotted with chestnut, measuring 1$\frac{1}{4}$ x $\frac{7}{8}$ in (31 x 22 mm). Eggs are laid on consecutive days, so the relatively large number of the clutch, compared to the small body size of this bird, is not complete for over a week.

ORDER	Gruiformes
FAMILY	Rallidae
BREEDING RANGE	Eastern North American coasts
BREEDING HABITAT	Freshwater and brackish marshes, rice paddies
NEST TYPE AND PLACEMENT	Ground nest on elevated platform, made from grasses and sedges, just above water level
CONSERVATION STATUS	Least concern
COLLECTION NO.	FMNH 5567

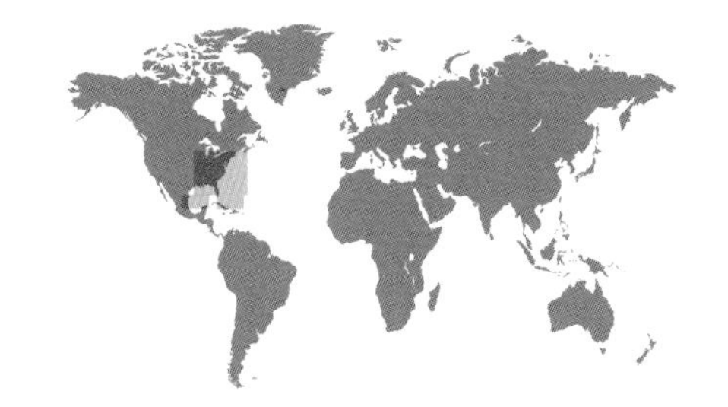

ADULT BIRD SIZE
15–19 in (38–48 cm)

INCUBATION
21–23 days

CLUTCH SIZE
6–14 eggs

RALLUS ELEGANS

KING RAIL

GRUIFORMES

This large North American rail is mostly active during the daytime, unlike many of its smaller, nocturnal relatives. It typically moves around shallow water, probing the mud for aquatic invertebrates. In the northern part of its ranges, habitat loss has severely impacted its populations, but in the southern states, freshwater and brackish swamps continue to support strong and viable populations.

Clutch

Both parents take turns incubating the eggs, and the chicks are fully feathered in down at hatching, ready to leave the nest. These young are still unable to feed themselves, however, and so they rely on both parents to hunt for small arthropod prey and feed these from beak to beak for up to six weeks after hatching.

Actual size

The egg of the King Rail is pale buff, with occasional brown spotting, and measures 1⅔ x 1⅛ in (41 x 30 mm) in size. The nest typically has a canopy of vegetation to hide the eggs from predators searching from above.

ORDER	Gruiformes
FAMILY	Rallidae
BREEDING RANGE	North America, from coast to coast
BREEDING HABITAT	Freshwater marshes
NEST TYPE AND PLACEMENT	Ground nest of matted-down vegetation and dry leaves, typically away from the water's edge
CONSERVATION STATUS	Least concern
COLLECTION NO.	FMNH 5610

ADULT BIRD SIZE
8½–10½ in (22–27 cm)

INCUBATION
20–22 days

CLUTCH SIZE
4–13 eggs

RALLUS LIMICOLA

VIRGINIA RAIL

GRUIFORMES

Clutch

This widespread rail is both secretive and resourceful in avoiding potential dangers. It hides from the view of predators by pushing its narrow body through dense marsh vegetation without peeking above it; its forehead feathers are so strong that they can withstand the wear-and-tear of plant leaves and stems rubbing continuously against them. This species also relies heavily on its feet to get away from danger, showing some of the highest leg-to-wing muscle ratios among all birds. When surprised in the water, it swims and even dives under the surface to propel itself to safety using its wings.

Parent birds show similar resourcefulness in their breeding strategies: the eggs are hidden by a canopy built over the nest; pairs also build dummy nests in their breeding territory to confuse potential predators. When these lines of defense are not enough, and the eggs or chicks are threatened directly, both adults vigorously attack the predator to keep the next generation safe.

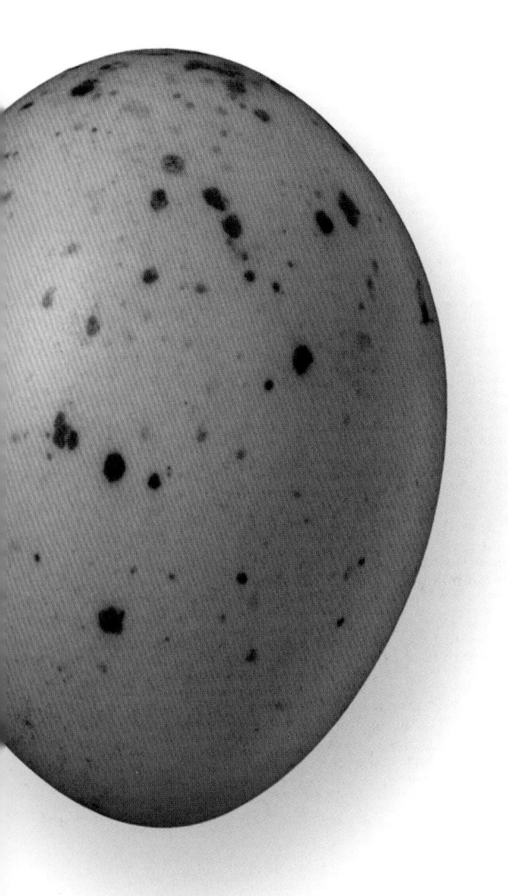

Actual size

The egg of the Virginia Rail is white or buff, with sparse gray or brown spotting, and 1⅓ x ⅞ in (32 x 24 mm) in dimensions. As the eggs are laid, both parents begin to take turns to incubate them; they also continue adding nesting material to further conceal and protect the nest.

ORDER	Gruiformes
FAMILY	Rallidae
BREEDING RANGE	Coastal North America, the Caribbean, and South America
BREEDING HABITAT	Brackish swamps, tidal marshes
NEST TYPE AND PLACEMENT	Ground nest of twigs and dead vegetation, placed on tussocks of grass and other live plants
CONSERVATION STATUS	Least concern
COLLECTION NO.	FMNH 5604

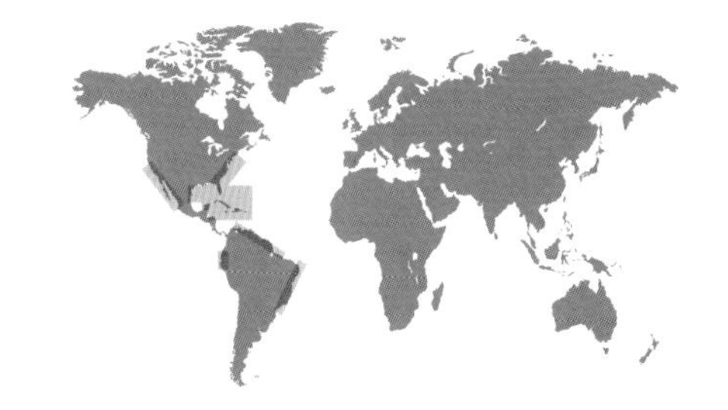

ADULT BIRD SIZE
13–16 in (32–41 cm)

INCUBATION
20–23 days

CLUTCH SIZE
7–10 eggs

RALLUS LONGIROSTRIS

CLAPPER RAIL

GRUIFORMES

Though geographically widespread and with a large total population, this coastal marsh specialist has, in some regions, become rare, and even endangered locally, due to loss of wetlands near heavily populated oceanic shores. Across its large distribution, populations can be migratory or sedentary (in the tropics). Some populations, including some of the smallest and most endangered, are isolated and are both genetically and morphologically distinctive, raising the issue of whether or not they should be considered separate species.

Clutch

Both male and female invest heavily in parental care, including incubating the eggs and feeding the young; chicks depend on the parents feeding them for many months. As a result, pairs can breed successfully only when both the mother and the father survive to take part in these critical breeding activities.

The egg of the Clapper Rail is creamy white in color, with irregular purple-brown speckling, and measures 1¾ x 1¼ in (44 x 31 mm) in size. The coloration camouflages the eggs during the brief periods when both parents are off the nest feeding before one of them returns to resume incubation.

Actual size

ORDER	Gruiformes
FAMILY	Gruidae
BREEDING RANGE	Central temperate Asia, from the Black Sea to Mongolia, and the Atlas Mountains in North Africa
BREEDING HABITAT	Grassy fields, typically near rivers or streams
NEST TYPE AND PLACEMENT	Ground nest, in a scrape within patches of taller grasses
CONSERVATION STATUS	Least concern
COLLECTION NO.	FMNH 21459

ADULT BIRD SIZE
33–39 in (85–100 cm)

INCUBATION
27–29 days

CLUTCH SIZE
1–3 eggs

ANTHROPOIDES VIRGO

DEMOISELLE CRANE

GRUIFORMES

Clutch

Though a widespread and common crane in most of its range, westernmost populations near the Black Sea and in the Atlas Mountains are declining and near extinction today. Habitat loss, through the conversion of grasslands into farmland, heavily impacts this crane, especially during the nesting period when it requires vegetation cover to keep the eggs safe from predators.

Demoiselles breed in central Asia and their migratory path takes them directly over the highest peaks of the Himalayan Mountains. The journey is extremely demanding, so arriving birds are exhausted and, despite standing up to 3 ft (90 cm) tall, vulnerable to predation by hawks and falcons. Once the adults arrive on the breeding ground, the female initiates courtship dances that include a vocal duet with the male. The resulting mate bond is strong, and both parents take turns to incubate the eggs and look after the young, but the male takes charge of protecting the nest and the chicks from intruders and predators.

Actual size

The egg of the Demoiselle Crane is rusty beige in color, with darker reddish brown blotches, elongated in shape, and 3 x 1¾ in (75 x 45 mm) in dimensions. The eggs are laid in a nest situated in grasses tall enough to conceal the incubating bird, but short enough that it can raise its neck to look out for predators.

ORDER	Gruiformes
FAMILY	Gruidae
BREEDING RANGE	Forested central regions of North America, today limited to two sites in Alberta, Canada and the state of Wisconsin
BREEDING HABITAT	Open swamps surrounded by forests
NEST TYPE AND PLACEMENT	Ground nest, placed in an elevated area surrounded by swamps
CONSERVATION STATUS	Endangered
COLLECTION NO.	FMNH 5542

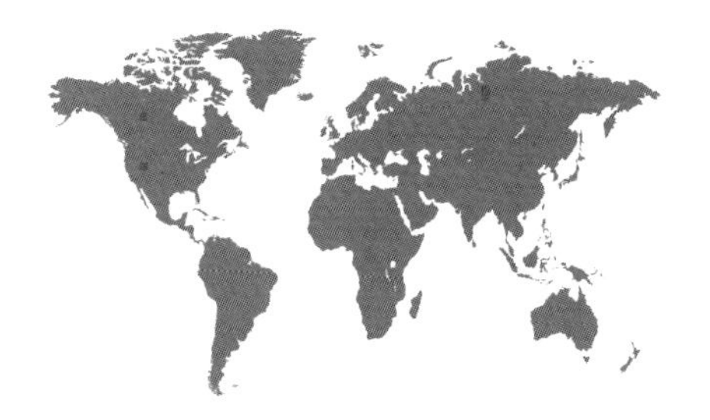

ADULT BIRD SIZE
52–60 in (132–150 cm)

INCUBATION
29–31 days

CLUTCH SIZE
2 eggs

GRUS AMERICANA

WHOOPING CRANE

GRUIFORMES

This is one of the rarest bird species in North America; a third of its global population occurs in captive breeding colonies. Wild populations are closely monitored and attempts have been made to start new populations by having Whooping Crane eggs raised by Sandhill Crane foster parents (see page 66).

Clutch

The breeding cycle lasts nearly a full year, from courtship displays to nest building through incubation and raising the young. Both parents attend and feed the chicks, who receive food for six to eight months after hatching. Accordingly, the family unit stays together not only on the breeding grounds, but also during the migratory journey and on the wintering grounds, with the parents serving as behavioral role models for the young.

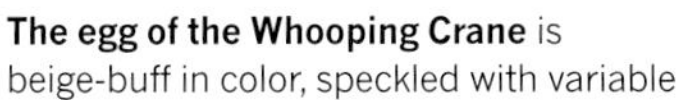

The egg of the Whooping Crane is beige-buff in color, speckled with variable dark large blotches, and measures 4 x 2⅓ in (100 x 60 mm) in size. The female invests more heavily in incubating the eggs and the male stands by protecting the nest site.

Actual size

ORDER	Gruiformes
FAMILY	Gruidae
BREEDING RANGE	North America, Cuba, and eastern Siberia
BREEDING HABITAT	Open marshes and bogs
NEST TYPE AND PLACEMENT	Ground nest of large mound of plant matter, floating or attached to standing vegetation
CONSERVATION STATUS	Least concern
COLLECTION NO.	FMNH 5548

ADULT BIRD SIZE
31–47 in (80–120 cm)

INCUBATION
28–30 days

CLUTCH SIZE
1–3 eggs

GRUS CANADENSIS

SANDHILL CRANE

GRUIFORMES

Clutch

Sandhill Cranes have a broad distribution and diverse migratory strategies: their Arctic and subarctic populations are strongly migratory to and from the northern breeding sites, whereas the southern populations are sedentary all year around. Members of the Idaho Sandhill population have served as step-parents for chicks of the endangered Whooping Crane (see page 65).

Bowing, leaping, and circling, while emitting duetting calls, Sandhill Crane pairs perform an elaborate and coordinated mating dance. The function of these intricate mating behaviors is perhaps to test each other's suitability for the prolonged and intensive duties required during the incubation and parental provisioning chores critical to raise the needy young to independence. For example, young cranes stay with their parents for 9–10 months after hatching, and even when capable of feeding on their own, they benefit from the parents' knowledge and experience to lead them to profitable feeding grounds and to protect them from attacks by competitors.

Actual size

The egg of the Sandhill Crane is pale brown, marked irregularly with darker brown spots. It measures 3⅔ x 2⅓ in (93 x 59 mm) in dimensions. Though the eggs are distinct in appearance, the parents do not recognize their own, and readily accept foreign eggs in the nest.

ORDER	Gruiformes
FAMILY	Aramidae
BREEDING RANGE	Temperate and tropical South and Central America, the Caribbean, Florida
BREEDING HABITAT	Open freshwater marshes, swamp forests, shores of rivers and ponds
NEST TYPE AND PLACEMENT	A platform of sticks and vines, anywhere from the ground to high on tree limbs
CONSERVATION STATUS	Least concern
COLLECTION NO.	FMNH 15956

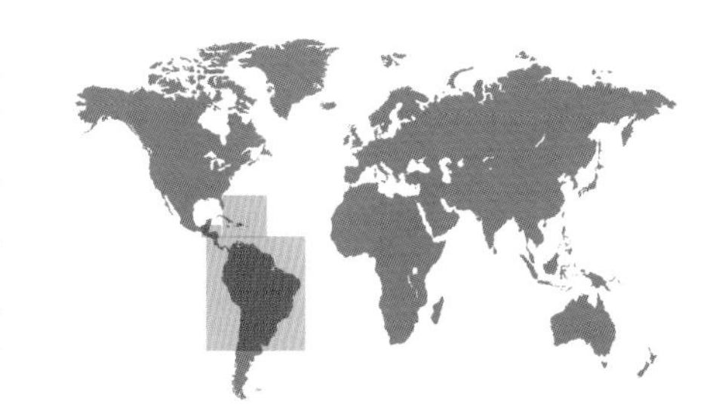

ARAMUS GUARAUNA

LIMPKIN

GRUIFORMES

ADULT BIRD SIZE
25–29 in (64–73 cm)

INCUBATION
26–28 days

CLUTCH SIZE
5–7 eggs

Although it resembles herons and ibises, this unique bird is more closely related to cranes. It feeds almost exclusively on large apple snails (*Pomacea* spp.), using its long curved beak to quickly (in 10–20 seconds) extract the flesh from the shell, without having to break it open. Once a common bird, habitat loss and modification has decimated some of its localized populations, including those in Southern Florida.

Clutch

A male provides nuptial gifts (usually extracted snail flesh) to a female visiting his display territory, in the hopes of convincing her to settle with him, instead of moving onto another male's display site. At the time of female visitations, the male's nest is already constructed and ready for the female to lay her eggs. The hatchlings are fully feathered, like crane chicks, and readily leave the nest on their own feet within a day of hatching , but they require provisioning from both parents for up to four months before independence.

The egg of the Limpkin is highly variable, from greenish to grayish in background color, with brownish or purplish streaks and blotches, and 2⅓ x 1¾ in (60 x 44 mm) in size. The parents take turns to incubate the eggs, but during the male's shift, he readily leaves the clutch to chase intruders away, so the female stays nearby to keep the eggs warm during territorial fights.

Actual size

ORDER	Podicipediformes
FAMILY	Podicipedidae
BREEDING RANGE	Western North America, central Mexico
BREEDING HABITAT	Freshwater marshes and swamps, with extensive open water
NEST TYPE AND PLACEMENT	Floating nest, anchored to emergent vegetation, built up of plant material in the shape of a mound
CONSERVATION STATUS	Least concern
COLLECTION NO.	FMNH 15452

ADULT BIRD SIZE
22–30 in (55–75 cm)

INCUBATION
24 days

CLUTCH SIZE
3–4 eggs

AECHMOPHORUS OCCIDENTALIS

WESTERN GREBE

PODICIPEDIFORMES

Clutch

Western Grebes are true aquatic specialists, with their streamlined body shaped to dive effectively under water in pursuit of fish. The feet of this and other species of grebes are positioned so far back on their torso that walking on land is slow and awkward. Most of the daily life of the Western Grebe thus takes place on or in the water, including its elaborate mating display where both sexes "rush" in unison, running side by side with each other across the surface of the water.

Nesting also takes place on the water, with the nest mound typically hidden in emergent vegetation, and anchored against floating away by plant stems. This species can form large nesting colonies, with hundreds to thousands of birds in close proximity sharing a single large lake.

Actual size

The egg of the Western Grebe is, initially, plain and pale bluish and free of spots, but it becomes stained rusty by the soiled plant material inside the nest and used to cover the eggs when the parents leaves the nest to feed. It measures 2¼ x 1½ in (58 x 39 mm) in dimensions.

ORDER	Podicipediformes
FAMILY	Podicipedidae
BREEDING RANGE	Northern Eurasia and northwestern North America
BREEDING HABITAT	Freshwater lakes and swamps with open water
NEST TYPE AND PLACEMENT	Open bowl of vegetation, floating or on rock emerging from the open water
CONSERVATION STATUS	Least concern
COLLECTION NO.	FMNH 1750

PODICEPS AURITUS

HORNED GREBE

PODICIPEDIFORMES

ADULT BIRD SIZE
12–15 in (31–38 cm)

INCUBATION
23–24 days

CLUTCH SIZE
3–8 eggs

A circumpolar species, this conspicuous grebe is familiar to birdwatchers in Europe, Asia, and North America. The "horns" after which the species is named are golden feather patches behind the ears, which are under muscular control so that they can be raised or lowered as long-distance visual displays and communication signals.

These grebes are dedicated parents, providing a "watertaxi" for their characteristically striped young chicks, which ride on the parents' backs. When the adult dives after its fish prey, the chicks can remain on its back, or they can jump off and swim on the surface until the parent emerges again.

Clutch

Actual size

The egg of the Horned Grebe is white to brownish or bluish green, clear of spotting, and measures 2¼ x 1½ in (58 x 39 mm) in size. The striped hatchlings are ready to leave the nest and dive behind the parents within a day, but typically remain on the nest mound for days, and become independent only two months later.

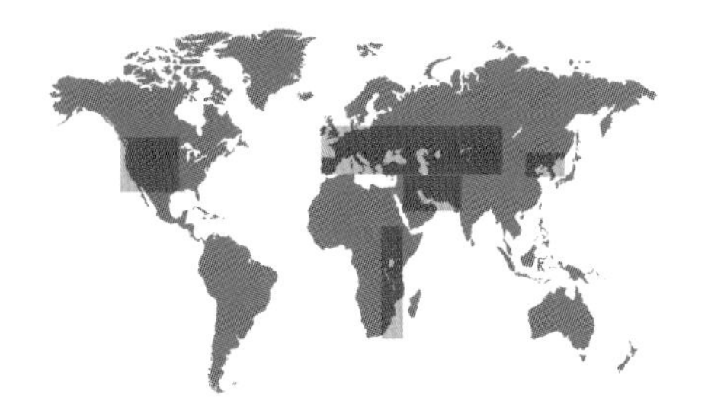

ORDER	Podicipediformes
FAMILY	Podicipedidae
BREEDING RANGE	Every continent, except Australia, South America, and Antarctica
BREEDING HABITAT	Swamps and lakes with dense vegetation
NEST TYPE AND PLACEMENT	A mound of plant matter on emergent plants in shallow water or directly in the lake shore vegetation
CONSERVATION STATUS	Least concern
COLLECTION NO.	FMNH 15247

ADULT BIRD SIZE
12–14 in (30–35 cm)

INCUBATION
21 days

CLUTCH SIZE
3–4 eggs

PODICEPS NIGRICOLLIS

EARED GREBE

PODICIPEDIFORMES

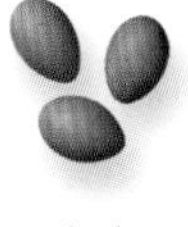

Clutch

Also known as the Black-necked Grebe, this species is socially monogamous, with both the female and male actively involved in building the nest, incubating the eggs, and looking after the young. Nesting in loose, but occasionally large, colonies of several hundred individuals, the family quickly leaves the nest after hatching. The chicks can swim and dive, but cannot fish at hatching, and so the parents provision their young by feeding them small fish from beak to beak.

At about ten days of age the parents may split the brood, and lead each half separately for the rest of the three-week period before the chicks finally fledge. Such "brood divisioning" occurs in many birds with dependent young, and is thought to reduce the risk of losing all the young due to predation of one or both of the parents or all the young if they remain together.

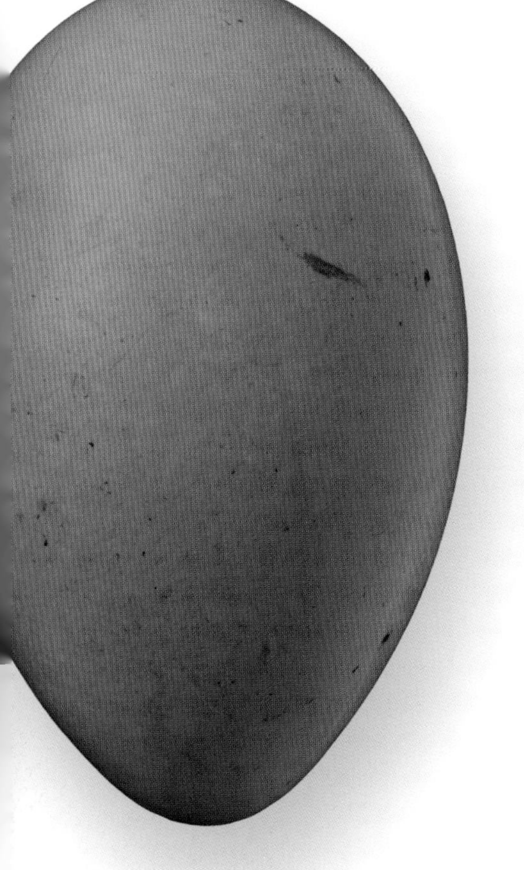

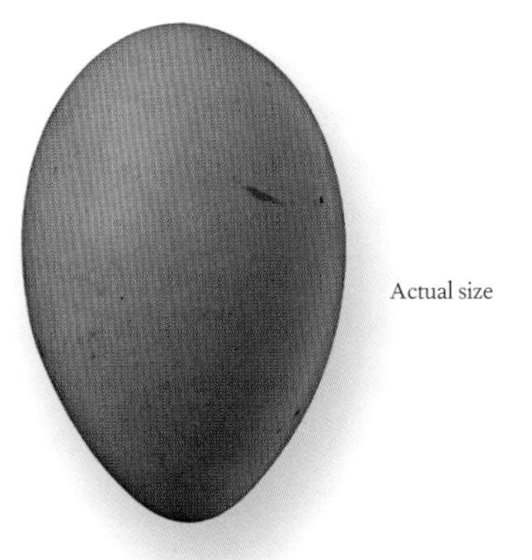

Actual size

The egg of the Eared Grebe is chalky bluish or greenish, immaculate, but quickly becomes stained during the incubation period by the plant material inside the nest cup. It measures 1¾ x 1⅛ in (45 x 30 mm) in size.

ORDER	Podicipediformes
FAMILY	Podicipedidae
BREEDING RANGE	Southern North America, the Caribbean, and Central and South America
BREEDING HABITAT	Temporary and permanent wetlands, ponds, and slow-moving rivers
NEST TYPE AND PLACEMENT	A pile of wet, decaying plant matter, anchored to emergent vegetation, floating in shallow water
CONSERVATION STATUS	Least concern
COLLECTION NO.	FMNH 4047

ADULT BIRD SIZE
9–11 in (22–27 cm)

INCUBATION
21–22 days

CLUTCH SIZE
3–7 eggs

TACHYBAPTUS DOMINICUS

LEAST GREBE

PODICIPEDIFORMES

This is the smallest and least well-understood grebe species in the Americas. It can be secretive and readily avoids detection or observation by quickly diving under water and, curiously, remaining there with just its beak breaking the surface to allow breathing. This small grebe is a comfortable hunter, both below and above the water surface, diving or pecking at immersed vegetation and small insect prey.

Clutch

Unlike many other grebes, Least Grebes remain in pairs during the breeding season, nesting at considerable distances from their nearest neighbor. Both parents share the responsibility of building the nest and incubating the eggs, as well as looking after the young. Upon hatching, the parents and chicks quickly vacate the nest and rely on swimming and diving in the open water to keep safe from predators.

Actual size

The egg of the Least Grebe is whitish to pale blue or green in color, immaculate, and measures 1⅓ x ⅞ in (34 x 23 mm) in size. Within 20 minutes of hatching the chicks are able to climb on the back of the parent and cling onto it during dives for up to 40 minutes.

ORDER	Phoenicopteriformes
FAMILY	Phoenicopteridae
BREEDING RANGE	Islands and continental shores of the Caribbean, Galapagos
BREEDING HABITAT	Mudflats, lagoons, coastal lakes
NEST TYPE AND PLACEMENT	Ground nest a crater of built-up mud, standing in shallow water
CONSERVATION STATUS	Least concern
COLLECTION NO.	FMNH 2170

ADULT BIRD SIZE
47–57 in (120–145 cm)

INCUBATION
28–32 days

CLUTCH SIZE
1 egg

PHOENICOPTERUS RUBER

AMERICAN FLAMINGO

PHOENICOPTERIFORMES

The only flamingo native to North America, this charismatic and conspicuous species relies on the safety of large flocks and hard-to-access breeding sites, rather than weaponry or aggression, to defend its nest. Crowded colonial life is so essential for the American Flamingo that zookeepers have been able to induce small groups of captive flamingos to nest and breed successfully by playing back sounds of large numbers of birds recorded at large colonies.

The success of the American Flamingo's strategy of feeding and breeding on remote mudflats and lagoons has allowed them to maintain population sizes despite laying just a single egg per year. Some individuals can survive to more than 40 years of age, assuring that at least a handful of those breeding attempts will yield a viable chick to recruit for the next generation.

Actual size

The egg of the American Flamingo is chalky white in color, immaculate, elongated in shape, and 3⅓ x 2 in (85 x 53 mm) in size. Occasionally, two eggs are laid in the same nest, apparently by the same female, and both parents take on the duties of incubation and provisioning the young.

ORDER	Anseriformes
FAMILY	Anhimidae
BREEDING RANGE	Tropical northern South America
BREEDING HABITAT	Marshes and swamps with dense vegetation
NEST TYPE AND PLACEMENT	A floating platform of swamp plants
CONSERVATION STATUS	Least concern
COLLECTION NO.	FMNH 2866

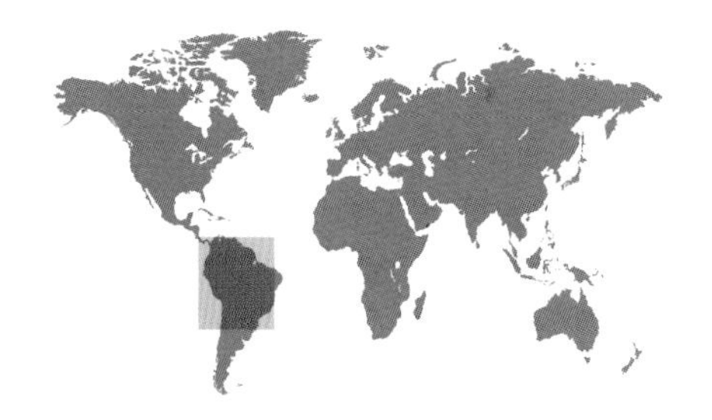

ADULT BIRD SIZE
33–37 in (84–95 cm)

INCUBATION
42–47 days

CLUTCH SIZE
2–7 eggs

ANHIMA CORNUTA

HORNED SCREAMER

ANSERIFORMES

Screamers are related to ducks and geese, but their heads and beaks, as well as their habits, are chickenlike. Adult Horned Screamers are ornamented with a pair of spurs on their legs and a long, thin, spiny horn on the top of the head that grows continually throughout life.

Unlike ducks and geese, screamers have partially webbed feet and feel as comfortable on water as in grassland and perching on tree branches. Both the male and female construct the nest, with the female incubating the eggs during the daytime, and the male typically warming the eggs at night. The hatchlings, like those of related species, leave the nest on their own feet within a day of hatching.

Clutch

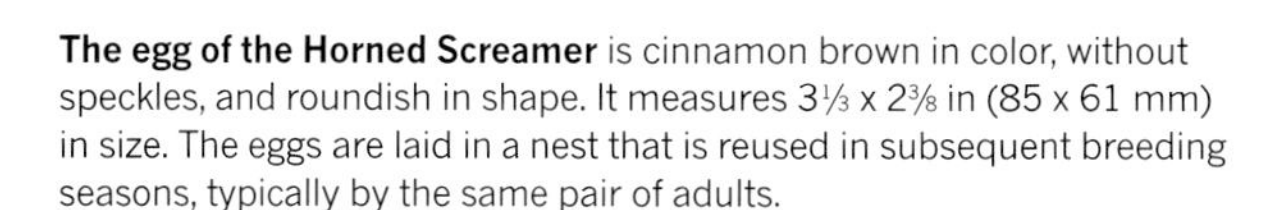

The egg of the Horned Screamer is cinnamon brown in color, without speckles, and roundish in shape. It measures 3⅓ x 2⅜ in (85 x 61 mm) in size. The eggs are laid in a nest that is reused in subsequent breeding seasons, typically by the same pair of adults.

Actual size

ORDER	Anseriformes
FAMILY	Anhimidae
BREEDING RANGE	Northwestern South America
BREEDING HABITAT	Lowland marshes, swamps, and banks of slow-moving rivers
NEST TYPE AND PLACEMENT	The bulky ground nest is made of sticks and marsh vegetation, sited near or in shallow water
CONSERVATION STATUS	Near threatened
COLLECTION NO.	FMNH 2374

ADULT BIRD SIZE
30–36 in (76–91 cm)

INCUBATION
40–47 days

CLUTCH SIZE
2–7 eggs

CHAUNA CHAVARIA

NORTHERN SCREAMER

ANSERIFORMES

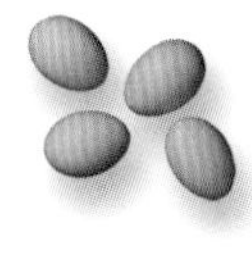

Clutch

This large, boldly patterned, gooselike species has the distinction of being one of the loudest bird species in the world. Unlike most other duck relatives, Northern Screamers readily roost on dead branches at the top of tall trees, and they call loudly from these prominent perching sites. Duets between females and males establish and reinforce the pair bond, which can last from across consecutive breeding seasons to a full lifetime of social monogamy. The parents fully collaborate in all aspects of parental care, including protecting the young by attacking birds of prey, snakes, and felids, and chasing them out of the breeding territory.

Already limited in its geographic distribution to coastal plains of Venezuela and Colombia, and perhaps because of its status as a unique species, its numbers have steadily declined to below 10,000 individuals. This is largely due to habitat destruction, modification, and pollution associated with oil pipelines in lowland swamplands, and also because of illegal hunting.

Actual size

The egg of the Northern Screamer is creamy white in color with pale spotting, and measures 3 x 2⅛ in (77 x 54 mm) in size. The eggs are laid at two-day intervals, and the chicks hatch fully downed and ready to leave the nest.

ORDER	Anseriformes
FAMILY	Anatidae
BREEDING RANGE	Arctic and subarctic tundra in Eurasia and North America
BREEDING HABITAT	Pools, ponds, and lakes across coastal plains
NEST TYPE AND PLACEMENT	The bulky nest is placed on the ground on a mound or ridge, built up from grasses and leaves, with little down lining
CONSERVATION STATUS	Least concern
COLLECTION NO.	FMNH 21118

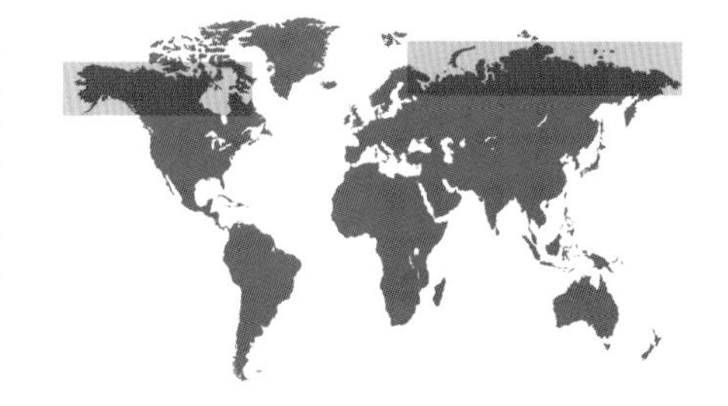

ADULT BIRD SIZE
47–59 in (120–150 cm)

INCUBATION
30–32 days

CLUTCH SIZE
3–5 eggs

CYGNUS COLUMBIANUS

TUNDRA SWAN

ANSERIFORMES

This is a large and common swan species; taxonomists currently consider that two races of Tundra Swan—the European Bewick's Swan and the American Whistling Swan—are separate from each other only at the subspecies level. Generally flocking during the winter and spending the nights floating on open water, in the breeding season these swans pair off, defend breeding ranges against other swans and most other species happening to approach the nest, and sleep on solid ground near their nest.

Clutch

Like many geese, but few duck species, both the female and the male are devoted to the prolonged duties and many costs of parental care; they only leave the nest when predators large enough to prey on the adults approach. These include wolves, bears, and of course, people.

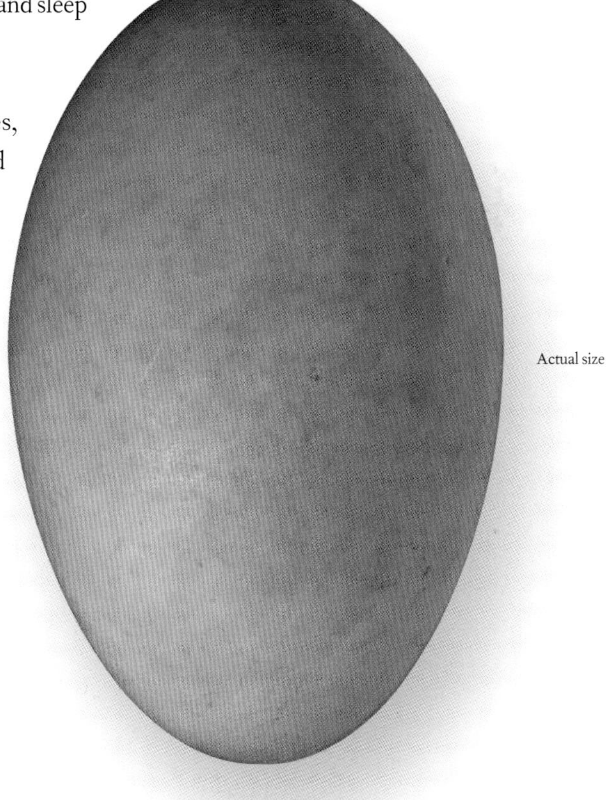

Actual size

The egg of the Tundra Swan is creamy to rusty white in color, free of prominent blotching, and measures 4⅛ x 2⅔ in (106 x 68 mm). The few bright eggs in the large bulky nests are vulnerable to predation, but when the parents are nearby, they defend the nest with vigor against predatory birds and small mammals.

ORDER	Anseriformes
FAMILY	Anatidae
BREEDING RANGE	Temperate North America, western Caribbean
BREEDING HABITAT	Marshes, swamps, and freshwater ponds
NEST TYPE AND PLACEMENT	Large cavities in tree trunks; readily uses nestboxes
CONSERVATION STATUS	Least concern
COLLECTION NO.	FMNH 5188

ADULT BIRD SIZE
17–20 in (43–51 cm)

INCUBATION
28–37 days

CLUTCH SIZE
6–16 eggs

AIX SPONSA

WOOD DUCK

ANSERIFORMES

Clutch

The gaudiest of North American ducks, Wood Ducks breed in coastal and inland ponds surrounded by old tree stands. In the absence of large tree cavities, Wood Ducks will readily use artificial nestboxes, and preferred nesting sites may be aggressively contested by several females. Eventually, one of these females wins and begins incubating her own eggs along with those of some of the others.

Hatchling Wood Ducks are protected by their mother, but not fed, and so they must depart for the nearest pond together by jumping out of the nest entrance hole and bouncing on the forest floor, cushioned by their down-covered bodies. Because several females may lay their eggs in the same nest, ducklings are not all full siblings, but they most likely benefit from avoiding predatory attacks by being part of a large cohort.

Actual size

The egg of the Wood Duck is glossy, creamy white to tan, 2⅛ x 1⅓ in (54 x 34 mm) in size, but may become discolored during incubation. Young females often return to their natal nestbox where their mother hatched them, and may lay eggs in the same cavity until one of them takes ownership and begins incubation of the extra-large clutch.

ORDER	Anseriformes
FAMILY	Anatidae
BREEDING RANGE	North America and Eurasia
BREEDING HABITAT	Lakes, ponds, swamps, and seasonal wetlands
NEST TYPE AND PLACEMENT	Ground nest of a scrape, lined with grass and feathers, in brush or tall grass, typically not near water
CONSERVATION STATUS	Least concern
COLLECTION NO.	FMNH 5182

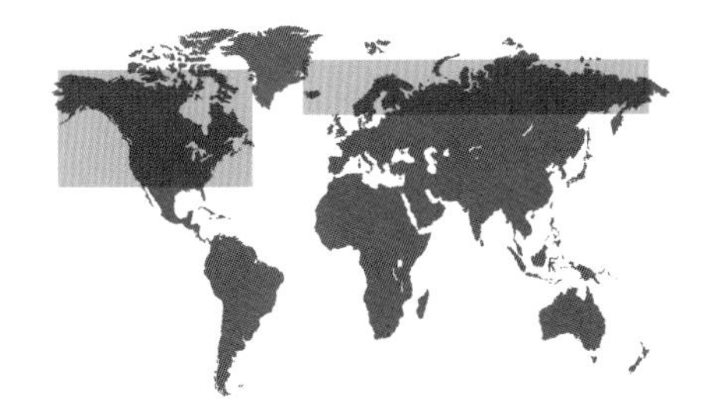

ANAS ACUTA

NORTHERN PINTAIL

ANSERIFORMES

ADULT BIRD SIZE
20–30 in (51–75 cm)

INCUBATION
22–25 days

CLUTCH SIZE
3–12 eggs

Like most birds, the plumage of male and female Northern Pintails differs markedly. Males have distinctive colors and a long tail, while females display camouflaged marbled brown tones. It is thus not surprising that the duties of incubation, during the long northern daylight hours, are carried out exclusively by the female. She incubates the eggs for over three weeks in a hidden nest on the ground.

Female pintails incubate their eggs for over 80 percent of the day, during which they have just three short breaks (called recesses) to leave the nest, drink, forage briefly, and preen their feathers. The more hidden the nest, the more recesses a female takes, but as the eggs reach the hatching stage she takes fewer of them.

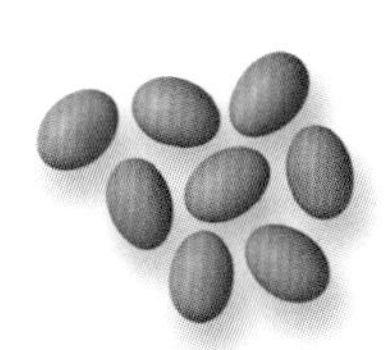

Clutch

The egg of the Northern Pintail is greenish buff to beige in color, clear of speckling, and measures 2 x 1½ in (53 x 38 mm) in size. This species is the earliest nesting duck in North America and the female begins to build the nest and lay the eggs shortly after the break-up of snow and ice cover on the breeding grounds.

Actual size

ORDER	Anseriformes
FAMILY	Anatidae
BREEDING RANGE	North America
BREEDING HABITAT	Shallow freshwater ponds, swamps, and wetlands
NEST TYPE AND PLACEMENT	A ground nest in the form of a shallow depression lined with grasses and down, in dense and brushy vegetation, often far from water
CONSERVATION STATUS	Least concern
COLLECTION NO.	FMNH 20005

ADULT BIRD SIZE
17–23 in (42–59 cm)

INCUBATION
22–25 days

CLUTCH SIZE
3–13 eggs

ANAS AMERICANA

AMERICAN WIGEON

ANSERIFORMES

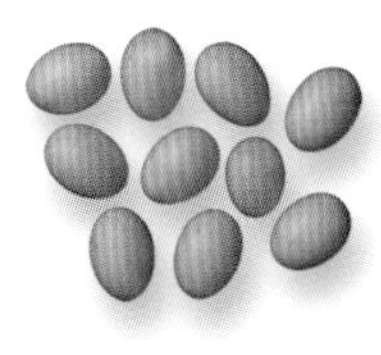

Clutch

The egg of the American Wigeon is creamy white in color, free of speckling, and measures 2 x 1½ in (53 x 37 mm) in size. The young emerge fully covered in down feathers that come to resemble the cryptic brownish plumage of the female at the time of fledging, about six weeks after hatching.

On its way to breed this duck is a common migratory species across North America from the Midwestern prairie pothole region to the Alaskan and Canadian Arctic. Together with the Mallard (see page 83) and the Wood Duck (see page 76), it is one of the most commonly harvested waterfowl. White bills and foreheads give the males a distinctive look and the nickname "baldpate."

As a type of "dabbling duck," wigeons seek food from the surface of open waters, often in mixed flocks with coots and loons, and occasionally may appear to be "snatching" some plant matter accidentally brought to the surface by those diving species. Wigeons also visit, and lead their young to, grassy areas to consume green leafy plants and, later in the year, fallen grains from agricultural areas.

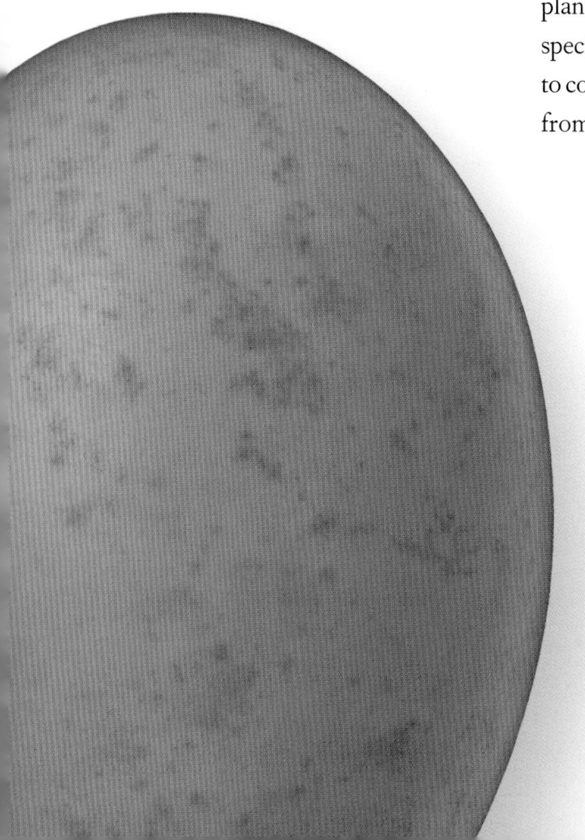

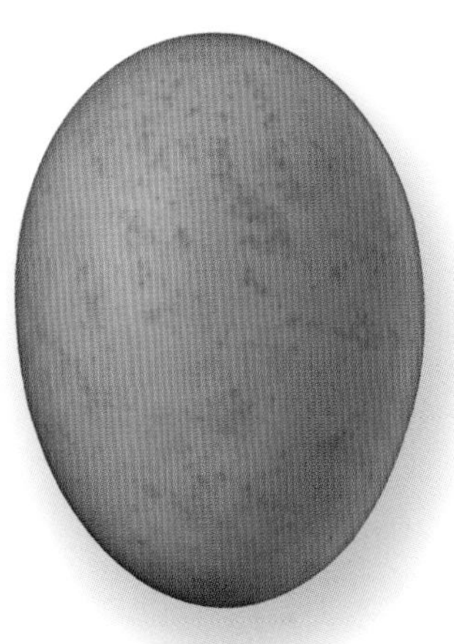

Actual size

ORDER	Anseriformes
FAMILY	Anatidae
BREEDING RANGE	Arctic and temperate North America and Eurasia
BREEDING HABITAT	Shallow grassy marshes, with thick layers of bottom mud
NEST TYPE AND PLACEMENT	Simple ground scrape, surrounded by vegetation on at least three sides, placed near open water
CONSERVATION STATUS	Least concern
COLLECTION NO.	FMNH 15543

ADULT BIRD SIZE
17–20 in (44–51 cm)

INCUBATION
24 days

CLUTCH SIZE
8–12 eggs

ANAS CLYPEATA

NORTHERN SHOVELER

ANSERIFORMES

This conspicuous and unique duck makes the most of its feeding specialization: a broad, spoon-shaped bill best suited to sifting through vast amounts of swamp water to filter and trap small aquatic invertebrates. Though laborious to collect, these foods provide more than sufficient protein supplies for the female to invest energy in forming and laying her eggs.

Clutch

The female relies on her own nutrient reserves to provision the eggs with sufficient fats to fuel embryonic development; with each gram of lipid deposited into an egg, a female loses 1/40 oz (0.75 g) of fat reserves stored during the wintering period. Because the males provide no energetic or material investment in the egg, their fat reserves remain stable during the breeding period. The female also bears the rest of the costs of parental care, both incubating the eggs and protecting the ducklings on her own.

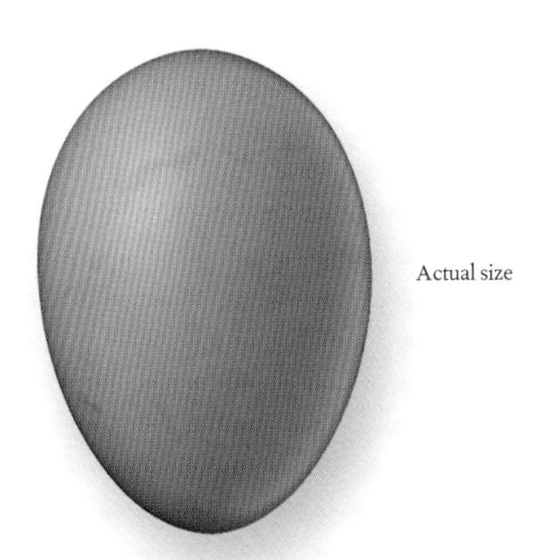

Actual size

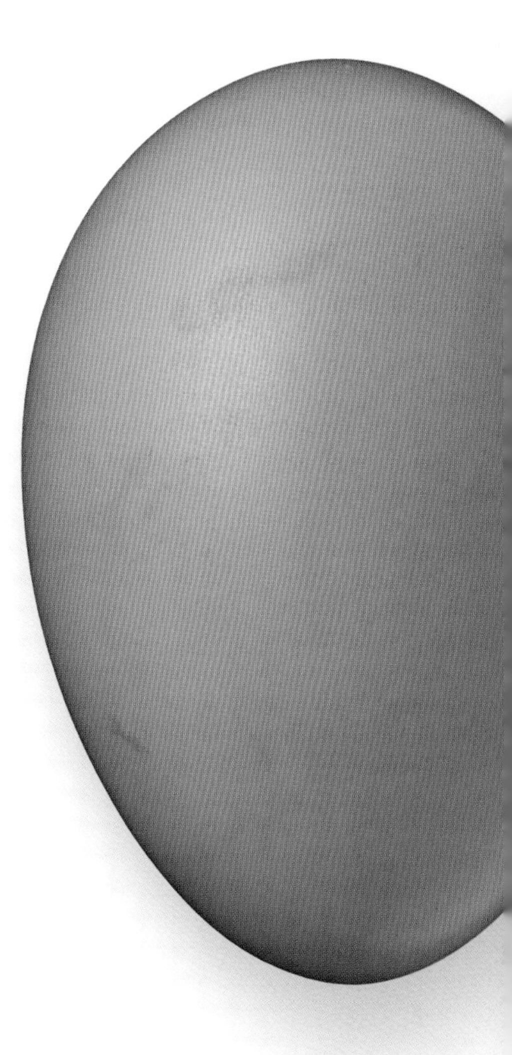

The egg of the Northern Shoveler is pale greenish gray to olive-beige in color, immaculate, and measures 1¾ x 1¼ in (45 x 33 mm) in size. When the female is flushed from the nest, she may defecate on the eggs, perhaps to mask their color and smell from sight- and scent-driven predators.

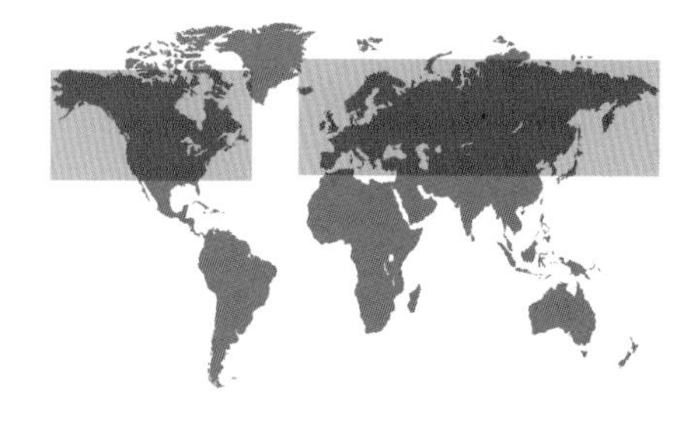

ORDER	Anseriformes
FAMILY	Anatidae
BREEDING RANGE	Arctic and temperate North America and Eurasia
BREEDING HABITAT	Shallow freshwater ponds, tidal creeks, and mudflats, with dense emergent vegetation
NEST TYPE AND PLACEMENT	Ground nest in the form of a shallow depression lined with grasses, feathers, and down
CONSERVATION STATUS	Least concern
COLLECTION NO.	FMNH 5162

ADULT BIRD SIZE
12–15 in (31–39 cm)

INCUBATION
21–23 days

CLUTCH SIZE
5–16 eggs

ANAS CRECCA

GREEN-WINGED TEAL

ANSERIFORMES

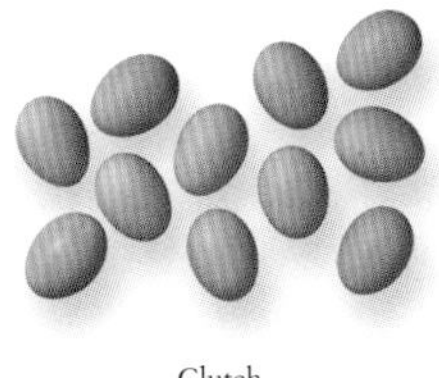

Clutch

Green-winged Teals, also known as Eurasian Teals, arrive earlier on their northern breeding grounds than most waterfowl; they land soon after the melting of heavy snow cover reveals permanent and seasonal wetlands. Despite the dangers of arriving to breed before the last of the winter storms, and the costly migratory journeys to and from the breeding grounds, the longevity record for this duck species is more than 20 years based on returns from official banding records.

The female and the male both become sexually mature in the winter after their birth, and establish a consortship, which may look like a traditional pair bond. However the males leave the females shortly after the onset of incubation, and she alone proceeds to provide parental care.

The egg of the Green-winged Teal is buff or white in color, immaculate, and measures 1¾ x 1⅜ in (46 x 36 mm) in size. The hatchlings are fully covered in down and fledge at five to six weeks of age—the fastest growth rate among all North American ducks.

Actual size

ORDER	Anseriformes
FAMILY	Anatidae
BREEDING RANGE	Western North, Central, and South America
BREEDING HABITAT	Seasonal and permanent wetlands, coastal swamps
NEST TYPE AND PLACEMENT	Ground depression, near water, lined with grasses and down feathers
CONSERVATION STATUS	Least concern
COLLECTION NO.	FMNH 15541

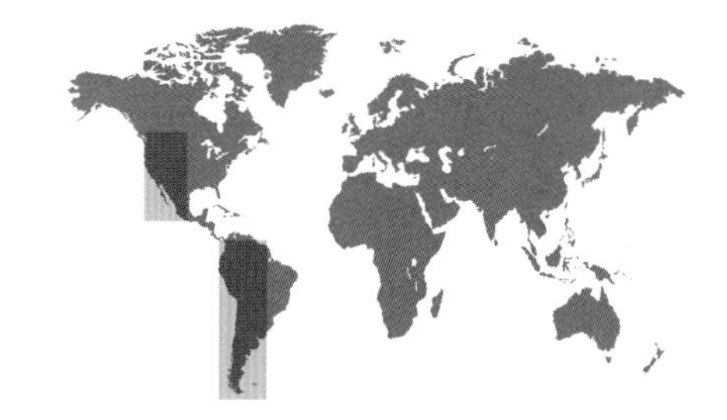

ADULT BIRD SIZE
14–17 in (36–43 cm)

INCUBATION
23 days

CLUTCH SIZE
8–10 eggs

ANAS CYANOPTERA

CINNAMON TEAL

ANSERIFORMES

This handsome duck is unlike most other North American species: it rarely breeds in the Midwestern prairie regions of the United States and Canada, and is more of a West Coast specialist. Interestingly, it migrates south from the northern hemisphere, but also maintains a large breeding population in South America.

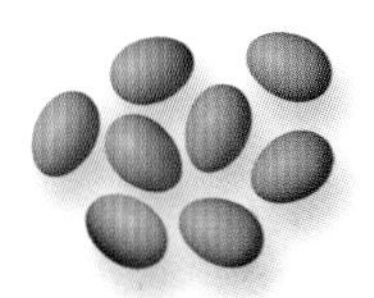

Clutch

The female builds her nest near the water, and hides it well under a mat of dried grasses and other vegetation. She approaches the nest furtively, through tunnels which she creates by pushing her body through the dense vegetation around the nest. Unlike most other ducks, the male and the female maintain their association through to the end of the incubation period.

The egg of the Cinnamon Teal is creamy white or beige, immaculate, and measures 1⅞ x 1⅓ in (47 x 34 mm) in dimensions. The hatchlings are fully downed in yellow colors with a prominent eye stripe. They leave the nest, following their mother, soon after hatching.

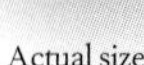

Actual size

ORDER	Anseriformes
FAMILY	Anatidae
BREEDING RANGE	Temperate and subarctic North America
BREEDING HABITAT	Vegetation near marshes, ponds, and other wetlands
NEST TYPE AND PLACEMENT	Ground scrape, lined with dried grasses, above the water's edge, covered by vegetation
CONSERVATION STATUS	Least concern
COLLECTION NO.	FMNH 2989

ADULT BIRD SIZE
14–16 in (36–41 cm)

INCUBATION
19–29 days

CLUTCH SIZE
6–14 eggs

ANAS DISCORS

BLUE-WINGED TEAL

ANSERIFORMES

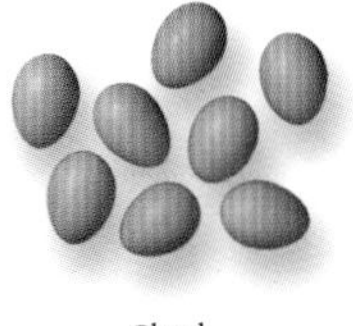

Clutch

The Blue-winged Teal is the second most common duck species in North America, surpassed only by the Mallard (see facing page). It is a long-distance migrant: females choose their mates in southern wintering grounds, and return to breed near the site where they themselves had hatched. Therefore, genetic diversity in this duck is maintained by the males' dispersal from natal to breeding sites.

Although female and male teals are often seen together, there is no traditional pair bond; once the eggs are fertilized and laid, the male leaves, while the female carries out all incubation duties. When she has to leave the nest temporarily, she covers it with thick vegetation to hide her bright white eggs from predators.

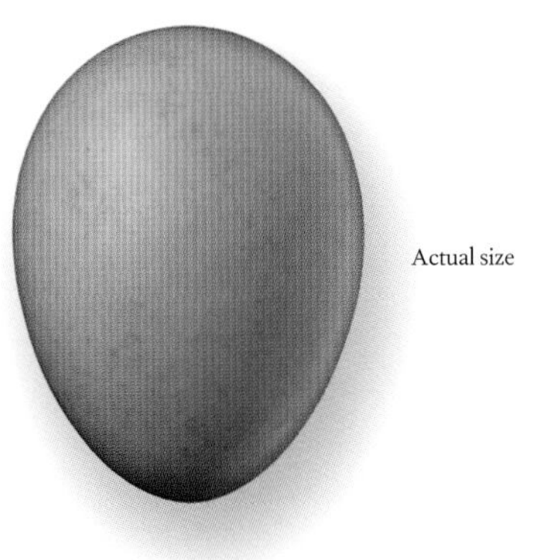

Actual size

The egg of the Blue-winged Teal is creamy white, and measures 1¾ x 1⅓ in (45 x 33 mm). Although it lacks maculation, as do all duck eggs, it may become stained by vegetation or feces during incubation.

ORDER	Anseriformes
FAMILY	Anatidae
BREEDING RANGE	Native to Eurasia, North Africa, the Americas, introduced to Australia and New Zealand
BREEDING HABITAT	Freshwater or saltwater wetlands, both natural and urban, from the Arctic to the subtropics
NEST TYPE AND PLACEMENT	Ground nest of leaf and other plant matter, near water, typically hidden under thickets of vegetation
CONSERVATION STATUS	Least concern
COLLECTION NO.	FMNH 5142

ANAS PLATYRHYNCHOS

MALLARD

ANSERIFORMES

ADULT BIRD SIZE
20–26 in (50–65 cm)

INCUBATION
23–30 days

CLUTCH SIZE
8–13 eggs

This is the ancestral wild species of most breeds of domesticated ducks. Genetic data suggests that the evolutionary origin of Mallards could have been Siberia. Fossils of Mallard populations in Europe appear abruptly, perhaps indicating early domestication by humans.

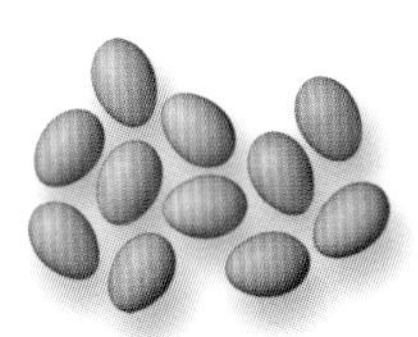

Clutch

In Australia and New Zealand, Mallards, also known as Wild Ducks, have been introduced and are considered invasive, competing for habitat with native species. In addition, Mallards readily interbreed with domestic ducks, and feral populations are often nonmigratory, especially in urban parks and ponds. These nonmigratory populations are also prone to hybridize with locally wintering native duck species, causing a genetic pollution of distinct species, and requiring conservation action to control Mallard populations.

The egg of the Mallard is creamy white to greenish buff, clear of speckles, and 2¼ x 1¼ in (58 x 32 mm) in size. Eggs are laid one day apart, and the female begins to incubate when the clutch is almost complete to allow the chicks to hatch and leave the nest in synchrony.

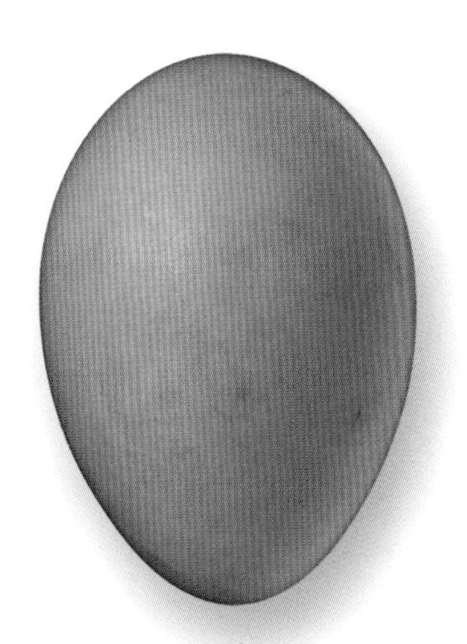

Actual size

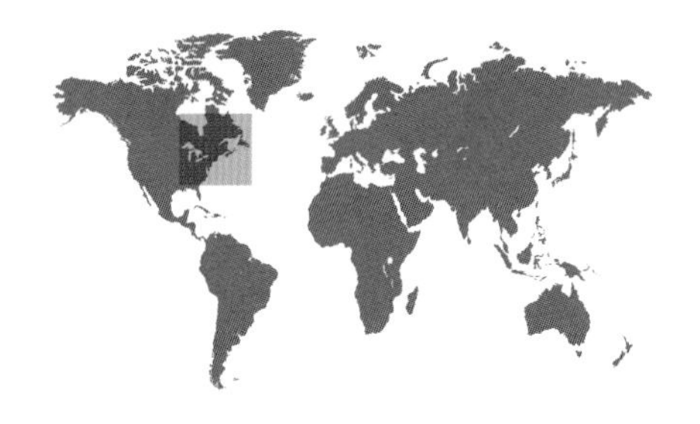

ORDER	Anseriformes
FAMILY	Anatidae
BREEDING RANGE	Eastern North America
BREEDING HABITAT	Shallow, freshwater wetlands and swamps
NEST TYPE AND PLACEMENT	Ground nest, in short vegetation near water, occasionally under bushes and trees
CONSERVATION STATUS	Least concern
COLLECTION NO.	FMNH 1974

ADULT BIRD SIZE
19–25 in (48–63 cm)

INCUBATION
28–32 days

CLUTCH SIZE
6–12 eggs

ANAS RUBRIPES

AMERICAN BLACK DUCK

ANSERIFORMES

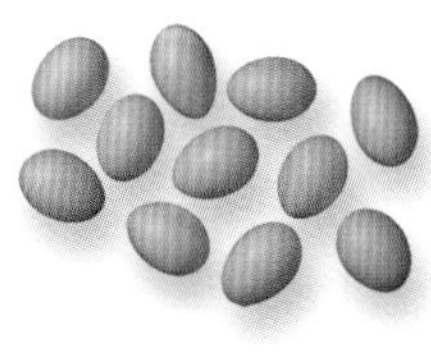

Clutch

Family life for American Black Ducks begins well before springtime, with a female making a decision in late fall or winter from among the many males courting her. Although the chosen male appears to follow the female faithfully, he only stays around until all the eggs are fertilized and laid, leaving incubation and parenting to the female.

Beginning in the last part of the twentieth century, hybridization with Mallards (see page 83) has been thought to cause major declines and genetic concerns for populations of this native species. But this decline seems to have slowed recently, with large flocks of American Black Ducks successfully overwintering along the eastern seaboard of the United States.

The egg of the American Black Duck is white to cream colored, occasionally pale green, and immaculate. It measures 2⅜ x 1⅔ in (61 x 43 mm) in size, similar in both shape and size to that of the Mallard.

Actual size

ORDER	Anseriformes
FAMILY	Anatidae
BREEDING RANGE	Native to North America, introduced into Europe, New Zealand
BREEDING HABITAT	Open pastures, grassy fields, near water
NEST TYPE AND PLACEMENT	Ground nest of bulky dried vegetation and downy feathers
CONSERVATION STATUS	Least concern
COLLECTION NO.	FMNH 21125

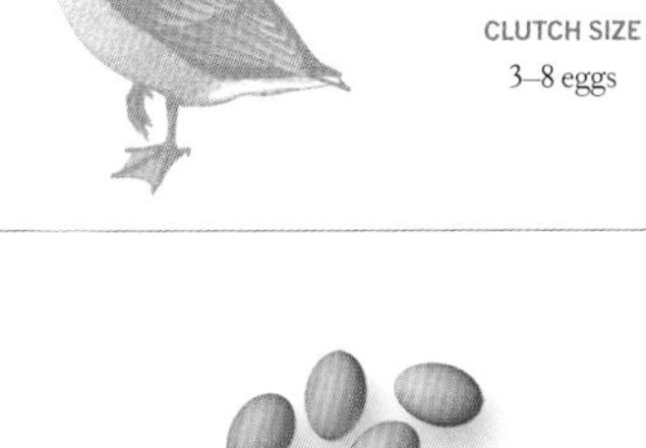

ADULT BIRD SIZE
30–43 in (75–110 cm)

INCUBATION
24–28 days

CLUTCH SIZE
3–8 eggs

BRANTA CANADENSIS

CANADA GOOSE

ANSERIFORMES

This is a highly adaptable goose that appears to be as happy nesting in parks and other human-modified habitats as it does on the tundra of the high Arctic. These qualities have allowed Canada Geese to spread from their native ranges into more southern and urbanized regions of North America, and also to establish themselves firmly as invasive exotics in Europe and New Zealand.

The close association of large populations of these birds with humans—and the threat that airborne birds may pose to aircraft taking off and landing—has led to efforts to control their numbers by methods that include culling and disruption of nesting by shaking the eggs so that they will not hatch.

Clutch

The egg of the Canada Goose is beige-white in color, immaculate, and 3¼ x 2¼ in (83 x 56 mm) in size. Only the female incubates the eggs, but the male stands guard nearby and is quick to defend the nest by attacking intruders, including dogs and people.

Actual size

ORDER	Anseriformes
FAMILY	Anatidae
BREEDING RANGE	Arctic North America and eastern Siberia
BREEDING HABITAT	Lowland marshes, coastal tundra wetlands
NEST TYPE AND PLACEMENT	Ground nest of deep bowl, lined with grass and down, placed at an elevated location, near lake shores
CONSERVATION STATUS	Least concern
COLLECTION NO.	FMNH 14997

ADULT BIRD SIZE
22–26 in (56–66 cm)

INCUBATION
23–26 days

CLUTCH SIZE
3–5 eggs

BRANTA BERNICLA

BRANT

ANSERIFORMES

Clutch

Population densities of the Brant or Brent Goose have increased in recent decades, partly because they now use agricultural fields and pastures on their wintering grounds, as well as coastal wetlands. However, in some summers flocks arriving at their tundra nesting grounds are finding the local food supplies sparse. In these years, many or all Brant females in a nesting area may forgo breeding, as they reabsorb the nutrients otherwise dedicated to forming the eggs.

In years when they do breed, the ground placement of the nest of the Brant makes the clutch vulnerable to predation by Arctic Foxes and other predators. However, the eggs hatch relatively quickly and the highly mobile and independently feeding goslings can fly within six short weeks of hatching. Females and males attend the young together, and pairs return year after year to the same breeding sites and lakes.

Actual size

The egg of the Brant is white and immaculate in appearance, measuring 2⅞ x 1⅞ in (73 x 47 mm) in size. The lipids, proteins, and calcium required for forming the egg are derived directly from the female's own stores built up at the wintering and migratory stopover sites.

ORDER	Anseriformes
FAMILY	Anatidae
BREEDING RANGE	Coastal areas of northern North America
BREEDING HABITAT	Open tundra, near lakes, streams, and shores
NEST TYPE AND PLACEMENT	Depression on high ground, lined with grasses, often reused from one year to the next
CONSERVATION STATUS	Least concern
COLLECTION NO.	FMNH 5953

ADULT BIRD SIZE
25–31 in (63–79 cm)

INCUBATION
22–25 days

CLUTCH SIZE
2–6 eggs

ANSER CAERULESCENS

SNOW GOOSE

ANSERIFORMES

Snow Geese are distinctly colored and appear in one of two main plumage morphs: white or blue gray. When it comes to choosing a mate, young geese rely on the appearance of their parents: those raised by blue-morph parents pair with blue mates, and those with white-morph parents pair with white mates. If the parents were different morphs, the offspring may choose either morph for mating.

Clutch

The pair bonds between adults, which may last a lifetime, form on the wintering grounds, and the pair migrates together to breed in the high Arctic, typically near the natal grounds of the female. Snow Geese nest colonially and often escape predation by Arctic Foxes and Great Skuas (see page 184) by settling near a resident pair of Snowy Owls (see page 300).

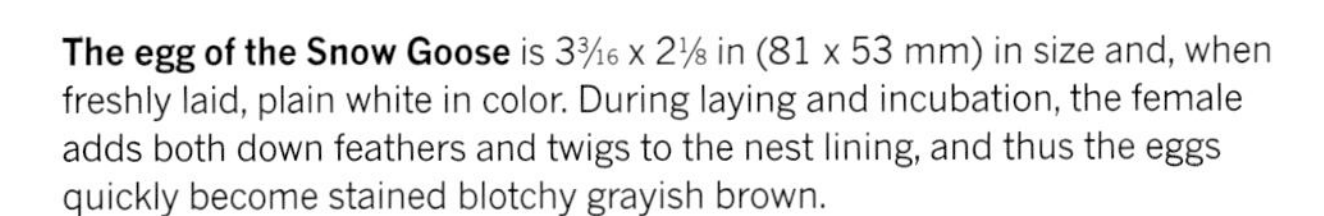

The egg of the Snow Goose is 3³⁄₁₆ x 2⅛ in (81 x 53 mm) in size and, when freshly laid, plain white in color. During laying and incubation, the female adds both down feathers and twigs to the nest lining, and thus the eggs quickly become stained blotchy grayish brown.

Actual size

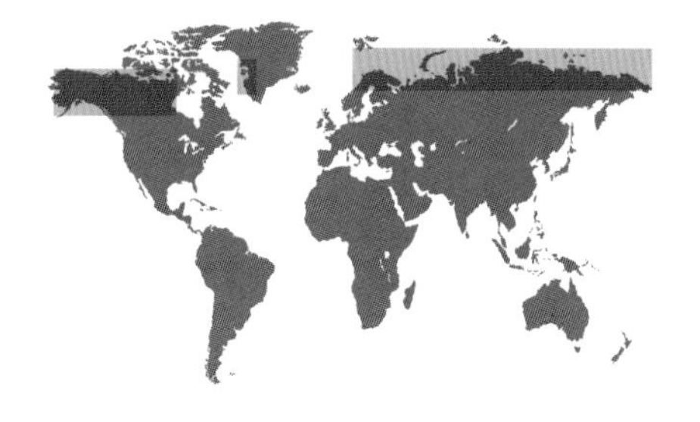

ORDER	Anseriformes
FAMILY	Anatidae
BREEDING RANGE	North American and Russian Arctic
BREEDING HABITAT	Edges of lakes, rivers, and wetlands, including bushy vegetation
NEST TYPE AND PLACEMENT	A scraped depression on the ground, lined by grasses and down feathers
CONSERVATION STATUS	Least concern
COLLECTION NO.	FMNH 21020

ADULT BIRD SIZE
25–32 in (64–81 cm)

INCUBATION
25–27 days

CLUTCH SIZE
3–6 eggs

ANSER ALBIFRONS

WHITE-FRONTED GOOSE

ANSERIFORMES

Clutch

Because this species has long been valued by hunters, the behavioral habits, ecological requirements, migratory movements, and population sizes are closely tracked by wildlife managers and hunters alike. The longest lived individual, a female which died in captivity at 47 years of age, was still laying eggs during its last year of life. In the wild, females and males form strong multiyear pair bonds, and nest in loose aggregations with other White-fronted Geese.

Parental care is intensive, as it encompasses both attending the goslings on the breeding ground and remaining in family units during the migratory and wintering periods. In some cases, parents and young return to the breeding grounds together and both generations breed; the original parents may then have the chance to look after both their grown offspring and their grand-offspring. This phenomenon is rare among waterfowl and is typically only seen in highly social vertebrates, including lions and humans.

The egg of the White-fronted Goose is yellowish or pinkish white in color, with no spotting, and measures 3 x 2⅛ in (76 x 54) mm in size. The female incubates the clutch while the male stands guard to look out for predators.

Actual size

ORDER	Anseriformes
FAMILY	Anatidae
BREEDING RANGE	Northwestern and Central North America
BREEDING HABITAT	Inland lakes and marshes
NEST TYPE AND PLACEMENT	Ground nest placed near lake shores, under thick vegetation of sages and brushes for cover
CONSERVATION STATUS	Least concern
COLLECTION NO.	FMNH 5205

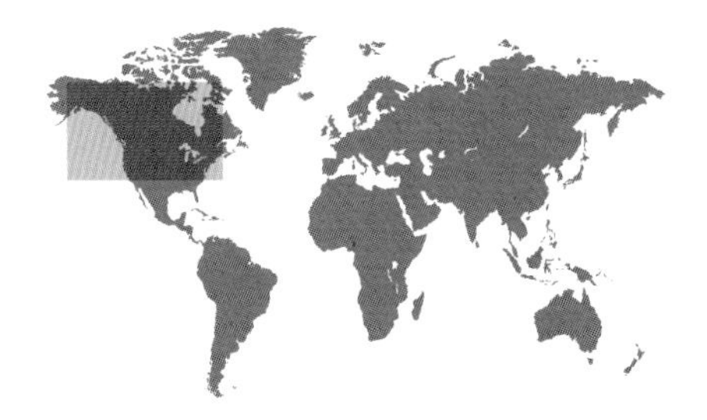

AYTHYA AFFINIS

LESSER SCAUP

ANSERIFORMES

ADULT BIRD SIZE
16½–17 in (42–43 cm)

INCUBATION
21–27 days

CLUTCH SIZE
8–10 eggs

Lesser Scaup are the most numerous diving-duck species in North America, but their population sizes have been consistently diminishing over the last three decades. This is surprising at first because Lesser Scaup are specialists on clams and other bivalves, and have switched their foraging habits to include a widespread freshwater-invasive organism, the Zebra Mussel (*Dreissena polymorpha*), in their diet on their migratory routes. However, Zebra Mussels readily accumulate toxins in polluted waters, and Scaup may suffer from consuming these toxic bivalves.

Male and female Scaup form consortships on the wintering grounds, but these cannot be called pair bonds, because the male does not participate in parental care after the eggs are fertilized. Nonetheless, consorting is beneficial for the female, because it reduces harassment from other males, allowing her to dive undisturbed to seek out bivalves and other foods to prepare for spring migration and the breeding season.

Clutch

The egg of the Lesser Scaup measures 2¼ x 1½ in (57 x 39 mm) in size, and is pale to dark olive or greenish buff. The female incubates the eggs and raises the young alone, while the male abandons her and joins other males to begin his molt.

Actual size

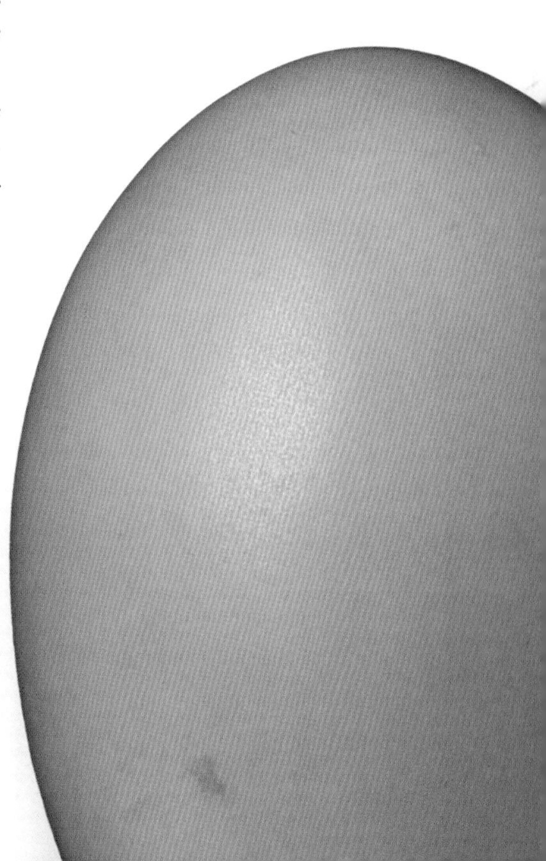

ORDER	Anseriformes
FAMILY	Anatidae
BREEDING RANGE	Central and western North America
BREEDING HABITAT	Prairie lakes and marshes, and foothill swamps
NEST TYPE AND PLACEMENT	Floating nests, in dense lakeshore or marsh vegetation; also lays eggs parasitically in other ducks' nests
CONSERVATION STATUS	Least concern
COLLECTION NO.	FMNH 5191

ADULT BIRD SIZE
16½–21¼ in (42–54 cm)

INCUBATION
24–28 days

CLUTCH SIZE
7–14 eggs

AYTHYA AMERICANA

REDHEAD

ANSERIFORMES

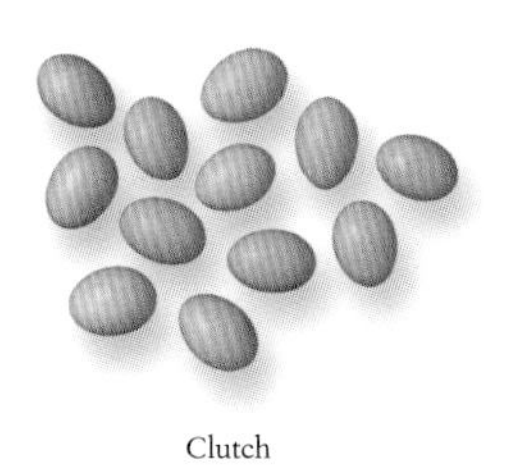

Clutch

This species is a brood parasite, which means that individual females will lay their eggs in nests of other individuals. Redhead females engage in both intraspecific parasitism: laying eggs in other Redhead nests, and interspecific parasitism: laying eggs in nests of other duck species, and occasionally those of coots or bitterns too. Parasitism enlarges the clutch size of the host female above the optimal levels determined by her nest size and parenting ability, and so ducklings from parasitized nests survive less well than those from nonparasitized nests.

Redhead ducklings raised by mothers of a different species face the conundrum of how to recognize their own species. Researchers suspect that genetically determined migratory paths to direct Redheads to the Gulf of Mexico and its most common host, the Canvasback, to the East Coast of the United States, provide young Redheads raised by Canvasbacks with the ability to pair eventually with other Redheads.

Actual size

The egg of the Redhead is cream colored, immaculate, and measures 2⅜–1⅓ in (61 x 43 mm) in size. Female Redheads attempting to lay eggs in Canvasback nests are resisted by the female, but neither the eggs nor the ducklings of the Redheads are discriminated against by the host female.

ORDER	Anseriformes
FAMILY	Anatidae
BREEDING RANGE	North America, Scandinavia, and Russia
BREEDING HABITAT	Arctic and subarctic taiga, evergreen and mixed-species forests
NEST TYPE AND PLACEMENT	High up in natural or woodpecker-carved tree cavities, near open-water lakes
CONSERVATION STATUS	Least concern
COLLECTION NO.	FMNH 15264

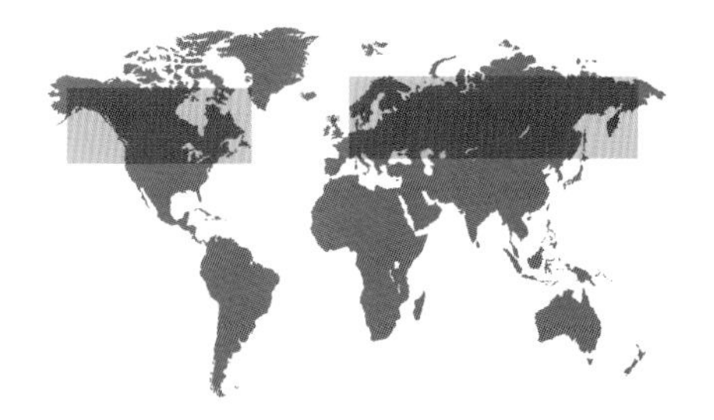

ADULT BIRD SIZE
15¾–20 in (40–51 cm)

INCUBATION
28–32 days

CLUTCH SIZE
5–16 eggs

BUCEPHALA CLANGULA

COMMON GOLDENEYE

ANSERIFORMES

Common Goldeneyes are both opportunistic parasites and dedicated parents. Limited by the availability of large, natural tree cavities for nesting, females typically return to the vicinity of their natal nest site for breeding. If the nest hole, or nestbox, is already occupied, they may lay one or more eggs parasitically, leaving the eggs to be incubated by the female using that nest, who is often a relative.

Once the ducklings hatch and make their way to a nearby lake, mothers typically fight for the ownership of the lake; the loser departs, leaving her young to be looked after by the victor. This process, called brood amalgamation, is beneficial to the winning mother, because some of the new young may be her nieces and nephews. Even if they are unrelated, more ducklings mean less predation pressure on her own young due to the dilution effect in larger flocks.

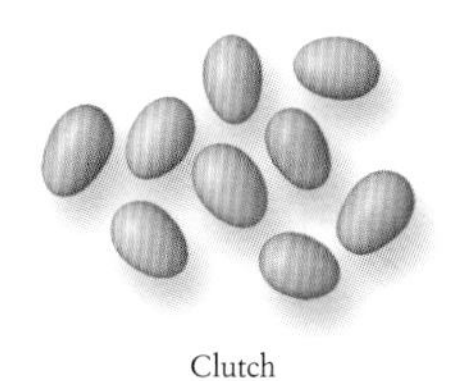

Clutch

The egg of the Common Goldeneye is glossy green and clear of spots; it measures 2⅓ x 1⅔ in (59 x 43 mm) in size. The ducklings remain in the nest for one or two days after hatching before heading to a nearby lake, following the calls of their mother.

Actual size

ORDER	Anseriformes
FAMILY	Anatidae
BREEDING RANGE	Northwestern North America, with some populations in east Canada and Iceland
BREEDING HABITAT	Arctic and wooded lakes, ponds, and parklands
NEST TYPE AND PLACEMENT	In a tree cavity, formed naturally or excavated by large woodpeckers; occasionally a ground hole, lined with down from the mother's breast
CONSERVATION STATUS	Least concern
COLLECTION NO.	FMNH 18290

ADULT BIRD SIZE
17–19 in (43–48 cm)

INCUBATION
29–31 days

CLUTCH SIZE
6–12 eggs

BUCEPHALA ISLANDICA

BARROW'S GOLDENEYE

ANSERIFORMES

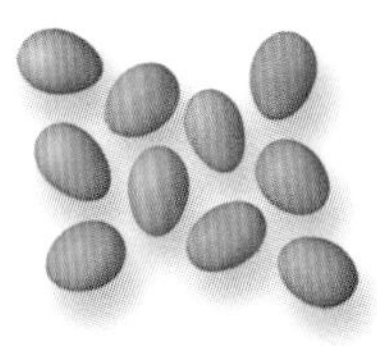

Clutch

The egg of the Barrow's Goldeneye is glossy green, immaculate, and 2⅜ x 1⅔ in (61 x 43 mm) in size. The ducklings are highly independent. Large groups can be attended by just a single female, as each young requires little brooding and feeds on its own, often away from the mother.

Barrow's Goldeneye breed in distant populations from Alaska's Pacific coast to Iceland, where they are called "house duck" due to their affiliation with human settlements. Despite these far-flung populations, the physical appearance of the species is uniform across its distribution, which rarely overlaps with that of the ecologically and visually similar sister species, the Common Goldeneye (see page 91).

As with Common Goldeneyes, many Barrow's Goldeneye females never establish or quickly lose guardianship of their own young, either because they initially lay eggs into available nests of other females, or because, once settling on a lake, mothers fight viciously until only one remains on the water to look after the young of both (or more) females.

Actual size

ORDER	Anseriformes
FAMILY	Anatidae
BREEDING RANGE	Mexico, Central and South America; with established feral populations in the United States, New Zealand, and Europe
BREEDING HABITAT	Forested swamps, lakes, streams, and nearby open fields
NEST TYPE AND PLACEMENT	In natural tree cavities or nestboxes, placed high up
CONSERVATION STATUS	Least concern
COLLECTION NO.	FMNH 2223

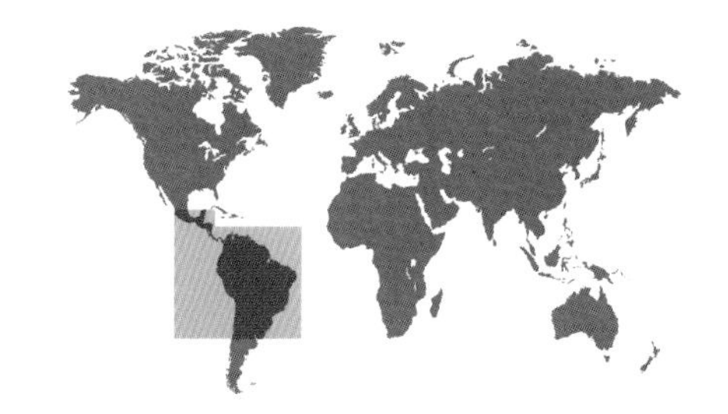

ADULT BIRD SIZE
25–34 in (64–86 cm)

INCUBATION
35 days

CLUTCH SIZE
8–16 eggs

CAIRINA MOSCHATA

MUSCOVY DUCK

ANSERIFORMES

Muscovy Ducks, known as Barbary Ducks in the culinary world, are a New World species with a long, shared history with humans, as they were domesticated in Mexico and South America prior to European settlement. The domesticated forms are genetically so distinct from the wild race that they have been elevated to their own subspecies status by systematic biologists. These subspecies also vary in size and flying ability, with the domesticated birds being larger and less prone to take to the wing.

Clutch

Despite their origin in the tropics, escaped and feral populations of the domesticated Muscovy Duck have adapted to colder climates and are considered invasive throughout the United States, New Zealand, and Europe. Their large size makes them a superior competitor for natural cavities used by native and endemic birds, including other duck species.

The egg of the Muscovy Duck is white and immaculate, and measures 2½ x 1⅞ in (64 x 47 mm) in dimensions. The female incubates the eggs alone, leaving the nest once a day for 20–30 minutes to feed, bathe, and preen.

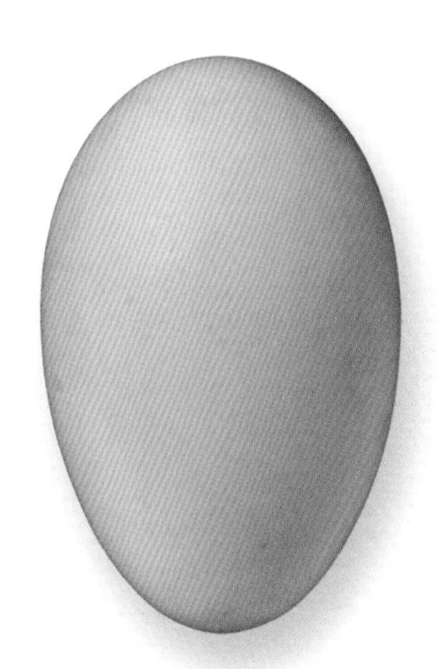

Actual size

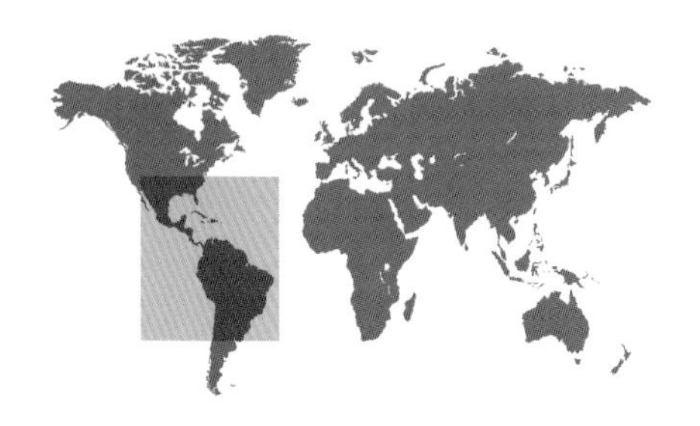

ORDER	Anseriformes
FAMILY	Anatidae
BREEDING RANGE	Southernmost United States, Central and South America
BREEDING HABITAT	Quiet lakes, swamps, ponds, often with forested shores
NEST TYPE AND PLACEMENT	Naturally hollow trees and other cavities, also nestboxes; occasionally nests on the ground
CONSERVATION STATUS	Least concern
COLLECTION NO.	FMNH 2192

ADULT BIRD SIZE
18½–22 in (47–56 cm)

INCUBATION
26–31 days

CLUTCH SIZE
12–16 eggs

DENDROCYGNA AUTUMNALIS

BLACK-BELLIED WHISTLING-DUCK

ANSERIFORMES

Clutch

This duck species has been characterized as the "most nonduck-like" species in North America. Its body plan, posture, coloration, foraging and reproductive behavior all set it apart from the typical diving or dabbling ducks. Most notably, female and male members of a pair are similar in size and coloration, do not migrate, and remain together year-round, often for many consecutive seasons.

Males and females also share equitably the extensive parental care duties, from incubation to hatching, and from leading the young to feed to protecting them from predators. When feeding in open fields and pastures on grain and other plant materials, these ducks may choose to forage during any part of the day, including extensive feeding periods during dark nights when they can avoid diurnal predators.

The egg of the Black-bellied Whistling-Duck is clear white and 2 x 1½ in (52 x 39 mm) in size. The eggs are laid on the bottom of the nest cavity without plant or feather lining; the only protection from breakage is the wood dust that accumulates naturally at the bottom of the nest hole.

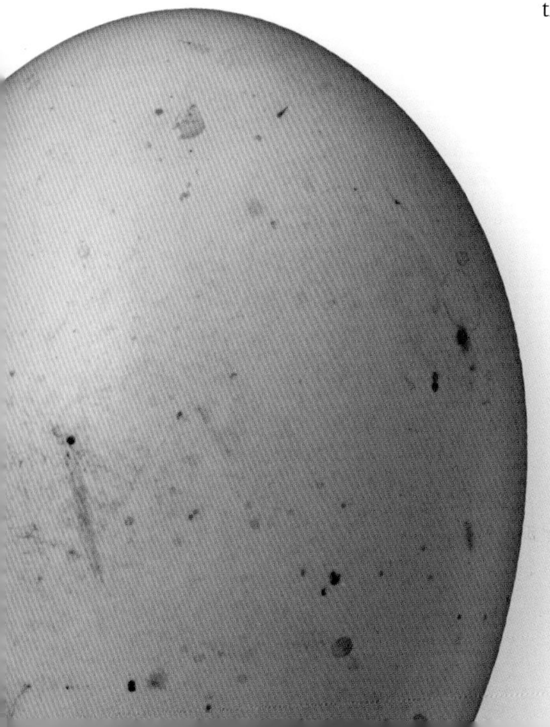

Actual size

ORDER	Anseriformes
FAMILY	Anatidae
BREEDING RANGE	Southern United States, Central and South America, the Caribbean, Africa, and Southeast Asia
BREEDING HABITAT	Freshwater ponds and lakes, including flooded plains managed for rice
NEST TYPE AND PLACEMENT	A simple grass-lined bowl positioned on top of sticks on a floating mat of vegetation or on flooded patches of swamp plants
CONSERVATION STATUS	Least concern
COLLECTION NO.	FMNH 5274

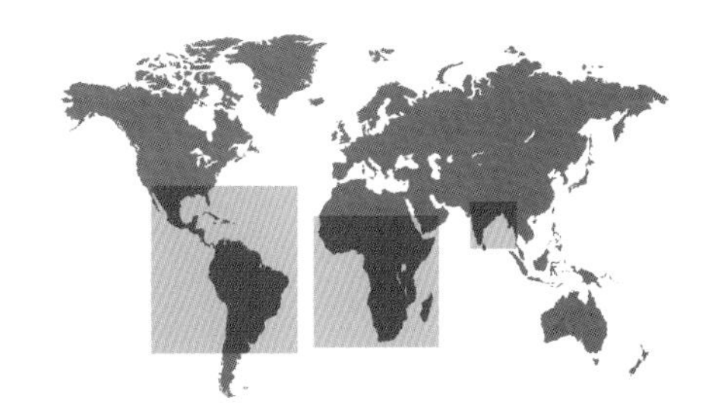

ADULT BIRD SIZE
19–21 in (48–53 cm)

INCUBATION
24–28 days

CLUTCH SIZE
8–12 eggs

DENDROCYGNA BICOLOR

FULVOUS WHISTLING-DUCK

ANSERIFORMES

This is a gregarious and noisy species, often associating with human-managed paddy fields and flooded pastures. Yet it is also wary, and readily takes flight when approached. The female and the male are similar in size and coloration. Surprisingly little is known about their courtship and mating displays; one possibility is that these displays are very limited compared to other duck species.

Clutch

Occurring in subtropical and tropical habitats, the Fulvous Whistling-Duck does not engage in long-distance migrations, although flocks of adult pairs and juveniles may move locally after the nesting season. Because of its association with agricultural fields and filter-feeding foraging, both adults and young are vulnerable to water pollution and pesticides.

The egg of the Fulvous Whistling-Duck is clear white to beige, and 2 x 1⅝ in (52 x 41 mm) in size. Once incubation begins, both parents attend the eggs and the hatchlings, providing equitable parental investment. In this respect, the species resembles swans and geese more than other ducks.

Actual size

ORDER	Anseriformes
FAMILY	Anatidae
BREEDING RANGE	Two distinct populations: Northwestern Asia and Alaska; eastern Canada, southern Greenland, and Iceland
BREEDING HABITAT	Cold and fast-moving streams and coastal cliffs
NEST TYPE AND PLACEMENT	Variable: well-concealed nests on ground, cliffs, and stumps, and in tree cavities near the water
CONSERVATION STATUS	Least concern overall; eastern North American population is endangered
COLLECTION NO.	FMNH 5229

ADULT BIRD SIZE
13–21 in (33–54 cm)

INCUBATION
25–30 days

CLUTCH SIZE
3–9 eggs

HISTRIONICUS HISTRIONICUS

HARLEQUIN DUCK

ANSERIFORMES

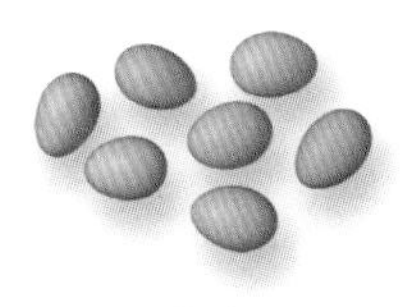

Clutch

The Harlequin Duck is not only remarkable for its rusty blue-white plumage pattern, but it also represents a monotypic genus: this means that it is evolutionarily distinct from other ducks. In North America the western populations of the species have remained stable, but on the Atlantic coast numbers have dwindled, likely due to the development of hydroelectric stations and dams on fast-moving rivers.

The species is well adapted to living in fast-moving, cold waters. A very dense plumage both insulates these small ducks from low temperatures and enables them to bounce and float high on the water after prolonged dives for feeding. The females are ready to attempt breeding at two years of age, and the males at three years, but they do not breed successfully until about five years after hatching.

Actual size

The egg of the Harlequin Duck is pale creamy to buff and measures 2 ¼ x 1⅝ in (58 x 41 mm) in size. The ducklings readily leave the nest after hatching, jumping after the mother into fast-moving streams and rocky coastal surfs, to begin dabbling and diving after crustaceans and other invertebrates to eat.

ORDER	Anseriformes
FAMILY	Anatidae
BREEDING RANGE	Arctic plains and coastal areas of North America, Europe, and Asia
BREEDING HABITAT	Marshes and pools in the tundra, but also along Arctic coastlines and mountain lakes
NEST TYPE AND PLACEMENT	Scrape in the ground, lined with leaves and padded with down feathers; in loose colonies near the water's edge, including on islands and peninsulas
CONSERVATION STATUS	Least concern
COLLECTION NO.	FMNH 164

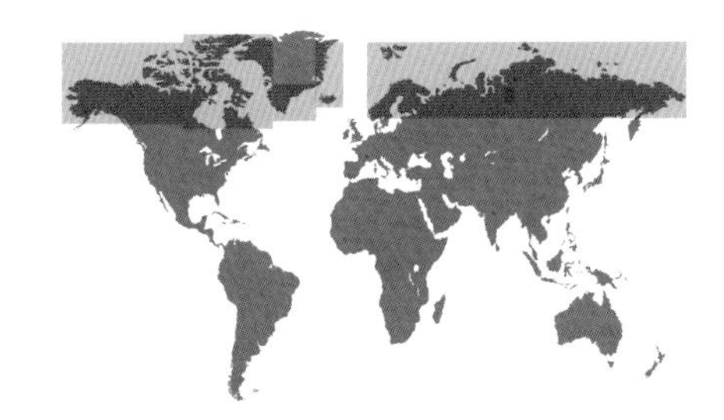

ADULT BIRD SIZE
15–23 in (38–58 cm)

INCUBATION
24–29 days

CLUTCH SIZE
5–10 eggs

CLANGULA HYEMALIS

LONG-TAILED DUCK

ANSERIFORMES

This is a unique duck, and not only because of the conspicuousness of its dark and light plumage patterns and the male's elongated central tail feathers. It is also the deepest-diving sea-duck, capable of spending more time in search of invertebrates and fish at a depth of 200 ft (60 m) than the time needed to catch its breath on the surface between the dives.

The consortship between female and male Long-tailed Ducks ends at the start of incubation, and the mother looks after the young alone. Within a day the hatchlings are capable of feeding and diving unassisted. Early on, females dive with the young, dislodging food items that are then quickly captured and consumed by the ducklings.

Clutch

The egg of the Long-tailed Duck is pale gray to olive green, immaculate, and 2⅛ x 1½ in (54 x 38 mm) in dimensions. During incubation, the initially white and clear eggs can become stained and soiled from wet leaves and other nesting materials, as is seen in a museum specimen shown above.

Actual size

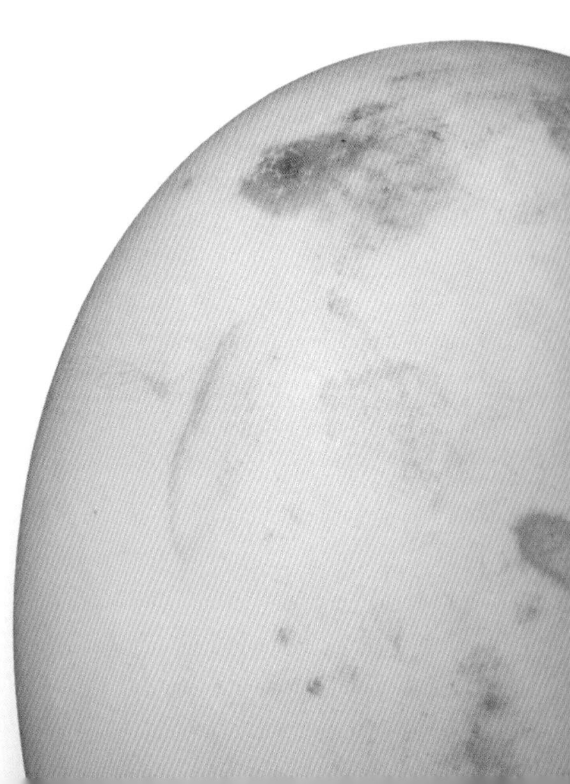

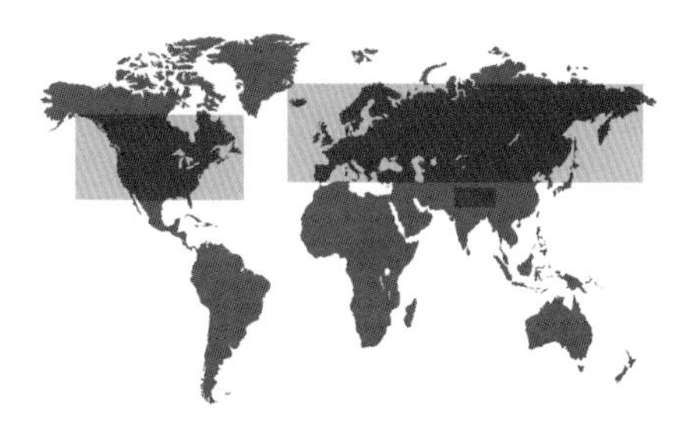

ORDER	Anseriformes
FAMILY	Anatidae
BREEDING RANGE	Forest lakes in northern North America, Europe, Siberia, and central to far eastern Asia
BREEDING HABITAT	Wetlands, grasslands interspersed with lakes and rivers
NEST TYPE AND PLACEMENT	Large cavities in mature trees; in areas void of trees, cliff and bank holes, often far from water
CONSERVATION STATUS	Least concern
COLLECTION NO.	FMNH 5131

ADULT BIRD SIZE
23–28 ½ in (58–72 cm)

INCUBATION
28–35 days

CLUTCH SIZE
8–12 eggs

MERGUS MERGANSER

COMMON MERGANSER

ANSERIFORMES

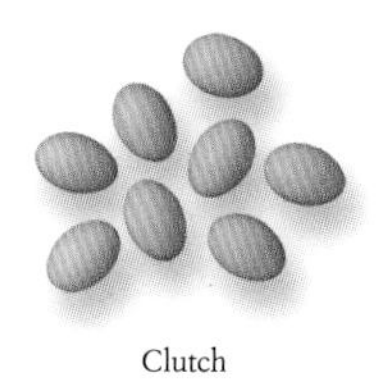

Clutch

Despite their distinct characteristics, including their loon-like body shape and gannet-like serrated beaks that help them to chase and grab fish prey firmly, mergansers (called goosanders in British English) are closely related to dabbling and diving ducks. After prolonged bouts of diving, Common Mergansers may settle on a rock above the stream water, and dry their wings in the warmth of direct sunlight.

Male and female Common Mergansers do not share parental care duties. Nest-hole selection, incubation, and protection of the eggs and ducklings are all undertaken by the female alone. Within a day or so of hatching, the ducklings are ready to leave the nest hole, and the female protects her mobile young but they feed on their own. After the breeding season, this species often forms flocks in the winter that engage in communal hunting for aquatic prey: they drive schools of fish toward the shallow waters near the shore, for easier capture by each member of the cooperating fishing party.

Actual size

The egg of the Common Merganser is white to yellowish and immaculate in color, and measures 2½ x 1¾ in (64 x 45 mm) in size. Where nest cavities are sparse, the female will build a bulky nest of twigs and rootlets, lined with down feathers plucked from her chest to protect the eggs.

ORDER	Anseriformes
FAMILY	Anatidae
BREEDING RANGE	Northern North America, Greenland, Europe, and Asia
BREEDING HABITAT	Freshwater lakes and rivers, far north in the tundra and taiga
NEST TYPE AND PLACEMENT	Depression in the ground, lined with down feathers. Hidden from above, typically under a boulder or dense brush
CONSERVATION STATUS	Least concern
COLLECTION NO.	FMNH 5133

MERGUS SERRATOR

RED-BREASTED MERGANSER

ANSERIFORMES

ADULT BIRD SIZE
20–25 in (51–64 cm)

INCUBATION
29–35 days

CLUTCH SIZE
7–12 eggs

This migratory merganser breeds further north and winters further south than any other merganser species. The adults feed almost exclusively on fish, while the ducklings consume mostly aquatic insects. Male–female consortships form in the winter and during spring migration, but the male abandons the female once the eggs are laid.

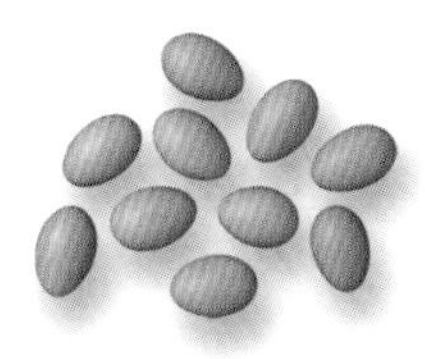

Clutch

Female Red-breasted Mergansers are quick to lead their hatchlings to open water, where the young dive to feed themselves independently of the mother. When two or more broods meet, they may combine into a crèche with one or more mothers attending the young. However, after a couple of weeks, the females depart and leave their flightless young to look after themselves, until they too become flighted. On their own, the young migrate to the southern wintering grounds near coastal waters of temperate seas and oceans.

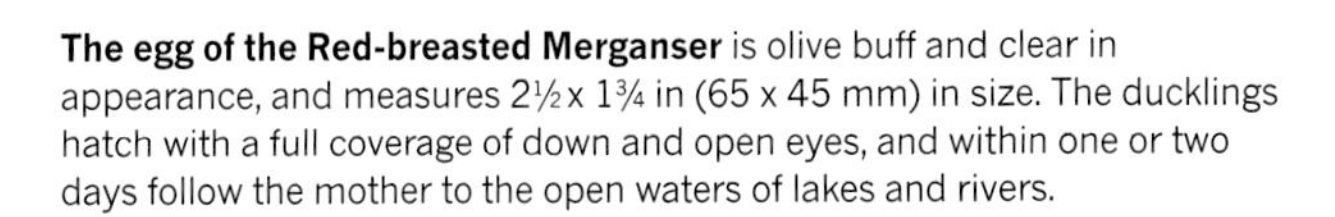

The egg of the Red-breasted Merganser is olive buff and clear in appearance, and measures 2½ x 1¾ in (65 x 45 mm) in size. The ducklings hatch with a full coverage of down and open eyes, and within one or two days follow the mother to the open waters of lakes and rivers.

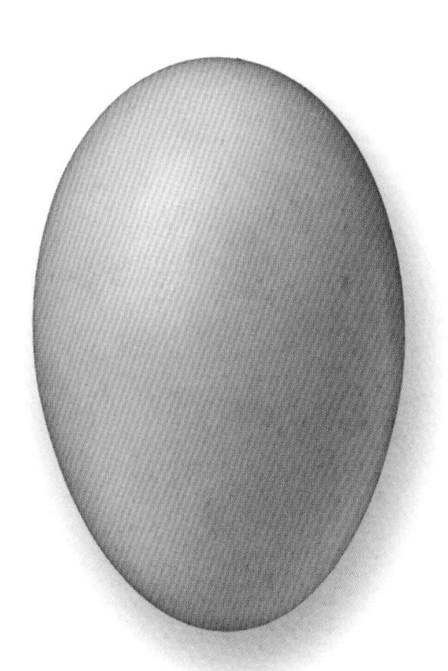

Actual size

ORDER	Anseriformes
FAMILY	Anatidae
BREEDING RANGE	Central and eastern Europe, central Asia, feral populations in Great Britain
BREEDING HABITAT	Deep, fresh or brackish lakes, rivers, and coastal lagoons
NEST TYPE AND PLACEMENT	Ground or floating nest, constructed of roots, twigs, and leaves, hidden in dense vegetation
CONSERVATION STATUS	Least concern
COLLECTION NO.	FMNH 5189

ADULT BIRD SIZE
21½ in (55 cm)

INCUBATION
26–28 days

CLUTCH SIZE
8–10 eggs

NETTA RUFINA

RED-CRESTED POCHARD

ANSERIFORMES

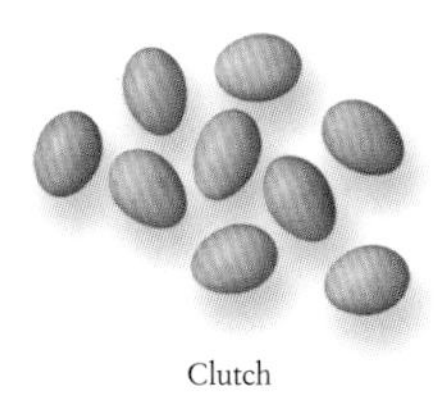

Clutch

The egg of the Red-crested Pochard is creamy to pale green, spotless, and measures 2¼ x 1⅝ in (58 x 41 mm) in size. Both mother and hatchlings feed mostly on aquatic plants, dabbling rather than diving for food, despite the closer systematic relationship of this species to diving, rather than dabbling, ducks.

The males of this duck are unique both in their appearance—with matching rusty brown head and bright red beak—and in their specialized mating behaviors. Courting males dive to retrieve plant and other food items to provide the female sitting on the water surface with offerings, called nuptial gifts. This extra food may help the female to gather and transfer the resources required for the formation and nourishment of the eggs and the developing embryos inside, but it also entices her to tolerate and accept the male as a breeding partner.

Later, and again unlike most other ducks, the male Red-crested Pochard remains with the female during incubation, providing her with more food to counter the energy costs and the lack of time to feed herself while incubating the eggs. He finally abandons the breeding female when the ducklings emerge, joining other males to complete their molt together, while the female attends to the young, protecting them from predators.

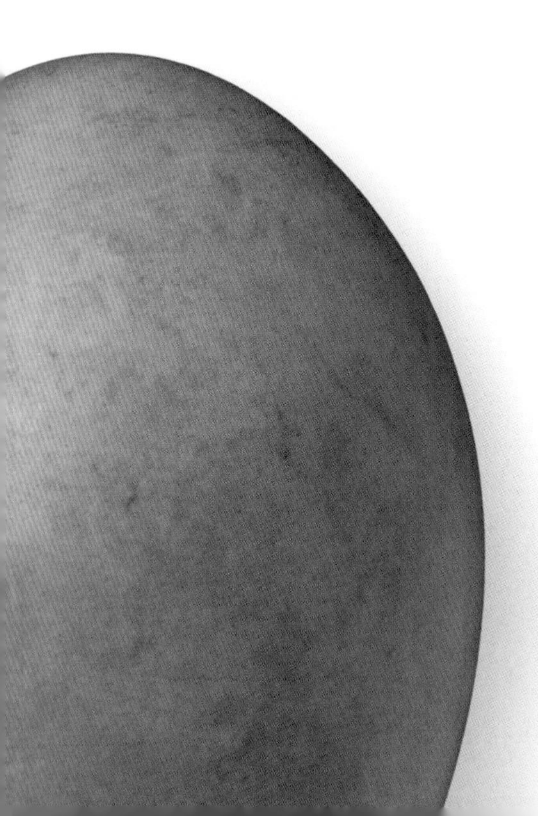

Actual size

ORDER	Anseriformes
FAMILY	Anatidae
BREEDING RANGE	North and western South America
BREEDING HABITAT	Swamps, marshes, and shallow lakes with dense vegetation
NEST TYPE AND PLACEMENT	Nests are built from grass often on floating mats of vegetation, and concealed by reeds and other plants from above
CONSERVATION STATUS	Least concern
COLLECTION NO.	FMNH 21373

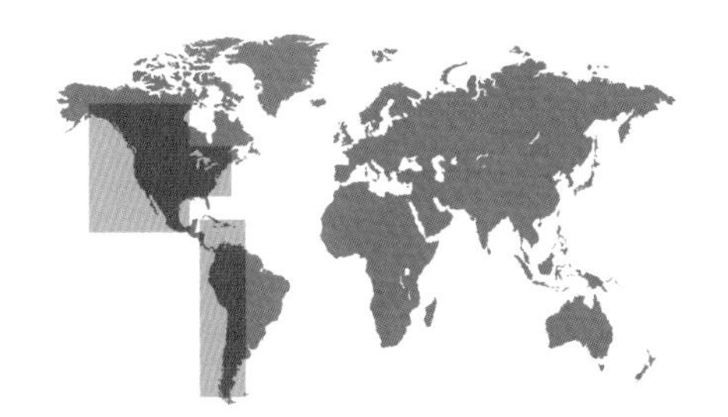

ADULT BIRD SIZE
14–17 in (35–43 cm)

INCUBATION
23–26 days

CLUTCH SIZE
5–15 eggs

OXYURA JAMAICENSIS

RUDDY DUCK

ANSERIFORMES

Ruddy Duck males are remarkably colored, with contrasting white, black, and rusty brown plumage and pale- to bright-blue beaks. The hue of the blue beak becomes more intense as the breeding season approaches, especially among the dominant males in larger social groups. Although most birds do not have external sexual organs, male waterfowl do, and Ruddy Ducks develop an extra-long phallus for mating, compared to other duck species. The phallus is shaped like a corkscrew, but interestingly the females' own reproductive organ, the vagina, curves in the opposite direction, making successful mating a cooperative rather than forcible interaction between the sexes.

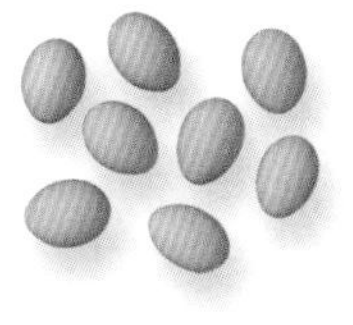

Clutch

Males do not help the females to nest, incubate, and protect the young, and some females lay their large eggs parasitically to reduce the cost of looking after too many eggs and ducklings. Although Ruddy Ducks are the second likeliest of North American waterfowl (after Redheads, see page 90) to lay eggs in other species' nests, scientists do not yet know how successful parasitically raised young are for this species.

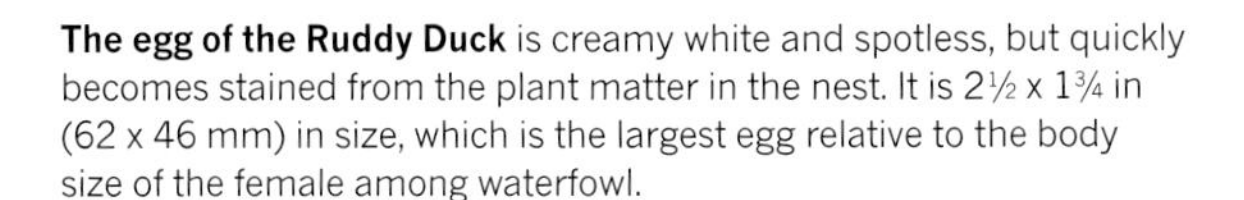

The egg of the Ruddy Duck is creamy white and spotless, but quickly becomes stained from the plant matter in the nest. It is 2½ x 1¾ in (62 x 46 mm) in size, which is the largest egg relative to the body size of the female among waterfowl.

Actual size

ORDER	Anseriformes
FAMILY	Anatidae
BREEDING RANGE	Northern North America, northern Europe, and eastern Siberia
BREEDING HABITAT	Coastal plains in the Arctic and subarctic
NEST TYPE AND PLACEMENT	A hollowed scrape in the ground, near the open water; lined thickly with eiderdown plucked from the female's own chest
CONSERVATION STATUS	Least concern
COLLECTION NO.	FMNH 5238

ADULT BIRD SIZE
19½–28 in (50–71 cm)

INCUBATION
24–26 days

CLUTCH SIZE
3–7 eggs

SOMATERIA MOLLISSIMA

COMMON EIDER

ANSERIFORMES

Clutch

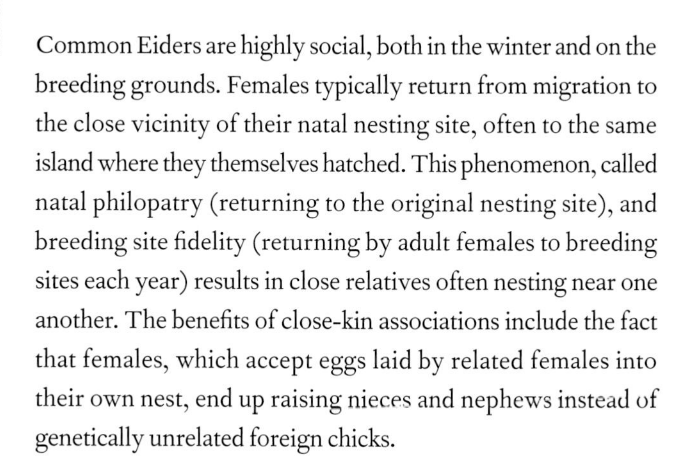

Common Eiders are highly social, both in the winter and on the breeding grounds. Females typically return from migration to the close vicinity of their natal nesting site, often to the same island where they themselves hatched. This phenomenon, called natal philopatry (returning to the original nesting site), and breeding site fidelity (returning by adult females to breeding sites each year) results in close relatives often nesting near one another. The benefits of close-kin associations include the fact that females, which accept eggs laid by related females into their own nest, end up raising nieces and nephews instead of genetically unrelated foreign chicks.

Perhaps the most famous colony of these eiders still exists in northern England, where in days long past traditional harvesting methods were having a severe impact on the adults, the eggs, and the nests. As a result, in 676 CE St. Cuthbert enacted the first-known bird-protection laws, and about 1,000 pairs still nest in the same area today. This is one of the earliest known examples of conservation.

Actual size

The egg of the Common Eider is olive or greenish, immaculate, and 3 x 2 in (78 x 51 mm) in size. The nest is lined densely with down to provide a thick and efficient insulation for the eggs during cold Arctic days. Harvesting of eiderdown from the nest sites remains a sustainable practice for many traditional communities, because it can take place after the eggs hatch and mother and ducklings have departed.

ORDER	Anseriformes
FAMILY	Anatidae
BREEDING RANGE	Coastal Alaska, northeastern Siberia
BREEDING HABITAT	Plains and flats, sloping toward the ocean shore; lake shores in the open tundra
NEST TYPE AND PLACEMENT	Ground nests, lined with down, situated on islands or lake peninsulas
CONSERVATION STATUS	Least concern overall; threatened in the United States
COLLECTION NO.	FMNH 14992

ADULT BIRD SIZE
20½–22½ in (52–57 cm)

INCUBATION
24–28 days

CLUTCH SIZE
3–9 eggs

SOMATERIA FISCHERI

SPECTACLED EIDER

ANSERIFORMES

Young Spectacled Eiders spend the first two to three years of their lives in open marine waters, and only then do they return to land to attempt breeding. Until the mid-1990s no one knew with scientific certainty where this species spent the winter. Since then, satellite tagging has established the location of extensive flocks in the open water between several islands in the Bering Sea, rendering this species an Arctic specialist during both the breeding and the non-breeding season. A deep-diving forager, these ducks consume bivalves and crustaceans picked off the bottom of the sea or lakes.

Clutch

Males sport a spectacular nuptial plumage in winter and pairs form at sea, prior to returning to the breeding grounds. The male chases rivals away from his mate, but later abandons the female after egg laying and departs from the breeding ground soon after, leaving the mothers to incubate the eggs and look after the ducklings on their own. The female protects her young and responds to predation threats from Arctic Foxes and mink by leading the young to open water. The mother and ducklings also hide from aerial predators, including Great Skuas (see page 184), by pushing their bodies into dense brush vegetation.

The egg of the Spectacled Eider is oval in shape and light olive green in color, and measures 2⅔ x 1¾ in (68 x 45 mm) in dimensions. The male not only abandons the female once the eggs are laid, but departs from the breeding grounds soon afterwards, leaving the mothers to incubate the eggs and look after the ducklings on their own.

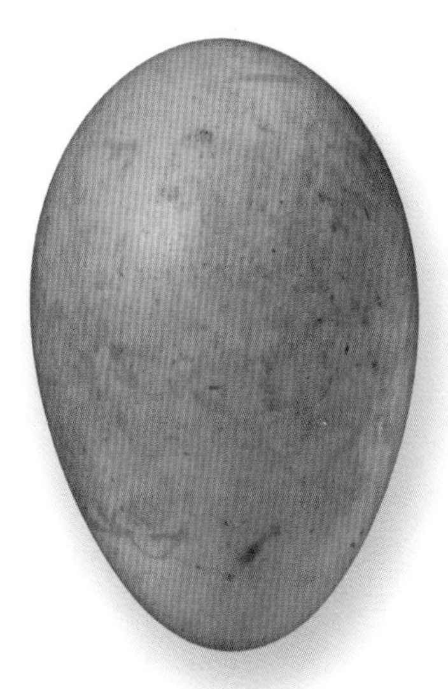

Actual size

ORDER	Anseriformes
FAMILY	Anatidae
BREEDING RANGE	Northern Africa, southern Europe, central to Southeast Asia
BREEDING HABITAT	Open country, steppe and plateaus with freshwater and brackish lakes or rivers
NEST TYPE AND PLACEMENT	Cliffs, rock and ground crevices, and tree holes, often far from water
CONSERVATION STATUS	Least concern
COLLECTION NO.	FMNH 2894

ADULT BIRD SIZE
23–27½ in (58–70 cm)

INCUBATION
28–30 days

CLUTCH SIZE
6–16 eggs

TADORNA FERRUGINEA

RUDDY SHELDUCK

ANSERIFORMES

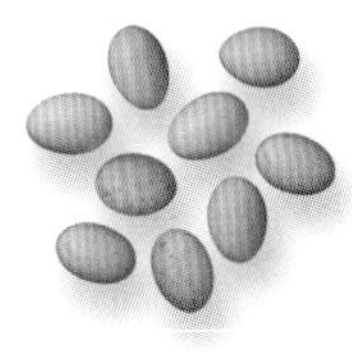

Clutch

The two sexes of this species appear similar from a distance, but the male carries a black neckband, and the female has a white patch around her beak. The pair bond of these shelducks is longer lasting than the temporary consortship of many other duck species, as both parents participate in raising the young. The female pulls out her down feathers to line the nest and the male stands guard over the incubating female and the brood of noisy ducklings.

This species is interspecifically territorial, which means that mating pairs become aggressive and attack intruding ducks and geese of other species, not just conspecifics. Pairs join others soon after the breeding season ends, and can be seen foraging in grassy fields by pulling and grazing on green shoots or dabbling in slow-moving water feeding on aquatic plants and invertebrates.

Actual size

The egg of the Ruddy Shelduck is a creamy white color, immaculate, and 2½ x 1⅞ in (67 x 48 mm) in size. The parents do not always lead the young to open water, and the family often feeds and hides from predators in open fields for prolonged periods away from lakes and streams.

ORDER	Ciconiiformes
FAMILY	Ciconiidae
BREEDING RANGE	Central and eastern Europe, western and central Asia; reintroduced to western Europe. Locally in South Africa
BREEDING HABITAT	Open grassy fields, marshes, and swamplands, often near human settlements
NEST TYPE AND PLACEMENT	Large, bulky nests built on top of trees, lamp poles, or chimneys, repaired and used repeatedly for years
CONSERVATION STATUS	Least concern
COLLECTION NO.	FMNH 21188

ADULT BIRD SIZE
3½–45½ in (100–115 cm)

INCUBATION
33–34 days

CLUTCH SIZE
3–4 eggs

CICONIA CICONIA

WHITE STORK

CICONIIFORMES

The White Stork in much of its distribution has become a strict commensalist of rural human settlements. These birds seek out nesting sites on top of chimneys and power poles, returning each year to the exact same site, to repair and reuse their nest, undisturbed by human neighbors and observers. Not only do these storks share their nesting habitat with people, the large, bulky stick nests also become homes for many other birds, including House Sparrows (see page 636), which may build nests among the sticks and branches in the base of the storks' nest.

The close association of pair-bonded White Storks, allied to their predictable annual return in spring and the conspicuous mating displays of the adults throwing their heads back while snapping their beaks loudly together, has turned this species into a popular subject of fairy tales and legends. Scientifically, however, White Storks also provided some of the earliest direct evidence of long migratory flights by birds. Specifically, during the spring of 1822, a stork returned to Europe with a traditional African hunting spear lodged firmly through its neck.

Clutch

The egg of the White Stork is oval, smooth, and of a slightly glossy white color. It measures 2⅞ x 2 in (72 x 52 mm) in size. Both members of the pair incubate the eggs, taking turns to brood them in cold weather, shade them in the heat, and provision the young with food and water.

Actual size

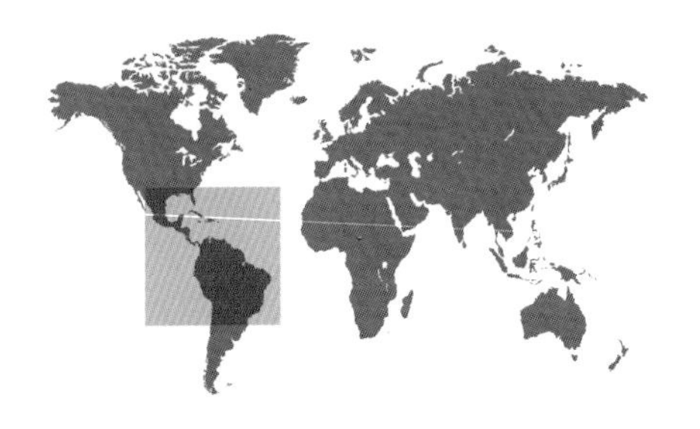

ORDER	Ciconiiformes
FAMILY	Ciconiidae
BREEDING RANGE	South and central America, Caribbean, southern United States
BREEDING HABITAT	Wet woodlands, swamps, and mangroves
NEST TYPE AND PLACEMENT	Colonial, builds large stick nests in the canopy of tall trees
CONSERVATION STATUS	Least concern; locally endangered in the United States
COLLECTION NO.	FMNH 5332

ADULT BIRD SIZE
33½–45½ in (85–115 cm)

INCUBATION
27–32 days

CLUTCH SIZE
3–5 eggs

MYCTERIA AMERICANA

WOOD STORK

CICONIIFORMES

Clutch

The Wood Stork is the heaviest wading bird in North America, and its large white body stands out conspicuously when walking in muddy shallow waters catching crustaceans, amphibians, and fish. A reflex snaps its beak shut when it touches a potential prey item. To sustain their own bodies as well as their breeding effort, the reproductive cycle of the Wood Stork is triggered by dropping water levels that allow these wading birds to catch enough fish to feed their young. Even then, food shortages may result in only the oldest chick(s) surviving to fledge.

Despite the large size of this stork, the nest is vulnerable to predation by grackles, crows, and vultures. In drought years, when the swamps underneath the breeding trees dry out, raccoons are able to climb and get at the nests, and all breeding attempts may fail within days due to mammalian egg and chick predation. The adults, by contrast, are safe from most predators, and return to the same sites year after year to attempt to breed.

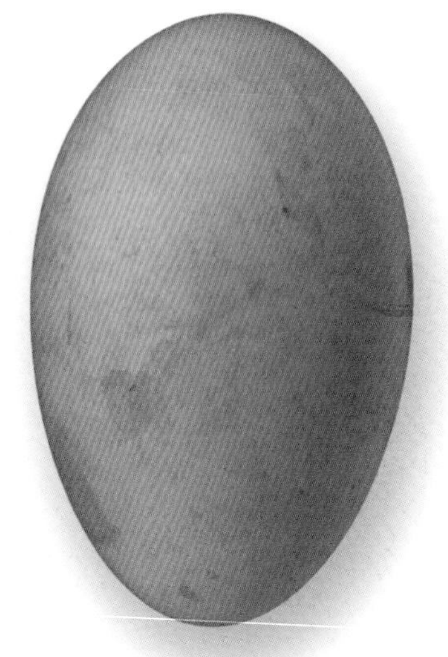

Actual size

The egg of the Wood Stork is cream colored and immaculate, and measures 2⅔ x 1¾ in (68 x 46 mm) in size. Like other storks, parents cool the chicks from the hot summer sun by bringing back water to the nest in their bills and dribbling it directly onto their young.

ORDER	Ciconiiformes
FAMILY	Ciconiidae
BREEDING RANGE	Sub-Saharan to South Africa, Madagascar
BREEDING HABITAT	Shallow swamps and rivers, near sandbanks and tree stands
NEST TYPE AND PLACEMENT	Constructed of twigs and branches, on small trees standing in water or tall trees on dry land
CONSERVATION STATUS	Least concern
COLLECTION NO.	FMNH 21165

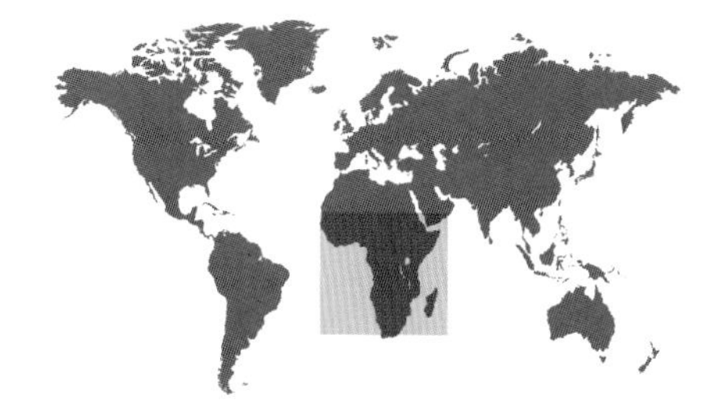

ADULT BIRD SIZE
35½–41½ in (90–105 cm)

INCUBATION
30–32 days

CLUTCH SIZE
2–3 eggs

MYCTERIA IBIS

YELLOW-BILLED STORK

CICONIIFORMES

Breeding for the Yellow-billed Stork is a colonial affair, and the female approaches groups of males who display at potential nesting sites. She is in charge of the final selection of the nest location, but both sexes gather sticks and other materials to build the bulky and messy nest in the course of just over a week. These storks make few sounds; one of them is the hollow noise made by the quick snap of the beak catching prey; another is the hissing sound made by nesting birds when neighbors accidentally come too close to one another; also, their nestlings produce a harsh begging call when demanding food from the parents.

Clutch

Unlike many wading birds, outside the breeding season these storks tend to fish and feed on their own, moving one foot around slowly and deliberately to flush out invertebrates, fish, and amphibians hiding in the mud. When detected, the prey is captured with one of the fastest reflexes in the avian world: a quick movement of the muscular neck and long beak with a curved tip. These impressive beaks allow the storks to capture diverse prey, both small and large.

The egg of the Yellow-billed Stork is dull white in color, and measures 3½ x 2⅔ in (88 x 67 mm) in size. The eggs are laid about two days apart, but incubation by both members of the pair begins on the first day of laying, and the chicks hatch asynchronously, establishing a natural order of size and dominance hierarchy in the nest.

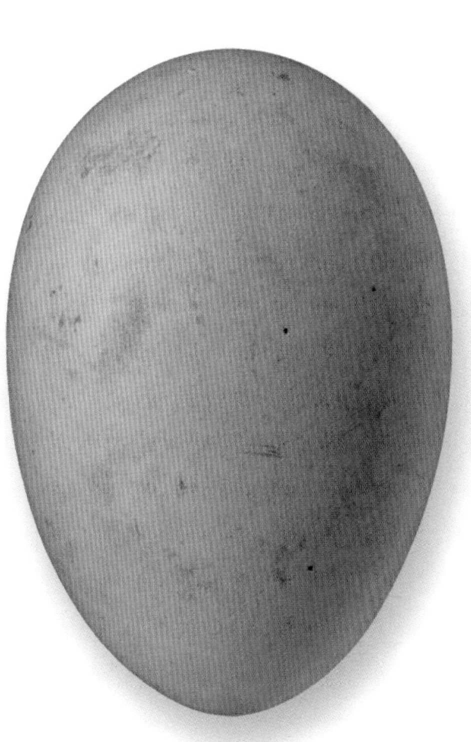

Actual size

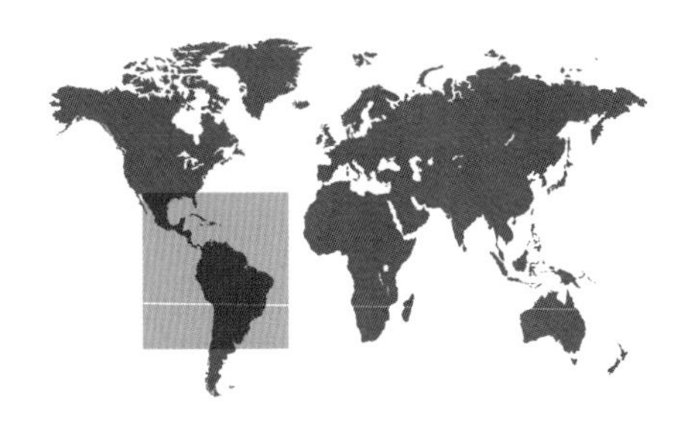

ORDER	Ciconiiformes
FAMILY	Threskiornithidae
BREEDING RANGE	Gulf of Mexico coastline, the West Indies, Central and South America
BREEDING HABITAT	Inland marshes, bays, and estuaries with stands of brushes and trees
NEST TYPE AND PLACEMENT	Platform nest of sticks, built in shrubs or trees, including mangroves
CONSERVATION STATUS	Least concern
COLLECTION NO.	FMNH 21161

ADULT BIRD SIZE
28–34 in (71–86 cm)

INCUBATION
22–24 days

CLUTCH SIZE
2–5 eggs

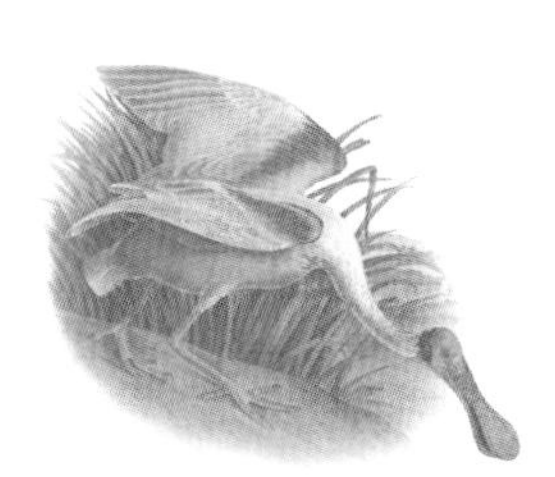

PLATALEA AJAJA

ROSEATE SPOONBILL

CICONIIFORMES

Clutch

Female and male Roseate Spoonbills are similar in size and coloration, leaving behavioral cues as the means of telling the sexes apart. The male attracts the female to a suitable nesting site by offering a piece of twig. Once chosen, the newly formed pair complete their nest together, take turns to incubate the eggs, and both members provision the chicks. Though they nest in loose colonies, often with other wading birds, these spoonbills are typically silent and solitary feeders during the breeding season.

The pink coloration of this spoonbill species is derived from consuming shrimp and other crustaceans, which in turn feed on carotenoid- (red pigment-) producing algae. The bright plumes, however, cost the birds dearly in past centuries; many spoonbills were shot to collect feathers as adornments for human fashion trends. Today, coastal habitat losses, and the fast-paced development of beachside communities, present the gravest conservation concern for this species.

Actual size

The egg of the Roseate Spoonbill is whitish in background, with brown markings of speckles and blotches, and measures 2 ½ x 1¾ in (65 x 44 mm) in size. The eggs are vulnerable to predation by raccoons and the chicks by invasive fire ants.

ORDER	Pelicaniformes
FAMILY	Threskiornithidae
BREEDING RANGE	Northern South America, and the Caribbean islands
BREEDING HABITAT	Wetlands, marshes, tropical grasslands (llanos), and coastal swamps and rainforests
NEST TYPE AND PLACEMENT	A loose aggregation of sticks, built high above the water in the canopy; trees growing on islands preferred
CONSERVATION STATUS	Least concern
COLLECTION NO.	FMNH 5318

ADULT BIRD SIZE
21½–25 in (55–63 cm)

INCUBATION
19–23 days

CLUTCH SIZE
2–4 eggs

EUDOCIMUS RUBER

SCARLET IBIS

PELICANIFORMES

Juvenile Scarlet Ibis are blotchy brown gray, but molt into a uniformly bright red color when mature, making this the only wading bird with deep-red coloration. Despite its dramatic and unique appearance, some behavioral ecologists and evolutionary geneticists have begun to consider that the Scarlet and White Ibises (see page 110) may belong to the same biological species. White Ibis readily accept and raise eggs and young of the Scarlet Ibis; the fledged birds, having imprinted on the foster-parents, will seek mates from their host species, and successfully hybridize across the species boundaries, generating the "pink ibis," which can be seen, rarely, both in nature and also in captivity.

Clutch

Scarlet Ibises are entrepreneurial in their foraging, using their long beaks to catch not only shrimps and crustaceans in shallow, muddy waters, but also scarab beetles in open grasslands. They have also been seen chasing other wading birds to steal prey, and trailing livestock or ducks moving through the grass to catch insects disturbed by these pedestrian neighbors.

The egg of the Scarlet Ibis is dull green, with brown speckles, and 2 x 1⅜ in (51 x 36 mm) in size. In large colonies, many females lay their eggs in synchrony, leading to waves of chicks hatching and fledging together, and so perhaps reducing relative predation pressure on any one egg, chick, or fledgling.

Actual size

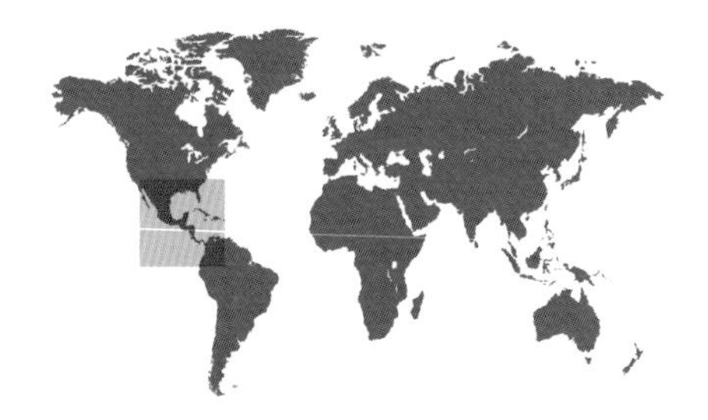

ORDER	Pelicaniformes
FAMILY	Threskiornithidae
BREEDING RANGE	Coastal Gulf of Mexico and southeastern United States, Central America and northwestern South America
BREEDING HABITAT	Coastal marshes and wetlands, mangroves
NEST TYPE AND PLACEMENT	Stick nests, on branches of shrubs or trees, typically over open water
CONSERVATION STATUS	Least concern
COLLECTION NO.	FMNH 21170

ADULT BIRD SIZE
22–27 in (56–68 cm)

INCUBATION
21–23 days

CLUTCH SIZE
2–3 eggs

EUDOCIMUS ALBUS

WHITE IBIS

PELICANIFORMES

Clutch

The egg of the White Ibis is pale greenish, with brown blotching, and measures 2¼ x 1½ in (58 x 39 mm) in size. The beak of the hatchling quickly begins to grow disproportionately long relative to its skeletal growth. Between the second and sixth weeks of life, nestling ibises have three dark bands on their beaks, which distinguish them from nestlings of other bird species breeding in the same mixed colony sites.

The White Ibis (also known as the American White Ibis to distinguish it from the Australian White Ibis, which belongs to a different genus) is a highly gregarious, colonially nesting wading bird. Males and females are similar in appearance and difficult to tell apart from a distance, but measurements of birds indicate that sexual size differences are consistent, and females measure and weigh about 20–25 percent less than males.

The female takes the initiative in nesting, selecting the tree and the branch where the nest is to be built, though both sexes gather material to build the nest platform, whose immediate vicinity is also defended by the pair. The species breeds in large, sometimes vast, colonies. But unlike the traditional nesting aggregations of many other wading and seabirds, White Ibis colonies are ephemeral, and may disband or move away to nearby nesting sites within one or two years of the initial settlement.

Actual size

ORDER	Pelicaniformes
FAMILY	Threskiornithidae
BREEDING RANGE	Eastern North America, coastal Caribbean, Europe, Southeast Asia, Africa, Pacific islands, Australia
BREEDING HABITAT	Marshes and wetlands
NEST TYPE AND PLACEMENT	Shallow stick and twig nest, lined with grasses, in low bushes and trees
CONSERVATION STATUS	Least concern
COLLECTION NO.	FMNH 5320

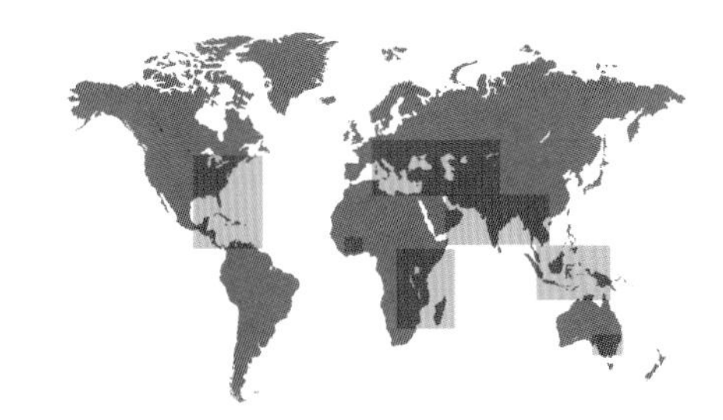

ADULT BIRD SIZE
19–26 in (48–66 cm)

INCUBATION
20–23 days

CLUTCH SIZE
3–4 eggs

PLEGADIS FALCINELLUS

GLOSSY IBIS

PELICANIFORMES

This is the world's most widely distributed ibis species, occurring on all continents in both hemispheres. They nest in colonies with ibis of the same and different species, as well as herons and egrets. However, around the nest's vicinity, these ibis are highly aggressive and territorial. Both sexes incubate the eggs and feed the nestlings, changing guard over the eggs and small chicks following prolonged vocal displays to one another.

Incubation is asynchronous, which means that by the time the last laid egg hatches, the new chick is typically younger and smaller than its nest mates. Parents feed the chicks by regurgitating recently captured food and putting it directly into the beaks of the chicks. However, unlike the case in many heron nests, this ibis's young do not directly fight with each other, perhaps because the parents appear to preferentially feed the smallest chick in the nest first. Thus, parental control over their progeny, it appears, is complete in the Glossy Ibis.

Clutch

The egg of the Glossy Ibis is pale blue or green, immaculate, and elliptical in shape, measuring 2 x 1½ in (52 x 37 mm) in size. The hatchlings grow rapidly and leave the nest after just one week, but do not become flighted until three weeks later. They leave the nesting colony with their parents at two months of age.

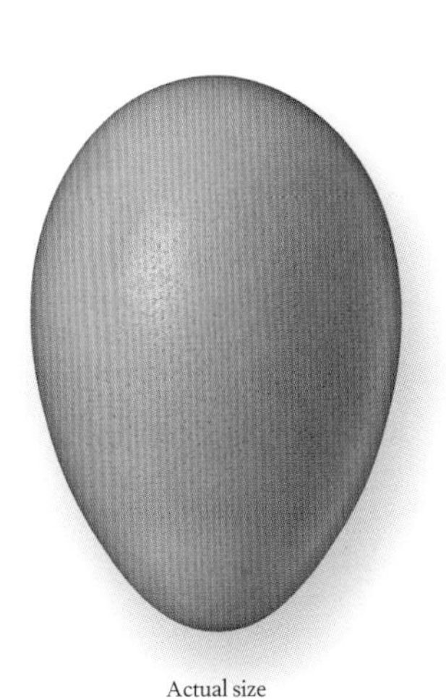

Actual size

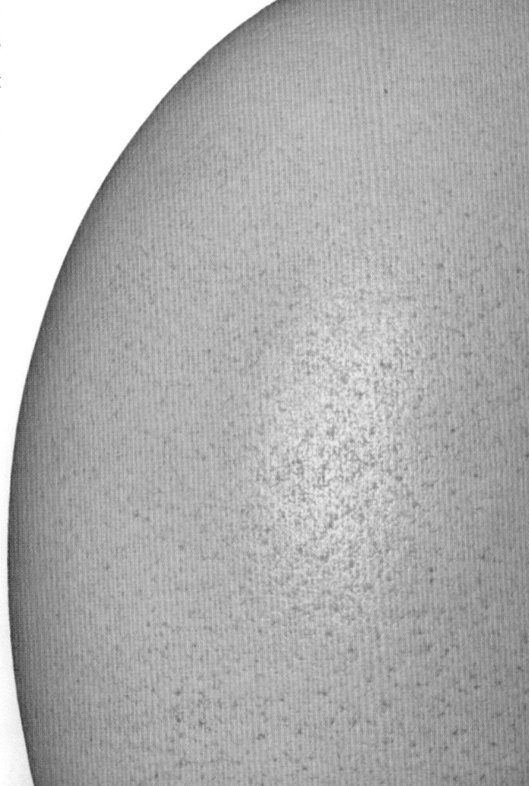

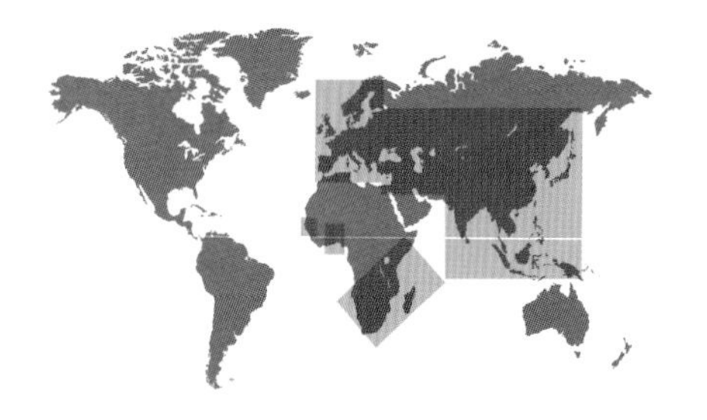

ORDER	Pelicaniformes
FAMILY	Ardeidae
BREEDING RANGE	Temperate Europe, central and south Asia, southern Africa
BREEDING HABITAT	Wetlands and flooded grasslands
NEST TYPE AND PLACEMENT	Bulky reed nest, typically in trees on lakes and seashores, occasionally directly on emergent marsh vegetation
CONSERVATION STATUS	Least concern
COLLECTION NO.	FMNH 5425

ADULT BIRD SIZE
33–40 in (84–102 cm)

INCUBATION
27–28 days

CLUTCH SIZE
3–4 eggs

ARDEA CINEREA

GRAY HERON

PELICANIFORMES

Clutch

Gray Herons breed in large colonies with other birds, including several other species of herons. The recent expansion of Great Cormorant (see page 132) populations has led to direct competition with herons, typically resulting in cormorant colonies replacing tree-based heronries. However, when tree-nesting habitat is no longer available, Gray Herons will successfully nest near the ground, building their bulky nests on emergent marsh vegetation, displaying much-needed flexibility to avoid loss of reproductive opportunities where nesting sites are limited.

In contrast to life in crowded breeding colonies, Gray Herons tend to seek out quiet lake shores and slow-moving streams to forage on their own. There, they take measured steps or wait for long periods motionless, until striking the water for fish and amphibians. They also might spear small mammals from the shore and hatchling birds from nests. In recent years, Gray Herons have invaded urban areas, including Amsterdam in the Netherlands, where they also feed on garbage and other food discarded in the busy city streets.

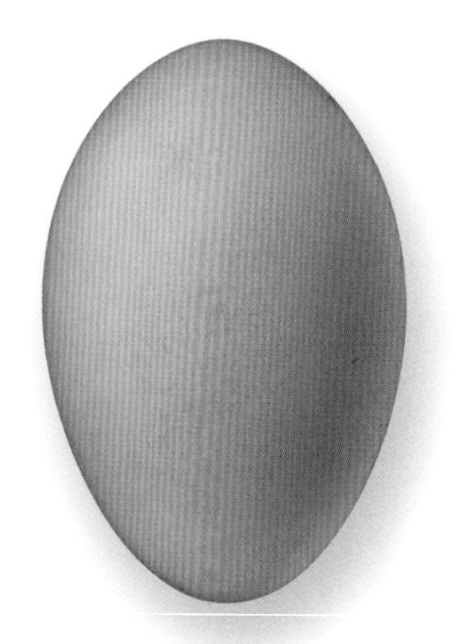

Actual size

The egg of the Gray Heron is pale greenish in color, clear of spots, and 2⅜ x 1⅔ in (61 x 43 mm) in dimensions. Due to pesticide exposure and accumulation in the 1960s, Gray Herons suffered reduced breeding success, with high embryo mortality and egg-shell thinning causing nesting failure. Today, populations are stable or growing.

ORDER	Pelicaniformes
FAMILY	Ardeidae
BREEDING RANGE	North and Central to South America, the Caribbean
BREEDING HABITAT	Diverse wetlands, from coastal marshes and mangroves to inland lakes and rivers
NEST TYPE AND PLACEMENT	Bulky stick nests on trees, typically in a heronry numbering up to several hundred conspecifics
CONSERVATION STATUS	Least concern
COLLECTION NO.	FMNH 5397

ARDEA HERODIAS

GREAT BLUE HERON

PELICANIFORMES

ADULT BIRD SIZE
38–54 in (97–137 cm)

INCUBATION
25–30 days

CLUTCH SIZE
3–6 eggs

Great Blue Herons are the largest herons in North America, and due to their size, they are able to use diverse aquatic resources, from flooded fields to deeper lakes and rivers. They usually feed alone or in small groups, occasionally far from the open water when in search of mice or lizards in meadows. The ability of this heron to forage in varied habitats allows it to remain during the colder seasons, making it a year-round resident throughout much of its range.

Parenting is done by both sexes, including incubation and feeding. Adult herons collect and store food for their young in their crop and regurgitate it to the nestlings. With a full brood waiting in the nest, the parents may collect four times the amount of food required to nourish themselves outside the breeding season.

Clutch

The egg of the Great Blue Heron is light blue and clear in coloration, and measures 2½ x 1¾ in (64 x 46 mm) in size. Human disturbance is particularly harmful during the early part of the breeding period, and entire colonies may be abandoned after intrusions by people, leaving behind eggs or young chicks that will be lost.

Actual size

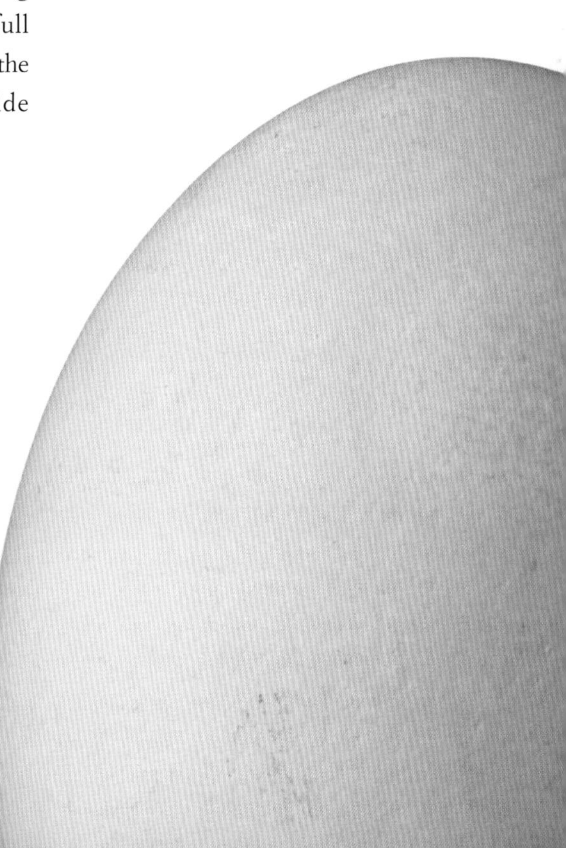

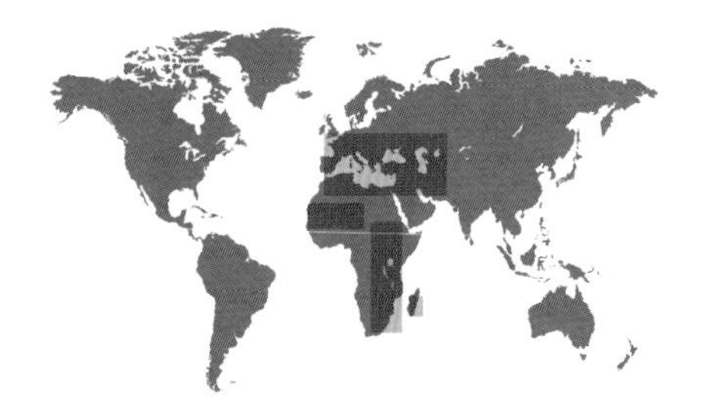

ORDER	Pelicaniformes
FAMILY	Ardeidae
BREEDING RANGE	Southern Europe, western Asia, the Mediterranean, Sub-Saharan Africa, Madagascar
BREEDING HABITAT	Wetlands, marshes
NEST TYPE AND PLACEMENT	Bulky tree nest overhanging the open water, or near ground and water, in reed beds
CONSERVATION STATUS	Least concern
COLLECTION NO.	FMNH 21218

ADULT BIRD SIZE
17–18½ in (43–47cm)

INCUBATION
20 days

CLUTCH SIZE
2–4 eggs

ARDEOLA RALLOIDES

SQUACCO HERON

PELICANIFORMES

Clutch

Breeding in large colonies of up to several thousand pairs, the Squacco Heron build platform nests that are occasionally used by other birds to lay their eggs in, a phenomenon called brood parasitism. Incubating foreign eggs and raising unrelated chicks is evolutionarily costly, and experiments show that this species deserts nests that are parasitized by larger eggs; but if the eggs are similar in size to their own, they will continue to incubate them. This implies that nest desertion is an adaptation to reduce the cost of interspecific brood parasitism by larger egrets and herons breeding in the same rookery.

The Squacco Heron has a large distribution and a similarly extensive total population size, but locally it can be rare, and is vulnerable to loss of habitat due to destruction of its preferred freshwater lake and pond habitats. In Nigeria it is hunted and its skin is sold in local markets for medicinal purposes.

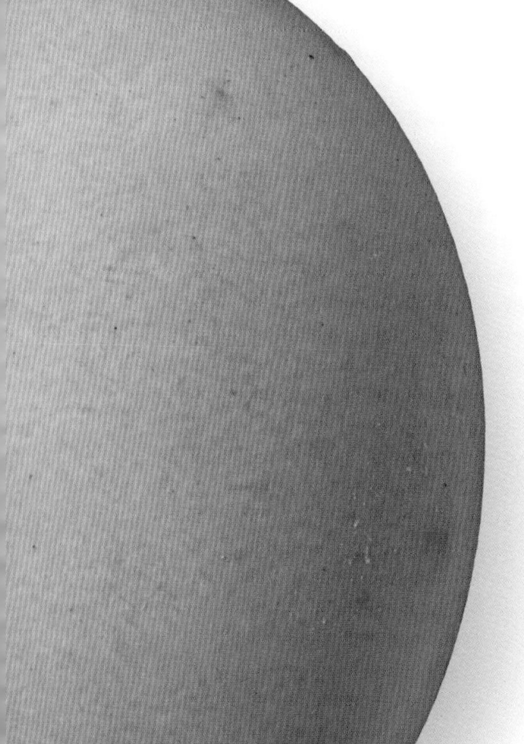

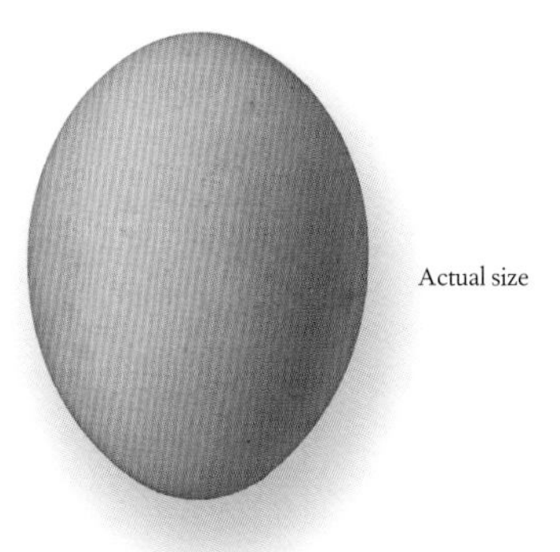

Actual size

The egg of the Squacco Heron is greenish blue and immaculate, and measures 1½ x 1⅛ in (39 x 29 mm) in dimensions. This is a crepuscular species, and so the parents are mostly active feeding themselves and their young during early evening hours.

ORDER	Pelicaniformes
FAMILY	Ardeidae
BREEDING RANGE	Temperate North America
BREEDING HABITAT	Wetlands, marshes, shallow ponds
NEST TYPE AND PLACEMENT	On emergent vegetation, directly over water, a platform of leaves lined with grass
CONSERVATION STATUS	Least concern
COLLECTION NO.	FMNH 5344

BOTAURUS LENTIGINOSUS

AMERICAN BITTERN

PELICANIFORMES

ADULT BIRD SIZE
23½–33½ in (60–85 cm)

INCUBATION
24–28 days

CLUTCH SIZE
2–3 eggs

The American Bittern combines behavior and plumage patterns to disappear from its predators through crypsis. Its solitary foraging and breeding habits, its crepuscular activity pattern, its well-camouflaged plumage, measured walking steps, and sudden freezing immobility allow it to blend in easily against a background of cattails, reed stems, and other marsh vegetation.

Clutch

In the spring, the males use their deep voices to call to females and to indicate their territorial boundaries; the booming quality of this sound is aided by a specialized chamber in the bittern's esophagus. When calling is not enough, competing males approach one another low to the ground, and display their otherwise hidden bright white shoulder feathers to establish dominance. The males will also display these white ruffs to a female to court her. Occasionally, females settle with a male whose territory is already occupied by another female, making the mating system of this species polygynous.

The egg of the American Bittern is elliptical, buff or olive to brown in color, clear of markings, and 2 x 1½ in (49 x 37 mm) in size. The pair nests solitarily, with the female building the nest platform and the male standing guard nearby.

Actual size

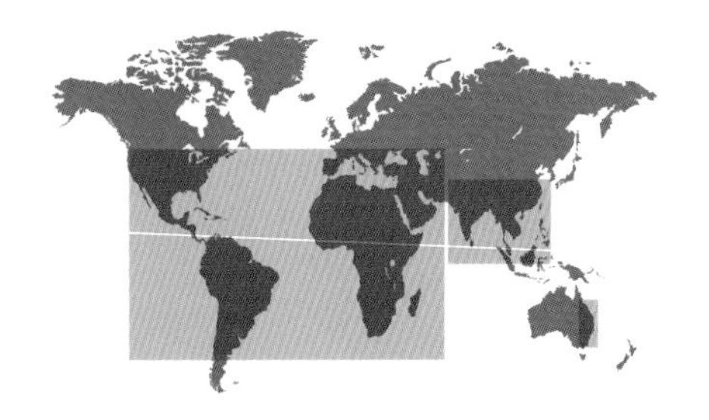

ORDER	Pelecaniformes
FAMILY	Ardeidae
BREEDING RANGE	Southeast Asia, southern Europe, Africa, recently expanded naturally into South, Central, and North America, and Australasia
BREEDING HABITAT	Woodlands near lakes, grasslands, and pastures with livestock
NEST TYPE AND PLACEMENT	Colonies in trees in water, often with other wading birds; loosely woven stick nest
CONSERVATION STATUS	Least concern
COLLECTION NO.	FMNH 21216

ADULT BIRD SIZE
34½–38 in (88–96 cm)

INCUBATION
23 days

CLUTCH SIZE
1–5 eggs

BUBULCUS IBIS

CATTLE EGRET

PELECANIFORMES

Clutch

Cattle Egrets have undergone one of the most widespread natural expansions of their distribution range, crossing open oceans and dispersing across several continents. Today, breeding birds are established on every continent except for Antarctica, but vagrant birds have already been seen on remote subantarctic islands. It has been suggested that the association of this species with human-maintained grazing practices has resulted in this dramatic range-growth over the past 100 years.

Cattle Egrets have become foraging specialists in grassfields near large mammals, both wild and domestic, while hunting for insects flushed by the grazing herds. As a result, the visual system of this heron is now more like that of a landbird, causing these egrets frequently to miss when hunting for prey underwater. This is unlike other herons, which rely on a sophisticated mental correction of the striking-angle of their beak between where the aquatic prey appears when viewed refracted through the water, and where it physically is in the water.

Actual size

The egg of the Cattle Egret is pale bluish white, clear of spotting, and 2 x 1¾ in (53 mm x 45 mm) in size. The eggs hatch asynchronously, and the later-hatched chicks often starve before fledging. Cattle Egrets are also brood parasitic, laying eggs both in con- and heterospecific nests, although many of these eggs fail to hatch.

ORDER	Pelecaniformes
FAMILY	Ardeidae
BREEDING RANGE	Temperate North America, Central America, northern South America, the Caribbean, South and Southeast Asia, and Australia
BREEDING HABITAT	Small wetlands, in coastal and low-lying areas
NEST TYPE AND PLACEMENT	Tree stands in swamps, typically high up, but occasionally on brushes or the ground; a basket-shaped stick nest
CONSERVATION STATUS	Least concern
COLLECTION NO.	FMNH 5498

ADULT BIRD SIZE
17½ in (44 cm)

INCUBATION
19–21 days

CLUTCH SIZE
3–5 eggs

BUTORIDES VIRESCENS

GREEN HERON

PELECANIFORMES

The Green Heron is part of what scientists call a species-complex, together with the similar but globally widespread Striated and the local-endemic Lava or Galapagos Herons. The high degree of mobility of Green and Striated Herons may result in ongoing genetic exchange between them, despite their geographic isolation in most of the breeding ranges.

Green Herons are solitary foragers, often standing still on a bank or perched on a branch directly above the water, waiting for prey to swim by. As one of the few tool-using bird species, these birds occasionally pick up sticks or plant stems, and drop them onto the water, looking for fish and insects to swim up to the water surface and reach striking distance so that the heron can feed upon them. The parents rely on their efficient hunting strategies to nourish the many chicks in the nest. The young leave the nest before being fully feathered and flighted, but continue to be attended and provisioned by both parents.

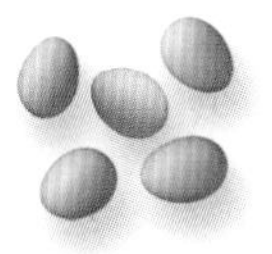

Clutch

The egg of the Green Heron is pale green, immaculate, and measures 1½ x 1³⁄₁₆ in (38 x 30 mm) in size. The female lays the eggs at two-day intervals, but incubation typically begins at clutch completion.

Actual size

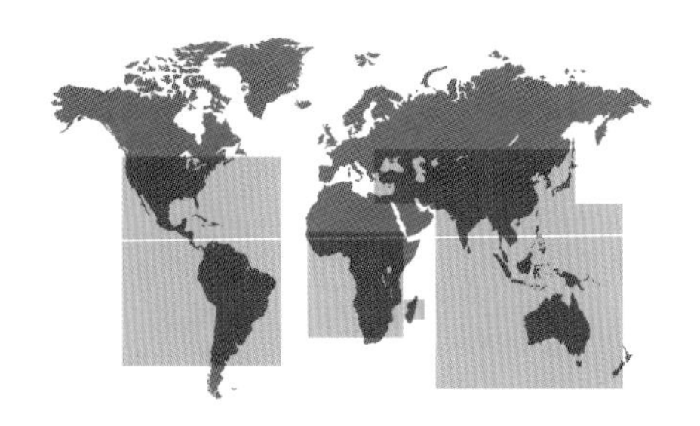

ORDER	Pelecaniformes
FAMILY	Ardeidae
BREEDING RANGE	North and South America, the Caribbean, Sub-Saharan Africa, Asia, Europe, and Australasia, including a single colony in New Zealand
BREEDING HABITAT	Tree stands or reed beds, near fresh or saltwater ponds and marshes
NEST TYPE AND PLACEMENT	Stick nest on top of trees, but also in reed stands; forming colonies with other herons, ibises, spoonbills, and cormorants
CONSERVATION STATUS	Least concern
COLLECTION NO.	FMNH 5432

ADULT BIRD SIZE
37–41 in (94–104 cm)

INCUBATION
23–27 days

CLUTCH SIZE
2–6 eggs

ARDEA ALBA

GREAT EGRET

PELECANIFORMES

Great Egrets are a cosmopolitan species, occurring and breeding on all continents, except Antarctica. It is an evolutionary conundrum that requires further research to understand whether and how these globally widespread populations maintain genetic linkage to one another, appearing morphologically and behaviorally similar between distant localities.

Following the initial nest building and display of the male near the top of a tree, the pair bond is established. The female contributes to the completion of the nest, and both parents provide parental care for the young. Despite the cost involved in building a large, bulky nest, but perhaps also because of the nest's critical role in the male's courtship and mating displays, nests are typically built anew and rarely reused from year to year.

Clutch

The egg of the Great Egret is pale blue in color, immaculate, and 2¼ x 1⅔ in (58 x 42 mm) in size. The large nest platform is lined with wet or fresh plant materials. These dry out, forming a stable cup to hold the eggs throughout the incubation by both parents.

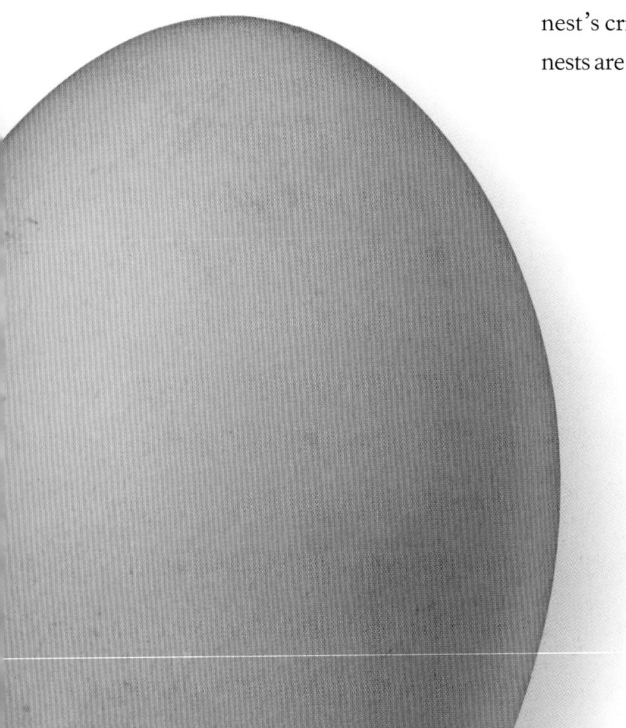

Actual size

ORDER	Pelecaniformes
FAMILY	Ardeidae
BREEDING RANGE	From Mexico through Central America and north-central South America
BREEDING HABITAT	Mangrove swamps
NEST TYPE AND PLACEMENT	Stick nest, in trees or bushes; solitary or in small groups
CONSERVATION STATUS	Least concern
COLLECTION NO.	FMNH 497

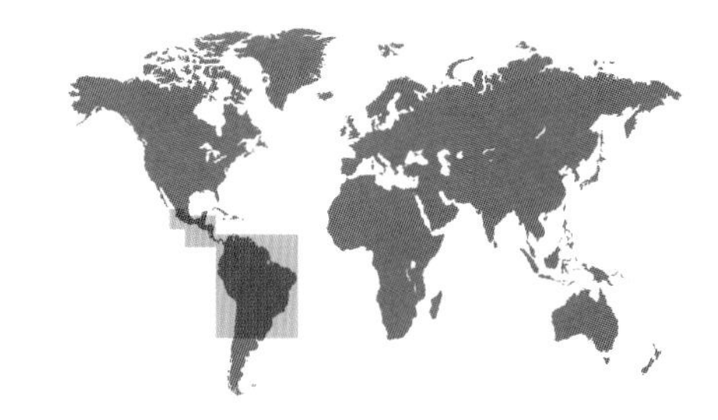

ADULT BIRD SIZE
18–21½ in (46–54 cm)

INCUBATION
21–26 days

CLUTCH SIZE
2–4 eggs

COCHLEARIUS COCHLEARIUS

BOAT-BILLED HERON

PELECANIFORMES

This is a remarkable-looking, distinctive species that escapes notice because of its nocturnal habits. The impressive boatlike bill is used to snap prey items under the water, and to take up large volumes of water and mud to capture prey lying motionless in the dark. To this end, the bill is extra-sensitive, automatically opening in response to the lightest touch.

Clutch

Both sexes build the nest, a shallow platform, together and they provide biparental care. At the onset of the mating season, typically in the rainy months in the neotropics, both sexes use their beaks to preen and clap loudly; the mates also copulate at or on the nest, unlike many other heron species. To keep their plumage dry from the swamp and rainwater, these birds continuously grow powder-down feathers; their structure quickly disintegrates, and the powder is spread by the beak and the feet throughout the rest of the feathers to keep them water-repellent and, thus, dry.

The egg of the Boat-billed Heron is bluish white in color and speckled with cinnamon spots, forming a ring near the blunt end; after incubation and in storage at museums, this maculation is often missing. The egg measures 2 x 1⅜ in (50 x 35 mm) in size.

Actual size

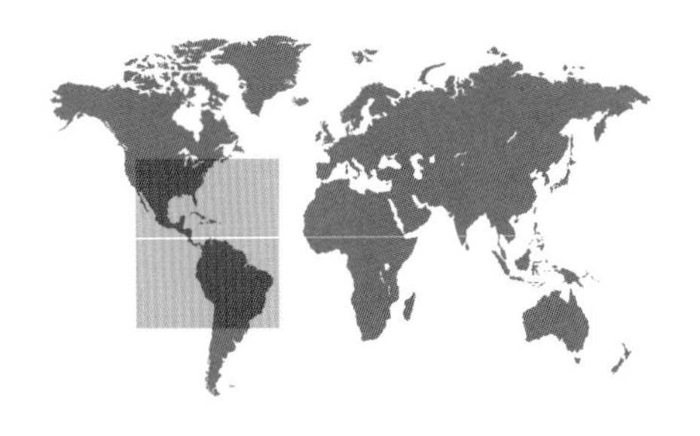

ORDER	Pelecaniformes
FAMILY	Ardeidae
BREEDING RANGE	Temperate North America, Central and South America, and the Caribbean
BREEDING HABITAT	Freshwater swamps, brackish lagoons, coastal thickets and islands
NEST TYPE AND PLACEMENT	Platform nest of sticks, reeds and grasses, in trees and bushes; in colonies of mixed heron species
CONSERVATION STATUS	Least concern
COLLECTION NO.	FMNH 5464

ADULT BIRD SIZE
22–29 in (56–74 cm)

INCUBATION
21 days

CLUTCH SIZE
3–6 eggs

EGRETTA CAERULEA

LITTLE BLUE HERON

PELECANIFORMES

Clutch

The egg of the Little Blue Heron is pale blue, clear, and $1\frac{3}{4} \times 1\frac{1}{3}$ in (46 mm x 34 mm) in dimensions. The male chooses the nesting site, and displays to the female, but both sexes build the nest, incubate the eggs, and feed the young.

Adult Little Blue Herons are dark blue, but young birds are pure white up to two years of age. Interestingly, the white plumage pattern of the young may have evolved and been maintained not just to reduce aggression from breeding adult conspecifics, but also to increase tolerance by other small, white wading birds, including Snowy Egrets (see facing page). White-colored Little Blue Herons are tolerated more and capture more fish when next to Snowy Egrets, compared to older, dark-feathered individuals. The white plumage may also protect the young herons from predation, especially when roosting or flying together with other white herons and egrets.

When not stalking prey by standing still in shallow waters, this species can also follow farmers and livestock in fields and pastures, capturing small insects flushed by the equipment and the movement of the herd.

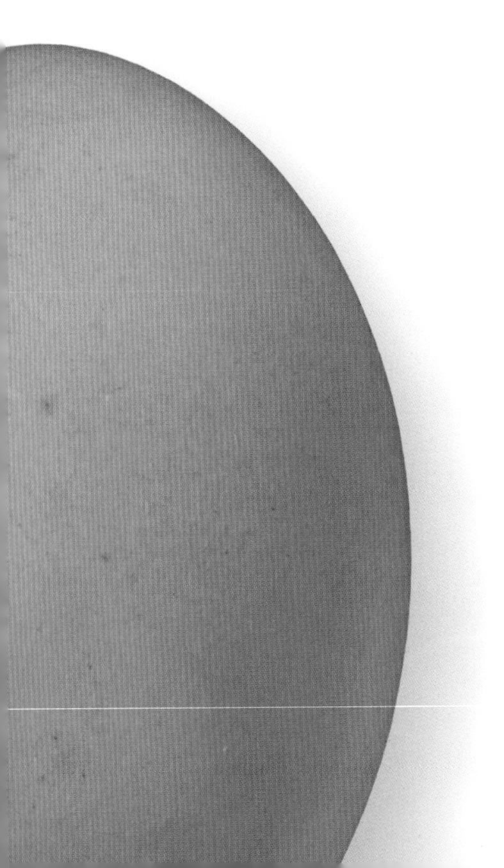

Actual size

ORDER	Pelecaniformes
FAMILY	Ardeidae
BREEDING RANGE	Temperate North America, Central and South America, and the Caribbean
BREEDING HABITAT	Tree stands in wetlands, shores and islands of inland lakes, and estuaries
NEST TYPE AND PLACEMENT	Platform of sticks, in trees and bushes, often in mixed-species colonies with other herons
CONSERVATION STATUS	Least concern
COLLECTION NO.	FMNH 5437

ADULT BIRD SIZE
22–26 in (56–66 cm)

INCUBATION
20–24 days

CLUTCH SIZE
3–5 eggs

EGRETTA THULA

SNOWY EGRET

PELECANIFORMES

Unlike many other widespread herons, which form a single species throughout their range, this species has a strictly New World distribution, with a similar Old World counterpart, the Little Egret, breeding in Eurasia. However, in recent decades, Little Egrets have themselves arrived in the Bahamas and the Caribbean, potentially providing the opportunity for the species to either hybridize or to establish overlapping distributions.

Clutch

The long lace-like chest, back, and neck plumes grown by Snowy Egrets during the breeding season were immensely popular as decorations for women's hats in the late 1880s. This led to excessive hunting and a decline of the species. But since this fashion madness ended, the species' numbers have recovered strongly. The male attracts the female with prominent movements involving his showy head plumes; if he is chosen, the female builds the nest, and both parents incubate the eggs and provision the hatchlings.

The egg of the Snowy Egret is oval in shape, pale green or blue in color and clear of spots, and measures 1⅔ x 1¼ in (43 x 32 mm) in dimensions. The blue tint of this egret's egg is caused by an avian pigment, biliverdin, which is deposited onto the eggshell in the female's oviduct during the night before the egg is laid.

Actual size

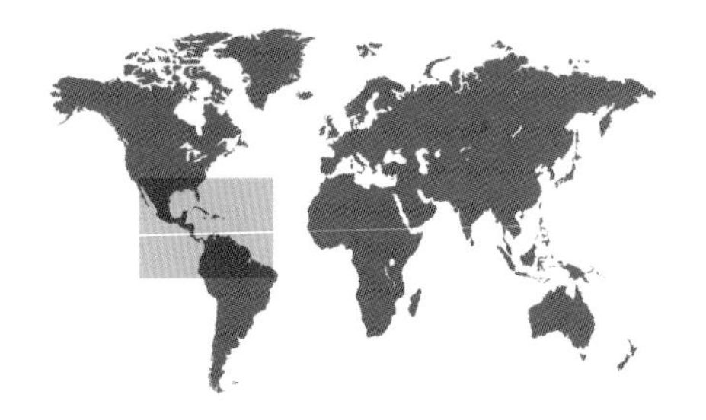

ORDER	Pelecaniformes
FAMILY	Ardeidae
BREEDING RANGE	Southern North America, the Caribbean, and northern South America
BREEDING HABITAT	Marshes, swamps, and lagoons
NEST TYPE AND PLACEMENT	Stick nest, built on reed stems, in a bush or on a low tree
CONSERVATION STATUS	Least concern
COLLECTION NO.	FMNH 516

ADULT BIRD SIZE
22–30 in (56–76 cm)

INCUBATION
21 days

CLUTCH SIZE
3–5 eggs

EGRETTA TRICOLOR

TRICOLORED HERON

PELECANIFORMES

Clutch

The egg of the Tricolored Heron is pale blue or green, clear of blotching, and 1¾ x 1¼ in (44 x 32 mm) in size. The male establishes the breeding site, and both sexes build the nest, incubate the eggs, and provision the young.

The Tricolored Heron, formerly known as the Louisiana Heron, is a colonially breeding wading bird, nesting in mixed-species colonies with other herons and egrets. It may be difficult to spot away from the colony, standing motionless in tall grasses or wading slowly in deep water, with its chest touching the surface. But breeding and feeding are interlinked, and most colonies are formed within 1–2 miles (2–3 km) of abundant foraging sites.

Ritualized foraging behavior is also an important part of this species' courtship display: the males perform it during the early mating season from a suitable nesting platform within the breeding colony. Once a female chooses a male, the display intensity grows, the mates copulate, and nest building begins at full steam. The pair bond remains stable throughout the breeding season, and occasionally, across years, too.

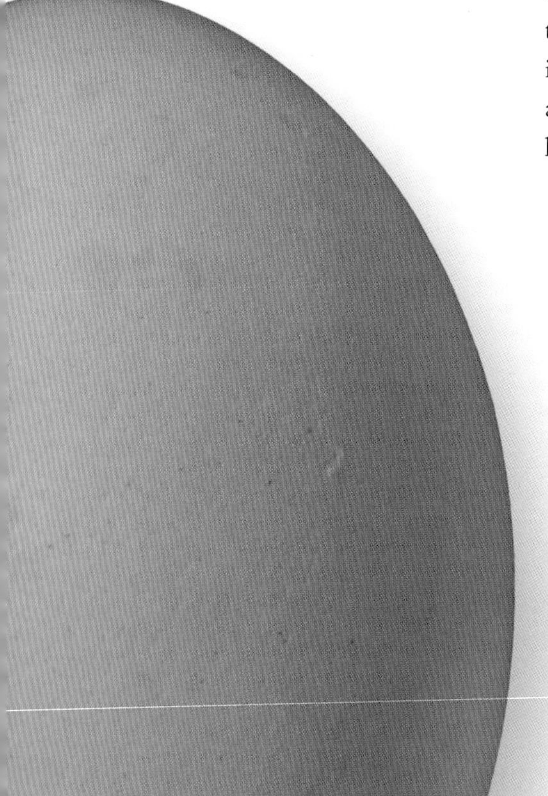

Actual size

ORDER	Pelecaniformes
FAMILY	Ardeidae
BREEDING RANGE	Temperate North America, the Caribbean, riparian and coastal South America
BREEDING HABITAT	Freshwater and salt marshes, with tall and dense vegetation
NEST TYPE AND PLACEMENT	Platform of vegetation, with a roof of pulled marsh plants, built above water
CONSERVATION STATUS	Least concern, but can be locally rare or endangered
COLLECTION NO.	FMNH 5364

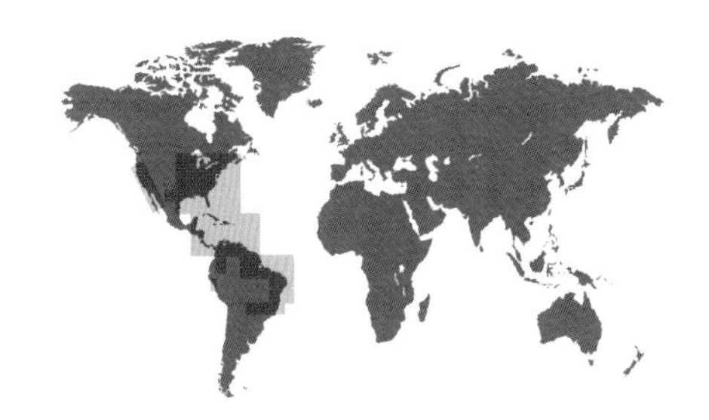

ADULT BIRD SIZE
11–14 in (28–36 cm)

INCUBATION
17–20 days

CLUTCH SIZE
2–5 eggs

IXOBRYCHUS EXILIS

LEAST BITTERN

PELECANIFORMES

The Least Bittern is a master of disguise. Small and cryptic in coloration, it responds to the approach of potential predators, including people, by freezing and raising its neck straight up, blending into the background of swaying thickets of marsh grasses. Rather than stepping out into the open, members of the pair communicate with each other using gender-specific vocal cues while they remain hidden in dense reed beds. There they also nest, incubate the eggs, and feed the young, until these climb out of the nest up to three weeks before they become flighted and independent from the parents.

Globally, this species' conservation status is of "least concern," because of the vast geographic distribution and its occasionally dense, loosely colonial, nesting populations. Yet, locally, it may suffer from habitat loss, fragmentation, and modification so that, for example, in Massachusetts this highly secretive species is listed as endangered.

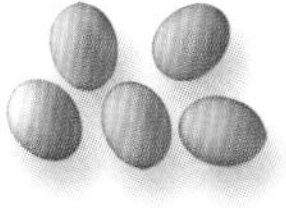

Clutch

Actual size

The egg of the Least Bittern is pale blue or green, matte, elliptical in shape, and 1¼ x 1 in (31 x 24 mm) in size. Nest predation by mammals and snakes, and wakes from boat traffic, are the most frequent causes of clutch failure, but changes in marsh vegetation structure, including the presence of invasive *Phragmites* reeds, also degrade the quality of this bird's nesting habitats.

ORDER	Pelecaniformes
FAMILY	Ardeidae
BREEDING RANGE	Temperate and tropical regions of North, Central, and South America, the Caribbean
BREEDING HABITAT	Swamps, marshes, coastal islands
NEST TYPE AND PLACEMENT	Loose stick nest, lined with grasses and leaves, often high up in tree canopy
CONSERVATION STATUS	Least concern
COLLECTION NO.	FMNH 5536

ADULT BIRD SIZE
24 in (61 cm)

INCUBATION
24–25 days

CLUTCH SIZE
3–5 eggs

NYCTANASSA VIOLACEA

YELLOW-CROWNED NIGHT-HERON

PELECANIFORMES

Clutch

This night-heron breeds in small colonies where predators may access the nests; larger colonies form in more remote locations, including lake and offshore islands with no tree-climbing mammalian predators. Nesting colonies remain in use year after year, with individual nests being refurbished and expanded by the same pair again and again. Rookeries are typically near the water to guarantee a short commuting distance to and from feeding sites in the dawn and dusk hours.

The initial nest construction is part of the courtship display between newly formed pairs: the female waits at the nest site, while the male delivers long bare branches, often as long as his own wingspan. Once the mate bond is firmly (re)established, both parents build up the nest, leaving just a small depression in the center in which to lay and incubate the eggs.

Actual size

The egg of the Yellow-crowned Night-Heron is pale blue and clear in color, and measures 2⅛ x 1¾ in (55 x 42 mm) in size. The chicks hatch with light down and are blind, but their eyes open in a day, and the young vigorously compete for parental provisions.

ORDER	Pelecaniformes
FAMILY	Ardeidae
BREEDING RANGE	Tropical and temperate regions of all continents, except Australia and Antarctica
BREEDING HABITAT	Fresh and saltwater marshes, swamps, lakes, and wooded creeks
NEST TYPE AND PLACEMENT	Flimsy stick-platform nest, in dense colonies, placed in foliage of thickets, bushes, and trees
CONSERVATION STATUS	Least concern
COLLECTION NO.	FMNH 5517

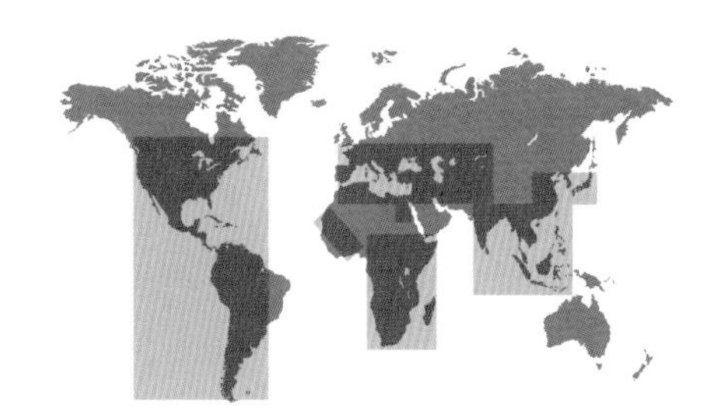

NYCTICORAX NYCTICORAX

BLACK-CROWNED NIGHT-HERON

PELECANIFORMES

ADULT BIRD SIZE
23–26 in (58–66 cm)

INCUBATION
24–26 days

CLUTCH SIZE
3–5 eggs

True to its name, this is a crepuscular species that becomes active outside the bright daylight hours. Although it breeds in noisy and busy single- and mixed-species colonies, on the foraging grounds this species is quiet, slow-moving, and highly defensive of its feeding site. It mostly captures small fish, amphibians, and invertebrates, but occasionally hunts small mammals and birds too.

In contrast to the sharp and unmistakable black and gray-white feathers of the adult, the juveniles are streaky brown and beige in color. They take two to three years to develop the full adult plumage patterns, even though they may become sexually mature in earlier years. This phenomenon is called delayed plumage maturation, and may benefit the younger birds by preventing aggression and attacks by older breeding adults.

Clutch

The egg of the Black-crowned Night-Heron is pale blue or green, clear of speckles, and 2 x 1½ in (52 x 37 mm) in dimensions. Both parents incubate the egg and feed the chicks, with most feeding visits only undertaken in the early morning and late evening hours.

Actual size

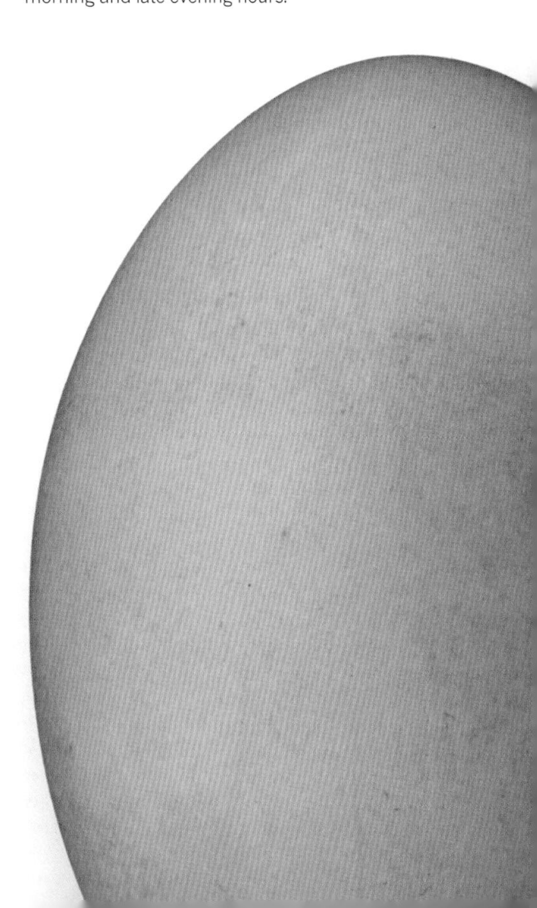

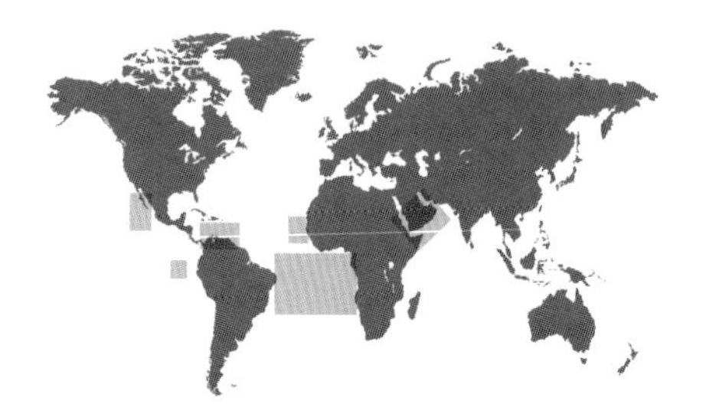

ORDER	Phaethontiformes
FAMILY	Phaethontidae
BREEDING RANGE	Tropical Atlantic, eastern Pacific, and Indian Oceans
BREEDING HABITAT	Isolated, oceanic islands
NEST TYPE AND PLACEMENT	A simple scrape, with the egg laid directly on the ground at a cliff face or in a crevice
CONSERVATION STATUS	Least concern
COLLECTION NO.	FMNH 15097

35½–41½ in (90–105 cm), 19 in (48 cm) without elongated tail feathers

INCUBATION
42–46 days

CLUTCH SIZE
1 egg

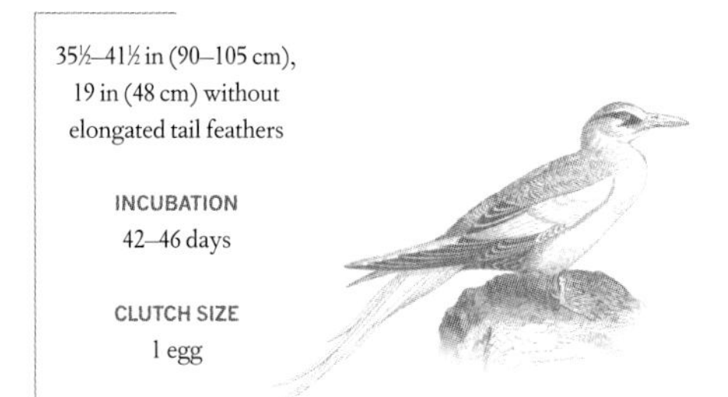

PHAETHON AETHEREUS

RED-BILLED TROPICBIRD

PHAETHONTIFORMES

The egg of the Red-billed Tropicbird is white buff to pale purple in color, with reddish brown maculation, and 2½ x 1½ in (60 x 40 mm) in size. If the single egg is broken or lost to weather or predators within days after laying, the female will lay a second, replacement egg and the pair continues to incubate it, and provision the hatchling together.

Red-billed Tropicbirds are impressive at flying, diving, and fishing, yet are poor swimmers for an open ocean bird. After a dive to catch fish, they quickly reappear at the surface of the ocean with their elongated white tail feathers pointing straight up. Then they take flight, because the plumage is poorly insulated and can quickly become waterlogged. Once in the air, the tail feathers are aligned with the rest of the body, generating the characteristic silhouette for this species and other tropicbirds.

Potential mates perform extensive, aerial courtship displays before settling on an isolated cliff face on a remote island to begin breeding. Despite the lack of a distinct nest structure, pairs may return to the same island and cliff face to breed across many years. Whereas the nestling tropicbird looks more like a ball of white fluffy bubbles, by the time it is a fledgling it resembles the adults (both sexes are similar to each other), with the exception of the young having a yellow, rather than red, beak.

Actual size

ORDER	Suliformes
FAMILY	Fregatidae
BREEDING RANGE	Coastal subtropical and tropical Atlantic and eastern Pacific Oceans
BREEDING HABITAT	Offshore islands and mangrove cays
NEST TYPE AND PLACEMENT	Shallow platform nest of branches and twigs, on top of trees and bushes
CONSERVATION STATUS	Least concern
COLLECTION NO.	FMNH 20864

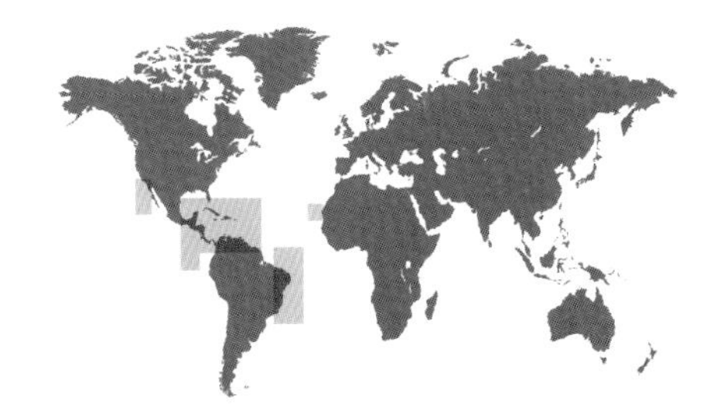

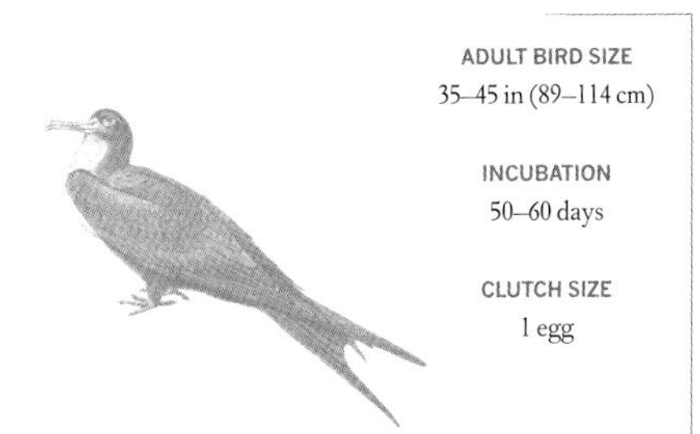

ADULT BIRD SIZE
35–45 in (89–114 cm)

INCUBATION
50–60 days

CLUTCH SIZE
1 egg

FREGATA MAGNIFICENS

MAGNIFICENT FRIGATEBIRD

SULIIFORMES

Magnificent Frigatebirds live and sleep on the wing when away from the nesting colonies, and cover thousands of miles and several oceans between breeding attempts. Thus most populations extensively share their genetic make-up with other distant breeders. Nevertheless, studies show that the Galapagos population is genetically distinct from other conspecifics, thereby influencing species' limits and regional conservation status.

The egg of the Magnificent Frigatebird is clear white, and measures 2⅔ x 1⅞ in (68 x 47 mm) in size. Both sexes incubate the egg, but the male abandons the chick soon after hatching, while the female remains to provision the young for nearly a year.

The species is a quintessential oceanic bird, but one that does not land on or dive in the water for food. It is a thief, or kleptoparasite, using its acrobatic flying skills to cover long distances to locate feeding frenzies of other seabirds. Once a successful target, such as a booby, is located, the Frigatebird pursues and pecks at the booby in mid-air, until it eventually regurgitates its freshly caught prey. The pirate-parasite Frigatebird then dives and catches the food before it hits the water.

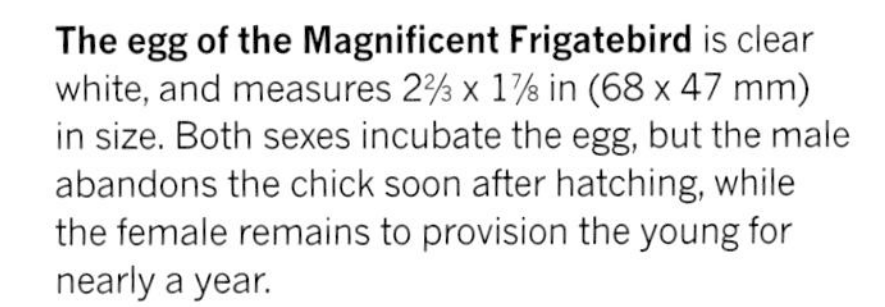

Actual size

ORDER	Pelecaniformes
FAMILY	Pelecanidae
BREEDING RANGE	North-central and western North America
BREEDING HABITAT	Isolated islands, typically in brackish or freshwater lakes
NEST TYPE AND PLACEMENT	Ground nest scrape, with twigs and branches, within large colonies
CONSERVATION STATUS	Least concern
COLLECTION NO.	FMNH 5109

ADULT BIRD SIZE
50–65 in (127–165 cm)

INCUBATION
30 days

CLUTCH SIZE
2–3 eggs

PELECANUS ERYTHRORHYNCHOS

AMERICAN WHITE PELICAN

PELECANIFORMES

Clutch

This pelican is a fishing expert, relying on shallow dives or quick plunges of its head to engulf prey underwater with the expandable pouch connected to its lower mandible. Often found in small groups, they can cooperate by driving schools of fish toward the shallow water, each member benefiting directly by more easily capturing more fish.

On the breeding grounds, these birds are clumsy walkers on solid ground. They rely on open water between their nests and any mammalian predators for safe and successful breeding. Direct disturbance by humans and dogs, or other terrestrial predators, can result in large-scale nest abandonment and failure. Both parents attend the eggs and provision the chicks by feeding them regurgitated prey. Once the chicks become mobile and begin to walk away from the natal nests, they form crèches, with the parents having to locate and identify their own young for feeding.

The egg of the American White Pelican is an immaculate white, and it measures 3½ x 2¼ in (90 x 57 mm) in dimensions. The eggs hatch on different days, and the younger chicks often die as a result of aggression and starvation.

Actual size

ORDER	Pelecaniformes
FAMILY	Pelecanidae
BREEDING RANGE	Coastal southern North America, the Caribbean, Central America, and coastal South America
BREEDING HABITAT	Coastal and estuarine marshes
NEST TYPE AND PLACEMENT	Large stick and grass platform in short trees, nests colonially
CONSERVATION STATUS	Least concern
COLLECTION NO.	FMNH 5113

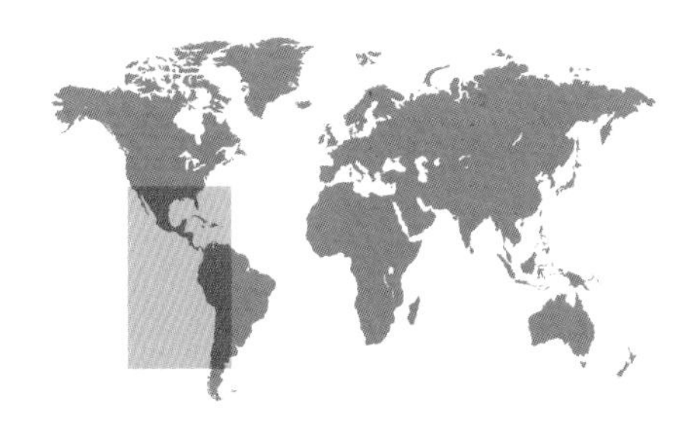

ADULT BIRD SIZE
39½–54 in (100–137 cm)

INCUBATION
29–32 days

CLUTCH SIZE
1–4 eggs

PELECANUS OCCIDENTALIS

BROWN PELICAN

PELECANIFORMES

Clutch

Brown Pelicans nest in large colonies in coastal swamps, a habitat that is vulnerable to hurricanes, oil spills, and other natural and human-caused catastrophes. There, even in typical years, they face varying levels of food supplies. As a result, Brown Pelicans engage in facultative siblicide: chicks tolerate each other only when food is plentiful, but in food-poor years they fight to the death over scraps of fish brought by the parents.

The Brown Pelican, along with the Peregrine Falcon (see page 282), is also a poster child for population recovery following the elimination of harmful pesticides in the larger ecosystem. Eggshell thinning caused by DDT led to the Brown Pelican being listed as endangered in the United States, even though its broad geographic distribution would have assured the survival of its populations elsewhere. Following the 1972 ban on DDT, its populations in the United States have made a dramatic recovery.

The egg of the Brown Pelican is chalky white, but quickly becomes soiled with mud and feces; it measures 2¾ x 1¾ in (70 x 45 mm) in size. The eggs are incubated from the time when they are laid, resulting in the asynchronous hatching of the young and a size hierarchy that favors the survival of older nestlings in food-poor years.

Actual size

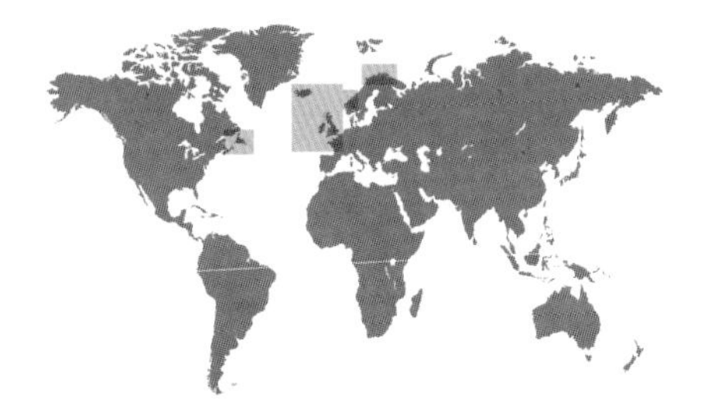

ORDER	Pelecaniformes
FAMILY	Sulidae
BREEDING RANGE	Eastern and western North Atlantic
BREEDING HABITAT	Coastal cliffs and plateaus, oceanic island
NEST TYPE AND PLACEMENT	Ground nest on cliffs and flat areas, built up into a small mound of dirt, mud, and feces
CONSERVATION STATUS	Least concern
COLLECTION NO.	FMNH 5040

ADULT BIRD SIZE
32–43½ in (81–110 cm)

INCUBATION
44 days

CLUTCH SIZE
1 egg

MORUS BASSANUS

NORTHERN GANNET

PELECANIFORMES

The predatory Northern Gannet is an aerial fish- and squid-specialist, locating its prey under the water surface, and diving on it at top speed. Once in the water, gannets can continue to propel themselves down in pursuit of the prey using their wings for movement and their eyes for guidance. Male and female Northern Gannets are similar in size, yet males dive deeper than females, and capture different types of prey.

In large and noisy breeding colonies, gannets locate their mates through individually distinct vocalizations. Once the pair bond is firmly established, both parents are actively involved in incubating the egg and feeding the young. This cooperation in breeding is absolutely critical for gannets, because during incubation, the parents rely on changing guard at the nest with one another, and leaving to refuel, every four to seven days.

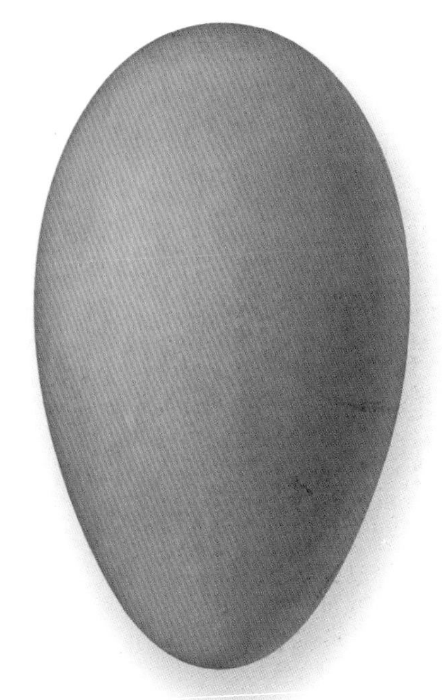

Actual size

The egg of the Northern Gannet is pale blue or green and immaculate in color, but it quickly becomes soiled with mud and feces. It measures 3¼ x 2 in (82 x 50 mm) in size. Gannets do not develop a brood patch, and instead both sexes use increased blood flow into their vascularized lobed feet to cradle and incubate the egg.

ORDER	Suliformes
FAMILY	Phalacrocoracidae
BREEDING RANGE	North America from western Alaska to Mexico and the Bahamas
BREEDING HABITAT	Wetlands and tree stands near coastal areas, and along inland rivers and lakes
NEST TYPE AND PLACEMENT	Large, chunky stick nest, on trees, cliffs, and occasionally on the ground of isolated islands
CONSERVATION STATUS	Least concern
COLLECTION NO.	FMNH 5063

ADULT BIRD SIZE
27½–35½ in (70–90 cm)

INCUBATION
25–28 days

CLUTCH SIZE
4–5 eggs

PHALACROCORAX AURITUS

DOUBLE-CRESTED CORMORANT

SULIFORMES

The Double-crested Cormorant has both suffered and benefited from human activities over the past 50 years. In the 1960s, due to the accumulation of DDT and other pesticides through the food chain, the breeding attempts of this species often failed, with fragile eggs breaking before the hatching date. Since the ban on DDT in the United States and Canada, these cormorants have significantly recovered in range and numbers.

Clutch

In turn, another series of anthropogenic events have helped this cormorant, including the accidental introduction of invasive fish species, like the Alewife, into the Great Lakes, and the spread of commercial catfish farming in southern portions of its range. The extra prey fish produced by these developments allowed this cormorant's breeding colonies to exceed pre-DDT sizes. Now, in response to sport-fishermen's complaints that the cormorants are eating too many fish, wildlife agencies have begun to allow the culling of Double-crested Cormorants on both the nesting and wintering grounds.

The egg of the Double-crested Cormorant is bluish white in coloration, has a chalky surface, and measures 2⅜ x 1½ in (61 x 39 mm) in size. These cormorants segregate together to form loud subcolonies at rookeries, and often successfully displace herons, ibises, and other waders also trying to nest at the same sites.

Actual size

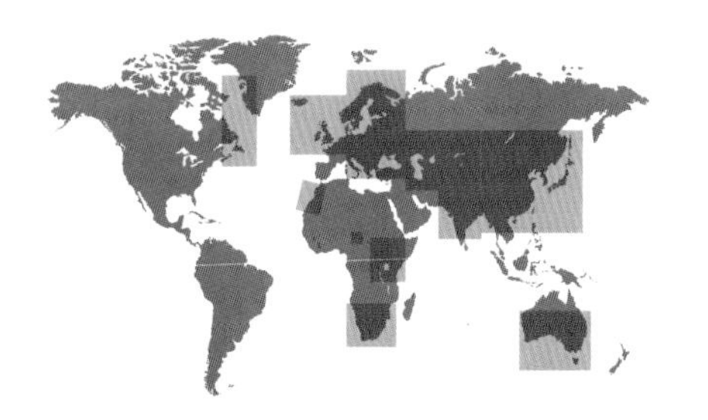

ORDER	Suliformes
FAMILY	Phalacrocoracidae
BREEDING RANGE	North America, central and Southeast Asia, Africa, Europe, and Australia
BREEDING HABITAT	Wetlands, lakes, rivers, sheltered inshore waters
NEST TYPE AND PLACEMENT	Stick nest in trees, on cliff ledges, and on the ground on rocky islands free of predators
CONSERVATION STATUS	Least concern
COLLECTION NO.	FMNH 5058

ADULT BIRD SIZE
33–35½ in (84–90 cm)

INCUBATION
28–31 days

CLUTCH SIZE
3–5 eggs

PHALACROCORAX CARBO

GREAT CORMORANT

SULIFORMES

Clutch

This cormorant has a broad distribution, breeding on all continents except for South America and Antarctica. The species is a fish-specialist, diving underwater to chase prey, to snap (but not spear) it firmly with its hooked beak. In parts of east Asia people capture and use these birds for traditional fishing, keeping the bird from swallowing the fish it grabs by placing a tight neck ring below the throat. To allow underwater diving and pursuit of prey, the eyes of this cormorant accommodate the dramatic change in the optical properties of air versus water, and allow the birds to see schools of fish and grab their prey from above the surface of the water.

Though mostly dark in plumage, during the breeding season the thighs and the heads of both males and females are adorned with white-tipped feathers. Pairs are faithful to their nests and colony sites, returning year after year to breed at the same location. The acidic guano of these birds rapidly kills off nesting trees; they become dead and bare, and this allows birders to distinguish trees with cormorant nests from trees with heron and egret nests.

Actual size

The egg of the Great Cormorant is pale bluish or green, clear of markings, but may be covered with a white chalky layer. It measures 2½ x 1⅝ in (63 x 41 mm) in size. In North America, the species often makes its nest in mixed-species colonies with other cormorants and gulls.

ORDER	Pelecaniformes
FAMILY	Anhingidae
BREEDING RANGE	Southern United States, Central America, and tropical South America
BREEDING HABITAT	Wetlands, slow-moving rivers, mangrove channels, and lakes
NEST TYPE AND PLACEMENT	Stick and twig nest, built in trees near shorelines of lakes and rivers
CONSERVATION STATUS	Least concern
COLLECTION NO.	FMNH 5932

ADULT BIRD SIZE
29½–37½ in (75–95 cm)

INCUBATION
25–30 days

CLUTCH SIZE
3–5 eggs

ANHINGA ANHINGA

ANHINGA

PELECANIFORMES

The word "anhinga" comes from the Brazilian Tupi language, denoting a devil- or snake-bird. This name comes from the typical appearance of Anhingas, with their curved necks and long beaks moving along above the water with the rest of the body submerged. The female and the male of the species are similar in size but can be told apart, even when swimming, by the lighter buff face and chest feathers of the female.

Once males and females arrive at the breeding sites, pairs form quickly, often within a few days. Nesting activities may become highly synchronized between different pairs, especially where this species forms loose colonies of dozens or hundreds of pairs. The female takes one or more days between laying subsequent eggs, and so the clutch is completed within about a week. During laying she remains on the nest, and relies on the male feeding her, to prevent theft of nesting materials by nearby pairs. Once incubation begins, both parents take turns to warm the eggs and, later, feed the young.

Clutch

The egg of the Anhinga is pale green in color, clear from speckles, and measures 2⅛ x 1⅓ in (53 x 35 mm). The eggs hatch asynchronously, and the chicks undergo several changes in color during their first weeks, from naked to tan and then to white down.

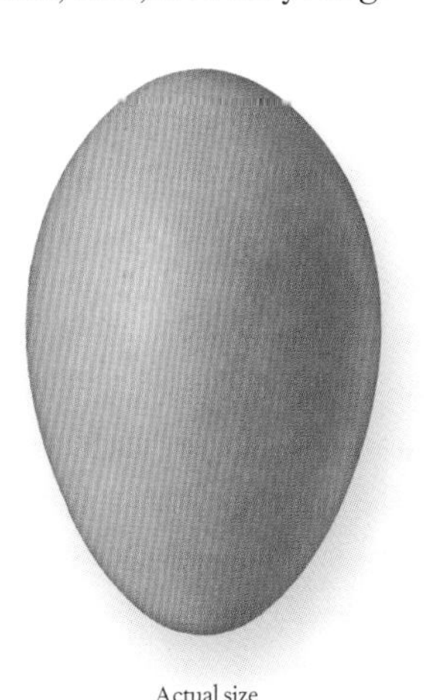

Actual size

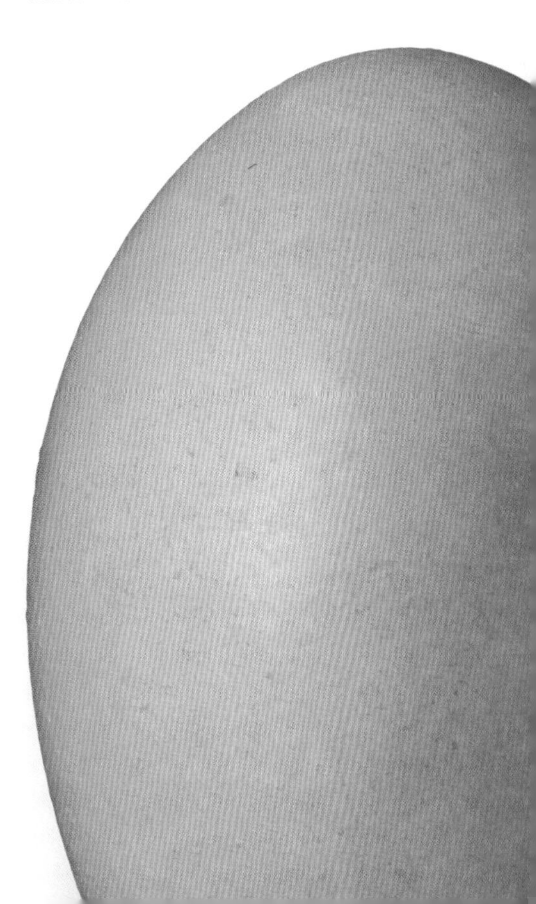

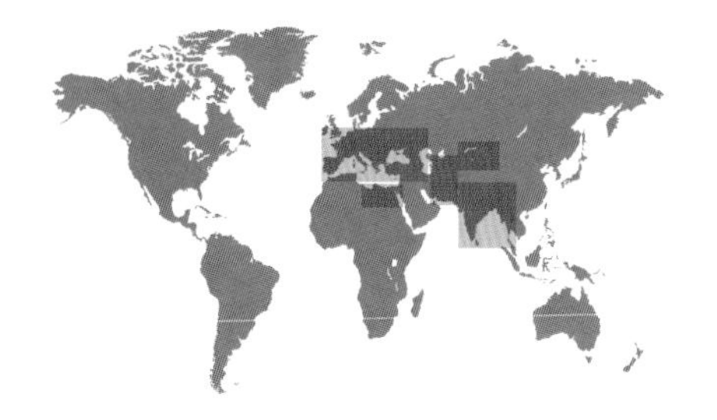

ORDER	Charadriiformes
FAMILY	Burhinidae
BREEDING RANGE	Central Europe, central and south Asia, north Africa
BREEDING HABITAT	Wetlands, open fields
NEST TYPE AND PLACEMENT	Minimal scraping on the ground, barely lined
CONSERVATION STATUS	Least concern
COLLECTION NO.	FMNH 21376

ADULT BIRD SIZE
15½–17½ in (40–44 cm)

INCUBATION
24–26 days

CLUTCH SIZE
2–3 eggs

BURHINUS OEDICNEMUS

EURASIAN THICK-KNEE

CHARADRIIFORMES

Clutch

The egg of the Eurasian Thick-knee is beige in background color, heavily blotched with brown markings, and 2⅛ x 1½ in (54 x 39 mm) in size. This species shows a fixed-action pattern of compulsively retrieving any egg displaced from the shallow nest scraping, thereby assuring incubation success of its entire clutch.

The Eurasian Thick-knee, or Stone-curlew, is a secretive and often hard-to-spot shorebird relative, that typically occurs in dry open habitats, far from permanent water shores. Its population is widely distributed across three continents, but it remains locally vulnerable to habitat loss due to human development and agricultural expansion. It feeds after sunset on invertebrates and small reptiles and mammals, relying on its large eyes, strong legs, and powerful feet to find and chase down its prey.

Both sexes incubate the eggs, and the highly cryptic plumage of the adults, and the precocial hatchlings' own cryptic down, as well as the species' crepuscular and nocturnal foraging behaviors, ensure that breeding families, adults, eggs, and chicks generally escape notice.

Actual size

ORDER	Charadriiformes
FAMILY	Charadriidae
BREEDING RANGE	South America
BREEDING HABITAT	Wet or seasonal grasslands, pastures
NEST TYPE AND PLACEMENT	Sparsely lined scraping, on ground in open fields
CONSERVATION STATUS	Least concern
COLLECTION NO.	FMNH 14754

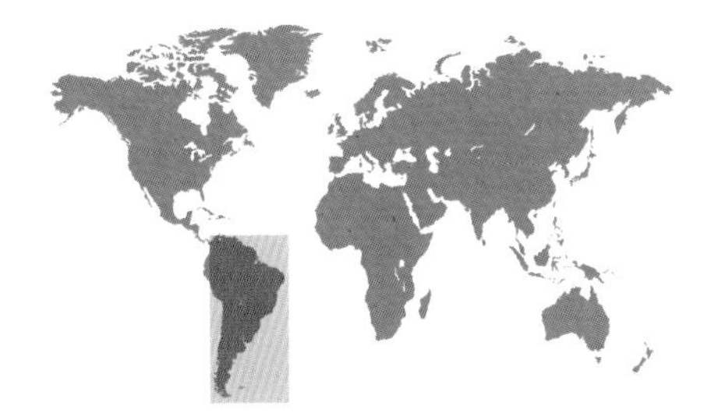

ADULT BIRD SIZE
12⅝–15 in (32–38 cm)

INCUBATION
27–30 days

CLUTCH SIZE
1–4, typically 3 eggs

VANELLUS CHILENSIS

SOUTHERN LAPWING

CHARADRIIFORMES

Southern Lapwings are conspicuous inhabitants of South American grasslands. They are both colorful in appearance and noisy in behavior. They may breed as pairs in loose colonies, with colony members alerting others to approaching danger and mobbing potential predators as a group. This way, the eggs in the highly vulnerable ground nests are kept safe for the duration of the incubation.

In addition to the adults mobbing predators to protect their nests, the eggs and the hatching chicks are also highly cryptic in coloration, blending into the background of dirt and dry vegetation. Once the chicks hatch and dry, they are fully independent after the first day, and feed and hide on their own, although their parents guide them to safety and continue to mob and keep predators away.

Clutch

The egg of the Southern Lapwing is 2 x 1¼ in (50 x 33 mm) in size, dark green to beige in background color and heavily maculated with brown spots of different sizes and distinct patterns. The clutch's overall appearance is such that it is well camouflaged against the sparse nest lining and its surroundings, and thus thoroughly hidden from visually oriented avian and mammalian predators.

Actual size

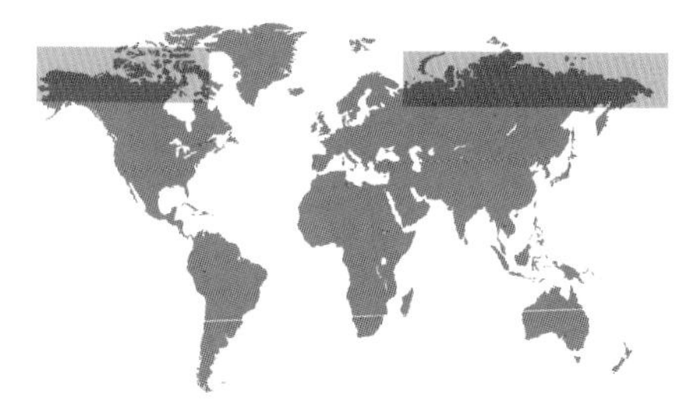

ORDER	Charadriiformes
FAMILY	Charadriidae
BREEDING RANGE	Arctic regions of North America and Eurasia
BREEDING HABITAT	Open grasslands, dry heath tundra
NEST TYPE AND PLACEMENT	Shallow scrape in gravel, lined with lichens, twigs, and pebbles
CONSERVATION STATUS	Least concern
COLLECTION NO.	FMNH 15062

ADULT BIRD SIZE
10½–12 in (27–30 cm)

INCUBATION
26–27 days

CLUTCH SIZE
4 eggs

PLUVIALIS SQUATAROLA

BLACK-BELLIED PLOVER

CHARADRIIFORMES

Clutch

The Black-bellied Plover, known as the Grey Plover in Europe, is one of the most widely dispersed shorebird species in the world. This is because it migrates long distances across continents and over open water from its Arctic breeding grounds to coastal wintering areas in temperate regions of the northern hemisphere, but also in coastal regions of South America, Africa, and Australasia. Because young of this species do not begin breeding until two years of age, these plovers can be seen throughout the year at any region of their worldwide distribution, which accounts for their cosmopolitan occurrence and familiarity to birdwatchers.

Black-bellied Plovers serve an important role as sentinels for foraging shorebirds, detecting approaching predators. They are the first to call the alarm and take flight, alerting the other species in the flock. Incubating adults are also quick to flush from the nest when predators approach, but once the danger is over, they reliably return, and do not abandon the clutch of eggs.

Actual size

The egg of the Black-bellied Plover is pinkish to brownish, with dark spotting heaviest around the blunt end. It measures 2⅛ x 1½ in (53 x 37 mm) in size. Both parents attend the nest and the chicks, before they become independent at five to six weeks of age.

ORDER	Charadriiformes
FAMILY	Charadriidae
BREEDING RANGE	Canadian and Alaskan Arctic
BREEDING HABITAT	Open, rocky and grassy hillside slopes in the tundra
NEST TYPE AND PLACEMENT	Scrape in the ground, lined with lichens, grass, and leaves
CONSERVATION STATUS	Least concern
COLLECTION NO.	FMNH 16021

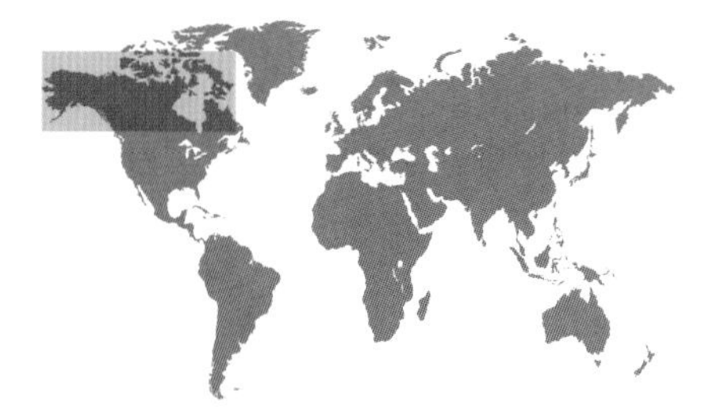

ADULT BIRD SIZE
9½–11 in (24–28 cm)

INCUBATION
26–27 days

CLUTCH SIZE
4 eggs

PLUVIALIS DOMINICA

AMERICAN GOLDEN PLOVER

CHARADRIIFORMES

American Golden Plovers are a conspicuous sight in their breeding grounds, where they nest in the open, dry Arctic tundra, and display aggressively to neighbors to protect their breeding territories. Some individuals are also territorial in the wintering grounds, far to the south in Patagonia, protecting a patch of the coastal mudflat where they spend most daylight hours foraging. On the breeding grounds, males settle near suitable nesting sites and scrape a hollow in the ground for the nest. They incubate the eggs during the day, while the females incubate at night.

This species is also conspicuous when it forms large flocks during migration, and moves directly south from the East Coast of North America to the eastern shores of South America, without a stopover. It has been said that Columbus himself may have noticed flocks of Golden Plovers flying south over the open ocean as his ships were approaching the Caribbean after spending 65 days at sea sailing from Europe.

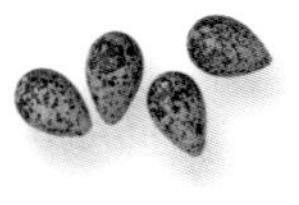

Clutch

The egg of the American Golden Plover is white to buff in background color, and heavily blotched with brown and black spots. It measures 1⅞ x 1⅓ in (48 x 33 mm) in size. The chicks hatch fully downed, leave the nest within hours, and can feed themselves within just one day.

Actual size

ORDER	Charadriiformes
FAMILY	Charadriidae
BREEDING RANGE	Southern and central Europe, Africa, and Asia
BREEDING HABITAT	Brackish inland lakes, sparsely vegetated sandy seacoasts
NEST TYPE AND PLACEMENT	Shallow scrape in the sand
CONSERVATION STATUS	Least concern, but locally endangered
COLLECTION NO.	FMNH 15319

ADULT BIRD SIZE
6–6½ in (15–17 cm)

INCUBATION
25–26 days

CLUTCH SIZE
3–5 eggs

CHARADRIUS ALEXANDRINUS

KENTISH PLOVER

CHARADRIIFORMES

Clutch

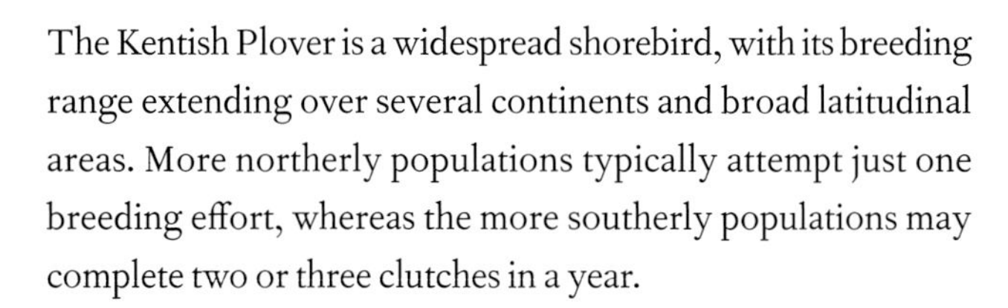

The Kentish Plover is a widespread shorebird, with its breeding range extending over several continents and broad latitudinal areas. More northerly populations typically attempt just one breeding effort, whereas the more southerly populations may complete two or three clutches in a year.

Kentish Plovers do not feed their young. But it is clear that parental care by both parents provides important benefits to the chicks, because the natural loss or experimental removal of a single parent (whether the female or the male) results in fewer surviving young. Surprisingly, however, some individual parents still abandon their growing young to establish a new pair and attempt to breed with another mate, leaving their original mate to protect the brood until independence. This brood desertion, involving both females and males, also means that Kentish Plovers can be both polygynous (one male, multiple females) and polyandrous (one female, multiple males) during each breeding season.

The egg of the Kentish Plover is buff to beige in background, and heavily maculated with blackish spots and lines. It is 1¼ x ⅞ in (31 x 23 mm) in size. The male scrapes the nest and incubates at night, with the female taking charge of the eggs during the daylight hours.

Actual size

ORDER	Charadriiformes
FAMILY	Charadriidae
BREEDING RANGE	Arctic Canada and Alaska
BREEDING HABITAT	Mossy tundra, gravel and sandy flats, wetlands
NEST TYPE AND PLACEMENT	Shallow scrape in the ground, lined with pebbles and shells
CONSERVATION STATUS	Least concern
COLLECTION NO.	FMNH 2255

ADULT BIRD SIZE
6½–7½ in (17–19 cm)

INCUBATION
26–31 days

CLUTCH SIZE
4 eggs

CHARADRIUS SEMIPALMATUS

SEMIPALMATED PLOVER

CHARADRIIFORMES

The Semipalmated Plover is a small but assertive shorebird: males vehemently protect and defend the breeding territory from intruding neighbors. If calling and running to patrol the breeding site are not enough, the male will often take flight and circle the territory repeatedly. He also uses flight, in a butterfly-like flutter, as a courtship display to attract the female's attention to form a season-long pair bond.

When not flying, displaying, or courting, these plovers glean small invertebrates from muddy flats, shallow waters, and grassy patches. If no prey is in sight, the birds may occasionally stir the mud by shaking one of their feet to dislodge prey, grabbing it quickly with their short beaks. This foraging strategy works efficiently both in the breeding grounds and on migratory stopover sites, where fast feeding on plentiful prey is essential for rapid refueling before the next long stretch of nonstop flight.

Clutch

Actual size

The egg of the Semipalmated Plover is buff in color, has spotting with dark brown and black speckles, and measures 1¼ x ⅞ in (32 x 23mm) in dimensions. Both parents incubate the eggs and attend the chicks, but the female often abandons the brood before the young are fully independent.

ORDER	Charadriiformes
FAMILY	Charadriidae
BREEDING RANGE	North America, Ecuador and Peru
BREEDING HABITAT	Flat, dry grassy areas, often mixed with rocks and pebbles, occasionally flat building roofs
NEST TYPE AND PLACEMENT	Shallow scrape on ground, amidst pebbles and rocks
CONSERVATION STATUS	Least concern
COLLECTION NO.	FMNH 8021

ADULT BIRD SIZE
8–11 in (20–28 cm)

INCUBATION
22–28 days

CLUTCH SIZE
4–6 eggs

CHARADRIUS VOCIFERUS

KILLDEER

CHARADRIIFORMES

Clutch

Although technically the Killdeer is a true plover and, thus, a shorebird, it frequently lives away from water. Its familiar, loud call of "killdeer" can be heard in dry grassy fields or even graveled parking lots. This species is an example of commensalism, as it extensively shares with people different parklands or more densely built, flat urban habitats, where it breeds successfully despite the dangers presented by people, cars, cats, and dogs.

The Killdeer is best known for its conspicuous broken-wing display in the breeding season. In the presence of real or potential danger, Killdeers pretend to be unable to fly, pulling their wings flatly behind them, while they lead predators away from the eggs or young chicks. When the parent reaches a safe distance, it simply flies up and returns to its progeny to continue guard duty.

Actual size

The egg of the Killdeer is buff to beige, with dark markings of brown and black speckles, and it is 1½ x 1 in (38 x 27 mm) in size. The parents may make several scrapings as potential nests, perhaps to confuse predators, before the female begins to lay her eggs at one of these sites.

ORDER	Charadriiformes
FAMILY	Recurvirostridae
BREEDING RANGE	Southern North America, Central and northeastern South America, the Caribbean, Hawaii
BREEDING HABITAT	Freshwater ponds and coastal wetlands
NEST TYPE AND PLACEMENT	A scraped hollow in the ground, lined sparsely with twigs and pebbles, or not at all
CONSERVATION STATUS	Least concern, endangered in Hawaii
COLLECTION NO.	FMNH 16162

ADULT BIRD SIZE
14–15½ in (35–39 cm)

INCUBATION
21–26 days

CLUTCH SIZE
4 eggs

HIMANTOPUS MEXICANUS

BLACK-NECKED STILT

CHARADRIIFORMES

Like all stilts, the Black-necked Stilt has a typically distinctive look that includes a long straight beak and a curved neck, a strongly patterned black-and-white plumage, and incredibly thin, but long, legs. In fact, stilts are only second to flamingos when it comes to the relative length of these legs compared to the rest of the body and beak. Wading through shallow water, these birds are adept at capturing aquatic insects and other prey under the surface.

While foraging in flocks, many birds use sentries to watch for danger while the rest of the flock feeds. Stilts have evolved another mechanism: left and right sidedness in visual functions. They use their right eye to detect and reach for prey, and their left eye to spot predators and potential competitors or mates. This lateralization, much like handedness in humans, is controlled by different sides of the brain.

Clutch

The egg of the Black-necked Stilt is beige to olive gray in background color, and heavily speckled with brown spots. It is 1¾ x 1³⁄₁₆ in (45 x 30 mm) in size. The nest is often sited on an elevated area near the shore or on small islands, amidst clumps of vegetation, to further protect the cryptic eggs from visually driven land predators, including many mammals.

Actual size

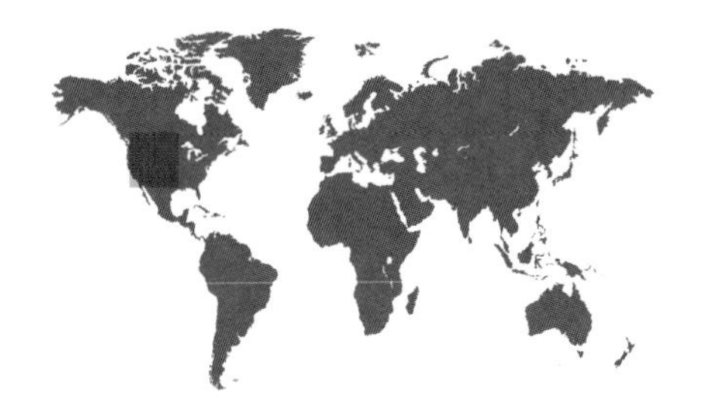

ORDER	Charadriiformes
FAMILY	Recurvirostridae
BREEDING RANGE	Temperate North America
BREEDING HABITAT	Wetlands, ponds, and coastal marshes
NEST TYPE AND PLACEMENT	Ground scraping, in the shape of a shallow saucer; may be lined with feathers, grasses, and twigs, or completely empty
CONSERVATION STATUS	Least concern
COLLECTION NO.	FMNH 16150

ADULT BIRD SIZE
17–18½ in (43–47 cm)

INCUBATION
22–29 days

CLUTCH SIZE
3–4 eggs

RECURVIROSTRA AMERICANA

AMERICAN AVOCET

CHARADRIIFORMES

Clutch

American Avocets breed in both saltwater and freshwater marshes and ponds and nest in small, loose colonies of between 10 and 15 pairs. They forage in shallow open waters, moving their heads back and forth just beneath the surface to catch crustaceans and insects with their gently upward-curving bill. Avocets often flock and forage together. While most individuals are occupied in scanning and sifting through the water for prey, a handful of others in the flock are keeping an eye out for predators and other dangers.

Visual displays are also important for the courtship of the American Avocet, as elaborate movements of the body and the neck are part of both nuptial and post-mating behavioral interactions between the pair-bonded mates. Once the eggs are laid, both parents incubate, and once the chicks are hatched, both adults attend the young, leading them to productive feeding areas and away from predators.

Actual size

The egg of the American Avocet is greenish brown in color, with irregular dark brown and black markings throughout. It is pointed in shape, and measures 2 x 1⅓ in (50 x 34 mm). The chicks will leave the nest within 24 hours of hatching.

ORDER	Charadriiformes
FAMILY	Jacanidae
BREEDING RANGE	South and Southeast Asia, Indonesian archipelago
BREEDING HABITAT	Slow-moving rivers and lakes, with floating vegetation
NEST TYPE AND PLACEMENT	A mound of twigs and leaves, positioned on top of floating mats and leaves of water plants
CONSERVATION STATUS	Least concern, endangered in Taiwan
COLLECTION NO.	FMNH 15312

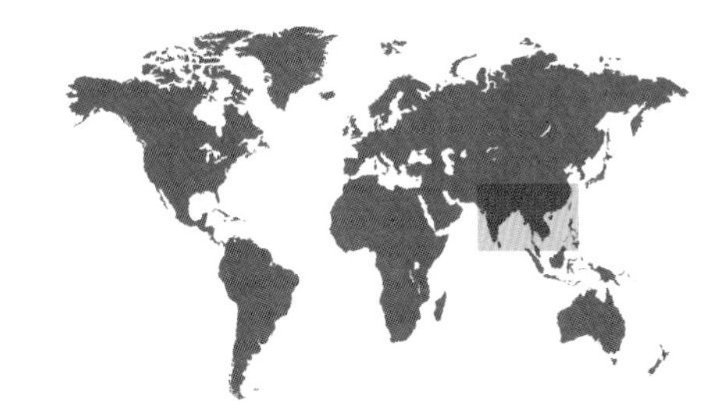

ADULT BIRD SIZE
11–12 in (28–31 cm)
(15½–19½ in / 40–50 cm with tail plumes)

INCUBATION
22–28 days

CLUTCH SIZE
4 eggs

HYDROPHASIANUS CHIRURGUS

PHEASANT-TAILED JACANA

CHARADRIIFORMES

The Pheasant-tailed Jacana is a conspicuous marsh bird, with its prominent long tail feathers fully grown only in the breeding season. It is the female, and not the male, however, who displays the longest and most distinctly colored plumes. In fact, the entire family system is driven and controlled by the female: she establishes and defends the territory from other females and she protects the nests and her mates and young from predators. In turn, the males build the nest, incubate the eggs, and attend the young almost exclusively. Females are often mate bonded to several males, up to four, rendering this mating system polyandrous (one female, several males).

Males are smaller than and subordinate to the female, but they do have one weapon to ensure that they are raising their own young and not those of another male: a male can commit infanticide by tossing the earlier laid eggs (which may not be his) from his floating nest into the water. The female then quickly mates with the male again and lays replacement eggs in his nest; this time, hopefully, they will have been fertilized by this male who will then provide paternal care.

Clutch

The egg of the Pheasant-tailed Jacana is olive green to deep bronze in color with small fine speckles, and 1⅜ x 1 in (36 x 26 mm) in size. The nest and the eggs are often destroyed by predators or floods, but the female, emancipated from the costs of caring for the young, can quickly lay replacement clutches.

Actual size

ORDER	Charadriiformes
FAMILY	Jacanidae
BREEDING RANGE	South and Southeast Asia
BREEDING HABITAT	Shallow ponds and lakes
NEST TYPE AND PLACEMENT	A floating platform, on top of emerging leaves and other vegetation
CONSERVATION STATUS	Least concern
COLLECTION NO.	FMNH 15310

ADULT BIRD SIZE
11–12 in (28–31 cm)

INCUBATION
22–24 days

CLUTCH SIZE
4 eggs

METOPIDIUS INDICUS

BRONZE-WINGED JACANA

CHARADRIIFORMES

Clutch

The Bronze-winged Jacana is predominantly polyandrous: a female defends large breeding territories from other females, and maintains a harem of males with whom she mates repeatedly. Repeated mating with a male ensures both that he does not depart from the harem, and that he will be readily available to incubate and raise the female's eggs and chicks.

In turn, the males are territorial against other, neighboring males, whether or not they belong to the same female's harem. These male territories are used for feeding and breeding exclusively by a male, the polyandrous female, and their young. The males are smaller and subordinate to the females, but among males, the relatively larger ones are able to secure better habitat patches, with more emergent and floating vegetation. This increases a male's chances of successful foraging and reproduction.

Actual size

The egg of the Bronze-winged Jacana is yellowish buff to dark, deep red, with long black lines scattered across it. It measures 1⅜ x 1 in (36 x 25 mm) in size. It is oval and elongated in shape, and the surface is bright and glossy.

ORDER	Charadriiformes
FAMILY	Jacanidae
BREEDING RANGE	Mexico and Central America, the Caribbean
BREEDING HABITAT	Wetlands, lakes, and rivers with floating vegetation
NEST TYPE AND PLACEMENT	A compact mound of pulled stems and leaves, formed into a floating mat
CONSERVATION STATUS	Least concern
COLLECTION NO.	FMNH 1721

ADULT BIRD SIZE
8½–9½ in (21–24 cm), females are twice as heavy as mates

INCUBATION
28 days

CLUTCH SIZE
4 eggs

JACANA SPINOSA

NORTHERN JACANA

CHARADRIIFORMES

In common with all other jacanas, a female Northern Jacana keeps a harem of males. She mates with each in turn and can breed all year long when water levels are stable. Each male incubates and tends to his own young. Polyandry, thus, frees the female from the cost of nest building, incubating, and protecting the young. She trades off parenting for increased egg production.

Jacanas live, feed, mate, and breed in marshes. They are poor fliers, and when they take to the wing, their flight is labored and their long legs and toes are simply dragged behind, dangling below the body. The Northern Jacana is called the "Jesus Bird" in Jamaica, because its long, thin toes and light body weight enable this bird to step across stems and leaves of floating vegetation, so appearing as if it can walk on water.

Clutch

Actual size

The egg of the Northern Jacana is brown in color with black markings, and measures 1³⁄₁₆ x ⅞ in (30 x 23 mm) in size. Jacana hatchlings are called "downies," and stay with the father bird for protection, although they are fully capable of feeding themselves independently.

ORDER	Charadriiformes
FAMILY	Scolopacidae
BREEDING RANGE	Arctic Europe and Asia
BREEDING HABITAT	Shores of rivers and lakes in taiga woodlands
NEST TYPE AND PLACEMENT	Shallow depression on the ground, near the shore, in the open, or near a tussock of grasses
CONSERVATION STATUS	Least concern
COLLECTION NO.	FMNH 18632

ADULT BIRD SIZE
8½–10 in (22–25cm)

INCUBATION
21–22 days

CLUTCH SIZE
3–4 eggs

XENUS CINEREUS

TEREK SANDPIPER

CHARADRIIFORMES

Clutch

The egg of the Terek Sandpiper is light gray in color, with brown to black speckles and blotches throughout. It is 1½ x 1⅛ in (38 x 28 mm) in dimensions. Both parents incubate the eggs, and the precocial young become flighted after just two weeks.

This medium-sized sandpiper runs rapidly, propelling its solid body over mudflats. Like an avocet, it has a longish, slightly upward-curving beak, with which it forages for insects and crustaceans by chasing them down directly and capturing them on the move. It often washes its food at the shoreline before eating it. The Terek Sandpiper is so distinctive evolutionarily, that it is placed in its own genus; its closest phylogenetic relatives are the larger shanks and phalaropes.

The breeding range of this species covers such remote areas of taiga forest and tundra across Siberia that changes in population trends and potential threats in the breeding ranges are difficult to monitor for conservation management. Habitat loss on migratory routes, especially in coastal Southeast Asia, and continued use of the pesticide DDT in southern India, may be negatively affecting populations.

Actual size

ORDER	Charadriiformes
FAMILY	Scolopacidae
BREEDING RANGE	Europe and Asia
BREEDING HABITAT	Lake and river shores, and wetlands in the Palearctic
NEST TYPE AND PLACEMENT	Shallow depression in the ground
CONSERVATION STATUS	Least concern
COLLECTION NO.	FMNH 15330

ADULT BIRD SIZE
7–8 in (18–20 cm)

INCUBATION
21–22 days

CLUTCH SIZE
4 eggs

ACTITIS HYPOLEUCOS

COMMON SANDPIPER

CHARADRIIFORMES

The Common Sandpiper is a solitary species generally found along freshwater streams and lakes, although small flocks form during migration. This sandpiper picks invertebrate prey from along the shore, often washing it before eating. In their wintering grounds in Africa, the birds will perch on the backs of hippopotamuses and crocodiles, feeding on insects and parasites found there.

Both father and mother incubate the eggs, but once these hatch, the female may leave the family behind before the young become independent. Once the young are flighted, they make their way to wintering grounds in Africa, South Asia, or Australia, where they may spend the next several months. They return to the breeding grounds in the northern hemisphere only in their second full year of life.

Clutch

Actual size

The egg of the Common Sandpiper is pink gray in background color, with fine brownish red maculation. It is pointed, and measures 1½ x 1 in (36 x 26 mm) in size. The eggs are cryptic already, but the nest can be further hidden among shrubs and trees for better protection from predators that hunt by sight.

ORDER	Charadriiformes
FAMILY	Scolopacidae
BREEDING RANGE	Arctic and north temperate North America
BREEDING HABITAT	In wooded areas, near wetlands and lakes
NEST TYPE AND PLACEMENT	In trees, usurping open cup nests built and used by songbirds
CONSERVATION STATUS	Least concern
COLLECTION NO.	FMNH 100

ADULT BIRD SIZE
7½–9 in (19–23 cm)

INCUBATION
23–24 days

CLUTCH SIZE
4–5 eggs

TRINGA SOLITARIA

SOLITARY SANDPIPER

CHARADRIIFORMES

Clutch

True to its name, the Solitary Sandpiper never flocks, whether in the breeding season, along its migratory routes, or on its wintering grounds in South America. Perhaps because it lives mostly on its own, it is one of the first species in mixed-species flocks to give high-pitched alarm calls in response to approaching predators.

The species was described by the American ornithologist, Alexander Wilson in 1813, yet its nest and eggs were only first documented 90 years later, in 1903. This is because its breeding range spans the remote northern coniferous forests of Canada and Alaska and because it nests in trees. The Solitary Sandpiper is one of only two sandpiper species that typically usurps songbird nests. When American Robin (see page 503) or Rusty Blackbird (see page 612) nests are not available, the female will build her own cup-shaped nest high up on a tree branch. It is not yet known how the hatchlings get down from the nest to find their way to nearby water to begin feeding.

The egg of the Solitary Sandpiper is olive grey in color, with brown markings, and measures 1½ x 1 in (36 x 26 mm) in size. The female refurbishes the songbird nest that the male locates, and they take turns to incubate the eggs.

Actual size

ORDER	Charadriiformes
FAMILY	Scolopacidae
BREEDING RANGE	Northern Europe and Asia
BREEDING HABITAT	Open boggy taiga, wet and swampy forest clearings
NEST TYPE AND PLACEMENT	A ground scrape in short vegetation
CONSERVATION STATUS	Least concern
COLLECTION NO.	FMNH 15325

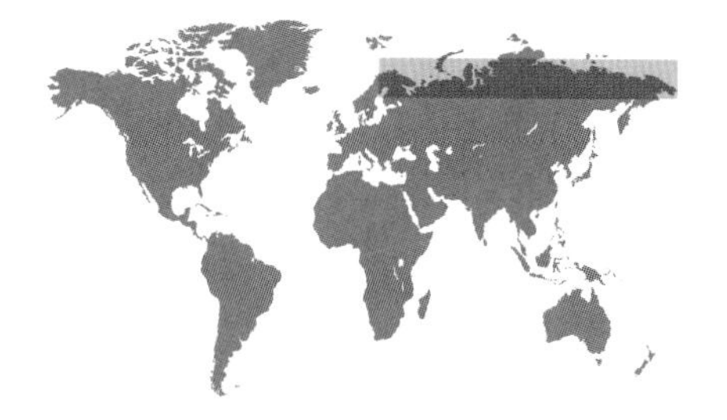

ADULT BIRD SIZE
$11\frac{1}{2}$–12 in (29–31 cm)

INCUBATION
23–24 days

CLUTCH SIZE
4 eggs

TRINGA ERYTHROPUS

SPOTTED REDSHANK

CHARADRIIFORMES

The Spotted Redshank is a long-legged shorebird that forages both on shore and in shallow water in search of insect and crustacean prey. Unlike most sandpipers, however, redshanks also occasionally feed on aquatic larvae by swimming out to the open water and submerging their heads fully to capture prey in the water column.

For nesting purposes, this species selects a tussock of grass or a shrub on the shore of a river or pond. The eggs are well hidden, both by their cryptic coloration and by the shadows of the vegetation cast onto the nest. Visual modeling of predator eyesight has revealed that camouflage of eggs is most effective when in shaded areas. Both parents incubate the eggs but the female typically deserts the clutch before hatching, leaving the male to complete incubation and to protect the precocial young until independence.

Clutch

The egg of the Spotted Redshank is greenish in color, and has large brown-black blotching; it measures $1\frac{7}{8}$ x $1\frac{1}{4}$ in (47 x 32 mm) in size. The parents use tree trunks and branches near the nest as lookout posts to watch for would-be predators.

Actual size

ORDER	Charadriiformes
FAMILY	Scolopacidae
BREEDING RANGE	Northern Europe and Asia
BREEDING HABITAT	Clearings in taiga woodlands, wet bogs, marshes, and lake shores
NEST TYPE AND PLACEMENT	Shallow scrape in the ground, lined sparsely with grasses
CONSERVATION STATUS	Least concern
COLLECTION NO.	FMNH 16479

ADULT BIRD SIZE
12–14 in (30–35 cm)

INCUBATION
24–27 days

CLUTCH SIZE
3–5 eggs

TRINGA NEBULARIA

COMMON GREENSHANK

CHARADRIIFORMES

Clutch

The largest species in its genus, the Common Greenshank is conspicuous both while feeding and when breeding. It is an active forager, chasing down prey on foot and taking sharp turns to capture the crustacean or insect it is after. Alternatively, it will wade belly-deep into water, swimming and walking while probing for prey.

On the breeding grounds, the males arrive before the females to establish breeding territories. Males display to interested females using loud calls and upward jumps and flights, occasionally turning and tumbling in mid-air on their way to land. The male also prepares several nesting scrapes, and the female chooses one in which she lays her eggs. The nest is typically made beside a fallen log or a tall hummock, providing cover for the clutch and a lookout post for the adults. Both parents incubate the eggs, taking turns, and jointly protect the young.

The egg of the Common Greenshank is gray to buff in color, has red spotting, and measures 2 x 1½ in (51 x 35 mm) in size. Some nests contain eight eggs, twice the typical clutch size, because the nest is used by two females mated to the same male.

Actual size

ORDER	Charadriiformes
FAMILY	Scolopacidae
BREEDING RANGE	Eastern and midwestern North America
BREEDING HABITAT	Coastal salt marshes and barrier islands, prairie freshwater wetlands, and grasslands
NEST TYPE AND PLACEMENT	Ground nest, hidden in grasses, often in a colony, lined with grasses and pebbles
CONSERVATION STATUS	Least concern
COLLECTION NO.	FMNH 16129

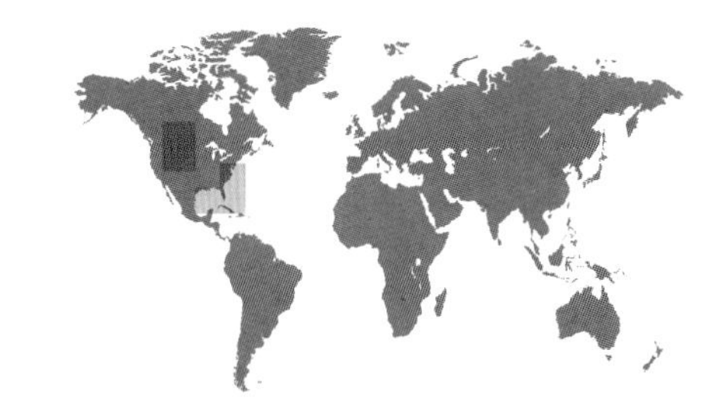

ADULT BIRD SIZE
13–16 in (33–41 cm)

INCUBATION
22–29 days

CLUTCH SIZE
4 eggs

TRINGA SEMIPALMATA

WILLET

CHARADRIIFORMES

Willets forage by touch, using their sensitive beak tips to locate and catch aquatic prey both by day and night. On the breeding ground, the male bird defends the area around the nest, and the nearby shoreline, against other Willets. He patrols the boundaries of the territory using a ritualized walk, which can escalate into a physical attack.

Members of a pair remain faithful to one another during the breeding season, and across the years. They often return to breed in the same area and pair up repeatedly with one another. The two sexes also inspect potential nest sites together, with the male making trial scrapes and the female deciding which one to use for the nest. Both parents incubate the eggs, but only the male spends the full night on the nest. Once the chicks hatch, it is the female that leaves the male and their brood behind, typically up to several weeks before the young become independent.

Clutch

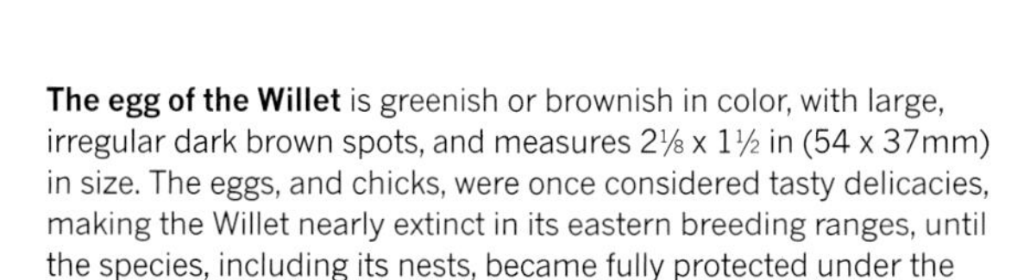

The egg of the Willet is greenish or brownish in color, with large, irregular dark brown spots, and measures 2⅛ x 1½ in (54 x 37mm) in size. The eggs, and chicks, were once considered tasty delicacies, making the Willet nearly extinct in its eastern breeding ranges, until the species, including its nests, became fully protected under the 1918 Migratory Bird Treaty.

Actual size

ORDER	Charadriiformes
FAMILY	Scolopacidae
BREEDING RANGE	Arctic and subarctic North America
BREEDING HABITAT	Boreal forest ponds and grassy lake shores
NEST TYPE AND PLACEMENT	Ground nest, in drier grassy areas some distance from water, lined with grasses, leaves, and pine needles
CONSERVATION STATUS	Least concern
COLLECTION NO.	FMNH 16139

ADULT BIRD SIZE
9–10 in (23–25 cm)

INCUBATION
22–23 days

CLUTCH SIZE
4 eggs

TRINGA FLAVIPES

LESSER YELLOWLEGS

CHARADRIIFORMES

Clutch

The Lesser Yellowlegs is a common and conspicuous breeder in the boreal evergreen forests of North America, occupying the shores of lakes, bogs, and grassy clearings. They call frequently and loudly, and can be seen running along the shore chasing prey items to eat. Relatively tame, they tolerate intruders at close range. The female builds the nest on higher, drier ground, perhaps on a low ridge near the lakeshore. She builds either on her own or within a loose concentration of other yellowlegs.

Despite looking very different from one another, the larger Willet (see page 151) and the smaller Lesser Yellowlegs are mutual closest relatives based on DNA sequence comparisons, a fact that highlights how evolution can lead to different appearances even in genetically close relatives.

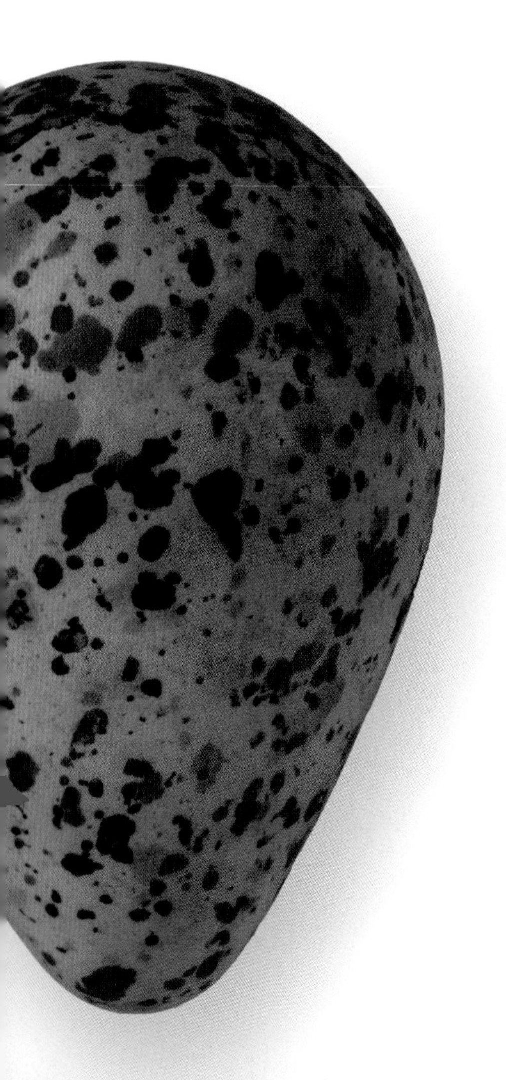

Actual size

The egg of the Lesser Yellowlegs is gray with brown markings, and 1⅔ x 1⅛ in (42 x 29 mm) in size. Both sexes incubate the eggs, and defend the hatchlings, but the female typically deserts the brood ten days before the chicks can fly.

ORDER	Charadriiformes
FAMILY	Scolopacidae
BREEDING RANGE	Temperate and subarctic North America
BREEDING HABITAT	Prairies, open grassland, and fields; sometimes airport lawns and cropland
NEST TYPE AND PLACEMENT	A shallow scrape on the ground, loosely lined with plant matter and hidden by grasses
CONSERVATION STATUS	Least concern
COLLECTION NO.	FMNH 5834

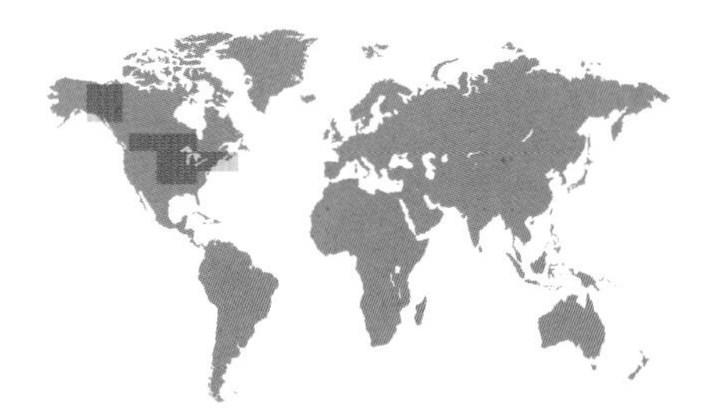

BARTRAMIA LONGICAUDA

UPLAND SANDPIPER

CHARADRIIFORMES

ADULT BIRD SIZE
11–12½ in (28–32 cm)

INCUBATION
23–24 days

CLUTCH SIZE
4 eggs

The numbers of Upland Sandpipers declined dramatically over their midwestern North American range in the late nineteenth century due to conversion of prairie grasslands to agriculture, as well as the same hunting pressures that led to the extinction of Eskimo Curlews and Passenger Pigeons. Today, many Upland Sandpipers make their homes and breed successfully on the grassy areas that are maintained around airports.

Upland Sandpipers often precede their mating by spectacular courtship displays, with birds soaring upward in wide circles while giving a distinctive whistle. Their nests, built on the ground in dense vegetation, are concealed by grasses pulled over the nest in an arch. Both parents share incubation duties and are hard to dislodge from the nest even when approached. During breeding, Upland Sandpipers may form loose colonies in which the timing of nesting and the hatching of young may be synchronized across broods.

Clutch

The egg of the Upland Sandpiper is cream in color, and finely maculated with brown spots; bigger blotches may be present on the blunt end of the egg. Glossy and smooth, the eggs are oval to subelliptical in shape, and measure 1¾ x 1¼ in (45 x 33 mm) in size.

Actual size

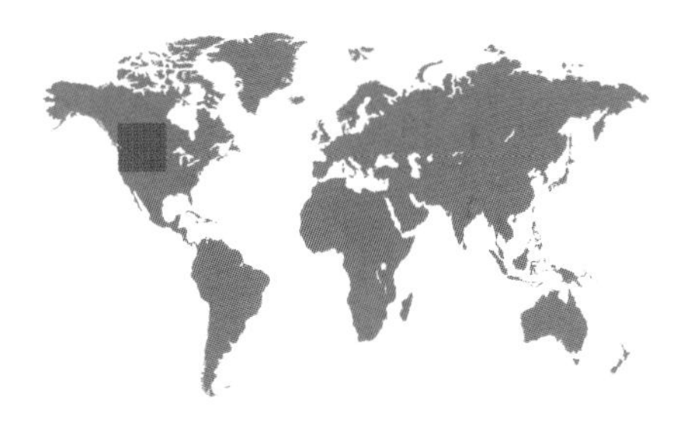

ORDER	Charadriiformes
FAMILY	Scolopacidae
BREEDING RANGE	Temperate central and western North America
BREEDING HABITAT	Open grasslands
NEST TYPE AND PLACEMENT	Shallow hollow, lined with grasses and weeds
CONSERVATION STATUS	Least concern
COLLECTION NO.	FMNH 16066

ADULT BIRD SIZE
19½–25½ in (50–65 cm)

INCUBATION
27–30 days

CLUTCH SIZE
4 eggs

NUMENIUS AMERICANUS

LONG-BILLED CURLEW

CHARADRIIFORMES

Clutch

The impressively long, downward-curving beak of this large shorebird allows it to hunt for earthworms, crabs, and crickets hiding in long burrows. But these birds also probe mud, catch grasshoppers off plants, and even devour small lizards and other birds' eggs. Historically, this was a popularly hunted species; today, habitat destruction of prairie grasslands presents an ongoing threat for long-term population viability and the health of the Long-billed Curlew.

Nesting begins with the male scraping a hollow in the ground, and the female using her breast and bill to widen it. She then lines it with pebbles, dirt, and plant matter. Both parents incubate the eggs, with the female taking her turn during the day and the male during the night. Once the eggs hatch, both parents protect the young from predators, but the female typically abandons her mate and brood before the young reach independence.

Actual size

The egg of the Long-billed Curlew is white to olive in color, has brown spots, and measures 2⅝ x 1⅞ in (66 x 47 mm) in size. The nest is often made near piles of dirt or manure; these are thought to provide shade and hide the eggs from predators.

ORDER	Charadriiformes
FAMILY	Scolopacidae
BREEDING RANGE	Northern Europe, northern Asia, and Alaska
BREEDING HABITAT	Coastal plains and open tundra
NEST TYPE AND PLACEMENT	Shallow cup in the mossy tundra, lined with grasses and leaves
CONSERVATION STATUS	Least concern
COLLECTION NO.	FMNH 15324

ADULT BIRD SIZE
14½–16 in (37–41 cm)

INCUBATION
20–22 days

CLUTCH SIZE
4 eggs

LIMOSA LAPPONICA

BAR-TAILED GODWIT

CHARADRIIFORMES

Female Bar-tailed Godwits are not only larger than males, but they also have longer beaks than the males', even relative to their body sizes. This allows the two sexes to forage in slightly, but consistently, different water depths and on different types of firmer or softer mudflats. The benefit of such foraging-related sexual dimorphism is that the pair can divide during the breeding season, and so more efficiently exploit the available food resources in their territory.

The Bar-tailed Godwit is also known as the record holder for the longest nonstop migratory flight; a female tagged with a satellite transmitter flew directly from Alaska to New Zealand, a distance of over 7,000 miles (11,000 km), in eight days. On their way back north, these birds leave New Zealand and fly to the South China Sea for a stopover to refuel, before taking flight again to arrive back in Alaska for the next breeding season.

Clutch

The egg of the Bar-tailed Godwit is olive to pale brown in color, and spotted with dark blotches. It is 2⅛ x 1½ in (54 x 37 mm) in size. Both parents incubate the eggs and look after the chicks, and adults and young move together to the coastline before flying south on migration.

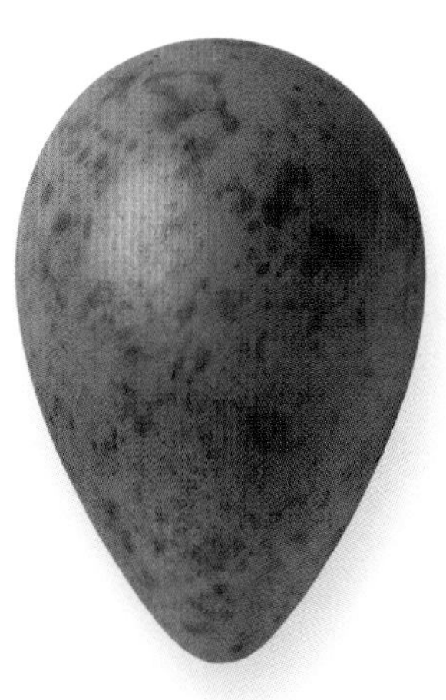

Actual size

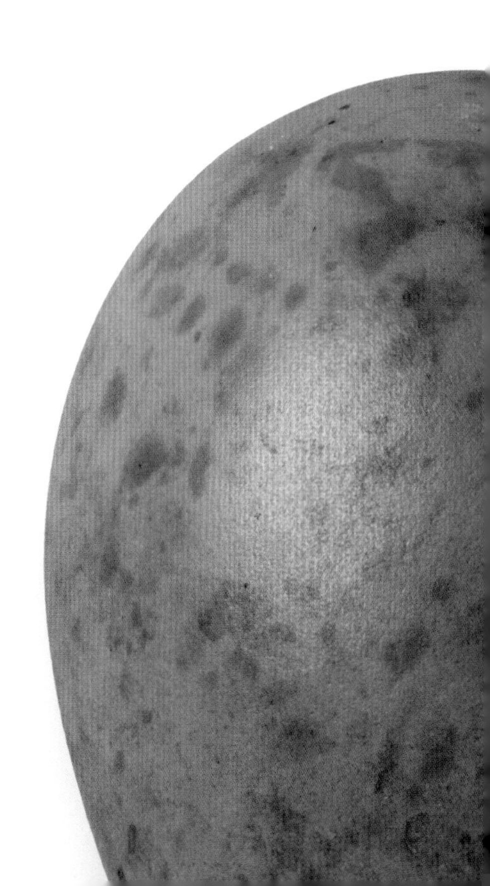

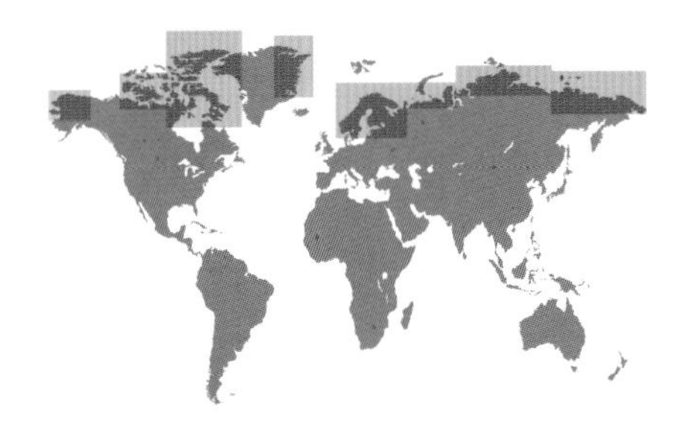

ORDER	Charadriiformes
FAMILY	Scolopacidae
BREEDING RANGE	Northern coastlines of North America, Europe, and Asia
BREEDING HABITAT	Grassland and wetlands near the coastline, rocky Arctic shores
NEST TYPE AND PLACEMENT	Shallow scrape in the ground, lined with grass
CONSERVATION STATUS	Least concern
COLLECTION NO.	FMNH 1196

ADULT BIRD SIZE
8½–10 in (21–26 cm)

INCUBATION
22–24 days

CLUTCH SIZE
4 eggs

ARENARIA INTERPRES

RUDDY TURNSTONE

CHARADRIIFORMES

Clutch

The egg of the Ruddy Turnstone is smooth, glossy, and oval in shape, pale green to brown in color, and has dark brown markings. It measures 1⅝ x 1⅛ in (41 x 29 mm) in size. Turnstones will attack and consume unattended eggs of other birds, including those of their own species.

True to its name, the turnstone uses its wedge-shaped bill to pry between rocks and to move pebbles around as it searches for insects and crustaceans hiding in crevices. It also inspects and picks prey from the top of the sand and the surface of driftwood. This feeding strategy is highly adaptable and successful globally, and allows these birds to migrate around the globe, from the Arctic through temperate and tropical shores, and all the way south to Australia and New Zealand.

The male Ruddy Turnstone establishes his breeding territory early in the season, and attracts a female by consorting with her throughout the area. He ceremonially scrapes nest-shaped patches in the ground, and the potential pair inspect these scrapes together. Eventually, when the female decides on a mate, she builds and lines her own nest, away from all the male scrapes made during the courtship.

Actual size

ORDER	Charadriiformes
FAMILY	Scolopacidae
BREEDING RANGE	Pacific coasts and islands of Arctic North America and Asia
BREEDING HABITAT	Rocky, mossy, or grassy fields in the coastal tundra
NEST TYPE AND PLACEMENT	Simple scrape on the ground, lined with grasses or leaves
CONSERVATION STATUS	Least concern
COLLECTION NO.	FMNH 15358

ADULT BIRD SIZE
9–15½ in (23–39 cm)

INCUBATION
20 days

CLUTCH SIZE
4 eggs

CALIDRIS PTILOCNEMIS

ROCK SANDPIPER

CHARADRIIFORMES

The Rock Sandpiper is a handsome, yet inconspicuous, bird; on the breeding grounds the bird's earthy tones, ranging from beige to reddish to brown and black, camouflage it well among the rocks and mosses of its coastal tundra habitat. The male establishes the breeding territory, and calls repeatedly to advertize the territory to prospective females. Once the pair bond is formed, both sexes provide some parental care. Male and female share incubation and lead the young away from harm. Yet, just days after hatching, the female typically deserts the brood and leaves the male to finish raising the young. Despite this, divorce is rare, and pairs established in one year remain together in the second year, as long as both members return to the same breeding site.

Rock Sandpipers feed by searching for small aquatic insects and crustaceans, often while standing breast deep or swimming in the coastal waters; during the summer they also consume berries, seeds, and moss.

Clutch

The egg of the Rock Sandpiper appears buff to olive in background, has brown markings, and measures 1½ x 1¹⁄₁₆ in (38 x 27 mm) in dimensions. The male makes several nest scrapes before the female selects one of them and lays in it.

Actual size

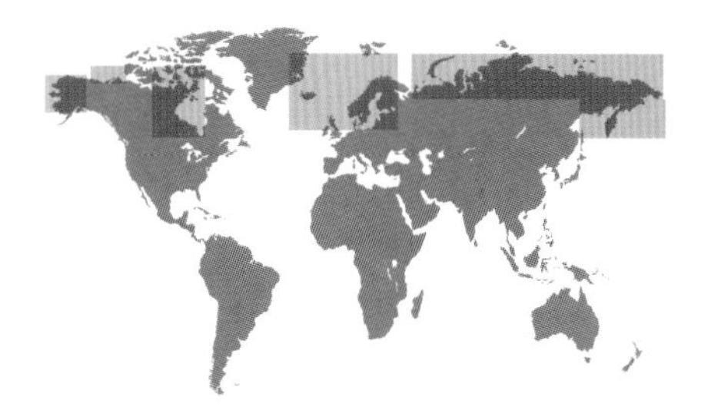

ORDER	Charadriiformes
FAMILY	Scolopacidae
BREEDING RANGE	Circumpolar Arctic regions of North America and Eurasia
BREEDING HABITAT	Moist, wet coastal tundra, grasslands
NEST TYPE AND PLACEMENT	Simple scrape in the ground, lined with grasses
CONSERVATION STATUS	Least concern
COLLECTION NO.	FMNH 18494

ADULT BIRD SIZE
6½–8½ in (16–22 cm)

INCUBATION
21–22 days

CLUTCH SIZE
4 eggs

CALIDRIS ALPINA

DUNLIN

CHARADRIIFORMES

Clutch

The Dunlin is one of the best-known shorebirds, as its migratory routes take it to the coastal areas of most major north-temperate ocean shores. Dunlins migrate in large flocks and frequently join mixed-species flocks at migratory stopover sites. Once they arrive on the breeding grounds, pairs quickly form, and mate repeatedly. Pairs from previous years typically are reestablished but in about 25 percent of pairs divorce occurs. By moving onto a new territory with a new mate, female divorcees can double their annual breeding success.

Living in groups provides safety for these small sandpipers, and they readily take flight in response to approaching danger, creating a loud noise caused by the air moving through the stiff flight feathers of the many individuals. In response to aerial predators, Dunlin can also dive under water, and, when approached from the shore, they readily head out to the open water, confidently swimming away from danger.

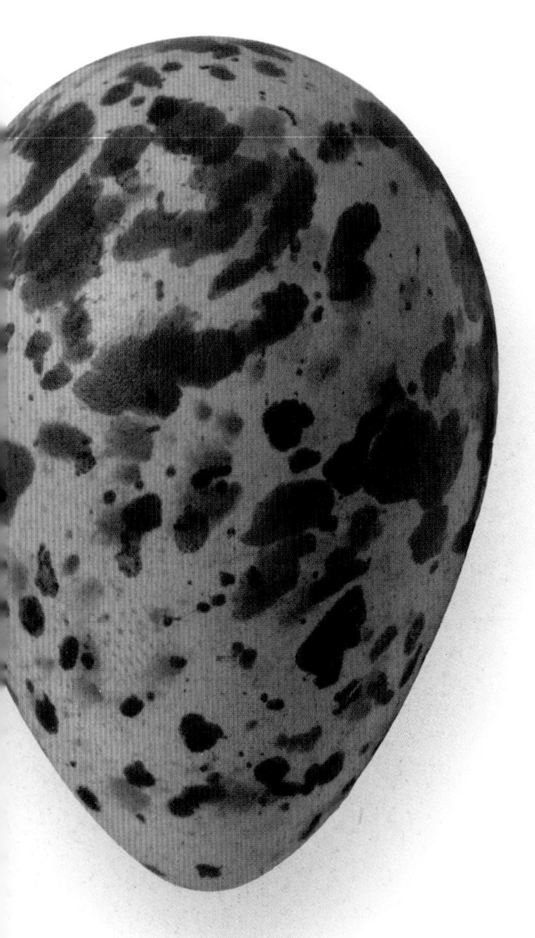

Actual size

The egg of the Dunlin is pale greenish to olive in coloration, with brown speckles, and its dimensions are 1⅜ x 1 in (36 x 25 mm). The eggs are incubated by both parents, but the female typically abandons the brood and leaves the nesting area soon after hatching.

ORDER	Charadriiformes
FAMILY	Scolopacidae
BREEDING RANGE	Pacific coasts and islands of Arctic North America and Asia
BREEDING HABITAT	Rocky, mossy, or grassy fields in the coastal tundra
NEST TYPE AND PLACEMENT	Simple scrape on the ground, lined with grasses or leaves
CONSERVATION STATUS	Least concern
COLLECTION NO.	FMNH 15358

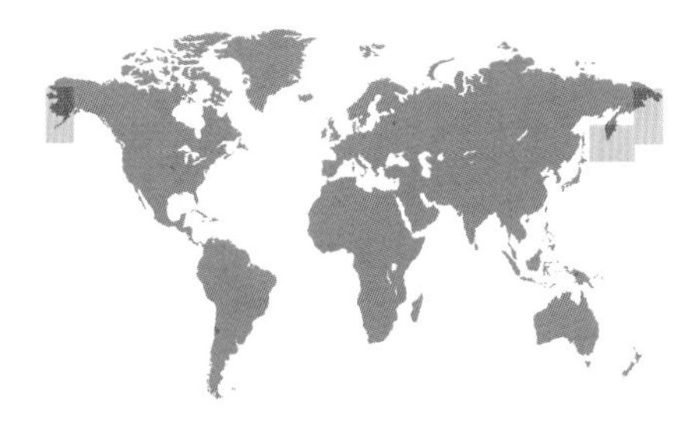

ADULT BIRD SIZE
9–15½ in (23–39 cm)

INCUBATION
20 days

CLUTCH SIZE
4 eggs

CALIDRIS PTILOCNEMIS

ROCK SANDPIPER

CHARADRIIFORMES

The Rock Sandpiper is a handsome, yet inconspicuous, bird; on the breeding grounds the bird's earthy tones, ranging from beige to reddish to brown and black, camouflage it well among the rocks and mosses of its coastal tundra habitat. The male establishes the breeding territory, and calls repeatedly to advertize the territory to prospective females. Once the pair bond is formed, both sexes provide some parental care. Male and female share incubation and lead the young away from harm. Yet, just days after hatching, the female typically deserts the brood and leaves the male to finish raising the young. Despite this, divorce is rare, and pairs established in one year remain together in the second year, as long as both members return to the same breeding site.

Rock Sandpipers feed by searching for small aquatic insects and crustaceans, often while standing breast deep or swimming in the coastal waters; during the summer they also consume berries, seeds, and moss.

Clutch

The egg of the Rock Sandpiper appears buff to olive in background, has brown markings, and measures 1½ x 1¹⁄₁₆ in (38 x 27 mm) in dimensions. The male makes several nest scrapes before the female selects one of them and lays in it.

Actual size

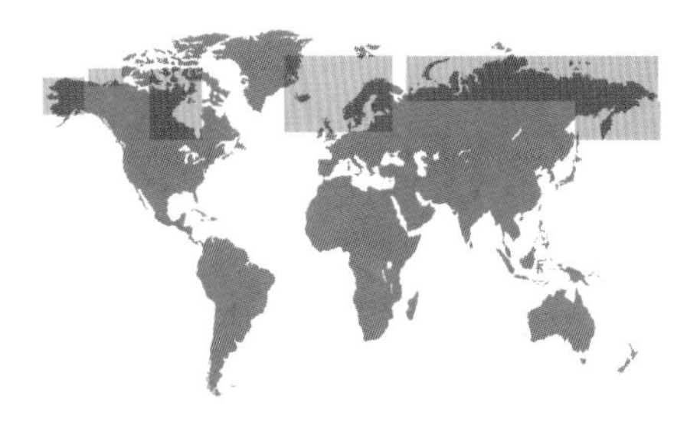

ORDER	Charadriiformes
FAMILY	Scolopacidae
BREEDING RANGE	Circumpolar Arctic regions of North America and Eurasia
BREEDING HABITAT	Moist, wet coastal tundra, grasslands
NEST TYPE AND PLACEMENT	Simple scrape in the ground, lined with grasses
CONSERVATION STATUS	Least concern
COLLECTION NO.	FMNH 18494

ADULT BIRD SIZE
6½–8½ in (16–22 cm)

INCUBATION
21–22 days

CLUTCH SIZE
4 eggs

CALIDRIS ALPINA

DUNLIN

CHARADRIIFORMES

Clutch

The Dunlin is one of the best-known shorebirds, as its migratory routes take it to the coastal areas of most major north-temperate ocean shores. Dunlins migrate in large flocks and frequently join mixed-species flocks at migratory stopover sites. Once they arrive on the breeding grounds, pairs quickly form, and mate repeatedly. Pairs from previous years typically are reestablished but in about 25 percent of pairs divorce occurs. By moving onto a new territory with a new mate, female divorcees can double their annual breeding success.

Living in groups provides safety for these small sandpipers, and they readily take flight in response to approaching danger, creating a loud noise caused by the air moving through the stiff flight feathers of the many individuals. In response to aerial predators, Dunlin can also dive under water, and, when approached from the shore, they readily head out to the open water, confidently swimming away from danger.

Actual size

The egg of the Dunlin is pale greenish to olive in coloration, with brown speckles, and its dimensions are 1⅜ x 1 in (36 x 25 mm). The eggs are incubated by both parents, but the female typically abandons the brood and leaves the nesting area soon after hatching.

ORDER	Charadriiformes
FAMILY	Scolopacidae
BREEDING RANGE	Arctic North America
BREEDING HABITAT	Grasslands, open sedge tundra, sometimes near taiga woodlands
NEST TYPE AND PLACEMENT	Ground scrape in dry areas, lined with grasses, occasionally near shrubs
CONSERVATION STATUS	Least concern
COLLECTION NO.	FMNH 15064

ADULT BIRD SIZE
8–9 in (20–23 cm)

INCUBATION
19–21 days

CLUTCH SIZE
3–4 eggs

CALIDRIS HIMANTOPUS

STILT SANDPIPER

CHARADRIIFORMES

The Stilt Sandpiper is an evolutionary enigma: extensive DNA sequences have not yet helped systematists to resolve this species' taxonomic affiliations. Is it a member of a typical sandpiper genus, a relative of another genetic odd-ball, the Curlew Sandpiper, or perhaps a sole member of its own genus? With the advance of more complete sequencing of full genomes, and the equally fast-moving improvements in analytical approaches, before long the phylogenetic affiliations of this species should be resolved.

Males engage in conspicuous courtship flights to attract females. Once the pair bond is established, Stilt Sandpipers maintain strict territorial boundaries and nest some distance away from others of their own species. However, they do place their nests in the general vicinity of other species, perhaps as an anti-predator defense mechanism to benefit from early detection and communal mobbing of approaching danger.

Clutch

Actual size

The egg of the Stilt Sandpiper is cream to olive green in color with brown spots, and 1⅜ x 1 in (36 x 26 mm) in size. Incubation and parental care are carried out by both parents.

ORDER	Charadriiformes
FAMILY	Scolopacidae
BREEDING RANGE	Northern Europe and Asia, Aleutian Islands
BREEDING HABITAT	Boreal wetlands, taiga bogs
NEST TYPE AND PLACEMENT	A scrape on the ground, lined with grasses
CONSERVATION STATUS	Least concern
COLLECTION NO.	FMNH 15338

ADULT BIRD SIZE
6–7 in (15–18 cm)

INCUBATION
21–22 days

CLUTCH SIZE
4 eggs

LIMICOLA FALCINELLUS

BROAD-BILLED SANDPIPER

CHARADRIIFORMES

Clutch

This sandpiper is the sole member of its genus. It resembles in appearance a typical small sandpiper such as the Dunlin (see page 158), but its genetic make-up points toward affinities with the larger ruffs. It is also different from other shorebirds in that this species is less gregarious on the breeding grounds, and even during migration, and occurs in relatively small flocks of around ten individuals.

At the onset of the breeding season, the male performs a conspicuous courtship flight to impress the female, circling the bog or other wetland repeatedly. Once the pair bond forms, the female and the male attend the eggs and the chicks together, taking turns in incubation and guarding. Eventually, the female leaves the family behind and the male attends to the young until they become independent.

Actual size

The egg of the Broad-billed Sandpiper is gray to brown in background color, with large dark speckles, and 1³⁄₁₆ x ⅞ in (30 x 22 mm) in size. The nest may be hidden underneath tufts of wetland grasses, Even when visible, the eggs are well camouflaged against the surrounding muddy ground.

ORDER	Charadriiformes
FAMILY	Scolopacidae
BREEDING RANGE	Northern Eurasia
BREEDING HABITAT	Wetlands, marshes, and river deltas
NEST TYPE AND PLACEMENT	Shallow scrape, lined with leaves and stems
CONSERVATION STATUS	Least concern
COLLECTION NO.	FMNH 5819

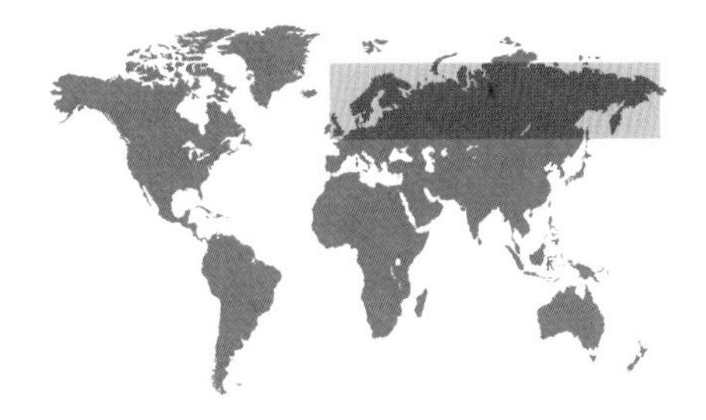

ADULT BIRD SIZE
11½–12½ in (29–32 cm)

INCUBATION
20–23 days

CLUTCH SIZE
3–4 eggs

PHILOMACHUS PUGNAX

RUFF

CHARADRIIFORMES

The Ruff has what is possibly the most complex set of gender roles among birds, with not two, not three, but four sex-specific plumage types: three types of males, and at least one type of female. Unlike most waders or shorebirds, nesting and parenting for the Ruff is done by the female only. Instead of providing paternal care, the males spend most of the breeding season displaying in social flocks, called leks, whose sole purpose is to attract females. Why should a male display next to many other males, some of whom may be more impressive and attractive to the females? With the Ruff, it is because females are more attracted to larger leks; displaying alone would yield few or no matings at all.

Clutch

The three types of male Ruffs include territorial individuals who carry ornamental collar feathers and do most of the displaying, satellite males who are less colorful and display on the edge of the lek, and female-mimic males. Female-mimics develop plumage that resembles a female's, allowing them to get into the lek on the sly and breed with some of the actual females. These male roles have been shown to be genetically inherited.

The egg of the Ruff is sandy buff or greenish gray in color with heavy reddish brown maculation, and with a second layer of spots in lavender. It measures 1¾ x 1¼ in (44 x 31 mm) in size. The nest is hidden under tall grasses and tussocks to avoid predation by visually hunting predators.

Actual size

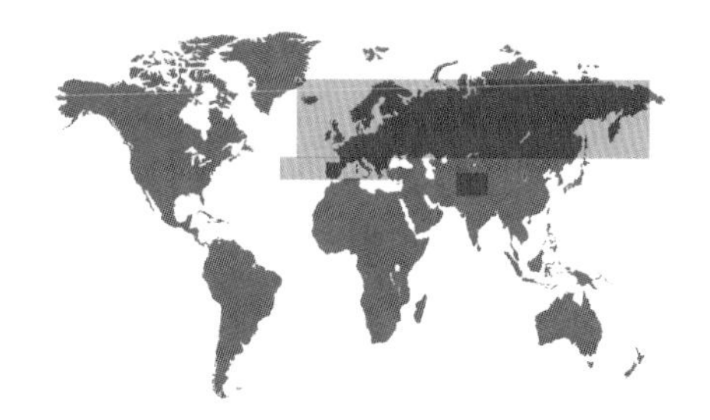

ORDER	Charadriiformes
FAMILY	Scolopacidae
BREEDING RANGE	Arctic and temperate montane Eurasia
BREEDING HABITAT	Marshes, bogs, tundra wetlands
NEST TYPE AND PLACEMENT	Grass-lined hollow in wet meadows , well hidden under grass leaves and stems
CONSERVATION STATUS	Least concern
COLLECTION NO.	FMNH 18718

ADULT BIRD SIZE
10–10½ in (25–27 cm)

INCUBATION
18–21 days

CLUTCH SIZE
4 eggs

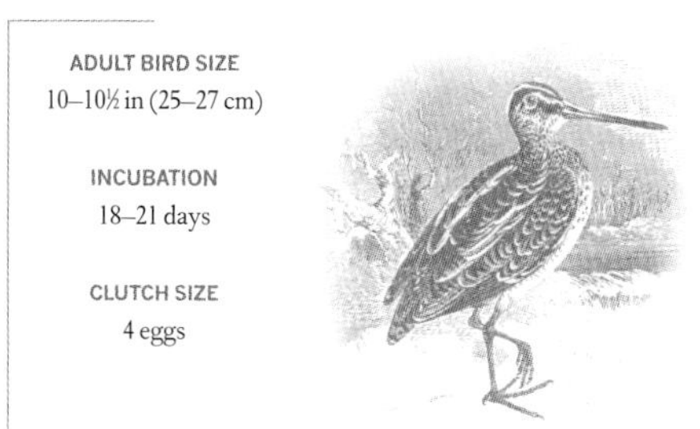

GALLINAGO GALLINAGO

COMMON SNIPE

CHARADRIIFORMES

Clutch

The egg of the Common Snipe is pale grayish to dark green in background color with rich brown spotting, and 1½ x 1⅛ in (39 x 28 mm) in size. The coloration of the eggs, hatchlings, and the attending parents all provide great camouflage in marsh vegetation. Both parents look after the young, but each takes half of the brood under their own, separate care.

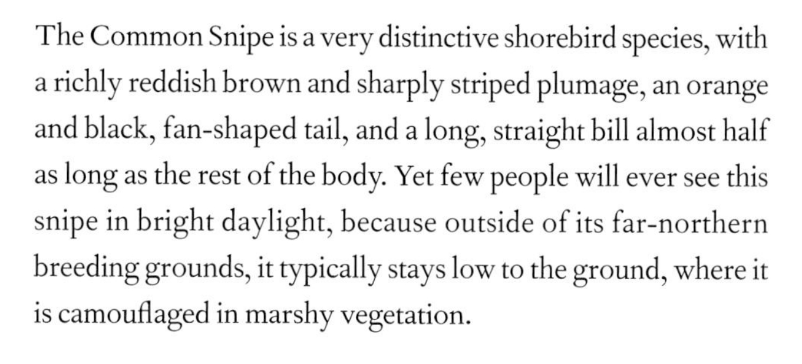

The Common Snipe is a very distinctive shorebird species, with a richly reddish brown and sharply striped plumage, an orange and black, fan-shaped tail, and a long, straight bill almost half as long as the rest of the body. Yet few people will ever see this snipe in bright daylight, because outside of its far-northern breeding grounds, it typically stays low to the ground, where it is camouflaged in marshy vegetation.

However, occasional conspicuousness is the means through which snipes escape predators and attract mates; when flushed by danger, these birds call out loudly and fly away in a confusing zigzag pattern. In turn, on the breeding grounds, males complete several circles on the wing above the display site from where the female might be watching. The males then make several shallow dives in which they spread their stiff tail feathers. The air rushing through the tail generates a loud sound. Because this sound is not generated by vocal chords, bioacousticians call this process of sound-making by feathers "sonation."

Actual size

ORDER	Charadriiformes
FAMILY	Scolopacidae
BREEDING RANGE	Eastern North America
BREEDING HABITAT	Clearings in forests, shrubby areas near water
NEST TYPE AND PLACEMENT	Shallow ground nest of twigs and leaves, with brushy or sapling cover
CONSERVATION STATUS	Least concern
COLLECTION NO.	FMNH 16003

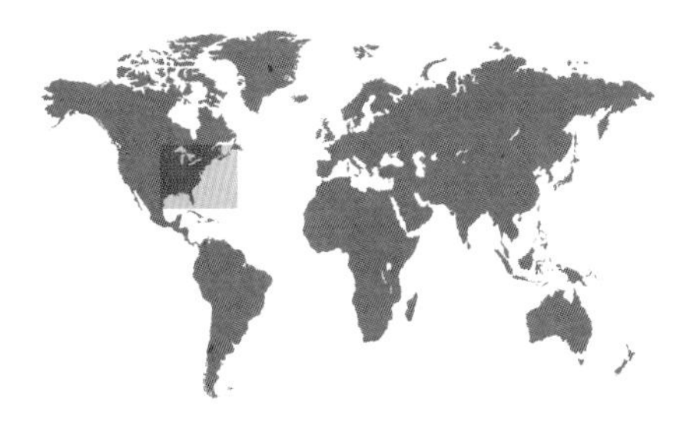

ADULT BIRD SIZE
10½–12 in (27–31 cm)

INCUBATION
20–22 days

CLUTCH SIZE
4 eggs

SCOLOPAX MINOR

AMERICAN WOODCOCK

CHARADRIIFORMES

The American Woodcock is a shorebird species which has taken to living in open woodland and adjacent fields in eastern North America. It is also one of the few wader-relatives that is still a commonly hunted game species; in the United States alone, around half a million woodcocks are killed each year.

Generally an inconspicuous species, male woodcocks put on quite a show to attract their mates each spring. They fly, zigzag, flutter, and make dives in midair, while calling and chirping to any nearby female on the ground. If the female likes this show, she flies directly in and lands nearby on the ground. The species is polygynous, and the male will continue his attempts to attract new females throughout the breeding season.

Clutch

Actual size

The egg of the American Woodcock is creamy buff in background color with brown spotting, and measures 1½ x 1⅛ in (38 x 29 mm) in size. The female attends the eggs and the chicks on her own, even if the nest is typically within 500 ft (150 m) of the male's singing grounds. Like their mother, the precocial chicks are superbly camouflaged as they feed on their own, probing the soft soil and leaf litter with their already elongated beaks.

ORDER	Charadriiformes
FAMILY	Scolopacidae
BREEDING RANGE	Western and central North America
BREEDING HABITAT	Wetlands and lakes in prairie grasslands
NEST TYPE AND PLACEMENT	A shallow scrape in the ground, lined with grass leaves
CONSERVATION STATUS	Least concern
COLLECTION NO.	FMNH 5696

ADULT BIRD SIZE
8½–9 in (22–23 cm)

INCUBATION
18–23 days

CLUTCH SIZE
4 eggs

PHALAROPUS TRICOLOR

WILSON'S PHALAROPE

CHARADRIIFORMES

Clutch

The egg of the Wilson's Phalarope is buff colored in background with large brown spots, and it measures 1⅝ x 1⅓ in (41 x 34 mm) in size. The female and the male start the nest together, but once egg laying is complete, the female leaves the breeding grounds, and the male completes the parental care duties of attending and protecting the young.

The Wilson's Phalarope is one of five species of birds named after Alexander Wilson, the nineteenth-century scientist often considered the father of American ornithology. It is the largest of the phalaropes, and shares many breeding characteristics with the other species, including reversed sexual dimorphism of size, plumage, and parental behaviors. The females are larger and brighter than the males: they protect the territory, while the males are the sole caretakers of the young.

Phalarope females often mate with more than one male, abandoning the first clutch of eggs to mate and nest with another male. Females do not incubate the eggs but they do provide protection from danger. When a predator approaches, the female positions herself to cover the eggs in a fake incubation posture, while the male does a broken-wing display to lead the predator away. Once the predator leaves, the female herself walks away from the nest.

Actual size

ORDER	Charadriiformes
FAMILY	Glareolidae
BREEDING RANGE	North Africa and southwest Asia
BREEDING HABITAT	Dry open country, semidesert
NEST TYPE AND PLACEMENT	Simple, unlined depression on the ground
CONSERVATION STATUS	Least concern
COLLECTION NO.	FMNH 15345

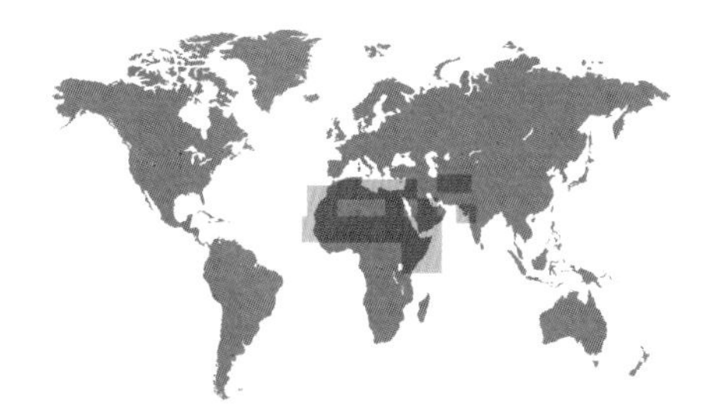

CURSORIUS CURSOR

CREAM-COLORED COURSER

CHARADRIIFORMES

ADULT BIRD SIZE
7½–8½ in (19–22 cm)

INCUBATION
18–19 days

CLUTCH SIZE
2 eggs

Coursers are long-legged like many other shorebirds, but their lifestyle has evolved to be independent from bodies of water for feeding and breeding. The Cream-colored Courser prefers to breed in dry and semidesert habitats throughout its wide distribution. Locally, this habitat preference has caused population declines, especially on the Canary Islands where tourism-related development has encroached on the flat and open breeding habitats favored by the species. In other regions, especially in the Arabian peninsula, this species has benefited from successful habitat-conservation efforts aimed at larger, historically hunted gamebird species.

Coursers show a consistent seasonality in breeding. Throughout their predominantly desert habitat, they begin to nest in the early spring, just after the brief winter rains have ended. But the species will also re-nest and lay a new clutch if the first breeding attempt fails. This flexibility in restarting the breeding cycle allows them to reproduce in response to unpredictable rains.

Clutch

The egg of the Cream-colored Courser is sandy brown in background, with brown maculation, and measures 1⅓ x 1 in (34 x 26 mm) in size. The chicks are nidifugous, meaning that they leave the nest soon after hatching. Nonetheless, they still require protection and provisioning by their parents before fledging.

Actual size

ORDER	Charadriiformes
FAMILY	Glareolidae
BREEDING RANGE	Southeast Europe, southwest Asia
BREEDING HABITAT	Open, short grass steppe, plowed fields, grazed meadows
NEST TYPE AND PLACEMENT	Shallow depression on ground, with small pieces of twigs and dried leaves
CONSERVATION STATUS	Near threatened
COLLECTION NO.	FMNH 20765

ADULT BIRD SIZE
9½–11 in (24–28 cm)

INCUBATION
35–42 days

CLUTCH SIZE
3–4 eggs

GLAREOLA NORDMANNI

BLACK-WINGED PRATINCOLE

CHARADRIIFORMES

Clutch

The egg of the Black-winged Pratincole is roundish, buff to olive in background color with heavy dark-brown maculation, and measures 1¹⁄₁₆ x ⅞ in (27 x 22 mm) in size. Both parents incubate the eggs, but the ground nests in open pastures are vulnerable to trampling by livestock.

Despite being genetically close to shorebirds, pratincoles are not at all like shorebirds in appearance and behavior. In shape and flight they are more like swifts and swallows. They catch flying prey on the wing, swooping around grazing herds of wild and domestic livestock. With the increasing pressure to control insects with pesticides, and the ongoing conversion of open fields into agricultural areas or managed pastures, the Black-winged Pratincole has experienced a severe decline of 50 percent of its estimated population size in just ten years of intensive monitoring in Europe and central Asia.

Despite its unique appearance, this pratincole shares several reproductive traits with many other shorebirds, including ground nesting and typical clutch sizes of four eggs. The species breeds in loose, small-to-medium-sized colonies of between five and 500 pairs. Larger colonies are often sited near wetter meadows and areas of water that provide plentiful insect prey for the adults and their chicks.

Actual size

ORDER	Charadriiformes
FAMILY	Rostratulidae
BREEDING RANGE	Sub-Saharan Africa, south and southwest Asia
BREEDING HABITAT	Wetlands, reed beds, near shores of ponds and streams
NEST TYPE AND PLACEMENT	Shallow scrape in soft ground, lined with stems and leaves, typically near water
CONSERVATION STATUS	Least concern
COLLECTION NO.	FMNH 20744

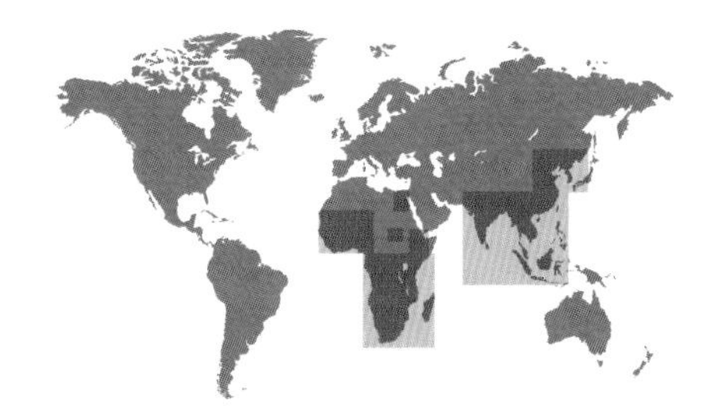

ADULT BIRD SIZE
9–11 in (23–28cm)

INCUBATION
15–21 days

CLUTCH SIZE
2–3 eggs

ROSTRATULA BENGHALENSIS

GREATER PAINTED-SNIPE

CHARADRIIFORMES

Painted-Snipes are not close relatives of the "real" snipes. Like snipes, they use their long beaks to probe for aquatic prey, but their appearance is more like that of a typical wader, with long legs. The female is one of the most colorful shorebird species, and sports a rich chocolate brown plumage and distinctive chest bands, compared to the smaller and plainer, brown and striped male. As might be expected from this "reversed" sexual dimorphism, this species is polyandrous, with the female initiating courtship with the male, before she lays the eggs for him. Then she moves on to mate with another male.

The male provides all parental care of the eggs and chicks. He will lead chicks to feeding sites, and away from predators into dense vegetation. When threatened, the male also carries the chicks directly, by tucking them under his wings as he moves away from danger.

Clutch

The egg of the Greater Painted-Snipe is beige in background color, has fine dark brown maculation, and measures 1½ x 1⅛ in (38 x 28 mm). Only the male incubates the eggs; his plumage is far less colorful than the female's and so it affords him camouflage to hide from predators during incubation.

Actual size

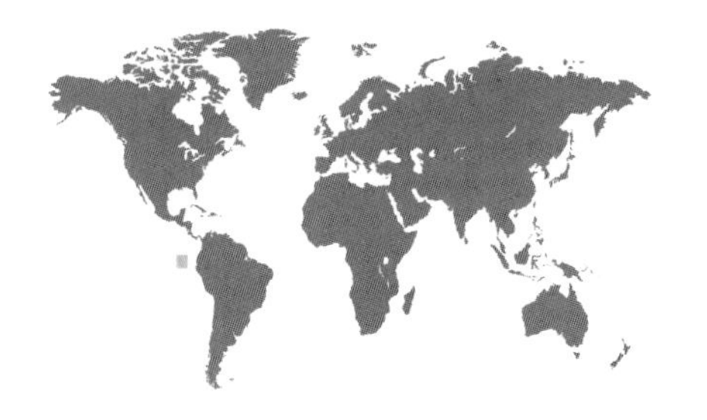

ORDER	Charadriiformes
FAMILY	Laridae
BREEDING RANGE	Galapagos Islands (Ecuador), and Malpelo Island (Colombia)
BREEDING HABITAT	Rocky cliffs and beaches
NEST TYPE AND PLACEMENT	Small platform of pebbles, sticks, sea urchin spines, and coral pieces, to prevent the egg from rolling away
CONSERVATION STATUS	Least concern
COLLECTION NO.	FMNH 18981

ADULT BIRD SIZE
26 in (66 cm)

INCUBATION
31–34 days

CLUTCH SIZE
1 egg

CREAGRUS FURCATUS

SWALLOW-TAILED GULL

CHARADRIIFORMES

This is a highly social species, and it nests and forages in large groups off the lava-rock beaches of the Galapagos Islands. It is a near-endemic of the Galapagos, with just a few pairs breeding on another island between the Galapagos and mainland South America. This species changes both its plumage and skin coloration to prepare for the breeding season, with its white head turning grayish black and its black fleshy eye ring turning bright red. Young birds often do not attempt to breed until about five years of age.

The Swallow-tailed Gull is the only strictly nighttime-hunting shorebird or gull species. It leaves at sunset in loud groups for foraging trips to catch nocturnal crabs. Two traits help with this behavior: its eyes are larger than expected for its body size, and each eye has an extra layer of reflective tissue opposite the lens, helping the bird to see better in minimal light.

Actual size

The egg of the Swallow-tailed Gull is off-white in background color with gray and dark brown maculation, and measures $2\frac{5}{8}$ x $1\frac{13}{16}$ in (66 x 46 mm) in size. During incubation the parent faces away from the ocean, a behavior typically seen in other gulls nesting on exposed shorelines.

ORDER	Charadriiformes
FAMILY	Laridae
BREEDING RANGE	Aleutian Islands, north Pacific
BREEDING HABITAT	Rock ledges on vertical sea cliffs
NEST TYPE AND PLACEMENT	Ground nests built up from mud, grass, seaweed, located on steep cliffs
CONSERVATION STATUS	Vulnerable
COLLECTION NO.	FMNH 15076

ADULT BIRD SIZE
14–15½ in (35–39 cm)

INCUBATION
23–32 days

CLUTCH SIZE
1–3 eggs

RISSA BREVIROSTRIS

RED-LEGGED KITTIWAKE

CHARADRIIFORMES

The Red-legged Kittiwake is a true "seagull" in the sense that it is one of the few gull species with marine-only breeding, wintering, and foraging ranges. Nesting on just a handful of Arctic islands in the Bering Sea, it feeds on small fish and crustaceans from the ocean. After the nesting season is over, this kittiwake is rarely seen near shore, spending the non-breeding season far out at sea.

The global population of this species has always been limited to just a handful of breeding localities on small islands, but in the last four decades alone its numbers have dropped by 50 percent. It has been suggested that lower marine productivity in the region, partly the result of human-caused overfishing, may be responsible for these consistent declines. It is also feared that increased commercial activity in the region will introduce Norway rats, a fierce nest predator, to the islands with kittiwake nesting colonies. Thus intensive monitoring and management of this vulnerable species and its breeding sites is needed.

Clutch

The egg of the Red-legged Kittiwake is brownish in background color with dark brown and lilac maculation, and measures 2¼ x 1⅝ in (58 x 41 mm) in size. Unlike most gull chicks, which begin to wander as they grow, the kittiwake's chick stays in the nest for nearly five weeks, probably to avoid falling off the steep cliffs used for nesting.

Actual size

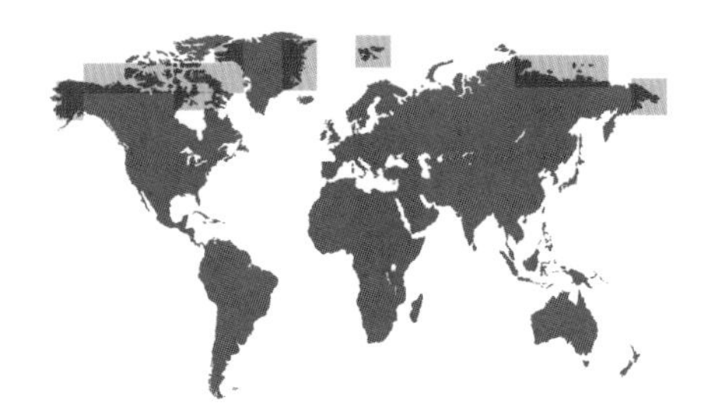

ORDER	Charadriiformes
FAMILY	Laridae
BREEDING RANGE	Northernmost parts of North America, Europe, and Asia
BREEDING HABITAT	Open fields, moist tundra grounds, typically near freshwater ponds
NEST TYPE AND PLACEMENT	Depression in vegetation, typically with no additional lining
CONSERVATION STATUS	Least concern
COLLECTION NO.	FMNH 15080

ADULT BIRD SIZE
10½–13 in (27–33 cm)

INCUBATION
25 days

CLUTCH SIZE
2–3 eggs

XEMA SABINI

SABINE'S GULL

CHARADRIIFORMES

Clutch

The egg of the Sabine's Gull is olive brown in background color with irregular, dark greenish brown maculation, and measures 1¾ x 1¼ in (45 x 32 mm) in dimensions. The hatchlings are relatively precocial; they may leave the nest within days of hatching, and the parents lead them to open water, where the chicks quickly start feeding themselves.

The Sabine's Gull is the only member of its genus. Genetically, its closest relative is another Arctic-nesting gull, the all-white Ivory Gull, which is also a monotypic, or single, member of its own genus. DNA studies show that the two species diverged from one another more than two million years ago, each evolving into species with distinct genetic and morphological traits.

The Sabine's Gull has a delicate flight more closely resembling that of terns than gulls. But like other gulls, it hovers over the water to spot and pick prey items from the water surface; it also forages by swimming around, by walking on mudflats, and, like a phalarope, by spinning in shallow water to draw up prey hidden in the mud below. During the breeding season on the tundra, both parents and chicks forage over and in freshwater ponds. In winter, the species becomes pelagic, and forages offshore in the ocean.

Actual size

ORDER	Charadriiformes
FAMILY	Laridae
BREEDING RANGE	Temperate North America
BREEDING HABITAT	Open fields near rivers, lakes, and coastal areas
NEST TYPE AND PLACEMENT	A ground scrape, lined with twigs, sticks, leaves, or lichens
CONSERVATION STATUS	Least concern
COLLECTION NO.	FMNH 372

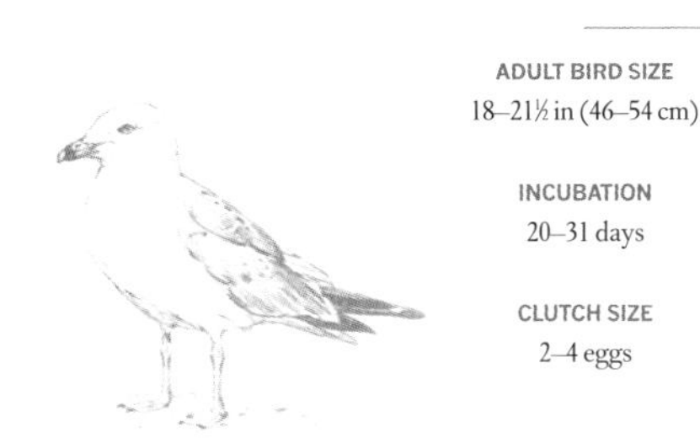

ADULT BIRD SIZE
18–21½ in (46–54 cm)

INCUBATION
20–31 days

CLUTCH SIZE
2–4 eggs

LARUS DELAWARENSIS

RING-BILLED GULL

CHARADRIIFORMES

Ring-billed Gulls breed further inland than most other gull species, and their large nesting colonies are typically associated with freshwater lakes, swamps, and streams. These gulls also are the most common species seen at municipal garbage-dumps, which have boosted local colony sizes and rates of reproduction. When the dumps close, or switch to indoor processing, Ring-billed Gull population densities plummet.

Nesting in a large colony, these birds rely on an acute sense of recognition of space and individuals: each must identify its own nest, mate, eggs, and chicks. After hatching, the nestlings stay around for just four to five days, and then they become mobile. This puts extra pressure on parents to recognize their own chicks by sound and sight. Mistakes happen, and some young are lost by their parents. But a few may also be adopted by foster-parents and survive to fledge successfully.

Clutch

The egg of the Ring-billed Gull is grayish to brown in background color with brown to lilac maculation, and measures 2⅓ x 1⅔ in (59 x 42 mm) in size. In some colonies, observers have also found large egg-shaped pebbles in the nest; these may provide distraction for egg-predators or represent overeager parents confusing rocks for their own eggs.

Actual size

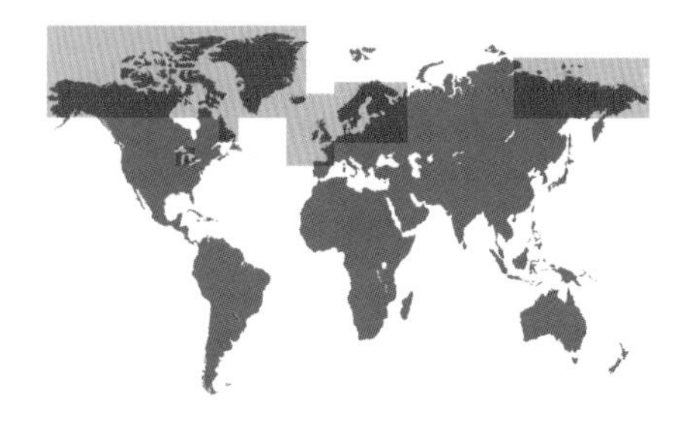

ORDER	Charadriiformes
FAMILY	Laridae
BREEDING RANGE	Northern North America, Europe, and Asia
BREEDING HABITAT	Islands, beaches, and wetlands
NEST TYPE AND PLACEMENT	Scraping in soft soil, sand, or short vegetation
CONSERVATION STATUS	Least concern
COLLECTION NO.	FMNH 2416

ADULT BIRD SIZE
23½–25 in (60–64 cm)

INCUBATION
28–30 days

CLUTCH SIZE
2–4 eggs

LARUS ARGENTATUS

HERRING GULL

CHARADRIIFORMES

Clutch

Herring Gulls are conspicuous, noisy, and densely distributed along most northern hemisphere coastlines throughout the year, making them one of the best-known species to local beachgoers. They readily feed not only on fish and crustaceans, but also on leftovers and other garbage on the beach or at garbage dumps. These gulls are also prominent predators of eggs and chicks of other colonially nesting waterbirds including smaller gulls and terns, and also shorebirds, shearwaters, petrels, and puffins.

The breeding season begins with prominent displays and courtship between the sexes: the male, settling on a display territory, is approached by an interested female. If he accepts, they call and raise their bodies in synchrony, and eventually copulate. The nest site is chosen by both members of the pair, and parental duties are also shared. If the nest is successful, and both mates return to the breeding site the following year, the pair bond is maintained. Failed nesting attempts are typically followed by divorce.

The egg of the Herring Gull is olive in color with a darker brown blotching. It measures 2½ x 1⅞ in (65 x 48 mm) in dimensions. Although these gulls are themselves egg predators of other smaller species, their own eggs are also vulnerable to predation by even larger gulls, corvids, herons, and mammals like raccoons.

Actual size

ORDER	Charadriiformes
FAMILY	Laridae
BREEDING RANGE	Tropical and subtropical coasts and oceanic islands around the world
BREEDING HABITAT	Isolated islands, offshore islets
NEST TYPE AND PLACEMENT	A platform nest of sticks and twigs, typically on cliffs, or in trees and bushes; occasionally lays the egg on bare ground
CONSERVATION STATUS	Least concern
COLLECTION NO.	FMNH 4837

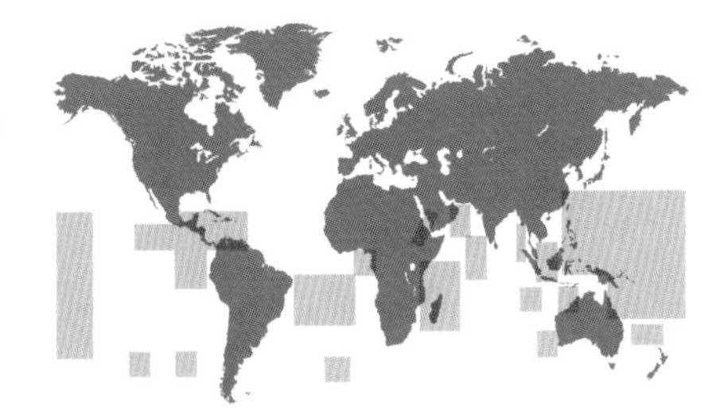

ADULT BIRD SIZE
15½–17½ in (40–45 cm)

INCUBATION
33–36 days

CLUTCH SIZE
1 egg

ANOUS STOLIDUS

BROWN NODDY

CHARADRIIFORMES

Breeding is a colonial affair for the Brown (or Common) Noddies. On suitable, remote islands, entire patches of forest, or cliffs, may be covered by thousands of noddies calling, displaying, incubating eggs, or feeding chicks. The female and the male bow and nod to one another in their nuptial displays, accompanied by courtship feeding and flights, during which the male transfers a small freshly caught fish to the female.

Incubation is carried out by both sexes; they take turns on the nest every one or two days while their mate is feeding at sea. Once the chick hatches, the parents are efficient providers, and the chick quickly reaches the weight of the parents in just three weeks. At the time of fledging, the chick can even weigh more than the parents, but once the chick takes flight, it quickly loses the excess weight, learning how to fend for itself while the parents feed it less and less.

The egg of the Brown Noddy is a pinky cream in background color with lilac and chestnut maculation, and measures 2 x 1⅜ in (52 x 35 mm) in dimensions. When these birds nest on trees, many eggs and chicks accidentally fall from high branches, leaving the ground underneath a veritable cemetery of past breeding seasons.

Actual size

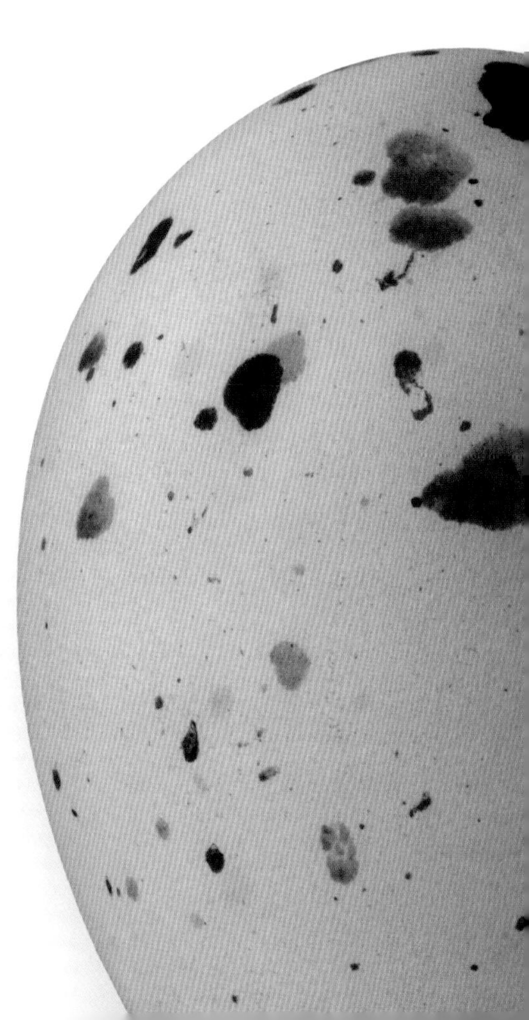

ORDER	Charadriiformes
FAMILY	Laridae
BREEDING RANGE	Tropical and subtropical islands in all three major oceans
BREEDING HABITAT	Sandy or rocky islands with tree stands to provide nesting sites
NEST TYPE AND PLACEMENT	A loose platform of twigs and leaves on a tree branch
CONSERVATION STATUS	Least concern
COLLECTION NO.	FMNH 14797

ADULT BIRD SIZE
13½–15½ in (35–40 cm)

INCUBATION
34 days

CLUTCH SIZE
1 egg

ANOUS MINUTUS

BLACK NODDY

CHARADRIIFORMES

The Black Noddy, which is also known as the White-capped Noddy, is most commonly seen offshore near remote oceanic islands, and in dense nesting colonies weighing down tree branches on sandy atolls and other low-lying islands. When at sea, these birds do not dive into the water after food, but instead rely on tuna and other schools of predatory fish to drive small fish to the surface, where the noddies can scoop them up in an act of kleptoparasitism.

Black Noddy parents consider their nest to be the home base; if a chick wanders along the branch far from the nest, or falls to the ground, it will not be recognized or attended by the parents, and it usually perishes. The adults will then wait until the next nesting season to attempt reproduction again. Pairs return to the same island, and often to the same branch of the same tree, to breed annually. It takes three to five years for juveniles to become fully mature and make their first breeding attempt.

Actual size

The egg of the Black Noddy is reddish yellow in background color with rusty brown maculation, and is $1\frac{11}{16} \times 1\frac{1}{4}$ in (43 x 31 mm) in size. The egg is laid on a platform of twigs, leaves, and dried fecal matter, with the nest often being reused over consecutive breeding seasons.

ORDER	Charadriiformes
FAMILY	Laridae
BREEDING RANGE	Southern coasts and inland major river systems of North America
BREEDING HABITAT	Gravel or sandy beaches and coastlines on the sea, lakes, or rivers
NEST TYPE AND PLACEMENT	Ground scraping in the sand or gravel, occasionally on rooftops
CONSERVATION STATUS	Least concern
COLLECTION NO.	FMNH 18322

STERNULA ANTILLARUM

LEAST TERN

CHARADRIIFORMES

ADULT BIRD SIZE
8½–9 in (21–23 cm)

INCUBATION
20–22 days

CLUTCH SIZE
2–3 eggs

Least Terns prefer low-lying, sandy shores for nesting and breeding, which makes their greatest competitors none other than people. Habitat modification has thus negatively impacted seashore breeding colonies of this small tern species. Similarly, the inland populations that settle on small alluvial islands of slow-flowing rivers in North America have also been hurt by management to artificially alter water levels that coincides with the breeding season, and in general by the building up of river banks and construction of dams. Both of these activities interfere with the formation of temporary sand banks, further reducing the available nesting habitats for this tern species.

The eggs are highly variable in color and markings, often within the same clutch, possibly to enhance camouflage and crypsis. Perhaps because of this variation, the eggs are not individually recognized and parent terns readily retrieve and adopt eggs of other pairs nesting nearby.

Clutch

The egg of the Least Tern is olive buff in background color with purple and brown streaks and maculation, and is 1¼ x ⅞ in (31 x 23 mm) in dimensions. Both sexes make and repeatedly visit several nest scrapes, but it is the female who settles on one site for the laying of the eggs.

Actual size

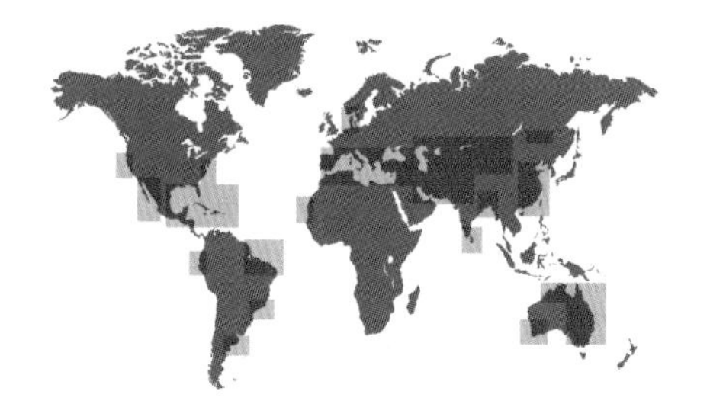

ORDER	Charadriiformes
FAMILY	Laridae
BREEDING RANGE	Southern Europe, temperate Asia, coastal North America, South America, and Australia
BREEDING HABITAT	Lakes, marshes, and seashores
NEST TYPE AND PLACEMENT	Shallow scrape in the ground, sometimes rimmed with sticks and other vegetation matter
CONSERVATION STATUS	Least concern
COLLECTION NO.	FMNH 4543

ADULT BIRD SIZE
13–15 in (33–38 cm)

INCUBATION
21–23 days

CLUTCH SIZE
3 eggs

GELOCHELIDON NILOTICA

GULL-BILLED TERN

CHARADRIIFORMES

Clutch

Gull-billed Terns belong to a monotypic genus of terns, which implies a degree of evolutionary uniqueness and isolation. Accordingly, the mandibles and the foraging strategies of this tern are unlike any of their relatives: they have a thick but stubby bill, and hunt by catching insects in midair, instead of diving into the water for fish.

Mating displays of this species also are atypical compared to most other terns. Instead of incorporating aerial elements, mating takes place solely on the ground or on sandy banks suitable for nesting. The female and male work together to establish the nest site, collect the few pieces of vegetation outlining the nest, incubate the eggs, and locate and provision the chicks once they leave the nest and mix in with the chicks from other nests in the small breeding colony.

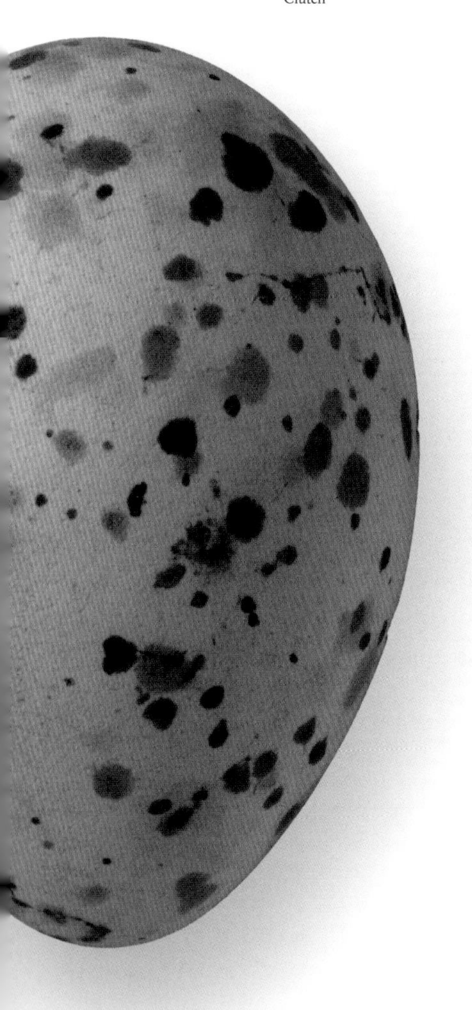

Actual size

The egg of the Gull-billed Tern is pinkish buff to ivory yellow in background color with dark brown maculation, and measures 1⅞ x 1⅓ in (47 x 34 mm). The eggs are well camouflaged and the chicks hatch with open eyes and are mobile within hours. Both these factors reduce the chances of predation.

ORDER	Charadriiformes
FAMILY	Laridae
BREEDING RANGE	Australasia, Asia, Africa, Europe, and North America
BREEDING HABITAT	Beaches and shorelines of large lakes, seas, and oceans
NEST TYPE AND PLACEMENT	A scrape in the ground, lined with dry vegetation, sticks, small pebbles, and broken shells
CONSERVATION STATUS	Least concern
COLLECTION NO.	FMNH 18312

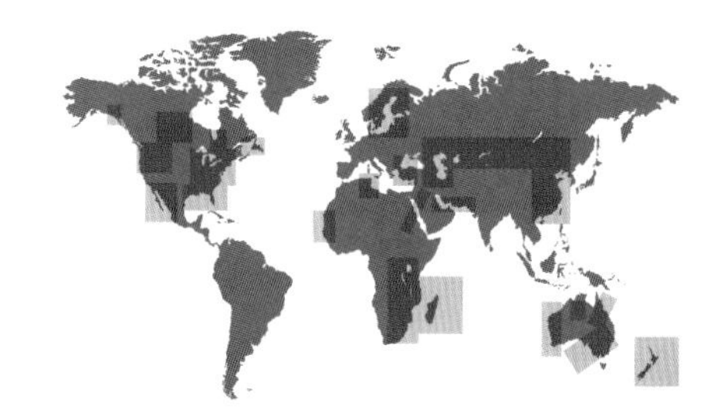

HYDROPROGNE CASPIA

CASPIAN TERN

CHARADRIIFORMES

ADULT BIRD SIZE
18½–21½ in (47–54 cm)

INCUBATION
26–28 days

CLUTCH SIZE
1–3 eggs

This conspicuous tern is the largest tern species in the world, with similar dimensions to some medium-sized gulls. It breeds in scattered colonies on the shores of lakes and other bodies of water across all continents except South America and Antarctica, making it an interesting species with respect to the genetic connectedness of such widely dispersed breeding populations. Based on morphology, all these populations are similar.

Clutch

Locally, Caspian Terns nest in small, loose colonies, often near other terns and gulls. The parents establish the nest and incubate the eggs together. They take turns in feeding the nestlings, diving for fish and other aquatic prey. The chicks stay in or around the nest for several days, and then they become mobile. The adults encourage the young to separate from their siblings, even if it means that the parents have to locate and recognize their own chicks as individuals (and not based on the location of the nest site alone). This separation reduces the chance of predators wiping out the whole brood at once.

The egg of the Caspian Tern is buff in background color with dark brown irregular blotching, and measures 2½ x 1¾ (65 x 45 mm) in dimensions. Nesting colonies have been increasingly established on dredged mounds and other habitats on shorelines either generated or modified by humans, thus locally increasing population sizes.

Actual size

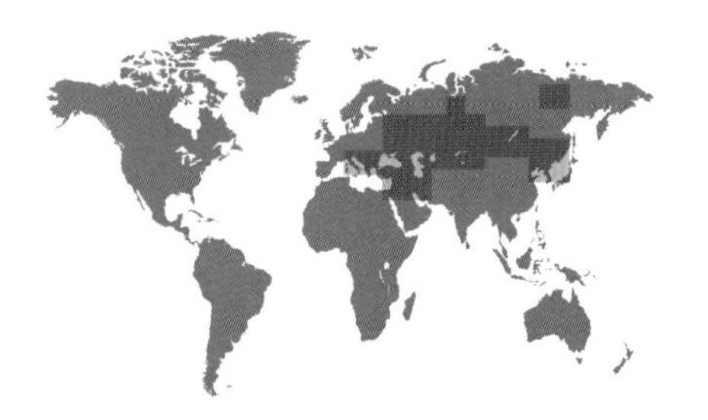

ORDER	Charadriiformes
FAMILY	Laridae
BREEDING RANGE	Temperate Eurasia
BREEDING HABITAT	Marshes, wetlands, lake shores
NEST TYPE AND PLACEMENT	A built-up platform of small reed stems and other vegetation, on the shore, or floating on the water
CONSERVATION STATUS	Least concern
COLLECTION NO.	FMNH 15082

ADULT BIRD SIZE
8½–10 in (22–25 cm)

INCUBATION
18–22 days

CLUTCH SIZE
2–4 eggs

CHLIDONIAS LEUCOPTERUS

WHITE-WINGED TERN

CHARADRIIFORMES

Clutch

White-winged Terns live and forage near water, but unlike terns with mostly white plumages (genus *Sterna*), these birds do not plunge-dive for fish. Instead they hunt for insects by flying above and near the water, or pick prey directly off the water's surface. In the nesting season, they stay near freshwater ponds, swamps, and marshes, but in the non-breeding season they wander to coastal areas, as well as to drier, inland, open fields.

On the breeding grounds, White-winged Terns maintain their own colonies, although in some European populations, apparently hybrid individuals with intermediate plumages between White-winged and Black Terns occur. Hybridization such as this allows scientists to assess the behavioral cues and the genetic consequences of recognizing one's own species and the mistakes that some individuals inevitably make in this recognition process.

Actual size

The egg of the White-winged Tern is buff to pale stone in background color with black and brown maculation, and measures 1⅜ x 1 in (35 x 25 mm) in size. Although individual pairs are fragile and too weak to attack egg predators, by their nesting in small, loose colonies of 20–40 pairs, the nests have a greater chance of success.

ORDER	Charadriiformes
FAMILY	Laridae
BREEDING RANGE	Temperate and subarctic Eurasia, central and eastern North America
BREEDING HABITAT	Flat, sandy, or rocky shores, beaches, and grassy patches
NEST TYPE AND PLACEMENT	Ground scrape in soil and sand, may be unlined or contain plant debris around the edges
CONSERVATION STATUS	Least concern
COLLECTION NO.	FMNH 4649

ADULT BIRD SIZE
12–15 in (31–38 cm)

INCUBATION
21–22 days

CLUTCH SIZE
2–3 eggs

STERNA HIRUNDO

COMMON TERN

CHARADRIIFORMES

Common Terns are thrifty and they have come to adapt to many human-induced changes to their shoreline-based habitats. They readily nest near humans, in places such as the edges of parking lots, or on abandoned rafts and other structures in or near the water. This species also shares colonies with several other terns, including the locally endangered Roseate Tern.

The Common Tern is a creature of habit. It returns to the same colony where it bred before, and may even occupy the same small patch of grassy area with the same mate. Males arrive and settle on the territory first, and then wait for several days for their mate to turn up at the colony. However, if the female is more than five days late, the male may pair up with a new female to begin the nesting attempt, ensuring enough time for incubation and chick development.

Clutch

The egg of the Common Tern is olive to buff in background color with dark brown maculation, and 1⅔ x 1⅛ in (42 x 30 mm) in size. Once hatched, the chicks begin to wander around the colony; unlike the situation in dense colonies of gulls, tern chicks are tolerated by other adults even away from their own nest.

Actual size

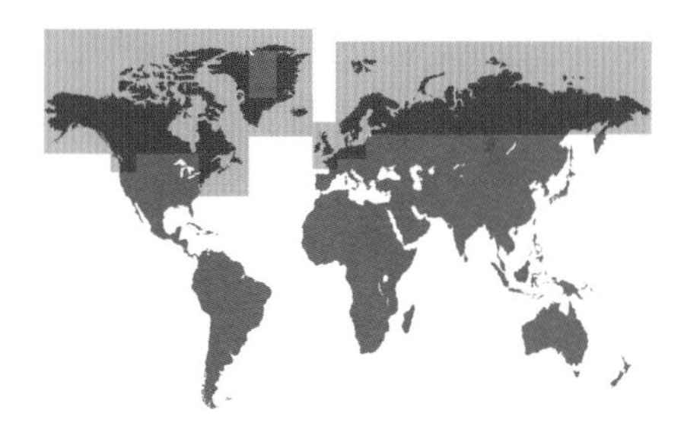

ORDER	Charadriiformes
FAMILY	Laridae
BREEDING RANGE	Arctic and subarctic North America and Eurasia
BREEDING HABITAT	Open tundra, clearings in boreal forests, rocky beaches, and islands
NEST TYPE AND PLACEMENT	Scrape in the open ground, or shallow depression in short vegetation
CONSERVATION STATUS	Least concern
COLLECTION NO.	FMNH 4673

ADULT BIRD SIZE
13–15½ in (33–39 cm)

INCUBATION
22–27 days

CLUTCH SIZE
2 eggs

STERNA PARADISAEA

ARCTIC TERN

CHARADRIIFORMES

Clutch

The egg of the Arctic Tern is olive to buff in background color with dark brown mottling, and measures 1⅝ x 1⅛ in (41 x 29 mm) in size. Both parents incubate the eggs and feed the mobile, but dependent, chicks after hatching.

The Arctic Tern is best known for its extraordinary annual migration, departing the Arctic breeding grounds and making its way to the Southern Ocean of the Antarctic; these journeys are undertaken both by adults and by each year's juveniles. Despite the energetic demands and the many dangers potentially encountered while criss-crossing half the globe, this species has a long lifespan, living up to 30 years of age.

On the breeding ground, the Arctic Tern forms colonies with other terns. It vigorously defends its nesting site against any intruders, no matter their size, even to the extent of occasionally drawing blood from researchers monitoring the colony. Fast- and agile-flying, these terns behave boldly and mob predators approaching their nests. Their aerial skills are such that they are able to outmaneuver and escape a predator should it decide to switch its attention to the parents.

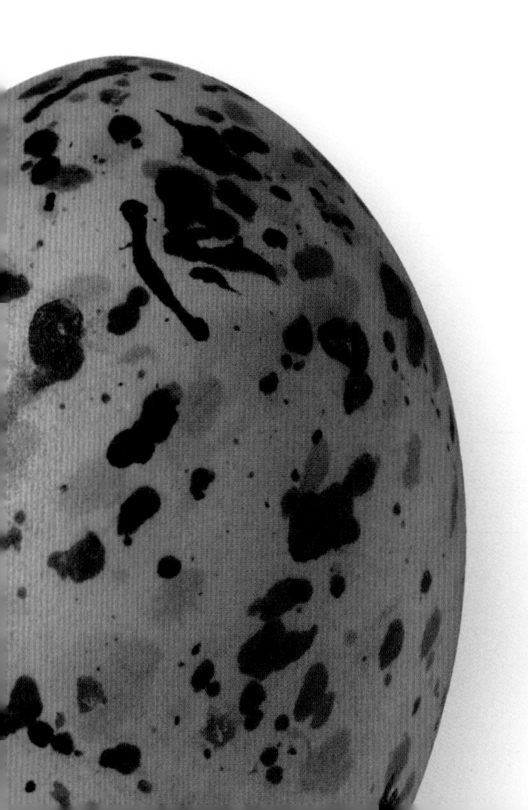

Actual size

ORDER	Charadriiformes
FAMILY	Laridae
BREEDING RANGE	Temperate North America
BREEDING HABITAT	Inland wetlands and marshes
NEST TYPE AND PLACEMENT	Simple scrape in the soil, or a raft of floating vegetation, or a cup formed of and lined with plant matter on top of a mound
CONSERVATION STATUS	Least concern
COLLECTION NO.	FMNH 4585

ADULT BIRD SIZE
13–14 in (33–36 cm)

INCUBATION
25–32 days

CLUTCH SIZE
2–3 eggs

STERNA FORSTERI

FORSTER'S TERN

CHARADRIIFORMES

Forster's Terns occupy a predominantly temperate breeding range in North America and subtropical and tropical wintering distribution further south. Small numbers of birds, however, regularly cross the Atlantic Ocean, and spend the cold winter months near the coast of the British Isles. At that time of the year, these are the only terns present, as most European species migrate further south for the non-breeding season.

On their inland breeding grounds, this North American species nests in large colonies, with individual parents vigorously defending their nest and eggs from intruders, irrespective of the size of that potential threat. The chicks hatch with down and open eyes, are able to walk, but typically remain in the nest for weeks to be provisioned by parents. When the chicks do wander, they may find their way to near the nest of another Forster's Tern or a Black Tern. As a result, these species occasionally feed each other's chicks.

Clutch

The egg of the Forster's Tern is olive to buff in background coloration, with dark brown maculation concentrated around the blunt pole, and measures 1⅔ x 1¼ in (43 x 31 mm). The midwestern inland wetland breeding grounds of this species in the U.S. have suffered from habitat modification, which has negatively impacted the nesting of this and many other waterbirds in the region.

Actual size

ORDER	Charadriiformes
FAMILY	Laridae
BREEDING RANGE	Southern California and northwestern Mexico
BREEDING HABITAT	Low, flat, and sandy islands
NEST TYPE AND PLACEMENT	Ground scrape, in high-density colonies
CONSERVATION STATUS	Near threatened
COLLECTION NO.	FMNH 4569

ADULT BIRD SIZE
15½–17 in (39–43 cm)

INCUBATION
25–26 days

CLUTCH SIZE
1–2 eggs

THALASSEUS ELEGANS

ELEGANT TERN

CHARADRIIFORMES

Clutch

The egg of the Elegant Tern is purplish white in background color with black and lilac spotting, and measures 2 x 1⅜ in (52 x 35 mm) in size. These birds are rarely aggressive toward predators near the colonies; instead, they rely on nesting near each other, and also near larger gull species, for protection.

The Elegant Tern has a globally restricted breeding range, with over 90 percent of its breeding population nesting on just a single island in the Gulf of California. Such a limited distribution automatically qualifies the species for a conservation alert and special consideration status. And although the Elegant Tern has expanded into more breeding areas in California in the last half-century, ongoing patterns of trampling, habitat modification, and predation by feral animals have negatively impacted the size and long-term prospects of several breeding colonies.

Before egg laying, the male catches and passes a fish from his bill to the female, as part of their courtship display. When selecting a nest site, these birds space themselves at just one body-length away from their nearest neighbor. Such high density serves as a passive, but highly efficient, mechanism of protection against predators, because in general only the edge-nesting pairs are then vulnerable to predation.

Actual size

ORDER	Charadriiformes
FAMILY	Laridae
BREEDING RANGE	Southern and eastern North America and lowland South America
BREEDING HABITAT	Sandy or gravel beaches
NEST TYPE AND PLACEMENT	Scrape nest in sand or gravel on shore, including artificial islands and rooftops
CONSERVATION STATUS	Least concern
COLLECTION NO.	FMNH 4871

ADULT BIRD SIZE
15½–19½ in (40–50 cm)

INCUBATION
21–25 days

CLUTCH SIZE
3–5 eggs

RYNCHOPS NIGER

BLACK SKIMMER

CHARADRIIFORMES

Black Skimmers are typically seen flying just above the water, with their knifelike lower mandible "skimming" through the water. When the beak touches a fish or other creature, the upper mandible snaps down automatically to catch the prey. These are colonial, social birds, both during the breeding season and at other times of the year. In North America, their preferred nesting and roosting habitat is on sandy beaches along warm-water coastlines, so skimmers and people compete for some of the best oceanfront beach views. The skimmers typically lose out.

When a nesting colony is established, pairs of Black Skimmers perform a ritualized display to scrape their nest. Standing upright, both the female and the male kick sand with their feet; although the nest is a simple scrape, the process of creating it is so much part of the courtship and nuptial display that it often takes a long time, and several attempts, to settle on a final nest site. The male scrapes and kicks more sand than the female, but both parents share duties to incubate the eggs and feed the mobile, but dependent, young.

Clutch

The egg of the Black Skimmer is bluish white or pale pink in background color, with brown to purple maculation, and measures 1¾ x 1¼ in (45 x 32 mm). Nesting in larger colonies assures safety and success in this species, and such colonies are maintained across years, whereas smaller, failed colonies may relocate annually.

Actual size

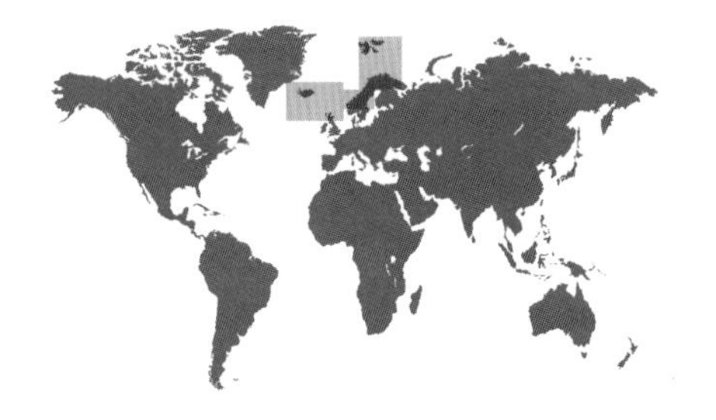

ORDER	Charadriiformes
FAMILY	Stercorariidae
BREEDING RANGE	Northern European coast and islands
BREEDING HABITAT	Coastal moorlands and rocky islets
NEST TYPE AND PLACEMENT	Ground nest on rocky coast or in fields, lined with grasses
CONSERVATION STATUS	Least concern
COLLECTION NO.	FMNH 4369

ADULT BIRD SIZE
19½–23 in (50–58 cm)

INCUBATION
29 days

CLUTCH SIZE
2 eggs

STERCORARIUS SKUA

GREAT SKUA

CHARADRIIFORMES

Clutch

The Great Skua is a large, bulky seabird, and it uses its size, rather than agility, to be successful on the breeding and wintering grounds. When it spots other birds—including gulls, terns, and even gannets—with freshly caught prey, it flies directly at them, pinches their wing in its beak, and follows the falling victim directly to the water, where it continues to attack it, until it regurgitates its prey. This advanced form of kleptoparasitism complements the other food sources of skuas, which typically include fish, the eggs and chicks of other birds, and rodents.

Individual Great Skuas may become specialists, feeding solely on stolen food or only on fish. Specialization seems to pay off for these individuals, compared to generalist feeders, because the specialists start nesting earlier in the season, lay larger eggs, and raise faster-growing chicks.

Actual size

The egg of the Great Skua is olive brown in background and with slightly darker brown maculation, and is 2⅞ x 2 in (73 x 50 mm) in size. Parents vigorously defend their eggs or young, flying toward and directly confronting the approaching danger, including people, instead of trying to hide or lead the intruder away from the nest.

ORDER	Charadriiformes
FAMILY	Stercorariidae
BREEDING RANGE	High Arctic of North America, Greenland, and Eurasia
BREEDING HABITAT	Dry, short grass or rocky tundra
NEST TYPE AND PLACEMENT	A shallow ground depression, often with pebbles
CONSERVATION STATUS	Least concern
COLLECTION NO.	FMNH 4391

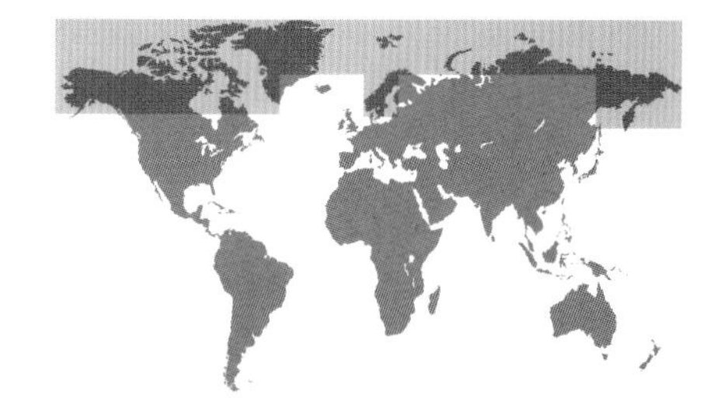

ADULT BIRD SIZE
15–23 in (38–58 cm)

INCUBATION
23–25 days

CLUTCH SIZE
2 eggs

STERCORARIUS LONGICAUDUS

LONG-TAILED JAEGER

CHARADRIIFORMES

Known as the Long-tailed Skua outside the Americas, this species is an active hunter of mammalian prey, including lemmings, in its Arctic breeding grounds; as the populations of lemmings fluctuate, so does the productivity of the jaegers' breeding attempts. However, this species also will revert to its relatives' most common foraging tactic: parasitism and stealing. By chasing after and outmaneuvering other seabirds, it catches their regurgitated prey, instead of spending prolonged periods fishing on its own.

Clutch

The two eggs are incubated from the day on which each is laid, perhaps as a necessity to avoid dangerously low temperatures in the cool Arctic breeding grounds. Because the eggs may be laid several days apart, the chicks hatch asynchronously. Both parents incubate the eggs and attend the chicks, which are semi-precocial: they are downy and mobile soon after they hatch, but they require prey to be delivered and fed to them, in small morsels, by the parents during the first weeks of life.

The egg of the Long-tailed Jaeger is olive brown in coloration with dark brown maculation, and it measures 2⅛ x 1½ in (55 x 38 mm) in size. Although both parents incubate the eggs, the female does so more often and for longer periods than the male does.

Actual size

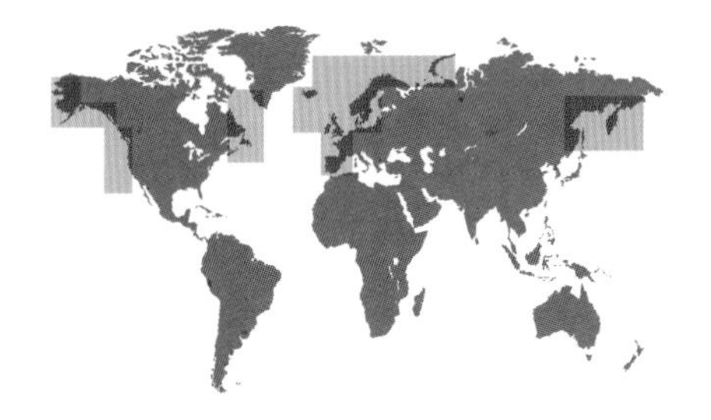

ORDER	Charadriiformes
FAMILY	Alcidae
BREEDING RANGE	Coastlines of the northern Pacific and Atlantic Oceans
BREEDING HABITAT	Rocky cliffs or flat areas on rocky headlands and islands, in full ocean view
NEST TYPE AND PLACEMENT	No nest, egg is typically placed on bare, rocky ledges
CONSERVATION STATUS	Least concern
COLLECTION NO.	FMNH 4244

ADULT BIRD SIZE
15–17 in (38–43 cm)

INCUBATION
30 days

CLUTCH SIZE
1 egg

URIA AALGE

COMMON MURRE

CHARADRIIFORMES

The Common Murre, also known as the Common Guillemot, breeds in colonies called "loomeries," which are so densely packed that individuals are often able to reach and touch their neighbors. Without a distinct nest structure or spatial isolation, it is paramount that murre parents can recognize their own egg as they come and go from the colony. It appears that the immense variation in the background coloration of eggs and the distinctiveness of their maculation assist with individual egg recognition. Upon returning to incubate the egg, the parent carefully examines it, probably to ensure that it has the right markings.

In the absence of a protective nest, the shape of the Common Murre's egg also functions to anchor it safely on the narrow cliff ledge where this bird breeds. Even when it is knocked by an adult taking flight, the egg rolls in a narrow circle back to its original position. The chick grows fast to fledge at just three weeks of age.

Actual size

The egg of the Common Murre can be a wide range of shades of greenish blue in background color, with blackish brown markings, and measures 3³⁄₁₆ x 2 in (81 x 50 mm) in size. Relative to body size, the egg is large and comprises 11 percent of the female's body weight.

ORDER	Charadriiformes
FAMILY	Alcidae
BREEDING RANGE	Arctic and subarctic coastlines of North America and Eurasia
BREEDING HABITAT	Cliff faces on oceanic shores and islands
NEST TYPE AND PLACEMENT	No nest, egg is laid on bare rock of ledges on steep cliffs
CONSERVATION STATUS	Least concern
COLLECTION NO.	FMNH 4317

ADULT BIRD SIZE
17–17½ in (43–45 cm)

INCUBATION
31–34 days

CLUTCH SIZE
1 egg

URIA LOMVIA

THICK-BILLED MURRE

CHARADRIIFORMES

When breeding, these birds form vast colonies, occasionally of over a million individuals. The single chick grows fast and require lots of parental trips; with each trip the parent brings just one food item back in its beak. Older birds are more successful at feeding their chicks more often, and also learn to nest in the more central, safer parts of the colony. Thus, experience imparts greater fitness to breeding murres. These birds breed in the densest of colonies, with each incubating adult requiring less than a square foot (0.1 m^2) of space.

Since hunting by humans led to the extinction of the flightless Great Auk (see page 189), the Thick-Billed Murre is the largest living species among all alcid seabirds. Although flighted, these murres spend most of their time at sea, diving and using their wings to propel themselves by essentially "flying underwater" in pursuit of fish prey.

The egg of the Thick-billed Murre is tan to dark green in color with black markings; it measures 3⅛ x 2 in (80 x 50 mm). The highly variable background color and maculation of the eggs ensures that each one is unique in appearance, meaning it can be recognized by its parents.

Actual size

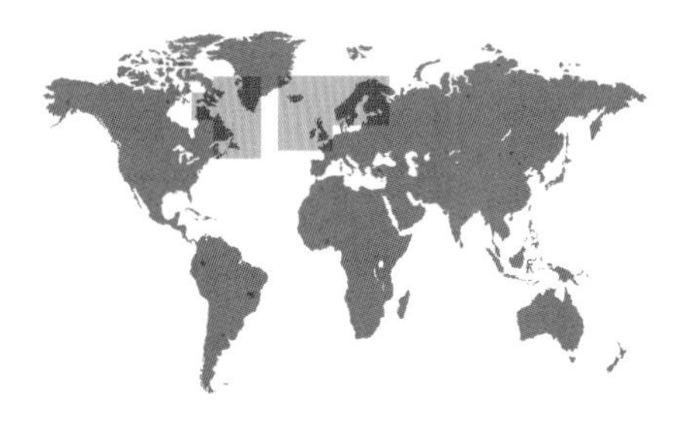

ORDER	Charadriiformes
FAMILY	Alcidae
BREEDING RANGE	Subarctic and boreal waters on both coasts of the Atlantic Ocean
BREEDING HABITAT	Coastal areas off continental-shelf waters, near boulders and caves, or narrow cliff ledges
NEST TYPE AND PLACEMENT	Egg is laid on barren rock, or in shallow bowl of rocks, shells, feathers, and vegetation
CONSERVATION STATUS	Least concern
COLLECTION NO.	FMNH 4360

ADULT BIRD SIZE
15–17 in (38–43 cm)

INCUBATION
35 days

CLUTCH SIZE
1 egg

ALCA TORDA

RAZORBILL

CHARADRIIFORMES

Compared to other birds, Razorbills plan for the long haul in their reproductive behaviors. Individuals do not breed until four to five years of age, and the female carefully chooses her social mate by approaching different males during their courtship displays, as if to incite them to compete for her favors. Once the pair bond is established, it is maintained for life. The birds return to the same breeding colony and the same nesting site, where they display and copulate for weeks before the single egg is laid.

Once the pair reaches older age, they may skip a breeding season between nesting attempts. Taking a season off ensures the parents are in better physical condition for the next breeding season. As she gets older, the female may copulate with several mates before laying her egg, to ensure that fertilization occurs. Previously exploited by hunters, today large nesting colonies are stable or even growing in size. The lifespan of this species is long, with the oldest recorded individual living for 42 years.

Actual size

The egg of the Razorbill is white to reddish brown in background coloration, with darker brown spots around the blunt end. It measures 3 x 1⅞ in (76 x 48 mm) in size. The parents take turns to incubate the egg; those laid in crevices are more likely to survive attacks from predators than eggs on open cliffs.

ORDER	Charadriiformes
FAMILY	Alcidae
BREEDING RANGE	Cold, north Atlantic coastlines, both in Europe and North America
BREEDING HABITAT	Rocky, barren offshore islands
NEST TYPE AND PLACEMENT	Egg laid directly on rock surface
CONSERVATION STATUS	Extinct
COLLECTION NO.	FMNH 2908

PINGUINUS IMPENNIS

GREAT AUK

CHARADRIIFORMES

ADULT BIRD SIZE
30–33½ in (75–85 cm)

INCUBATION
42 days

CLUTCH SIZE
1 egg

Great Auks were flightless, a trait they shared with other extinct species in the alcid family. However, the Great Auk's closest relatives, the Razorbill (see previous page) and the Dovekie, can fly. This suggests that flightlessness evolved independently in both Great Auk and other extinct flightless alcids. The colonial and breeding behaviors of the species were similar to those of most other extant relatives, probably with a single monogamous pair incubating a single egg and raising the dependent chick within a large and densely packed breeding colony.

The Great Auk was the largest member of its family, and its colonial habits made it vulnerable to human exploitation through hunting, and predation by introduced rats and other predatory mammals on islands with long-standing colonies. The last individuals known to be nesting were killed in 1844. The species has never been studied by a trained scientist; knowledge of the Great Auk's behavior has been distilled from native people's legends and whalers' accounts.

The egg of the Great Auk was white with a bluish tint, and patterned with dark brown blotches and streaks. It measured 4¾ x 3 in (120 x 75 mm). Fewer than 100 eggs of the Great Auk remain, and so most collections obtain a handpainted, realistic cast for their display, as shown below.

Actual size

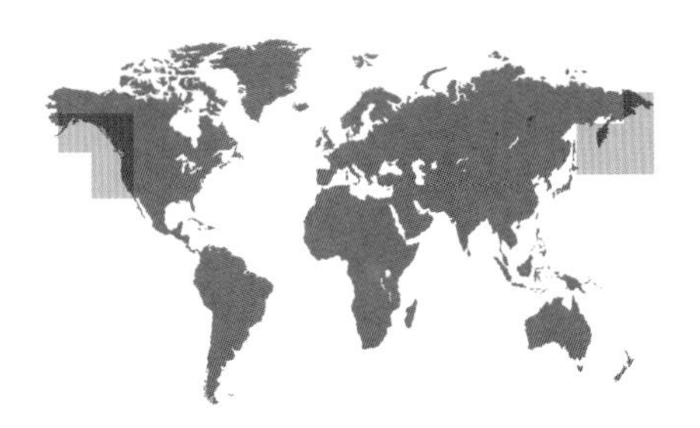

ORDER	Charadriiformes
FAMILY	Alcidae
BREEDING RANGE	North Pacific Ocean, along both Asian and North American coasts
BREEDING HABITAT	Rocky shores, cliffs, and islands
NEST TYPE AND PLACEMENT	Shallow scrape in soil or gravel, inside a cavity or boulder cavity
CONSERVATION STATUS	Least concern
COLLECTION NO.	FMNH 4215

ADULT BIRD SIZE
12–14 in (30–35 cm)

INCUBATION
30–32 days

CLUTCH SIZE
1–2 eggs

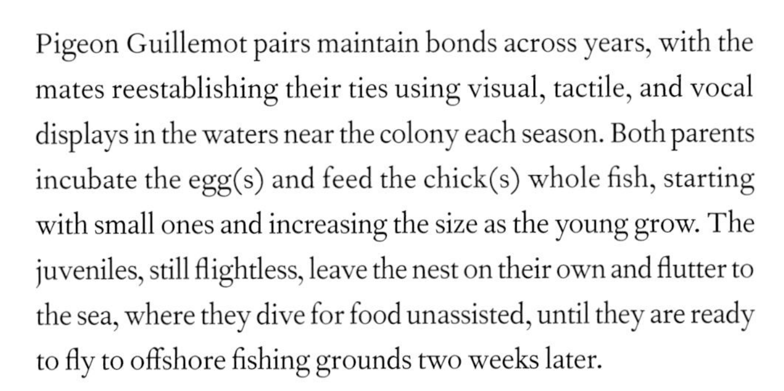

CEPPHUS COLUMBA

PIGEON GUILLEMOT

CHARADRIIFORMES

Clutch

The egg of the Pigeon Guillemot is greenish to bluish cream in background color and has dark brown to black and lavender blotches; it measures 2⅜ x 1⅝ in (60 x 41 mm) in size. Where each egg hatches is important, because the adult birds return to their own natal colony for breeding.

Pigeon Guillemot pairs maintain bonds across years, with the mates reestablishing their ties using visual, tactile, and vocal displays in the waters near the colony each season. Both parents incubate the egg(s) and feed the chick(s) whole fish, starting with small ones and increasing the size as the young grow. The juveniles, still flightless, leave the nest on their own and flutter to the sea, where they dive for food unassisted, until they are ready to fly to offshore fishing grounds two weeks later.

The Pigeon Guillemot, a handsome, mostly black seabird with prominently red feet, spends most of its life at sea, coming to land only to nest and to raise the young. It is a loosely colonial species, with breeding aggregations ranging from a couple of pairs to hundreds. Adults can hop and crawl high onto near-vertical cliff faces to reach their nesting crevice, using their clawed webbed feet and intensive wing-flapping.

Actual size

ORDER	Charadriiformes
FAMILY	Alcidae
BREEDING RANGE	Alaska, possibly also eastern Siberia
BREEDING HABITAT	Inland, mountain-top slopes, above the tree line
NEST TYPE AND PLACEMENT	Egg laid on barren ground or in crevice, on south-facing slopes
CONSERVATION STATUS	Critically endangered
COLLECTION NO.	FMNH 18533

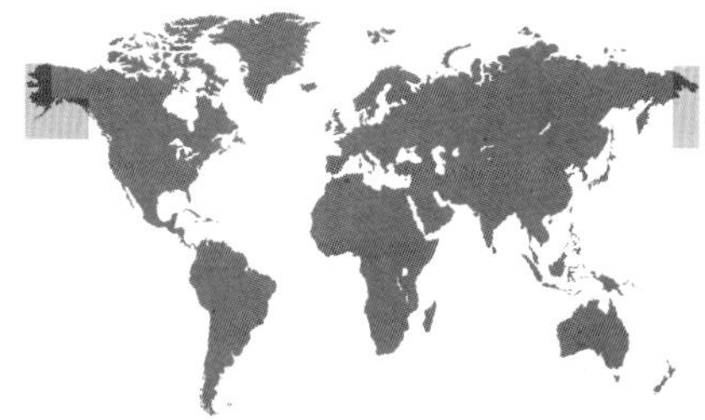

ADULT BIRD SIZE
7½–9 in (19–23 cm)

INCUBATION
Unknown (!), estimated at 30 days

CLUTCH SIZE
1 egg

BRACHYRAMPHUS BREVIROSTRIS

KITTLITZ'S MURRELET

CHARADRIIFORMES

This murrelet is one of the rarest and least-known North American seabirds. Unusually for a seabird, it nests solitarily on high mountain tops inland from the Alaskan coastline, near glaciers and close to the snow line. The nesting grounds are remote and difficult to locate, so the breeding biology of the species is poorly known, including the duration of its incubation period. The chicks are covered in a dense and warm down layer until just before departing the nesting site to protect them from the high elevation and exposed position of the breeding sites. We have yet to discover how the fledglings make their way to the coastal waters after the adults cease parental care.

Ongoing habitat changes, including the retreat of glaciers near the breeding grounds and at sea level where the adults forage, have paralleled a dramatic decrease in the population sizes of this species. Most of the monitored nesting localities have seen a 30–80 percent decline in breeding pair numbers over the last two decades. It is also estimated that 5–15 percent of the global population was killed off in the *Exxon Valdez* oil spill in 1989.

The egg of the Kittlitz's Murrelet is pale olive to olive green in background color with brown maculation, and it measures 2⅜ x 1⁷⁄₁₆ in (60 x 37 mm) in size. Typical maculation for this species consists of small speckles, larger spots, and thin scrawls, especially near the blunt end of the egg. This specimen from the Field Museum, collected in 1886 from the Aleutian Islands, represents an unusual record for the species in that it is immaculate.

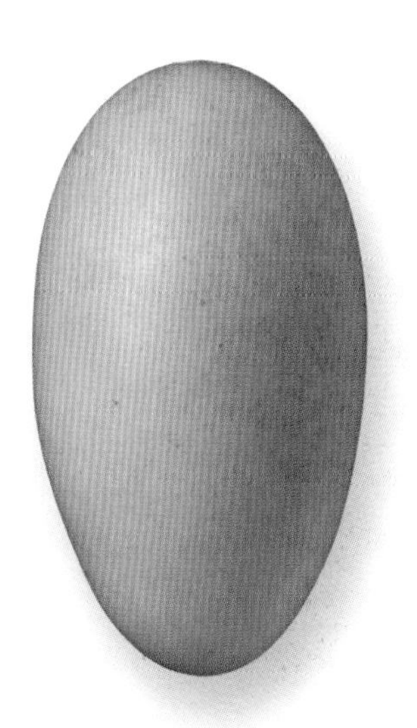

Actual size

ORDER	Charadriiformes
FAMILY	Alcidae
BREEDING RANGE	Northern Pacific coastline, both in Asia and in North America
BREEDING HABITAT	Rock outcrops and grassy patches on islands
NEST TYPE AND PLACEMENT	Along rocky seacoasts in crevices, under rocks, and in burrows in the ground
CONSERVATION STATUS	Least concern, but locally a species of "special concern"
COLLECTION NO.	FMNH 4174

ADULT BIRD SIZE
8–9½ in (20–24 cm)

INCUBATION
23–36 days

CLUTCH SIZE
1–2 eggs

SYNTHLIBORAMPHUS ANTIQUUS

ANCIENT MURRELET

CHARADRIIFORMES

Clutch

Ancient Murrelets are small seabirds and, while incubating, arrive and depart their burrow-nest under darkness, presumably to avoid predators. The parents do not feed their young in the nest chambers, perhaps to reduce exposure to land-dwelling predators. Instead they leave the nest and fly toward the ocean shore, calling repeatedly to lead the chicks to the sea. Under the cover of the night sky, the one-to-three-day-old chicks crawl out of the burrows and head to the shore. Once in the water, the families locate their members by their calls, and then directly swim offshore for at least 12 more hours. The chicks are then fed in the open ocean by the parents for a month or more, until they can feed and fly on their own.

The burrow-nests of Ancient Murrelets make them vulnerable to mammalian predators, including foxes and raccoons. Invasion of colony islands by these predators is frequent, requiring continued conservation monitoring effort, and repeated mammal-removal programs, to ensure these birds' breeding success in the long run.

Actual size

The egg of the Ancient Murrelet is pale green in coloration, with light and dark brown maculation; it measures 2⅓ x 1½ in (59 x 38 mm) in size. Occasionally, 3–4 eggs are found together, but the differences in the maculation suggest that these clutches are the product of two or more females laying together.

ORDER	Charadriiformes
FAMILY	Alcidae
BREEDING RANGE	Pacific coastline of temperate and Arctic North America
BREEDING HABITAT	Offshore islands
NEST TYPE AND PLACEMENT	Short burrows in the ground, also crevices and holes in buildings
CONSERVATION STATUS	Least concern
COLLECTION NO.	FMNH 4165

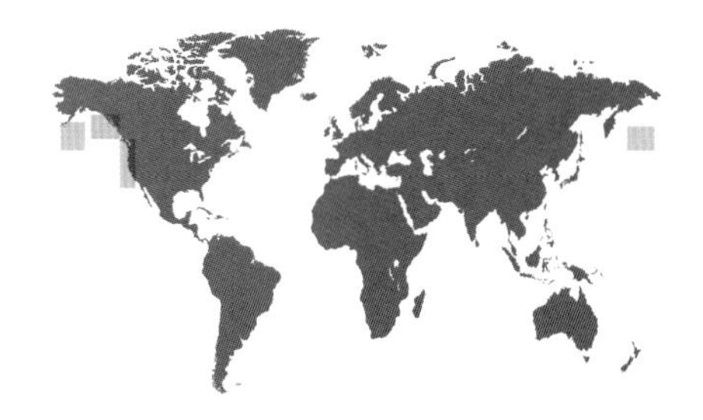

PTYCHORAMPHUS ALEUTICUS

CASSIN'S AUKLET

CHARADRIIFORMES

ADULT BIRD SIZE
8–9 in (20–23 cm)

INCUBATION
40 days

CLUTCH SIZE
1 egg

The Cassin's Auklet, another diminutive seabird, is unusual among its relatives in that adults maintain the same dark head plumes throughout the year, instead of molting into an ornamental, nuptial plumage prior to mating. In fact, individuals from some southern populations remain around the nesting colony all year round, instead of moving offshore to spend the non-breeding season on the open sea.

Breeding is a well-coordinated cooperative affair between the female and the male; the mates take turns in incubating the egg, changing guard every three days or so, typically during the nighttime. The Cassin's Auklet is one of the few alcids that can complete two nesting attempts per breeding season, especially in the southern populations. To feed the young, parents provide the chick with nutrient-rich half-digested krill and other planktonic prey items, stored and transported in a special gular pouch, just under the tongue.

The egg of the Cassin's Auklet is white or whitish green in background color, immaculate, and 1¾ x 1¼ in (44 x 32 mm) in size. The eggs are variable in size, and quickly become darker grey or soiled as incubation progresses.

Actual size

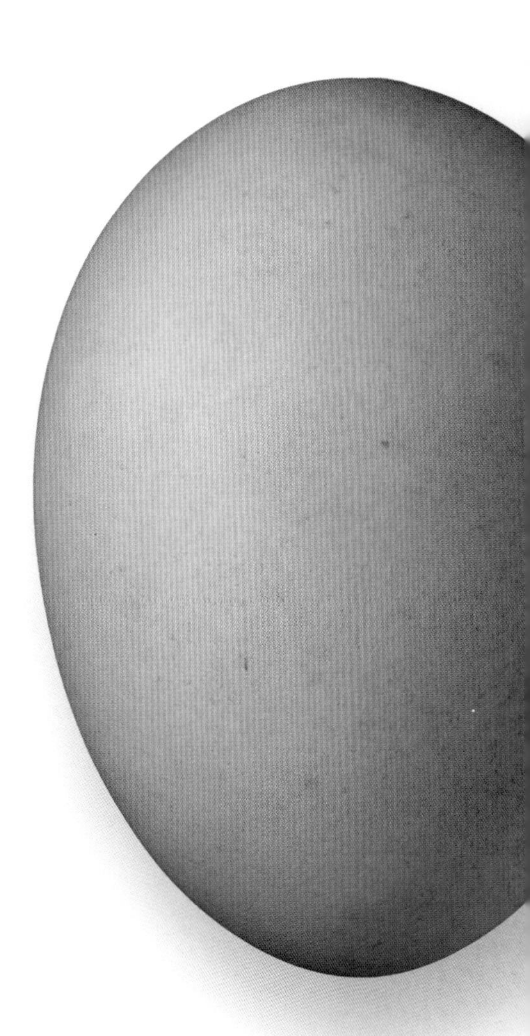

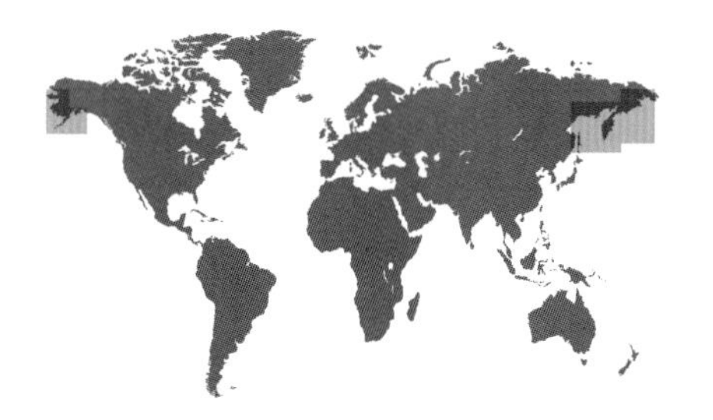

ORDER	Charadriiformes
FAMILY	Alcidae
BREEDING RANGE	Arctic Pacific coasts and islands of North America and Asia
BREEDING HABITAT	Rocky cliffs on the mainland and islands
NEST TYPE AND PLACEMENT	Bare ground in rock crevices, or in burrows
CONSERVATION STATUS	Least concern
COLLECTION NO.	FMNH 1938

ADULT BIRD SIZE
9–10 in (23–26 cm)

INCUBATION
35–36 days

CLUTCH SIZE
1 egg

AETHIA PSITTACULA

PARAKEET AUKLET

CHARADRIIFORMES

The egg of the Parakeet Auklet is chalk white to bluish white in background color, has no speckling, and measures 2⅛ x 1½ in (54 x 37 mm) in size. The parents take turns to incubate the egg, typically spending one day at sea, followed by a day in the nest chamber.

Parakeet Auklets are dedicated parents during the breeding season, with both sexes equitably sharing incubation of the egg and provisioning of the nestling. Once the chick can fly, the parents stop feeding it. The weight profile of the chick reflects this parental strategy: first the chick grows rapidly, even surpassing the mass of its parents. Once the adults stop provisioning, the chick loses weight and departs the nest, flying toward the open ocean soon afterwards.

This auklet is a specialist of the Arctic seas in the north Pacific Ocean; it winters at sea, foraging on invertebrate zooplankton by diving to depths of 100 ft (30 m), and only returns to land to nest and breed. Its unusual, tubular beak is probably an adaptation to handle soft gelatinous prey, including jellyfish.

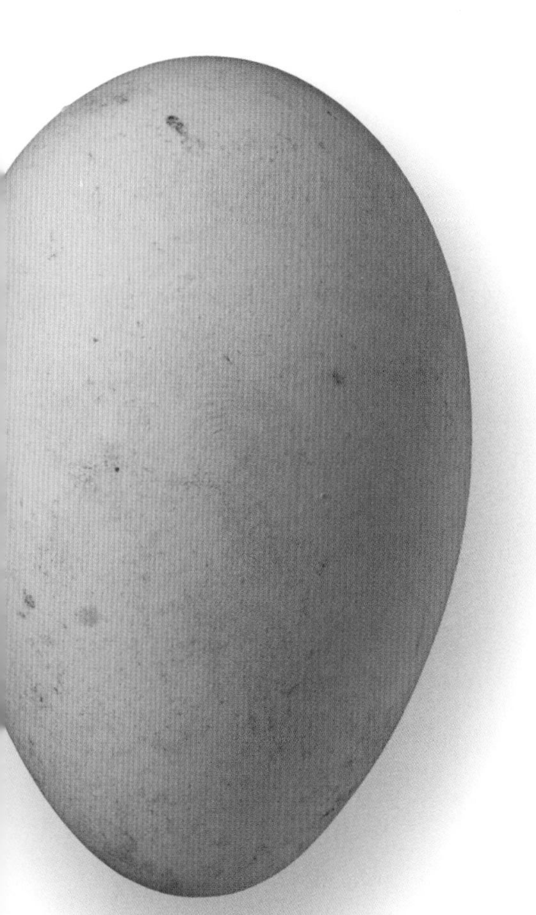

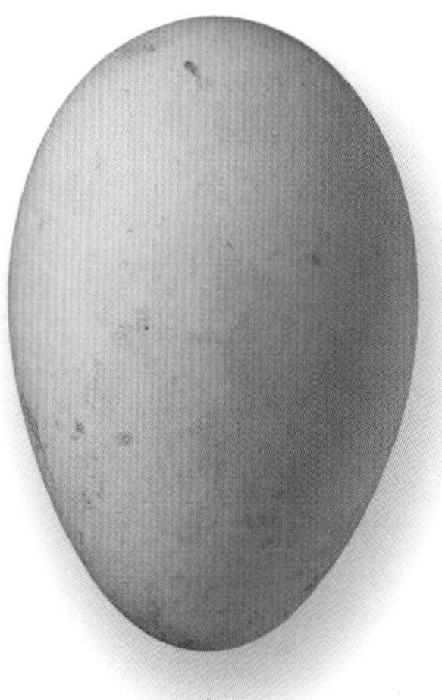

Actual size

ORDER	Charadriiformes
FAMILY	Alcidae
BREEDING RANGE	Arctic and temperate north Pacific Ocean, islands along both North American and Asian coastlines
BREEDING HABITAT	Grassy fields sloping toward the open ocean, also rocky outcroppings
NEST TYPE AND PLACEMENT	Digs its own ground burrows or uses rock crevices
CONSERVATION STATUS	Least concern
COLLECTION NO.	FMNH 4160

ADULT BIRD SIZE
11–11½ in (28–29cm)

INCUBATION
39–52 days

CLUTCH SIZE
1 egg

CERORHINCA MONOCERATA

RHINOCEROS AUKLET

CHARADRIIFORMES

This species is named after the horn-like vertical white sheath at the base of its beak, which is displayed only during the breeding season. It nests on remote islands in both forested and grassy habitats, in colonies of thousands of individuals. When it digs its own burrow in the soil, this can be long, with several forks and dead-ends, before reaching the nesting chamber. The chick is semiprecocial, covered in down and can readily walk; yet it remains in the nest for over a month and a half. During this time the adults arrive and depart under cover of darkness. This reduces both predation on the adults as well as kleptoparasitism by large gulls who try to steal the fish adult auklets bring back to feed their nestling.

Despite being called an auklet, this species resembles the puffins in appearance, behavior, and DNA. For example, it typically lines up several fish prey items in its beak before flying back to the colony to feed the nestling, a behavior more akin to puffins than other auklets. Returning with more than one prey item is important for these birds, because their feeding sites can be up to 30 miles (50 km) away from the breeding site.

The egg of the Rhinoceros Auklet is off-white in background color, immaculate, and 2¾ x 1¾ in (69 x 46 mm) in size. On several islands without large colonies of gulls, the parents attend the nest throughout all hours of the day, to reduce the risk of losing their egg to avian nest predators.

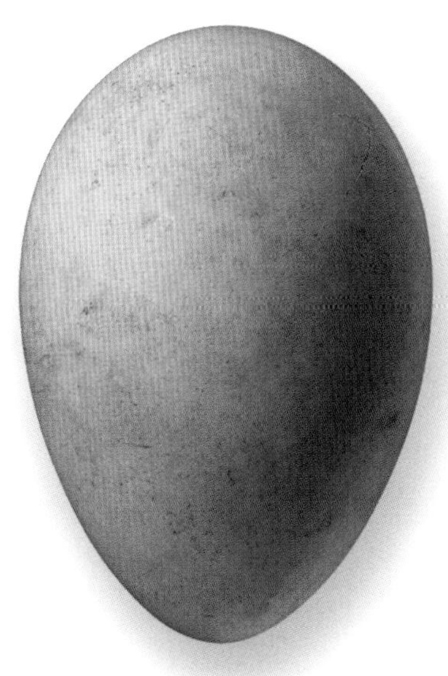

Actual size

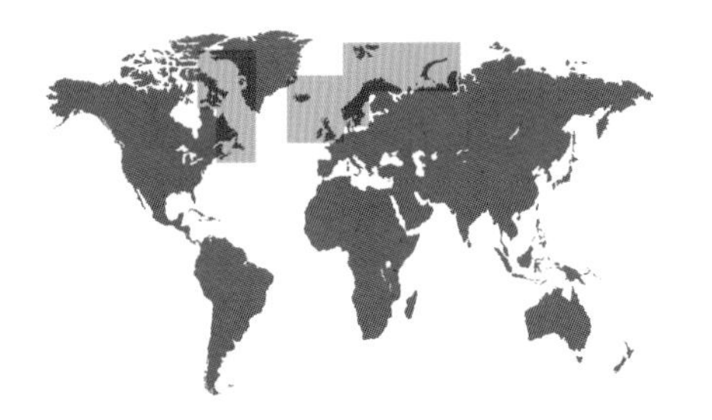

ORDER	Charadriiformes
FAMILY	Alcidae
BREEDING RANGE	North Atlantic coasts and islands off Europe, Greenland, and North America
BREEDING HABITAT	Rocky seacoasts with cliff faces and grassy fields
NEST TYPE AND PLACEMENT	Excavated burrows in grassy slopes facing the ocean, occasionally in rocky crevices
CONSERVATION STATUS	Least concern
COLLECTION NO.	FMNH 4126

ADULT BIRD SIZE
10–11½ in (26–29 cm)

INCUBATION
39–45 days

CLUTCH SIZE
1 egg

FRATERCULA ARCTICA

ATLANTIC PUFFIN

CHARADRIIFORMES

Atlantic Puffins breed in large colonies, often mixed with other seabirds, and frequently on offshore islands unreachable by land mammals. It takes four to five years for a puffin to sexually mature, and then it begins breeding by establishing a pair bond which lasts across several breeding seasons. The male is larger, more boldly colored, and invests more heavily in excavating the nesting burrow, but both sexes take turns incubating the egg and provisioning the chick. After the nestling departs the burrow, it heads directly to sea, where the parents continue to feed it until it reaches independence.

When swimming in the ocean, the Atlantic Puffin, also known as the Common Puffin, displays the most conspicuous of its traits, the boldly patterned head plumes and the colorful beak. This allows for a suite of social and reproductive decisions to be undertaken in the water, including both pair bonding and mating. These birds only truly return to land to nest.

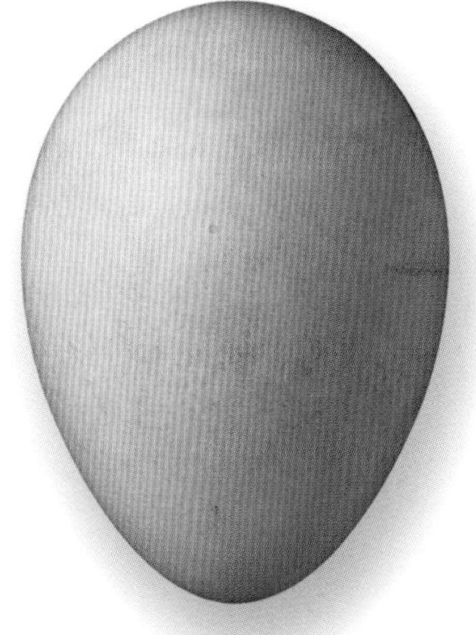

Actual size

The egg of the Atlantic Puffin is white with a green or beige tint in background color; it has brown maculation, and measures 2½ x 1¾ in (63 x 45 mm) in size. The most common predators of puffin eggs are other birds, including large gulls and skuas.

ORDER	Charadriiformes
FAMILY	Alcidae
BREEDING RANGE	Isolated islands in the north Pacific off North America and Asia
BREEDING HABITAT	Coastal slopes, vegetation, and rocky outcroppings
NEST TYPE AND PLACEMENT	Simple burrow or a crevice between rocks, well lined with plant materials and feathers
CONSERVATION STATUS	Least concern
COLLECTION NO.	FMNH 2980

ADULT BIRD SIZE
14–15½ in (36–40 cm)

INCUBATION
45 days

CLUTCH SIZE
1 egg

LUNDA CIRRHATA

TUFTED PUFFIN

CHARADRIIFORMES

The Tufted (or Crested) Puffin is a uniquely colored seabird that breeds in dense colonies on offshore locations. The single egg is incubated by both parents taking turns, and the pair also cooperate in provisioning the nestling. It takes a lot to feed the fast-growing chick, and puffin parents are often seen to return to the breeding colony with several intact fish lodged between their mandibles.

This puffin is the preferred prey species for Arctic Foxes, even when nesting in large, mixed-species breeding aggregations. To escape this predation, the puffins dig nesting burrows or select crevices that are especially hard to approach from land. The strategy also makes the species hard for researchers and conservation managers to monitor. Nesting habitat protection, including the removal of mammalian invaders on the few islands where the largest colonies are established, is a conservation priority to maintain puffin populations.

The egg of the Tufted Puffin is off-white in background color, immaculate, and 2¾ x 1⅞ in (72 x 48 mm) in size. Nesting on steep slopes and rock faces, and on remote islands, allows these puffins, and their eggs, to foil the attempts of many of their mammalian predators.

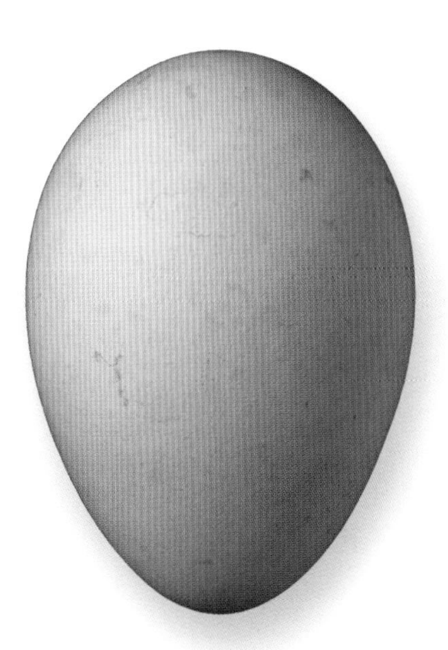

Actual size

LARGE NON-PASSERINE LAND BIRDS

This chapter starts with ancient flightless land birds (ostriches, emus, rheas, and kiwis; collectively called ratites) and their relatives, the flighted tinamous; included are also the birds of prey, the gamebirds, the parrots, and the hornbills. Ratites and tinamous are restricted largely to the southern hemisphere, but hawks and eagles, owls as well as pheasants, grouse and ptarmigan have adapted well to living on most continents throughout the seasons, including the cold northern hemisphere winters. The eggs of these birds vary in size and shape, as does the size of a clutch. The large flightless birds, tinamous, and gamebirds produce precocial young that are ready to leave the nest soon after hatching; most of these nest and feed on the ground. In contrast, the birds of prey, parrots, and hornbills often build nests high above the ground and produce altricial young that rely on the parents to secure food for them during development.

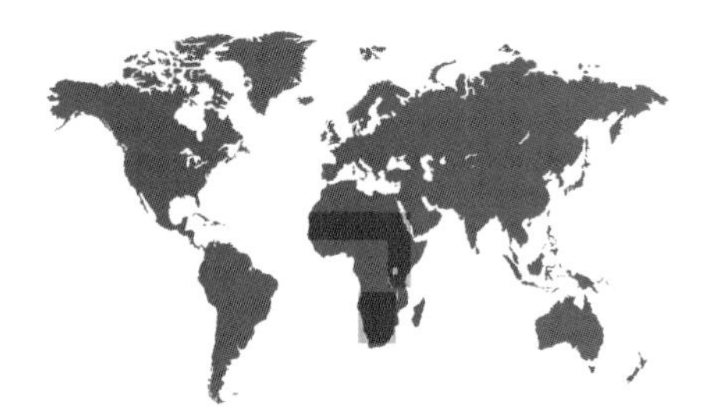

ORDER	Struthioniformes
FAMILY	Struthionidae
BREEDING RANGE	Sub-Saharan Africa, recently extinct from the Arabian peninsula
BREEDING HABITAT	Desert, short grassland, savanna
NEST TYPE AND PLACEMENT	Simple, shallow pit in the ground
CONSERVATION STATUS	Least concern
COLLECTION NO.	FMNH 2195

ADULT BIRD SIZE
6 ft 10 in–8 ft 10 in
(2.1–2.7 m)

INCUBATION
35–45 days

CLUTCH SIZE
10–12 eggs per female; 20 eggs on average per nest

STRUTHIO CAMELUS

OSTRICH

STRUTHIONIFORMES

Clutch

The Ostrich is well known for the distinction between the sexes both in terms of plumage and behavior: the males' black-and-white plumes once were highly sought after by people as fashion and clothing accessories. The male incubates the eggs at night, when its plumage patterns provide for strong camouflage. In turn, the female's brown-gray plumage allows her to be cryptic while sitting on the eggs during the day.

Nesting is a communal affair in ostriches; several females lay in the same nest. But when egg laying is complete, the dominant female culls some of the eggs by moving them out of the nest scrape, to reduce the clutch size to a manageable 20 or so eggs. These are incubated by her and a single male. The Ostrich is the largest of all living birds, and hence, when first fertilized, the egg is the largest single cell on earth.

The egg of the Ostrich is off-white to cream in background color and immaculate with a glossy sheen. It measures 7 x 5½ in (178 x 140 mm) in size. When the eggs are exposed to direct sun, their thick shell and pale color screen the embryo from harmful ultraviolet radiation.

Actual size

ORDER	Rheiformes
FAMILY	Rheidae
BREEDING RANGE	Subtropical and temperate eastern South America; a self-sustaining feral population is now established in Germany
BREEDING HABITAT	Grasslands and pastures
NEST TYPE AND PLACEMENT	Shallow ground scrape, with the ground cleared of sticks and branches around it
CONSERVATION STATUS	Near threatened
COLLECTION NO.	FMNH 21156

ADULT BIRD SIZE
35½–59 in (90–150 cm)

INCUBATION
29–43 days

CLUTCH SIZE
5–10 eggs per female, 26 eggs on average per nest

RHEA AMERICANA

GREATER RHEA

RHEIFORMES

The social structure of Greater Rhea communities follows a strong seasonal pattern; in the winter, flocks of adults of both sexes and varied ages are maintained. As the spring approaches, the males become increasingly aggressive and solitary, while the females break off into small groups.

Clutch

Greater Rhea males court and mate with several females in a flock, and often remain in the vicinity of their eventual nesting site. In contrast, after flocks of females mate with a male, they often move between the distant nesting sites of several different males.The eggs are also laid by many females in each nest, on different days: up to 80 eggs have been counted in a single nest. As they develop, the chicks call to others from inside the egg, and through this communication, and a flexible incubation period, the eggs end up hatching within two days of each other, even if they have been laid nearly two weeks apart.

The egg of the Greater Rhea is greenish to bluish yellow, and fades to dull cream in background color when exposed to light. It is immaculate, and measures 5⅛ x 3½ in (130 x 90 mm) in dimensions.

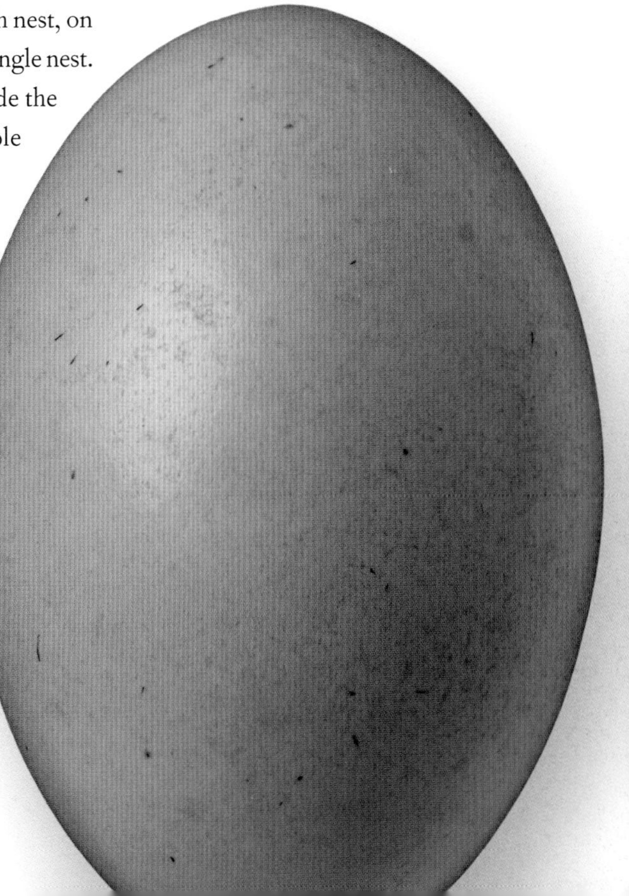

Actual size

ORDER	Casuariiformes
FAMILY	Casuariidae
BREEDING RANGE	Northeastern Australia, New Guinea, and nearby Indonesian islands
BREEDING HABITAT	Lowland rainforests
NEST TYPE AND PLACEMENT	Large ground nest built up from leaves, twigs, and other plant materials
CONSERVATION STATUS	Vulnerable
COLLECTION NO.	FMNH 2203

ADULT BIRD SIZE
4 ft 3 in–5 ft 7 in (1.3–1.7 m)

INCUBATION
50 days

CLUTCH SIZE
3–5 eggs

CASUARIUS CASUARIUS

SOUTHERN CASSOWARY

CASUARIIFORMES

Clutch

The Southern Cassowary is the largest of the three extant cassowary species, and, by weight, it is the second largest living bird species, second only to the Ostrich (see page 200). The female is larger and brighter than the male, and she is also more dominant in territorial disputes between neighbors. Yet it is the father who becomes a true threat to people during the breeding season; he alone incubates the eggs and attends the young. When threatened, he storms and uses his sharp inner toe-claws to kick up and slice through the enemy's skin and flesh.

The Southern Cassowary is conspicuously colored when viewed in a zoo or encountered at one of several camping grounds in tropical Queensland, Australia, where it has become tame. Yet, in the low light of its native dense rainforest, its black plumage, grayish casque on the top of the head, and blue-red wattle hanging below, appear more cryptic than contrasting.

The egg of the Southern Cassowary is pale green in background color, and measures 5½ x 3¾ in (139 x 95 mm) in size. The nest is a built-up mattress, which allows moisture to drain away and so keeps the eggs dry after the frequent heavy rainfalls.

Actual size

ORDER	Casuariiformes
FAMILY	Dromaiidae
BREEDING RANGE	Australia
BREEDING HABITAT	Forests, wooded savanna, and grassland; absent from rainforests and extremely arid lands
NEST TYPE AND PLACEMENT	Shallow depression in the ground, lined with leaves, twigs, bark, and grass
CONSERVATION STATUS	Least concern
COLLECTION NO.	FMNH 2955

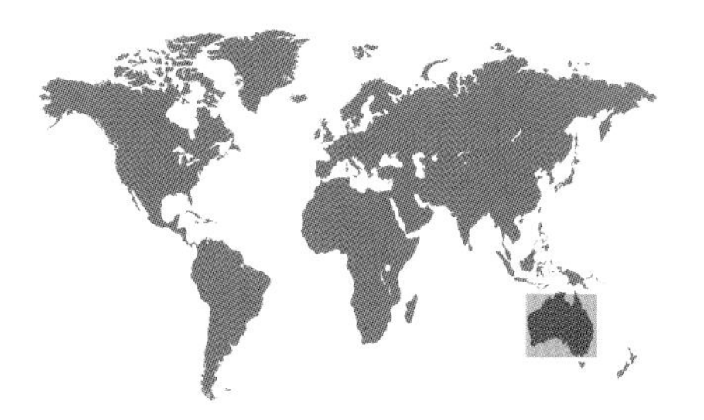

ADULT BIRD SIZE
4 ft 7 in–5 ft 3 in (1.4–1.6 m)

INCUBATION
56 days

CLUTCH SIZE
11–20 eggs

DROMAIUS NOVAEHOLLANDIAE

EMU

CASUARIIFORMES

Emus have well-defined parental roles. Females range widely while males undertake all the incubation. Remaining on the nest for almost the full 24 hours of each day, the male rises only to turn the eggs; eating and drinking nothing, he can lose up to 20 percent of his body weight before the eggs hatch. The male also tends the chicks on his own. Females often mate with more than one male; the eggs in a single nest may derive from multiple mothers and fathers.

Clutch

Although common throughout Australia, some Emu populations are at risk locally from hunting and nest predation. Several island populations have now become extinct, following the fate of other large flightless ratites on islands, among them the great Elephantbirds of Madagascar (see page 205) and the Moas of New Zealand.

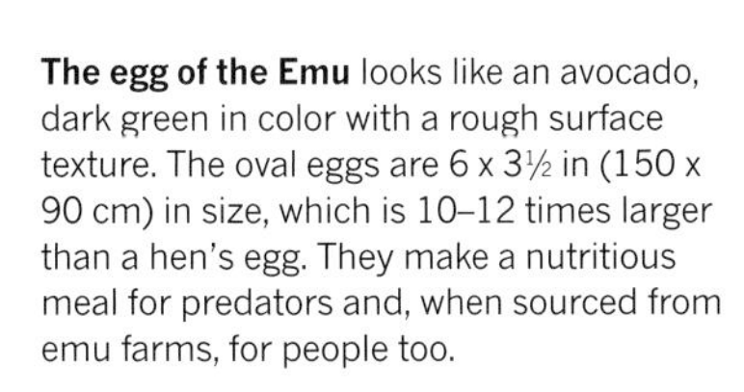

The egg of the Emu looks like an avocado, dark green in color with a rough surface texture. The oval eggs are 6 x 3½ in (150 x 90 cm) in size, which is 10–12 times larger than a hen's egg. They make a nutritious meal for predators and, when sourced from emu farms, for people too.

Actual size

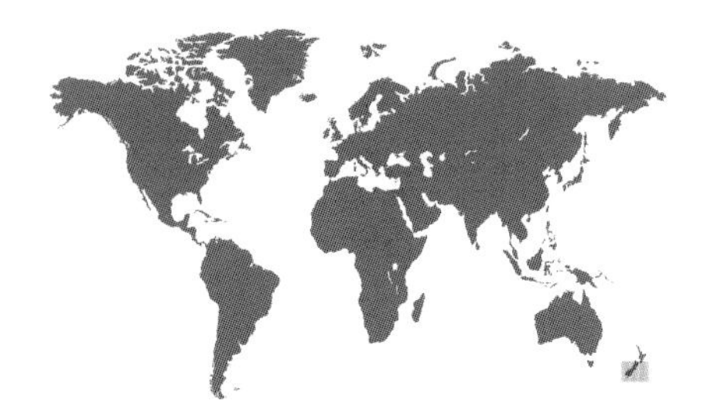

ORDER	Apterygiformes
FAMILY	Apterygidae
BREEDING RANGE	New Zealand
BREEDING HABITAT	Forest understory
NEST TYPE AND PLACEMENT	Underground burrows below vegetation or between rootlets
CONSERVATION STATUS	Vulnerable
COLLECTION NO.	FMNH 22383

ADULT BIRD SIZE
18–22 in (45–55 cm)

INCUBATION
74–84 days

CLUTCH SIZE
1–2 eggs

APTERYX AUSTRALIS

SOUTHERN BROWN KIWI

APTERYGIFORMES

Clutch

The female Southern Brown Kiwi invests nearly 40 percent of her body weight into laying the immense, single egg. It is no surprise that she takes several weeks if she lays a second egg. In turn, mostly the male is in charge of incubating the egg(s). Adult kiwis can successfully defend themselves and their eggs against stoats and rats, but the hatchlings are vulnerable until they reach about 26 oz (750 g) in weight. Therefore, when a kiwi's nesting burrow is located, conservation managers typically collect the egg, and send it to be incubated artificially at one of several captive hatching facilities in New Zealand.

On Stewart Island, where its population densities have remained high, these typically nocturnal birds frequently forage during daytime hours, too. Still, as a species, it suffers from the same fate and population trend as all other kiwis: its numbers are dwindling as introduced mammals devastate the native flora and fauna in New Zealand. Conservation biologists predict its swift demise unless rapid nationwide protective actions are taken.

The egg of the Southern Brown Kiwi is white to beige, with a bluish tint, and clear of spots. It measures 4¾ x 3⅛ in (120 x 80 mm) in size. The kiwi egg is nearly as large as a single rhea or six hen eggs, yet its shell is thin; not much thicker than the chicken egg.

Actual size

ORDER	Aepyornithiformes
FAMILY	Aepyornithidae
BREEDING RANGE	Madagascar
BREEDING HABITAT	Probably in wet forests, with fruit-bearing trees and bushes
NEST TYPE AND PLACEMENT	Unknown; eggshells are found buried in the sand on beaches, dunes, and river banks
CONSERVATION STATUS	Extinct
COLLECTION NO.	FMNH P14545

AEPYORNIS MAXIMUS

GREAT ELEPHANTBIRD

AEPYORNITHIFORMES

ADULT BIRD SIZE
10 ft (3 m) or greater in height

INCUBATION
Unknown

CLUTCH SIZE
Unknown

Elephantbirds belong to an ancient group of flightless birds, called ratites, which also includes ostriches, rheas, emus, and kiwis. Elephantbirds were the tallest and some of the heaviest birds known to walk the Earth: perhaps only one of the now extinct emu relatives in Australia weighed as much as the largest elephantbird.

These giants were once common and hunted by local people in Madagascar. Even early European visitors and colonists had the chance to sight them, observers including Marco Polo in the twelfth century, and the seventeenth-century French governor of the island nation. However, these observations recorded little with respect to the breeding, social, and foraging behaviors of elephantbirds. The last of the species died out soon afterward, around the end of the seventeenth century or early eighteenth century, probably due to a combination of hunting pressure and habitat changes brought about by humans.

The egg of the Great Elephantbird was off white to peach in background color, and measured 13½ x 9½ in (342 mm x 241 mm) in size; this equals 100–150 chicken eggs in volume. The specimen shown here is a rare whole sub-fossil egg from the Field Museum's paleontology collection.

Actual size

ORDER	Tinamiformes
FAMILY	Tinamidae
BREEDING RANGE	Disjunct populations along the Andes, South America
BREEDING HABITAT	Humid low- and mid-elevation forests
NEST TYPE AND PLACEMENT	Leafy depression at the base of a tree
CONSERVATION STATUS	Vulnerable
COLLECTION NO.	FMNH 2856

ADULT BIRD SIZE
15½–18 in (40–46 cm)

INCUBATION
Unknown

CLUTCH SIZE
2+ eggs

TINAMUS OSGOODI

BLACK TINAMOU

TINAMIFORMES

Clutch

The egg of the Black Tinamou is glossy blue in background color and free of speckling, and 2½ x 2⅓ in (64 x 59 mm) in size. The egg shown here was reconstructed from a broken egg sample housed in the Field Museum's collection, one of the few such samples known.

While all tinamous are secretive, when it comes to the Black Tinamou, scientists and conservation managers have discovered very little about this species. This may be due to its disjointed distributional range in the montane forests of the Andes. Few eggs and nests have ever been seen in nature, and none has been monitored, and so hardly anything is known about the breeding biology of this species.

It is suspected that the small and distant breeding populations in Colombia and in Peru represent a recent restriction in the overall distribution of the species, due to ongoing hunting and extensive deforestation. However, with advances in remote monitoring, using nighttime motion-activated cameras, recent discoveries have also placed this species in eastern Ecuador, implying a more continuous distribution prime for new discoveries and future studies.

Actual size

ORDER	Tinamiformes
FAMILY	Tinamidae
BREEDING RANGE	Central America and tropical South America
BREEDING HABITAT	Humid to dry forests, including rainforests, cloud forests, and dry seasonal woodlands
NEST TYPE AND PLACEMENT	Scraped ground nest between tree buttresses
CONSERVATION STATUS	Near threatened
COLLECTION NO.	FMNH 2474

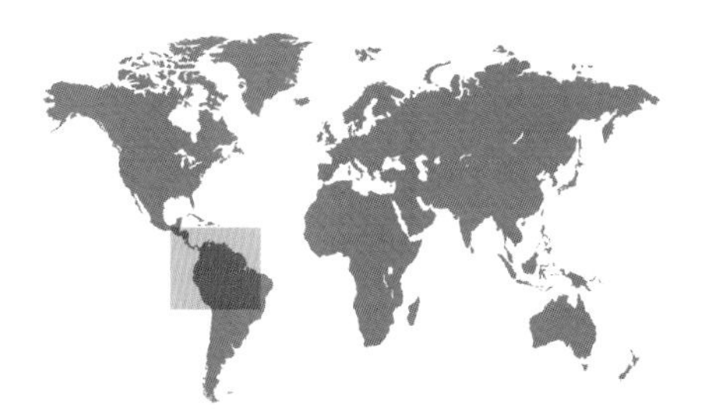

ADULT BIRD SIZE
15½–17½ in (40–45 cm)

INCUBATION
19–20 days

CLUTCH SIZE
3–5

TINAMUS MAJOR

GREAT TINAMOU

TINAMIFORMES

Tinamous lay the most brilliantly colored eggs of all birds and the Great Tinamou lays some of the brightest eggs among its relatives. Together with the ratites, tinamous belong to the ancient group of birds called paleognaths. Scientists used to think that the color of the ancestral bird egg might have been the pure white color of the calcite ($CaCO_3$) which makes up the avian eggshell matrix. However, the colorful eggs of the ratites and tinamous may contradict this theory.

How can the conspicuous blue egg of the Great Tinamou survive in the predator-rich tropical habitats? The answer may lie in the intensity of paternal care. Females move in groups and lay eggs into several nests, while the male dedicates all his time to incubating a set of eggs, hiding them underneath his own cryptic plumage. The male sits so tightly on the nest that a researcher can pluck one of his feathers for paternity tests before he flushes off the nest.

Clutch

The egg of the Great Tinamou is glossy turquoise in background color and clear of speckles. It measures 2¼ x 1⅞ in (58 x 48 mm) in size. The eggs, incubated atop the wet leaf litter on the forest floor, can become soiled prior to hatching, as seen in this museum specimen.

Actual size

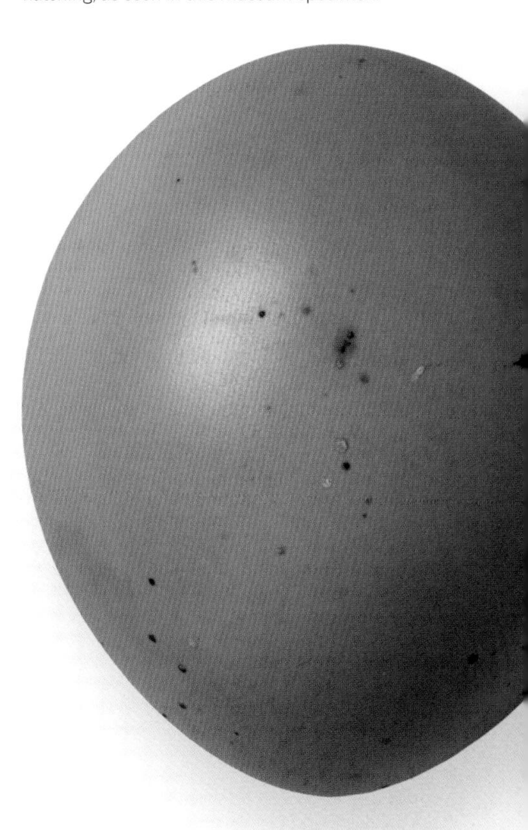

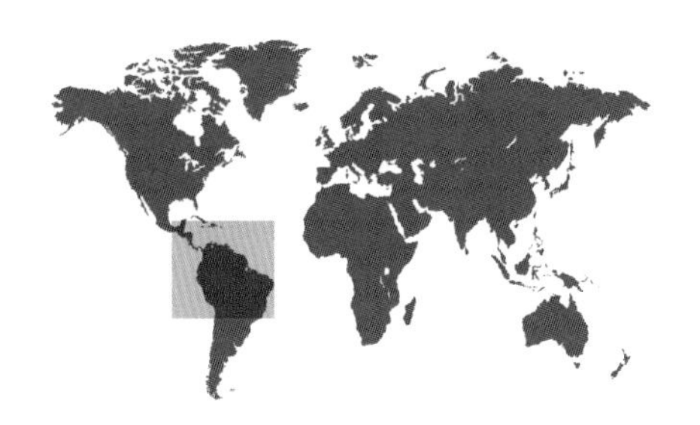

ORDER	Tinamiformes
FAMILY	Tinamidae
BREEDING RANGE	Central America and northern South America
BREEDING HABITAT	Lowland humid forests, riparian corridors, and shrubland
NEST TYPE AND PLACEMENT	Small depression in forest floor, often near the base of a tree
CONSERVATION STATUS	Least concern
COLLECTION NO.	FMNH 2475

ADULT BIRD SIZE
8½–9½ in (22–24 cm)

INCUBATION
16–19 days

CLUTCH SIZE
1–2 eggs

CRYPTURELLUS SOUI

LITTLE TINAMOU

TINAMIFORMES

Clutch

The egg of the Little Tinamou is reddish clay or purplish drab in background color, and immaculate. It measures 1½ x 1 in (37 x 26 mm) in size. The small clutch size suggests that only a single female lays in the nest.

The Little Tinamou is a secretive ground-dwelling bird that can be easily confused with a quail or partridge in the forest understory. It is rarely seen, even though it often occurs near human settlements and in modified habitats. When it is sighted, it is typically on its own, irrespective of the season. This implies a solitary lifestyle for both the female and the male, and a lack of the female alliances observed in other tinamou species, where multiple females seek out males to lay eggs communally in the same nest.

In Costa Rica and Trinidad the species breeds all year round, yet little is known about its social, nesting, and parental behaviors. Only a single bird attends the nest, but no one has confirmed to date that it is the male, as found in most other tinamou species. The incubating adult sits so tightly on the nest that it can be touched with a stick, and even marked with a dab of paint for subsequent identification of the same individual from a distance.

Actual size

ORDER	Tinamiformes
FAMILY	Tinamidae
BREEDING RANGE	Tropical and subtropical lowland South America
BREEDING HABITAT	Wooded habitats, from wet rainforest to seasonally dry cerrados
NEST TYPE AND PLACEMENT	Small depression in the ground
CONSERVATION STATUS	Least concern
COLLECTION NO.	FMNH 2200

ADULT BIRD SIZE
11–12 in (28–30 cm)

INCUBATION
17 days

CLUTCH SIZE
4–5 eggs

CRYPTURELLUS UNDULATUS

UNDULATED TINAMOU

TINAMIFORMES

The Undulated Tinamou lives in the dense understory of wet and dry forests, remaining hidden from the eyes of competitors and predators alike. It can occasionally be seen crossing clearings and roads that cut through these forests, but even then these birds quickly run away from approaching people and vehicles rather than taking flight.

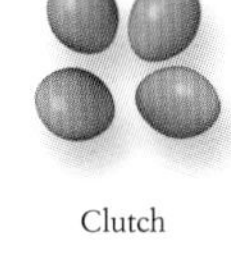

Clutch

The most conspicuous aspect of this species' behavior is its incessant calling: two whistles followed by a raspy note. It can be heard repeatedly throughout all daylight hours. When mimicked by a playback or a whistle, the male can be attracted to the source; this suggests that these calls are used as male-to-male communication displays. However, the exact functions of the call, as well as many aspects of the behavior of this species, remain to be discovered.

The egg of the Undulated Tinamou is glossy reddish pink or light gray in background color, and free of maculation. It measures 2 x 1½ in (50 x 40 mm) in size. The male alone looks after the eggs and the chicks, relying on his rippled ("undulated") plumage to remain cryptic and stay hidden from danger.

Actual size

ORDER	Tinamiformes
FAMILY	Tinamidae
BREEDING RANGE	Subtropical and tropical Mexico, along both coastlines
BREEDING HABITAT	Moist lowland forests and secondary growths
NEST TYPE AND PLACEMENT	Shallow ground scrape
CONSERVATION STATUS	Least concern
COLLECTION NO.	FMNH 2153

ADULT BIRD SIZE
10–12 in (25–30 cm)

INCUBATION
16 days

CLUTCH SIZE
2–3 eggs

CRYPTURELLUS CINNAMOMEUS

THICKET TINAMOU

TINAMIFORMES

Clutch

The egg of the Thicket Tinamou is bronze purple to porcelain pink in background color, clear of spots, and 1½ x 1¼ in (40 x 33 mm) in dimensions.

The Thicket Tinamou is a relatively large, ground-dwelling bird. Its earth-toned plumage, and a habit of moving through the dense undergrowth of terrestrial bromeliads and sapling trees in the wet coastal forests, make it hard to spot. When approached by a potential threat, it prefers to run or freeze until it is safe to resume walking and feeding again.

This species' presence becomes apparent during the breeding season, when the male needs to attract females for mating. He produces a hollow, two- or three-syllable-long whistle that concludes in an upward slur at the end of the calling bout. The simple, tonal quality of the call travels well in the dense forest, and it can be heard for long distances. This allows the female both to locate the source and assess the male's potential as a mate. The nest may contain up to seven eggs, laid by several different females but always incubated by just one male.

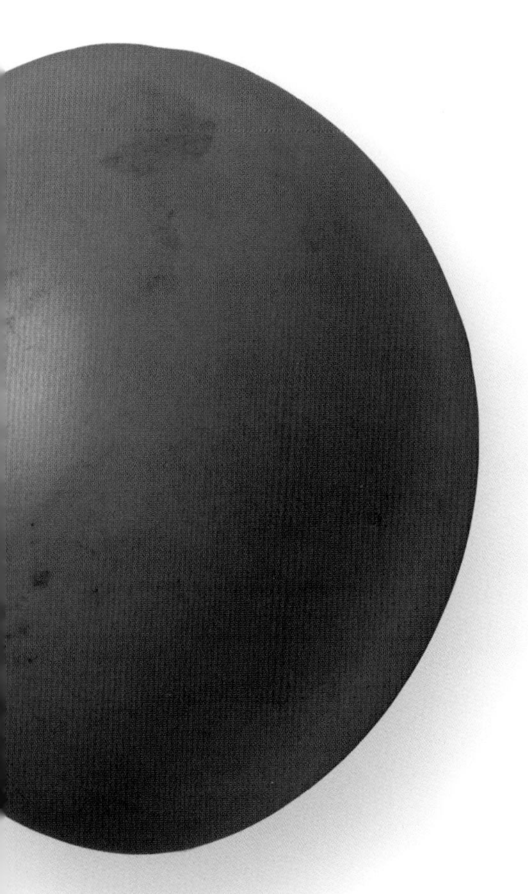

Actual size

ORDER	Tinamiformes
FAMILY	Tinamidae
BREEDING RANGE	Central and eastern South America
BREEDING HABITAT	Tropical, subtropical, and high-elevation grassland
NEST TYPE AND PLACEMENT	Simple scrape on ground
CONSERVATION STATUS	Least concern
COLLECTION NO.	FMNH 2466

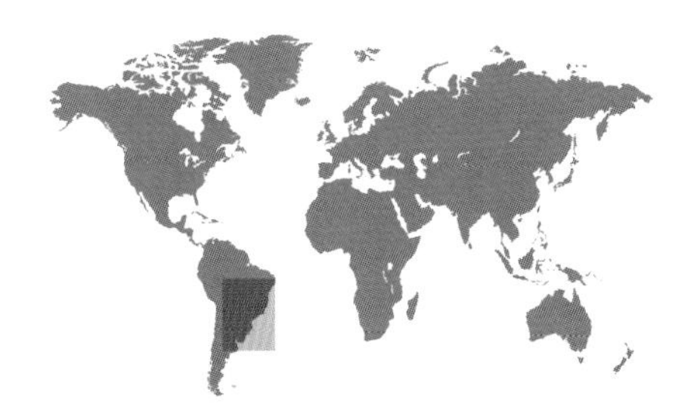

ADULT BIRD SIZE
15½–16 in (40–41 cm)

INCUBATION
19–21 days

CLUTCH SIZE
4–8 eggs

RHYNCHOTUS RUFESCENS

RED-WINGED TINAMOU

TINAMIFORMES

The Red-winged Tinamou is a secretive bird, hiding and moving in dense grass fields. It only becomes conspicuous when it takes flight to escape predators. Unlike many other tinamous that are usually active at low light, this species is mostly on the move during the hottest hours in the middle of the day.

Clutch

The female does not vocalize, but selects a male based on both his calls and his behaviors. Instead of drawing the female directly to the nest site, the male engages in a ritual called "forage-following," which means that he trails behind the female as she seeks out seeds, insects, and fruits to feed on. When she shows interest in him, he leads her to the nest scrape, they mate, and she deposits an egg. She then moves on for good and he turns his attention to attract other females. The result is a clutch containing eggs laid by several different females. Only the male incubates the eggs and chaperones the chicks after they hatch. It is unclear, however, if the male actually fathers all the eggs that are laid in his nest.

The egg of the Red-winged Tinamou is red to purple in color, has no speckles, and is 2⅛ x 1⅝ in (55 x 41 mm) in size. It is covered by a bright glaze, almost resembling the texture and color of a painted or ceramic Easter egg.

Actual size

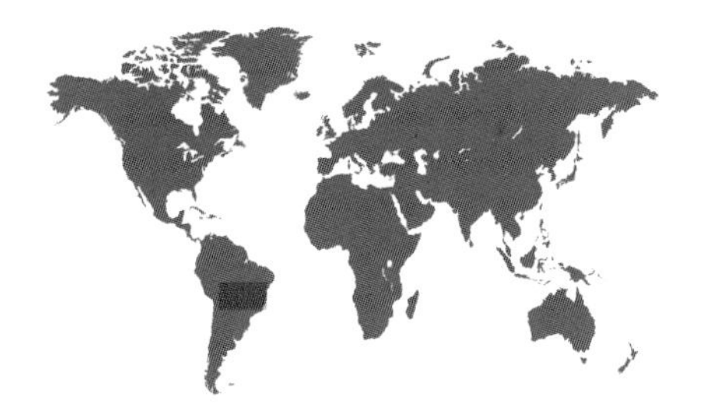

ORDER	Tinamiformes
FAMILY	Tinamidae
BREEDING RANGE	Inland South America
BREEDING HABITAT	Grassy shrubland, pasture, and savanna
NEST TYPE AND PLACEMENT	On ground, in dense vegetation, or between tree roots and buttresses
CONSERVATION STATUS	Least concern
COLLECTION NO.	FMNH 2471

ADULT BIRD SIZE
10–10½ in (25–27 cm)

INCUBATION
17–34 days for tinamous in general, unknown for this species

CLUTCH SIZE
3–5 eggs

NOTHURA BORAQUIRA

WHITE-BELLIED NOTHURA

TINAMIFORMES

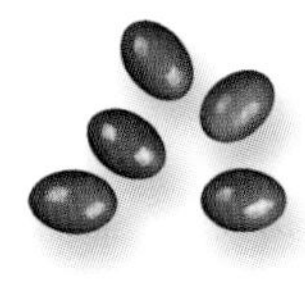

Clutch

The egg of the White-bellied Nothura is rich chocolate brown in background color and immaculate, and it measures 1¾ x 1¼ in (43 x 31 mm) in size.

Actual size

A white-washed underbelly and bright yellow legs are characteristic of both sexes of the White-bellied Nothura but because of its generally cryptic upper plumage and ground-dwelling lifestyle, it is often called a 'quail' in local languages. This tinamou species becomes distinctive during the mating season when the male calls to attract the females, or when it is foraging for seeds and fruits, frequently crossing dirt roads cutting across pastures and grassy fields. Up to four different females may lay in the same nest, but only the male takes care of the eggs and the young.

The White-bellied Nothura has a disjunct distribution, with separate populations breeding in eastern Brazil and also in Bolivia and Paraguay. Where it occurs, it is known to be an elusive and secretive bird; when approached by predators or people, it often crouches rather than flying off. When immediately threatened, it may hide in burrows made by other animals, including armadillos.

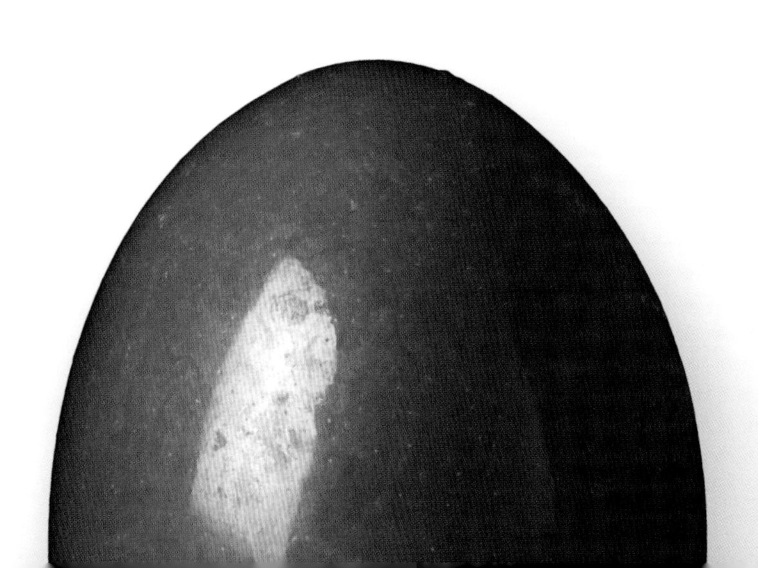

ORDER	Tinamiformes
FAMILY	Tinamidae
BREEDING RANGE	Southeastern South America
BREEDING HABITAT	Grassland, farmland, dry brushland
NEST TYPE AND PLACEMENT	Ground scrape in dense bush
CONSERVATION STATUS	Least concern
COLLECTION NO.	FMNH 2201

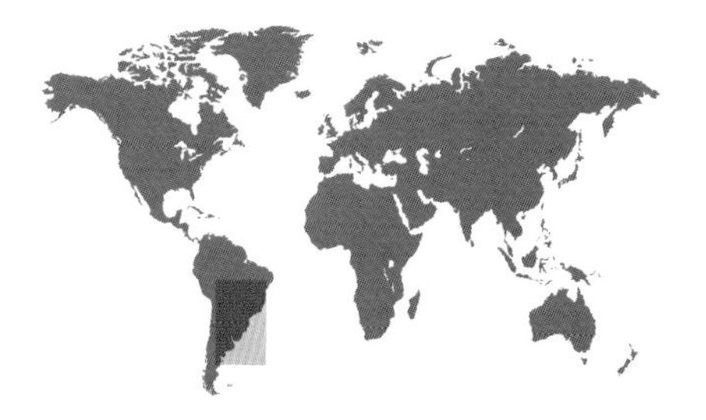

ADULT BIRD SIZE
9–10 in (23–26 cm)

INCUBATION
34 days

CLUTCH SIZE
4–6 eggs

NOTHURA MACULOSA

SPOTTED NOTHURA

TINAMIFORMES

In South America, tinamous are a popular gamebird, hunted for their meat throughout the continent. The Spotted Nothura is no exception, but unlike some of the other tinamou species, in most of its range strict seasonal bans and other hunting regulations assure that its populations are not exploited and remain stable. This is because the natural history of the species is such that just a few months of a hunting ban are sufficient for reproduction to begin rapidly and populations to grow in size.

Females of this species grow extremely fast and become sexually mature at as early as two to three months of age, after which they are able to lay four to six clutches of eggs per year. By contrast, the male takes longer to mature, and also needs more time than a female to become ready to breed after each attempt.

Clutch

Actual size

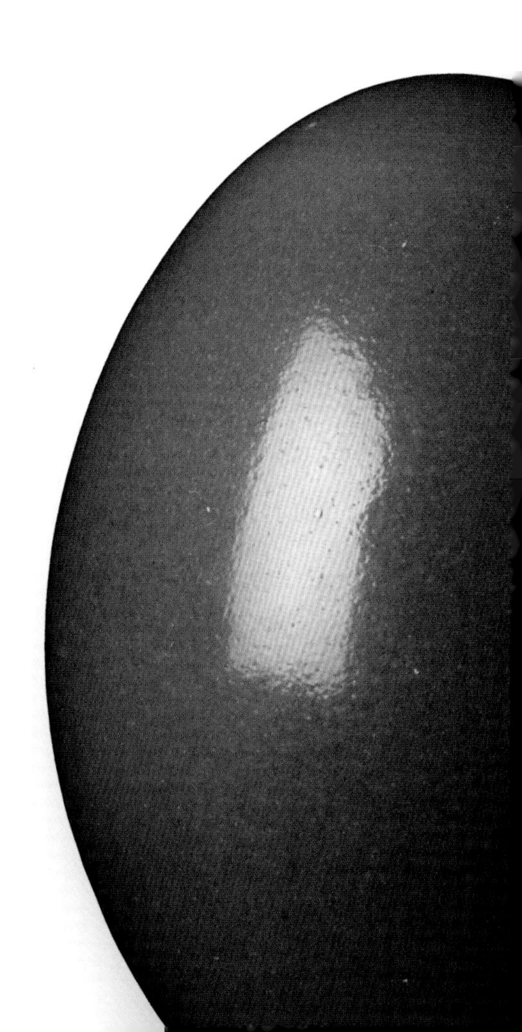

The egg of the Spotted Nothura is rich maroon to chocolate brown in background color, free of speckles with a gloss like porcelain. It measures 1⅝ x 1⅛ in (40 x 29 mm) in size. The male alone incubates the eggs and attends the highly mobile, precocial chicks.

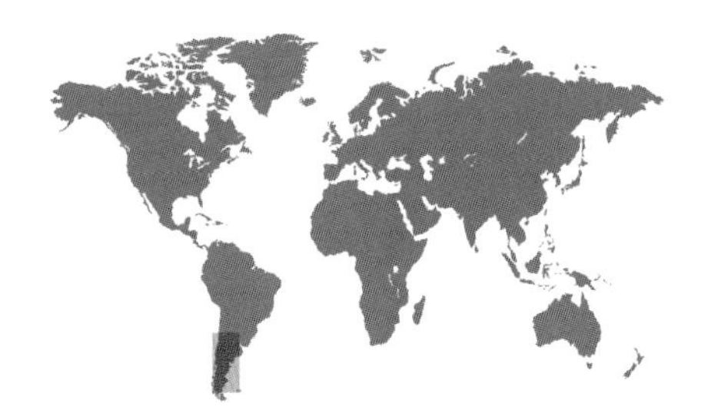

ORDER	Tinamiformes
FAMILY	Tinamidae
BREEDING RANGE	Southern South America
BREEDING HABITAT	Dry shrubland and farmland
NEST TYPE AND PLACEMENT	Scrape in the ground, lined with some vegetation, near low bushes
CONSERVATION STATUS	Least concern
COLLECTION NO.	FMNH 2151

ADULT BIRD SIZE
15–16 in (38–41 cm)

INCUBATION
20–21 days

CLUTCH SIZE
5–6 eggs

EUDROMIA ELEGANS

ELEGANT CRESTED-TINAMOU

TINAMIFORMES

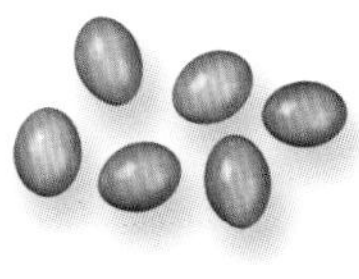

Clutch

The Elegant Crested-Tinamou is a predominantly terrestrial bird, preferring to crouch tightly to the ground when threatened, with its neck and crest erected, rather than taking flight. After the female has laid the eggs, she quickly abandons them, leaving all parental duties to the male, while she seeks another male's nest to lay in. The bright eggs are well hidden under the cryptic plumage of the male, and the nest is located in dense vegetation. However, if a nest is lost due to predation, a female will return to the previous male and produce a replacement clutch to avoid a total reproductive failure for that breeding season.

This is a highly social species, and roosts in flocks numbering 50 to 100 individuals in the non-breeding season. During the day, however, the birds break into smaller groups to forage. Breeders become territorial, with the males using a loud, sad, whistlelike voice to demarcate their territorial boundaries.

Actual size

The egg of the Elegant Crested-Tinamou is glossy emerald green in background color, and clear of maculation, and it measures 2 x 1½ in (53 x 39 mm) in size. During decades of storage in museum collections, even if in a dry and dark place, egg colors fade and lose their original tint, as seen in this faintly lime-colored egg.

ORDER	Galliformes
FAMILY	Megapodiidae
BREEDING RANGE	Sulawesi, Indonesia
BREEDING HABITAT	Sandy areas near lakes, rivers, and beaches
NEST TYPE AND PLACEMENT	Deep burrow in sand, warmed by solar or geothermal heat
CONSERVATION STATUS	Endangered
COLLECTION NO.	FMNH 2954

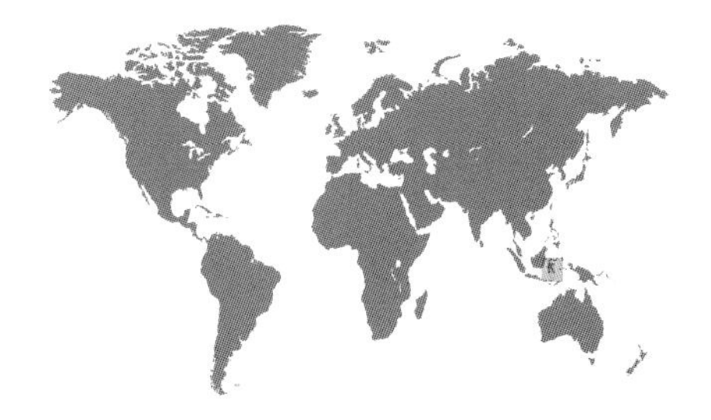

ADULT BIRD SIZE
21½–23in (55–60 cm)

INCUBATION
60 days

CLUTCH SIZE
8–12 eggs

MACROCEPHALON MALEO

MALEO

GALLIFORMES

Members of the family Megapodiidae carefully select their egg-laying site, often defended as a valuable resource by a male, but the females provide no maternal care for their eggs or young. Female Maleos, for example, migrate long distances, mostly on foot, from their uphill rainforest habitats to coastal communal breeding sites. Once there, they do not build nests. Instead females lay each egg at the end of a long tunnel dug into soft sand. There, over the course of two to three months, the eggs remain while they are warmed by the sun, or by geothermal heat generated by nearby volcanic activity. Upon hatching, the young Maleo is capable not only of running but also of hiding, flying, and fending for itself.

The eggs, vulnerable on their own, benefit from the presence of many other eggs laid close to one another to avoid predation by birds, snakes, and pigs. However, as suitable habitat is destroyed, the efficacy of such "safety in numbers" is greatly reduced.

Clutch

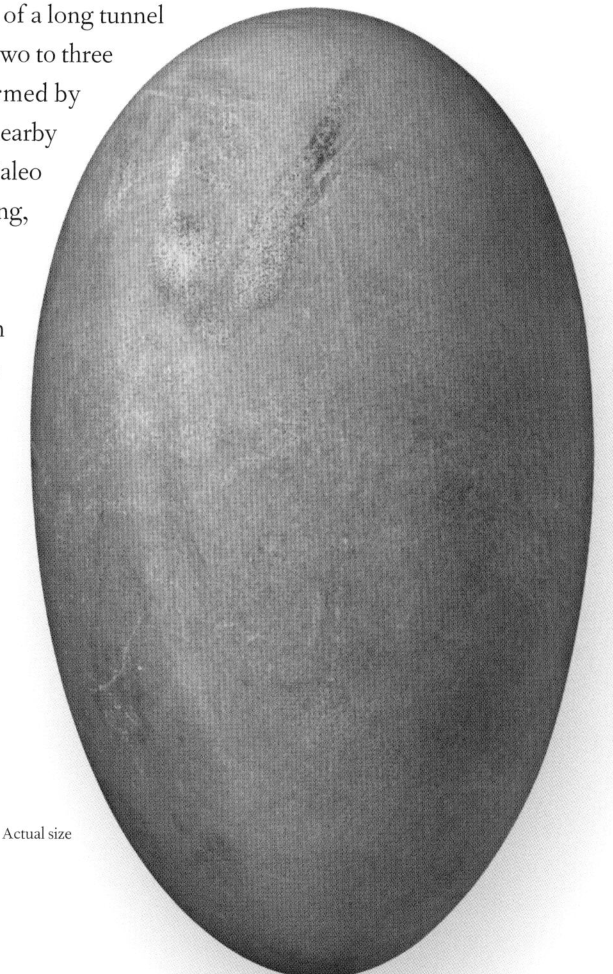

The egg of the Maleo is clear and white, with a pinkish tint, and it measures 5 x 3⅛ in (130 x 79 mm) in size. It is about five times the size of a chicken egg, and is consumed locally as a delicacy.

Actual size

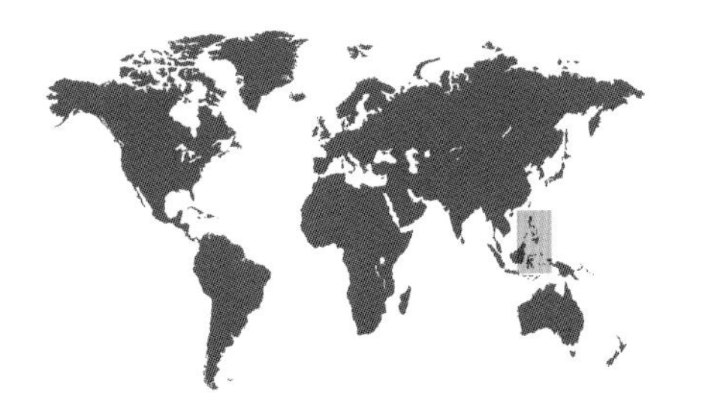

ORDER	Galliformes
FAMILY	Megapodiidae
BREEDING RANGE	Indonesia, Malaysia, and the Philippines
BREEDING HABITAT	Tropical lowland and montane forests
NEST TYPE AND PLACEMENT	Burrow among decaying roots of dead trees
CONSERVATION STATUS	Least concern
COLLECTION NO.	FMNH 2911

ADULT BIRD SIZE
14–15 in (36–38 cm)

INCUBATION
70 days

CLUTCH SIZE
10 eggs

MEGAPODIUS CUMINGII

TABON SCRUBFOWL

GALLIFORMES

Clutch

The Tabon or Philippine Scrubfowl, like all other megapodes, has reverted to a reptilian breeding strategy: it relies on environmental heat sources to provide warmth to induce and maintain embryonic growth and development. Eggs are laid by several females into heaps of moist sand mixed with wet leaf litter, where the eggs lie untouched. Eventually, the chick digs itself out of the heap of rotting vegetation that provides all the heat required to incubate the egg successfully.

The lack of parental brooding is paralleled by a prolonged incubation period for the Tabon Scrubfowl; this is due to the low intensity of externally generated heat sources compared to the direct transfer of parental body warmth to the egg. To withstand its long incubation period, the megapode embryo is well nourished by large and fatty yolk reserves. This makes megapode eggs highly vulnerable to ground-dwelling predators, for whom the large, yolk-rich eggs are easy prey.

The egg of the Tabon Scrubfowl is creamy to pale pink in background color, spotless, elongated in shape, and 3½ x 2 in (90 x 53 mm) in size. Unlike those of all other birds, megapode eggs are not rotated on a daily basis, and yet the embryo matures and hatches successfully.

Actual size

ORDER	Galliformes
FAMILY	Cracidae
BREEDING RANGE	Mexico, south Texas, Central America
BREEDING HABITAT	Thickets and dense, semi-arid scrubland
NEST TYPE AND PLACEMENT	Shallow saucer of twigs and plant fibers, lined with moss and leaves, on tree branches
CONSERVATION STATUS	Least concern
COLLECTION NO.	FMNH 6020

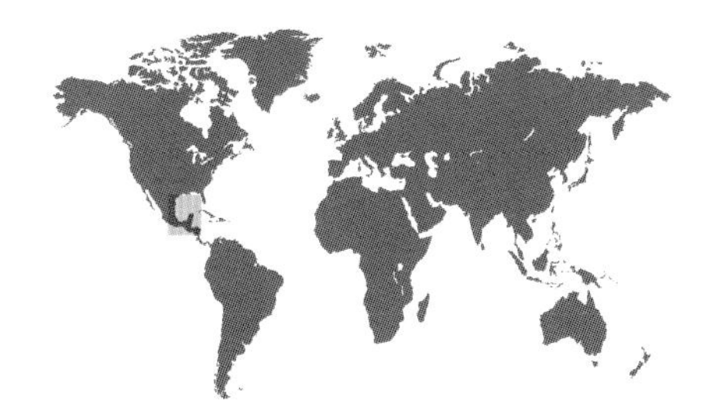

ADULT BIRD SIZE
21½–22 in (55–56 cm)

INCUBATION
21–28 days

CLUTCH SIZE
2–4 eggs

ORTALIS VETULA

PLAIN CHACHALACA

GALLIFORMES

While feeding, these birds move around in loose, often noisy flocks of four to six individuals. In contrast, nesting and parenting are strictly a female-only affair; males are promiscuous and do not assist the females. The female alone builds the flimsy nest, incubates the eggs, and leads the chicks away from the nest after hatching, toward safety and feeding sites. The nest is arboreal, and the chicks follow the female on their own feet by climbing and clinging to tree branches.

The Plain Chachalaca is the only member of the family Cracidae that reaches the United States in its distribution. Its Texan nickname, the tree pheasant, aptly describes the fowl-like aspects of both its appearance and its habit of roosting in trees at night, but also its non-chickenlike habit of spending most of the daytime in the tree canopy, walking, gliding, and feeding among thick and thin branches alike.

Clutch

The egg of the Plain Chachalaca is cream white in background color, clear of spots, and measures 2⅓ x 1⅔ in (59 x 42 mm) in size. The texture of the eggshell is unusually rough, compared to most other, smoother fowl eggs.

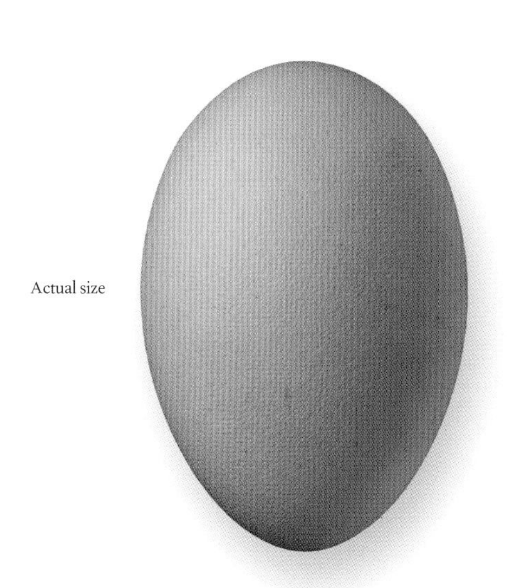

Actual size

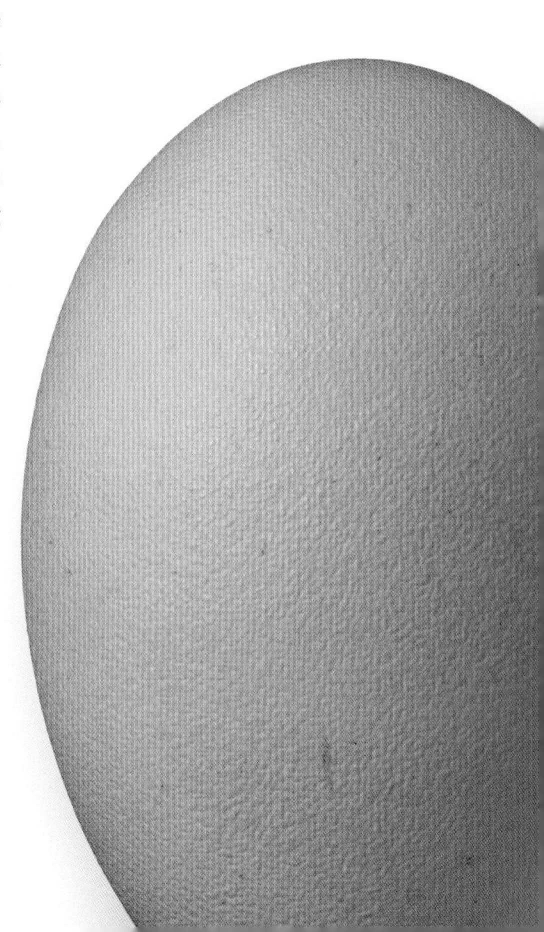

ORDER	Galliformes
FAMILY	Cracidae
BREEDING RANGE	Northern South America
BREEDING HABITAT	Humid forests, thickets along rivers
NEST TYPE AND PLACEMENT	Stick nest in trees, lined with leaves and bark
CONSERVATION STATUS	Least concern
COLLECTION NO.	FMNH 2390

ADULT BIRD SIZE
33½–37½ in (85–95 cm)

INCUBATION
30–32 days

CLUTCH SIZE
1–3 eggs

CRAX ALECTOR

BLACK CURASSOW

GALLIFORMES

Clutch

This species spends about equal time among tree branches and on the ground, in thick riparian scrub, foraging for seeds, fruits, and invertebrates. The Black Curassow is the only species in its genus, and the female and the male display the same uniformly black plumage coloration, making members of the monogamous pair bond difficult to tell apart. Similarity in plumage in this species is also complemented by shared parental duties when it comes to leading the young to food and protecting them from predators. Only the female incubates the eggs, but both parents provision the chicks, both before and after they leave the nest within days of hatching.

Curassows, as well as guans and chachalacas, are popular gamebirds, and are frequently hunted by native peoples for sustenance, especially in the Amazon. Ongoing hunting pressure on these large birds has led to increasing concern about future population trends for the species.

The egg of the Black Curassow is buff to tan in background color, with dark brown maculation, and measures 3 x 2¼ in (78 x 57 mm) in size. Female curassows are wary and will stop laying a second egg in the nest if the first egg goes missing.

Actual size

ORDER	Galliformes
FAMILY	Numididae
BREEDING RANGE	Sub-Saharan Africa, introduced to the West Indies, Brazil, Australia, and France
BREEDING HABITAT	Low shrubs and trees, open savanna, and farmland
NEST TYPE AND PLACEMENT	Scraped bare nest on ground
CONSERVATION STATUS	Least concern
COLLECTION NO.	FMNH 14770

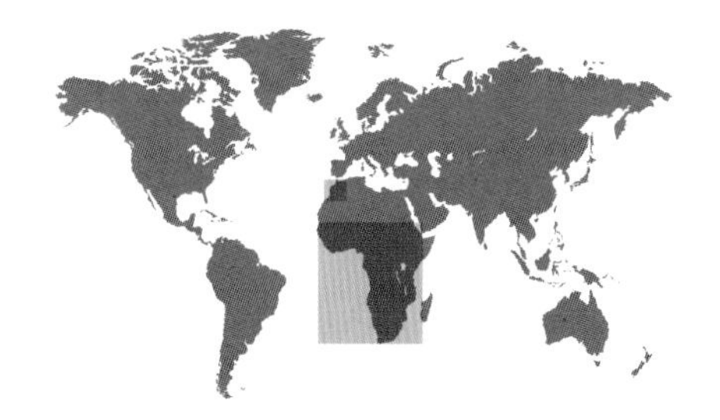

ADULT BIRD SIZE
21–23 in (53–58 cm)

INCUBATION
26–28 days

CLUTCH SIZE
8–15 eggs

NUMIDA MELEAGRIS

HELMETED GUINEAFOWL

GALLIFORMES

During the breeding season, these otherwise flocking fowl become aggressive, with the males frequently engaging in physical fights with each other that lead to bloody cuts and scars. Only the females scrape their nest and incubate the eggs. Laying one egg per day they begin incubation only once all the eggs are laid in the well-hidden nest. Once the young hatch, the females often return to their flock, bringing along the precocial and mobile young with them.

An ancestor of domesticated guineafowl, this species is both colorful and sociable in its native and introduced ranges. In recent years, Helmeted Guineafowl have also expanded their habitat into suburban areas, switching their nightly roosts from tree branches to rooftops and high fences, successfully fighting off cats, and avoiding direct confrontation with dogs.

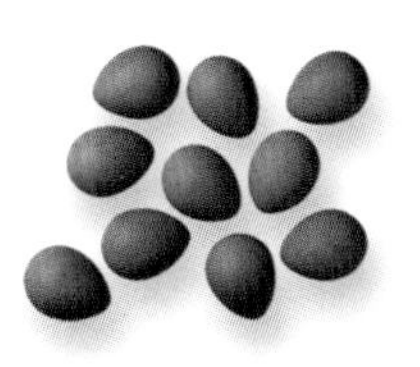

Clutch

The egg of the Helmeted Guineafowl is buff in background color with light, small, dark brown maculation. It measures 2 x 1½ in (50 x 37 mm) in dimensions. The eggshell is so thick relative to its size that the hatchlings—called keets—end up breaking the shell into many small fragments just to work their way out of it.

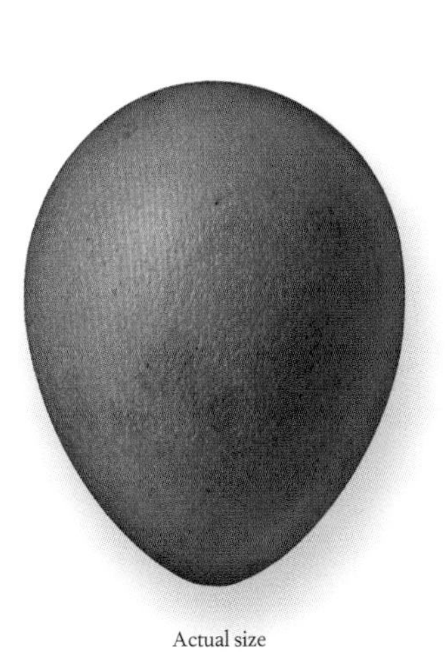

Actual size

ORDER	Galliformes
FAMILY	Odontophoridae
BREEDING RANGE	Western North America
BREEDING HABITAT	Dry shrubland and chaparral, up to high elevations
NEST TYPE AND PLACEMENT	Simple scrape on ground
CONSERVATION STATUS	Least concern
COLLECTION NO.	FMNH 15930

ADULT BIRD SIZE
10–12 in (26–31 cm)

INCUBATION
21–25 days

CLUTCH SIZE
9–10 eggs

OREORTYX PICTUS

MOUNTAIN QUAIL

GALLIFORMES

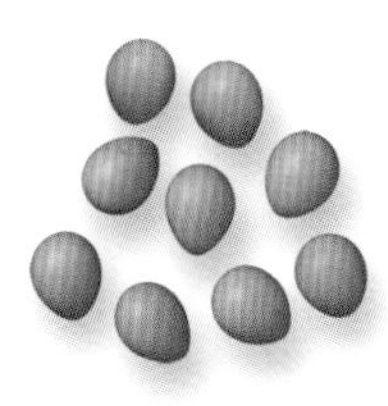

Clutch

Mountain Quail are sociable birds, and can be found in flocks, called coveys. During the breeding season, however, pairs break off from the covey and display to each other by tossing twigs and leaves, as if gathering nesting materials, even though the nest is typically a shallow scrape in the understory with little or no lining. Once the pair bond is established, the female and the male closely cooperate in preparing to raise their young, with both sexes involved in choosing the nest site, incubating the eggs, and protecting the chicks.

Such biparental care is made possible by the development of a brood patch, a bare area of skin on the belly of both the female and the male, so they can take turns incubating the eggs. Later, when the eggs hatch, the female may leave the male to look after the first brood of young, while she lays a second clutch that she alone takes care of.

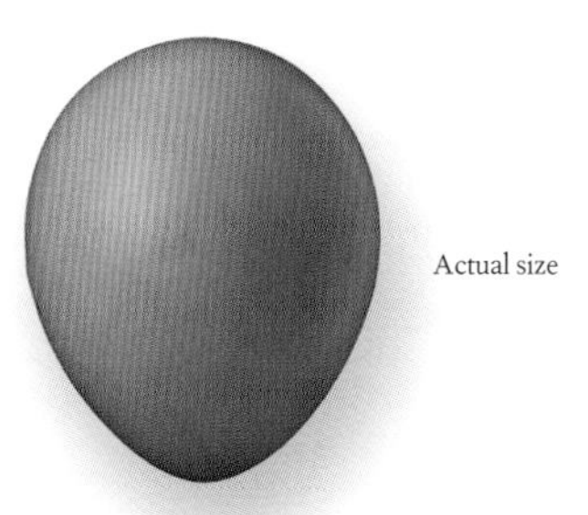

Actual size

The egg of the Mountain Quail is white or pale pink in background color and immaculate, and it measures 1⅓ x 1 in (34 x 25 mm) in size. The eggs are hidden from predators in a nest concealed in vegetation, at the base of a tree or shrub stem.

ORDER	Galliformes
FAMILY	Odontophoridae
BREEDING RANGE	Western North America, introduced to South America and several Pacific Islands
BREEDING HABITAT	Coastal sagebrush, chaparral, and desert shrub
NEST TYPE AND PLACEMENT	Shallow depression lined with stems and grasses
CONSERVATION STATUS	Least concern
COLLECTION NO.	FMNH 9070

ADULT BIRD SIZE
9½–10½ in (24–27 cm)

INCUBATION
22–23 days

CLUTCH SIZE
12–16 eggs

CALLIPEPLA CALIFORNICA

CALIFORNIA QUAIL

GALLIFORMES

Raising young is a group affair for these quail, with broods of several different pairs amalgamated into one large group, which is looked after by all the parents. Group parenting is beneficial, because more adults means more vigilance against predators; also, more chicks means there is less chance of any one chick being picked off by a predator. However, in flocks where both California and Gambel's Quail parents attend the young, the chicks may accidentally imprint on the wrong species, leading to mating between the two species.

Clutch

The California Quail is a well-known and distinctive bird, familiar to people living in coastal California cities and suburbs, and popularized in many Hollywood movies and Disney cartoons. Outside the mating season, the species occurs in large flocks, called quail coveys.

Actual size

The egg of the California Quail is white to cream in background color, has variable brown maculation, and measures 1¼ x ½ in (32 x 25 mm). Occasionally, females will lay their eggs into existing nests, typically of other quail, but some have been found even in the nest of a Wild Turkey (see page 247).

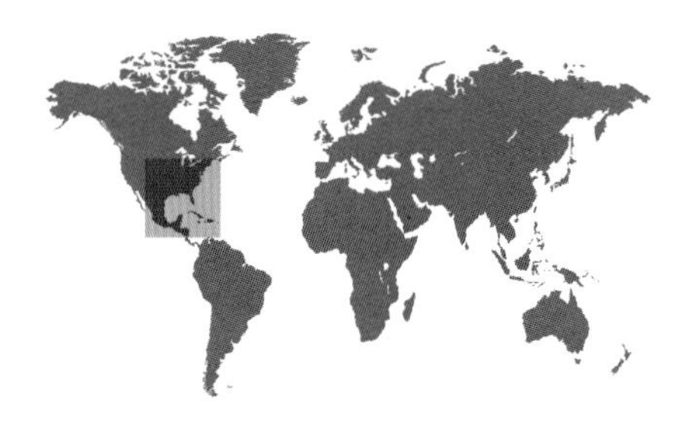

ORDER	Galliformes
FAMILY	Odontophoridae
BREEDING RANGE	Eastern and central North America, Mexico, the Caribbean
BREEDING HABITAT	Grasslands
NEST TYPE AND PLACEMENT	Simple scrape in the ground
CONSERVATION STATUS	Least concern, but some subspecies are endangered
COLLECTION NO.	FMNH 9050

ADULT BIRD SIZE
9½–11 in (24–28 cm)

INCUBATION
23–24 days

CLUTCH SIZE
12–16 eggs

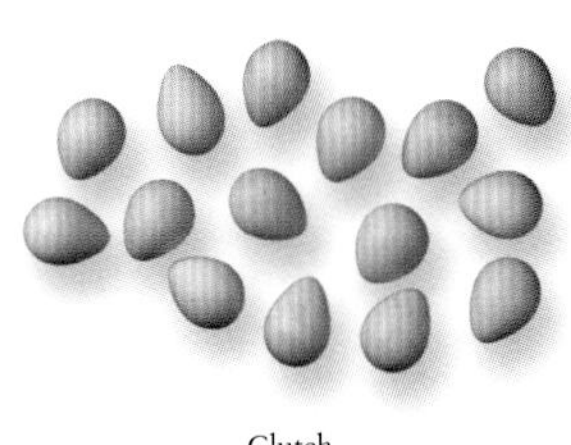

COLINUS VIRGINIANUS

NORTHERN BOBWHITE

GALLIFORMES

Clutch

The Northern Bobwhite is polyandrous, in that up to half of the females in a population mate with and lay eggs for two or more males. Both females and males incubate the nest, but a female will often abandon her first mate in search of a second, and a third, with whom she begins new clutches. This flexible and fast-paced clutch-initiation strategy may allow the species to recover after years of heavy predation, low food availability, cold weather, and intense hunting pressure, all of which are taking their toll in many regions.

The Northern Bobwhite has over 20 different subspecies. They are connected by the relative similarity of the females' plumage coloration across distant geographic areas, and separated by the different facial patterns of males. Despite their terrestrial and secretive habits, the distinctive call of the male ("bob white!") advertizes the presence of this species throughout its range.

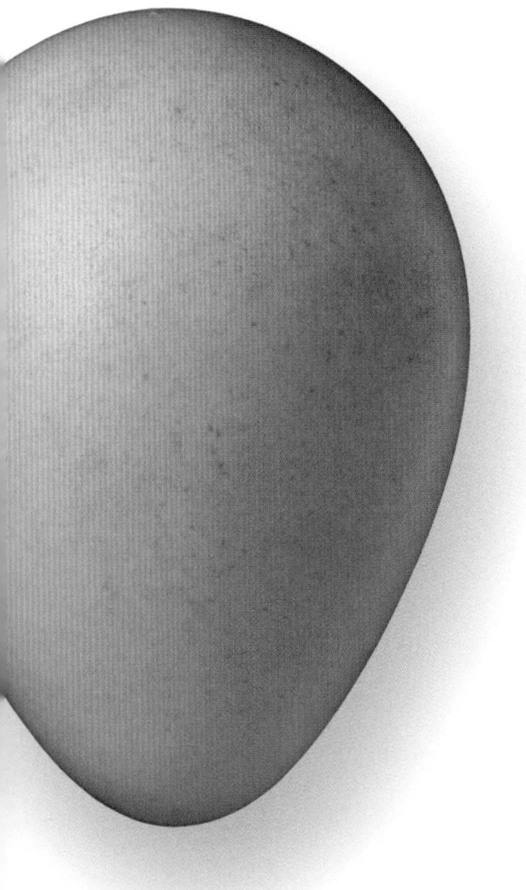

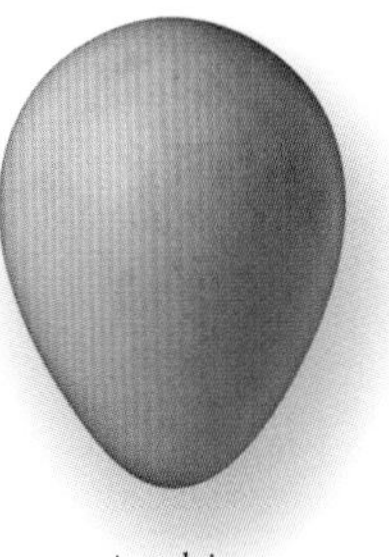

Actual size

The egg of the Northern Bobwhite is white with some gloss, immaculate, and 1³⁄₁₆ x 1 in (30 x 25 mm) in dimensions. If the eggs are taken by predators, the female can readily lay a second clutch.

ORDER	Galliformes
FAMILY	Odontophoridae
BREEDING RANGE	Mexico and southwestern USA
BREEDING HABITAT	Open woodlands of oaks and pines
NEST TYPE AND PLACEMENT	Built-up ground nest, made of grasses and other dry vegetation under dense shrub canopy
CONSERVATION STATUS	Least concern
COLLECTION NO.	FMNH 15931

CYRTONYX MONTEZUMAE

MONTEZUMA QUAIL

GALLIFORMES

ADULT BIRD SIZE
8–9 in (20–23 cm)

INCUBATION
23–27 days

CLUTCH SIZE
6–16 eggs

This small and distinctively colored New World quail maintains year-round territories. It occurs in small coveys of four to eight individuals in the non-breeding season, yet pairs become highly territorial and quite distantly spaced for such a small species in the breeding season. Before nesting commences the male calls loudly to his mate for several months. Nesting coincides with the onset of the unpredictable summer monsoon rains where the species resides.

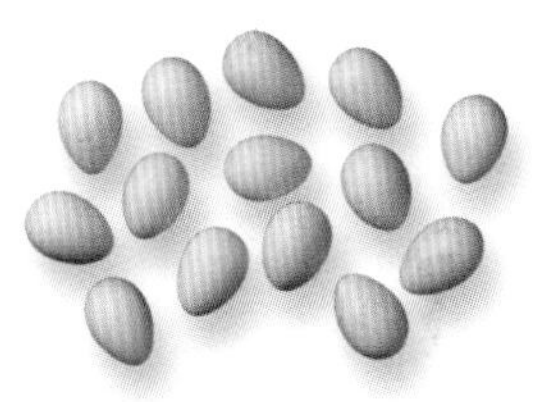

Clutch

As a ground-dwelling species, these quail rely on camouflage and immobility to remain undetected by predators. When approached too closely, they jump high and then take flight in an explosive escape response. During incubation, the male is in charge of standing guard to protect the nest while the female sits on the nest; the male only rarely incubates the eggs himself.

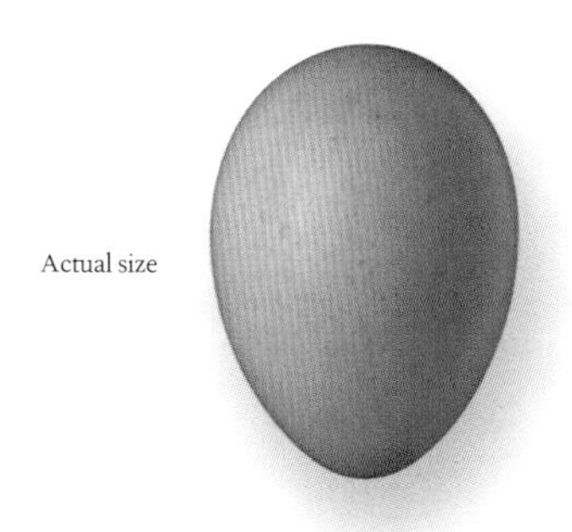

Actual size

The egg of the Montezuma Quail is chalk white to light buff in background color, with no maculation, and it measures 1⅓ x 1¹⁄₁₆ in (34 x 27 mm). The white eggs become stained from the decaying nesting materials as the incubation progresses, especially in years with heavy monsoon rains.

ORDER	Galliformes
FAMILY	Phasianidae
BREEDING RANGE	Southeastern Europe, introduced into many other temperate regions as a gamebird
BREEDING HABITAT	Rocky slopes, and lightly wooded montane areas
NEST TYPE AND PLACEMENT	Ground scrape lined with twigs and leaves
CONSERVATION STATUS	Near threatened
COLLECTION NO.	FMNH 20894

ADULT BIRD SIZE
13½–14 in (34–35 cm)

INCUBATION
24–26 days

CLUTCH SIZE
8–14 eggs

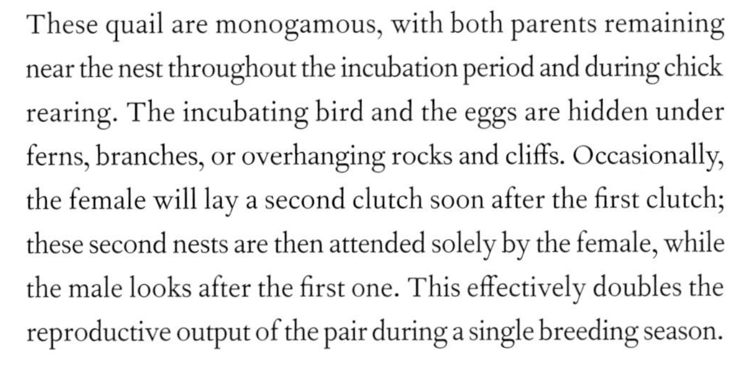

ALECTORIS GRAECA

ROCK PARTRIDGE

GALLIFORMES

Clutch

The egg of the Rock Partridge is pale cream to brown in background color with fine red maculation, and measures 1½ x 1⅓ in (40 x 34 mm).

These quail are monogamous, with both parents remaining near the nest throughout the incubation period and during chick rearing. The incubating bird and the eggs are hidden under ferns, branches, or overhanging rocks and cliffs. Occasionally, the female will lay a second clutch soon after the first clutch; these second nests are then attended solely by the female, while the male looks after the first one. This effectively doubles the reproductive output of the pair during a single breeding season.

The Rock Partridge is a conspicuous and distinctively colored species, which inhabits rocky and brushy high elevations in several Mediterranean countries. With climate shifts, habitat destruction, and hunting pressure, this species is continuing to decline in numbers in its native range. By contrast, throughout much of its introduced range, the species is doing quite well, having become naturalized on several other continents.

Actual size

ORDER	Galliformes
FAMILY	Phasianidae
BREEDING RANGE	Central to western Asia
BREEDING HABITAT	Grasslands and scrub, typically near water
NEST TYPE AND PLACEMENT	Ground scrape, lined with leaves and twigs
CONSERVATION STATUS	Least concern
COLLECTION NO.	FMNH 21445

ADULT BIRD SIZE
13–14 in (33–36 cm)

INCUBATION
18–19 days

CLUTCH SIZE
8–12 eggs

FRANCOLINUS FRANCOLINUS

BLACK FRANCOLIN

GALLIFORMES

The male Black Francolin is both loud in voice and conspicuous in plumage coloration and pattern, whereas the quieter female is drabber and cryptic in appearance. Moving and feeding in flocks during the non-breeding season, with the onset of mating these birds break away into pairs, and the males aggressively defend their territories from nearby competitors.

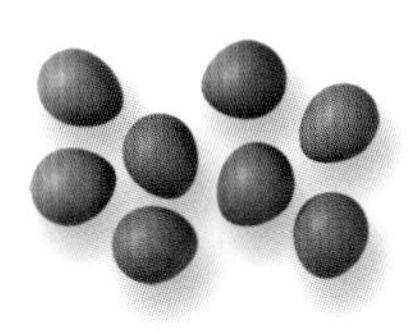

Clutch

The morphological and behavioral sex differences well complement the specific reproductive roles of the pair: the male uses his voice and displays to attract a female for pair bonding, providing her with insects as nuptial gifts. Once the pair bond is settled, the female quietly takes charge of nesting and incubating. The nest and the eggs are well hidden from view by dense overhanging vegetation. Both parents provide warmth or shade, and protection from danger for the precocial young.

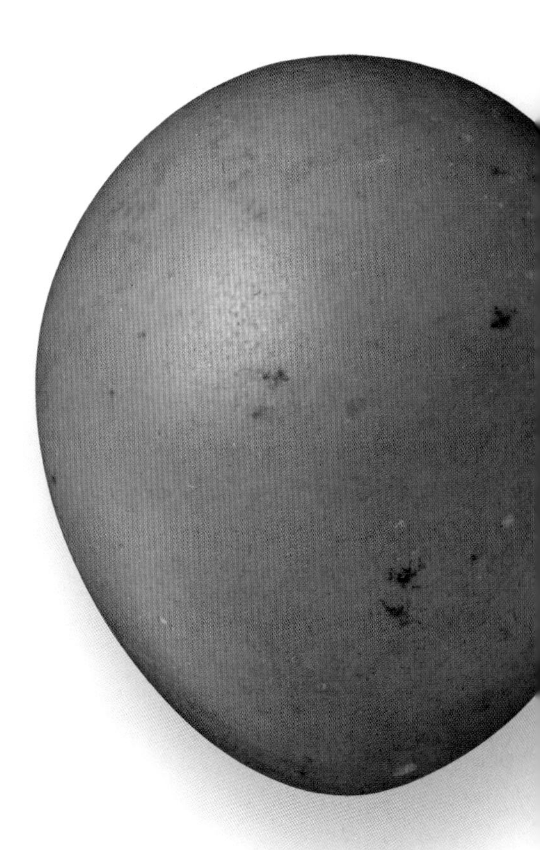

Actual size

The egg of the Black Francolin is olive to pale buff in background color, round in shape with small white spots near the pole, and 1¾ x 1½ in (46 x 37 mm) in size.

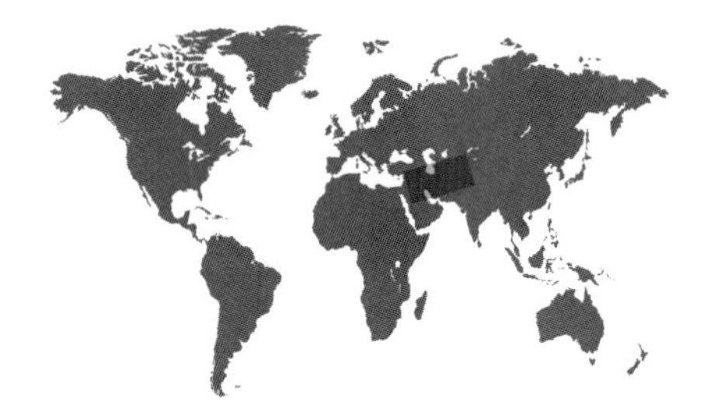

ORDER	Galliformes
FAMILY	Phasianidae
BREEDING RANGE	Western Asia
BREEDING HABITAT	Dry, open scrub and grasslands, semi-deserts
NEST TYPE AND PLACEMENT	Barren or sparsely lined ground scrape
CONSERVATION STATUS	Least concern
COLLECTION NO.	FMNH 18614

ADULT BIRD SIZE
8½–10 in (22–25 cm)

INCUBATION
23–25 days

CLUTCH SIZE
8–16 eggs

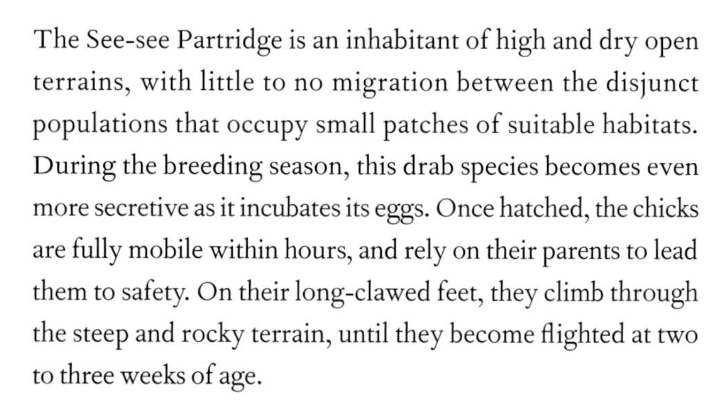

AMMOPERDIX GRISEOGULARIS

SEE-SEE PARTRIDGE

GALLIFORMES

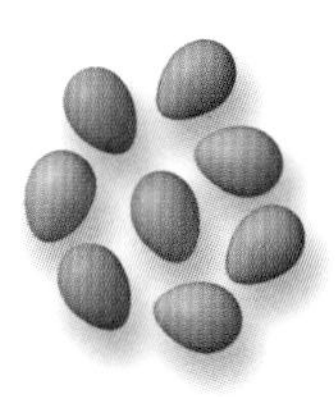

Clutch

The egg of the See-see Partridge is dull-white to pale-cream in coloration, clear of speckles, and measures 1½ x 1 1/16 in (36 x 27 mm) in size. These birds have been bred in captivity, and it is hoped that individuals may be reintroduced throughout the native range to augment or reestablish populations of the species.

The See-see Partridge is an inhabitant of high and dry open terrains, with little to no migration between the disjunct populations that occupy small patches of suitable habitats. During the breeding season, this drab species becomes even more secretive as it incubates its eggs. Once hatched, the chicks are fully mobile within hours, and rely on their parents to lead them to safety. On their long-clawed feet, they climb through the steep and rocky terrain, until they become flighted at two to three weeks of age.

The lack of individuals' movement between even neighboring populations has resulted in strongly patterned genetic structure even across short geographic distances. This provides a mechanism to maintain both high genetic variation across the whole species, and a great degree of local adaptation within each population. However, it is also a headache for conservation scientists, who want to identify the significant population units for management and protection.

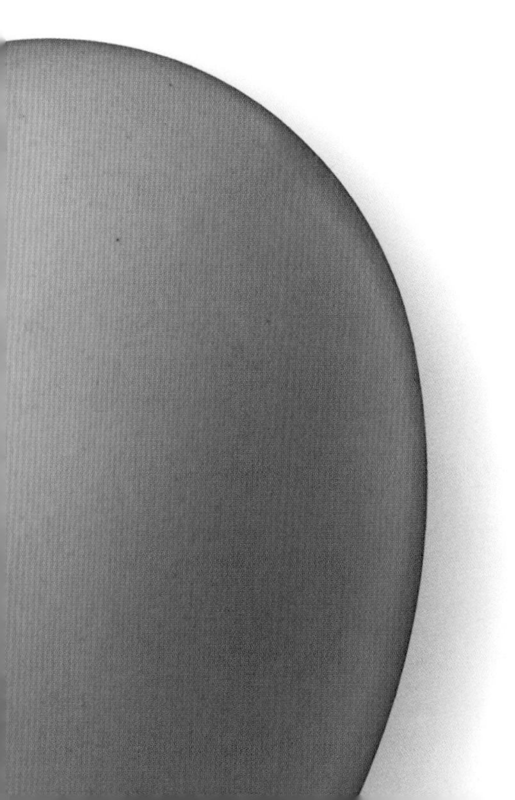

Actual size

ORDER	Galliformes
FAMILY	Phasianidae
BREEDING RANGE	Central and east Asia
BREEDING HABITAT	Farmland, open woodlands
NEST TYPE AND PLACEMENT	Depression in ground, lined with leaves and grasses
CONSERVATION STATUS	Least concern
COLLECTION NO.	FMNH 20886

ADULT BIRD SIZE
11–12 in (28–30 cm)

INCUBATION
25 days

CLUTCH SIZE
18–20 eggs

PERDIX DAURICA

DAURIAN PARTRIDGE

GALLIFORMES

Behaviorally, the Daurian Partridge is a sedentary species that forms coveys of both sexes during the non-breeding season. Several months prior to nesting, these flocks divide into pairs. Although this species is monogamous and females and males engage in elaborate mating displays, individuals often switch pair bonds prior to nesting. But once breeding has begun, the female is fully dedicated to nesting and incubating. The nest is located under a cover of grasses and brush to keep it safe from predators and the male guards it from nearby, providing an equitable share of care for the young chicks.

The Daurian Partridge was long considered to be an east Asian subspecies of the Gray Partridge. Recent genetic data have suggested that this widespread Asian group of species, called a species-complex, should be split into three distinct lines, with the Daurian Partridge occurring on the Tibetan Plateau, the Gray Partridge in Europe and northwestern Asia, and the Himalayan Partridge in the Himalayas.

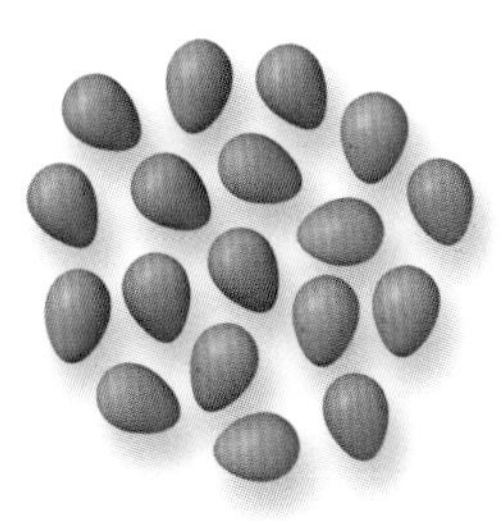

Clutch

The egg of the Daurian Partridge is pale to olive brown in background color, with no maculation, and measures 1⅓ x 1 in (34 x 25 mm) in size.

Actual size

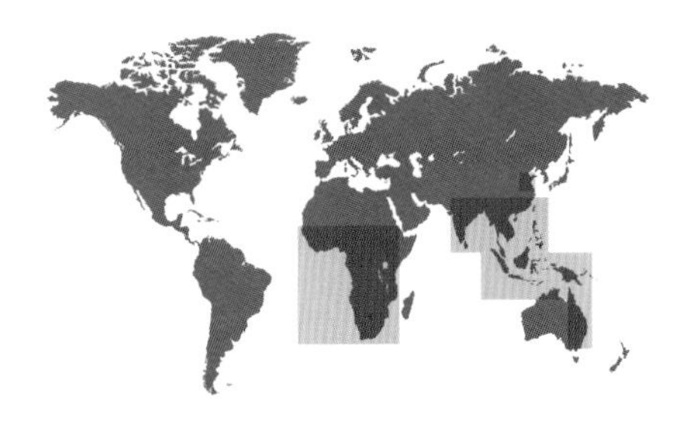

ORDER	Galliformes
FAMILY	Phasianidae
BREEDING RANGE	Sub-Saharan Africa, south and southeastern Asia, Australia
BREEDING HABITAT	Wet grasslands, damp woodlands
NEST TYPE AND PLACEMENT	Shallow, well-lined scrape, hidden under tufts of grass
CONSERVATION STATUS	Least concern
COLLECTION NO.	FMNH 15296

ADULT BIRD SIZE
4½–6 in (12–15 cm)

INCUBATION
18–19 days

CLUTCH SIZE
4–8 eggs

COTURNIX CHINENSIS

BLUE-BREASTED QUAIL

GALLIFORMES

Clutch

The Blue-breasted Quail is distributed throughout three continents, yet locally its populations are sparse, secretive, and hard to pinpoint. In the wild, the species forms transient pair bonds, as the female alone incubates the eggs; the male often stands guard nearby. Once the eggs hatch, both parents attend and protect the young. In some cases, the female leaves the brood and lays another full clutch of eggs that are incubated and attended solely by her. This pattern of rapid second clutching allows for successful reproduction even in the face of heavy predation on the eggs and young.

The Blue-breasted Quail is commonly kept in captivity, so much of its behavior is known from domesticated birds. In captivity they tolerate each other even at close quarters, with both sexes keeping minimal personal space around their nesting site. However, males will still defend the nesting site with low-intensity vocal contests.

Actual size

The egg of the Blue-breasted Quail is olive brown, buff, or reddish brown in color with finely freckled dark brown maculation. It measures 1 x ¾ in (25 x 19 mm) in size. The chicks hatch explosively from the egg, often in under five minutes after pipping begins.

ORDER	Galliformes
FAMILY	Phasianidae
BREEDING RANGE	Endemic to China
BREEDING HABITAT	Mature forests, with dense understory
NEST TYPE AND PLACEMENT	Nest is on branches of tall trees, made of leaves and mosses, or a scraped depression on the ground
CONSERVATION STATUS	Vulnerable
COLLECTION NO.	FMNH 18724

ADULT BIRD SIZE
19½–24 in (50–61 cm)

INCUBATION
28 days

CLUTCH SIZE
2–6 eggs

TRAGOPAN CABOTI

CABOT'S TRAGOPAN

GALLIFORMES

With its colorful plumage and patches of bare orange skin, there are few more spectacular sights in nature than a male Cabot's Tragopan displaying to a female. The courting male inflates the blue and red wattle (called a lappet) hanging from his throat, and erects a set of flashy black horns to frame his face, confronting the female with a fleshy explosion of color. The female carefully inspects the displaying male, and may decide to mate with him right there and then, or move on to another male's territory and watch his display.

The males demarcate their breeding territories through calling and attacking intruder males, whereas the females roam widely between territories. Only the females are in charge of the nesting, incubation, and chick protection. Inside the egg, the embryos mature fast, and the hatchlings are able to fly as soon as they emerge.

Clutch

The egg of the Cabot's Tragopan is buff in background color with pale brown maculation, and measures 1⅞ x 1⅝ in (49 x 41 mm) in size. The rusty beige tones of the egg blend in well against the colors of the forest floor in this species' native habitats, littered with dried pine needles and other leaf matter.

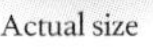

Actual size

ORDER	Galliformes
FAMILY	Phasianidae
BREEDING RANGE	Endemic to Taiwan
BREEDING HABITAT	Mature broadleaf and secondary forests
NEST TYPE AND PLACEMENT	On the ground, under cover
CONSERVATION STATUS	Near threatened
COLLECTION NO.	FMNH 18727

ADULT BIRD SIZE
21½–28½ in (55–72 cm)

INCUBATION
25–28 days

CLUTCH SIZE
2–6 eggs

LOPHURA SWINHOII

SWINHOE'S PHEASANT

GALLIFORMES

Clutch

This pheasant is a creature of habit, as it follows a preset path on a daily basis during its foraging. Active at dusk and dawn, and in dense fog, it is rarely seen, despite the deep blue and bright white plumage of the male. Solitary during most of the year, consortship between females and males occurs during the early part of the breeding season: the females soon leave the males and nest on their own. The eggs are kept safe from sight through the cryptic plumage of the female in a nest that is typically positioned under a large shelter, such as a fallen log.

A species commonly kept by pheasant fanciers, this bird was once near extinction, with a total population of just 200 individuals 50 years ago. As an endemic to an island heavily altered by human development, this population crash was nearly the end of these birds in the wild. However, multiple conservation actions have increased numbers to some 10,000 or so individuals.

Actual size

The egg of the Swinhoe's Pheasant is pale pink to pale yellow in background color and clear of speckles. It measures 2⅛ x 1⅞ in (53 x 49 mm) in size. Exposure to the elements, dirt, and the female's claws leaves otherwise immaculate eggs often discolored and scratch-marked.

ORDER	Galliformes
FAMILY	Phasianidae
BREEDING RANGE	Endemic to China
BREEDING HABITAT	Montane, mature forests
NEST TYPE AND PLACEMENT	On ground, in a shallow hollow or under a fallen log or tall grass
CONSERVATION STATUS	Vulnerable
COLLECTION NO.	FMNH 18739

ADULT BIRD SIZE
38–39½ in (96–100 cm)

INCUBATION
28 days

CLUTCH SIZE
4–14 eggs

CROSSOPTILON MANTCHURICUM

BROWN EARED-PHEASANT

GALLIFORMES

Unlike most ornamental pheasants, the Brown Eared-Pheasant and its close relatives are brightly colored and conspicuously patterned; yet, unlike most other pheasants where only the males are colorful, this species' distinctive plumage, including the mustachelike, stiff ear-cover feathers, is shared by both sexes. It is not known whether plumage ornamentation serves a role in mutual mate choice, both by the female when choosing her mate, and by the male when pairing with the female.

During the breeding season, these birds break away from larger flocks, and form socially monogamous couples. The female is in charge of directing the couple's movements, from foraging to inspecting potential nest sites. In turn, the male performs mate-guarding behaviors, including remaining in close proximity to the female, pecking at the female's feet to dissuade her from paying attention to another male, and displaying to other males from a distance even when these are already part of another pair bond.

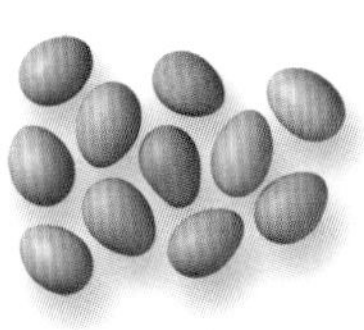

Clutch

The egg of the Brown Eared-Pheasant is pale stone green in background color, immaculate, and 2¼ x 1⅝ in (56 x 41 mm in size). Nests often fail because they are disturbed by people harvesting mushrooms in the forest undergrowth.

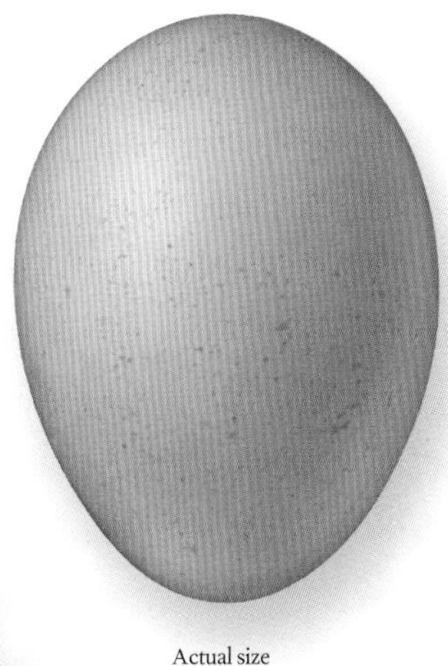

Actual size

ORDER	Galliformes
FAMILY	Phasianidae
BREEDING RANGE	Central Asia, the western Himalayas
BREEDING HABITAT	High-elevation grassland and scrubland, near steep cliffs
NEST TYPE AND PLACEMENT	Ground nest, hidden in vegetation, usually at the foot of an outcropping
CONSERVATION STATUS	Vulnerable
COLLECTION NO.	FMNH 18741

ADULT BIRD SIZE
23½–43½ in (60–110 cm)

INCUBATION
26 days

CLUTCH SIZE
9–12 eggs

CATREUS WALLICHII

CHEER PHEASANT

GALLIFORMES

Clutch

The egg of the Cheer Pheasant is pale yellowish gray in background color with reddish brown speckles of maculation, and measures 2¹⁄₁₆ x 1⅓ in (53 x 40 mm) in dimensions.

During the mating season, Cheer Pheasants break away from their loose flocks and disperse in pairs. Nesting and incubating is only undertaken by the female, but the male remains close by to distract or threaten approaching predators. Both parents stay with the chicks and each performs the broken-wing display when the young are under threat. The drab plumage of the adults and chicks is quite inconspicuous, and they keep low in dense grasses and shrubs to remain hidden from predators, including hunting humans.

This species both benefits and suffers from its association with human settlements. Its preferred montane habitats include scrub- and grasslands kept low by grazing livestock and crop cultivation, but this large gamebird is also frequently hunted for its meat by local villagers. The declining numbers and increasing isolation between populations warrant concerns about the conservation status of this pheasant.

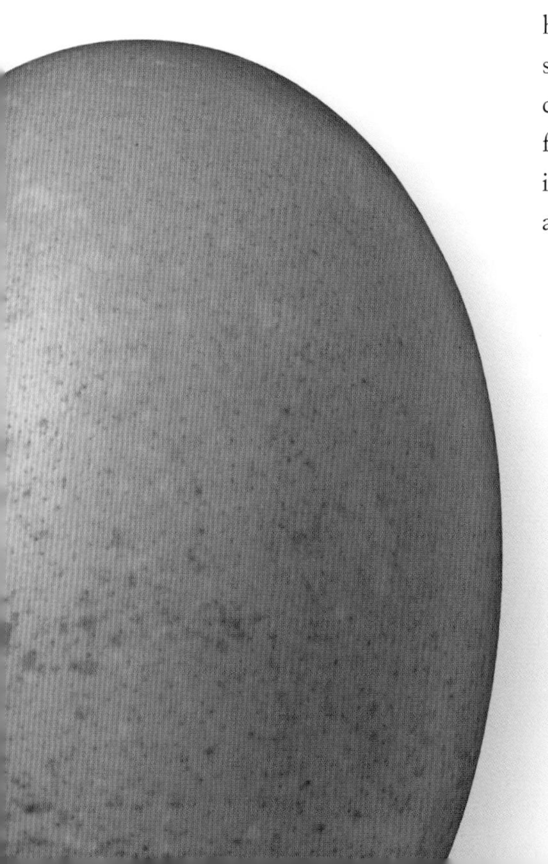

Actual size

ORDER	Galliformes
FAMILY	Phasianidae
BREEDING RANGE	Southeastern China
BREEDING HABITAT	Montane, evergreen forests and scrubs
NEST TYPE AND PLACEMENT	Scrape on ground
CONSERVATION STATUS	Near threatened
COLLECTION NO.	FMNH 18721

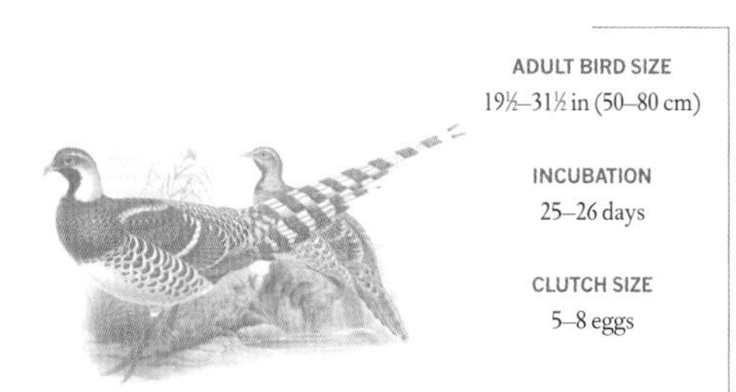

ADULT BIRD SIZE
19½–31½ in (50–80 cm)

INCUBATION
25–26 days

CLUTCH SIZE
5–8 eggs

SYRMATICUS ELLIOTI

ELLIOT'S PHEASANT

GALLIFORMES

The Elliot's Pheasant is a large, forest-dwelling galliform, which has suffered from the deforestation of its endemic habitats in China. Female and male Elliot's Pheasants do not form partnerships for parental duties. After mating, the male may become aggressive toward the female, either by trying to mate with her repeatedly or by attempting to chase her away in favor of attracting new females. The hatchlings are looked after by the female alone, and they grow fast, with differences between female and male feathers showing as early as a few weeks of age.

Due to its prominent coloration, this is a popular breed among bird fanciers, and today there are 10 to 15 times as many captive individuals as the number estimated to be in the wild (around 10,000). Like many other galliforms, this species readily hybridizes in captivity with other pheasants both within and outside its genus, making many captive stocks of dubious value for potential reintroduction projects.

Clutch

The egg of the Elliot's Pheasant is cream (as shown) to white in background color with no maculation, and measures 2 x 1½ in (50 x 38 mm) in size.

Actual size

ORDER	Galliformes
FAMILY	Phasianidae
BREEDING RANGE	Asia; introduced and naturalized in many other parts of the world
BREEDING HABITAT	Woodlands, scrub, farms, and grasslands
NEST TYPE AND PLACEMENT	Ground nest concealed in grass
CONSERVATION STATUS	Least concern
COLLECTION NO.	FMNH 21415

ADULT BIRD SIZE
19½–27½ in (50–70 cm)

INCUBATION
23–26 days

CLUTCH SIZE
8–10 eggs

PHASIANUS COLCHICUS

RING-NECKED PHEASANT

GALLIFORMES

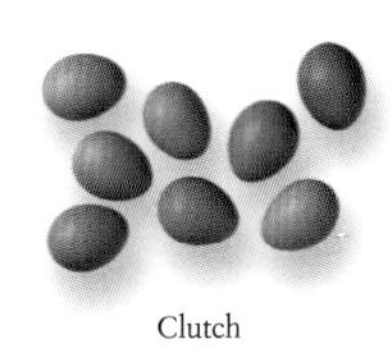

Clutch

The egg of the Ring-necked Pheasant is brownish olive in background color, clear of heavy maculation, and measures 1¾–1⅜ in (45 x 36 mm) in size.

Female Ring-necked Pheasants choose mates based on their physical attributes, including the length of tail feathers, the brightness of black ear-tufts, and, perhaps most importantly, the size of their leg spurs. These traits are thought to be related to the quality of offspring that the male would sire and this is what females are assessing. The female deposits varying amounts of steroid hormones, including testosterone, into the egg yolk, which in turn affect the sexual attractiveness and behaviors of her adult sons and daughters. The male mates with several females and the hens alone undertake all incubating and rearing duties. The chicks are independent within 12–14 days of hatching.

The Ring-necked, or Common, Pheasant, is historically one of the earliest introduced and naturalized species in Europe, having established itself from released stock over hundreds of years, so that in many places it is treated as a native species. This is also the case in the United States, where after its relatively recent introduction following European settlement, the pheasant became the official state bird of South Dakota.

Actual size

ORDER	Galliformes
FAMILY	Phasianidae
BREEDING RANGE	Malay Peninsula, Sumatra, and Borneo
BREEDING HABITAT	Jungle, rainforest
NEST TYPE AND PLACEMENT	Ground nest, scraped under dense vegetation
CONSERVATION STATUS	Near threatened
COLLECTION NO.	FMNH 18628

ARGUSIANUS ARGUS

GREAT ARGUS

GALLIFORMES

ADULT BIRD SIZE
29½–71 in (75–180 cm)

INCUBATION
24–26 days

CLUTCH SIZE
2 eggs

This large and distinctive species lacks bright colors, but makes up for it by the extraordinary length and the elaborate patterning of its plumage. The long secondary wing feathers are fully covered in eyelike patterns, or ocelli. The females resemble the males, but are drabber in their colors and have much shorter plumes.

Clutch

During the mating season, the males form loose aggregations, called leks, in the forest understory. Each male clears a small patch, from where he calls out loudly and displays to any approaching females. The females visit several males, observe them closely during their dance, and eventually choose males based on their plumes and the intensity of their displays. After mating has occurred, the female alone scrapes the nest, incubates the eggs, and looks after hatchlings, while the male continues displaying to attract more females. Although Great Argus chicks depart the nest soon after hatching, they are slow to achieve independence from parental care.

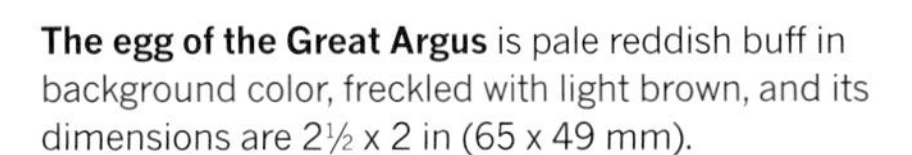

The egg of the Great Argus is pale reddish buff in background color, freckled with light brown, and its dimensions are 2½ x 2 in (65 x 49 mm).

Actual size

ORDER	Galliformes
FAMILY	Phasianidae
BREEDING RANGE	Remnant populations throughout Southeast Asia
BREEDING HABITAT	Tropical dry and seasonal forests
NEST TYPE AND PLACEMENT	Scrape nest on the ground
CONSERVATION STATUS	Endangered
COLLECTION NO.	FMNH 2500

ADULT BIRD SIZE
3 ft 3 in–7 ft 4 in
(100–224 cm)

INCUBATION
26–28 days

CLUTCH SIZE
3–6 eggs

PAVO MUTICUS

GREEN PEAFOWL

GALLIFORMES

Clutch

The Green Peafowl looks familiar, as it resembles the Indian Peafowl (or peacock) in its coloration and the male's decorated train. But there are notable differences. For instance, the species' bright green and blue plumage coloration is shared by both the male and the female, making it sexually monochromatic. However, the male weighs three to four times as much as the female, which marks it out as one of the most size-dimorphic of modern bird species.

The mating behavior of the Green Peafowl is also distinctive. Instead of forming leks, like Indian Peafowl, the Green Peafowl maintains distinct male territories, and strong affiliations with one or more females in that territory. This is similar to the haremlike mating system of wild Jungle Fowl, and its domesticated form, the chicken. The parents continue their association after the eggs hatch, and so entire family groups of parents and growing chicks can be seen roosting together at night, in trees, for safety from ground predators.

The egg of the Green Peafowl is pale pink in background color, clear of distinct spots, and is 3 x 2 in (75 x 53 mm) in size. Within two weeks of hatching the young peafowl are able to fly.

Actual size

ORDER	Galliformes
FAMILY	Phasianidae
BREEDING RANGE	Northern Europe and western and central Asia
BREEDING HABITAT	Mature coniferous woodlands
NEST TYPE AND PLACEMENT	Ground nest lined with vegetation
CONSERVATION STATUS	Least concern, but subspecies are locally endangered or extinct
COLLECTION NO.	FMNH 2443

TETRAO UROGALLUS

EURASIAN CAPERCAILLIE

GALLIFORMES

ADULT BIRD SIZE
23½–31½ in (60–80 cm)

INCUBATION
26–29 days

CLUTCH SIZE
5–12 eggs

The life of the Eurasian Capercaillie balances time spent on the ground foraging, mating, and nesting, with time spent in trees for displays and safety. During the mating season, the males gather near each other, and begin their displays by calling loudly from tree branches. This aggregation of males, or lek, is a central place for females to find potential mates. Once the females arrive at a lek, the dominant male in the group continues his courtship on the ground, and eventually he alone mates with females. Nest building and incubating are chores undertaken solely by the females. The precocial young are ready to leave the nest, guided by only their mother, just one day after hatching.

Clutch

As a large bird with a heavy flight, this species escapes most danger on foot. When the threat is imminent, these birds take off accompanied by a loud noise generated by the wings and tail feathers; this sound is thought to deter predators from pursuing their prey any further.

The egg of the Eurasian Capercaillie is buff yellow in background color with sparse brown spotting, and it measures 2¼ x 1⅔ in (57 x 42 mm) in size. The eggs are hidden from view, by dense vegetation above the nest, where the cryptically feathered female spends 23 hours a day incubating them.

Actual size

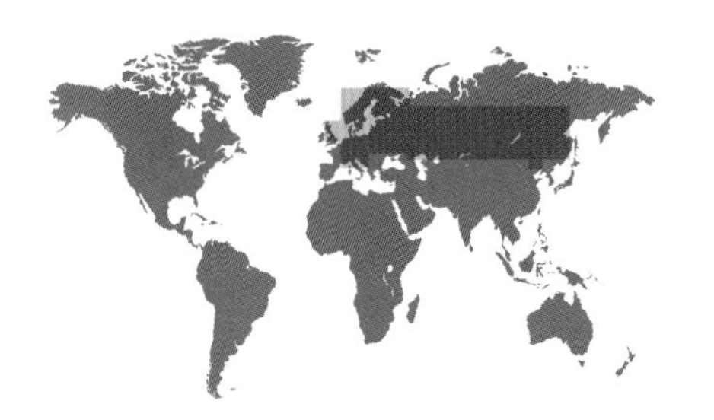

ORDER	Galliformes
FAMILY	Phasianidae
BREEDING RANGE	Northern and central Eurasia
BREEDING HABITAT	Grasslands, bogs, and young forest stands
NEST TYPE AND PLACEMENT	Ground depression, lined with leaves and feathers
CONSERVATION STATUS	Least concern
COLLECTION NO.	FMNH 1216

ADULT BIRD SIZE
15¾–22 in (40–55 cm)

INCUBATION
25 days

CLUTCH SIZE
7–9 eggs

TETRAO TETRIX

BLACK GROUSE

GALLIFORMES

Clutch

The Black Grouse is a formidable and distinctively colored gamebird, occurring throughout much of the suitable habitat even in densely populated European countries. The nesting success of Black Grouse has been steadily declining over the past several decades, and locally some populations have gone extinct or become small enough to be considered endangered. The reasons for this are not known.

The male Black Grouse is polygynous and his role in breeding stops at fertilizing the eggs. The female selects the nest site, collects the nesting materials, incubates the eggs, and attends the chicks. Her preference is to nest in younger, regenerating forests, instead of older, mature stands. The nest is well hidden under the canopy of trees and bushes, and the eggs are concealed by the cryptic plumage of the female herself. Eggs in ground nests are vulnerable to predation, and inexperienced females are less successful in breeding than those who have bred before, suggesting that, with age, females may learn how better to conceal their nests.

Actual size

The egg of the Black Grouse is brownish buff in background color, with blotchy reddish brown maculation, and measures 2 x 1⅜ in (51 x 36 mm) in size.

ORDER	Galliformes
FAMILY	Phasianidae
BREEDING RANGE	Northern and central Eurasia
BREEDING HABITAT	Dense, mixed deciduous-coniferous woodlands
NEST TYPE AND PLACEMENT	Ground cup, lined with fallen leaves
CONSERVATION STATUS	Least concern
COLLECTION NO.	FMNH 15302

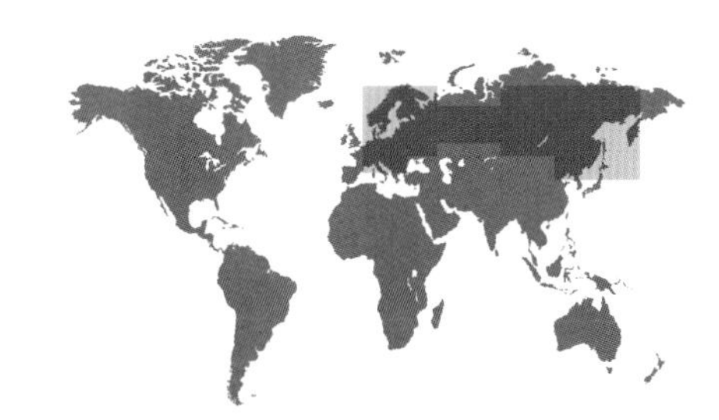

ADULT BIRD SIZE
14–15½ in (35–39 cm)

INCUBATION
23–27 days

CLUTCH SIZE
3–6 eggs

BONASA BONASIA

HAZEL GROUSE

GALLIFORMES

The female Hazel Grouse relies on a dense forest understory to keep her eggs safely hidden from visual predators during the incubation period. However, in common with the Black Grouse (see facing page), the population sizes and the nesting success of these birds have been slowly but steadily declining. This may be due to habitat loss, including the clear-cutting of boreal forests, as well as habitat modification resulting from the selective removal of dense understory vegetation in managed woodlands, which would otherwise have functioned to protect the nest. Younger and older females have similar nesting success, with predation being the most important cause of nest failure.

The Hazel Grouse is a nonmigratory gamebird, able to weather the harsh winters of its mostly boreal distribution. In some areas of the species' range, grouse-hunting by local peoples is an important cultural and economic activity, whereas in other areas, hunting has been limited to the autumnal non-breeding season or banned altogether.

Clutch

The egg of the Hazel Grouse is creamy white in background color, with brown maculation, and is 1½ x 1⅛ in (38 x 29 mm) in size.

Actual size

ORDER	Galliformes
FAMILY	Phasianidae
BREEDING RANGE	Eastern and northern North America
BREEDING HABITAT	Mixed and deciduous forests
NEST TYPE AND PLACEMENT	Ground bowl, at base of a tree, lined with leaves
CONSERVATION STATUS	Least concern
COLLECTION NO.	FMNH 5955

ADULT BIRD SIZE
17–19½ in (43–50 cm)

INCUBATION
24 days

CLUTCH SIZE
9–14 eggs

BONASA UMBELLUS

RUFFED GROUSE

GALLIFORMES

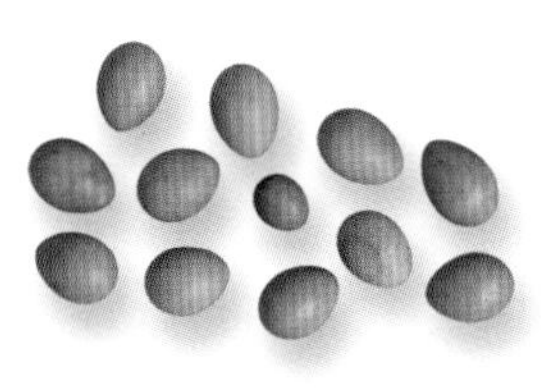

Clutch

The egg of the Ruffed Grouse is light brownish buff in background color, with no or few sparse red speckles, and measures 1⅝ x 1³⁄₁₆ in (40 x 30 mm) in dimensions. In some clutches, females produce and lay one or more smaller eggs (runts) which often lack yolk and fail to hatch.

Being non-migratory has its advantages for Ruffed Grouse, as it allows both females and males to become familiar with their breeding territories. For the male, this means selecting a large log from which it displays, producing drumming-like noises in the spring to attract females. For the female, this means that she frequently hears and is able to visit several males before deciding to mate with the one whose display she finds most impressive. The female alone then easily finds a safe and covered spot to build her nest and incubate the eggs, typically at the foot of a trunk or underneath a bush in the undergrowth.

The Ruffed Grouse is a forest specialist bird; it is a year-long resident, with special adaptations to survive the winter. For example, it grows projections on its toes that create "snowshoes" that allow it to forage about without breaking through the upper crust of snow. At night, it dives into deep, insulating stands of snow to keep warm.

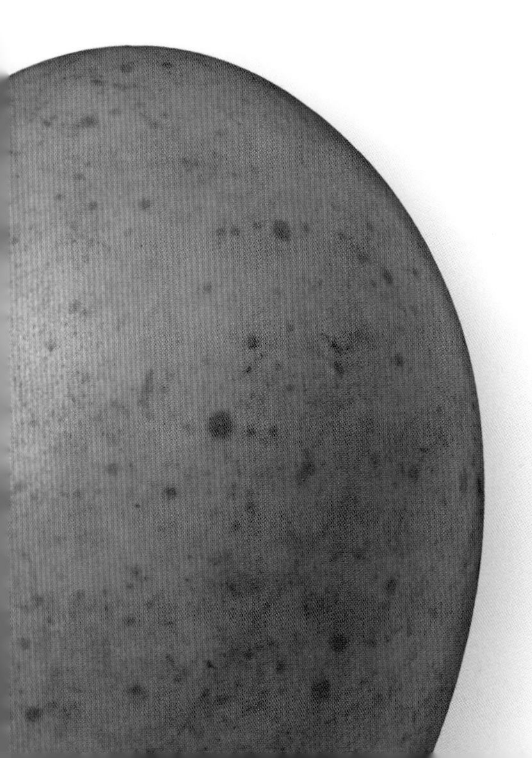

Actual size

ORDER	Galliformes
FAMILY	Phasianidae
BREEDING RANGE	Central-western North America
BREEDING HABITAT	Sparse sage-brush and open grazing fields
NEST TYPE AND PLACEMENT	Ground nest, under a sagebush
CONSERVATION STATUS	Near threatened
COLLECTION NO.	FMNH 6063

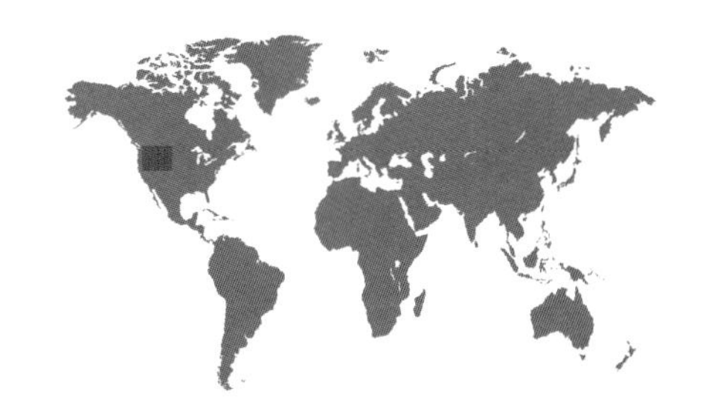

CENTROCERCUS UROPHASIANUS

GREATER SAGE-GROUSE

GALLIFORMES

ADULT BIRD SIZE
19½–29½ in (50–75 cm)

INCUBATION
25–29 days

CLUTCH SIZE
6–13 eggs

The extravagant breeding displays of the Greater Sage-Grouse are matched by few other species. Gathering in groups, called leks, on an open plateau clear of tall vegetation, male sage-grouse strut, dance, call, and pop, using highly conspicuous visible and audible displays. Researchers have built self-propelled robots costumed as female sage-grouse to understand the dynamic nature of male displays. The results showed that once a female enters the lek, the males reposition and readjust their mating displays in order to beam their sounds most loudly in the direction of the visiting female.

Clutch

In addition to habitat loss due to the conversion of sage fields into crops and grazing land, increasing sources of anthropogenic noise, such as roads and mining, have further impacted this sage-grouse's population health. Such sound pollution reduces the lek-site attendance of males at otherwise suitable breeding habitats. Those birds that do decide to remain on noisy leks show elevated levels of stress hormones.

The egg of the Greater Sage-Grouse is buff in background color, and measures 2⅛ x 1½ in (55 x 38 mm) in size. As in most other grouse species, the female mates with males at the display site, but she alone takes care of all parental duties. Her cryptic plumage, compared to that of the male, enhances clutch survival because the eggs are less likely to be detected.

Actual size

ORDER	Galliformes
FAMILY	Phasianidae
BREEDING RANGE	Endemic to the United States' Rocky Mountain region
BREEDING HABITAT	Open, sage-brush habitat
NEST TYPE AND PLACEMENT	Scrape on the ground
CONSERVATION STATUS	Endangered
COLLECTION NO.	FMNH 22381

ADULT BIRD SIZE
18–22 in (46–56 cm)

INCUBATION
25–29 days

CLUTCH SIZE
6–8 eggs

CENTROCERCUS MINIMUS

GUNNISON SAGE-GROUSE

GALLIFORMES

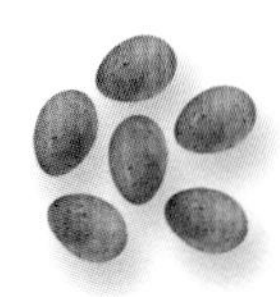

Clutch

This is the first new bird species described from the United States since the nineteenth century; it was formally accepted as a full species in 2000, following a detailed description of the morphology, genetic structure, and the behavior of several distinct populations of the former Greater Sage-Grouse (see previous page). Specifically, the mating behaviors and displays of the Gunnison Sage-Grouse are highly distinctive, and include a unique use of rhythms, plumes, and sounds to attract females to the males' communal display grounds.

The grouse is an obligate sage-specialist: it relies on a mature and stable sage ecosystem to feed and breed. This habitat is under threat from grazing and habitat conversion. The strutting sites for mating are located on a bare and flat area where each male's dancing site can be inspected by females from a distance. The female moves between the different males to inspect their plumage and listen to their calls, and eventually settles to mate with one of the more dominant males in the center of the display site. When disturbed by other animals, including humans, females are quick to desert a full clutch of eggs. Many other nests are also directly destroyed by predators.

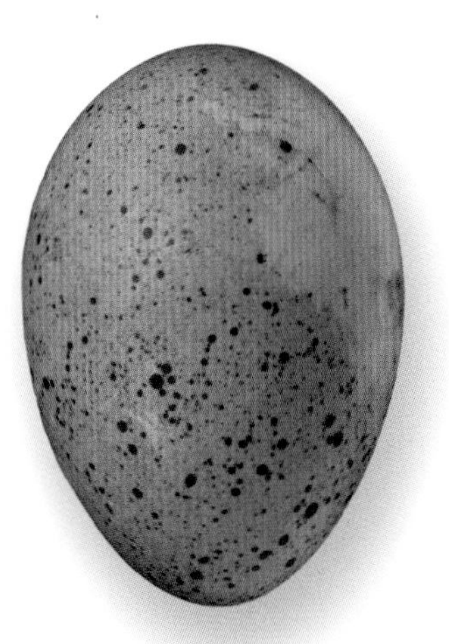

Actual size

The egg of the Gunnison Sage-Grouse is buff olive in background color, has sparse brown markings, and measures 2¼ x 1⅜ in (55 x 35 mm) in size.

ORDER	Galliformes
FAMILY	Phasianidae
BREEDING RANGE	Boreal and montane North America
BREEDING HABITAT	Boreal forests
NEST TYPE AND PLACEMENT	On the ground, concealed by a leafy bush or other vegetation
CONSERVATION STATUS	Least concern
COLLECTION NO.	FMNH 15891

ADULT BIRD SIZE
15–21 in (38–53 cm)

INCUBATION
24 days

CLUTCH SIZE
4–7 eggs

FALCIPENNIS CANADENSIS

SPRUCE GROUSE

GALLIFORMES

Mating is a brief affair for the Spruce Grouse, as the males provide no parental duties; the females alone nest, incubate, and attend the young. Females become ready to breed in their second year of life, but males often delay establishing their first territories until they are in their third year. Incubation begins with the last egg laid, so that the chicks hatch and leave the nest at the same time. The chicks grow fast and are able to fly from the ground onto low-lying branches one week after hatching. The female stays close to the chicks, brooding them for warmth when they call during cold days and nights.

The main food staple of adult Spruce Grouse, a specialist of the coniferous forests of Canada and Alaska, is freshly clipped pine needles, stored and digested overnight in their crops. Hatchlings, by contrast, feed on insects and berries throughout the spring and summer, and switch to needles only in the fall and winter.

Clutch

Actual size

The egg of the Spruce Grouse is buff in background color, with mottled brown maculation, and is 1⅔ x 1¼ in (43 x 32 mm) in dimensions.

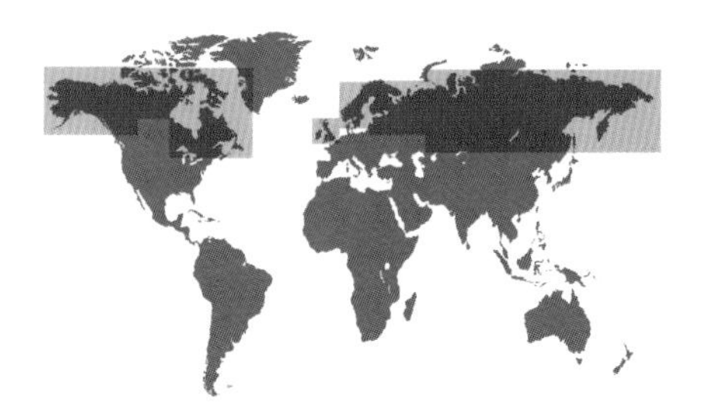

ORDER	Galliformes
FAMILY	Phasianidae
BREEDING RANGE	Arctic regions of North America, Europe, and Asia
BREEDING HABITAT	Open and shrubby tundra
NEST TYPE AND PLACEMENT	Scrape on the ground, typically under brush, near a clearing
CONSERVATION STATUS	Least concern
COLLECTION NO.	FMNH 5969

ADULT BIRD SIZE
14–17½ in (35–44 cm)

INCUBATION
21–22 days

CLUTCH SIZE
4–10 eggs

LAGOPUS LAGOPUS

WILLOW PTARMIGAN

GALLIFORMES

Clutch

The egg of the Willow Ptarmigan is buff in background, has dark brown to black maculation, and measures 1¾ x 1¼ in (44 x 32 mm) in size.

The Willow Ptarmigan is highly territorial and mating is centered around breeding territories defended by solitary males, who arrive at the mating grounds up to a month before the females. Most males are monogamous and closely follow females both during the pre-laying and the nesting period. During these times 97 percent of females are in the presence of a male! Mate-guarding is not to protect the female from predators but to assure paternity of the embryos, both during the first nesting attempt, and during any replacement attempts if the first clutch is lost to predation. The female alone incubates the eggs but both parents attend the chicks, leading them to food and safety.

This species has a wide distribution, occurring in large numbers throughout the polar regions of all three northern continents. Adult plumage changes dramatically between rusty red in the summer to almost pure white in the winter, except for the subspecies in the British Isles, called the Red Grouse.

Actual size

ORDER	Galliformes
FAMILY	Phasianidae
BREEDING RANGE	Central and northern North America
BREEDING HABITAT	Open grasslands, with herbaceous cover and occasional shrubs
NEST TYPE AND PLACEMENT	Shallow depression on the ground, lined with feathers, ferns, or grass
CONSERVATION STATUS	Least concern
COLLECTION NO.	FMNH 5999

ADULT BIRD SIZE
16–18½ in (41–47 cm)

INCUBATION
23–25 days

CLUTCH SIZE
6–15 eggs

TYMPANUCHUS PHASIANELLUS

SHARP-TAILED GROUSE

GALLIFORMES

Breeding for this species is a test of male dominance, endurance, and skill. Males gather in their historic dancing sites up to one month before the females arrive, often when the site is still covered in snow. They establish dominance hierarchies: older and larger males typically occupy more central display sites. Females begin to visit the dancing sites when the snow cover has thinned, and may revisit the same site repeatedly, while also visiting nearby sites. Eventually, mating takes place and the female leaves to take care of the next generation on her own. Around each dancing ground, the eggs in most nests are typically sired by the most dominant of males.

The Sharp-tailed Grouse is a quintessential inhabitant of grassland prairies, with both foraging and breeding tied closely to short grass and open habitats. The loss of natural grasslands and the conversion of grazing fields into croplands have resulted in dramatic declines in populations across the species range.

Clutch

The egg of the Sharp-tailed Grouse is buff drab in coloration, with small brown maculation, and is 1⅔ x 1¼ in (43 x 32 mm) in size.

Actual size

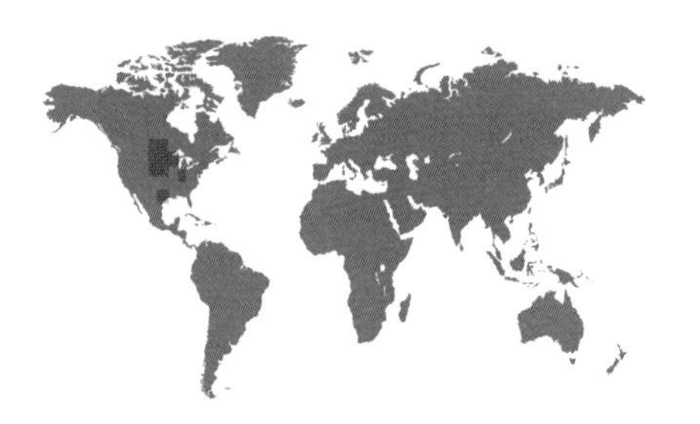

ORDER	Galliformes
FAMILY	Phasianidae
BREEDING RANGE	Middle North America
BREEDING HABITAT	Open grasslands, oak savanna
NEST TYPE AND PLACEMENT	Ground scrape, lined with vegetation
CONSERVATION STATUS	Vulnerable; its eastern subspecies, the Heath Hen, went extinct in 1932, and its Texan subspecies is endangered
COLLECTION NO.	FMNH 5991

ADULT BIRD SIZE
15½–18 in (40–46 cm)

INCUBATION
23–25 days

CLUTCH SIZE
10–15 eggs

TYMPANUCHUS CUPIDO

GREATER PRAIRIE-CHICKEN

GALLIFORMES

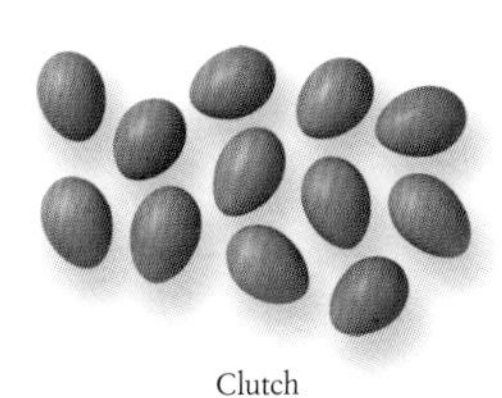

Clutch

Prior to mating, males return and establish their individual display sites, or leks, which are typically reused across years. There they dance, puff, and boom for long hours and weeks while females visit and inspect the males. Females visit, watch, and listen at multiple sites and observe multiple males at each site. After mating, however, the male provides no more parental care, and all duties of nesting, incubating, and attending the young fall onto the female. When Ring-necked Pheasants (see page 234) lay parasitically into a Prairie-Chicken nest, these foreign eggs hatch earlier, causing the female Prairie-Chicken to lead the pheasant chicks away and leave her own unhatched eggs to perish.

For this species of open grassland, habitat conversion of the prairie into agricultural fields has been the greatest cause of population decline and conservation concern. Once common throughout many of the Midwestern regions of the United States, today the species persists in small and isolated populations. Because of the birds' sedentary nature, genetic diversity has declined within each population, necessitating conservation management action that increasingly relies on translocating individuals between sites to generate artificial gene flow for viability and diversity.

Actual size

The egg of the Greater Prairie-Chicken is buff to olive in color, has dark brown maculation, and measures 1⅔ x 1¼ in (42 x 32 mm). These eggs are indistinguishable from the handful of remaining museum-held eggs of the now-extinct Heath Hen, which was a subspecies of the Greater Prairie-Chicken.

ORDER	Galliformes
FAMILY	Phasianidae
BREEDING RANGE	Temperate North America
BREEDING HABITAT	Forests and clearings
NEST TYPE AND PLACEMENT	Ground depression, lined with leaves
CONSERVATION STATUS	Least concern
COLLECTION NO.	FMNH 6008

ADULT BIRD SIZE
29½–49 in (75–125 cm)

INCUBATION
28 days

CLUTCH SIZE
10–15 eggs

MELEAGRIS GALLOPAVO

WILD TURKEY

GALLIFORMES

The breeding behavior of the Wild Turkey is a classic example of cooperation driven by kin selection. Related groups of males stay together during their lifespans, and strut to display and attract females as a team. Larger groups are apparently more attractive to females than solitary males. However, only the alpha male mates and fertilizes the eggs. What, then, can be the benefit of communal display to subordinate males? The answer is relatedness: teams of turkey males are close relatives, typically brothers. By helping a brother to reproduce, subordinate males benefit from indirect reproduction, with many nieces and nephews passing on some of their genes.

Clutch

The Wild Turkey and the Muscovy Duck (see page 93) are the two endemic bird species native to the New World that have been domesticated by humans. Although the domesticated stock is derived from a Mexican subspecies, there have also been many accidental releases, and reintroductions of wild populations following local extinctions, so that today's Wild Turkey populations represent a genetic mix from many sources.

The egg of the Wild Turkey is cream to tan in background color with reddish or pink spots, and 2½ x 1⅞ in (62 x 47 mm) in size. Most eggs are sired by just a single male, so that most chicks, called poults, are full siblings of each other.

Actual size

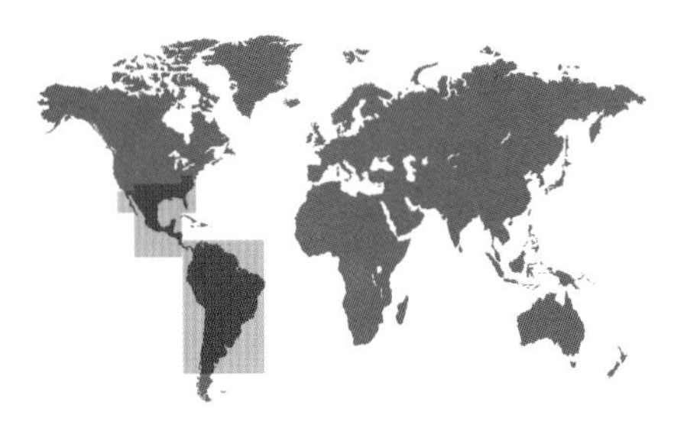

ORDER	Accipitriformes
FAMILY	Cathartidae
BREEDING RANGE	Southern North America, Central and South America
BREEDING HABITAT	Open areas interspersed with woods and brush; also larger stands of lowland wet forests
NEST TYPE AND PLACEMENT	Ground scrape with no lining; nearby area may be decorated with brightly colored cloth, glass, or plastic
CONSERVATION STATUS	Least concern
COLLECTION NO.	FMNH 20979

ADULT BIRD SIZE
23½–27 in (60–68 cm)

INCUBATION
28–41 days

CLUTCH SIZE
2 eggs

CORAGYPS ATRATUS

BLACK VULTURE

ACCIPITRIFORMES

Clutch

These vultures form pair bonds after a prolonged display, where multiple males walk and hop around the ground, with their wings held open, while a female inspects them. Once the pair is formed, they settle on a nesting spot, and both parents incubate the eggs and provision the young until independence. The eggs are typically laid directly on the ground, but remain concealed under logs or in a rock crevice. Despite the birds being familiar to people, especially near garbage dumps of tropical areas, few can say that they have actually seen the nest or the chicks of Black Vultures. This is because nesting takes place in dense woodlands, and nests are hidden under logs or in shallow cavities.

Black Vultures are conspicuous birds as they soar above forests and open fields in search of carrion. They scan for aggregations of other vultures that reveal the location of dead animals. Vultures are specialized to consume and digest rotten meat, and they feed their young with half-digested, regurgitated morsels, instead of carrying pieces of the carrion to their nest.

Actual size

The egg of the Black Vulture is creamy white to gray blue in background color, with dark to reddish brown blotching, and measures 3 x 2 in (76 x 51 mm) in size.

ORDER	Accipitriformes
FAMILY	Cathartidae
BREEDING RANGE	Temperate North America, Central and South America, the Caribbean
BREEDING HABITAT	Forests, shrubland, mixed open fields, urban areas
NEST TYPE AND PLACEMENT	Bare ground, inside crevices or caves
CONSERVATION STATUS	Least concern
COLLECTION NO.	FMNH 6171

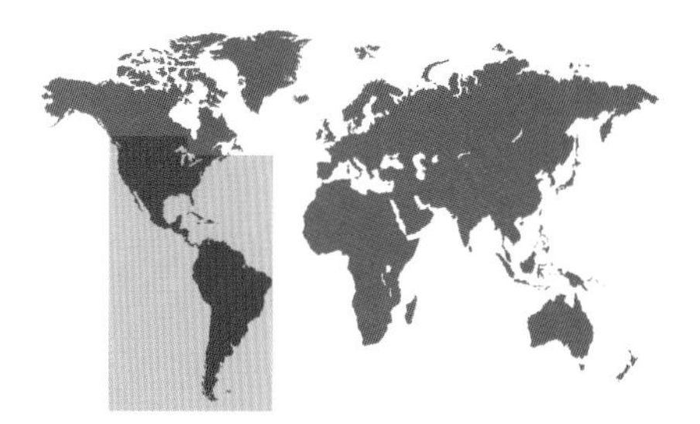

ADULT BIRD SIZE
25–32 in (64–81 cm)

INCUBATION
28–40 days

CLUTCH SIZE
2 eggs

CATHARTES AURA

TURKEY VULTURE

ACCIPITRIFORMES

The Turkey Vulture is the most common and widespread carrion-eater in the New World—its population size is estimated at 4.5 million across the Americas. It finds carcasses to eat using its keen sense of smell. A social and gregarious species, it is often seen around road-kills, and at roosting sites throughout the day. By contrast, nesting is a solitary affair, with single pairs occupying a cave, crevice, or hollows.

Clutch

Whereas few predators attack the adult Turkey Vulture, the eggs and the young are vulnerable to other birds and mammals. To protect themselves, the hatchling vultures hiss like snakes and regurgitate acidic, half-digested meat to spew around the nest and directly onto the intruder. The nestlings are also attended by both parents, and, in response to danger, the large and formidable adults put up their own fight to deter or mislead potential predators.

The egg of the Turkey Vulture is cream white tinged with gray in the background color; it has purple to brown maculation, and is 2¾ x 1⅞ in (70 x 48 mm) in size.

Actual size

ORDER	Accipitriformes
FAMILY	Cathartidae
BREEDING RANGE	Reintroduced populations in southwestern United States
BREEDING HABITAT	From scrubby chaparral to forested montane areas
NEST TYPE AND PLACEMENT	Platform of loose debris on cliff edges
CONSERVATION STATUS	Critically endangered
COLLECTION NO.	FMNH 15033

ADULT BIRD SIZE
46–53 in (117–134 cm)

INCUBATION
53–60 days

CLUTCH SIZE
1 egg

GYMNOGYPS CALIFORNIANUS

CALIFORNIA CONDOR

ACCIPITRIFORMES

Male California Condors become reproductively mature at six years of age. In the wild, the male initiates mating with a display flight. In captivity, pairing is prescribed by managers, and such pair bonds often fail to be established. Once mated, a female lays just one egg per nesting attempt. If successful, she will skip a full breeding season before nesting again two years later. This is because incubation lasts for two months, the chick takes six to seven months to fledge, and the juvenile remains with the parents well into its second year of life, before becoming independent.

This well-known species is a flagship example of conservation efforts using captive breeding programs. Historically widespread throughout much of the south and west of North America, the species became all but extinct in the wild. The last remaining wild birds were brought into captivity where breeding efforts have built up the population enough to allow reintroduction in several parts of the former distribution.

Actual size

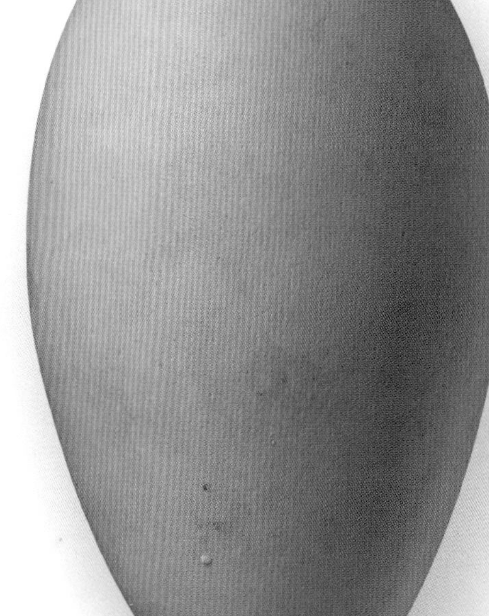

The egg of the California Condor is pale blue green in background color, clear of speckling, and measures 4¼ x 2⅝ in (108 x 66 mm) in dimensions. Unassisted by the parents, it takes several days for the hatchling to emerge completely from the egg.

ORDER	Accipitriformes
FAMILY	Accipitridae
BREEDING RANGE	All continents, except for South America and Antarctica
BREEDING HABITAT	In areas near any type of open body of water, from oceans to rivers and ponds
NEST TYPE AND PLACEMENT	Large platform on trees, stumps, cliffs, and power-poles, built of sticks and lined with bark, sod, grasses, vines, and algae
CONSERVATION STATUS	Least concern
COLLECTION NO.	FMNH 7039

PANDION HALIAETUS

OSPREY

ACCIPITRIFORMES

ADULT BIRD SIZE
19½–25½ in (50–65 cm)

INCUBATION
36–42 days

CLUTCH SIZE
1–4 eggs

Adult Ospreys are reproductively mature at three years of age, but when tall nesting structures are sparse and already occupied, individuals may wait until five to six years of age to begin nesting. The pair builds or refurbishes the nest together, but the initial nesting site is chosen by the males, before the females return from migration. It is easy to spot an unmated male with a suitable nesting site: he displays to attract nearby females through a sky-dance, as he flies high and plunges below with a fish or a branch held tightly in the talons. Nests built by younger birds are smaller, but with years of reuse and repair, these nests grow in size, becoming safer for both eggs and nestlings.

Ospreys breed throughout most continents in the world. They move about in small flocks in the winter, and form loose nesting aggregations in the breeding season. By nesting around other Ospreys, pairs benefit by being able to observe the flight direction of neighboring pairs and so assess where the best fishing areas are.

Clutch

The egg of the Osprey is creamy to pinkish cinnamon in background color, with reddish brown maculation, and measures 2⅓ x 1¾ in (60 x 45 mm).

Actual size

ORDER	Accipitriformes
FAMILY	Accipitridae
BREEDING RANGE	Temperate Europe and western Asia
BREEDING HABITAT	Deciduous and mixed forests
NEST TYPE AND PLACEMENT	Platform of twigs and green, leafy branches, also lined with leaves, high up on a tree
CONSERVATION STATUS	Least concern
COLLECTION NO.	FMNH 21391

ADULT BIRD SIZE
20½–23½ in (52–60cm)

INCUBATION
30–35 days

CLUTCH SIZE
2 eggs

PERNIS APIVORUS

EUROPEAN HONEY-BUZZARD

ACCIPITRIFORMES

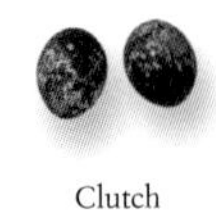

Clutch

During the breeding season, the male protects his territory from other males, and attracts a female by conspicuous flights above the tree canopy, complete with snapping sounds produced by his wings. The Honey-Buzzard is particularly sensitive to human disturbances near the nest, and will readily abandon its eggs midway through incubation. Both parents feed the chicks in the nest, occasionally delivering not only a cropful of wasps, but pieces of honeycomb filled with bee larvae for the young to feed on.

This bird of prey is a specialist feeding on stinging insects: wasps, hornets, and bees. With its elongated beak and feather-covered feet it breaks into wasp nests, while receiving few stings on exposed skin. Compared to other birds of prey, the species has relatively weak defenses, so to protect itself from predatory goshawks, it is thought that the plumage of the juvenile Honey-Buzzard has evolved to mimic that of the juvenile Common Buzzard, which is stronger and better able to defend itself.

Actual size

The egg of the European Honey-Buzzard is cream in background color with different shades of large brown spotting. It is 2 x 1⅝ in (50 x 41 mm) in size.

ORDER	Accipitriformes
FAMILY	Accipitridae
BREEDING RANGE	Southeastern North America, Central and South America
BREEDING HABITAT	Woodlands, riparian corridors, and forested wetlands
NEST TYPE AND PLACEMENT	Platform of branches and twigs, lined with mosses, high up in trees
CONSERVATION STATUS	Least concern
COLLECTION NO.	FMNH 15593

ADULT BIRD SIZE
20½–25 in (50–64 cm)

INCUBATION
28 days

CLUTCH SIZE
1–2 eggs

ELANOIDES FORFICATUS

SWALLOW-TAILED KITE

ACCIPITRIFORMES

These migratory birds return annually to their northern breeding grounds where they form loose colonies of five to ten nesting pairs settling near one another. Nest building and incubation are carried out by both parents, and they can be seen changing guard at the nest both day and night. But the female alone broods and protects the young chicks. Meanwhile the male delivers a staple of large flying insects and occasional snakes or birds to the female, who then provides meals in portions to each of the chicks.

In some Central American populations, each nest produces just one chick; this is because the earlier-hatching and larger egg yields an older nest mate which acts aggressively toward the younger chick, until that one perishes. The immediate cause of such siblicide is unknown, but it does not appear to be hunger, because even well-fed older siblings attack their younger nest mates.

Clutch

The egg of the Swallow-tailed Kite is dull white in background color with reddish to burgundy maculation, and it is 1⅞ x 1 7/16 in (47 x 37 mm) in dimensions. The eggs and young chicks, in the exposed nest in a tree top, are vulnerable to predation by other birds of prey and arboreal snakes.

Actual size

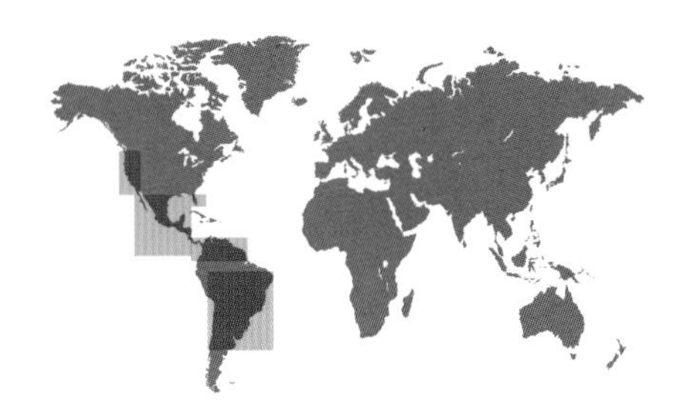

ORDER	Accipitriformes
FAMILY	Accipitridae
BREEDING RANGE	Western and southern North America, Central and South America
BREEDING HABITAT	Grasslands, savanna, open woodlands, and wetlands
NEST TYPE AND PLACEMENT	Platform nest in trees, lined with grass and leaves
CONSERVATION STATUS	Least concern
COLLECTION NO.	FMNH 15589

ADULT BIRD SIZE
12½–15 in (32–38 cm)

INCUBATION
30–32 days

CLUTCH SIZE
4 eggs

ELANUS LEUCURUS

WHITE-TAILED KITE

ACCIPITRIFORMES

Clutch

The egg of the White-tailed Kite is dull white in color, with dark brown maculation, and it measures 1⅔ x 1⅓ in (43 x 33 mm) in size.

During the breeding season, several weeks before the eggs are laid, pairs begin the prolonged task of nest building. First the sticks of the nest are collected and aligned, and then the nest cup is padded with weeds and leaves. While nesting, the pairs maintain an all-purpose territory for feeding and breeding against other conspecifics. Egg collecting in the early part of the twentieth century contributed to severe declines in the species' numbers, but now these birds are becoming more common in many areas. Current conservation efforts focus on providing safe habitats for breeding and undeveloped open habitats for feeding.

White-tailed Kites are specialist hunters of open grasslands and marshes, catching their mammalian and insect prey by hovering above the ground and plunging to strike. During the non-breeding season, these kites can be seen in loose social groups of 10 to 50 individuals. By day they disperse to hunt in individually defended territories. At night they come together to roost communally in trees.

Actual size

ORDER	Accipitriformes
FAMILY	Accipitridae
BREEDING RANGE	Southern North America and the Caribbean, Central and South America
BREEDING HABITAT	Wetlands, marshes
NEST TYPE AND PLACEMENT	Nest in swamp vegetation, on bushes or reed stumps
CONSERVATION STATUS	Least concern; Florida subspecies is endangered
COLLECTION NO.	FMNH 15034

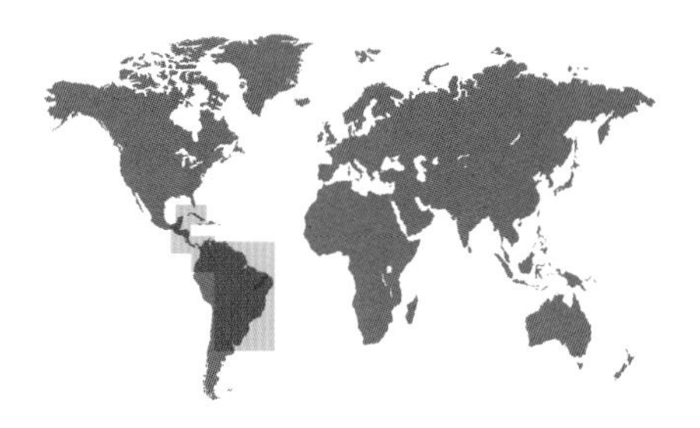

ADULT BIRD SIZE
14–15½ in (36–39 cm)

INCUBATION
26–28 days

CLUTCH SIZE
2–4 eggs

ROSTRHAMUS SOCIABILIS

SNAIL KITE

ACCIPITRIFORMES

This species is an extreme foraging specialist; it preys predominantly on aquatic snails, and occasionally on insects or small rodents. Its sharply curved beak is a clear adaptation to remove snails efficiently from their shells. The nestlings require more than the usual intensity of parental care; feeding involves not only bringing the snails to the nest, but also extracting them from their shells before they can be consumed. With wetland habitat transformation and losses, both snails and kites have declined in their North American ranges.

As a wetland-breeding species, these birds nest in loose colonies, both with conspecifics and with other marsh-breeding bird species, including herons and egrets. Mixed-species colonies are beneficial in not only providing safety in numbers, but also increasing overall vigilance, as different species have different activity patterns and can alert colony members to danger at different times of the day and night.

Clutch

The egg of the Snail Kite is buff in coloration with brown maculation, and is 1¾ x 1⅜ in (45 x 36 mm) in dimensions. When laid in marsh vegetation, the eggs are often lost due to sudden natural or managed changes in water levels.

Actual size

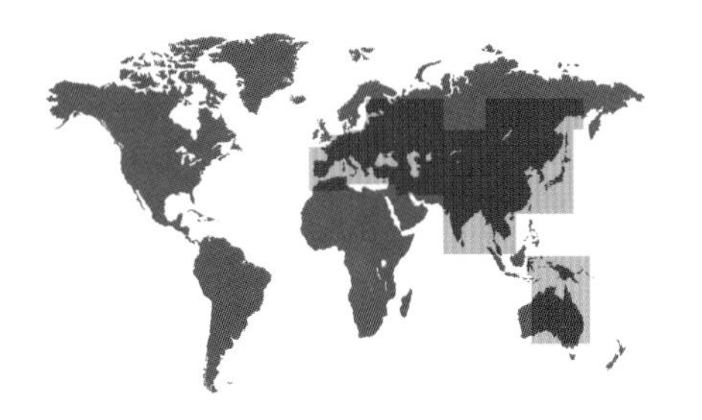

ORDER	Accipitriformes
FAMILY	Accipitridae
BREEDING RANGE	Temperate Eurasia, North Africa, and Australasia
BREEDING HABITAT	Open and mixed-forest areas, wetlands; also urban regions
NEST TYPE AND PLACEMENT	Roughly constructed platform of twigs in a tree
CONSERVATION STATUS	Least concern
COLLECTION NO.	FMNH 14766

ADULT BIRD SIZE
22–24 in (55–60 cm)

INCUBATION
30–34 days

CLUTCH SIZE
2–3 eggs

MILVUS MIGRANS

BLACK KITE

ACCIPITRIFORMES

Clutch

The breeding effort of Black Kites is shared between members of the pair, and, as is typical for other raptors, the smaller male is in charge of delivering food to the female and the chicks. The female remains on the nest, brooding the chicks for warmth and protecting them from predators, including other, cannibalistic Black Kites. An opportunist throughout its range, the Black Kite will prey on a large number of different food sources, from carrion to nestlings, and from urban garbage to insects and mammals fleeing wildfires.

The Black Kite's opportunistic foraging behavior also translates into flexibility in breeding ranges and nest site choices. In India, for example, this species is most commonly seen in large urban areas, soaring above busy streets and parks in search of food, or roosting in large flocks. When this kite breeds in cities, its presence and large numbers often become a nuisance. In other areas it is a bird of open brushland.

Actual size

The egg of the Black Kite is cream in background color with dark and light brown maculation, and measures 2 x 1½ in (51 x 39 mm) in dimensions. The safety of the eggs is often increased by nesting near other kites, as well as bustling colonies of herons and cormorants.

ORDER	Accipitriformes
FAMILY	Accipitridae
BREEDING RANGE	Central Asia
BREEDING HABITAT	Seacoast, river and lake shores
NEST TYPE AND PLACEMENT	Large intricate nests in tree tops, lined with sticks, grasses, hay, and green leaves
CONSERVATION STATUS	Vulnerable
COLLECTION NO.	FMNH 1262

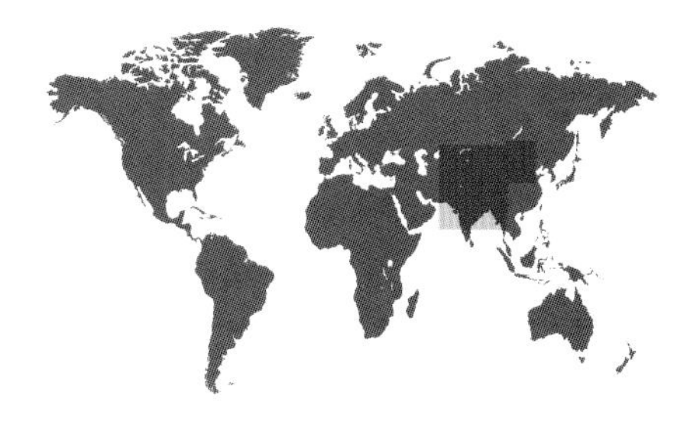

ADULT BIRD SIZE
28½–33 in (72–84 cm)

INCUBATION
40 days

CLUTCH SIZE
1–3 eggs

HALIAEETUS LEUCORYPHUS

PALLAS'S SEA-EAGLE

ACCIPITRIFORMES

The nesting and the parental care duties are undertaken by both sexes of Pallas's Sea-Eagle; together they build a bulky platform-nest, and they take turns in incubating the eggs and then feeding the young. Typically a canopy nester, occasionally the eggs are laid in a ground nest in the more arid, treeless regions of this species' wide distribution. Providing enough food for the growing chicks is a tall task; the last-laid egg yields the last-hatched chick, which typically starves due to competition for limited provisions with its older and larger nest mates.

Clutch

This large sea-eagle has a broad distribution, but modifications of its habitat and competition with people for its preferred prey of freshwater fish have led to the declines of many local populations, and of the species overall. In India, for example, invasive water hyacinths obscure the open waters in many ponds and rivers, which prevents these eagles from spotting fish at the water's surface.

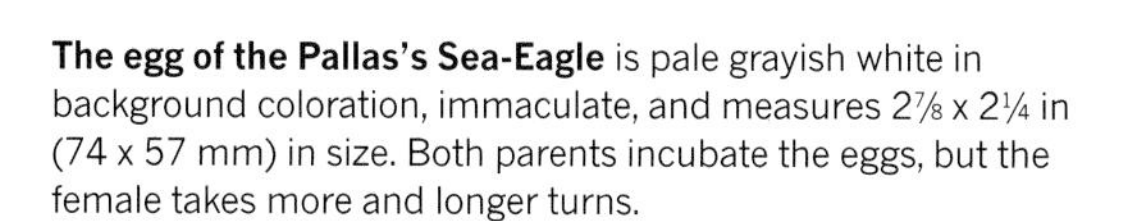

The egg of the Pallas's Sea-Eagle is pale grayish white in background coloration, immaculate, and measures 2⅞ x 2¼ in (74 x 57 mm) in size. Both parents incubate the eggs, but the female takes more and longer turns.

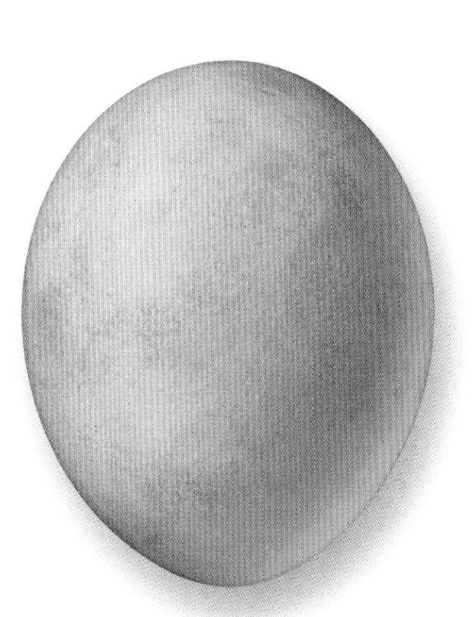

Actual size

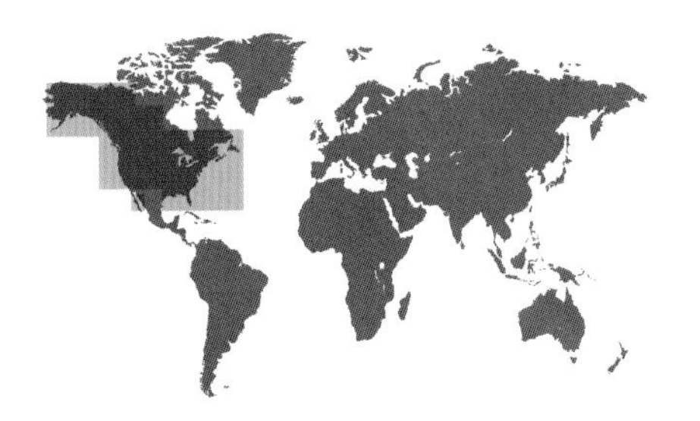

ORDER	Accipitriformes
FAMILY	Accipitridae
BREEDING RANGE	Temperate and Arctic North America
BREEDING HABITAT	Mature forests, especially along lakes and rivers, and the seacoast
NEST TYPE AND PLACEMENT	Large stick platform and a lined cup, in tall trees or on cliffs
CONSERVATION STATUS	Least concern; threatened status in continental United States
COLLECTION NO.	FMNH 6801

ADULT BIRD SIZE
28–38 in (71–96 cm)

INCUBATION
34–36 days

CLUTCH SIZE
1–3 eggs

HALIAEETUS LEUCOCEPHALUS

BALD EAGLE

ACCIPITRIFORMES

Clutch

Notable changes in both plumage and ranging behaviors mark the onset of the reproductive stage of Bald Eagles. At around the age of four to five years, individuals acquire the well-known white-headed and white-tailed plumage, and return to their natal region to establish a territory and find a mate. Once mated, members of the pair remain together across years, unless they repeatedly fail to fledge a brood or the mate dies. These cases result in divorce and local dispersal, or re-pairing and remaining at the original pair's territory.

The Bald Eagle's massive nest, called an aerie, is the result of a prolonged and intensive construction process, lasting several weeks within each season and the nest typically survives many years of use, repair, and reuse. If a nest falls due to winds and storms, these eagles will typically rebuild it in the same or a nearby tree inside their breeding territory.

Actual size

The egg of the Bald Eagle is dull white in background color, immaculate, and 2¾ x 2³⁄₁₆ in (70 x 55 mm) in dimensions. During incubation, the nest linings, including pine needles, often become moist or wet, and stain the eggs, which become increasingly discolored as the hatching date nears.

ORDER	Accipitriformes
FAMILY	Accipitridae
BREEDING RANGE	Southwestern Europe, northern and equatorial Africa, the Middle East, and south Asia
BREEDING HABITAT	Dry plains and lowland forests; often near human settlements, including large cities
NEST TYPE AND PLACEMENT	Cliffs or buildings, made of sticks, leaves, and garbage
CONSERVATION STATUS	Endangered; despite the large distribution, local populations have suffered 30–75 percent declines in recent decades
COLLECTION NO.	FMNH 1255

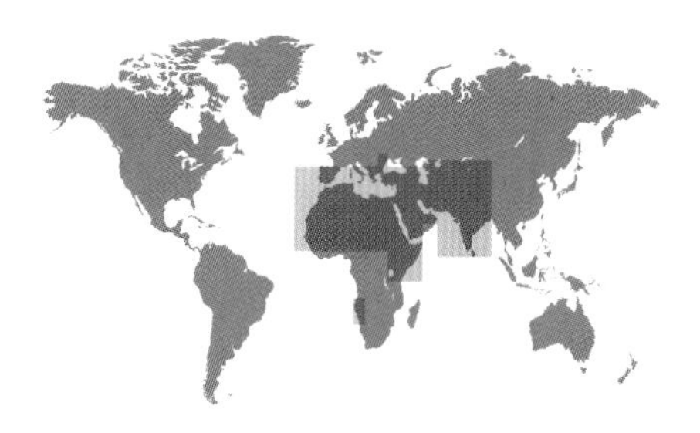

ADULT BIRD SIZE
18½–27½ in (47–70 cm)

INCUBATION
42 days

CLUTCH SIZE
2 eggs

NEOPHRON PERCNOPTERUS

EGYPTIAN VULTURE

ACCIPITRIFORMES

Egyptian vultures are typically seen alone or in pairs during the day, but aggregate in groups at night for roosting in trees or on building roofs. Breeding takes place in a haphazardly constructed stick nest, or in a previously used but refurbished eagle's nest. Occasionally, females mate with two males, both of whom may attend and help raise the young in a single nest. With the decline of freely available mammalian carrion, due to declining large wildlife stocks as well as modern practices of swiftly removing carcasses, these vultures have suffered severe population losses.

Clutch

The Egyptian Vulture is one of the most resourceful species of birds, making innovative use of diverse materials available to them in nature, and especially around human settlements. For example, these vultures use small rocks to drop and smash against the thick shells of ostrich eggs to access the nutrient- and energy-rich contents. Birds have been seen holding sticks in their beaks to rake up wool left behind from sheep shearing. They carry the wool to their nests where it provides a soft lining.

The egg of the Egyptian Vulture is red clay in background color with red, brown, and black spotting, and measures 2½ x 2⅛ in (65 x 55 mm) in size. Both sexes incubate the eggs and provision the young.

Actual size

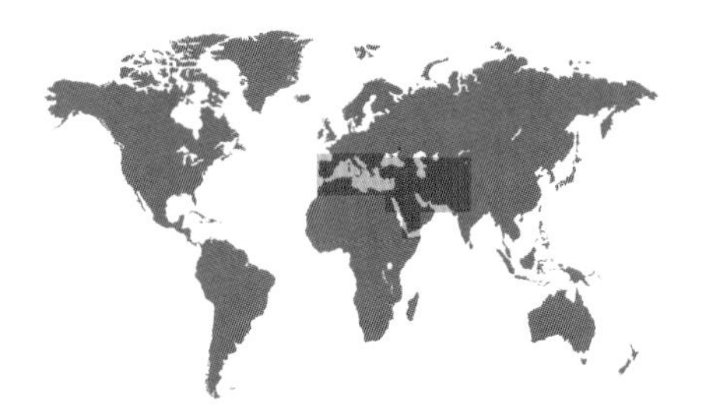

ORDER	Accipitriformes
FAMILY	Accipitridae
BREEDING RANGE	Mediterranean Europe, north Africa, and southwestern Asia
BREEDING HABITAT	Open montane fields, with steep cliffs
NEST TYPE AND PLACEMENT	Ground nest on rock or cliff ledges
CONSERVATION STATUS	Least concern
COLLECTION NO.	FMNH 1315

ADULT BIRD SIZE
37½–41½ in (95–105 cm)

INCUBATION
50–58 days

CLUTCH SIZE
1 egg

GYPS FULVUS

EURASIAN GRIFFON

ACCIPITRIFORMES

The egg of the Eurasian Griffon is a dull white in background color, has reddish brown maculation, and measures 3⅝ x 2¾ in (91 x 71 mm) in dimensions. The hatching success of eggs in the wild can be as low as 35 percent, but in captive programs for conservation projects, around 90 percent of artificially incubated eggs can be hatched.

Nesting in colonies of 15 to 20 pairs, Griffon mates share parental duties equitably; they build the nest together, and both incubate the eggs and provision the young. In areas with few cliffs or caves, these large birds easily usurp the nests of other smaller raptors to lay their eggs and raise their young. This large and powerful vulture is a specialist on carrion and never catches or consumes live prey at all. Being a carrion specialist in heavily populated regions poses an indirect health hazard for both adults and the growing vulture chick: namely, the accumulation in their bodies of toxic heavy metals, especially lead ingested from the carcasses of shot animals.

Despite protective measures around its habitat and, especially, its nesting sites throughout southern Europe, governmental decisions to remove all cattle and sheep carcasses from fields to reduce the spread of mad-cow disease has had a devastating effect on the food sources for these birds. Today, vulture "cafes," where carrion is regularly provided by conservation managers near Griffon breeding colonies, provide an artificial way to reduce food scarcity for this species.

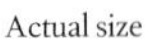
Actual size

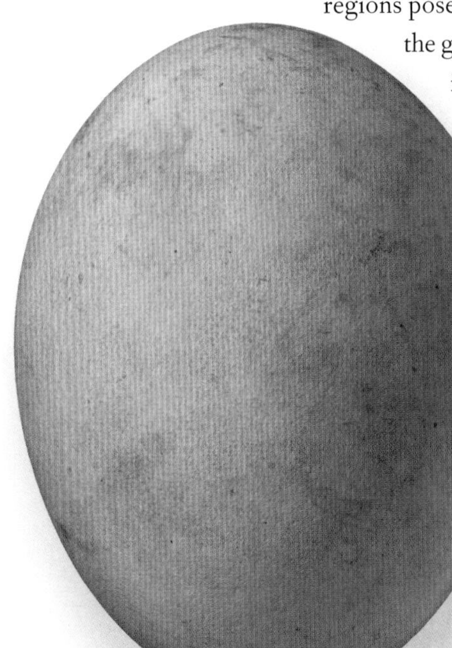

ORDER	Accipitriformes
FAMILY	Accipitridae
BREEDING RANGE	Montane areas of Europe, a central band in Asia from west to east
BREEDING HABITAT	Upland forests, open steppe, and grazing areas
NEST TYPE AND PLACEMENT	Platform nest in trees or on cliff ledges
CONSERVATION STATUS	Near threatened
COLLECTION NO.	FMNH 1250

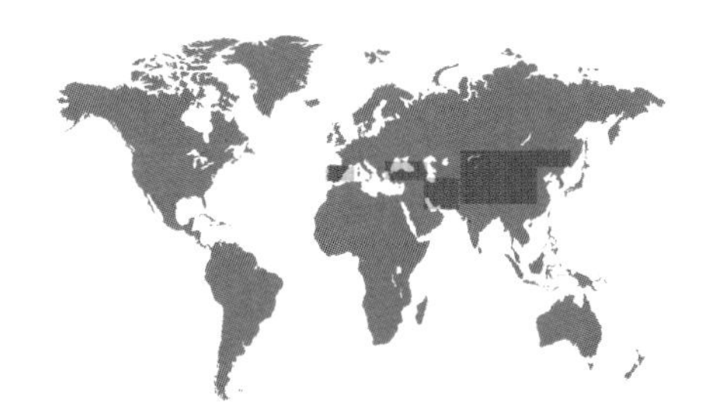

ADULT BIRD SIZE
38½–47 in (98–120 cm)

INCUBATION
50–56 days

CLUTCH SIZE
1 egg

AEGYPIUS MONACHUS

CINEREOUS VULTURE

ACCIPITRIFORMES

The Cinereous Vulture appears to have a low reproductive rate, laying one or rarely two eggs in the nest. However, the fledging success is substantial, at over 50 percent. It takes over half a year from laying the egg to fledging the young, and requires the full cooperation of both parents all the way through. The demise of the availability of large-bodied mammal carcasses, and other changes to high-elevation montane habitats, have raised conservation concerns over these vultures. Adult numbers are decreasing, and the species now requires captive breeding and translocation to reestablish locally extinct populations.

The Cinereous Vulture is the largest bird of prey living in the Old World, only outsized by the condors of the New World. Biologists have long recognized that among widespread groups of related species, the largest species are frequently those that live furthest from the equator (called Bergmann's rule). The Cinereous Vulture fits this pattern, being both the largest Old World and the furthest-north-occurring vulture.

The egg of the Cinereous Vulture is off-white in background color with reddish lilac maculation; it measures 3½ x 2¾ in (90 x 70 mm) in size. This species boasts a nearly 90 percent hatching success of the egg, which is among the highest in all birds.

Actual size

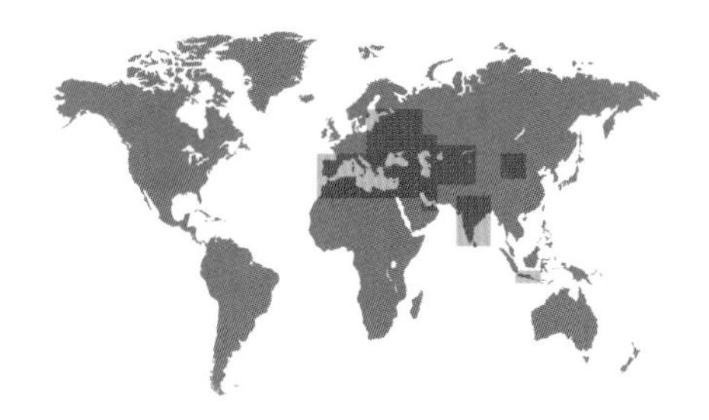

ORDER	Accipitriformes
FAMILY	Accipitridae
BREEDING RANGE	Temperate Europe, and central-western and south Asia
BREEDING HABITAT	Open cultivated plains, arid deciduous scrub, and semi-deserts
NEST TYPE AND PLACEMENT	Stick platform in trees, with a leaf-lined cup
CONSERVATION STATUS	Least concern
COLLECTION NO.	FMNH 20866

ADULT BIRD SIZE
24–27 in (61–68 cm)

INCUBATION
45–47 days

CLUTCH SIZE
1 egg

CIRCAETUS GALLICUS

SHORT-TOED SNAKE EAGLE

ACCIPITRIFORMES

Breeding is a slow and long-term affair for the Short-toed Snake Eagle; it takes five years for the young to become mature and nest on their own. Although pairs typically return to the same forested area year after year, they do not always reuse the nest built in the previous year. Once the chick hatches, it requires both parents to attend and provision it until it becomes fully flighted more than two months later.

This species is a specialist hunter, feeding mostly on reptiles, but also on small mammals, birds, and large insects. When carrying a large snake that is still alive, the air-borne bird may find itself entangled and fall to the ground until it can finally finish off its prey with a strong bite. Modern agricultural practices, and the reduction in numbers of reptiles in general, have resulted in the loss of this top predator's preferred foods, and population sizes have declined.

The egg of the Short-toed Snake Eagle is white in background color, immaculate, and measures 2¾ x 2¼ in (71 x 57 mm) in size. The female alone incubates the egg, while the male nourishes her with snakes and lizards.

Actual size

ORDER	Accipitriformes
FAMILY	Accipitridae
BREEDING RANGE	Temperate Europe, north Africa, and central Asia
BREEDING HABITAT	Swamps and other wetlands
NEST TYPE AND PLACEMENT	Ground nest in reed beds, comprised of sticks, stems, and grass
CONSERVATION STATUS	Least concern
COLLECTION NO.	FMNH 2454

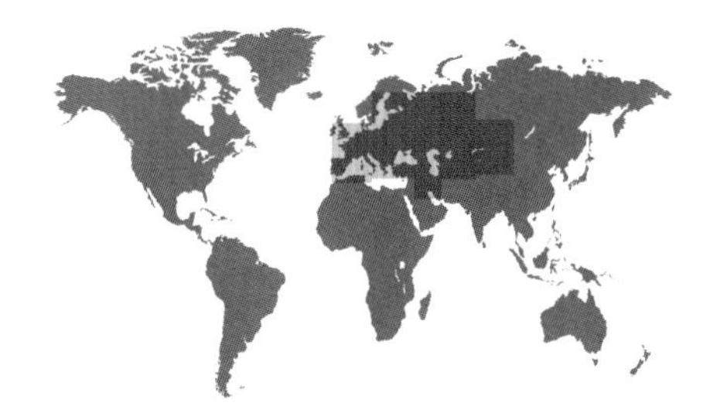

ADULT BIRD SIZE
16½–22 in (42–56 cm)

INCUBATION
31–38 days

CLUTCH SIZE
3–8 eggs

CIRCUS AERUGINOSUS

EURASIAN MARSH-HARRIER

ACCIPITRIFORMES

The Eurasian Marsh-Harrier is a common inhabitant of wetlands, often seen flying low over open water and reed beds. It preys on diverse aquatic life such as fish, frogs, and waterbirds, including sick shorebirds and helpless nestlings, leisurely carrying the captured prey in its talons to a dry spot to process and consume it. During the breeding season most, if not all, of the food supplied to the female and the chicks is provided by the male who meets the female near their nest and may transfer the prey to her in midair.

When breeding, these birds are territorial in that the breeding pairs keep other marsh-harriers away from the immediate proximity of their own nests. At other times, they form loose foraging and roosting flocks, patrolling marshlands and spending the night together in groups of 5 to 20 individuals.

Clutch

The egg of the Eurasian Marsh-Harrier is white, with a blue to green tint in background color, may be speckled with dark spots, and measures 2 x 1⅝ in (51 x 40 mm) in size. The eggs are laid at intervals of 2–5 days, and the chicks hatch on a series of different days.

Actual size

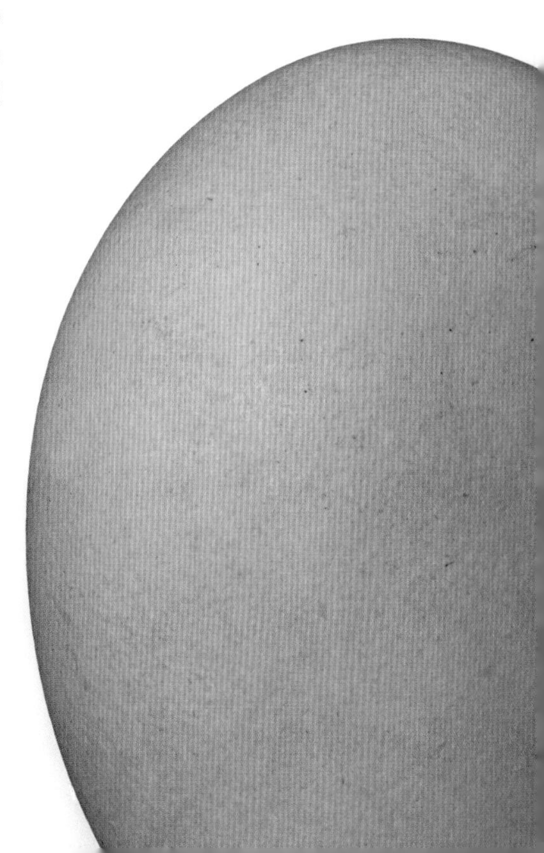

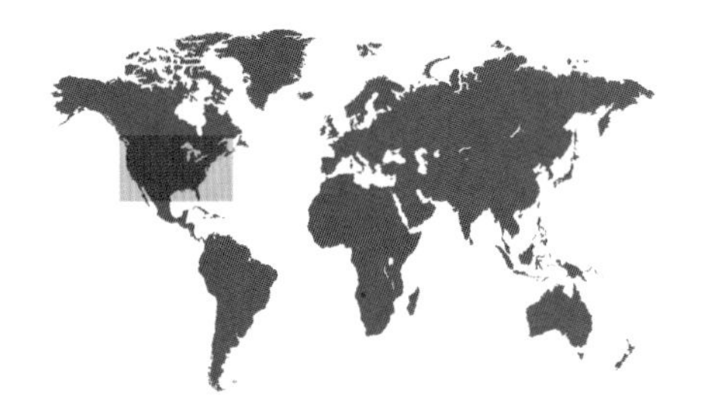

ORDER	Accipitriformes
FAMILY	Accipitridae
BREEDING RANGE	Temperate North America
BREEDING HABITAT	Forests, and wooded parklands, including suburbs and cities
NEST TYPE AND PLACEMENT	Nest made of sticks and branches, high in the canopy in dense woodlands
CONSERVATION STATUS	Least concern
COLLECTION NO.	FMNH 6294

ADULT BIRD SIZE
14½–15½ in (37–39 cm)

INCUBATION
30–36 days

CLUTCH SIZE
3–4 eggs

ACCIPITER COOPERII

COOPER'S HAWK

ACCIPITRIFORMES

Clutch

The breeding territory of the Cooper's Hawk is established by the male, who then defends it from competitors and intruders. He also attracts a female by engaging in courtship flights over the forest canopy. Once a pair bond has formed, the male maintains it through courtship feeding, handing over to the female small prey items plucked clean of feathers or fur. The nest is also constructed mostly by the male, typically by repairing an existing nest or building in a tree near the old nest. Then, it is the female's turn to be in charge: she incubates the eggs and tears prey items into smaller morsels to feed the newly hatched young.

This hawk is a specialized bird-hunter. It typically leaves small songbirds alone and instead focuses on larger, woodpecker- to pigeon-sized prey, which it catches in midair or plucks off tree branches in the forest. Cooper's Hawks have become a common species in some suburbs and cities where they often hunt birds at birdfeeders.

Actual size

The egg of the Cooper's Hawk is pale blue to bluish white in background color, and measures 1⅞ x 1½ in (47 x 37 mm) in size. To keep the nest and the eggs clean, the female departs briefly and defecates at a nearby perch before resuming her incubation duties.

ORDER	Accipitriformes
FAMILY	Accipitridae
BREEDING RANGE	North and Central America and the Caribbean
BREEDING HABITAT	Dense forests with closed canopy
NEST TYPE AND PLACEMENT	Platform nest of twigs and branches, near the top of a conifer tree
CONSERVATION STATUS	Least concern
COLLECTION NO.	FMNH 15609

ACCIPITER STRIATUS

SHARP-SHINNED HAWK

ACCIPITRIFORMES

ADULT BIRD SIZE
9½–13½ in (24–34 cm)

INCUBATION
30–35 days

CLUTCH SIZE
3–8 eggs

Both members of Sharp-Shinned Hawk pairs collect nesting materials, including branches, twigs, and bark, which are then arranged by the female to form a nest structure. As with most predatory birds, the female is about a third larger than the male; she tends to catch larger prey, which diversifies the types of creature eaten by the young. During the early hatchling stage, when the female remains with the nestlings, the male brings in mostly small birds as food. Once the chicks grow larger, the female can leave them and join her mate in bringing their young larger items.

This is the smallest hawk in North America, and it hunts for small perching birds in the forest and near wooded edges of mixed habitats. It requires dense forest stands for nesting, but during spring and fall migration, large numbers can be seen over open terrain following mountain ridges and other geological formations along a north–south direction.

Clutch

The egg of the Sharp-shinned Hawk is pale blue in background color with brown to violet maculation, and is 1½ x 1³⁄₁₆ in (38 x 30 mm) in dimensions. Without a prominent nest bowl to contain them, the eggs sometimes get knocked around and may fall off the large nest platform.

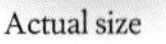

Actual size

ORDER	Accipitriformes
FAMILY	Accipitridae
BREEDING RANGE	Temperate and subarctic Eurasia and North America
BREEDING HABITAT	Deciduous or coniferous forests
NEST TYPE AND PLACEMENT	A bulky bowl of sticks lined with bark and leaves, placed high up in trees
CONSERVATION STATUS	Least concern
COLLECTION NO.	FMNH 15274

ADULT BIRD SIZE
20–25½ in (51–65 cm)

INCUBATION
39–42 days

CLUTCH SIZE
2–4 eggs

ACCIPITER GENTILIS

NORTHERN GOSHAWK

ACCIPITRIFORMES

Clutch

A large and agile predator, this species hunts not only for birds, but for many other types of prey, including squirrels and other mammals inhabiting dense woodlands. Hunting diverse prey is essential for breeding success: once the nest is completed, the female takes charge of incubating the eggs and brooding the hatchlings, while the male provides her and the next generation with a steady supply of varied meals. When two or more prey items are caught in short succession, the adults may store the prey by leaving them hidden among leaves of forking tree branches for one or two days, and then recovering and delivering the prey to the nestlings.

In traditional falconry, Goshawks are known as "cook's birds." In contrast to falcons, which typically hunt in open areas, the Goshawks' rounded wings allow them to seek larger bird and mammal prey in a variety of habitats. This made them helpful as hunting birds for cooks looking to prepare a diverse range of freshly caught game for an evening's meal.

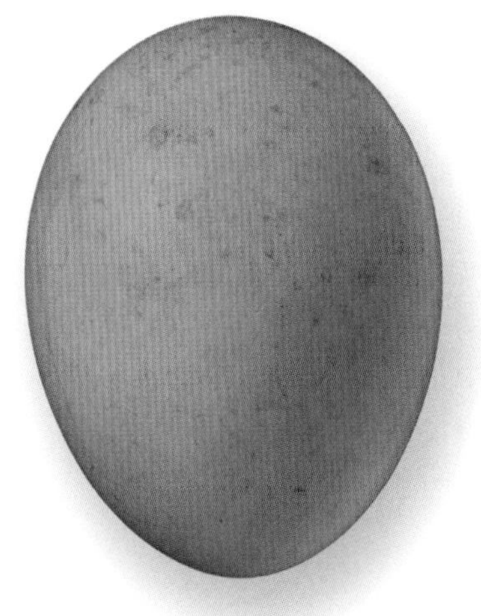

Actual size

The egg of the Northern Goshawk is bluish white in background color, and measures 2⅓ x 1¾ in (58 x 45 mm) in size. Both members of the pair vigorously defend the nest from intruders such as raccoons and other mammalian predators that can easily climb trees.

ORDER	Accipitriformes
FAMILY	Accipitridae
BREEDING RANGE	Coastal subtropical and tropical South, Central, and North America, and the Caribbean
BREEDING HABITAT	Wooded areas along rivers and coasts, including mangrove stands
NEST TYPE AND PLACEMENT	Platform nest of sticks and branches, placed in tree canopy
CONSERVATION STATUS	Least concern
COLLECTION NO.	FMNH 1310

ADULT BIRD SIZE
17–22 in (43–56 cm)

INCUBATION
38 days

CLUTCH SIZE
1–3 eggs

BUTEOGALLUS ANTHRACINUS

COMMON BLACK-HAWK

ACCIPITRIFORMES

Breeding is a coastal affair for the Common Black-Hawk; the nest is typically constructed within sight of standing or moving water. Following a series of undulating courtship flights, the pair bond is established firmly, and can last through most of the pair's lifespans. Both sexes collect nesting materials, including twigs and mistletoe branches. The pair copulates up to four times a day prior to egg laying to ensure fertilization. Unlike many raptors, both sexes partake in incubation, and they also share parental duties of brooding and feeding the hatchlings.

An adaptable species, this hawk frequents coastal areas, woodlands, parklands, and urban garbage dumps throughout its range. The opportunistic foraging behavior of Black-Hawks includes preying on land crabs near popular Caribbean beaches, and feeding these to chicks at nests in nearby tree groves.

Clutch

The egg of the Common Black-Hawk is grayish white in background color, with brown maculation, and is 2⅓ x 1¾ in (58 x 45 mm) in size. The clutch is well protected in the bulky nest structure, which is often reused and added to between breeding attempts.

Actual size

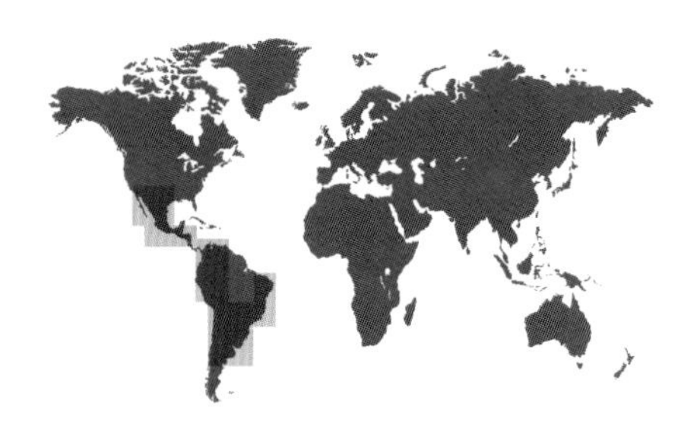

ORDER	Accipitriformes
FAMILY	Accipitridae
BREEDING RANGE	Patchy distribution throughout southern North America, Central America, and South America
BREEDING HABITAT	Desert scrub, including urban areas of desert cities, arid grassland, and wooded savanna
NEST TYPE AND PLACEMENT	Nest platform in a tree, made of sticks and twigs, and lined with soft mosses
CONSERVATION STATUS	Least concern
COLLECTION NO.	FMNH 6348

ADULT BIRD SIZE
18–23 in (46–59 cm)

INCUBATION
31–36 days

CLUTCH SIZE
2–4 eggs

PARABUTEO UNICINCTUS

HARRIS'S HAWK

ACCIPITRIFORMES

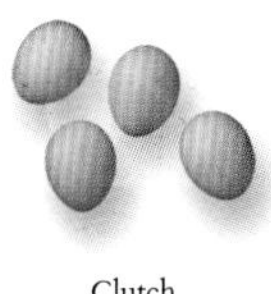

Clutch

Breeding is a social activity in this species. In some places (for example, Texas) where there are roughly equal numbers of males and females, a central breeding pair is assisted by several adult helpers during hunts; these helpers are not allowed near the eggs or nestlings, but do feed the fledglings after they have left the nest. In other areas (including Arizona) where the sex ratio is biased toward males, a central female will take on two or more males as breeding mates, and practice simultaneous polyandry where the males not only fertilize the eggs, but also assist in nesting, incubation, and provisioning the offspring.

The social life of the Harris's Hawk has several tiers of complexity. It is the only North American raptor known to engage in group hunting, with teams of two to seven birds moving across the habitat to flush and capture prey, or to continue pursuing the same prey through a relay of different hawk individuals in the group, until the prey is caught. The most dominant individual is typically the largest female in the team.

Actual size

The egg of the Harris's Hawk is white with a blue tint in background color, with buff to gray maculation, and is $2\frac{1}{8}$ x $1\frac{2}{3}$ in (53 x 42 mm) in dimensions. Females are reproductively active throughout the year, and can lay full clutches three or four times in years when prey is plentiful.

ORDER	Accipitriformes
FAMILY	Accipitridae
BREEDING RANGE	Eastern North America, also California
BREEDING HABITAT	Deciduous woodlands, riparian corridors, and forested wetlands
NEST TYPE AND PLACEMENT	Platform of sticks and branches, high up in the crotch of a tree
CONSERVATION STATUS	Least concern
COLLECTION NO.	FMNH 326

ADULT BIRD SIZE
17–24 in (43–61 cm)

INCUBATION
28–33 days

CLUTCH SIZE
2–4 eggs

BUTEO LINEATUS

RED-SHOULDERED HAWK

ACCIPITRIFORMES

Nesting is a long-term investment for this hawk, with both members of the pair returning to and refurbishing the same nest with branches and fresh leaves, year after year. One nest was recorded as being in use for 16 years in a row. To spot an active nest, it is sensible to look for signs of whitewash on the ground near tall trees; both the parents and the older chicks projectile-defecate from the nest to keep it hygienic during the long nestling period.

Typically these hawks inhabit dense forests, often near ponds or rivers, where they hunt rodents, reptiles, and amphibians. Recently, they have begun a slow but steady track toward sharing habitats with humans. They are increasingly seen in suburban wood lots, and their nests are often found in large trees in backyards and town parks. Where living near people, they become tolerant of traffic and pedestrians.

Clutch

The egg of the Red-shouldered Hawk is brown to lavender in background color with dark brown markings, and measures 2³⁄₁₆ x 1¾ in (55 x 44 mm) in size. With most nests built in the canopy, the eggs are vulnerable to predation by owls and crows.

Actual size

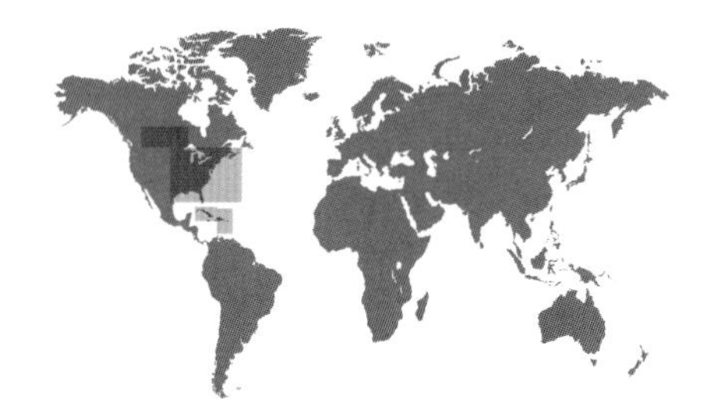

ORDER	Accipitriformes
FAMILY	Accipitridae
BREEDING RANGE	Eastern and central temperate North America, the Caribbean
BREEDING HABITAT	Large tracts of deciduous or mixed-conifer forests
NEST TYPE AND PLACEMENT	Large bowl of sticks, in a tree, lined with bark chips, often furnished with green twigs and leaves
CONSERVATION STATUS	Least concern
COLLECTION NO.	FMNH 6674

ADULT BIRD SIZE
13½–17½ in (34–44 cm)

INCUBATION
28–31 days

CLUTCH SIZE
2–3 eggs

BUTEO PLATYPTERUS

BROAD-WINGED HAWK

ACCIPITRIFORMES

Clutch

The egg of the Broad-winged Hawk is white with a bluish tint, has brownish speckles, and measures 1⅞ x 1½ in (49 x 39 mm). Young adults, both females and males, keep out of the breeding pool, with most females laying their first eggs at only two years of age.

This species is one of five raptors in North America that completely abandon their summer breeding range: it migrates in winter to tropical forests in South America. Large flocks, or "kettles," of these hawks can be seen during spring and fall migration. Satellite tracking has revealed that migrants can cover 60–120 miles (100–200 km) per day. On the wintering grounds, individuals spread out across the forests of the Amazon Basin.

When the Broad-winged Hawk returns north to breed, it keeps a low and secretive profile in its preferred habitat of dense and contiguous forests. There, a pair maintains a well-defended territory by attacking and chasing away intruder Broad-winged Hawks. The male attracts the female with an aerial and vocal display. The female alone incubates the eggs while the male provisions her and the hatchlings with small mammals and reptiles as food. They rarely reuse their own nest from previous years; instead they often renovate an existing crow or hawk nest to provide a fresh base into which to lay the eggs.

Actual size

ORDER	Accipitriformes
FAMILY	Accipitridae
BREEDING RANGE	North, central, and western North America
BREEDING HABITAT	Open grasslands and prairies with sparse groves of trees and riparian corridors
NEST TYPE AND PLACEMENT	Stick platform, with a bowl lined with leaves, weeds, mosses, and wool, placed in a solitary tree or in small groves
CONSERVATION STATUS	Least concern
COLLECTION NO.	FMNH 6616

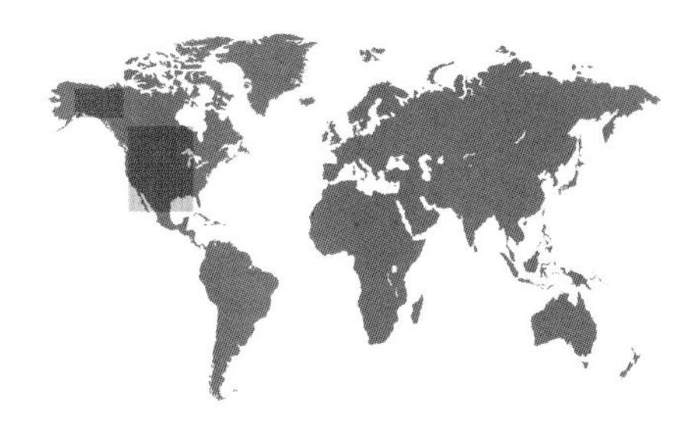

ADULT BIRD SIZE
19–20 in (48–51 cm)

INCUBATION
34–35 days

CLUTCH SIZE
2–3 eggs

BUTEO SWAINSONI

SWAINSON'S HAWK

ACCIPITRIFORMES

The Swainson's Hawk is highly gregarious, spending much of the year in large groups, migrating conspicuously to and from their wintering grounds in the pampas of South America. When returning to the breeding grounds, many individuals, especially younger birds, opt not to breed and instead remain in flocks of dozens to hundreds of individuals, to soar and roost together in open areas.

Clutch

By contrast, nesting pairs become territorial and defend a breeding site aggressively against conspecifics. When suitable habitat is limited, territory defense may be restricted to the immediate vicinity of the nest, with the birds flying dozens of miles away in search of prey-rich hunting grounds. Most Swainson's Hawks breed as monogamous pairs, but a small percentage of individuals consistently engage in polyandry, as one female mates and cooperates with two or more males to sire and provision the young.

The egg of the Swainson's Hawk is dull white in background color, plain or with brown maculation. It measures 2¼ x 1¾ in (57 x 44 mm) in dimensions. The female alone incubates the eggs for much of the day, but is relieved briefly by the male while she feeds and stretches her wings.

Actual size

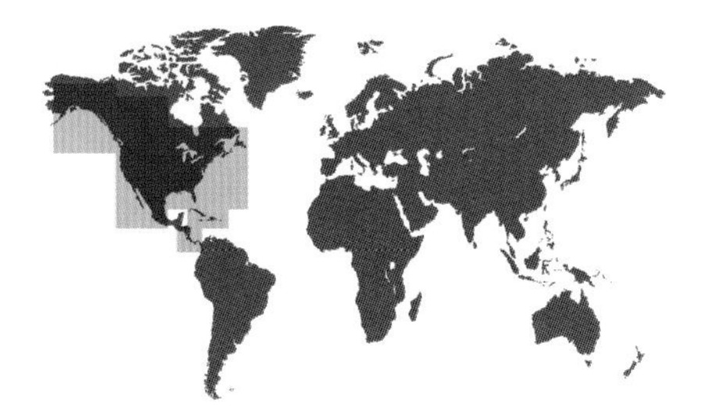

ORDER	Accipitriformes
FAMILY	Accipitridae
BREEDING RANGE	North America, Central America, and the Caribbean
BREEDING HABITAT	Open areas, including grasslands, semi-deserts, agricultural fields, and urban parklands
NEST TYPE AND PLACEMENT	Stick platform, lined with leaves and bark, placed high in the canopy, on cliffs, or on building ledges
CONSERVATION STATUS	Least concern
COLLECTION NO.	FMNH 6420

ADULT BIRD SIZE
17½–25½ in (45–65 cm)

INCUBATION
28–35 days

CLUTCH SIZE
2–3 eggs

BUTEO JAMAICENSIS

RED-TAILED HAWK

ACCIPITRIFORMES

Clutch

Red-tailed Hawks are the most common birds of prey across North America. Mated pairs return to the previous year's breeding site, and confirm their commitment to the pair bond in nuptial flights: they fly together in circles, perform high-speed chases, and tumble downward, clasping each other's feet. Instead of building a new nest, which can take up to a week, they typically refurbish the old one to promptly begin egg laying. During the breeding season, hunting by the pair as a coordinated unit can increase prey capture rates and foraging success.

Red-tailed Hawks are familiar in both rural and urban habitats, including New York City's Central Park. Their loud shrill calls can bc hcard casily cvcn whcn a bird is sitting dccp within a forest canopy; these calls are frequently used as the typical "spine-chilling wildlife call" in soundtracks of Hollywood movies, irrespective of the actual species depicted on screen. These birds have swiftly adapted to city life by switching their prey from ground-dwelling mice, voles, and others rodents, to slow-moving city squirrels and pigeons.

Actual size

The egg of the Red-tailed Hawk is buff in background color with brown or purple spotting, and it measures $2\frac{3}{16}$ x $1\frac{7}{8}$ in (56 x 47 mm) in dimensions. The female relies on her mate for food during the incubation, and mated pairs typically stay together and become more successful with each passing breeding season.

ORDER	Accipitriformes
FAMILY	Accipitridae
BREEDING RANGE	Arctic and subarctic Eurasia and North America
BREEDING HABITAT	Open tundra, grasslands near boreal forests
NEST TYPE AND PLACEMENT	Large bowl of sticks, lined with grasses and feathers, placed high on cliffs, buffs, and trees
CONSERVATION STATUS	Least concern
COLLECTION NO.	FMNH 6721

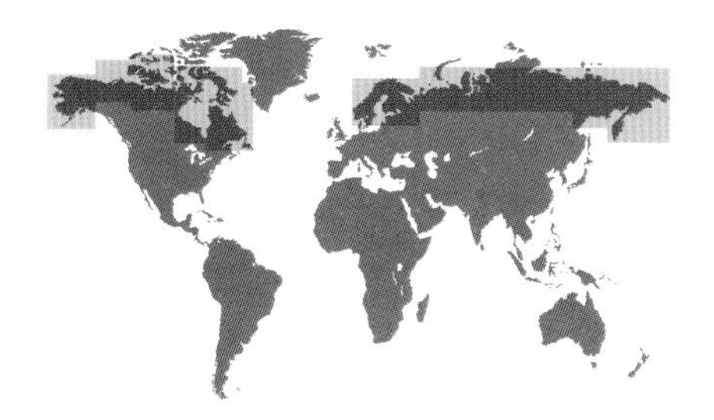

ADULT BIRD SIZE
18½–20½ in (47–52 cm)

INCUBATION
31–37 days

CLUTCH SIZE
3–5 eggs

BUTEO LAGOPUS

ROUGH-LEGGED HAWK

ACCIPITRIFORMES

Rough-legged Hawks are less aggressive toward members of their own species than other hawks are during the nesting season. When both members of the previous year's pair return to the breeding grounds, they reestablish the monogamous pair bond, and build a brand new nest together on a high cliff or rocky outcropping. The home hunting range is typically centered around the current year's nesting site and its size depends on local prey density and availability.

The Rough-legged Hawk is a specialist of open country where, similar to kestrels and kites, it is one of a handful of raptors that search for prey while hovering in midair. This species also completes a full annual migration, abandoning its Arctic and subarctic breeding sites in the winter to move to central North America and Eurasia.

Clutch

The egg of the Rough-legged Hawk is off-white in background color with brown maculation, and measures 2³⁄₁₆ x 1⅞ in (56 x 45 mm) in size. The female alone incubates the eggs, while the male feeds her and protects the vicinity of the nest from potential predators, including gyrfalcons and Great Skuas (see page 184).

Actual size

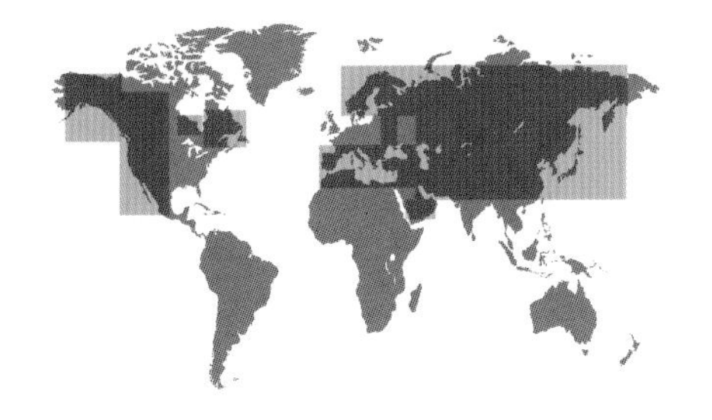

ORDER	Accipitriformes
FAMILY	Accipitridae
BREEDING RANGE	Temperate and subarctic North America, Eurasia, and North Africa
BREEDING HABITAT	Grasslands, tundra, and other mostly open habitats
NEST TYPE AND PLACEMENT	Large and bulky stick nest lined with leaves and mosses, placed in trees and on cliffs, occasionally on human structures
CONSERVATION STATUS	Least concern
COLLECTION NO.	FMNH 14933

ADULT BIRD SIZE
27½–33 in (70–84 cm)

INCUBATION
40–45 days

CLUTCH SIZE
2–3 eggs

AQUILA CHRYSAETOS

GOLDEN EAGLE

ACCIPITRIFORMES

Clutch

Golden Eagles form a long-lasting pair bond and cooperation is also essential for successful breeding. Mates maintain a vast and exclusive all-purpose territory: they both breed and forage on it, and they attack intruding eagles to keep the prey-base to themselves. The nests are vast structures, with the pair returning and reusing the previous year's nest, while continually building and expanding it year after year. These huge raptors have few predators, so there is no selective pressure to hide the nest and eggs.

The Golden Eagle is one of the largest birds of prey, occurring throughout the northern hemisphere where it feeds on larger ground-dwelling mammals and birds or seeks out carrion to consume. Occasionally, mated pairs hunt cooperatively, with one bird distracting the fleeing prey while the second bird approaches to catch it from the other side.

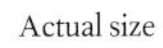

Actual size

The egg of the Golden Eagle is creamy white in background color, with cinnamon to brown marking, and 3 x 2¼ in (75 x 58 mm) in size. The high placement of the nest typically allows the incubating female to command a clear view of the breeding territory and any potential intruders in it.

ORDER	Accipitriformes
FAMILY	Accipitridae
BREEDING RANGE	Northern and southern tip of Africa, central Asia, southern and eastern Europe
BREEDING HABITAT	Forests with clearings, dry scrublands
NEST TYPE AND PLACEMENT	Bulky platform of twigs, lined with leaves, in trees, rarely on cliffs
CONSERVATION STATUS	Least concern
COLLECTION NO.	FMNH 1286

ADULT BIRD SIZE
16½–21 in (42–53 cm)

INCUBATION
37–40 days

CLUTCH SIZE
1–3 eggs

HIERAAETUS PENNATUS

BOOTED EAGLE

ACCIPITRIFORMES

During the breeding season, the female and the male Booted Eagle spend much of their time engaged in conspicuous ritualized courtship flights—they complete vertical loops of powered flight upward, then drop down in a free fall, often while locking their talons and facing each other. After the pair bond is established, they build a new, or repair an existing, nest structure, sometimes usurping the nest of another raptor. The female engages in most of the incubation while the male provides her and the young with small mammal, bird, and insect prey to feed on.

Clutch

Booted Eagles are small and stocky eagles, whose plumage is dimorphic, not between the sexes or between different geographic areas, but between two different color types occurring in overlapping ranges. Most of these eagles are dark brown, whereas a few individuals are pale gray. This type of color-hue polymorphism occurs in many birds, including other birds of prey.

The egg of the Booted Eagle is buff in background color with darker maculation; it measures 2³⁄₁₆ x 1¾ in (56 x 45 mm) in size. Despite prominent hatching asynchrony of several days, if more than one egg hatches, typically each chick survives to fledging.

Actual size

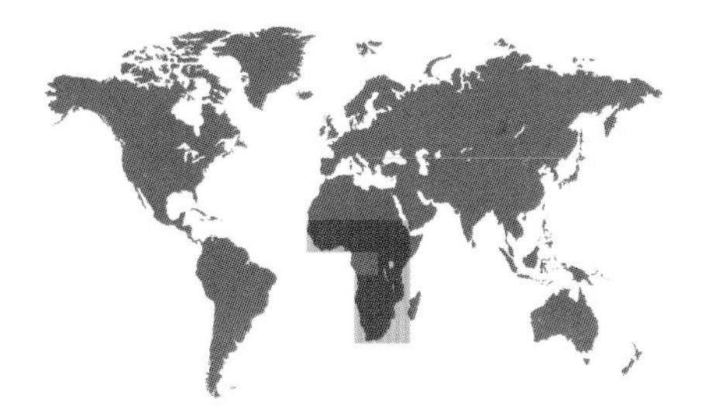

ORDER	Accipitriformes
FAMILY	Sagittariidae
BREEDING RANGE	Sub-Saharan Africa
BREEDING HABITAT	Open grasslands and mixed savanna
NEST TYPE AND PLACEMENT	Large, bulky nest of sticks, twigs, leaves, fur, and dung, placed in a tree
CONSERVATION STATUS	Vulnerable
COLLECTION NO.	FMNH 1094

ADULT BIRD SIZE
47–59 in (120–150 cm)

INCUBATION
45 days

CLUTCH SIZE
2–3 eggs

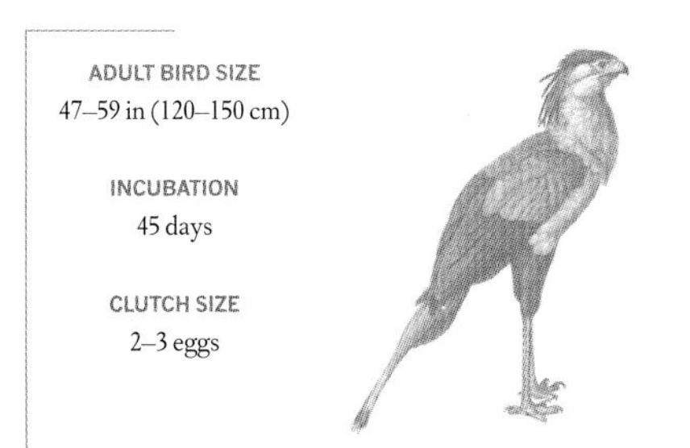

SAGITTARIUS SERPENTARIUS

SECRETARY-BIRD

ACCIPITRIFORMES

Clutch

Nesting is an arboreal affair for the Secretary-bird. The pair constructs a large and conspicuous nest high up in an acacia tree, which they maintain across breeding seasons. The female is in charge of most of the incubation, whereas both parents provision the chicks by feeding them with regurgitated, half-digested insect and small vertebrate prey. Fledglings require instruction from the parents on how to hunt on foot. Once they have acquired these skills, they become independent, and leave in search of their own home range.

Together with caracaras (see facing page), Secretary-birds are one of the few raptors that find most of their prey on foot, foraging on the ground for rodents, birds, and reptiles, including venomous snakes. For most of the day, pairs of these long-legged birds patrol their territory for prey, stomping it with their feet, or grabbing it with their bills. At night, the pair retreats to a roosting spot high up on a tree, out of the reach of nocturnal ground predators of the African savanna.

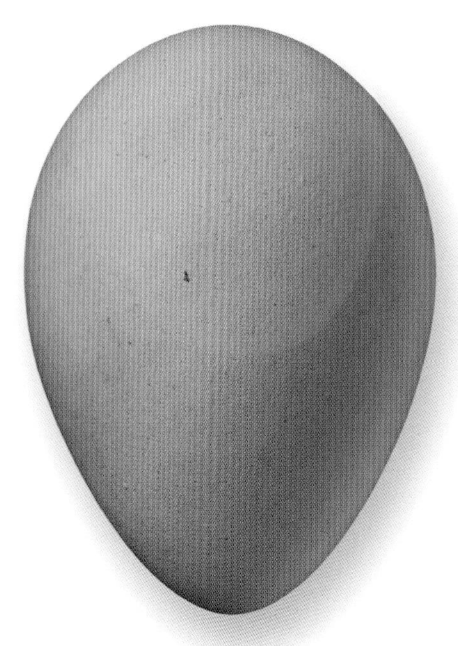

Actual size

The egg of the Secretary-bird is pale green or olive in background color, and measures 3⅛ x 2³⁄₁₆ in (79 x 56 mm) in size. The eggs become soiled during the prolonged incubation period, and hatch a few days apart, in the order in which they were laid.

ORDER	Falconiformes
FAMILY	Falconidae
BREEDING RANGE	Southern North America, Central America, Cuba, and northern South America
BREEDING HABITAT	Open grasslands, pastures, shrubs, plantations, and coastal forests
NEST TYPE AND PLACEMENT	Bowl of vines and sticks, constructed in a shrub or palm tree
CONSERVATION STATUS	Least concern
COLLECTION NO.	FMNH 6994

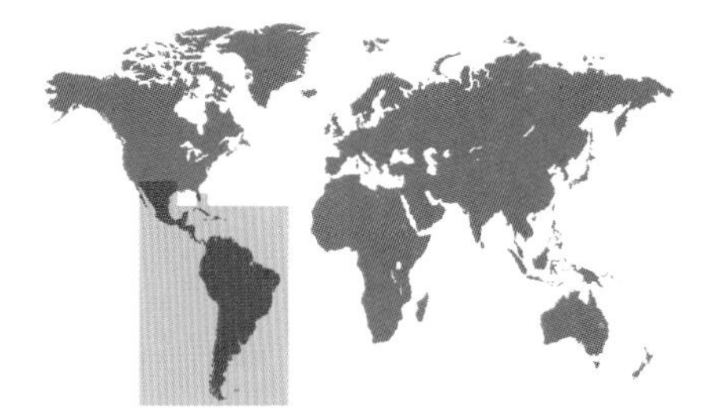

ADULT BIRD SIZE
19½–23 in (49–58 cm)

INCUBATION
28–32 days

CLUTCH SIZE
1–4 eggs

CARACARA CHERIWAY

CRESTED CARACARA

FALCONIFORMES

Caracaras maintain a home territory throughout the year. During the nesting season, the territory becomes smaller and centered around the nest. Suitable habitat for breeding is typically densely occupied by pairs of caracaras, which fight to keep out other caracaras, including recently matured birds attempting to take ownership of a territory to begin breeding themselves. Some of the excluded, young, non-breeding birds continue following older, breeding birds, and may eventually inherit the breeding site of a pair when one or both members dies or leaves the area.

Despite their close genetic affiliations with falcons, caracaras appear and act more like vultures or eagles. Their flight is slow as they seek potential prey on the ground, rather than catching it on the wing. In fact, the Crested Caracara typically forages on foot, often feeding on carrion or immobile prey, including sick birds and mammals, or helpless young.

Clutch

The egg of the Crested Caracara is a pink buff in background color with heavy dark spotting, and it measures 2⅓ x 1⅞ in (59 x 47 mm) in size. Both parents build the nest and incubate the eggs by taking turns.

Actual size

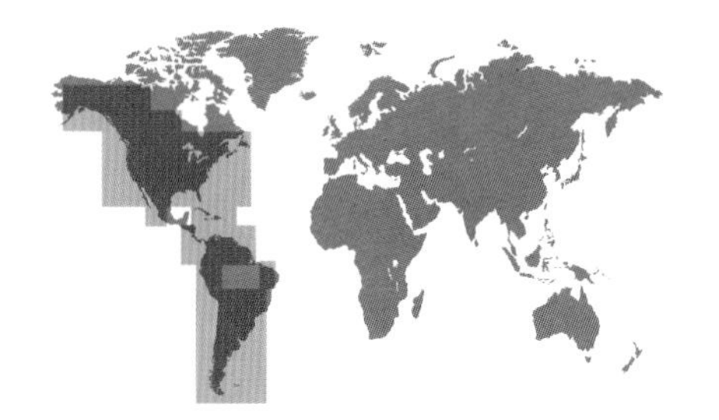

ORDER	Falconiformes
FAMILY	Falconidae
BREEDING RANGE	Temperate regions of most of the Americas
BREEDING HABITAT	Open areas, including grassland, semi-deserts and cultivated land; also urban areas
NEST TYPE AND PLACEMENT	Cavity nest in trees, cliffs, nestboxes, and buildings
CONSERVATION STATUS	Least Concern
COLLECTION NO.	FMNH 15844

ADULT BIRD SIZE
8½–12 in (22–31 cm)

INCUBATION
28–35 days

CLUTCH SIZE
3–7 eggs

FALCO SPARVERIUS

AMERICAN KESTREL

FALCONIFORMES

Clutch

The egg of the American Kestrel is white to cream in background color with brown or gray blotching, and 1⅓ x 1⅛ in (35 x 28 mm) in dimensions. The size of the egg is 10 percent larger than what is predicted for a typical bird species based on the female's size.

The smallest raptor in North America, American Kestrels typically feed on insects and occasionally small birds and mammals. At the same time, this species often becomes the prey itself for larger falcons and hawks. Female kestrels are bigger and more aggressive than the males. During the winter, the females occupy more open and more profitable habitats, while the subordinate males are constrained to hunt in woodlands where it is harder to find prey using the kestrel's typical hover-and-search hunting technique.

Incubation starts from the day the first egg is laid, resulting in a hierarchy of the age and size of the nestlings. The later-laid eggs' yolk contains increasing amounts of testosterone deposited by the mother, through which she manipulates the embryos' and the chicks' growth rate depending on clutch size, laying order, and food availability. The incubating and brooding female defecates while inside the nest, spraying her liquid feces on the wall of the nest cavity. With the growth of the chicks, the nest becomes increasingly smelly. The scent of the fecal matter is compounded by putrid remains of insects and mammals brought by the parents to feed the young.

Actual size

ORDER	Falconiformes
FAMILY	Falconidae
BREEDING RANGE	Islands of the Mediterranean, Canary Islands
BREEDING HABITAT	Cliffs facing the open sea
NEST TYPE AND PLACEMENT	Egg laid on bare ground of coastal cliffs, often in colonies
CONSERVATION STATUS	Least concern, but locally vulnerable
COLLECTION NO.	FMNH 18856

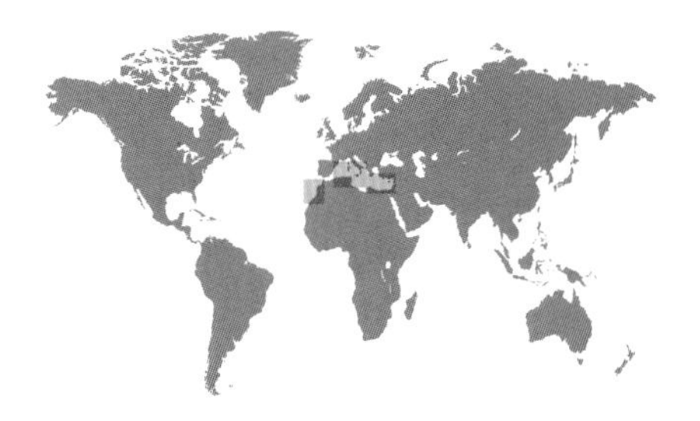

ADULT BIRD SIZE
14–16½ in (36–42 cm)

INCUBATION
30–33 days

CLUTCH SIZE
2–3 eggs

FALCO ELEONORAE

ELEONORA'S FALCON

FALCONIFORMES

The Eleonora's Falcon is both a habitat and foraging specialist, which influences its breeding biology as well. During much of the year, these birds feed on large flying insects, including dragonflies, that they capture in their talons in midflight, and consume without landing. During the breeding season, in late summer and early fall, the falcons shift their focus to catching migratory songbirds crossing the Mediterranean from their respective European nesting grounds to their African wintering grounds. The bountiful prey caught from among millions of migrating birds provides all the food needed to successfully raise this falcon's brood.

Nesting in colonies on vertical cliffs, these birds prefer to avoid human activities, including even the sight of lighthouse beams and other signs of settlements. A single remote island in the Greek archipelagos, Tilos, houses over 10 percent of the world's breeding population, making it vulnerable to losses due to human interference or weather events.

Clutch

The egg of the Eleonora's Falcon is reddish brown in background color, with dense darker maculation, and is 1⅔ x 1⅓ in (42 x 34 mm) in size. The color varies from pale brown with small dark spots for the first-laid egg to dark brown with many spots near the blunt pole for those laid later.

Actual size

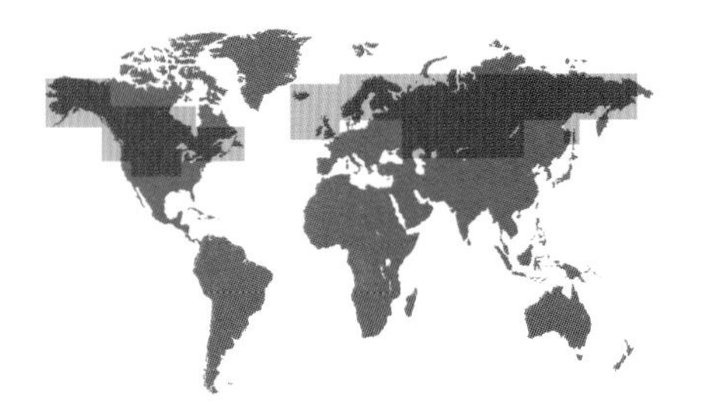

ORDER	Falconiformes
FAMILY	Falconidae
BREEDING RANGE	Temperate and subarctic regions of North America and Eurasia
BREEDING HABITAT	Open country and scrub to forests
NEST TYPE AND PLACEMENT	Usurps crow or hawk nests; may also place eggs on ground, cliffs, or building ledges
CONSERVATION STATUS	Least concern
COLLECTION NO.	FMNH 15840

ADULT BIRD SIZE
9½–12 in (24–30 cm)

INCUBATION
28–32 days

CLUTCH SIZE
4–5 eggs

FALCO COLUMBARIUS

MERLIN

FALCONIFORMES

Clutch

The Merlin, also known as the Pigeon Hawk, is a falcon occuring widely in the northern hemisphere. This species has a monogamous mating system: a female forms a pair bond with just one male. Wherever the nesting site is out of reach of typical urban predators, including cats and rats, the breeding attempt remains safe and is often very productive in this species, with most eggs hatching and most chicks fledging successfully. The female alone incubates the eggs, while the male provides her and the young chicks with provisions. But overwinter survival of juveniles is often very low; only a third of the fledglings typically return to begin breeding.

Known to many people throughout history and literature, this adaptable bird was one of the species most often used for falconry during the Middle Ages in Europe. Today, with its populations under protection throughout its range, the species also benefits from its adaptability to successfully live, feed, and nest in increasingly urban areas and modified habitats.

Actual size

The egg of the Merlin is rusty brown in background color with chestnut maculation, and measures 1⅝ x 1¼ in (40 x 32 mm).

ORDER	Falconiformes
FAMILY	Falconidae
BREEDING RANGE	Southern North America, South America
BREEDING HABITAT	Savanna, grassland, and open country
NEST TYPE AND PLACEMENT	Usurps stick nests in trees, made by other birds
CONSERVATION STATUS	Least concern; endangered in Texas and Mexico
COLLECTION NO.	FMNH 15830

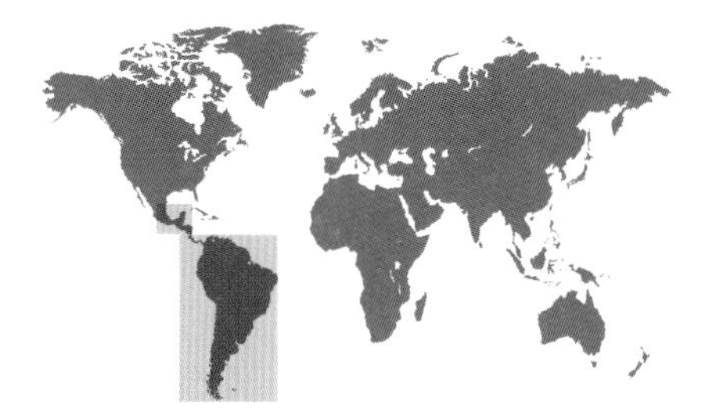

ADULT BIRD SIZE
15–17½ in (38–45 cm)

INCUBATION
32 days

CLUTCH SIZE
2–3 eggs

FALCO FEMORALIS

APLOMADO FALCON

FALCONIFORMES

Aplomado Falcon pairs rely on close cooperation for feeding and breeding. To capture prey, the female flies low and close to shrubs and bushes to flush small birds; the male, following closely behind, then swoops in to seize one. This species readily lays eggs into nests built by other bird species, as well as on artificial nest platforms placed in suitable habitat by conservation managers. During the breeding season, each member of the pair not only forages for itself, but returns to the nest with prey up to 20 to 30 times a day, feeding each chick frequently.

The Aplomado Falcon has a large distribution, spanning both hemispheres in the Americas, and so its global population faces no dangers. Yet, locally, especially in its northern range, the species has suffered tremendously, presumably from habitat alterations and losses to its prey base. To alleviate the threat of extinction in its native northern range, conservation management has focused heavily on breeding these birds in captivity, and releasing young falcons into the wild.

Clutch

The egg of the Aplomado Falcon is white to light pink in background color, with heavy brown maculation, and measures 1¾ x 1⅓ in (45 x 35 mm) in dimensions.

Actual size

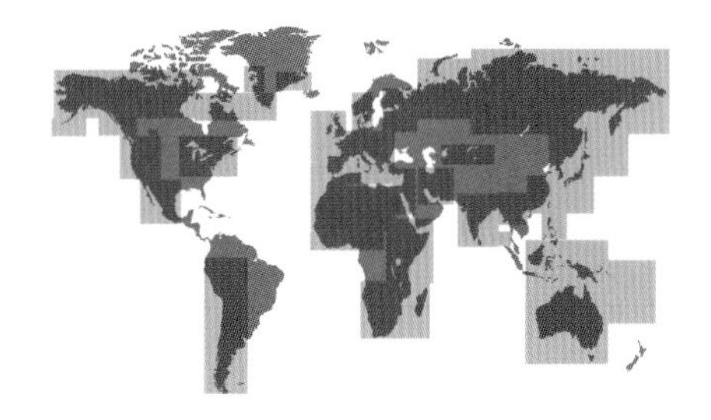

ORDER	Falconiformes
FAMILY	Falconidae
BREEDING RANGE	Worldwide; found on all continents except Antarctica
BREEDING HABITAT	A broad range of habitats; generally needs cliffs or tall buildings for nesting and open areas for hunting and provisioning
NEST TYPE AND PLACEMENT	A shallow scrape, with no lining materials added. Placed on a cliff or building ledge
CONSERVATION STATUS	Least concern
COLLECTION NO.	FMNH 6847

ADULT BIRD SIZE
16–20 in (41–51 cm)

INCUBATION
28–32 days

CLUTCH SIZE
2–6 eggs

FALCO PEREGRINUS

PEREGRINE FALCON

FALCONIFORMES

Clutch

The egg of the Peregrine Falcon is a cream or buff color, with strong, irregular markings of rust-brown. It is elliptical in shape, lacks gloss, and measures 2 x 1¾ in (51 x 44 mm). Incubation is done primarily by the female while the male provisions her.

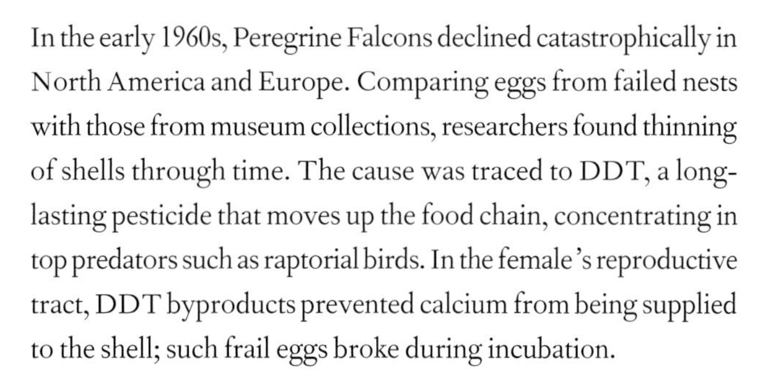

In the early 1960s, Peregrine Falcons declined catastrophically in North America and Europe. Comparing eggs from failed nests with those from museum collections, researchers found thinning of shells through time. The cause was traced to DDT, a long-lasting pesticide that moves up the food chain, concentrating in top predators such as raptorial birds. In the female's reproductive tract, DDT byproducts prevented calcium from being supplied to the shell; such frail eggs broke during incubation.

Peregrines became endangered in the United States, but a DDT ban, plus intensive breeding and reintroduction efforts, resulted in a strong recovery and, ultimately, their removal from the endangered species list. A key factor was the Peregrine's ability to become urbanized and adapt to nesting on tall buildings. Fast fliers, Peregrine Falcons can accelerate to 120 mph (200 km/h) during steep hunting dives for their bird prey, including city pigeons.

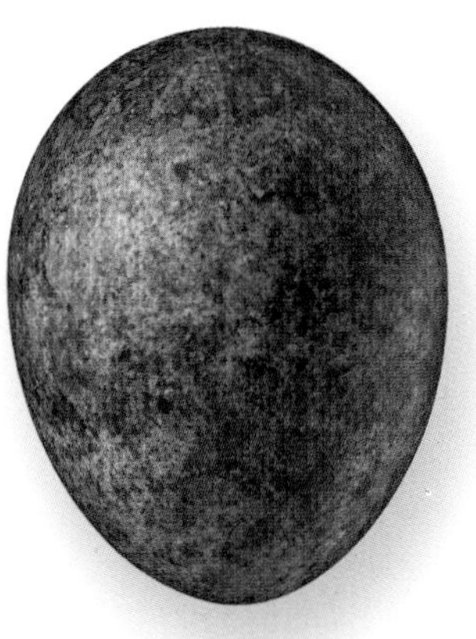

Actual size

ORDER	Falconiformes
FAMILY	Falconidae
BREEDING RANGE	Western North America
BREEDING HABITAT	Arid, open grasslands and semi-deserts
NEST TYPE AND PLACEMENT	On cliffs, in a shallow depression, without any nest lining; occasionally usurps nests of other raptors
CONSERVATION STATUS	Least concern
COLLECTION NO.	FMNH 15817

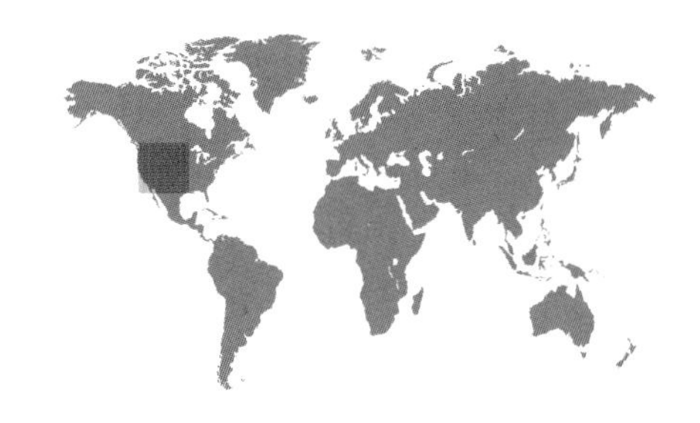

FALCO MEXICANUS

PRAIRIE FALCON

FALCONIFORMES

ADULT BIRD SIZE
14½–18½ in (37–47 cm)

INCUBATION
29–39 days

CLUTCH SIZE
3–6 eggs

Clutch

During the breeding season, Prairie Falcons may nest in the proximity of other cliff-dwelling species, including ravens, but they defend their territories vigorously from other Prairie Falcons. Mated pairs cooperate closely: the female is in charge of incubation and brooding the hatchlings, while the male supplies her and the young chicks with food. After the first two weeks, the female also heads out to hunt for prey to provision the growing and increasingly demanding offspring.

A large and efficient hunter, the Prairie Falcon is a specialist on rodents, including squirrels, and different birds, such as Horned Larks (see page 445) and Western Meadowlarks (see page 610), but also the introduced Common Starlings (see page 515), especially in the food-sparse winter period. Because of its arid breeding habitats, located away from most intense agricultural practices, the eggshells of this species were less affected by pesticides, such as DDT, during the middle of the twentieth century. Today, by being able to feed on common birds, this falcon's populations are increasing in size.

The egg of the Prairie Falcon is creamy, pink, or red-brown in color with brown maculation, and measures 2 x 1⅝ in (52 x 40 mm) in size. Incubation begins on the day when the first egg is laid, resulting in asynchronous hatching of the chicks.

Actual size

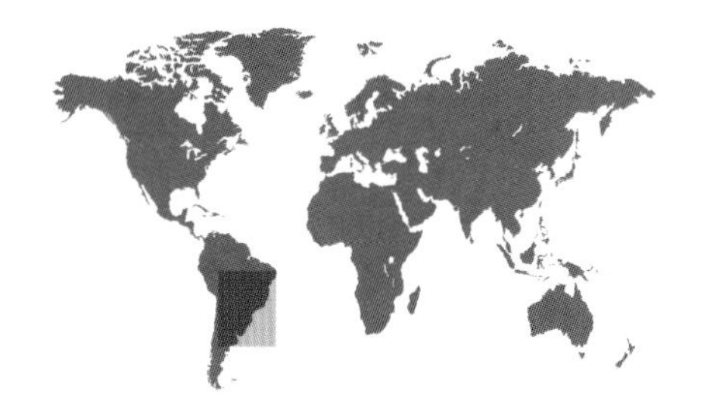

ORDER	Cariamiformes
FAMILY	Cariamidae
BREEDING RANGE	Southern South America
BREEDING HABITAT	Lowland grassland and savanna, sparse forests
NEST TYPE AND PLACEMENT	Typically a ground nest, occasionally in a bush or tree, made of branches and sticks, and lined with leaves and mud
CONSERVATION STATUS	Least concern
COLLECTION NO.	FMNH 14758

ADULT BIRD SIZE
29½–35½ in (75–90 cm)

INCUBATION
25–30 days

CLUTCH SIZE
2–3 eggs

CARIAMA CRISTATA

RED-LEGGED SERIEMA

CARIAMIFORMES

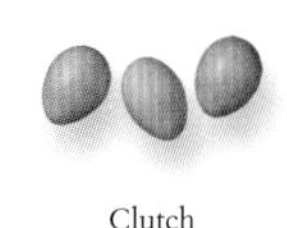

Clutch

Seriemas are an enigmatic group of birds, resembling in appearance and habits the Secretary-birds of Africa (see page 276): their long legs and hooked bill make them well suited to take lizards, snakes, and other terrestrial vertebrates that they catch on the ground. When night comes, they clamber into branches to roost. Red-legged Seriemas prefer to run from danger rather than fly. Their sharp central claw is a formidable weapon, used in territorial disputes against other seriemas.

During courtship, the long-necked male lowers his head and displays his prominent crown-feathers to attract the female. He also calls repeatedly during mate attraction. These seriemas are well known for their loud, barklike noises, produced by both sexes as well as by young chicks. Prior to breeding, both parents help to build the nest, and it can take up to a month to construct the bulky stick structure.

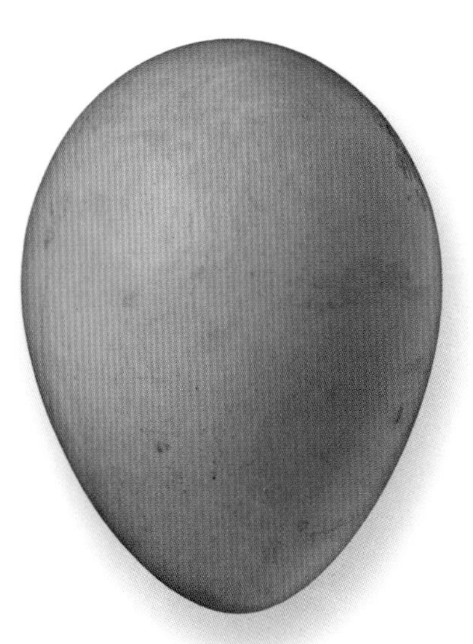

Actual size

The egg of the Red-legged Seriema is pink to dull white in background, with reddish-brown maculation, and it measures 2½ x 1⅞ in (63 x 48 mm). Both parents incubate the eggs and feed the chicks, until they are ready to hop to the ground from the nest, well before they are able to fly.

ORDER	Otidiformes
FAMILY	Otididae
BREEDING RANGE	Central and southern Europe, northwestern Africa, central Asia
BREEDING HABITAT	Open grasslands
NEST TYPE AND PLACEMENT	Shallow scrape on ground, hidden under vegetation
CONSERVATION STATUS	Vulnerable
COLLECTION NO.	FMNH 1688

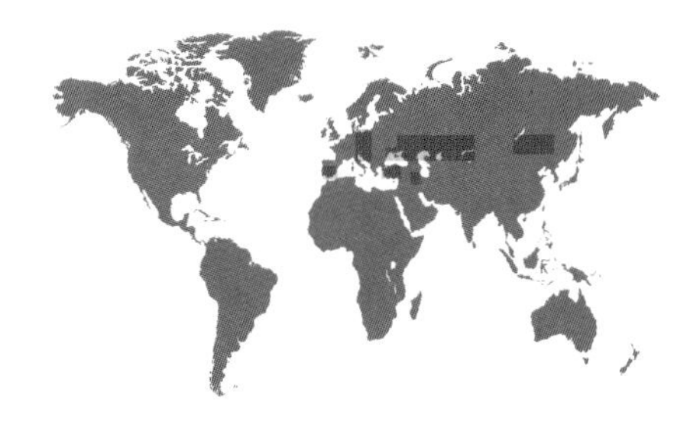

ADULT BIRD SIZE
31½–45½ in (80–115 cm)

INCUBATION
21–28 days

CLUTCH SIZE
1–3 eggs

OTIS TARDA

GREAT BUSTARD

OTIDIFORMES

Clutch

Although other birds may appear bigger, Great Bustards are probably the heaviest flying animal living on Earth today, weighing in at up to 46 lb (21 kg). Males and females are distinct in plumage pattern, size, and behavior. Males are larger and spend the winter fighting with others to establish dominance for the breeding season. Females are more cryptic, and their smaller size and drabber plumage make them well camouflaged for the incubation period. Nonetheless, agricultural activity, for instance mowing or harvesting during the incubation period, is uniformly detrimental to the nesting success of this species, resulting in many breeding attempts being lost prior to hatching.

The chicks of the Great Bustard leave the nest soon after hatching, and closely follow their mother for up to a year. Even when fully independent, male Great Bustards take several years to grow and become ready to deal with the intense and prolonged costs of fighting with other males before displaying for females. Outside the mating season, these large and noticeable birds widely roam the open countryside in groups.

The egg of the Great Bustard is glossy olive or tan in background color, has long brown blotches, and measures 3⅓ x 2¼ in (79 x 57 mm). The female alone incubates, but she is wary of human activities and many nests fail because of farming practices that flush the female or destroy the nest.

Actual size

ORDER	Otidiformes
FAMILY	Otididae
BREEDING RANGE	Europe, northwestern Africa, western and central Asia
BREEDING HABITAT	Dry grasslands, low-intensity agricultural fields
NEST TYPE AND PLACEMENT	Shallow depression in the ground, under grassy cover
CONSERVATION STATUS	Near threatened
COLLECTION NO.	FMNH 21360

ADULT BIRD SIZE
16½–17½ in (42–45 cm)

INCUBATION
20–22 days

CLUTCH SIZE
3–5 eggs

TETRAX TETRAX

LITTLE BUSTARD

OTIDIFORMES

Clutch

The egg of the Little Bustard is buff brown to deep olive green in background, with pale reddish spotting; it measures 2³⁄₁₆ x 1⅝ in (55 x 41 mm). The nest is so well concealed in dense grasses that the eggs are rarely seen and are only discovered by chance.

Little Bustards are habitat specialists, occurring in open grassland. Their populations are threatened throughout much of their range because of the expansion of agricultural fields. Living in open country is critical for these birds, because courtship behaviors rely mostly on visual signals given by the males, displaying in loose aggregations, to attract the females. Once mated, the female receives no further help from the male, and undertakes all nesting and parental duties on her own.

Male Little Bustards invest most of their time in the spring selecting and protecting their display sites. Older males tend to dominate younger males in confrontations that allow them to settle on more central patches. Interested females tend to visit and mate with older and more dominant males. This type of mating system, where males concentrate in an area to display competitively for females, is called lekking. It is seen in many other lineages, including galliformes.

Actual size

ORDER	Otidiformes
FAMILY	Otididae
BREEDING RANGE	Central and southern Europe, northwestern Africa, central Asia
BREEDING HABITAT	Open grasslands
NEST TYPE AND PLACEMENT	Shallow scrape on ground, hidden under vegetation
CONSERVATION STATUS	Vulnerable
COLLECTION NO.	FMNH 1688

ADULT BIRD SIZE
31½–45½ in (80–115 cm)

INCUBATION
21–28 days

CLUTCH SIZE
1–3 eggs

OTIS TARDA

GREAT BUSTARD

OTIDIFORMES

Clutch

Although other birds may appear bigger, Great Bustards are probably the heaviest flying animal living on Earth today, weighing in at up to 46 lb (21 kg). Males and females are distinct in plumage pattern, size, and behavior. Males are larger and spend the winter fighting with others to establish dominance for the breeding season. Females are more cryptic, and their smaller size and drabber plumage make them well camouflaged for the incubation period. Nonetheless, agricultural activity, for instance mowing or harvesting during the incubation period, is uniformly detrimental to the nesting success of this species, resulting in many breeding attempts being lost prior to hatching.

The chicks of the Great Bustard leave the nest soon after hatching, and closely follow their mother for up to a year. Even when fully independent, male Great Bustards take several years to grow and become ready to deal with the intense and prolonged costs of fighting with other males before displaying for females. Outside the mating season, these large and noticeable birds widely roam the open countryside in groups.

The egg of the Great Bustard is glossy olive or tan in background color, has long brown blotches, and measures 3⅓ x 2¼ in (79 x 57 mm). The female alone incubates, but she is wary of human activities and many nests fail because of farming practices that flush the female or destroy the nest.

Actual size

ORDER	Otidiformes
FAMILY	Otididae
BREEDING RANGE	Europe, northwestern Africa, western and central Asia
BREEDING HABITAT	Dry grasslands, low-intensity agricultural fields
NEST TYPE AND PLACEMENT	Shallow depression in the ground, under grassy cover
CONSERVATION STATUS	Near threatened
COLLECTION NO.	FMNH 21360

ADULT BIRD SIZE
16½–17½ in (42–45 cm)

INCUBATION
20–22 days

CLUTCH SIZE
3–5 eggs

TETRAX TETRAX

LITTLE BUSTARD

OTIDIFORMES

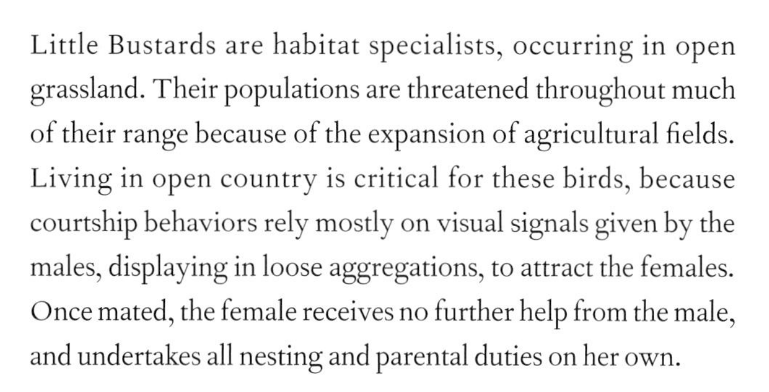

Clutch

The egg of the Little Bustard is buff brown to deep olive green in background, with pale reddish spotting; it measures 2 3/16 x 1 5/8 in (55 x 41 mm). The nest is so well concealed in dense grasses that the eggs are rarely seen and are only discovered by chance.

Little Bustards are habitat specialists, occurring in open grassland. Their populations are threatened throughout much of their range because of the expansion of agricultural fields. Living in open country is critical for these birds, because courtship behaviors rely mostly on visual signals given by the males, displaying in loose aggregations, to attract the females. Once mated, the female receives no further help from the male, and undertakes all nesting and parental duties on her own.

Male Little Bustards invest most of their time in the spring selecting and protecting their display sites. Older males tend to dominate younger males in confrontations that allow them to settle on more central patches. Interested females tend to visit and mate with older and more dominant males. This type of mating system, where males concentrate in an area to display competitively for females, is called lekking. It is seen in many other lineages, including galliformes.

Actual size

ORDER	Eurypygiformes
FAMILY	Rhynochetidae
BREEDING RANGE	New Caledonia
BREEDING HABITAT	Dense forests and shrubland
NEST TYPE AND PLACEMENT	Ground scrape or a simple nest of sticks or leaves
CONSERVATION STATUS	Endangered
COLLECTION NO.	WFVZ 155774

ADULT BIRD SIZE
20¼–21½ in (51–55 cm)

INCUBATION
33–37 days

CLUTCH SIZE
1 egg

RHYNOCHETOS JUBATUS

KAGU

EURYPYGIFORMES

The Kagu is a unique bird in anatomy and behavior, restricted to forested habitats throughout New Caledonia, in the western Pacific Ocean. This island lacked any ground-dwelling mammals until the recent arrival of humans, who introduced rats and dogs. The Kagu is well adapted to life without such ground threats: it rarely flies, and feeds and breeds on the ground. As a result, predation by rats on the eggs, and especially by dogs on the adults, has recently caused the extinction of entire local populations throughout the island.

While pairs defend their territory against other pairs all year round, including outside the breeding season, members of the pair may still forage in separate areas. However, once the egg hatches and the chick becomes independent, it often remains within the territory of the parents. Although these young do not help with breeding in the next year, they do contribute to territorial defense, and so they indirectly increase the success of their own parents' next breeding attempt.

The egg of the Kagu is gray brown, with darker blotches, and measures 2½ x 1⅞ in (63 x 47 mm) in dimensions. The egg is typically concealed near the base of a tree or under bushes, and both parents take turns in incubation.

Actual size

ORDER	Pterocliformes
FAMILY	Pteroclidae
BREEDING RANGE	Central Asia
BREEDING HABITAT	Steppes, semi-deserts
NEST TYPE AND PLACEMENT	Ground scrape, with no lining
CONSERVATION STATUS	Least concern
COLLECTION NO.	FMNH 20891

ADULT BIRD SIZE
12–16 in (30–41 cm)

INCUBATION
22–27 days

CLUTCH SIZE
2–3 eggs

SYRRHAPTES PARADOXUS

PALLAS'S SANDGROUSE

PTEROCLIFORMES

Clutch

The egg of the Pallas's Sandgrouse is cream to buff in background color, with darker spots and blotches, and measures 1⅔ x 1³⁄₁₆ in (42 x 30 mm) in size. Little is known about this species in the wild. In captivity, only the female incubates, while the male remains near the nest; both parents provision the chicks.

This species has flocking and nesting behaviors that resemble the habits of both shorebirds and galliformes; yet anatomical and genetic studies show that it belongs in its own, distinct taxonomic group. For example, despite terrestrial feeding and breeding patterns, its long and narrow wings allow it to fly long distances rapidly from the nesting grounds to the nearest source of water. Consuming water is essential because of a diet based predominantly on dry legume seeds.

The Pallas's Sandgrouse is a typical inhabitant of vast dry grasslands, the Asian steppes. Yet in recent history, in a phenomenon called eruptive dispersal and breeding, it has been known to reproduce sporadically as far west as the British Isles. This behavior is thought to be related to food scarcity in the birds' home range; once the nesting season is over, flocks of adults and newly produced young slowly return to the east.

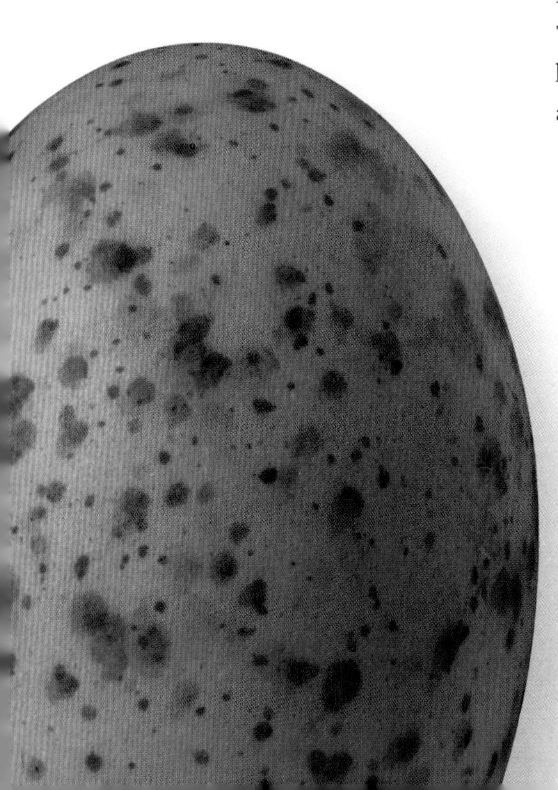

Actual size

ORDER	Pterocliformes
FAMILY	Pteroclidae
BREEDING RANGE	North Africa, southern Europe, western and central Asia
BREEDING HABITAT	Dry, open plains, short-grass fields
NEST TYPE AND PLACEMENT	Shallow, unlined hollow on the ground
CONSERVATION STATUS	Least concern
COLLECTION NO.	FMNH 1215

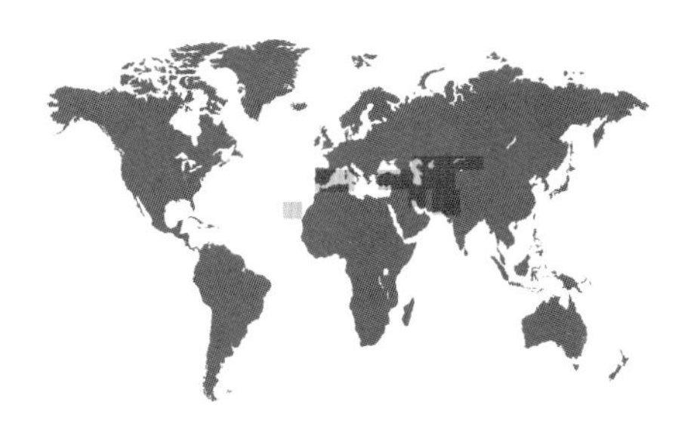

ADULT BIRD SIZE
13–15½ in (33–39 cm)

INCUBATION
23–28 days

CLUTCH SIZE
2–3 eggs

PTEROCLES ORIENTALIS

BLACK-BELLIED SANDGROUSE

PTEROCLIFORMES

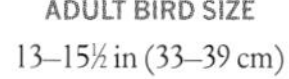

Black-bellied Sandgrouse are able to breed successfully in very dry areas, as long as there is some ground vegetation in which to hide the already cryptic eggs. Like other sandgrouse species, a critical habitat requirement for this grain-eating species is that the nesting sites have some open water sources nearby. Sandgrouse commute in flocks at dawn or dusk to drink water, often 20–30 miles (30–50 km) away. Making trips back and forth to water sources is also critical while raising the young; the fathers often carry water to their chicks by wetting their chest feathers at a pond or stream, then returning to the nest to let the chicks drink the water dripping from these feathers.

Clutch

This is a broadly distributed species, but local populations, especially on the Canary Islands and in Spain, have undergone a significant decline due to the use of irrigation that has turned open native grasslands into agricultural areas. Attempts to take wild-laid eggs into captivity to hand-raise the chicks, and to breed captive stocks in zoos, have been successful, assuring that captive-propagation, followed by reintroduction into former habitats, will be a viable conservation plan for this species. Elsewhere in its large range, however, the species maintains productive populations.

The egg of the Black-bellied Sandgrouse is creamy buff to pale green, with brown lines, and is 1⅞ x 1¼ in (47 x 32 mm) in size. Both sexes incubate the eggs, but only the males carry water back to the chicks.

Actual size

ORDER	Psittaciformes
FAMILY	Cacatuidae
BREEDING RANGE	Australia, except central desert, New Guinea, and neighboring islands
BREEDING HABITAT	Open woodlands, scrub, and suburban parks, often near rivers and canals
NEST TYPE AND PLACEMENT	Cavity in a tree, cliff, or termite mound; lined with wood shavings or dust
CONSERVATION STATUS	Least concern
COLLECTION NO.	FMNH 2885

ADULT BIRD SIZE
14–16 in (35–41 cm)

INCUBATION
25 days

CLUTCH SIZE
2–4 eggs

CACATUA SANGUINEA

LITTLE CORELLA

PSITTACIFORMES

Clutch

The egg of the Little Corella, like all parrot eggs, appears white in color, to the human observer, and immaculate. It measures 1½ x 1⅛ in (39 x 29 mm) in size. Both parents take turns to incubate the eggs and to provision the hatchlings.

Despite its name, the Little Corella is a large and conspicuous white cockatoo species that occurs throughout much of Australia. With the spread of irrigation channels and crop fields, flocks of these birds have become more common and also more destructive, descending on cereal fields and wreaking havoc by consuming vast amounts of grain.

Breeding is also part of the flocking habits of this species: colonies of corellas typically settle in tree stands with many natural cavities, often with more than one pair breeding in separate holes in the same tree. Little Corellas form close pair bonds, and the mates return year after year to reclaim the same nesting hollow for breeding. The eggs are laid several days apart, and incubation begins on the first day, so that hatching takes place across several days; older siblings and unhatched eggs can thus both be present in the nest at the same time.

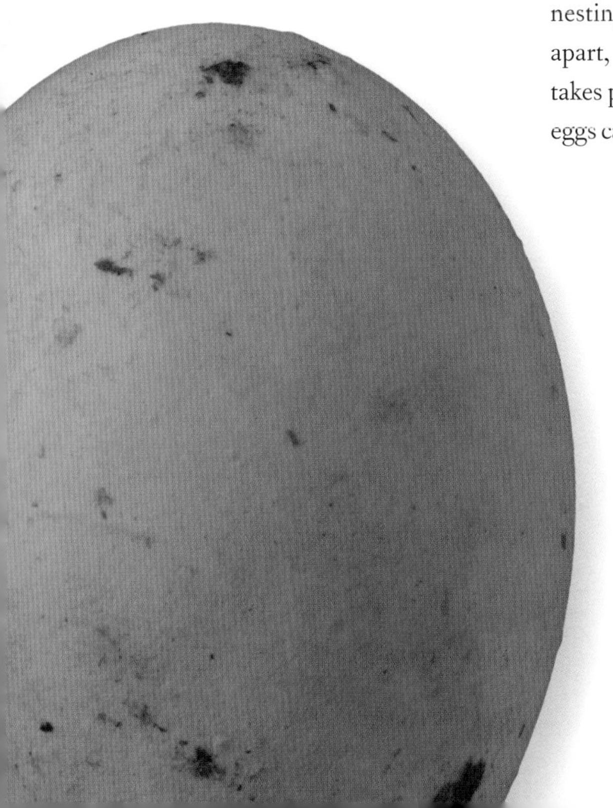

Actual size

ORDER	Psittaciformes
FAMILY	Psittacidae
BREEDING RANGE	Australia
BREEDING HABITAT	Dry forests, scrub, and grasslands
NEST TYPE AND PLACEMENT	Tree cavities, hollow fenceposts or logs, holes in buildings
CONSERVATION STATUS	Least concern
COLLECTION NO.	FMNH 2877

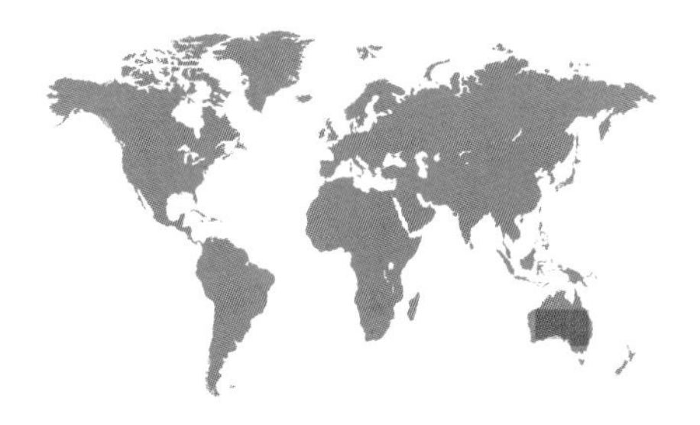

ADULT BIRD SIZE
7–8 in (18–20 cm)

INCUBATION
18–21 days

CLUTCH SIZE
4–6 eggs

MELOPSITTACUS UNDULATUS

BUDGERIGAR

PSITTACIFORMES

This small parakeet has been domesticated for over 150 years, making it the most popular parrot species available for purchase in pet stores. While much is known about the breeding behavior of captive budgies (a common name for the species), inferring information from captive birds to apply to wild stocks is not fully justified. Domestication has already altered the behavior and physiology of this species, including growth rates: domesticated birds are nearly double the size of wild forms.

During nesting, the female lays her eggs on alternate days, and she alone engages in incubation. She leaves the nest only to defecate, stretch, and forage briefly. Toward the end of the incubation period, and during the early nestling stage, she remains in the nest cavity, coming to the entrance hole to be fed exclusively by the male who regurgitates seeds and other plant matter to her.

Clutch

Actual size

The egg of the Budgerigar is pearl white in color, immaculate, and measures ⅔ x ⅝ in (18 x 15 mm) in size. The female keeps the male out of the nest during incubation, but allows him to enter when both parents begin to feed the chicks.

ORDER	Psittaciformes
FAMILY	Psittacidae
BREEDING RANGE	Eastern and northern Australia, Papua New Guinea
BREEDING HABITAT	Forests near rivers, mixed woodlands and scrub, and wooded farmland
NEST TYPE AND PLACEMENT	Large hollow in a tree
CONSERVATION STATUS	Least concern
COLLECTION NO.	FMNH 14721

ADULT BIRD SIZE
12–13 in (30–33 cm)

INCUBATION
21 days

CLUTCH SIZE
3–6 eggs

APROSMICTUS ERYTHROPTERUS

RED-WINGED PARROT

PSITTACIFORMES

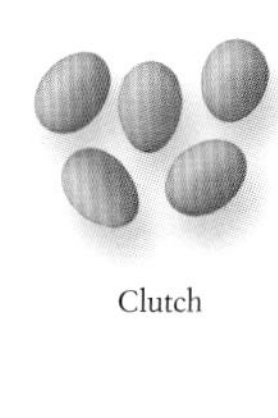

Clutch

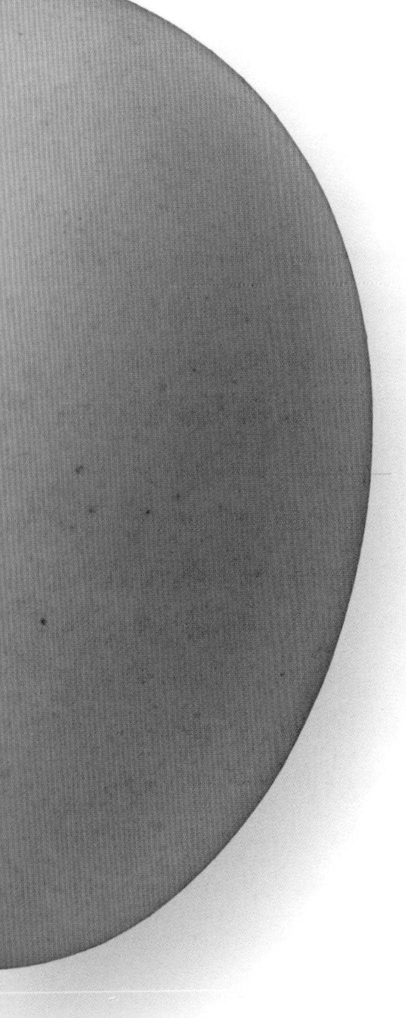

Australia is a hot-spot for parrot diversity, and most species prefer to nest in tree cavities. However, tree hollows are a precious resource on this continent: since there are no woodpeckers native to Australia to serve as primary nest excavators, competition for tree cavities is intense. Previously used cavities are often reused across breeding attempts, while active holes may be usurped by competitors. Red-winged Parrots occasionally compete with King Parrots for nest sites; in a few cases, these two species have also formed mixed pairs and produced viable, hybrid offspring.

This parrot is a highly social species, with pairs and family groups aggregating into flocks of 20 or more individuals to feed on fruits and search for water. They spend nearly all their time in the canopy, only landing on the ground for drinking or retrieving fallen food items.

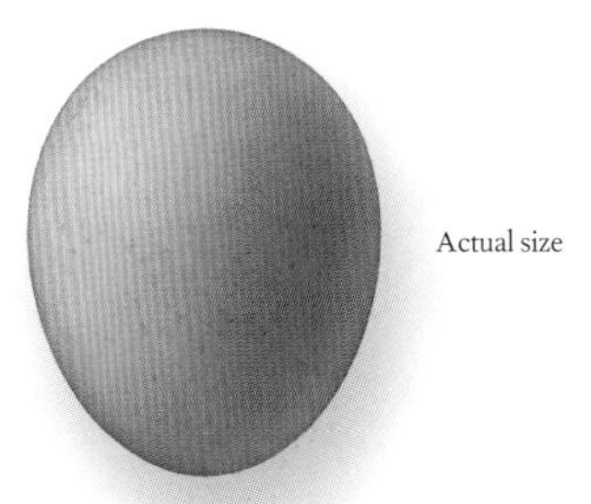

Actual size

The egg of the Red-winged Parrot is pale cream in background color, clear of spots, and 1¼ x 1 in (31 x 26 mm) in size. Only the female incubates the eggs, while the male searches for food and regurgitates it to his mate.

ORDER	Psittaciformes
FAMILY	Psittacidae
BREEDING RANGE	Southwestern Africa; feral in Arizona
BREEDING HABITAT	Arid regions
NEST TYPE AND PLACEMENT	In a rock crevice, under roofs, or in an empty compartment of large communal weaverbird nests
CONSERVATION STATUS	Least concern
COLLECTION NO.	FMNH 20799

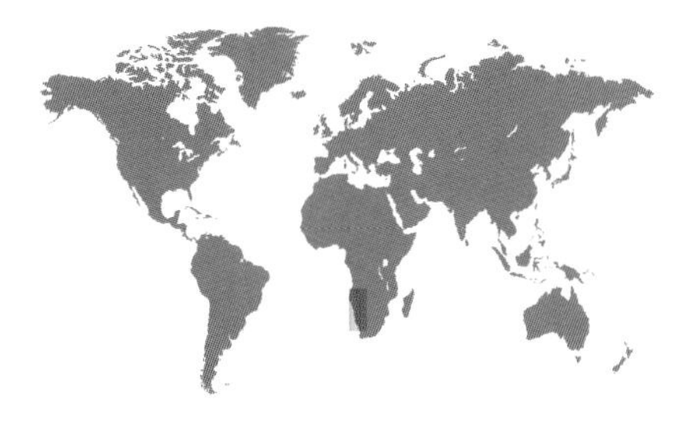

ADULT BIRD SIZE
6½–7 in (17–18 cm)

INCUBATION
23 days

CLUTCH SIZE
4–6 eggs

AGAPORNIS ROSEICOLLIS

ROSY-FACED LOVEBIRD

PSITTACIFORMES

Pair formation is a prolonged process for Rosy-faced Lovebirds, and there is no guarantee that any two adult individuals of the opposite sex will get along with one another, either in the wild or in captivity. If the pair bond is not established within days, these birds can become very aggressive; otherwise, the mates' relationship can last across many years. As a consequence of the dry landscapes they inhabit, breeding typically begins following rains, because the embryos inside the eggs require elevated moisture levels to develop normally. In recent decades, the breeding range of this species has expanded due to the establishment of more permanent water sources through agriculture and irrigation.

This is a highly gregarious species, always occurring in flocks composed of members of different mated pairs. Sociality often makes a species very suitable for the pet trade, because strong social bonding can also occur between the pet and its owner, but it also makes wild populations, including eggs and young, vulnerable to poaching and exportation.

Clutch

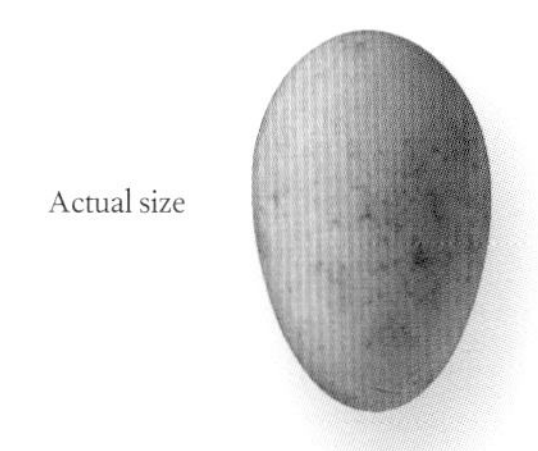

Actual size

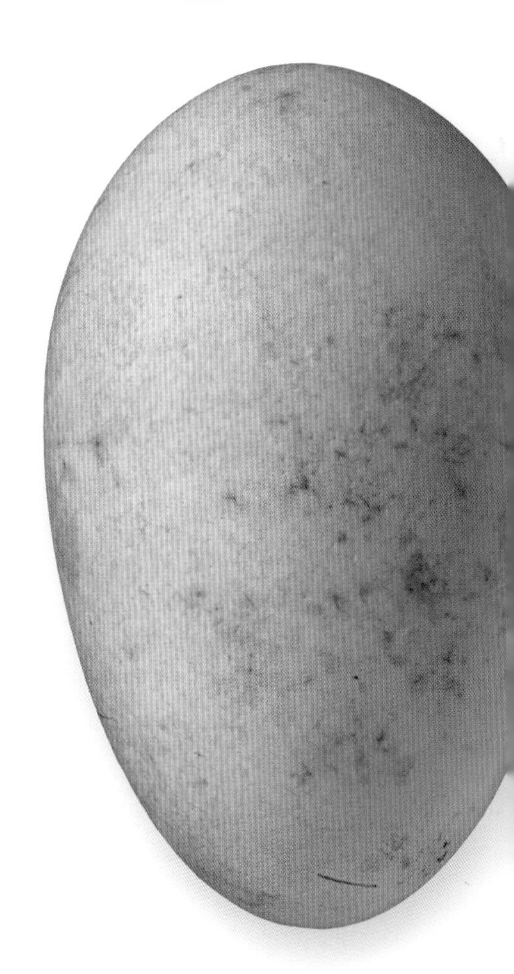

The egg of the Rosy-faced Lovebird is pale buff white in background, has no speckles, and its dimensions are ⅞ x ⅔ in (24 x 17 mm). The female alone incubates the eggs, and the male feeds her while she is in the nest.

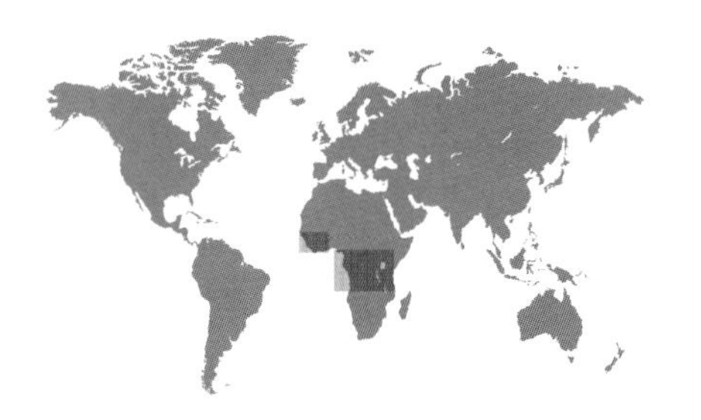

ORDER	Psittaciformes
FAMILY	Psittacidae
BREEDING RANGE	Equatorial Africa
BREEDING HABITAT	Primary and regrowing rainforests
NEST TYPE AND PLACEMENT	Natural tree hollows
CONSERVATION STATUS	Vulnerable
COLLECTION NO.	FMNH 20787

ADULT BIRD SIZE
11–13 in (28–33 cm)

INCUBATION
24–30 days

CLUTCH SIZE
2–3 eggs

PSITTACUS ERITHACUS

GRAY PARROT

PSITTACIFORMES

Clutch

The egg of the Gray Parrot is pure glossy white, and its dimensions are 1½ x 1⅛ in (39 x 28 mm). The female alone is in charge of incubating the eggs, relying on the male to feed her in the nest with regurgitated food.

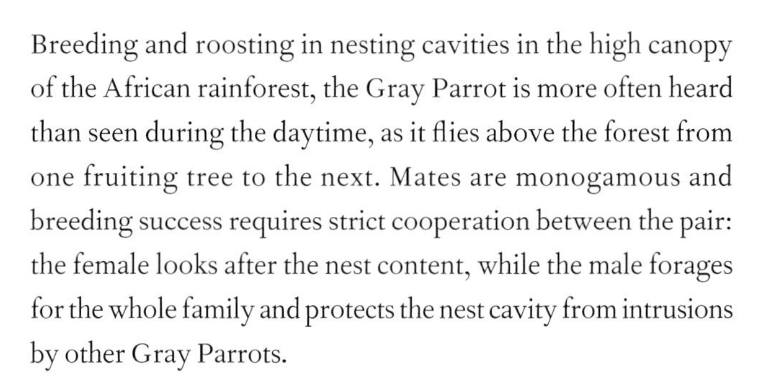

Breeding and roosting in nesting cavities in the high canopy of the African rainforest, the Gray Parrot is more often heard than seen during the daytime, as it flies above the forest from one fruiting tree to the next. Mates are monogamous and breeding success requires strict cooperation between the pair: the female looks after the nest content, while the male forages for the whole family and protects the nest cavity from intrusions by other Gray Parrots.

Admired for its intelligence by bird fanciers and scientists alike, the Gray Parrot has benefited from extensive accounts in the scientific and popular press about its cognitive abilities, including its propensity to mimic not only the voice, but also the meaning of their owner's speech. It is a large and slow-breeding parrot, which is prized highly in the pet trade. Collecting of wild birds for intercontinental sale has devastated this species' population numbers in the wild; today, it is under strict protection by CITES, the international agency that regulates trade in endangered wildlife.

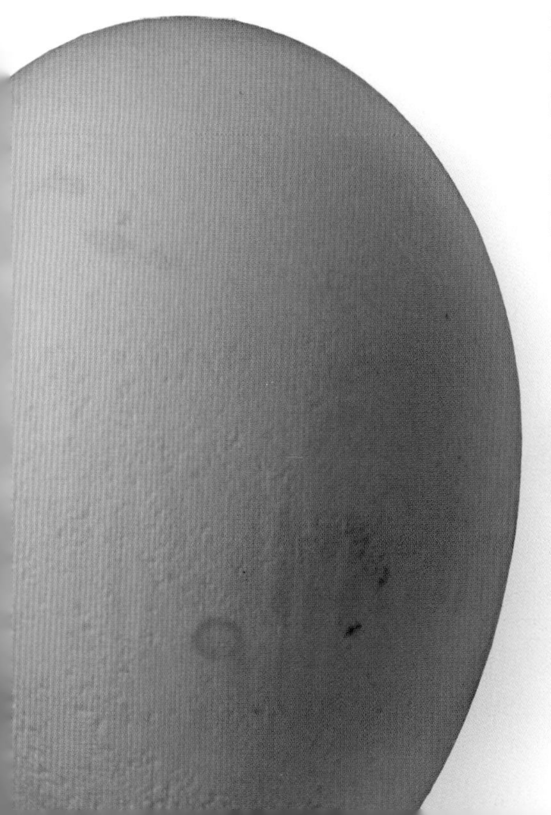

Actual size

ORDER	Psittaciformes
FAMILY	Psittacidae
BREEDING RANGE	Central South America
BREEDING HABITAT	Mixed riparian woodlands and open areas, including pastures
NEST TYPE AND PLACEMENT	Inside tree cavities or in burrows on vertical cliff faces
CONSERVATION STATUS	Endangered
COLLECTION NO.	FMNH 2480

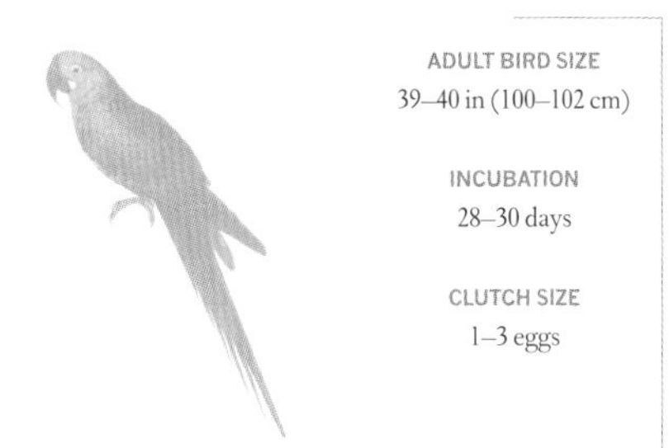

ADULT BIRD SIZE
39–40 in (100–102 cm)

INCUBATION
28–30 days

CLUTCH SIZE
1–3 eggs

ANODORHYNCHUS HYACINTHINUS

HYACINTH MACAW

PSITTACIFORMES

Sociality runs through all aspects of the life history of the Hyacinth Macaw. These birds move, feed, drink, and roost in flocks composed of established pairs. Among the pairs in the same flock, a clear dominance hierarchy develops, determining which pair gets first access to water and ripening palm nuts. The onset of reproduction is also synchronized within these flocks, with the pairs beginning to display courtship behaviors simultaneously, before finally breaking off into breeding units in search of a suitable nesting cavity.

One of the largest parrots, this beautiful, long-tailed bird was brought to the edge of extinction by uncontrolled capture for the pet trade during the 1980s. Estimates put the number of birds collected into the tens of thousands. Today, less than a tenth of that number persists in the wild in Brazil's Pantanal region, and in other small isolated populations in Brazil and Paraguay.

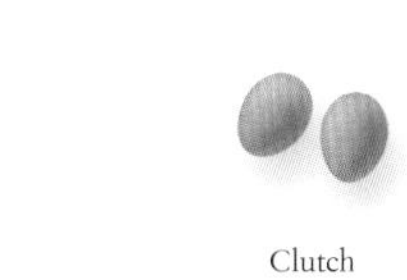

Clutch

The egg of the Hyacinth Macaw is white in background color with no maculation, and measures 1¾ x 1⅓ in (45 x 34 mm) in size. Conservation efforts to establish captive breeding colonies, to produce young birds for release back into the wild, have been hampered because many eggs are infertile.

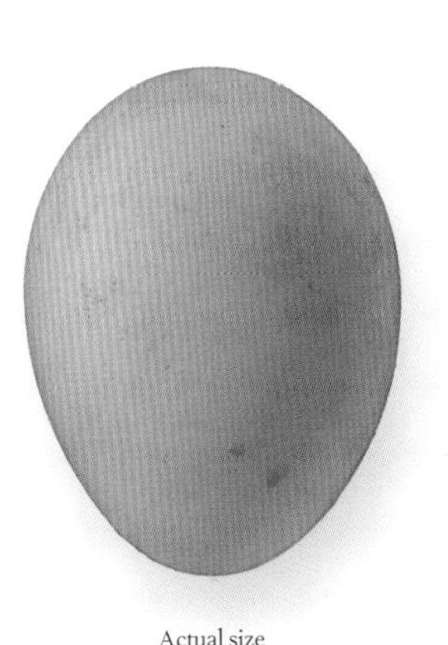

Actual size

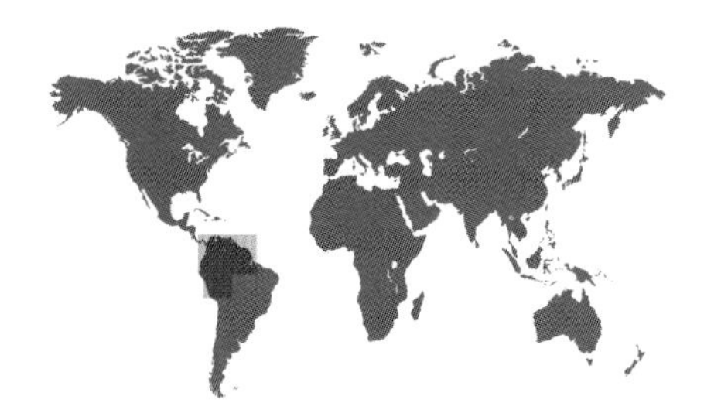

ORDER	Psittaciformes
FAMILY	Psittacidae
BREEDING RANGE	Tropical South America
BREEDING HABITAT	Lowland tropical forests, both humid and dry; also in savanna and some urban parks
NEST TYPE AND PLACEMENT	Cavity nest in a tree or palm hollow, or inside arboreal termite nests
CONSERVATION STATUS	Least concern
COLLECTION NO.	FMNH 879

ADULT BIRD SIZE
13–15 in (33–38 cm)

INCUBATION
26 days

CLUTCH SIZE
2–3 eggs

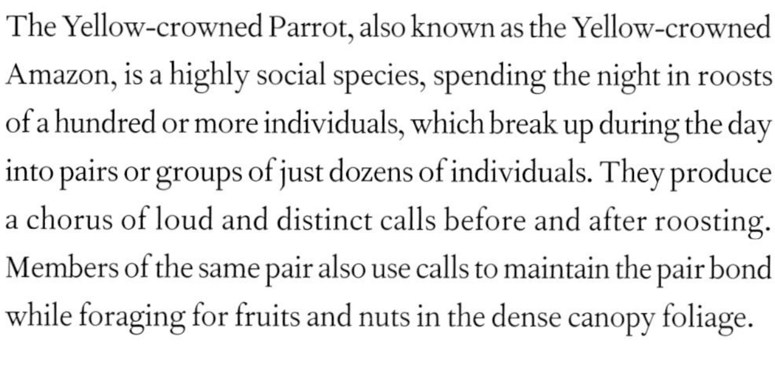

AMAZONA OCHROCEPHALA

YELLOW-CROWNED PARROT

PSITTACIFORMES

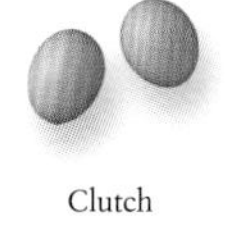
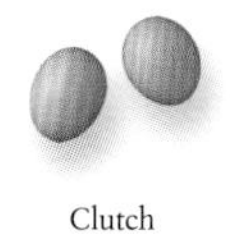

Clutch

The Yellow-crowned Parrot, also known as the Yellow-crowned Amazon, is a highly social species, spending the night in roosts of a hundred or more individuals, which break up during the day into pairs or groups of just dozens of individuals. They produce a chorus of loud and distinct calls before and after roosting. Members of the same pair also use calls to maintain the pair bond while foraging for fruits and nuts in the dense canopy foliage.

During the breeding season, pairs settle on a preferred and structurally sturdy nest cavity, in a tree trunk or a termite mound hanging off a branch. These nest sites must also be safe from tree-climbing snakes that prey on both the adult and the nestlings. In the cavity, the female alone incubates the eggs, while the male provisions her and the young chicks with regurgitated food pulp.

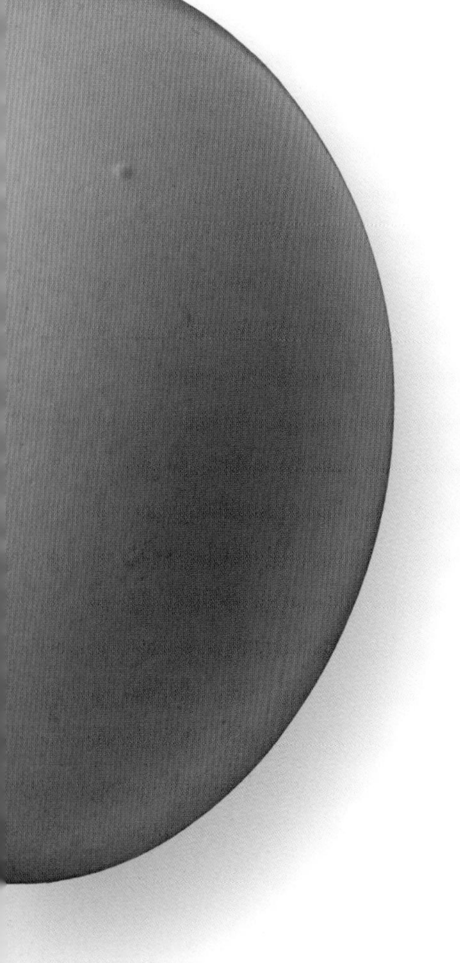

Actual size

The egg of the Yellow-crowned Parrot is clear white in color, and $1\frac{3}{8}$ x $1\frac{1}{16}$ in (35 x 27 mm) in size. The female only leaves the eggs briefly, typically at sunrise and sunset to forage and preen.

ORDER	Strigiformes
FAMILY	Tytonidae
BREEDING RANGE	All continents, except for Antarctica
BREEDING HABITAT	Open areas, including pastures and farmland near tree stands, old buildings, or cliffs
NEST TYPE AND PLACEMENT	Tree cavities, small caves; also attics in buildings and large holes in walls
CONSERVATION STATUS	Least concern
COLLECTION NO.	FMNH 7045

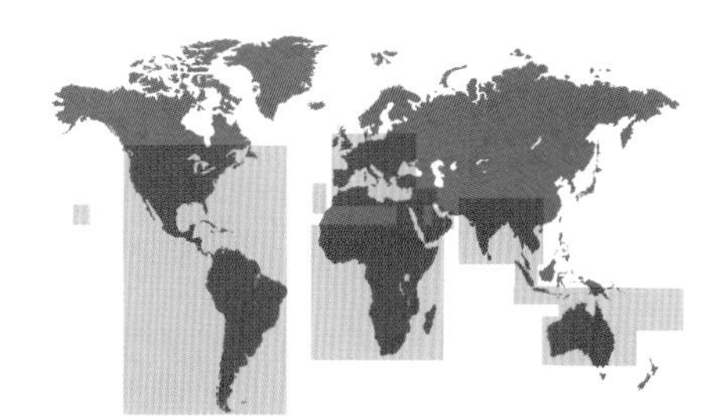

ADULT BIRD SIZE
12½–15½ in (32–40 cm)

INCUBATION
29–34 days

CLUTCH SIZE
4–7 eggs

TYTO ALBA

BARN OWL

STRIGIFORMES

The Barn Owl is the most widespread species of owl, having established a breeding population on the most remote of large landmasses, the New Zealand archipelago, during the last ten years. When breeding begins, the female takes two days between laying each egg; she alone incubates, while the male provisions her. The chicks and the fledglings are fed by both parents, including when they are fully grown but not yet skillful enough to capture prey on their own.

Common in rural and urban areas, this owl is seen by few people, because it begins hunting for mice, its preferred prey, only when night has fallen. The Barn Owl has also become a common laboratory species, used to study the details of its fine-tuned auditory system. Its hearing has evolved to accurately locate the direction and distance of mice based on the rustling noises they make as they move along the ground.

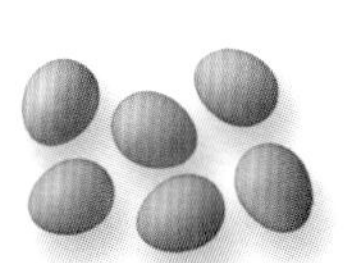

Clutch

The egg of the Barn Owl is 1¾ x 1¼ in (44 x 33 mm) in dimensions. Clutches of the white, immaculate eggs are typically laid in response to local food availabilities, which means that the timing of the breeding season varies across the species' vast range.

Actual size

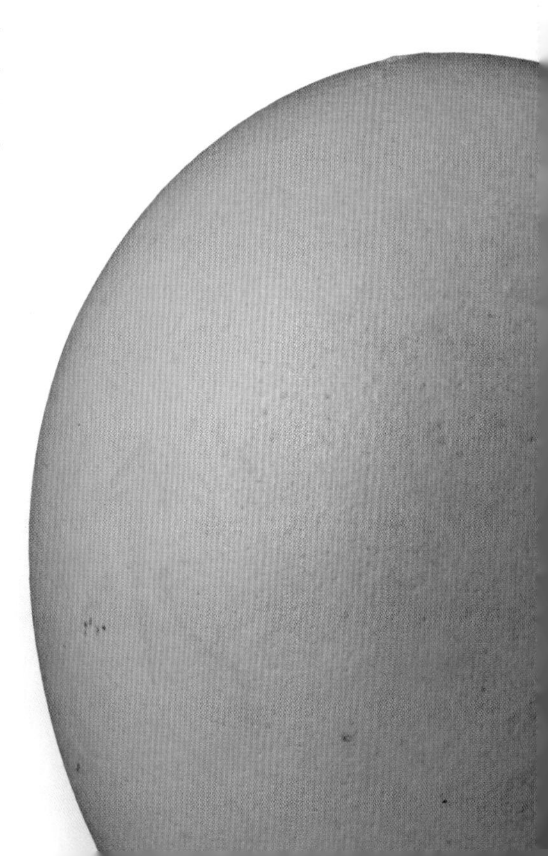

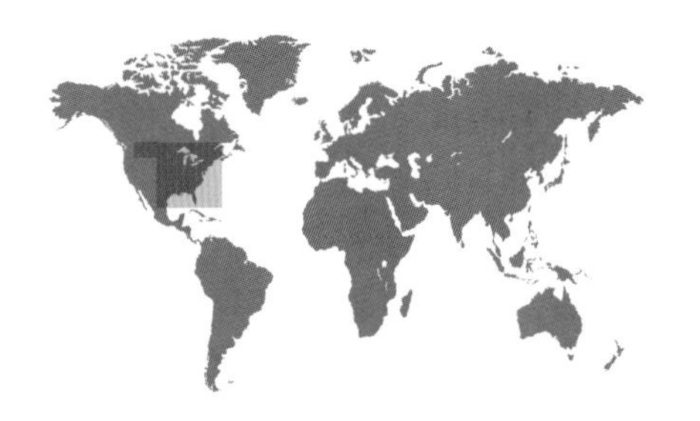

ORDER	Strigiformes
FAMILY	Strigidae
BREEDING RANGE	Eastern North America
BREEDING HABITAT	Forests, suburban and city parks
NEST TYPE AND PLACEMENT	Woodpecker or natural tree hole, no lining
CONSERVATION STATUS	Least concern
COLLECTION NO.	FMNH 7108

ADULT BIRD SIZE
6½–10 in (16–25 cm)

INCUBATION
27–34 days

CLUTCH SIZE
2–6 eggs

MEGASCOPS ASIO

EASTERN SCREECH OWL

STRIGIFORMES

Clutch

The egg of the Eastern Screech Owl is white in background color, and measures 1⅜ x 1³⁄₁₆ in (36 x 30 mm) in size. When a male mates with another female after egg laying has begun, this second female often displaces the first, lays her own eggs, and incubates both clutches.

This small, woodland owl is difficult to spot during the day because of its supremely camouflaged grey or rufous plumage. As with many owls and other birds of prey, the female is larger and flies more slowly, whereas the male is smaller and more agile. In contrast to most birds, in which the larger of the two sexes has a deeper call, in screech owls the smaller male has a lower-pitched voice.

During breeding, the whole family, including the female and the hatchlings, relies on the male's hunting skills to provision the brood. Even when food supplies are abundant, chicks in the nest fight viciously, and occasionally kill younger and smaller nest mates. Such siblicide is thought to be adaptive by allowing the larger and stronger nestlings to grow faster and fledge in a better condition to face new perils outside the nest.

Actual size

ORDER	Strigiformes
FAMILY	Strigidae
BREEDING RANGE	North, Central, and South America
BREEDING HABITAT	Forests and parklands, wooded areas near agricultural fields
NEST TYPE AND PLACEMENT	Natural cavities, burrows or tree holes excavated by other animals, abandoned squirrel nests; no lining
CONSERVATION STATUS	Least concern
COLLECTION NO.	FMNH 7159

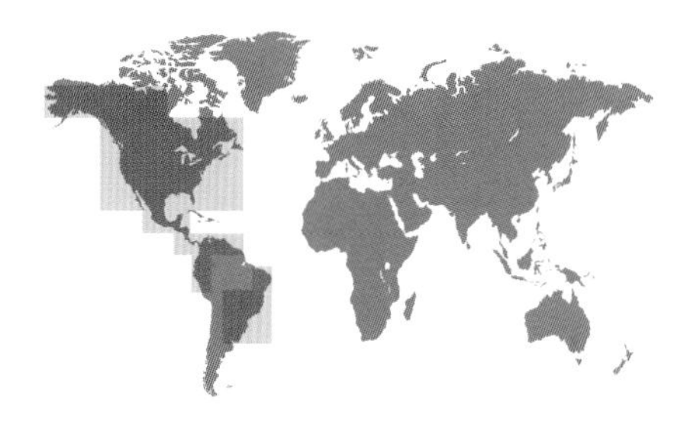

ADULT BIRD SIZE
18–25 in (46–63 cm)

INCUBATION
30–35 days

CLUTCH SIZE
2–4 eggs

BUBO VIRGINIANUS

GREAT HORNED OWL

STRIGIFORMES

Clutch

Male Great Horned Owls call throughout the year, uttering their characteristic deep hoots, while females only call during the mating season. Once vocal contact is established, the potential mates move closer to one another and display a series of bobs and bows as they continue their mutual assessment. When food is abundant, the male provisions the female during incubation and the early stages of chick rearing, whereas in years of food scarcity, the female leaves the nest soon after the eggs hatch to help catch food for the nestlings.

The Great Horned Owl, also known as the Tiger Owl, is a widespread, well-known, and powerful nocturnal predator; it commonly takes large prey, including skunks and osprey chicks. Nighttime nest predation by this owl was in fact a weighty hindrance in the reintroduction efforts of the Peregrine Falcon in some regions.

The egg of the Great Horned Owl is white in background color, and measures 2³⁄₁₆ x 1⅞ in (56 x 47 mm) in size. The number of eggs in a nest is typically higher when the food availability is greater in the weeks prior to egg laying.

Actual size

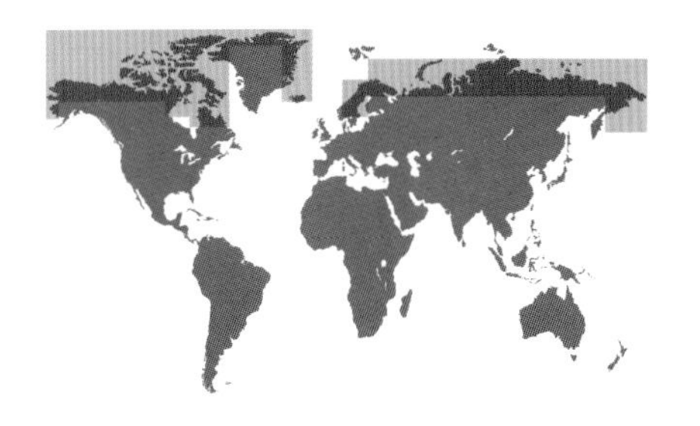

ORDER	Strigiformes
FAMILY	Strigidae
BREEDING RANGE	Arctic Eurasia and North America
BREEDING HABITAT	Tundra, and open coastal dunes
NEST TYPE AND PLACEMENT	Shallow hollow, scraped in the ground
CONSERVATION STATUS	Least concern
COLLECTION NO.	FMNH 7225

ADULT BIRD SIZE
20½–28 in (52–71 cm)

INCUBATION
32 days

CLUTCH SIZE
3–11 eggs

BUBO SCANDIACUS

SNOWY OWL

STRIGIFORMES

Clutch

The Snowy Owl is superbly adapted to life in a landscape without prominent perching sites: it is a tall predator whose white plumage makes it less conspicuous during the constant Arctic daylight. It uses small mounds in the open tundra for nesting and for surveying the landscape to sight potential danger and suitable prey. Flying low above the landscape, this owl also searches for small rodents and birds to hunt, and consumes its prey by swallowing it whole.

The female incubates the eggs while the male stands guard nearby; both sexes vigorously attack any predators that approach the nest. However, although this species is one of the largest predators in the tundra, its nest is still vulnerable to attacks by Great Skuas (see page 184), eagles, and Arctic Foxes. Nonetheless, because of aggressive defense of their clutch, other birds, including Snow Geese (see page 87), are often found nesting near Snowy Owls. The safety of these birds' eggs benefits because of their proximity to the Snowy Owls' nest sites.

Actual size

The egg of the Snowy Owl measures 2¼ x 1¾ in (57 x 45 mm) in size. Like all owl eggs, the Snowy Owl's are white and immaculate. Over the course of incubation, the white eggs may become soiled in the shallow ground nest.

ORDER	Strigiformes
FAMILY	Strigidae
BREEDING RANGE	Southwestern North America, Central America, and South America
BREEDING HABITAT	From moist forests to dry forests and cactus deserts
NEST TYPE AND PLACEMENT	Cavities in trees, stumps, or cactuses
CONSERVATION STATUS	Least concern
COLLECTION NO.	FMNH 7252

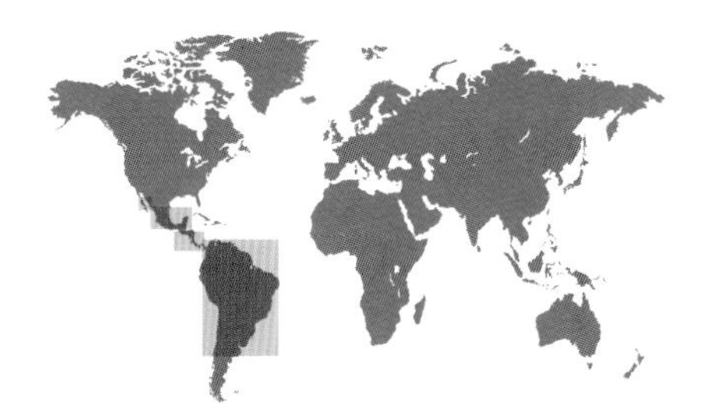

ADULT BIRD SIZE
5½–7 in (14–18 cm)

INCUBATION
23–28 days

CLUTCH SIZE
3–5 eggs

GLAUCIDIUM BRASILIANUM

FERRUGINOUS PYGMY-OWL

STRIGIFORMES

These owls spend most of the year alone, and only pair up during the breeding season. The male maintains the territory by calling frequently to keep out intruding competitors. The female alone incubates the eggs and broods the chicks. The male hunts to feed the female and the young chicks; after he transfers the food, the female tears it into small pieces and feeds it to the young. Just before fledging, the female joins the male in hunting to meet the growing nutritional demands of the brood.

This owl is a crepuscular hunter, active mostly around sunrise and sunset, but it can also be seen during the daytime, slowly walking along branches or flying with rapid wingbeats between the trees. It hunts for small rodents, lizards, and insects, and usually decapitates its prey before consuming the whole body or the soft parts.

Clutch

Actual size

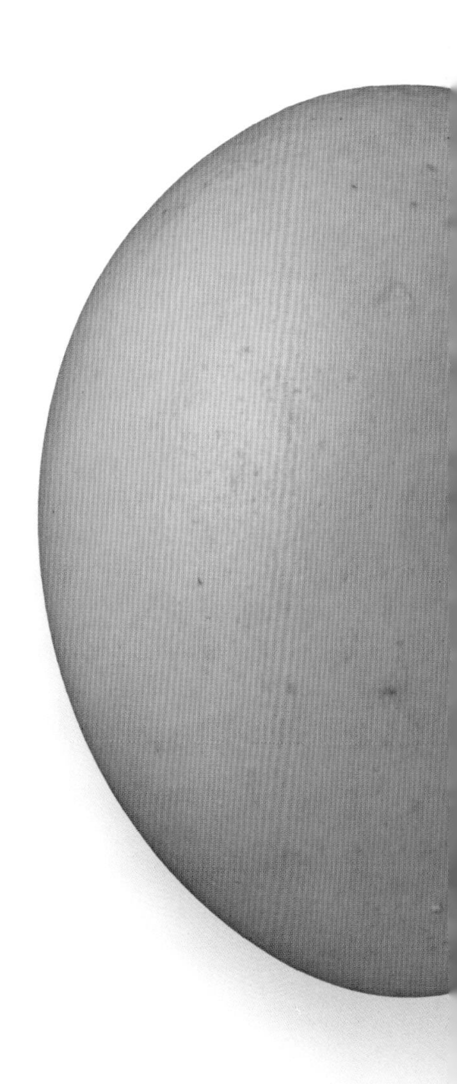

The egg of the Ferruginous Pygmy-Owl is white in background color, and measures 1⅛ x ⅞ in (29 x 23 mm) in size. As a small species, its nest is vulnerable to predation by birds and mammals. When threatened, the nestlings spread their wings and puff themselves up to appear larger.

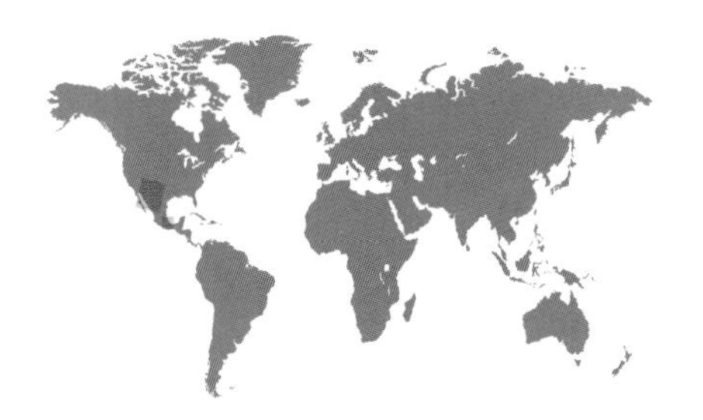

ORDER	Strigiformes
FAMILY	Strigidae
BREEDING RANGE	Southwestern North America
BREEDING HABITAT	Chaparral, cactus stands
NEST TYPE AND PLACEMENT	Cavity nest, often in saguaro cactus, originally excavated by woodpeckers
CONSERVATION STATUS	Least concern
COLLECTION NO.	FMNH 16343

ADULT BIRD SIZE
4½–5½ in (12–14 cm)

INCUBATION
20–23 days

CLUTCH SIZE
2–4 eggs

MICRATHENE WHITNEYI

ELF OWL

STRIGIFORMES

Clutch

The egg of the Elf Owl is white in background color, has no maculation, and is round in shape. It measures 1 1/16 x 7/8 in (27 x 23 mm) in size. The eggs are laid and incubated on the bare base of the nest cavity, even if it takes the female to remove all materials collected by another species previously using that same cavity.

Comparable in size to the House Sparrow and weighing only 1⅜ oz (40 g), this is the world's lightest owl. It feeds mostly on insects, many of which are attracted to flowering plants in the desert, including agave and cactus blooms. Elf Owls chase after these flying insects in the night, much like flycatchers do during the day. Its long wings, coupled with a short tail and compact, light body, make this bird an efficient flyer and a successful aerial hunter. To escape food shortages in the cold desert nights, the Elf Owl migrates to southern Mexico in winter.

With such a small stature, reproductive success for this owl depends on the safe location of the nest, typically high up from the ground, protected from predators by the natural defenses of its home plant, most frequently the 2-inch (5-cm) long sharp thorns of the aborescent saguaro cactus. The female remains in the nest during the period of incubation and brooding, while the male feeds her and the chicks.

Actual size

ORDER	Strigiformes
FAMILY	Strigidae
BREEDING RANGE	North, Central, and South America
BREEDING HABITAT	Grasslands, prairies, and pampas; avoids forests
NEST TYPE AND PLACEMENT	In ground burrows, including those excavated by prairie dogs or ground squirrels
CONSERVATION STATUS	Least concern
COLLECTION NO.	FMNH 7245

ADULT BIRD SIZE
7½–10 in (19–25 cm)

INCUBATION
21–28 days

CLUTCH SIZE
4–12 eggs

ATHENE CUNICULARIA

BURROWING OWL

STRIGIFORMES

The Burrowing Owl decorates the entrance of its nest with bits of mammal dung. Moisture evaporating from the dung in the hot sun serves to control the nest's microclimate. The scent of the dung also attracts insects, which are then caught by the parents and fed to the chicks. After hatching, the young owls appear at the nest entrance as early as two weeks of age, but remain in and around the nest for another month before fully fledging. After fledging, the male and the young may use one of several inactive nest burrows near the active site for roosting at night.

This is a diurnal owl, spending most of the daytime hours guarding its nest burrow or surveying its territory from the burrow entrance or a nearby fencepost. It feeds on insects and small rodents. Burrowing Owls often usurp ground burrows dug by prairie dogs or ground squirrels.

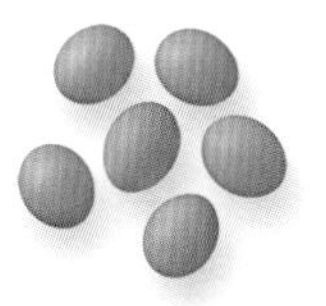

Clutch

Actual size

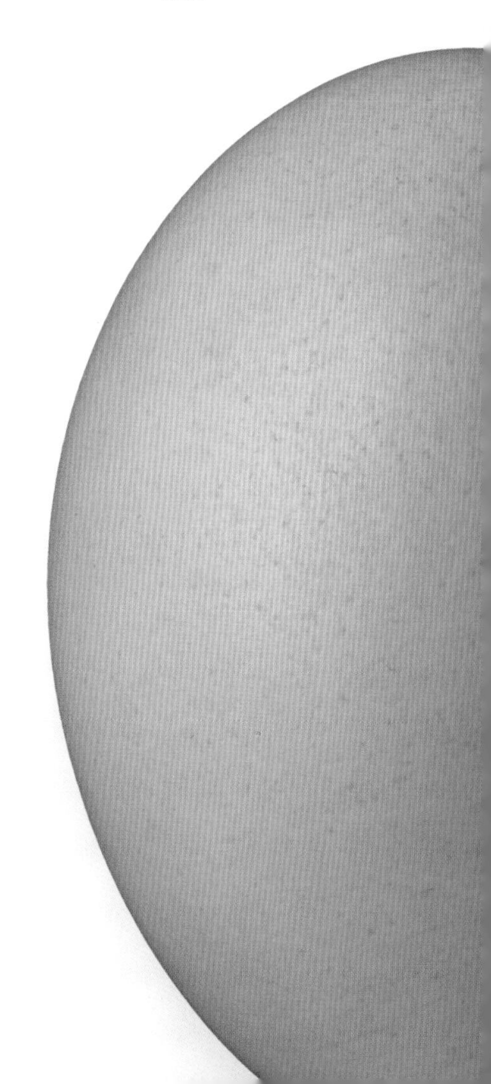

The egg of the Burrowing Owl is white in background color, and measures 1⅛ x 1 in (30 x 24 mm) in size. The female alone carries out incubation while the male feeds her, and the young chicks. Later, the female joins him to provision the full brood until fledging.

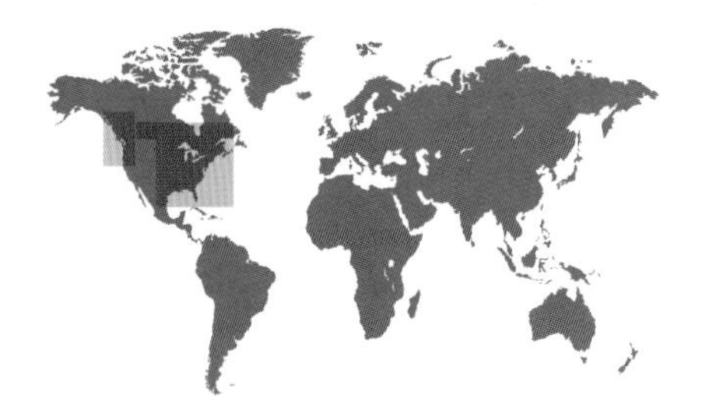

ORDER	Strigiformes
FAMILY	Strigidae
BREEDING RANGE	Northern and eastern North America, northern Central America
BREEDING HABITAT	Dense woodlands, including lowland riparian forests; recently also suburban parks
NEST TYPE AND PLACEMENT	Tree cavity, abandoned hawk, crow, or squirrel nest
CONSERVATION STATUS	Least concern
COLLECTION NO.	FMNH 16348

ADULT BIRD SIZE
17–19½ in (43–50 cm)

INCUBATION
28–33 days

CLUTCH SIZE
2–4 eggs

STRIX VARIA

BARRED OWL

STRIGIFORMES

Clutch

The Barred Owl is a large and successful forest predator, feared and mobbed by many diurnal birds, such as crows and jays, when its roosting branch is discovered. Yet its most dangerous enemy is a related species, the Great Horned Owl (see page 299), which can prey on nesting adults, chicks, and fledglings alike. When the breeding and foraging territories of these two species overlap, the Barred Owl moves away from its nemesis.

Colloquially also known as the Hoot Owl or the Eight Hooter for its distinctive call, prior to breeding the male Barred Owl hoots and displays frequently, while the female calls and approaches the courting male repeatedly. Only the female incubates the eggs, while the male provisions her. The chicks leave the nest while still flightless, probably to avoid simultaneous predation of the whole brood. The parents feed them in nearby trees until they are ready to fly and hunt independently.

Actual size

The egg of the Barred Owl is pure white, almost perfectly round in shape, and 2 x 1⅔ in (51 x 43 mm) in dimensions. The nest is often chosen up to a year before egg laying, but it is unknown whether the female or the male, or maybe both together, make the selection.

ORDER	Strigiformes
FAMILY	Strigidae
BREEDING RANGE	Temperate North America, Europe, and Asia
BREEDING HABITAT	Mixed grasslands and shrubland, open forests, parks
NEST TYPE AND PLACEMENT	Usurps the stick nests in trees built by hawks, crows, and other species
CONSERVATION STATUS	Least concern
COLLECTION NO.	FMNH 2450

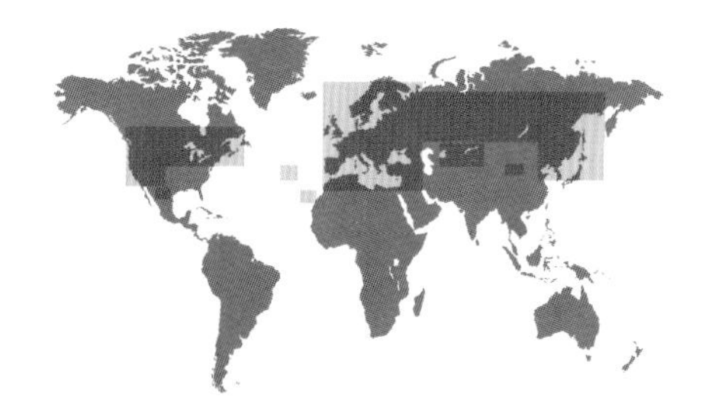

ASIO OTUS

LONG-EARED OWL

STRIGIFORMES

ADULT BIRD SIZE
14–15½ in (35–40 cm)

INCUBATION
25–30 days

CLUTCH SIZE
3–6 eggs

Unlike most owls, this species is quite sociable, forming daytime roosting aggregations during the non-breeding season. Pair bonds may form during these winter roosts, but during the breeding season, pairs break off from the group, search for an abandoned stick nest, and begin to protect the vicinity of that site. The male makes conspicuous zigzag flights as part of courtship and produces hooting calls to keep intruders away.

Once breeding begins, the female lays a variable number of eggs, with clutch sizes often directly tracking prey abundance—the more available prey, the more eggs. The chicks leave the nest and head into dense bushes nearby before they are flighted. Here, the female feeds them for just a few weeks before she departs from the breeding territory. The male stays and feeds the fledglings for several weeks more.

Clutch

The egg of the Long-eared Owl is white in background color, and measures 1⅝ x 1⅛ in (41 x 30 mm) in size. Only the female incubates the eggs, while the male provisions her, and the hatchlings too.

Actual size

ORDER	Strigiformes
FAMILY	Strigidae
BREEDING RANGE	The Americas and Eurasia; oceanic islands, including the Hawaiian and Galapagos archipelagos
BREEDING HABITAT	Open grasslands, marshes, sparse shrubland
NEST TYPE AND PLACEMENT	Ground scrape, lined with grasses
CONSERVATION STATUS	Least concern
COLLECTION NO.	FMNH 7072

ADULT BIRD SIZE
13½–17 in (34–43 cm)

INCUBATION
21–37 days

CLUTCH SIZE
4–7 eggs

ASIO FLAMMEUS

SHORT-EARED OWL

STRIGIFORMES

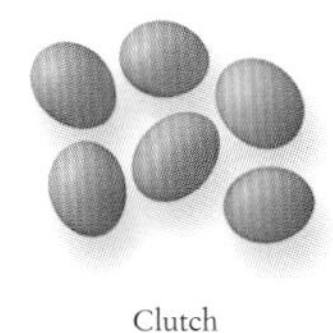

Clutch

At the onset of the breeding season, the male Short-eared Owl flies and swoops above the nest site repeatedly, in an attempt to attract an available female. The pair mate monogamously, and the female takes charge of incubating the eggs. The male defends the vicinity of the nest, and hunts for mammals and birds to provision the mother and the young.

The egg of the Short-eared Owl is cream white in background color, and measures 1½ x 1¼ in (39 x 31 mm) in size. In years with high vole densities, clutch sizes of up to 12 eggs have been recorded.

This owl is nearly as broadly distributed as the Barn Owl (see page 297), and also competes with that species for various rodent prey species. In some areas where nestboxes have been used to increase the number of breeding Barn Owls, Short-eared Owls often lose out. This is surprising because the two species are seemingly not in direct competition: Barn Owls forage only at night, while Short-eared Owls are crepuscular and often hunt even in bright daylight.

Actual size

ORDER	Strigiformes
FAMILY	Strigidae
BREEDING RANGE	Subarctic Eurasia and North America
BREEDING HABITAT	Dense coniferous forests, known as the taiga
NEST TYPE AND PLACEMENT	Tree cavity, usually an old woodpecker hole; no nest lining
CONSERVATION STATUS	Least concern; locally vulnerable
COLLECTION NO.	FMNH 7103

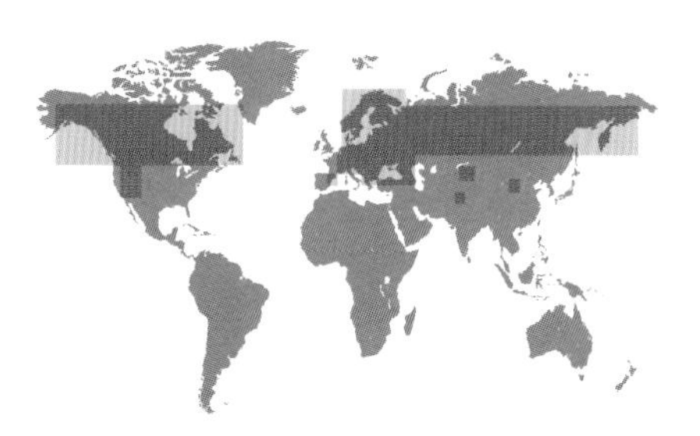

ADULT BIRD SIZE
8–11 in (20–28 cm)

INCUBATION
26–28 days

CLUTCH SIZE
3–6 eggs

AEGOLIUS FUNEREUS

BOREAL OWL

STRIGIFORMES

A sociable, nocturnal owl, this species, which also goes by the name Tengmalm's Owl, is most often identified by the male's mating calls given to attract the female in early spring. At other times, Boreal Owls sit quietly on perches and listen for rodent prey moving on the forest floor. Their ears are positioned asymmetrically behind their facial disk, which allows them to localize sound sources better in three dimensions. Once prey is localized, the birds swoop from their perches to capture it in their talons.

During most years, when mouse and vole densities are typical, one female and one male form a pair bond, to look after the eggs and the chicks. In years of high rodent abundances, a female may mate with two or three males, and lay eggs in several cavities, while the males may mate with more than one female. Such a promiscuous mating system is feasible in years when the abundance of food allows a single parent to feed and raise the whole brood successfully to fledging.

Clutch

Actual size

The egg of the Boreal Owl is creamy white in background color, and 1¼ x 1$\frac{1}{16}$ in (32 x 27 mm) in size. In the face of declining populations, as well as fewer woodpeckers excavating cavities in mature trees, conservation managers encourage breeding by placing nestboxes in suitable habitats.

ORDER	Strigiformes
FAMILY	Strigidae
BREEDING RANGE	Temperate North America
BREEDING HABITAT	Forests, both deciduous and evergreen
NEST TYPE AND PLACEMENT	Tree cavities excavated by woodpeckers; also uses nestboxes; does not line the bottom of the nest
CONSERVATION STATUS	Least concern
COLLECTION NO.	FMNH 7105

ADULT BIRD SIZE
7–8½ in (18–21 cm)

INCUBATION
20–23 days

CLUTCH SIZE
5–6 eggs

AEGOLIUS ACADICUS

NORTHERN SAW-WHET OWL

STRIGIFORMES

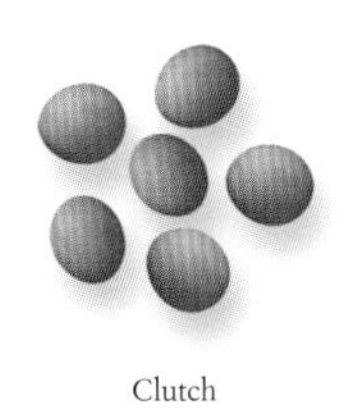

Clutch

This small woodland-dwelling owl breeds throughout North American forests; during the winter months, it expands its range, and roams widely into more open habitats, including urban areas and city parks. There, perching on a branch near the trunk, it waits steadily and turns its head to watch passers-by, making it possible for people to see this otherwise strictly nocturnal species during the day.

During breeding, the female owl is in full charge of incubating the eggs and brooding the hatchlings, while her mate feeds her and the young chicks frequent meals, mainly a diet of native wood mice. She tries to keep the nest clean and tidy, but by two weeks after the chicks hatch, the cavity is full of fecal matter and rotting prey remains. The female then leaves the chicks and begins to roost in another cavity, although she continues to return to the nest to feed them.

Actual size

The egg of the Northern Saw-whet Owl is white in background color, and 1³⁄₁₆ x 1 in (30 x 25 mm) in size. The female alone incubates the eggs, leaving only briefly to defecate and to cough up a pellet of hair and bone bits.

ORDER	Coraciiformes
FAMILY	Bucerotidae
BREEDING RANGE	Tropical southern Africa
BREEDING HABITAT	Woodlands and forested savanna
NEST TYPE AND PLACEMENT	Cavity nest, sealed with a mud wall
CONSERVATION STATUS	Least concern
COLLECTION NO.	FMNH 14771

ADULT BIRD SIZE
19–20 in (48–51 cm)

INCUBATION
25–32 days

CLUTCH SIZE
4–5 eggs

TOCKUS PALLIDIROSTRIS

PALE-BILLED HORNBILL

CORACIIFORMES

Breeding is a monogamous affair for the Pale-billed Hornbill, and the female and male must cooperate throughout the different breeding stages to be successful. Once an existing tree cavity is found and deemed suitable for nesting, the female builds a wall of mud, fruit pulp, and feces to protect it from predators and competitors; she will eventually seal herself inside, leaving just a narrow slit as an opening. The female alone incubates the eggs and attends the young chicks, while the male feeds his family fruits and insects through the opening in the mud wall of the nest entrance. As the chicks grow, the female breaks out of the cavity and reseals it, assisting the male in feeding the young until they are ready to fledge.

The Pale-billed Hornbill is familiar to many people in and outside of Africa, mostly because of its similarity to a cartoon character in Disney's *The Lion King*. Ecologically similar to toucans from the New World, hornbills stand alone in both their appearance and their genetic affiliations. The peculiar breeding behaviors of most tree-dwelling hornbills, which involve sealing the female inside the nest cavity, fully illustrate this uniqueness.

Clutch

The egg of the Pale-billed Hornbill is white in background color, and measures 1⅝ x 1¹⁄₁₆ in (40 x 27 mm) in size. During incubation, the female undergoes a wing-molt inside the nest, rendering her flightless during this period.

Actual size

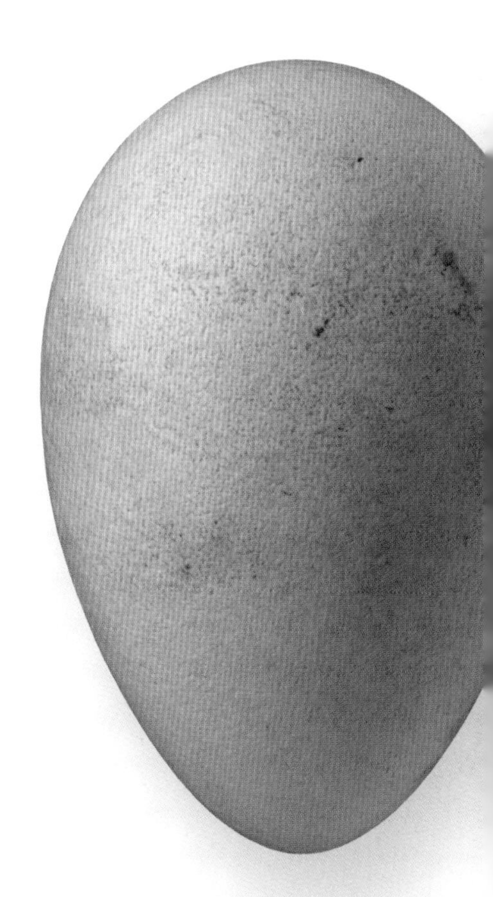

SMALL NON-PASSERINE LAND BIRDS

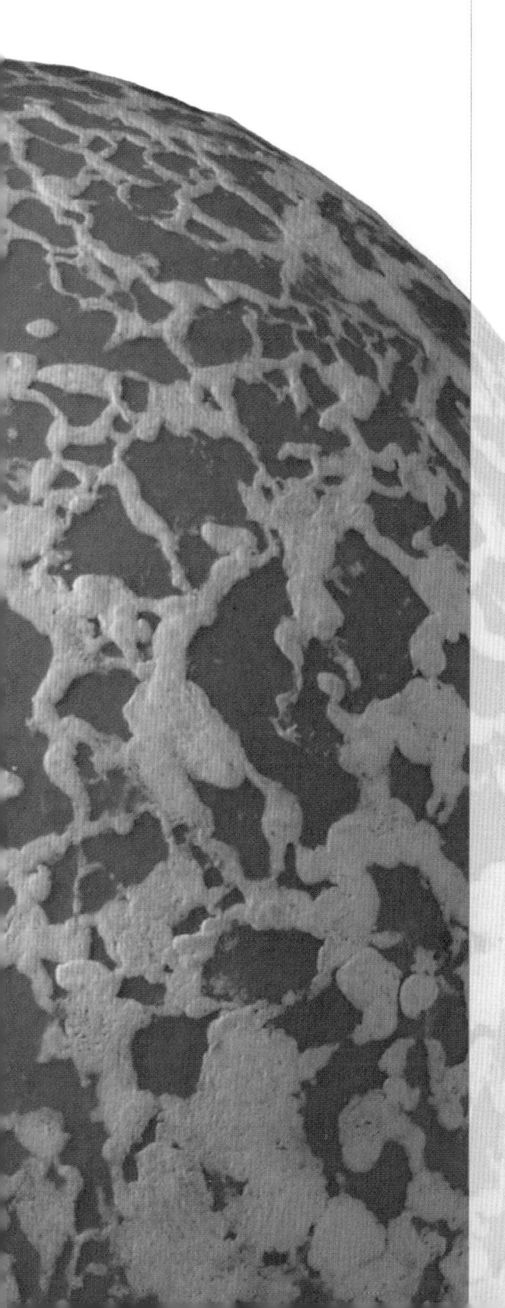

In this chapter, birds are grouped by their smaller size, and the fact that they are not Passerines (the perching birds that are featured in the following chapter.) Their diverse lineages have evolved a wide variety of feeding strategies; there are fish-, insect-, grain-, fruit-, or nectar-eaters, often with diverse body and bill shapes to match. They occur on every continent except Antarctica. Some, like woodpeckers, pigeons, kingfishers, and cuckoos have species in both the tropical and the temperate zones, whereas other groups, like jacamars, bee-eaters, rollers, barbets, motmots, todies, trogons and toucans, are largely confined to the tropics. Among the most behaviorally and morphologically specialized groups are the nocturnal nightjars, the aerially-living swifts, the parasitic cuckoos, and the incredible hummingbirds. Many, but not all these groups lay white eggs, and almost all these groups produce altricial young that develop in the nest, begging to solicit frequent visits for brooding and feeding by their parents.

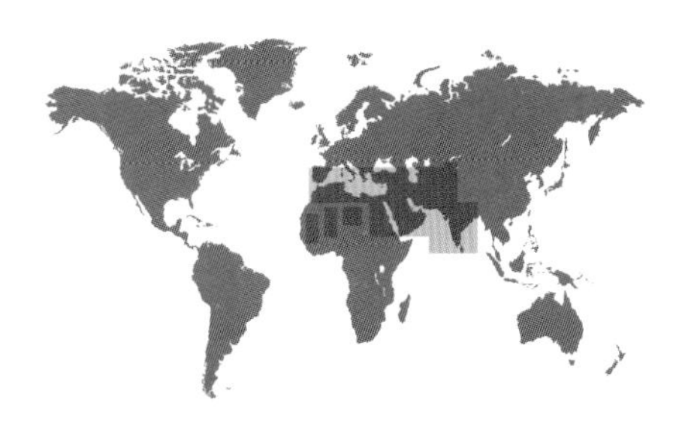

ORDER	Columbiformes
FAMILY	Columbidae
BREEDING RANGE	Native to Africa, Europe, and Asia; introduced on all other continents, except Antarctica
BREEDING HABITAT	Urban and suburban areas, farmland, rocky cliffs
NEST TYPE AND PLACEMENT	Flimsy platform of sticks, placed on buildings or cliff ledges with overhangs, or in dense ivy running up walls
CONSERVATION STATUS	Least concern
COLLECTION NO.	FMNH 20041

ADULT BIRD SIZE
12–14 in (30–36 cm)

INCUBATION
17–19 days

CLUTCH SIZE
2 eggs

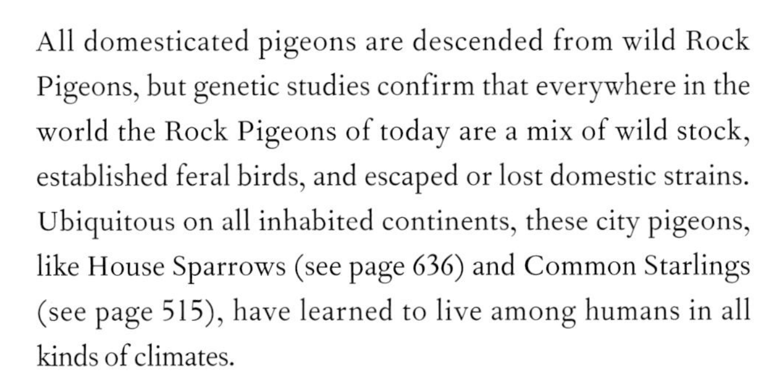

COLUMBA LIVIA

ROCK PIGEON

COLUMBIFORMES

Clutch

The egg of the Rock Pigeon measures 1½ x 1⅛ in (39 x 29 mm) in size. Like those of all doves and pigeons, the eggs are white and have no markings.

All domesticated pigeons are descended from wild Rock Pigeons, but genetic studies confirm that everywhere in the world the Rock Pigeons of today are a mix of wild stock, established feral birds, and escaped or lost domestic strains. Ubiquitous on all inhabited continents, these city pigeons, like House Sparrows (see page 636) and Common Starlings (see page 515), have learned to live among humans in all kinds of climates.

Breeding is not seasonal in this species, with freshly laid eggs and young squabs (chicks) present in nests throughout the year. Although the eggs of the Rock Pigeon are bright white and provide little camouflage, few predators (or people) ever spot a pigeon nest. This is due to its hidden placement, typically under an overhanging cliff or building ledge; and a dedicated incubation schedule: parents incubate the eggs constantly, only leaving the nest when one of them is replaced by the mate, who immediately continues the incubation.

Actual size

ORDER	Columbiformes
FAMILY	Columbidae
BREEDING RANGE	Caribbean islands and the Florida Keys
BREEDING HABITAT	Coastal woodlands, including mangroves
NEST TYPE AND PLACEMENT	Tree nest, often placed directly above water; a platform of sticks and branches, lined with smaller twigs
CONSERVATION STATUS	Near threatened
COLLECTION NO.	FMNH 6047

ADULT BIRD SIZE
13–14 in (33–35 cm)

INCUBATION
13–14 days

CLUTCH SIZE
1–3 eggs

PATAGIOENAS LEUCOCEPHALA

WHITE-CROWNED PIGEON

COLUMBIFORMES

The White-crowned Pigeon requires two distinct types of habitats for breeding and feeding: coastal mangroves for nesting, which are increasingly being destroyed to make space for coastal developments, and inland forests with mature fruit-bearing trees, which are frequently cut down to plant sugar cane and other crops on Caribbean islands. To cover the often formidable distance between nesting and feeding sites, this pigeon has a fast-paced flight, and can commute 12–30 miles (20–50 km) daily between suitable habitats, even on small islands. When food supplies are good, birds may breed up to four times in one year.

Unlike the Rock Pigeon and its descendants, the wild White-crowned Pigeon does not benefit from associating with people. Hunting and car collisions represent major causes of mortality for White-crowned Pigeons, to the extent that today this species has become a conservation concern.

Clutch

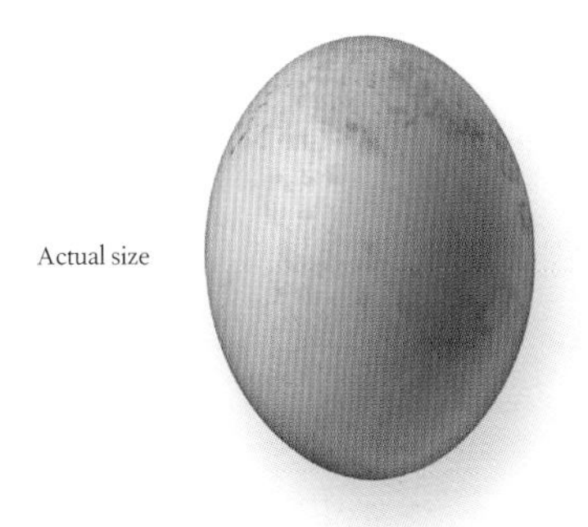

Actual size

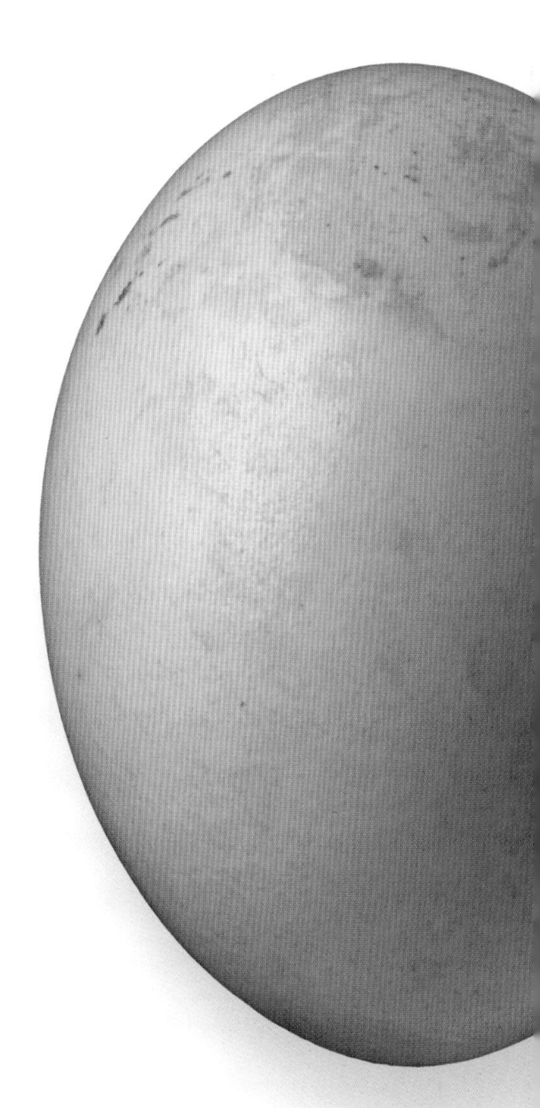

The egg of the White-crowned Pigeon is pale white in background color, and measures 1⅜ x 1 in (36 x 26 mm) in size. A skittish bird on the nest, it frequently abandons its eggs when disturbed by humans or predators.

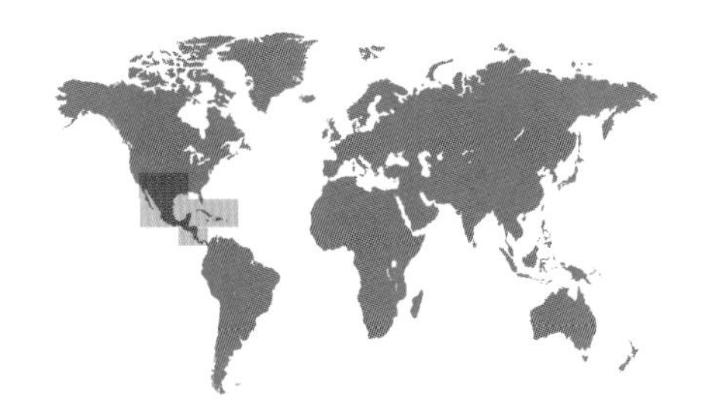

ORDER	Columbiformes
FAMILY	Columbidae
BREEDING RANGE	Southwestern North America, Central America, the Caribbean
BREEDING HABITAT	Dense thorny woodlands, riparian corridors, cactus forests, and suburban parks
NEST TYPE AND PLACEMENT	Twig nest, lined with grasses and mosses, placed on a tree branch with heavy foliage
CONSERVATION STATUS	Least concern
COLLECTION NO.	FMNH 16282

ADULT BIRD SIZE
11–12½ in (28–31 cm)

INCUBATION
14–20 days

CLUTCH SIZE
1–2 eggs

ZENAIDA ASIATICA

WHITE-WINGED DOVE

COLUMBIFORMES

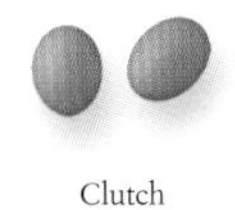

Clutch

The egg of the White-winged Dove is pale, dull white in background color, clear of spots, and measures 1 3/16 x 7/8 in (30 x 22 mm) in size. When a predator threatens the eggs, the incubating adult will feign a broken wing as a distraction.

The mating season of the White-winged Dove begins with the conspicuous courtship flight of the male, who spirals high up into the sky and returns with a stiff-winged descent to a cooing perch. Once the pair bond is established, the male provides the female with a selection of twigs and branches, which she then arranges to build up the nesting platform, lined with bark, grasses, and mosses. In the wild, the nest is built in dense tree stands; in city parks, this dove settles in tall ornamental trees.

Once commonly hunted, with its populations plummeting from 12 million to 1 million in Texas alone, today this species is on the move. It has recently begun a rapid expansion of its breeding range to include urban areas, such as town and city parks. Outside the breeding season, the White-winged Dove ranges from Canada to the east coast of North America.

Actual size

ORDER	Columbiformes
FAMILY	Columbidae
BREEDING RANGE	Temperate North America
BREEDING HABITAT	Open woodlands, also scrub and semi-deserts, city parks
NEST TYPE AND PLACEMENT	Twig, pine needle, and grass platform, no lining; placed on tree branches, but also on the ground or on building ledges
CONSERVATION STATUS	Least concern
COLLECTION NO.	FMNH 16273

ADULT BIRD SIZE
9–13½ in (23–34 cm)

INCUBATION
14 days

CLUTCH SIZE
2 eggs

ZENAIDA MACROURA

MOURNING DOVE

COLUMBIFORMES

The Mourning Dove is a common, widespread, and adaptable bird; it can successfully reproduce in urban areas, open forests, and the deserts of the American West. In arid regions, the adults are able to drink slightly saline spring water, without getting dehydrated. These mid-sized doves consume seeds all day long, ingesting and digesting 20 to 30 percent of their body weight on a daily basis.

The mating season begins with conspicuous flights and chases by both members of the eventual pair, as well as with intruding males trying to take over the resident male's territory. The male collects most of the nesting materials, which are then woven together, albeit loosely, by the female; the nest is so loose that the eggs are sometimes visible from underneath. The two eggs are bright white, but they are never uncovered because the two sexes take turns to incubate them. Instead of regurgitating seeds, both parents provision the young with nutritious and energy-rich "crop-milk," a secretion from the bird's neck pouch.

Clutch

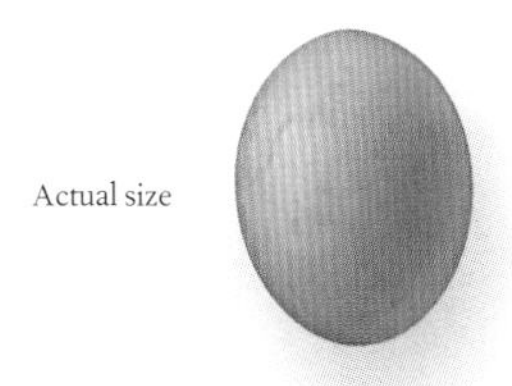

Actual size

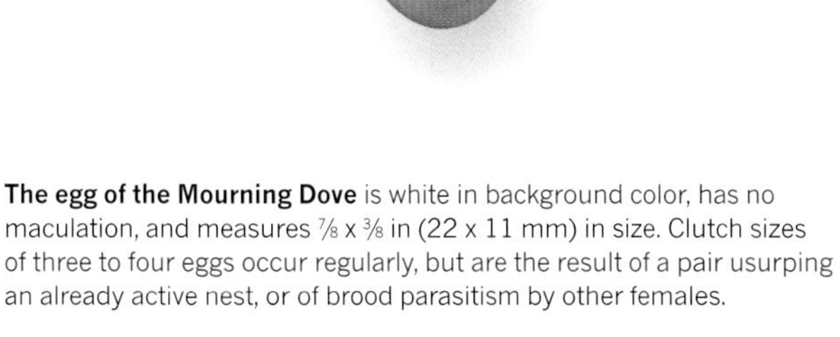

The egg of the Mourning Dove is white in background color, has no maculation, and measures ⅞ x ⅜ in (22 x 11 mm) in size. Clutch sizes of three to four eggs occur regularly, but are the result of a pair usurping an already active nest, or of brood parasitism by other females.

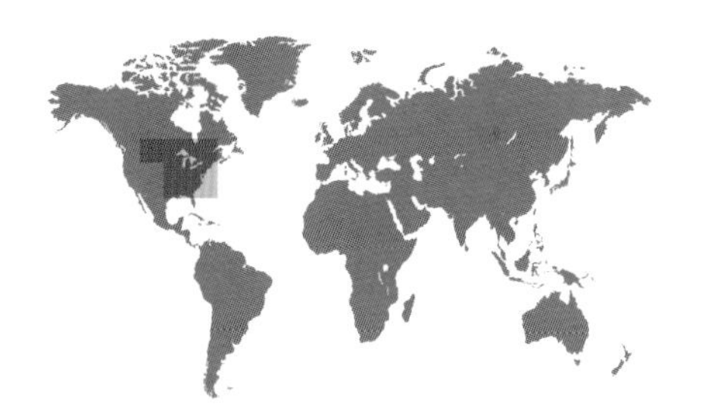

ORDER	Columbiformes
FAMILY	Columbidae
BREEDING RANGE	Formerly eastern North America
BREEDING HABITAT	Contiguous deciduous forests
NEST TYPE AND PLACEMENT	Colonial nesting in large stands of beeches and oaks
CONSERVATION STATUS	Extinct
COLLECTION NO.	FMNH 21499

ADULT BIRD SIZE
14–16 in (35–41 cm)

INCUBATION
12–14 days

CLUTCH SIZE
1 egg

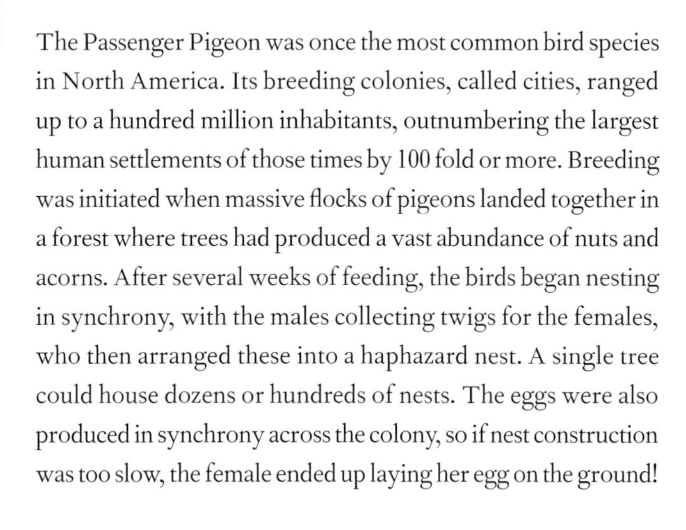

ECTOPISTES MIGRATORIUS

PASSENGER PIGEON

COLUMBIFORMES

The egg of the Passenger Pigeon was pale white to buff in background color, and 1½ x 1¹⁄₁₆ in (38 x 27 mm) in dimensions. If an egg was lost to a storm or predation, the female laid another within one week.

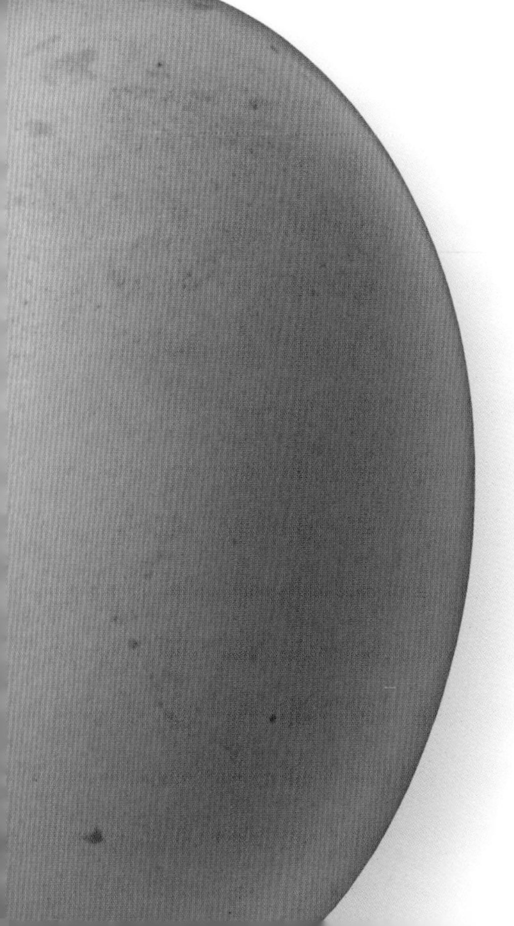

The Passenger Pigeon was once the most common bird species in North America. Its breeding colonies, called cities, ranged up to a hundred million inhabitants, outnumbering the largest human settlements of those times by 100 fold or more. Breeding was initiated when massive flocks of pigeons landed together in a forest where trees had produced a vast abundance of nuts and acorns. After several weeks of feeding, the birds began nesting in synchrony, with the males collecting twigs for the females, who then arranged these into a haphazard nest. A single tree could house dozens or hundreds of nests. The eggs were also produced in synchrony across the colony, so if nest construction was too slow, the female ended up laying her egg on the ground!

Both the female and the male took turns in incubating the egg and feeding the hatchling with crop-milk; after several days, they switched to softened food pulp regurgitated into the beak of the young. After two weeks, the parents abandoned the chick; in a few more days, the chick became flighted and joined one of the many flocks on its own.

Actual size

ORDER	Columbiformes
FAMILY	Columbidae
BREEDING RANGE	Southern United States, Central America, and northern South America
BREEDING HABITAT	Open forests, parklands, semi-arid areas, irrigated fields, residential neighborhoods
NEST TYPE AND PLACEMENT	Ground scrape lined with grasses, or a loose twig platform, lined with mosses, on branch
CONSERVATION STATUS	Least concern
COLLECTION NO.	FMNH 2393

ADULT BIRD SIZE
6–7 in (15–18 cm)

INCUBATION
12–14 days

CLUTCH SIZE
1–3 eggs

COLUMBINA PASSERINA

COMMON GROUND-DOVE

COLUMBIFORMES

This tiny dove species is about the size of a House Sparrow. As it forages throughout the day on the ground in search of seeds and grains, it is constantly under the threat of predation by both raptors and mammalian predators. Feeding is a full-time job for these birds, because they need to consume thousands of seeds during the day to meet their daily energy needs. Feeding is also part of the courtship behaviors; interested females first allow a future mate to follow them closely during foraging and in flight, and eventually allow him to "allofeed" her some regurgitated seeds to cement the pair bond.

Clutch

For nesting, these birds take just a couple of days to construct a flimsy structure, often on the ground. There, the bright white eggs are well hidden; the cryptically plumaged parents take turns to cover and incubate the eggs at all times, keeping them out of sight of any predators.

Actual size

The egg of the Common Ground-Dove is immaculate and white in color, and ⅞ x ⅔ in (22 x 16 mm) in size. Egg laying can occur throughout the year, depending on the local pattern of food abundance.

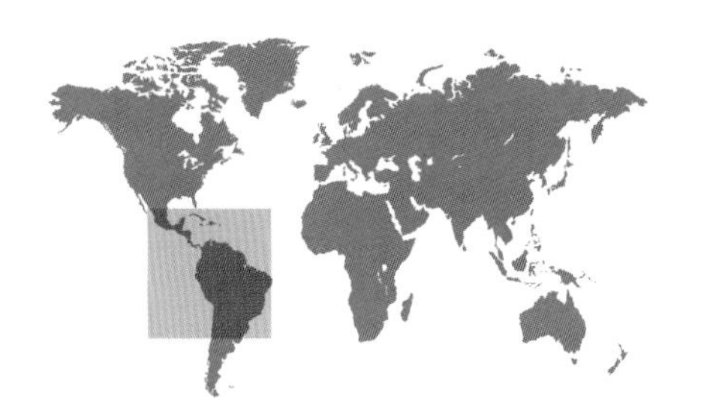

ORDER	Columbiformes
FAMILY	Columbidae
BREEDING RANGE	Central America, the Caribbean, tropical South America
BREEDING HABITAT	Woodlands, forested scrub, coffee plantations
NEST TYPE AND PLACEMENT	Platform nest built on a shrub, lined with dry leaves
CONSERVATION STATUS	Least concern
COLLECTION NO.	FMNH 20793

ADULT BIRD SIZE
7½–11 in (19–28 cm)

INCUBATION
10–11 days

CLUTCH SIZE
2 eggs

GEOTRYGON MONTANA

RUDDY QUAIL-DOVE

COLUMBIFORMES

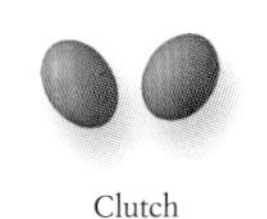

Clutch

This mid-sized, secretive dove spends most of its time on the ground, in search of seeds and small invertebrates. It is usually seen alone or in pairs, and is known to maintain a territory against conspecific intruders. In most areas, food supplies are relatively constant, enabling these doves to live in the same patch of habitat and breed for long parts of the year. However, recent evidence from Amazonia suggests that some populations move irregularly between years, being common at sites some years and rare or absent in others.

The nest is constructed at some height above the ground, typically on a stump or under dense foliage. Unlike most bird species, which lay all their eggs during the same part of the day, the first egg of the Ruddy Quail-Dove is laid in the morning of one day, and the second egg is laid in the afternoon of the next day. The parents take turns to incubate the clutch, with the male sitting on the eggs during morning hours and the female doing the same in the afternoons and at night.

Actual size

The egg of the Ruddy Quail-Dove is cream buff in background color, and measures 1⅛ x ¾ in (28 x 20 mm) in size. To keep the nest clean, and to reduce conspicuousness, soon after hatching the parents swallow the broken pieces of the eggshells, which are bright white on the inside.

ORDER	Cuculiformes
FAMILY	Musophagidae
BREEDING RANGE	Southeastern Africa
BREEDING HABITAT	Large tracts of dry forest
NEST TYPE AND PLACEMENT	Flimsy platform nest of twigs and branches, in tree canopy
CONSERVATION STATUS	Least concern
COLLECTION NO.	WFVZ 158458

ADULT BIRD SIZE
15½–17½ in (40–45 cm)

INCUBATION
22–23 days

CLUTCH SIZE
2–3 eggs

TAURACO PORPHYREOLOPHUS

VIOLET-CRESTED TURACO

CUCULIFORMES

This turaco spends most of its adult life in a pair bond. The pairs establish territories, then vocally advertize and aggressively defend them from neighbors and intruders. This is beneficial because the territory serves as the sole foraging range for this species. Together, the pair run and hop along branches to pick fresh leaves or reach the nearest fruiting tree as part of their fully vegetarian diet.

During the mating season, pairs reinforce their bond using crowing contact calls, often as a duet. Both sexes construct the nest and incubate the eggs by taking turns. Hatching is a prolonged process for the chicks, and may take up to two days. The hatchlings possess wing-claws, which make them resemble young dinosaurs. These claws, however, disappear before the chicks leave the nest. They crawl away from the nest canopy to await being fed by the parents; within several weeks they are fully fledged and become independent.

Clutch

The egg of the Violet-crested Turaco is creamy white in background color, has light yellow speckling, and measures 1⅝ x 1½ in (40 x 39 mm) in dimensions. Adult birds typically walk along branches to reach the nest, rather than flying in directly.

Actual size

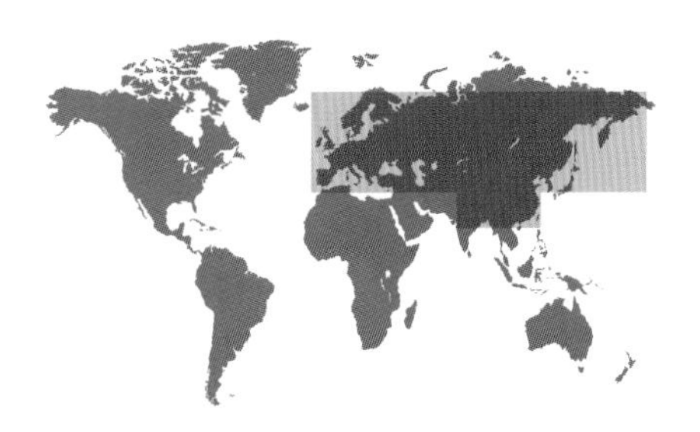

ORDER	Cuculiformes
FAMILY	Cuculidae
BREEDING RANGE	Temperate Eurasia
BREEDING HABITAT	Forests, open woodlands, reed beds
NEST TYPE AND PLACEMENT	Obligate brood parasite, lays eggs in other species' nests
CONSERVATION STATUS	Least concern
COLLECTION NO.	WFVZ 151110 (host species Eurasian Reed Warbler, *Acrocephalus scirpaceus*)

ADULT BIRD SIZE
12½–13½ in (32–34 cm)

INCUBATION
11–13 days

CLUTCH SIZE
1–2 eggs per host nest

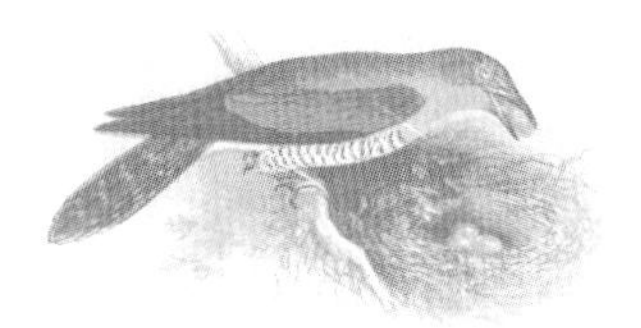

CUCULUS CANORUS

COMMON CUCKOO

CUCULIFORMES

Clutch

The egg of the Common Cuckoo is highly variable, from clear blue to beige and brown speckled, but each female lays a consistently colored and patterned egg type. The cuckoo egg's size is also variable but typically only slightly larger than the size of the host species' eggs; here, those of the Eurasian Reed Warbler.

This species is the world's best-studied obligate brood parasitic bird: the female cuckoo seeks out suitable and active nests of small songbirds in which to lay her eggs, to be cared for by the unwitting host. Host females typically lay some of their eggs in the early morning, and then leave the nest to feed themselves in order to nourish the next morning's egg. Capitalizing on their absence, the cuckoo female typically lays her egg in the mid-afternoon, so avoiding detection by the resident, soon-to-be foster-parents of the cuckoo chick. Once the cuckoo chick hatches, it quickly evicts all other eggs and nest mates.

To reduce the chances of detection by the host, cuckoos have evolved egg-color specific races, called gentes: females of the same gens lay eggs that mimic the color and maculation patterns of the egg of a host species. Similarities in light reflectance (human-visible and avian-perceivable ultraviolet range) between the cuckoo and the host eggshells lower the rate of rejection of the parasitic egg by the nest owners.

Actual size

ORDER	Cuculiformes
FAMILY	Cuculidae
BREEDING RANGE	Australia, New Zealand
BREEDING HABITAT	Forests and open woodlands
NEST TYPE AND PLACEMENT	Obligate brood parasite, lays eggs in other species' nests
CONSERVATION STATUS	Least concern
COLLECTION NO.	WFVZ 151112 (host species: Brown Thornbill, *Acanthiza pusilla*)

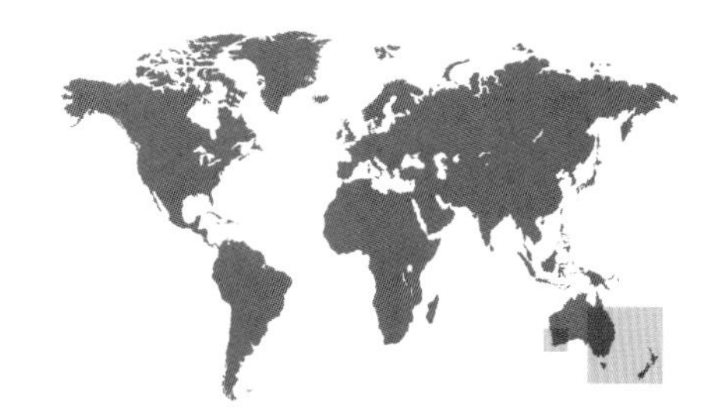

ADULT BIRD SIZE
6–6½ in (15–17 cm)

INCUBATION
12–15 days

CLUTCH SIZE
1 egg per host nest

CHALCITES LUCIDUS

SHINING BRONZE-CUCKOO

CUCULIFORMES

This is a small, emerald green cuckoo species, whose plumage hides it well in the tree foliage where it spends most of its time. Most populations of this cuckoo are migratory, while their hosts are sedentary. When they return from their wintering ground, these cuckoos will attack and destroy the contents of host nests if the host species have already begun breeding. This initiates a series of new nesting attempts by the host, which then become available for the eggs of the parasitic cuckoos.

The dark egg of the cuckoo is difficult to see in the enclosed, poorly illuminated nest of its host species; so rather than mimicking the beige and spotted host clutch, this cuckoo's egg's color makes it cryptic. Once the cuckoo egg hatches, the chick displaces all remaining eggs and nest mates to monopolize all foster-parental care. Remarkably, as a result of evolutionary host-specificity, the young parasite produces begging calls that match exactly what the host chicks would have sounded like, had they been alive.

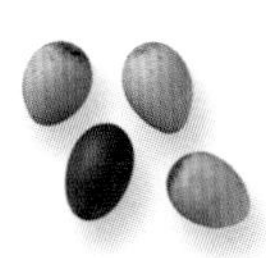

Clutch

The egg of the Shining Bronze-Cuckoo is dark brown to pale olive in background color, immaculate, and measures ¾ x ½ in (20 x 14 mm), which is comparable in size but distinct in coloration from the eggs of its host species; here, the Brown Thornbill.

Actual size

ORDER	Cuculiformes
FAMILY	Cuculidae
BREEDING RANGE	Temperate North America and the Caribbean
BREEDING HABITAT	Forests with clearings, open shrubland
NEST TYPE AND PLACEMENT	Loose platform of sticks and branches, lined with leaves and bark; on a low tree or a shrub branch
CONSERVATION STATUS	Least concern
COLLECTION NO.	FMNH 14984

ADULT BIRD SIZE
10–12 in (26–30 cm)

INCUBATION
12–14 days

CLUTCH SIZE
3–4 eggs

COCCYZUS AMERICANUS

YELLOW-BILLED CUCKOO

CUCULIFORMES

Clutch

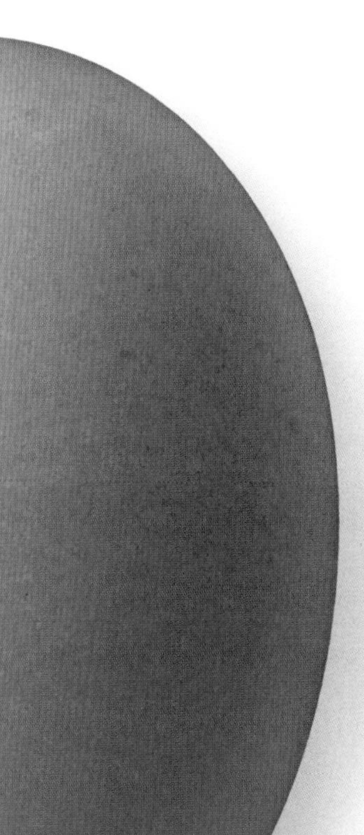

The Yellow-billed Cuckoo has long been considered an interspecific brood parasite, but in reality it usually builds its own nest and provisions its own young. In a few instances, however, this cuckoo's eggs have been discovered in the nests of other cuckoos and songbirds, most of which also lay bluish eggs; there also are reports of this species' chicks being raised by different host species.

Rare anecdotal reports and observations such as these are a critical part of the scientific discovery process, and helpful to establish the types of behavioral diversity that may exist in a species. But they can also be misleading. It took a systematic survey of over 10,000 songbird nests within the North American distribution of this cuckoo to reveal no consistent patterns of cross-species parasitism, even at low rates. The scientific process has now corrected years of misconception and it is clear that the primary breeding tactic of the Yellow-billed Cuckoo is to raise its own young.

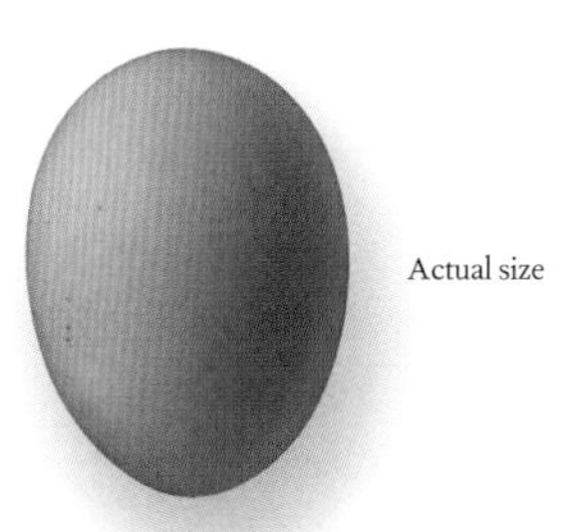

Actual size

The egg of the Yellow-billed Cuckoo is bluish green in background color, with no maculation, and measures 1¼ x ⅞ in (31 x 23 mm) in size. It was once thought that this cuckoo's egg color mimics that of the American Robin (see page 503), but this appears to be a simple convergence of egg colors in two bird species nesting, living, and breeding in same general habitat.

ORDER	Cuculiformes
FAMILY	Cuculidae
BREEDING RANGE	Tropical and temperate South America
BREEDING HABITAT	Pampas and mixed scrub forest, including farmland and suburbs
NEST TYPE AND PLACEMENT	Large cup nest, typically in thorny trees
CONSERVATION STATUS	Least concern
COLLECTION NO.	FMNH 2267

ADULT BIRD SIZE
13–15 in (32–38 cm)

INCUBATION
10–15 days

CLUTCH SIZE
1–3 eggs per female, 4–20 eggs per nest

GUIRA GUIRA

GUIRA CUCKOO

CUCULIFORMES

Guira Cuckoos are never alone, and most pairs join others to live in large groups. Members of these groups repair and reuse long-established nests, and multiple females lay their eggs in the same nest. Because of this communal behavior, Guira Cuckoos have become a model for the study of the evolution of sociality, cooperation, and group living in birds and other vertebrates.

Yet conflict is abundant in the daily life of Guira Cuckoo groups. Surprisingly, the eggs pay heavy costs by being pecked and tossed from the nest before incubation begins. Females also give their own chicks a competitive advantage by depositing more androgens into the yolks of their eggs. These steroid hormones, including testosterone, enhance the muscular development of the embryo, enabling the chicks to beg with greater vigor and to secure better chances of receiving parental provisions.

Clutch

The egg of the Guira Cuckoo is truly unique: its greenish blue background is covered by a chalky white lattice; the egg measures $1\frac{2}{3} \times 1\frac{1}{4}$ in (43 x 32 mm) in size. Many eggs are tossed by members of the breeding group and found broken under the nesting tree.

Actual size

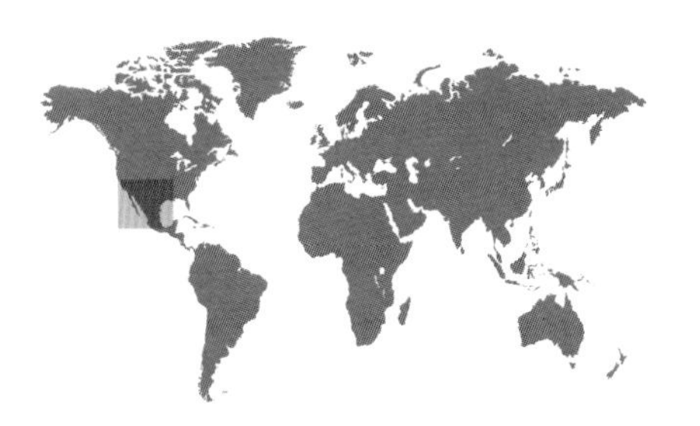

ORDER	Cuculiformes
FAMILY	Cuculidae
BREEDING RANGE	Southwestern North America
BREEDING HABITAT	Open scrubland, semi-arid fields, and deserts
NEST TYPE AND PLACEMENT	Shallow platform of thorny branches and sticks, lined with twigs, grasses, and snake skins
CONSERVATION STATUS	Least concern
COLLECTION NO.	FMNH 279

ADULT BIRD SIZE
20½–24½ in (52–62 cm)

INCUBATION
18–20 days

CLUTCH SIZE
3–6 eggs

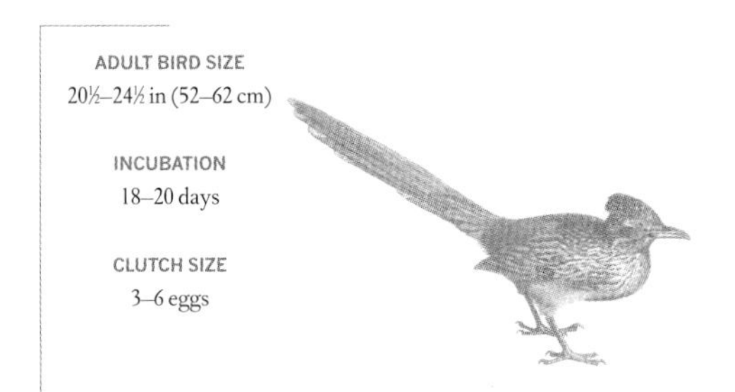

GEOCOCCYX CALIFORNIANUS

GREATER ROADRUNNER

CUCULIFORMES

Clutch

The egg of the Greater Roadrunner is chalky white in background color with a yellowish tint but otherwise clear, and measures 1⅝ x 1³⁄₁₆ in (40 x 30 mm). Some females engage in brood parasitism, and lay their eggs in nests of other roadrunners, and also of ravens or mockingbirds.

The Greater Roadrunner, immortalized by its popular Warner Brothers cartoon depiction, is a relative of the parasitic European Cuckoos. But unlike its Old World cousins, this species is parental, usually building its own nests and looking after its own young. A strong and agile predator, this bird is a successful hunter of anything smaller than itself, from venomous snakes to lizards, flying birds to nestlings, and arthropods of all sizes.

Pairs maintain an all-purpose territory for both foraging and breeding, and reside there all year long. Feeding also forms a part of the dynamic courtship display, which includes soft calls, coordinated dancelike movements, and courtship feeding, leading to copulation. The pair builds the nest together and incubates the eggs by taking turns. Similar to other cuckoos, the chicks also grow at an astonishing rate; they leave the nest at just 10 to 11 days of age, and begin foraging on their own only 16 days after hatching.

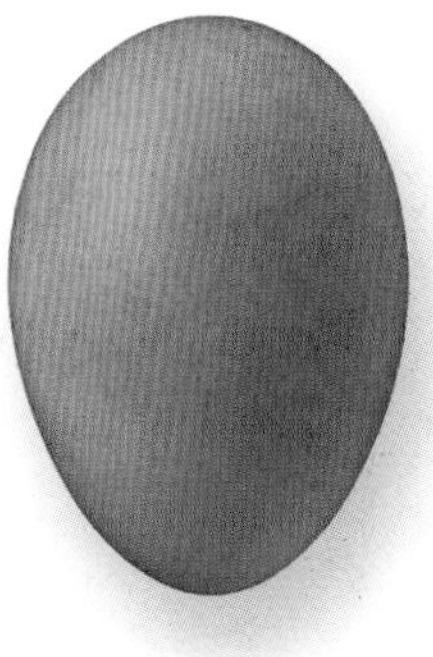

Actual size

ORDER	Cuculiformes
FAMILY	Cuculidae
BREEDING RANGE	Tropical and subtropical Americas, including the Caribbean; introduced to the Galapagos Islands
BREEDING HABITAT	Mixed grassland and forest, often near human settlements
NEST TYPE AND PLACEMENT	Large bulky cup nest, in trees
CONSERVATION STATUS	Least concern
COLLECTION NO.	FMNH 1332

ADULT BIRD SIZE
12–14 in (30–36 cm)

INCUBATION
13–21 days

CLUTCH SIZE
3–7 eggs per female, up to 22 eggs per nest

CROTOPHAGA ANI

SMOOTH-BILLED ANI

CUCULIFORMES

Anis are conspicuous, large, and noisy birds found throughout the neotropics, and they make it a group affair to build nests and provide care for their young. Breeding in groups is not unusual among birds, but the mating system of anis is anything but typical: multiple pairs of Smooth-billed Anis lay up to 22 eggs in the same nest, and individuals from different pairs take turns in incubating all the eggs and caring for the nestlings equitably.

Clutch

Surprisingly, despite the group's effort to look after the shared nest as a cooperative unit, the typical fate of this ani species' egg is failure. Many eggs are tossed out or pushed into the lining of the communal nest. This is because a breeding female tries to remove or bury the eggs that were laid before she arrived to lay her own. Thus, even in a communal effort, individual birds act selfishly and look after their own genetic interests.

Actual size

The egg of the Smooth-billed Ani is covered by a thick, white, chalky outer layer, and measures 1⅓ x 1 in (35 x 26 mm). As incubation begins, the chalky layer gradually wears off, revealing a pale blue background color.

ORDER	Opisthocomiformes
FAMILY	Opisthocomidae
BREEDING RANGE	Lowland tropical South America
BREEDING HABITAT	Rainforest, swamps, and mangrove stands
NEST TYPE AND PLACEMENT	Flimsy stick nest, placed in trees, often above water or marshes
CONSERVATION STATUS	Least concern
COLLECTION NO.	FMNH 2266

ADULT BIRD SIZE
25–26 in (63–66 cm)

INCUBATION
28 days

CLUTCH SIZE
2–3 eggs

OPISTHOCOMUS HOAZIN

HOATZIN

OPISTHOCOMIFORMES

Clutch

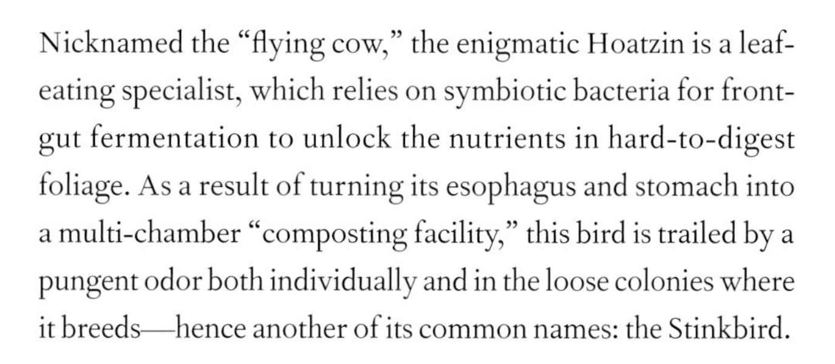

Nicknamed the "flying cow," the enigmatic Hoatzin is a leaf-eating specialist, which relies on symbiotic bacteria for front-gut fermentation to unlock the nutrients in hard-to-digest foliage. As a result of turning its esophagus and stomach into a multi-chamber "composting facility," this bird is trailed by a pungent odor both individually and in the loose colonies where it breeds—hence another of its common names: the Stinkbird.

The development of the Hoatzin reveals several other unique traits: the hatchlings possess double claws on each of their wings, and they use these to climb out into the nearby foliage from the loosely woven nest soon after hatching. The chicks are also quick to abandon their nesting tree whenever danger approaches; they drop into the water below, and swim underwater until they reach the next tree, where they again use their claws to climb up to safety.

The egg of the Hoatzin is creamy white in background color, has brown spotting, and measures 1⅔ x 1⅓ in (42 x 34 mm). The eggs in a Hoatzin colony are relatively safe from attack because the incubating adults rise in large numbers from their nests to confront intruding hawks and other predators.

Actual size

ORDER	Caprimulgiformes
FAMILY	Caprimulgidae
BREEDING RANGE	North and Central America, and the Caribbean
BREEDING HABITAT	Prairies, open scrubland, semi-deserts, burnt forests, and urban parks and rooftops
NEST TYPE AND PLACEMENT	No nest; eggs are laid directly on the ground, in a sheltered location near a log, bush, or boulder
CONSERVATION STATUS	Least concern
COLLECTION NO.	FMNH 7738

ADULT BIRD SIZE
$8\frac{1}{2}$–$9\frac{1}{2}$ in (22–24 cm)

INCUBATION
16–20 days

CLUTCH SIZE
2 eggs

CHORDEILES MINOR

COMMON NIGHTHAWK

CAPRIMULGIFORMES

A commonly seen and heard bird of the evening sky in North America, this nighthawk specializes in high-flying insect prey, which it catches on the wing. Throughout the summer, its typical "peernt" call is complemented by the males' booming noise, produced by the air rushing through their wing feathers while they dive to impress the females. When a potential mate approaches, the male drops to the open ground, spreads his tail and wings, and displays his bright white feather patches in the hope of convincing the female to mate and bond with him.

After mate bond is established, the female selects a secluded spot on the ground or on a gravel rooftop, and lays her highly cryptic eggs without any nest construction. During incubation, she gently settles down on the eggs with her breast feathers. But because the eggs are on flat ground, she also inadvertently rolls them to a new spot. Each time she departs for a feeding trip, she must relearn the location of the eggs so that she can accurately return and resume incubation.

Clutch

The egg of the Common Nighthawk is creamy white to pale olive gray in color, with heavy gray, brown, and black maculation. It measures $1\frac{3}{16}$ x $1\frac{3}{4}$ in (30 x 21 mm) in size. Only the female incubates the eggs, but both parents nourish the chicks.

Actual size

ORDER	Caprimulgiformes
FAMILY	Caprimulgidae
BREEDING RANGE	Temperate and subtropical western North America
BREEDING HABITAT	Grassland, pastures, and arid, open shrubs
NEST TYPE AND PLACEMENT	No nest, eggs are laid directly on the ground
CONSERVATION STATUS	Least concern
COLLECTION NO.	FMNH 7640

ADULT BIRD SIZE
7½–8½ in (19–21 cm)

INCUBATION
21–22 days

CLUTCH SIZE
2 eggs

PHALAENOPTILUS NUTTALLII

COMMON POORWILL

CAPRIMULGIFORMES

Clutch

The egg of the Common Poorwill is pale pink or buff in background color, has light purple markings, and measures 1 x ¾ in (26 x 20 mm). Although the eggs are highly cryptic against the bare ground, the female typically lays her egg near rocks or short vegetation for some additional cover.

This nightjar is a widespread occupant of vast arid grasslands and semidesert landscapes. Due to its occurrence into high mountain ranges, where summer nights and winters can be brutally cold, this species has evolved a unique ability to enter into days or weeks of a hibernation-like state known as torpor; in this state, the body temperature of the adult drops and all metabolic activities slow down.

If inclement weather threatens food sources during incubation, the attending parent, which can be either the male or the female, may enter torpor and cover the eggs while its own life slows down. When the weather improves and the adult's body temperature goes up again, incubation resumes. During incubation, the eggs are often rolled some distance away on a daily basis, perhaps to control exposure to sun and warmth, or to avoid detection by local predators retracing their previous night's steps.

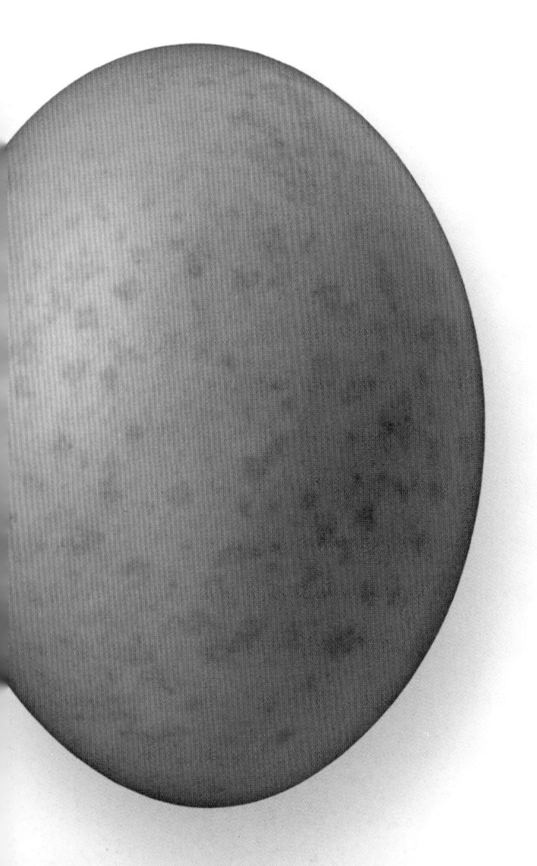

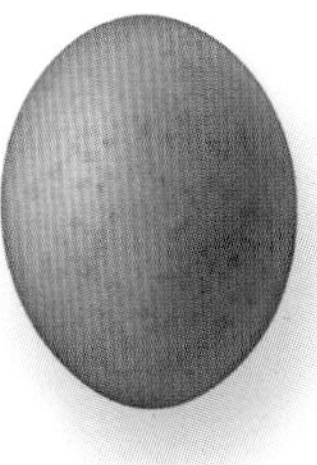

Actual size

ORDER	Caprimulgiformes
FAMILY	Caprimulgidae
BREEDING RANGE	Southeastern North America
BREEDING HABITAT	Pine, oak, or dry mixed forests
NEST TYPE AND PLACEMENT	No nest, eggs are laid on dried leaves or barren ground
CONSERVATION STATUS	Least concern
COLLECTION NO.	FMNH 18375

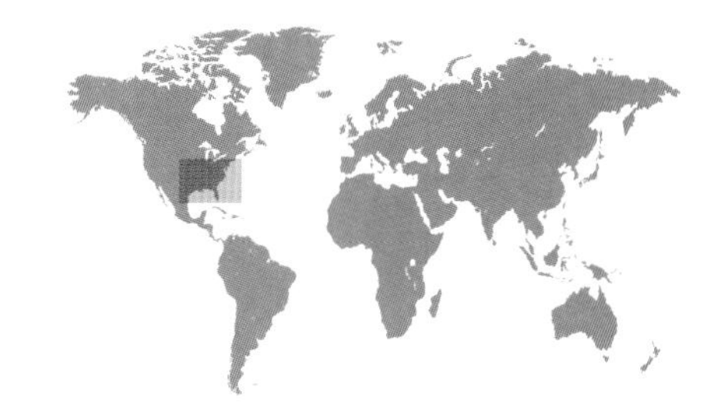

ADULT BIRD SIZE
11–12½ in (28–32 cm)

INCUBATION
20–21 days

CLUTCH SIZE
2 eggs

CAPRIMULGUS CAROLINENSIS

CHUCK-WILL'S-WIDOW

CAPRIMULGIFORMES

Birds of the dusk and moonlit nights, these large nightjars call incessantly in flight ("Chuck-will's-widow") while foraging for aerial insects. They capture prey with their wide-open beaks. Special funnel-like stiff feathers around the beak, called rictal bristles, expand the coverage for scooping insects from the air.

These birds also use vocalizations for breeding-related behaviors. For example, a male maintains a territory, often along an open section of a road or trail, by chasing intruders away while growling at them. Similarly, loud calls accompany the courtship display of the male to attract a female: he settles on a clear patch of the ground, often in the middle of a road, spreads his wings and tail, puffs his feathers, and begins vocalizing. Once a pair bond is cemented, only the female develops an incubation patch, but the male also sits on the eggs.

Clutch

The egg of the Chuck-will's-widow has a dull pink background color, is speckled with brown and lavender spots, and measures 1⅜ x 1 in (36 x 26 mm) in size. The eggs are already cryptic, but the incubating parents provide additional cover by blending in with the ground cover around them.

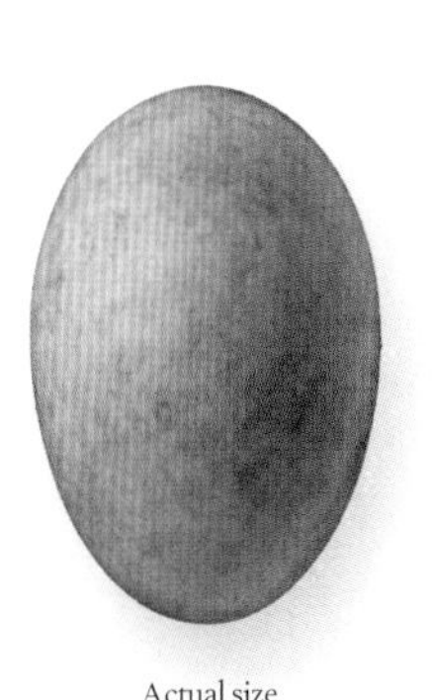

Actual size

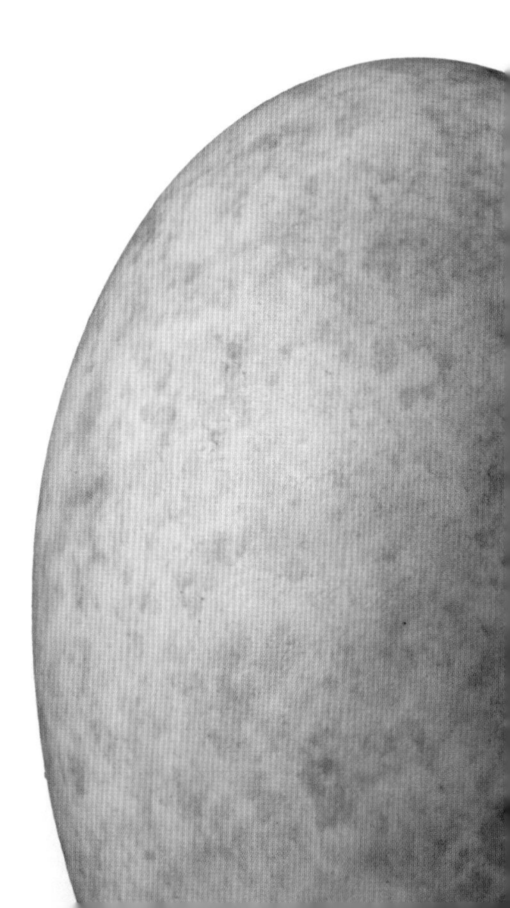

ORDER	Caprimulgiformes
FAMILY	Caprimulgidae
BREEDING RANGE	Southwestern Europe and northwestern Africa
BREEDING HABITAT	Open scrubland, sparse arid forests
NEST TYPE AND PLACEMENT	No nest, eggs are laid directly on the ground
CONSERVATION STATUS	Least concern
COLLECTION NO.	FMNH 1356

ADULT BIRD SIZE
12–13 in (30–34 cm)

INCUBATION
18 days

CLUTCH SIZE
2 eggs

CAPRIMULGUS RUFICOLLIS

RED-NECKED NIGHTJAR

CAPRIMULGIFORMES

Clutch

A local but relatively common nightjar species, this bird occasionally breeds in loose aggregations, with nesting territories much closer together than is typical of other nightjars, ranging from dozens of yards, rather than hundreds of yards, from each other.

Living near others of the same species definitely has benefits that may outweigh the costs of increased aggression between neighbors, greater competition for food, and elevated conspicuousness for predators. For example, studies have shown that when neighboring Red-necked Nightjars nest closer to one another, individual territorial defense is reduced, but communal nest defense, by mobbing of intruders, is stronger. Similarly, nesting synchrony is tighter in denser breeding groups; the greater number of individuals who have eggs and chicks at the same time, the less predation pressure they suffer per capita.

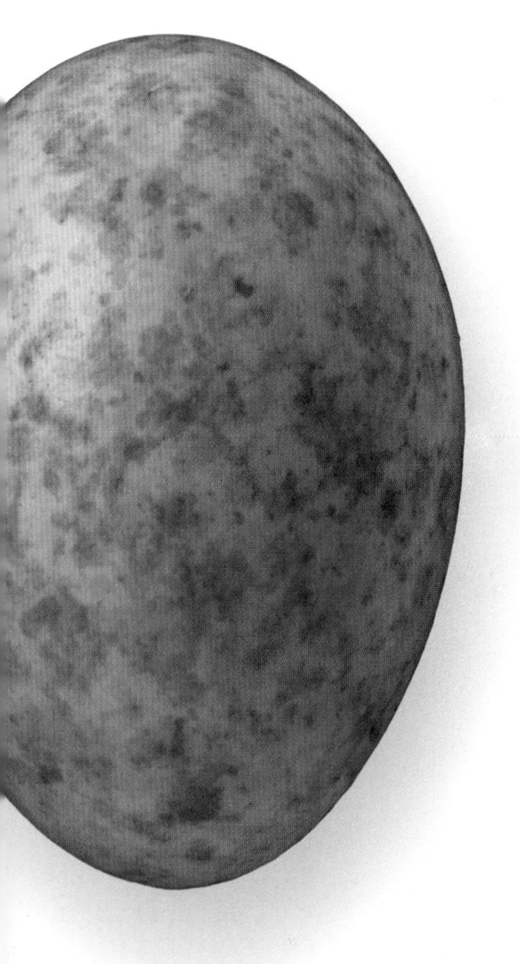

Actual size

The egg of the Red-necked Nightjar is creamy white in background color, with lilac and brownish black markings, and 1³⁄₁₆ x ⅞ in (30 x 22 mm) in size. Both sexes incubate the eggs, and an adult covers the eggs during the day, thereby providing visual protection from predators through its cryptic plumage patterns.

ORDER	Caprimulgiformes
FAMILY	Caprimulgidae
BREEDING RANGE	Eastern North America
BREEDING HABITAT	Deciduous or mixed conifer forests, with sparse undergrowth
NEST TYPE AND PLACEMENT	Shallow hollow, in leaf litter, or bare ground
CONSERVATION STATUS	Least concern
COLLECTION NO.	FMNH 19285

ADULT BIRD SIZE
8½–10 in (22–26 cm)

INCUBATION
19–21 days

CLUTCH SIZE
2 eggs

CAPRIMULGUS VOCIFERUS

EASTERN WHIP-POOR-WILL

CAPRIMULGIFORMES

A loud and common nightjar species, this species calls "whip-poor-will," like its name. To establish a pair bond, the female and the male perform a courtship display that involves following and dancing, calls and short flights, both on the ground and in the air, in forest clearings or on roadways. Once the mates begin to breed, the eggs are laid on the forest floor, and incubation begins with the first egg. Egg laying is influenced by the lunar cycle, so that moonlit nights usually follow after hatching, allowing the parents to see well enough to forage productively, providing enough food for themselves and the hatchlings.

One week after the eggs hatch, the female may leave the male to look after these chicks, while she lays another clutch of eggs in a second breeding attempt elsewhere. This is important because despite the near-perfect visual camouflage of the eggs and the chicks, and the mobility of the chicks to leave the nest and separate from each other, ground-dwelling predators, using olfactory cues, are still a common cause of brood loss.

Clutch

The egg of the Eastern Whip-poor-will is cream-colored or grayish white in coloration, has lavender, yellow, and brown maculation, and measures 1⅛ x ¾ in (29 x 21 mm).

Actual size

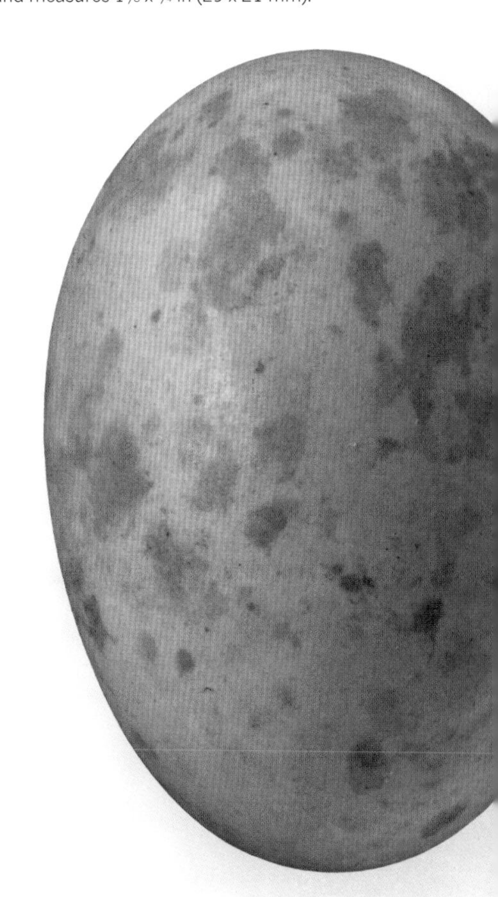

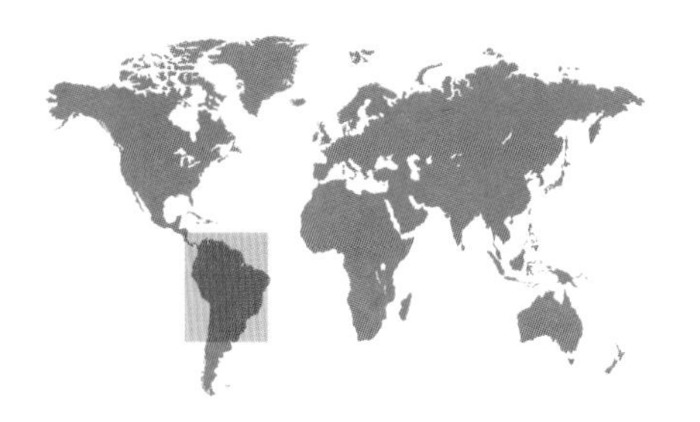

ORDER	Caprimulgiformes
FAMILY	Nyctibiidae
BREEDING RANGE	Central America and South America
BREEDING HABITAT	Open forests, and savanna
NEST TYPE AND PLACEMENT	Egg is laid into a depression in a log or tree trunk
CONSERVATION STATUS	Least concern
COLLECTION NO.	WFVZ 172910

ADULT BIRD SIZE
13–15 in (33–38 cm)

INCUBATION
30–33 days

CLUTCH SIZE
1 egg

NYCTIBIUS GRISEUS

COMMON POTOO

CAPRIMULGIFORMES

The egg of the Common Potoo is white in background color, with purple patterning, and 1⅔ x 1¼ in (42 x 31 mm) in dimensions. No nest structure is built, but the eggs are well protected through the camouflage provided by the incubating adult's plumage.

This large tropical relative of the nightjars spends its days in plain sunlight; yet it is rarely seen by hawks, falcons, or people. With its bark-toned plumage pattern and stretched-out neck and beak, its colors and shapes blend into the substrate, making the bird look like an extension of a broken tree stump. Unlike the nightjars, which hunt for nocturnal insects on the wing, the Common Potoo (also called the Gray or Lesser Potoo) uses its high perching spot to its advantage while foraging for insects, spotting them from above and swooping down to catch them.

The breeding season begins with the male calling loudly and repeatedly in a haunting four-note song for the female. Once the mates have settled with each other, a single egg is laid, and the lone nestling is provisioned by both parents during nighttime hours only. Infrared recordings have revealed that while the chick grows, the parents attend it less often, and it occasionally catches nearby flies while still waiting for the parents to provide some food.

Actual size

ORDER	Caprimulgiformes
FAMILY	Steatornithidae
BREEDING RANGE	Northern South America
BREEDING HABITAT	Caves in or near forests
NEST TYPE AND PLACEMENT	Placed on ledges inside caves; platform made of regurgitated fruit pulp and droppings
CONSERVATION STATUS	Least concern
COLLECTION NO.	FMNH 2496

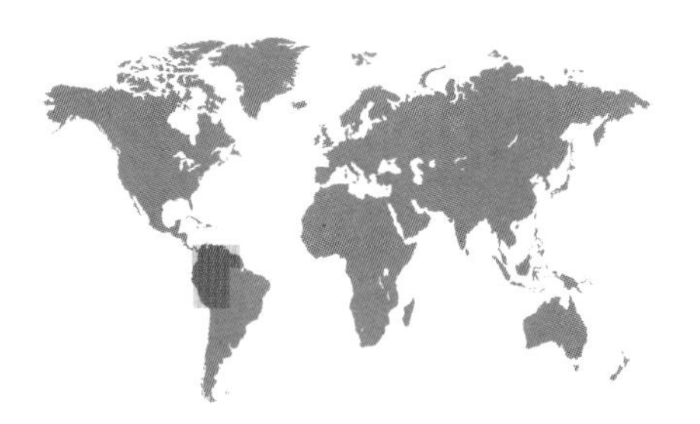

ADULT BIRD SIZE
16–19½ in (41–49 cm)

INCUBATION
32–35 days

CLUTCH SIZE
2–4 eggs

STEATORNIS CARIPENSIS

OILBIRD

CAPRIMULGIFORMES

Clutch

The Oilbird is a highly gregarious and colonial inhabitant of the neotropics; it nests in permanent colonies, and forages in large groups, making its presence known through its repeated clicking calls. This vocalization is its highly specialized navigation system. Through convergent evolution, similar to bats and dolphins, these birds use echolocation to find their way through the dark caves where their nests are located, as well as around the nighttime forest to land on fruiting trees. The only other birds with echolocation are some Asian swiftlets.

Throughout the year, the Oilbird resides in the same cave, and maintains a strong pair bond even outside the breeding season. The birds reuse the same nests year after year, which grow in height due to new material, especially droppings, added during the incubation and nesting stages. The chicks become very heavy before they fledge, weighing up to 50 percent more than the parents themselves—it is at this stage that local peoples have historically harvested oilbird squabs for their fat and meat.

The egg of the Oilbird is glossy and clear white in color, and measures 1⅔ x 1⅓ in (42 x 33 mm) in dimensions. The eggs become increasingly brown-stained during incubation as they roll around the nest made of fruit pulp and fecal matter.

Actual size

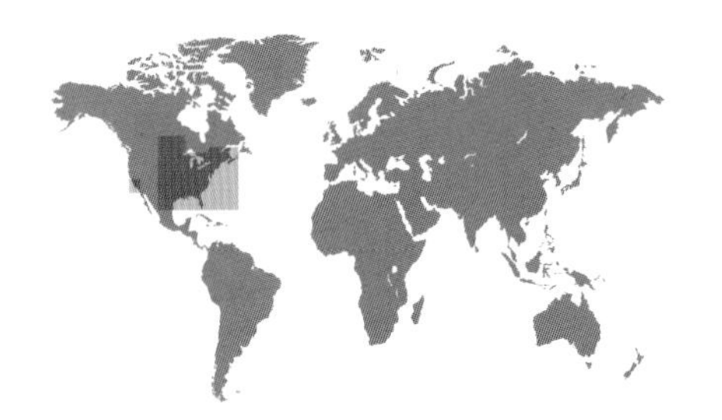

ORDER	Apodiformes
FAMILY	Apodidae
BREEDING RANGE	Eastern and central North America
BREEDING HABITAT	Historically, woodlands; today, built-up urban areas
NEST TYPE AND PLACEMENT	Historically, on vertical walls of large tree holes; today, mostly chimneys, airshafts, wells, or other interior spaces; a cup nest made of short sticks, glued to the wall with hardened saliva
CONSERVATION STATUS	Near threatened
COLLECTION NO.	FMNH 7785

ADULT BIRD SIZE
5–6 in (12–15 cm)

INCUBATION
19–21 days

CLUTCH SIZE
4–5 eggs

CHAETURA PELAGICA

CHIMNEY SWIFT

APODIFORMES

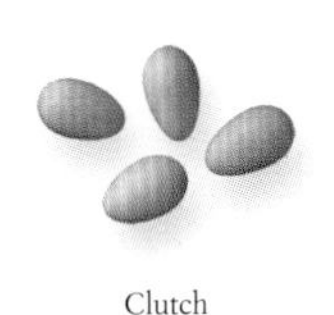

Clutch

Despite the popular notion that swifts do everything in the air, the Chimney Swift spends many critical stages of its reproductive cycle on a solid surface: it lands to collect twigs and branches used for the nest structure, it mates on the vertical walls of its nesting space, and both sexes take turns in incubating the eggs. The chicks are very tolerant of temperature fluctuations, and during short periods of cold, when flying insects to feed to the chicks are scarce, the chicks can go into a hibernation-like state, called torpor, to reduce metabolic rates and to save energy.

A vocal and social aerial insectivore, the Chimney Swift has found new breeding habitat in major North American cities. It was historically a common species in the industrialized, urban skies of eastern and central North America during the breeding season; in winter it migrates south to South America where the whereabouts of the entire species remains a mystery. Without complete knowledge of its annual cycle, it is also unclear why recent decades have seen a continuous decline in numbers of this once very common species.

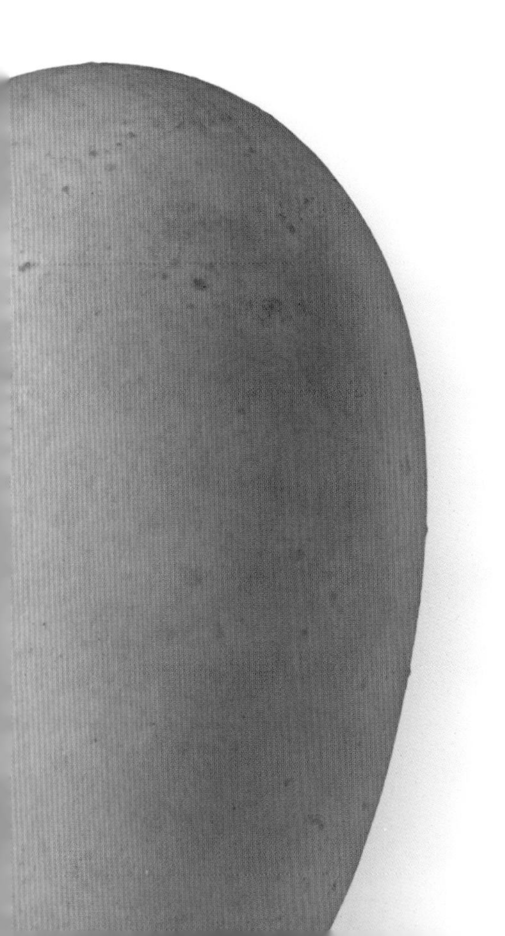

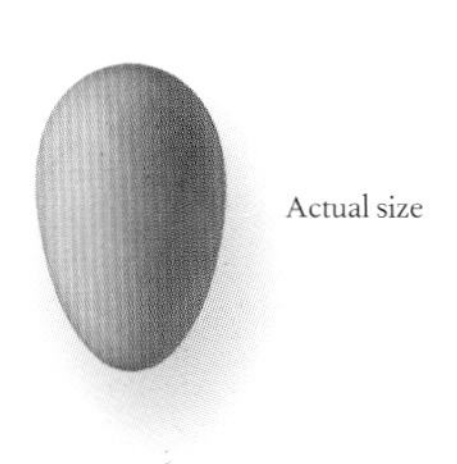

Actual size

The egg of the Chimney Swift is glossy white in background color, and is ⅞ x ½ in (21 x 14 mm) in size. Each egg weighs in at nearly 10 percent of the female's body weight, so producing a clutch of eggs represents a major reproductive effort for the female.

ORDER	Apodiformes
FAMILY	Apodidae
BREEDING RANGE	Temperate Europe, north Africa, and central Asia
BREEDING HABITAT	Forests, and built-up areas, mixed with open spaces
NEST TYPE AND PLACEMENT	Inside tree cavities, and on cliff walls; also under eaves and in holes in walls; a cup nest made of airborne plant fibers, spider silk, feathers, and saliva
CONSERVATION STATUS	Least concern
COLLECTION NO.	FMNH 2404

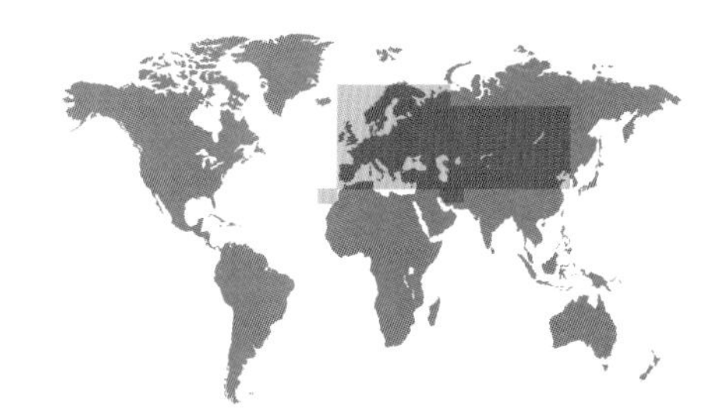

ADULT BIRD SIZE
6½–7 in (16–17 cm)

INCUBATION
20–22 days

CLUTCH SIZE
2–3 eggs

APUS APUS

COMMON SWIFT

APODIFORMES

An aerial specialist, this swift truly lives on the wing. It moves, feeds, sleeps, and copulates in midair; it even collects all nesting materials while flying. These are then glued together to construct the nest structure with lots of saliva; the salivary glands become enlarged during the breeding season. The nest is a valuable asset and is typically refurbished and reused, rather than rebuilt, on a yearly basis.

Clutch

Prior to mating, the male swift selects a suitable nesting site, as part of a smallish colony of several dozen other males. At first, females respond to courtship with aggressive chases of the male, but when a female accepts an invitation to inspect the site, she enters and then also allows the male to preen her face feathers. Pairs form monogamous bonds, and typically return to breed at the same site each spring after spending more than half of the year in the tropical wintering grounds.

Actual size

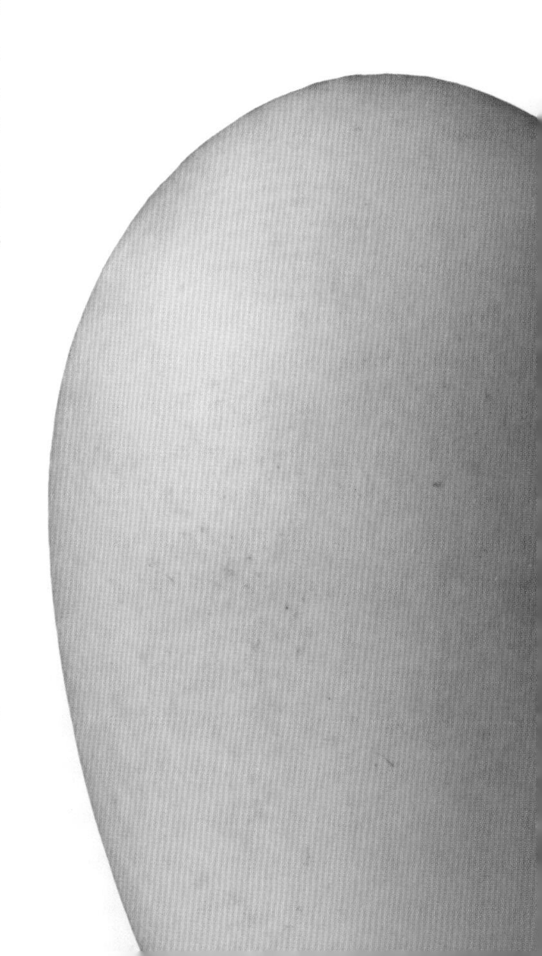

The egg of the Common Swift is pure white, and its dimensions are ⅞ x ⅔ in (24 x 16 mm). Both parents take turns to incubate the eggs, and successful parents remain together for several years.

ORDER	Apodiformes
FAMILY	Trochilidae
BREEDING RANGE	Northwestern South America
BREEDING HABITAT	Tropical humid forests on both slopes of the Andes
NEST TYPE AND PLACEMENT	A small hanging nest, made of mosses and stems, attached to tree branches in the high canopy
CONSERVATION STATUS	Least concern
COLLECTION NO.	FMNH 2944

ADULT BIRD SIZE
3½–8 in (9–20 cm)

INCUBATION
15–17 days

CLUTCH SIZE
2 eggs

AGLAIOCERCUS KINGI

LONG-TAILED SYLPH

APODIFORMES

Clutch

This hummingbird is a prime example of sexual selection by female choice for exaggerated male ornaments. Other than attracting the females, the elongated tail feathers of the male serve no practical purp≠≠ose. In fact, carrying and flying with them make foraging, escaping, and protecting display and feeding territories only possible for males who are in the best condition. Their reward is female preference for long tails, which eventually translates into fertilization and siring success of the next generation.

The female Long-tailed Sylph is charged with all aspects of parental care, including nest building and incubation. To nourish the fast-growing chicks, she must not only feed on energy-rich nectar, but also capture insects in flight or glean them off leaves and branches, to provide sufficient protein content for herself and her young.

Actual size

The egg of the Long-tailed Sylph is clear white in color, and ½ x ¼ in (13 x 7 mm) in dimensions. There is no guarantee that an occupied nest contains an egg—outside of the breeding season, females may build nests for the sake of nocturnal roosting.

ORDER	Apodiformes
FAMILY	Trochilidae
BREEDING RANGE	Northern South America
BREEDING HABITAT	Montane forest edges, open shrubby fields, thickets
NEST TYPE AND PLACEMENT	Woven, open cup nest, attached to a thin branch in a low shrub or tree
CONSERVATION STATUS	Near threatened
COLLECTION NO.	FMNH 2928

ADULT BIRD SIZE
4½–5¼ in (11–13 cm)

INCUBATION
Unknown

CLUTCH SIZE
2 eggs

ERIOCNEMIS CUPREOVENTRIS

COPPERY-BELLIED PUFFLEG

APODIFORMES

This hummingbird is a small specialist of montane forests and thickets, where it preferentially settles in sparse and edge habitats of stunted forests and feeds on nectar from long-tubed flowers. These forests have been intensively cleared and used for agriculture or grazing in much of the already limited and patchily occupied range of this species, resulting in its conservation status being elevated to near threatened. This is an interspecifically aggressive species, which means that it does not tolerate other hummingbirds that attempt to feed or breed on its own territory.

Both the male and the female exhibit a pouchlike bundle of white feathers at the base of their legs (hence the name "puffleg" for all the species in this genus). Like most other hummingbirds, the male is more brightly colored with iridescent patches of feathers. Because so little is known about the breeding biology and the natural history of this species, a specimen of a melanistic (all black) plumaged individual was considered for a while as a fully separate species.

Clutch

Actual size

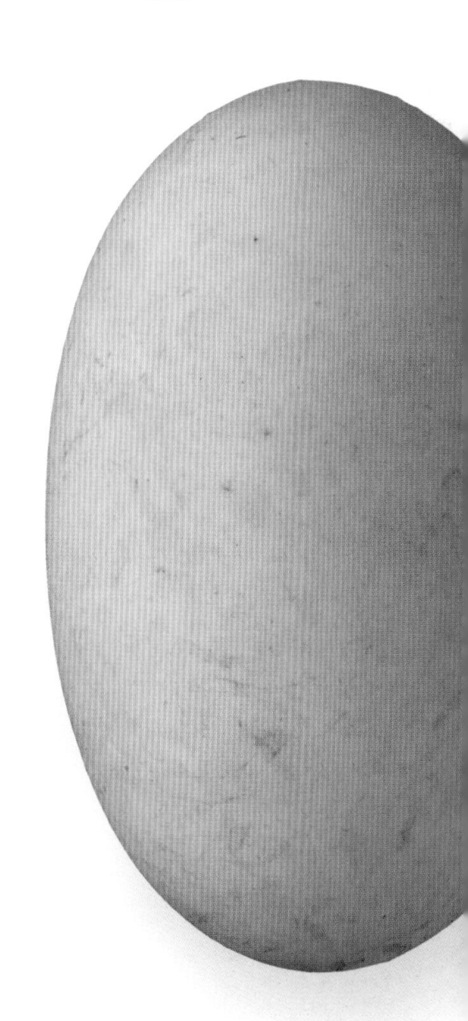

The egg of the Coppery-bellied Puffleg is pure white and immaculate in coloration, and ½ x ⅓ in (13 x 8 mm) in size. There is no detailed knowledge of the nesting and parental behavior of this species; such information is important for conservation initiatives to assess risk and to make progress with management plans.

ORDER	Apodiformes
FAMILY	Trochilidae
BREEDING RANGE	Eastern North America
BREEDING HABITAT	Deciduous woodlands and fields, including orchards and suburbs
NEST TYPE AND PLACEMENT	A small, walnut-sized cup attached firmly onto a tree branch
CONSERVATION STATUS	Least concern
COLLECTION NO.	FMNH 2061

ADULT BIRD SIZE
3–3½ in (7–9 cm)

INCUBATION
12–14 days

CLUTCH SIZE
2 eggs

ARCHILOCHUS COLUBRIS

RUBY-THROATED HUMMINGBIRD

APODIFORMES

Clutch

The Ruby-throated Hummingbird is the only hummingbird species that breeds in eastern North America. It also has the largest breeding range of any northern hemisphere hummingbird and flies thousands of miles during its annual migration. In the spring, following a spectacular courtship display involving sweeping flights like a pendulum and flashing his bright red gorge (throat) to the observing female, the male forgoes all parental duties. He does not build the nest, incubate the eggs, or provision the chicks.

The nest is built on top of a tree branch, often on a downward-sloping one; it is made of fluffy plant fibers held together by spider silk, and camouflaged with lichens. Once the eggs hatch, the deep nest houses the chicks securely. Often just a pair of thin long beaks sticking out of it reveal that the nest is in active use.

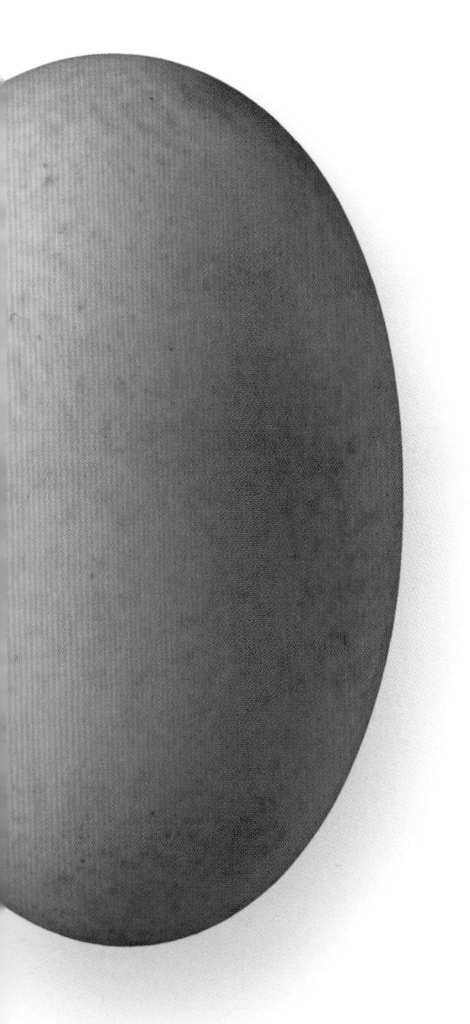

Actual size

The eggs of the Ruby-throated Hummingbird in the nest look like a pair of pearls in a small cup; the color is pure white, and non-glossy. They are pea-sized—½ x ⅓ in (13 x 9 mm)—and each egg weighs less than a paperclip.

ORDER	Apodiformes
FAMILY	Trochilidae
BREEDING RANGE	Western North America
BREEDING HABITAT	Riparian forests, tall scrubland, also in high deserts
NEST TYPE AND PLACEMENT	Deep cup nest on a tree branch, made of spider silk and plant down
CONSERVATION STATUS	Least concern
COLLECTION NO.	FMNH 134

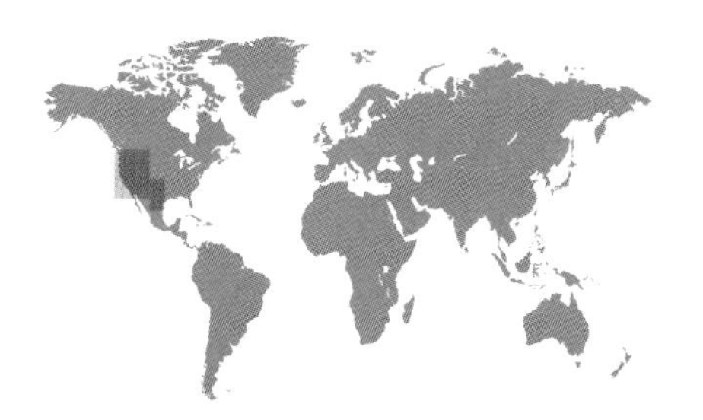

ADULT BIRD SIZE
3¼–3¾ in (8.5–9.5 cm)

INCUBATION
12–16 days

CLUTCH SIZE
2 eggs

ARCHILOCHUS ALEXANDRI

BLACK-CHINNED HUMMINGBIRD

APODIFORMES

This species is an inconspicuously ornamented and small hummingbird, but it may be one of the best-known species in the American West. This is because it is highly adaptable for life in a changing world, and has expanded its breeding habitats from increasingly sparse riverine forests, tall tree stands, and undisturbed scrub, to suburban backyards, and city parks. The male vigorously defends a well-defined mating territory by chasing away intruders; he also displays to approaching females by performing spectacular aerial dives.

Once the eggs are fertilized, the male turns his attention to attracting other females, and does not participate in any parental duties. The female alone builds the nest by collecting spider silk, cocoon fibers, and plant filaments; she incubates the eggs, and provisions the hatchlings by transferring to them a mix of energy-rich nectar and protein-rich small insects.

Clutch

Actual size

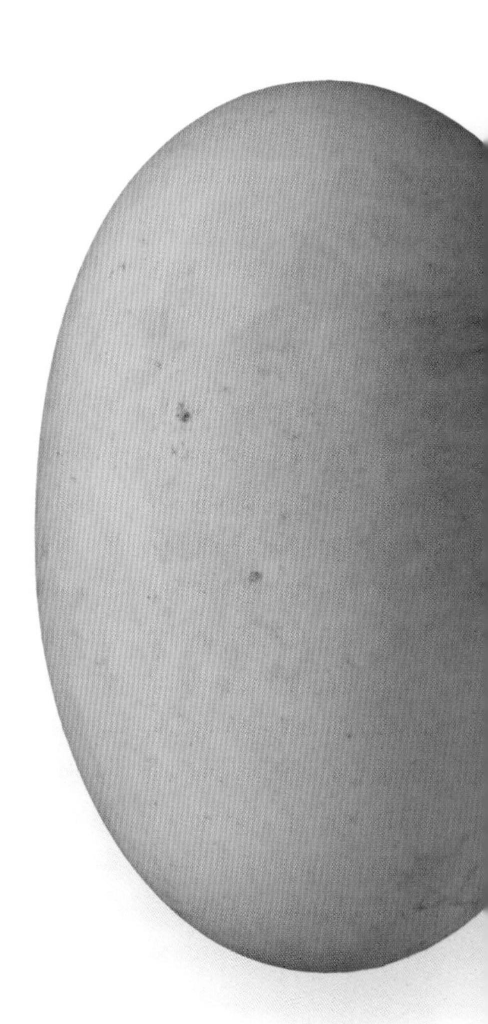

The egg of the Black-chinned Hummingbird is pure white in color, and measures ½ x ⅓ in (13 x 9 mm). Once the eggs hatch and the chicks develop, the nest cup, which is made of a flexible, natural mesh, expands to fit the growing size of the family.

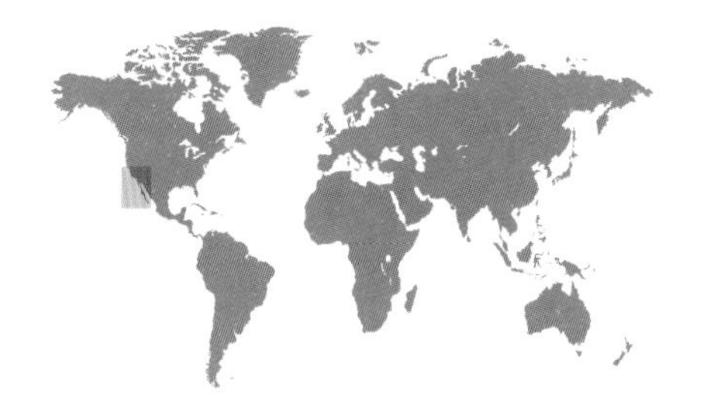

ORDER	Apodiformes
FAMILY	Trochilidae
BREEDING RANGE	Southwestern North America
BREEDING HABITAT	Desert and semi-desert; chaparral
NEST TYPE AND PLACEMENT	Cup nest on branches of shrubs and trees, composed of vegetation and spider silk
CONSERVATION STATUS	Least concern
COLLECTION NO.	FMNH 16408

ADULT BIRD SIZE
3½–4 in (9–10 cm)

INCUBATION
15–18 days

CLUTCH SIZE
2–3 eggs

CALYPTE COSTAE

COSTA'S HUMMINGBIRD

APODIFORMES

Clutch

This hummingbird is an inhabitant of the dry Sonoran desert in the southwestern United States. Some populations are migratory into coastal Mexico, but others stay in place year-round. If an adult experiences a severely cold night in the desert, it may go into torpor by dropping its heart rate from 500 to 50 beats per minute, so saving critical energy for its next morning's activity.

Males engage in aerial displays and battle for breeding grounds for one to three weeks before the females arrive. When they do arrive, the males compete with neighboring males for the attention of mates. Once sperm-transfer is complete, however, the female is in full charge of breeding, from nesting to fledging. She builds a nest by constructing a loosely knit cup of plant matter and spider silk on a branch, which she then finishes to form a cup shape by running her chin along the rim of the nest while sitting inside it.

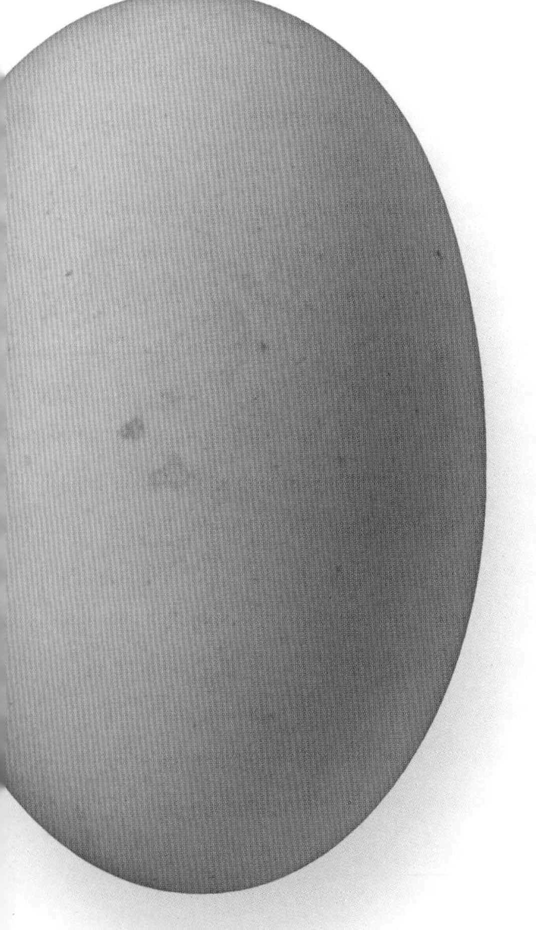

Actual size

The egg of the Costa's Hummingbird is clear and white in color, and measures ½ x ⅓ in (12 x 8 mm) in size. Following nest construction, the female delays laying for two to three days, presumably to save energy for subsequent care of the eggs.

ORDER	Apodiformes
FAMILY	Trochilidae
BREEDING RANGE	Northwestern North America, including Alaska
BREEDING HABITAT	Open woodlands, secondary growth
NEST TYPE AND PLACEMENT	Cup nest, in coniferous or deciduous trees, hidden by drooping branches; may form nesting aggregations of 20 to 30 individuals
CONSERVATION STATUS	Least concern
COLLECTION NO.	FMNH 3008

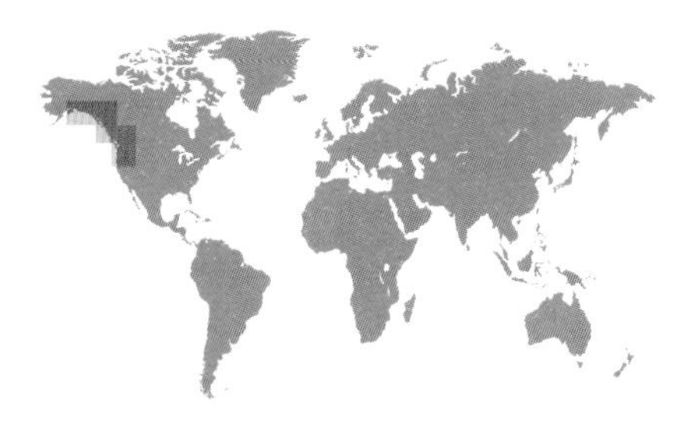

ADULT BIRD SIZE
3–3½ in (7–9 cm)

INCUBATION
15–17 days

CLUTCH SIZE
2–3 eggs

SELASPHORUS RUFUS

RUFOUS HUMMINGBIRD

APODIFORMES

From its wintering grounds in South America, the Rufous Hummingbird migrates further north than any other hummingbird, reaching the 61st parallel in Alaska. Factoring in its small size, it migrates further than any other bird species. This species has the shortest reproductive season of any hummingbird—it breeds in the Pacific Northwest where summers are fleeting. However, since the warmer days are also relatively long, there is plenty of time to hunt for insect prey, which allows the young to grow very quickly in the brief time available.

Despite the well-known migratory journey of this species, information about their breeding biology is sparse. The female is more cryptically colored than the male and she alone builds the nest and feeds the young by pumping food from bill to bill. Just before sunset, the rate of food delivery increases; presumably to satiate the chicks before nighttime, when the mother can no longer capture and deliver food.

Clutch

Actual size

The egg of the Rufous Hummingbird is white and immaculate, and ½ x ⅜ in (13 x 9 mm) in dimensions. The eggs typically hatch one or two days apart, suggesting that incubation begins on the day when the first egg is laid.

ORDER	Apodiformes
FAMILY	Trochilidae
BREEDING RANGE	Central America, and northern South America
BREEDING HABITAT	Open forests, second growths, coffee plantations, and gardens
NEST TYPE AND PLACEMENT	Cup of plant down and cobwebs, covered with lichen on the outside; placed on an external twig or branch, in a tree
CONSERVATION STATUS	Least concern
COLLECTION NO.	FMNH 2742

ADULT BIRD SIZE
3–4 in (7–10 cm)

INCUBATION
15–17 days

CLUTCH SIZE
2–3 eggs

AMAZILIA SAUCERROTTEI

STEELY-VENTED HUMMINGBIRD

APODIFORMES

Clutch

The Steely-vented Hummingbird is a conspicuous and noisy tropical species, common in many kinds of human-modified habitats, including forest edges, gardens, and parks. Its daily activities revolve around foraging, as birds begin their day searching out and aggressively defending nectar-rich flowers on a specific clump of blooming trees. Both females and males defend their feeding territories aggressively against all others; only during the courtship season does a male allow females to approach, at which time he performs his mating display. There is no pair bond, and the female alone is in full charge of nest building, incubation, and parental care for the next generation.

Based on the diversity of their vocalizations, some researchers have suggested that the Central American populations might be a different species from the South American populations. However, this awaits further study, since song divergence may not equate to genetic divergence. Hummingbirds, like parrots and songbirds, learn their songs by imitating others, thus passing on their vocal traditions from one generation to the next through cultural evolution.

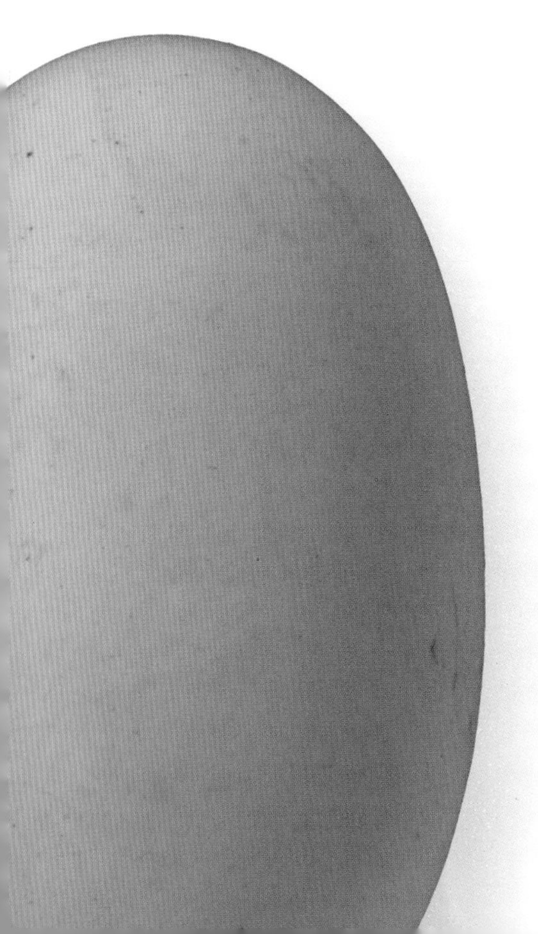

Actual size

The egg of the Steely-vented Hummingbird is white in background color, immaculate, and ½ x ⅜ in (14 x 9 mm) in dimensions. Despite their bright white color, the eggs, placed at the bottom of the deep cup of the tiny nest, are well hidden from predators at most angles.

ORDER	Trogoniformes
FAMILY	Trogonidae
BREEDING RANGE	Southern North America and Central America
BREEDING HABITAT	Dry woodlands, thorn brush, riparian forests
NEST TYPE AND PLACEMENT	Unlined nest in tree hole
CONSERVATION STATUS	Least concern
COLLECTION NO.	FMNH 3129

ADULT BIRD SIZE
11–12 in (28–30 cm)

INCUBATION
17–19 days

CLUTCH SIZE
2–4 eggs

TROGON ELEGANS

ELEGANT TROGON

TROGONIFORMES

The Elegant Trogon is an adaptable species that does not specialize on any particular habitat; instead it adjusts its foraging and breeding behavior to local opportunities, from lowland dry forests to highland conifer stands. It is an obligate cavity nester, but cannot excavate a nest hole on its own. It relies mostly on finding old woodpecker holes for breeding. Competition for cavities can be fierce in the forest, and trogons are known to lose their nest sites to more aggressive birds such as flycatchers and owls.

Nest sites are selected by both members of the pair. The male typically chooses a potential nesting cavity first, then calls repeatedly to the female while turning his head inside the hole again and again. If the female does not accept, he moves on and starts the process at the next suitable cavity entrance. The pair might inspect and try out several available cavities before settling and the female starting to lay eggs.

Clutch

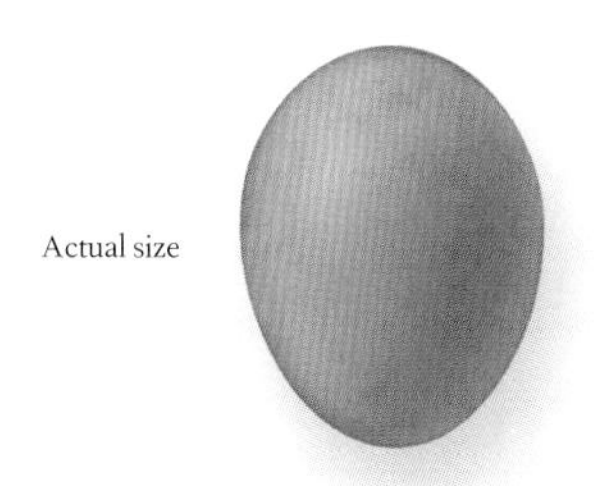

Actual size

The egg of the Elegant Trogon is white, dull in appearance, with a blue tint in background color, has no maculation, and measures 1 1⁄16 x 7⁄8 in (27 x 22 mm). Both parents incubate the eggs, each contributing about the same amount of time.

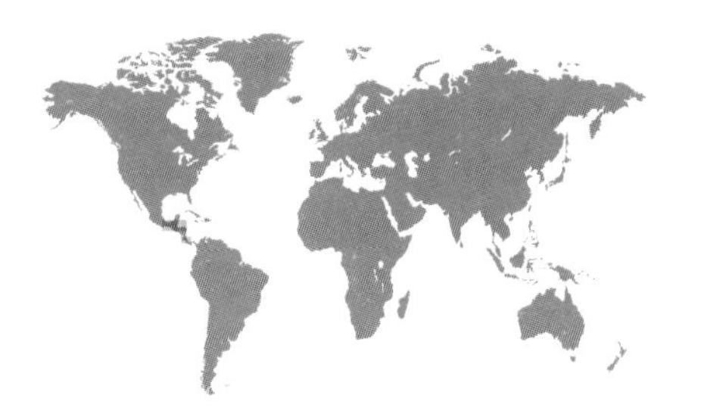

ORDER	Trogoniformes
FAMILY	Trogonidae
BREEDING RANGE	Central America
BREEDING HABITAT	Montane cloud forest
NEST TYPE AND PLACEMENT	Cavity nest, carved into a rotting tree trunk
CONSERVATION STATUS	Near threatened
COLLECTION NO.	WFVZ 24329

ADULT BIRD SIZE
14–41½ in (36–105 cm)

INCUBATION
18 days

CLUTCH SIZE
2 eggs

PHAROMACHRUS MOCINNO

RESPLENDENT QUETZAL

TROGONIFORMES

Clutch

Revered by both ancient and modern cultures in Central America, the Resplendent Quetzal has paid the cost of having brilliantly colored plumage by becoming a frequent trophy for hunters and feather collectors. The male has a pair of tail streamers nearly 2 ft (60 cm) long, which barely fit into the nest during incubation. Often they are folded over his back, and stick out of the nest hole as if they were branches of a tree fern.

This quetzal breeds in monogamous pairs, and the mates work together to excavate the nest cavity in the rotting trunk of a mature tree. The same nest can be reused across years. Once the eggs hatch, both parents continue to provide parental care for the young, feeding them fruits, berries, and insects. At around the time of fledging, the female often abandons the family unit, leaving the male to provide food for the fledglings until they reach independence.

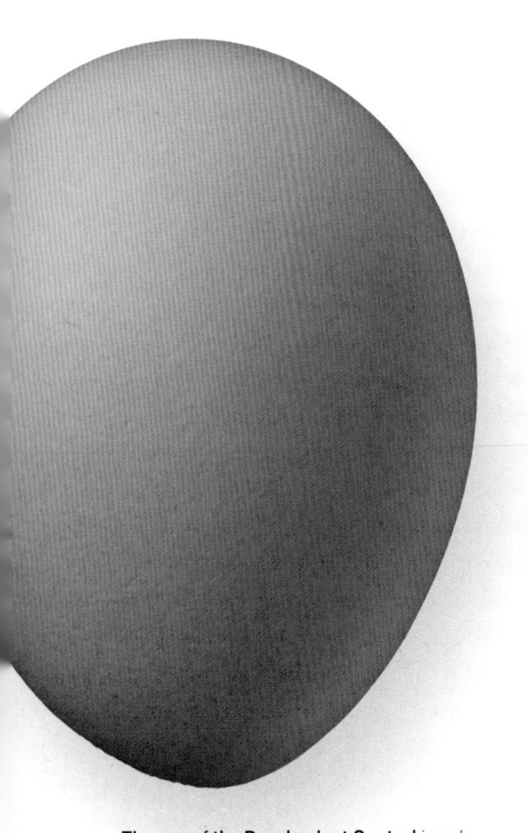

The egg of the Resplendent Quetzal is pale blue in background color, clear of spotting, and 1⅜ x 1⅓ in (36 x 33 mm) in dimensions. The nest site in a rotting tree limb exposes the eggs to weather damage or loss by predation, as intruders can easily access the eggs through the decayed nest walls.

Actual size

ORDER	Coraciiformes
FAMILY	Todidae
BREEDING RANGE	Hispaniola
BREEDING HABITAT	Low-elevation dry forest, and scrub
NEST TYPE AND PLACEMENT	Tunnel and nesting chamber, dug into sand or soil bank
CONSERVATION STATUS	Least concern
COLLECTION NO.	FMNH 19142

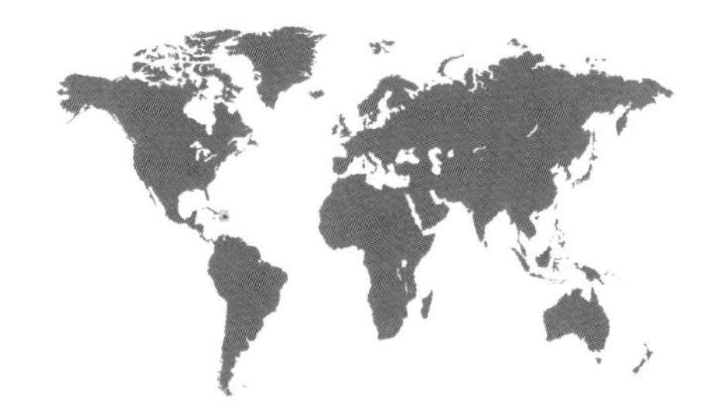

ADULT BIRD SIZE
4½–4¾ in (11–12 cm)

INCUBATION
16–20 days

CLUTCH SIZE
3–4 eggs

TODUS SUBULATUS

BROAD-BILLED TODY

CORACIIFORMES

The Broad-billed Tody is a representative of a uniquely Caribbean family of birds, and is one of two species of todies on the island of Hispaniola. The two species occur in different elevations and forest types; on all other Greater Antillean islands, only a single endemic species of tody is found. These tiny birds occupy a unique niche on these islands, foraging for insects and nesting in burrows.

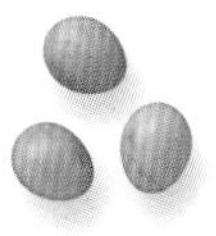

Clutch

Prior to the mating season, todies live alone, and can be seen perched at a prominent spot, scanning the air and the nearby ground for moving insects. They form pairs through an elaborate ritual of prolonged aerial displays, mutual flights and chases, with rattling sounds produced by stiff wing feathers. This species is monogamous, and both sexes contribute to the excavation of the nesting tunnel and chamber, incubation of the eggs, and feeding the chicks.

Actual size

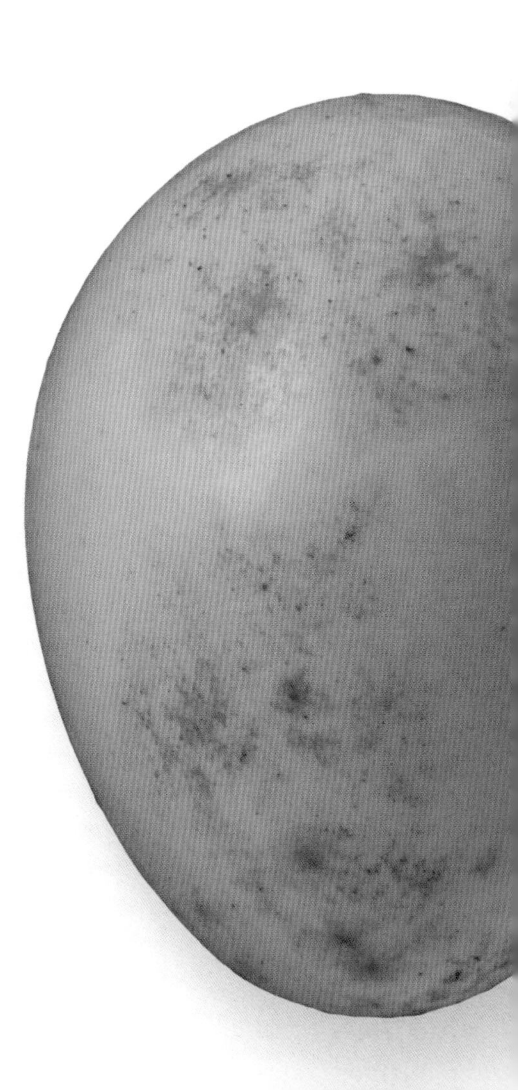

The egg of the Broad-billed Tody is clear and glossy white in color, and measures ⅔ x ½ in (17 x 13 mm) in dimensions. The eggs become stained as incubation proceeds by picking up the reddish color of the soil in the nesting chamber.

ORDER	Coraciiformes
FAMILY	Momotidae
BREEDING RANGE	Tropical, lowland South America
BREEDING HABITAT	Vertical drops
NEST TYPE AND PLACEMENT	Tunnel carved into vertical dirt or sand banks
CONSERVATION STATUS	Least concern
COLLECTION NO.	FMNH 237

ADULT BIRD SIZE
18–19 in (46–48 cm)

INCUBATION
17–22 days

CLUTCH SIZE
3–4 eggs

MOMOTUS MOMOTA

AMAZONIAN MOTMOT

CORACIIFORMES

Clutch

The egg of the Amazonian Motmot is immaculate white in color, and 1⅛ x 1 in (29 x 26 mm) in dimensions. Both parents incubate the eggs, with nighttime shifts typically carried out by the female.

This motmot shares the characteristic green-and-blue plumage of most other members of the motmot family, as well as the racket-shaped pair of tail feathers. These distinctive feathers are displayed by both sexes, but are slightly longer in the males. Their function may be to increase survival and, among the males, to attract mates. When approached by predators, both male and female motmots intensely wag their tails, thus potentially signaling to the predator that they can no longer launch a surprise attack. In the context of mating, females also may choose males with longer tails.

Once a pair is established, both members dig a nesting tunnel that is up to 5 ft (1.5 m) long, using their strong beaks and feet to dig into the moist soil. Amazonian Motmots nest solitarily, finding walls of sand or dirt in which to dig their nest hole; many nest entrances are hidden in larger primary cavities, such as caves, or behind dense foliage.

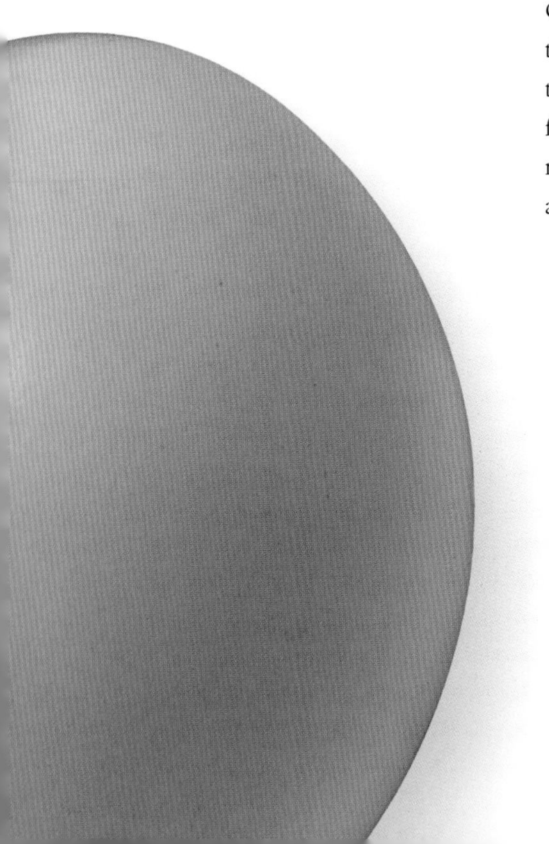

Actual size

ORDER	Coraciiformes
FAMILY	Alcedinidae
BREEDING RANGE	Temperate and southern Eurasia
BREEDING HABITAT	Lake and river shores with dense vegetation
NEST TYPE AND PLACEMENT	Cavity nest, dug into river bank or quarry wall
CONSERVATION STATUS	Least concern
COLLECTION NO.	FMNH 1358

ADULT BIRD SIZE
6½–6¾ in (16–17 cm)

INCUBATION
19–21 days

CLUTCH SIZE
5–7 eggs

ALCEDO ATTHIS

COMMON KINGFISHER

CORACIIFORMES

This kingfisher is territorial; the members of a pair will defend their own adjacent territories much of the time, but combine feeding territories in the early spring when nesting begins. Nest building, incubation, and chick raising are shared by the two mates; the female takes most of the night shifts, while the male changes guard with her during the day. As the chicks grow, they approach the cavity entrance to feed, leaving the rest of the nesting tunnel and chamber foul smelling with fecal matter, remains of uneaten prey, and regurgitated food pellets.

This is a strictly aquatic feeder, and so the presence of the Common Kingfisher is often taken as an indicator of high-quality, clear waters in a habitat. This is because the species requires a transparent water column to locate and catch fish by plunge diving after hovering in the air or sallying from a prominent perching site.

Clutch

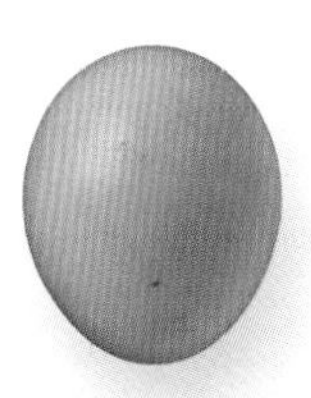

Actual size

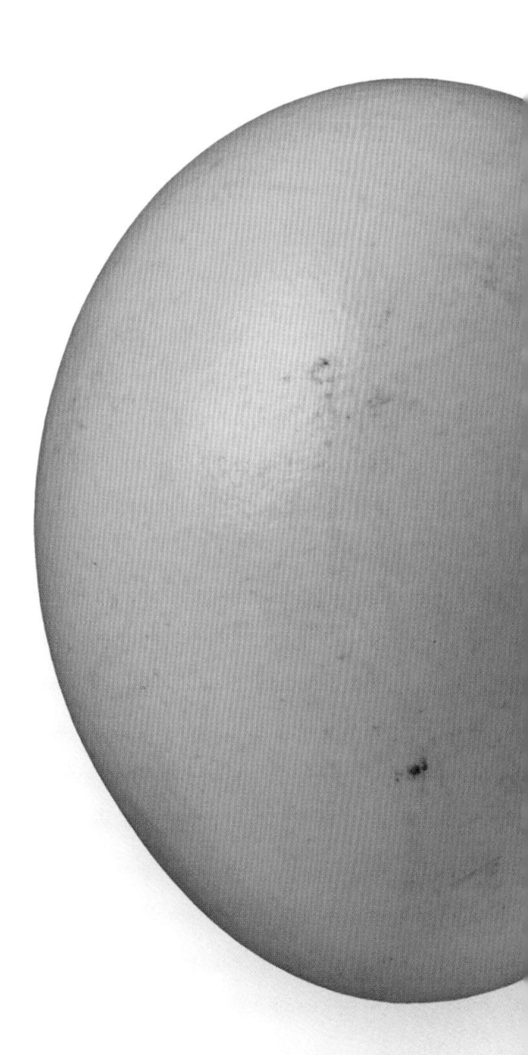

The egg of the Common Kingfisher is white with a pink tint in color, has no speckles, and measures ¾ x ⅔ in (20 x 18 mm). The tunnel leading to the nesting chamber has a slight upward slope, to keep rain or floodwater from reaching the eggs.

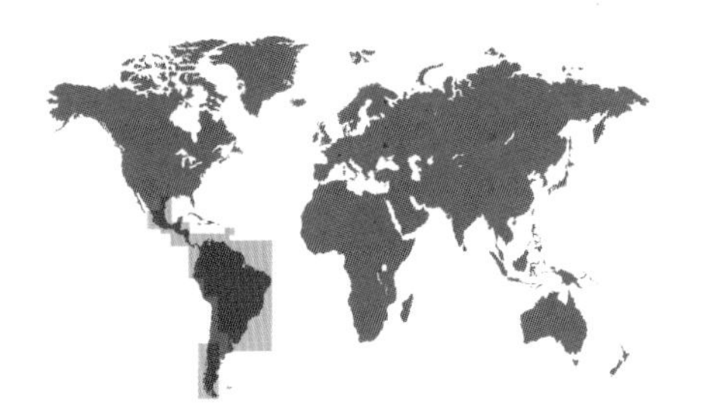

ORDER	Coraciiformes
FAMILY	Alcedinidae
BREEDING RANGE	Southern North America, Central, and South America
BREEDING HABITAT	Lakes and rivers, typically with forested shores
NEST TYPE AND PLACEMENT	Tunnel and nesting chamber, dug in river bank
CONSERVATION STATUS	Least concern
COLLECTION NO.	FMNH 21472

ADULT BIRD SIZE
15½–16 in (40–41 cm)

INCUBATION
22–24 days

CLUTCH SIZE
3–6 eggs

MEGACERYLE TORQUATA

RINGED KINGFISHER

CORACIIFORMES

Clutch

The Ringed Kingfisher is a large and brightly colored fishing specialist, which occurs from the southern tip of the United States to the southern tip of Patagonia. It requires large bodies of water for feeding and nests in higher banks by the shore, excavating its own nesting chamber on a yearly basis. Both adults in a pair participate in nest building: they loosen the soil and dirt with their strong beaks, and then use their feet to kick the sand back. Nest building is done in the dry season, when the water levels are lower so that more of the sand bank is exposed and there is less risk of flooding. The lower water levels also concentrate their fish prey.

The parents take turns in incubating the eggs and, later, provisioning the chicks. These parental duties are shared equitably between the sexes; for example, incubation is carried out in 24-hour bouts, with the changing of guard taking place swiftly in the early morning hours.

Actual size

The egg of the Ringed Kingfisher is glossy white in color, clear of spots, and is 1¾ x 1⅓ in (44 x 34 mm) in size. The incubating parents cough up pellets of fish scales and bones during each of their shifts, which attract flies to the nest entrance even prior to hatching.

ORDER	Coraciiformes
FAMILY	Alcedinidae
BREEDING RANGE	Temperate and subarctic North America
BREEDING HABITAT	Rivers and ponds inland, estuaries and marine bays on the coast
NEST TYPE AND PLACEMENT	Ground burrow, at the end of a tunnel, carved into a vertical river or lake bank
CONSERVATION STATUS	Least concern
COLLECTION NO.	FMNH 7333

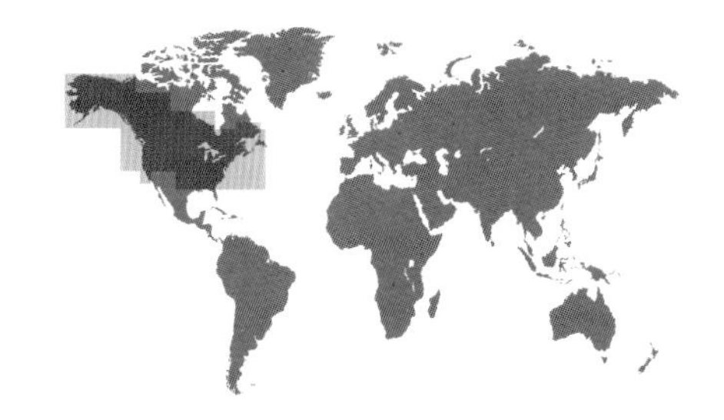

MEGACERYLE ALCYON

BELTED KINGFISHER

CORACIIFORMES

ADULT BIRD SIZE
11–14 in (28–35 cm)

INCUBATION
22–24 days

CLUTCH SIZE
5–8 eggs

Where the northern breeding range of the Ringed Kingfisher (see facing page) ends, the Belted Kingfisher takes over. The two species provide a nearly perfect coverage of all American landmasses. This North American species also uses deep and stationary or slow-moving bodies of water for feeding, diving for fish and invertebrates from a prominent perch on the shore.

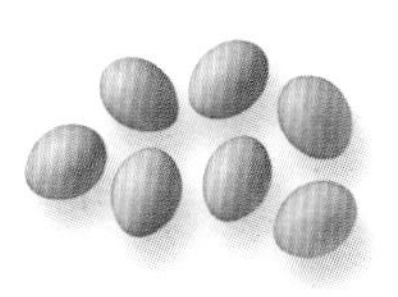

Clutch

Breeding pairs maintain an exclusive territory and keep all other competitors away to avoid competition for both food and nesting sites. This strong pattern of aggression is only lifted when the local breeding pair becomes "too busy" when feeding large nestlings and fledglings. At this time, previously homeless, typically younger, pairs may enter the territory and excavate a new hole and start to lay eggs, resulting in a wave of late-hatching clutches and late-fledging broods. Survival rates of the late-summer chicks is lower, but this is still better for these pairs than not breeding at all.

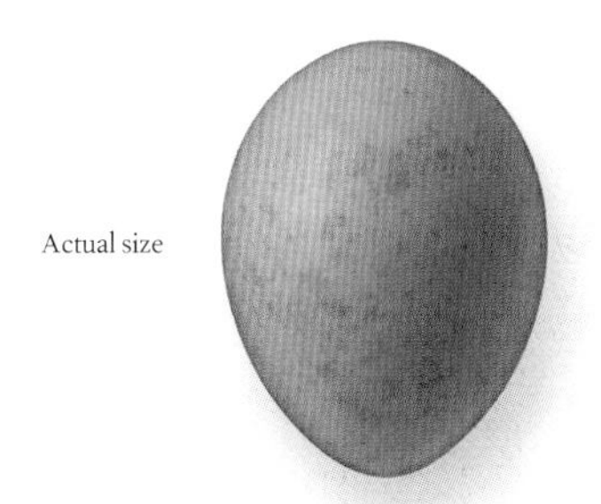

Actual size

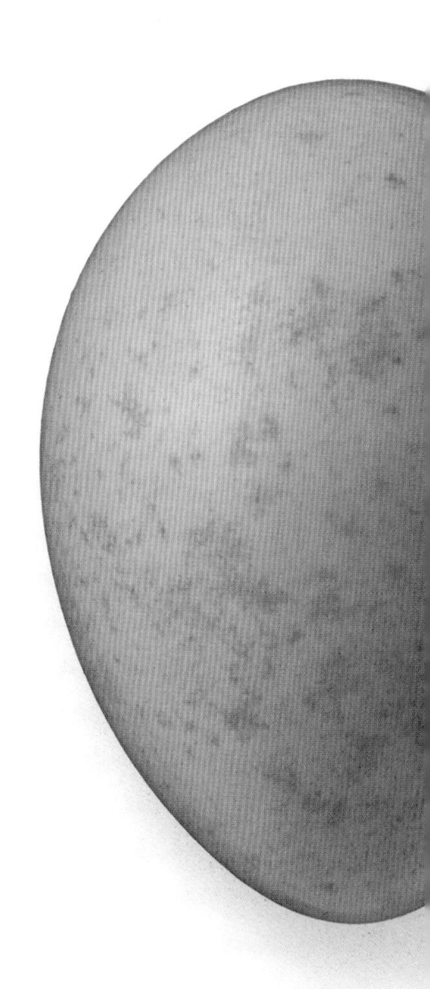

The egg of the Belted Kingfisher is pure white and glossy in appearance, and measures 1⅓ x 1 1/16 in (34 x 27 mm) in size. The incubating adult can recognize its mate based on sound alone when it calls outside of the nesting burrow as shown by a sudden change in heart rate following playback experiments of the mate's and a stranger's vocalizations.

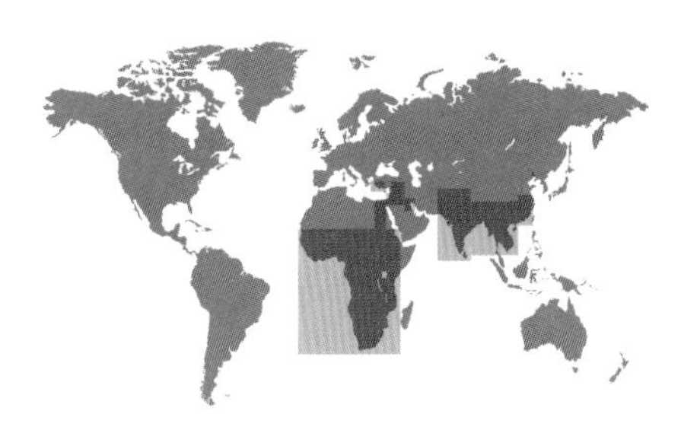

ORDER	Coraciiformes
FAMILY	Alcedinidae
BREEDING RANGE	Sub-Saharan Africa, southern and east Asia, eastern Mediterranean
BREEDING HABITAT	Rivers, lakes, flooded grasslands, and estuaries
NEST TYPE AND PLACEMENT	Long tunnel ending in a chamber, excavated into mud bank, usually above water's edge
CONSERVATION STATUS	Least concern
COLLECTION NO.	FMNH 20452

ADULT BIRD SIZE
9½–10 in (24–26 cm)

INCUBATION
17–19 days

CLUTCH SIZE
3–6 eggs

CERYLE RUDIS

PIED KINGFISHER

CORACIIFORMES

Clutch

The egg of the Pied Kingfisher is pure white, and 1⅛ x 1 in (29 x 24 mm) in size. Several pairs may nest in the same mud bank near each other, generating a loosely colonial breeding site.

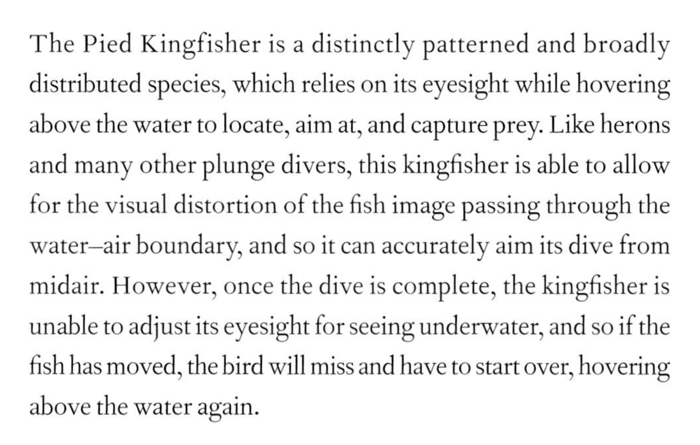

The Pied Kingfisher is a distinctly patterned and broadly distributed species, which relies on its eyesight while hovering above the water to locate, aim at, and capture prey. Like herons and many other plunge divers, this kingfisher is able to allow for the visual distortion of the fish image passing through the water–air boundary, and so it can accurately aim its dive from midair. However, once the dive is complete, the kingfisher is unable to adjust its eyesight for seeing underwater, and so if the fish has moved, the bird will miss and have to start over, hovering above the water again.

This species is solitary in its feeding, but gregarious at night, and roosts in large groups. Breeding is also often a communal affair, with both related and unrelated adults occasionally assisting the mated pair to feed the nestlings. Among such "helpers at the nest," relatives feed chicks more often than do non-relatives; but non-relatives often stay around for several breeding seasons, and eventually may inherit the nesting territory when a member of the resident pair dies.

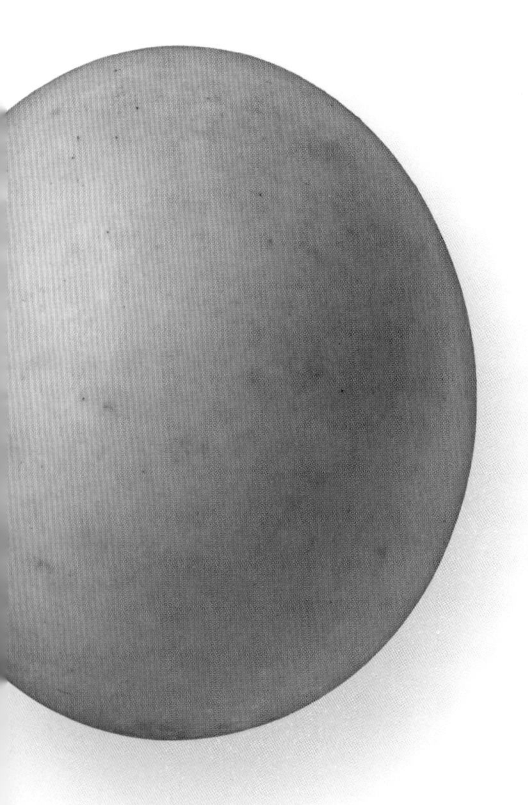

Actual size

ORDER	Coraciiformes
FAMILY	Alcedinidae
BREEDING RANGE	Southern North America, Central, and South America
BREEDING HABITAT	Forests and mangroves along streams and rivers
NEST TYPE AND PLACEMENT	Long tunnel, carved into a vertical river bank
CONSERVATION STATUS	Least concern
COLLECTION NO.	FMNH 15195

ADULT BIRD SIZE
7½–9½ in (19–24 cm)

INCUBATION
19–21 days

CLUTCH SIZE
3–4 eggs

CHLOROCERYLE AMERICANA

GREEN KINGFISHER

CORACIIFORMES

To the human eye, a female Green Kingfisher appears to be more colorful than the male, because of its rufous chest. However, having a fully rufous chest, compared to the rusty and white chest of the male, may not be seen in the same way by the birds themselves. Birds have a dramatically different visual system from humans, and bright whites, especially if these are complemented with ultraviolet undertones, may appear to other kingfishers to have more, rather than fewer, hues.

Once the sexually dichromatic female and male have established their nesting territory, they must find a suitable mud or sand bank into which they can carve a brand new nest burrow each season. Both sexes assist in nest excavation, using their beaks to break into the soil, and their partially fused front toes to shovel the sand behind them to dig deeper into the bank.

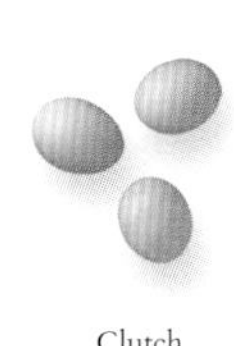

Clutch

The egg of the Green Kingfisher is pale white in color, immaculate, and 1 x ¾ in (24 x 19 mm) in size. Both parents keep the eggs warm, the female typically taking the night shift as well as several daytime shifts.

Actual size

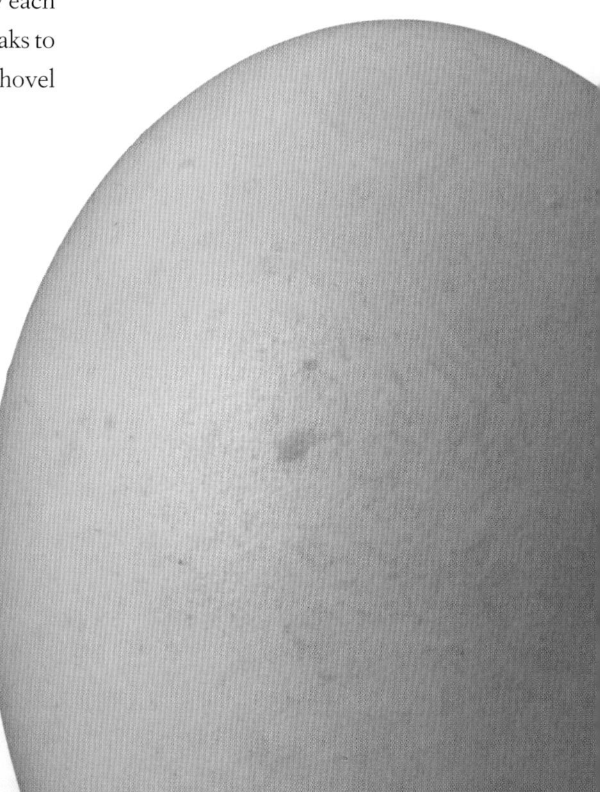

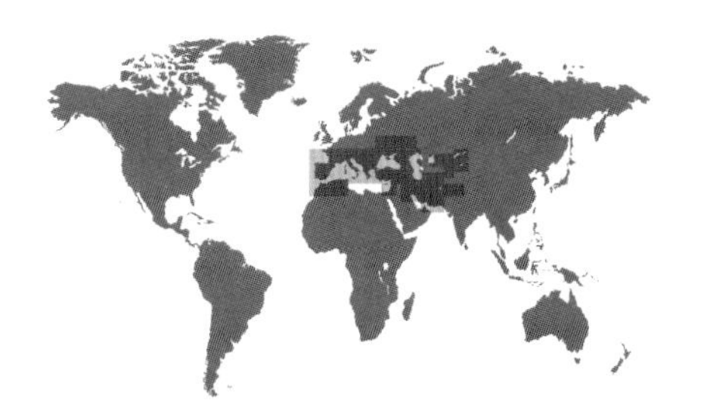

ORDER	Coraciiformes
FAMILY	Meropidae
BREEDING RANGE	Central and southern Europe, north Africa, and western Asia
BREEDING HABITAT	Grassland, savanna, and open forests
NEST TYPE AND PLACEMENT	Burrows in sand banks, or abandoned mines, preferably near rivers
CONSERVATION STATUS	Least concern, locally threatened
COLLECTION NO.	FMNH 21509

ADULT BIRD SIZE
10½–11½ in (27–29 cm)

INCUBATION
20–22 days

CLUTCH SIZE
5–8 eggs

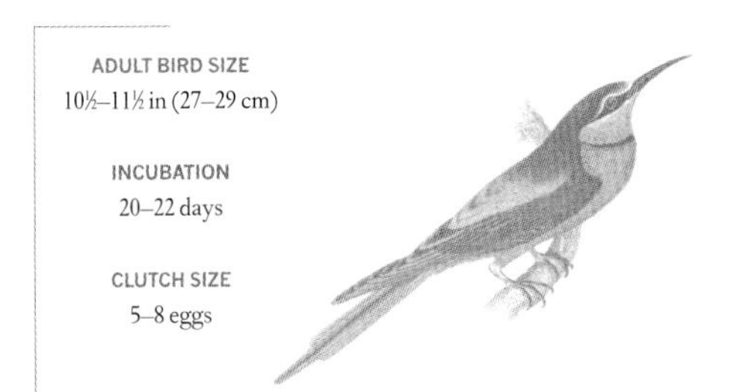

MEROPS APIASTER

EUROPEAN BEE-EATER

CORACIIFORMES

Clutch

A richly colored species, this bee-eater is a brief resident of its temperate breeding areas, arriving in May and departing by early September; it spends most of its year in the tropical wintering sites in southern Africa and Asia. The female and the male are identical in their appearance, and they also share the major aspects of the breeding cycle. Together they excavate new or repair previously used nesting chambers, incubate the eggs, and provision the offspring. Unlike many of its subtropical and tropical relatives, which are cooperative breeders with adult helpers feeding the chicks, European Bee-eater pairs typically do not have helpers at the nest.

This is a gregarious species, which feeds, roosts, and nests in colonies. During the last 30 years, however, its colony sizes have declined, and the once common colonies of 100 or more pairs nesting together are today smaller breeding colonies of barely 20 pairs.

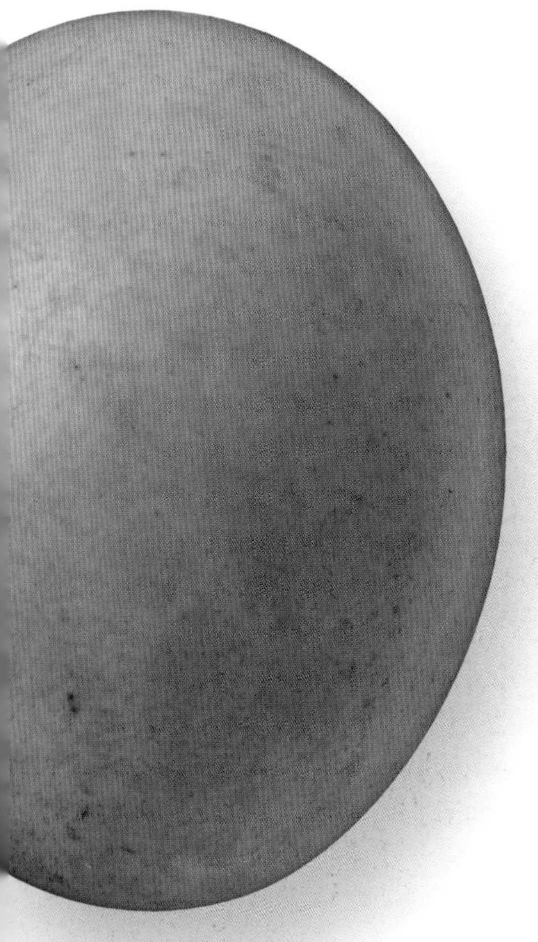

Actual size

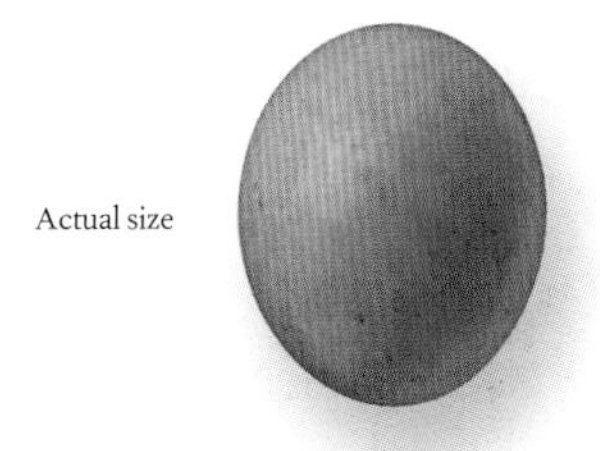

The egg of the European Bee-eater is spherical, white in background color, has no maculation, and measures 1 x ⅞ in (26 x 22 mm). The eggs are laid a day at a time, and incubated from the day when the first is laid, and so the chicks hatch on different days.

ORDER	Coraciiformes
FAMILY	Meropidae
BREEDING RANGE	Northern Africa, western Asia
BREEDING HABITAT	Dry savanna, semi-deserts
NEST TYPE AND PLACEMENT	Burrow nest, excavated in a river bank
CONSERVATION STATUS	Least concern
COLLECTION NO.	FMNH 21505

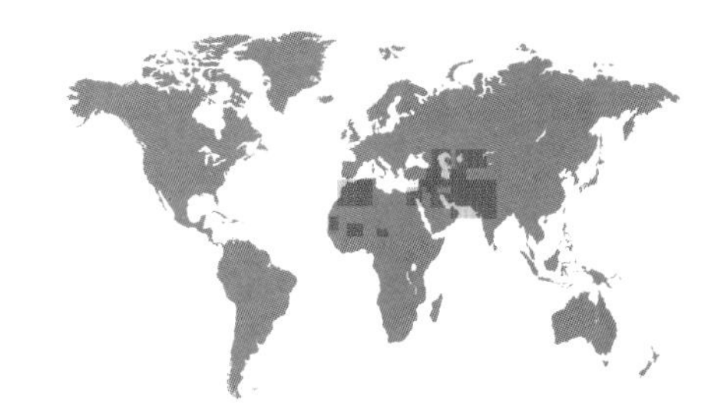

ADULT BIRD SIZE
6½–15½ in (17–40 cm)

INCUBATION
23–26 days

CLUTCH SIZE
4–8 eggs

MEROPS PERSICUS

BLUE-CHEEKED BEE-EATER

CORACIIFORMES

The breeding grounds of the Blue-cheeked Bee-eater, also known as the Madagascar Bee-eater, are in arid open areas punctuated by distant tree stands, with most pairs dispersed into lone nesting sites, or small loose colonies occasionally near other, more sociable bee-eaters. In the non-breeding season, the species forms large flocks that migrate to sub-Saharan Africa where they can often be found in open, grassy fields, along large rivers and in coastal mangroves on both the east and west coasts of Africa.

The two parents together excavate the nest tunnel and they also share feeding the young equitably and without the assistance of adult helpers. Like all bee-eaters, the adults frequently feed nestlings bees and wasps; they catch the insects in their long sharp bills and bang them against a tree branch or some other substrate to remove the stingers before the chicks eat them.

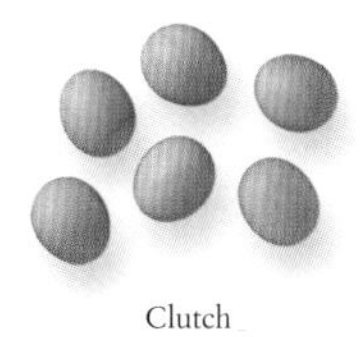

Clutch

Actual size

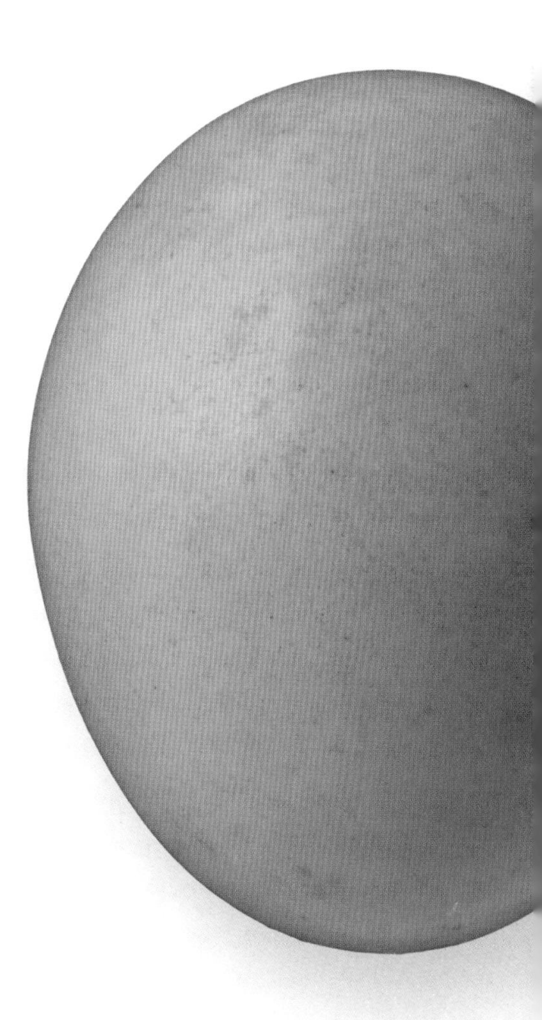

The egg of the Blue-cheeked Bee-eater is immaculate white in color, and measures 1 x ¾ in (24 x 20 mm). The eggs are laid on the barren ground of the nest chamber, and incubated by both parents, with only the female spending the night on the eggs.

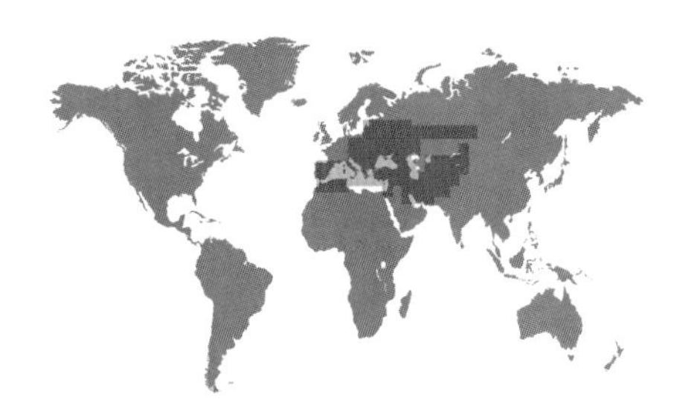

ORDER	Coraciiformes
FAMILY	Coraciidae
BREEDING RANGE	Europe, north Africa, and western Asia
BREEDING HABITAT	Open forests, wooded savanna
NEST TYPE AND PLACEMENT	Cavity in a tree or cliff, no nest lining
CONSERVATION STATUS	Near threatened
COLLECTION NO.	FMNH 21428

ADULT BIRD SIZE
11½–12½ in (29–32 cm)

INCUBATION
17–19 days

CLUTCH SIZE
3–6 eggs

CORACIAS GARRULUS

EUROPEAN ROLLER

CORACIIFORMES

Clutch

The egg of the European Roller is pure white in color, and 1⅜ x 1⅛ in (36 x 28 mm) in dimensions. The nestlings produce a foul-smelling excretion that repels scent-driven nest predators, typically small mammals but also snakes.

Though it breeds in the temperate regions, this roller migrates to tropical Africa for the winter and faces many risks during its long migration across the Mediterranean and Middle East, including being trapped or hunted for food by humans. Population trends have shown dramatic declines in recent decades, even though breeding populations are under strict protection in most European countries.

Many artificial nestboxes have been added by local conservation groups to augment the increasingly rare, naturally suitable nesting sites of this species, which are typically holes excavated by large woodpecker species, and other cavities in mature trees. Nestboxes provide new and, apparently, safe breeding opportunities for these rollers, with some surveys indicating a nearly 90 percent nesting success—from egg laying to fledging—a high success rate for a threatened species.

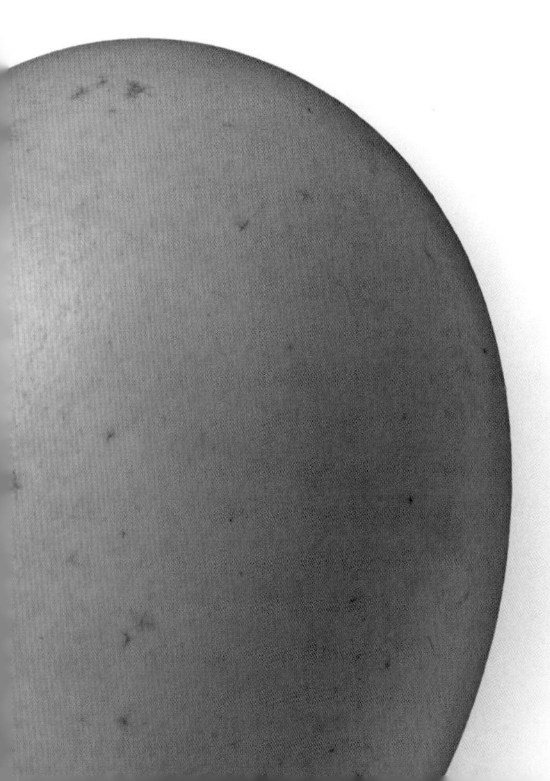

Actual size

ORDER	Coraciiformes
FAMILY	Coraciidae
BREEDING RANGE	East and Southeast Asia, northern and eastern Australia
BREEDING HABITAT	Forests, woodlands with mature trees, suburban parks
NEST TYPE AND PLACEMENT	Natural hollows in trees or coconut palms; burrows in sand banks
CONSERVATION STATUS	Least concern
COLLECTION NO.	FMNH 18859

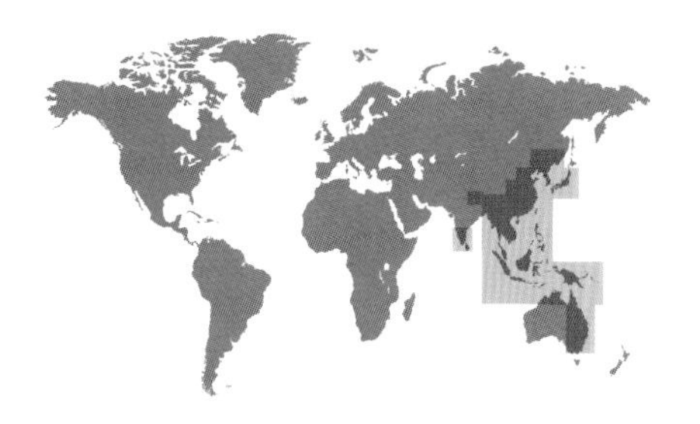

ADULT BIRD SIZE
10–12 in (25–30 cm)

INCUBATION
17–20 days

CLUTCH SIZE
3–4 eggs

EURYSTOMUS ORIENTALIS

DOLLARBIRD

CORACIIFORMES

The Dollarbird has a wide latitudinal distribution, spanning the tropics on either side of the equator, and both northern and southern temperate regions of the western Pacific Ocean. During the breeding season, pairs of these distinctive, blue-hued birds perch near each other at prominent spots, and sally out to capture flying insects for prey. They also frequently perform their undulating courtship flight together, before the onset of nesting and egg laying.

In Australia, where woodpeckers are absent and do not excavate new holes, the few available tree cavities are a resource highly valued by parrots and other burrow-nesting birds. Thus, Dollarbirds must defend their nest holes vigorously. At safe nesting sites, where eggs hatch and chicks fledge successfully, the cavities are often reused across different breeding seasons.

Clutch

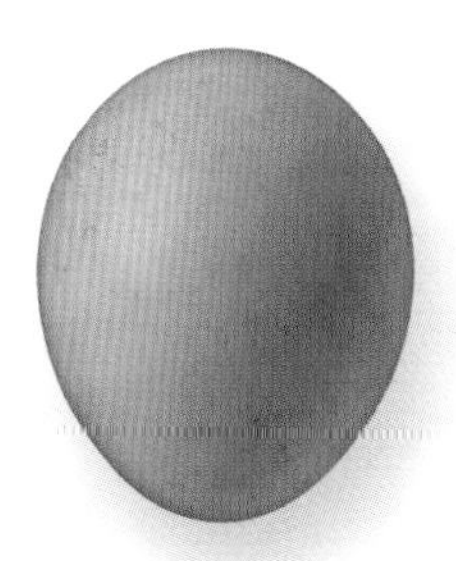

Actual size

The egg of the Dollarbird is pointed at one tip, dull white in background color, clear of any maculation, and 1⅓ x 1⅛ in (34 x 29 mm) in size. Both parents take turns in incubating the eggs and provide food for the nestlings.

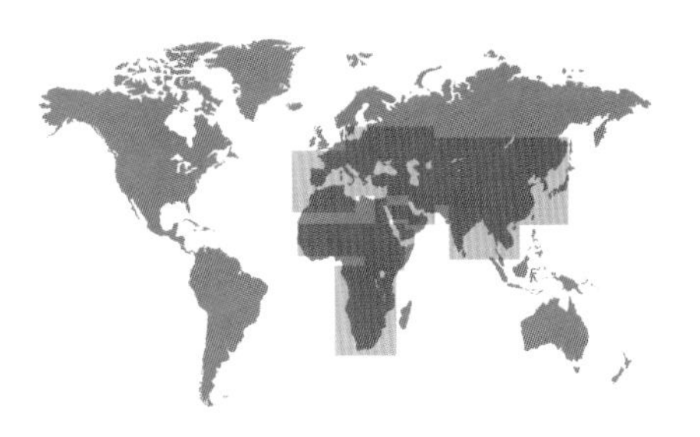

ORDER	Coraciiformes
FAMILY	Upupidae
BREEDING RANGE	Europe, North and sub-Saharan Africa, Central and South Asia
BREEDING HABITAT	Open areas, heathland, steppes, savanna and grasslands, forest clearings
NEST TYPE AND PLACEMENT	Tree holes, burrows in sand banks or the ground
CONSERVATION STATUS	Least concern
COLLECTION NO.	FMNH 20487

ADULT BIRD SIZE
10–12½ in (25–32 cm)

INCUBATION
15–18 days

CLUTCH SIZE
7–12 eggs

UPUPA EPOPS

EURASIAN HOOPOE

CORACIIFORMES

Clutch

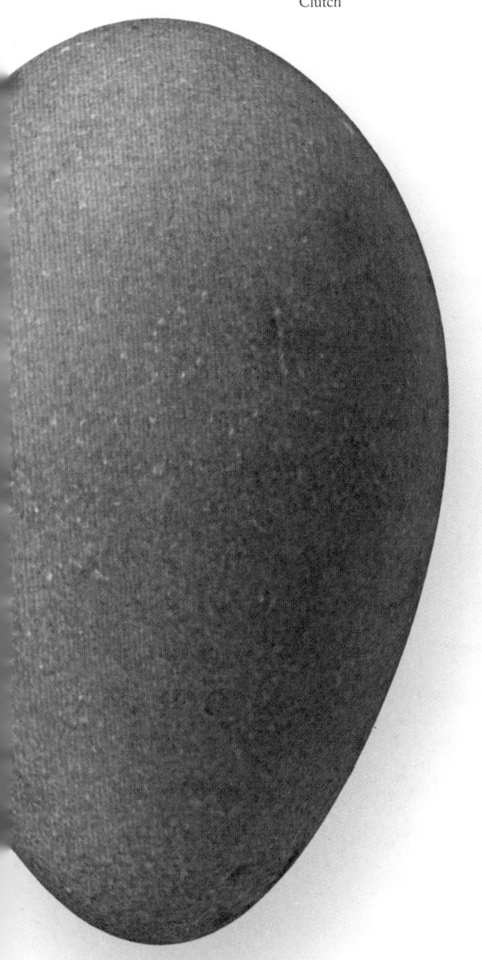

These birds are named after the sound of their calls. Hoopoes are highly distinctive grassland birds: their cinnamon-, black-, and white-patterned plumage, their undulating flight, long curved beaks, and the foul stench of their nests and chicks, have made these birds well known to people throughout their broad breeding range, and yielded one of their local names, the "stinky hoopoe."

The penetrating odor of the nest of the Eurasian Hoopoe is part of a suite of adaptations to nesting in potentially vulnerable ground burrows. The female, in charge of incubation on her own, develops a preen-gland secretion soon after nesting, which is transferred to the plumage and makes the nest smell like rotten meat. The hatchlings, too, produce this stinky gland product, but they are also able to generate a stream of fecal matter, to stab with their wings or beak, and to call with a snake-like hiss, all aimed at the face and senses of a nest-intruder.

Actual size

The egg of the Eurasian Hoopoe is milky blue in background color, and measures 1 x ⅔ in (25 x 18 mm) in size. With the onset of incubation, the eggs take on a dramatically different dark brown coloration (exhibited here) as they are stained by the nesting substrate and fecal matter.

ORDER	Coraciiformes
FAMILY	Phoeniculidae
BREEDING RANGE	Tropical sub-Saharan Africa, on either side of the equator
BREEDING HABITAT	Dry, wooded savanna, thorny bushland
NEST TYPE AND PLACEMENT	Inside a tree cavity, no lining
CONSERVATION STATUS	Least concern
COLLECTION NO.	FMNH 14772

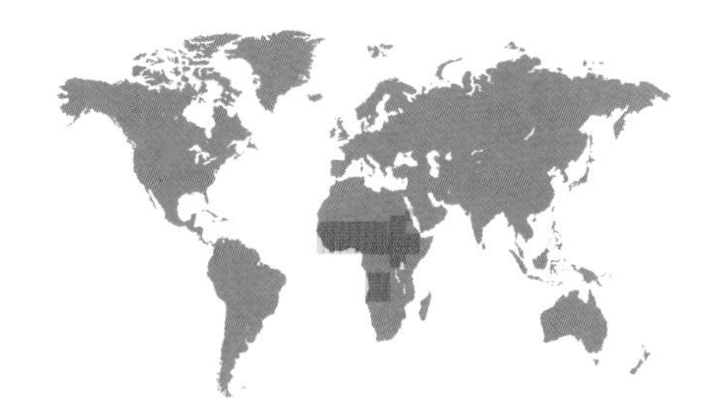

ADULT BIRD SIZE
10–12 in (26–30 cm)

INCUBATION
17–19 days

CLUTCH SIZE
2–3 eggs

RHINOPOMASTUS ATERRIMUS

BLACK SCIMITARBILL

CORACIIFORMES

The Black Scimitarbill is named for its long curved bill. Pairs move together through the trees, probing into the bark along larger branches for their insect prey. This behavior, along with their long tail and black plumage, make them a distinctive species in the dry forests where they live.

Scimitarbills often occur in the same areas as their similarly shaped and colored relatives, the Green Woodhoopoes (an alternative name for the Black Scimitarbill is the Black Wood Hoopoe). Though there is much resemblance in overall appearance and ecology, the two species are behaviorally distinct. The reproductive unit of the Black Scimitarbill is a single pair, whereas woodhoopoes live in larger family groups, with helpers assisting at the nest. Scimitarbills face competition for nest holes from other birds, and also experience high rates of predation at the nest, so they must frequently restart breeding attempts to complete a successful nesting cycle.

Clutch

Actual size

The egg of the Black Scimitarbill is white and clear in color, and measures ⅞ x ⅔ in (23 x 18 mm) in dimensions. Both members of the pair look after the eggs by taking turns, and both parents also brood and feed the nestlings.

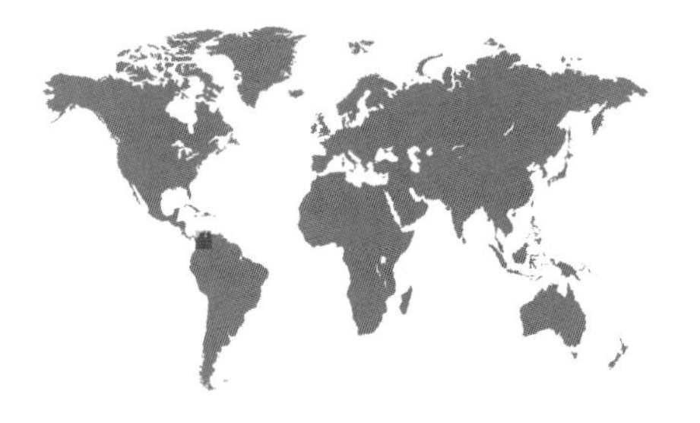

ORDER	Galbuliformes
FAMILY	Bucconidae
BREEDING RANGE	Northern South America
BREEDING HABITAT	Dry scrub, coastal and lowland forests
NEST TYPE AND PLACEMENT	Burrow nest, carved into an arboreal termite colony
CONSERVATION STATUS	Least concern
COLLECTION NO.	WFVZ 154603

ADULT BIRD SIZE
8–8½ in (20–22 cm)

INCUBATION
Unknown

CLUTCH SIZE
2–3 eggs

HYPNELUS RUFICOLLIS

RUSSET-THROATED PUFFBIRD

GALBULIFORMES

Clutch

This puffbird is a socially monogamous species, with pairs using vocalizations both prior to mate bonding and also as a cooperative territorial defense signal during nesting: they produce a series of notes, which are often given in synchrony by the female and the male. Such duetting behavior is displayed much more frequently in tropical and southern hemisphere birds than in northern temperate bird species.

Puffbirds are a strictly neotropical family; they are related to jacamars, but they have a more compact shape, stronger bills, and are more cryptically colored. The Russet-throated Puffbird is a distinctively patterned species, and feeds as a specialist perch-based insectivore: it is most often seen as it surveys open areas near tree stands for insects or as it tunnels into arboreal termitaria, constructed by tree-dwelling termites. With its hooked beak, it will also catch small lizards.

Actual size

The egg of the Russet-throated Puffbird is clear white in color, and ⅞ x ¾ in (23 x 19 mm) in size. Both parents incubate the eggs, but like too many tropical species little else is known about the breeding biology.

ORDER	Galbuliformes
FAMILY	Galbulidae
BREEDING RANGE	Central America and tropical South America
BREEDING HABITAT	Dry or wet forests, open woodlands
NEST TYPE AND PLACEMENT	Burrow nest, dug into river banks or termite mounds
CONSERVATION STATUS	Least concern
COLLECTION NO.	WFVZ 143371

ADULT BIRD SIZE
8½–10 in (22–25 cm)

INCUBATION
19–26 days

CLUTCH SIZE
2–4 eggs

GALBULA RUFICAUDA

RUFOUS-TAILED JACAMAR

GALBULIFORMES

The Rufous-tailed Jacamar is an efficient aerial insectivore. It searches for butterflies, moths, and other flying insects from prominent perch-sites, snaps them up in midair, and then returns to the perch to smash the insect against a branch before consuming it. These jacamars are the selective agents for the evolution of mimetic patterns between toxic and edible butterflies. Specifically, when young birds capture and taste a toxic butterfly, they quickly learn its specific wing patterns, and later they will not take any other butterflies that match the same pattern. Thus any butterflies that evolve patterns to mimic the toxic butterflies may be relatively safe from attack.

Both parents must be efficient predators of flying insects to raise a full brood of young, which require vast numbers of insects. Carrying the prey in their beaks, females and males complete frequent feeding trips to the nest. Housekeeping duties are ignored, so the nest of a developing brood of chicks is stinky and filthy with feces and decomposing insect body parts.

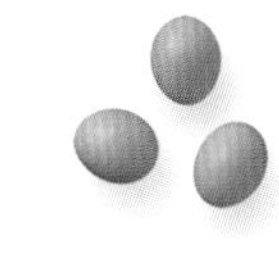

Clutch

Actual size

The egg of the Rufous-tailed Jacamar is white in background coloration, clear of maculation, and ⅞ x ¾ in (23 x 19 mm) in size. The sexes take turns to incubate the eggs, with the female engaged in night shifts.

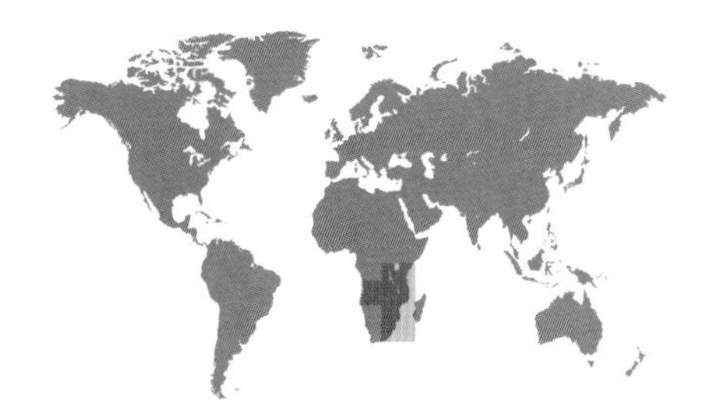

ORDER	Piciformes
FAMILY	Lybiidae
BREEDING RANGE	Subequatorial Africa
BREEDING HABITAT	Open grassland, gardens, forests, trees, open farmland with trees
NEST TYPE AND PLACEMENT	Cavity nest in a tree
CONSERVATION STATUS	Least concern
COLLECTION NO.	WFVZ 164035

ADULT BIRD SIZE
7½–8 in (19–20 cm)

INCUBATION
13–15 days

CLUTCH SIZE
2–5 eggs

LYBIUS TORQUATUS

BLACK-COLLARED BARBET

PICIFORMES

Clutch

The egg of the Black-collared Barbet is white and glossy in background color, clear of speckles, and measures 1 x ⅔ in (24 x 18 mm) in dimensions. Barbet nests can be parasitized by Lesser Honeyguides, and the honeyguide hatchling will kill its barbet nest mates using a sharp hooked bill-tooth.

Birds called "barbets" are found on three different continents, but despite the shared common names and appearances of American barbets, modern taxonomists have determined that these are closest relatives with toucans, whereas the Old World barbets deserve separate families for the African and Asian lineages. The Black-collared Barbet is a particularly gregarious and loud species within the African lineage, with groups foraging and calling together, and defending themselves and their young by mobbing predators. Outside the breeding season, nest cavities are used for communal nighttime roosting by up to a dozen adults.

During the mating season, these barbets perform loosely structured duets, with one member of the pair maintaining a strong rhythmic structure, while the other follows and engages in its own calls arrhythmically. The exact function of these duets is unknown, but they may be involved both in pair-communication and territorial defense against other pairs.

Actual size

ORDER	Piciformes
FAMILY	Ramphastidae
BREEDING RANGE	Tropical South America
BREEDING HABITAT	Humid forests, riverine woodlands, palm groves
NEST TYPE AND PLACEMENT	Unlined cavity nest; in rotting tree or in abandoned woodpecker cavity
CONSERVATION STATUS	Least concern
COLLECTION NO.	WFVZ 24477

ADULT BIRD SIZE
21–24 in (53–61 cm)

INCUBATION
16–20 days

CLUTCH SIZE
2–4 eggs

RAMPHASTOS TUCANUS

WHITE-THROATED TOUCAN

PICIFORMES

The White-throated Toucan is a conspicuous and colorful neotropical bird. It feeds on ripe fruit and berries, but also consumes fleshy insects, lizards, and chicks out of the nests of other birds when the opportunities arise. The two sexes are similar in appearance, with the male somewhat larger than the female. The sex roles, however, are equitable, with the mates spending much of the day foraging together. They also select the nesting cavity, incubate the eggs, and feed the chicks together, or by taking turns.

Nestlings of this toucan species have a thick layer of skin on the heel to protect them from the rough structure of the barren tree cavity floor. The hatchlings are naked and blind, and rely on their parents for brooding and feeding for several weeks in the nest. Even when the fledglings leave the nest, they need several weeks to become fully independent, and they take flight for short, undulating stretches between tall fruit-bearing trees.

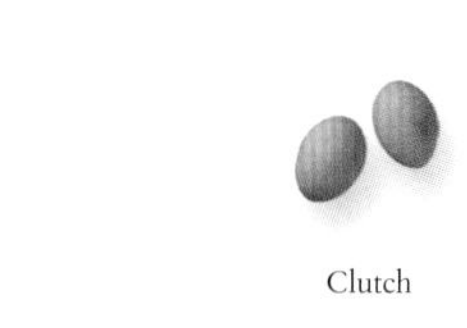

Clutch

The egg of the White-throated Toucan is pure white in color, and 1¾ x 1¼ in (46 x 32 mm) in dimensions. The eggs may become stained from the wood in the unlined tree cavity serving as the nest.

Actual size

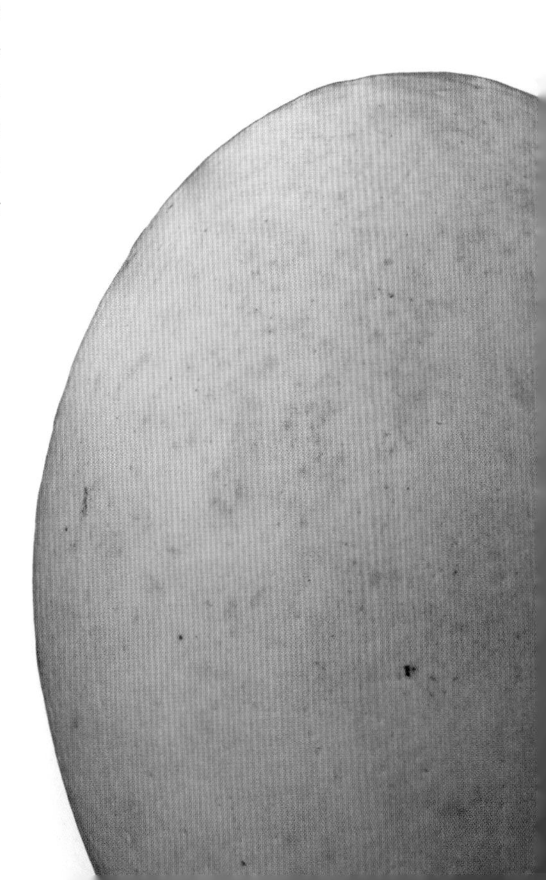

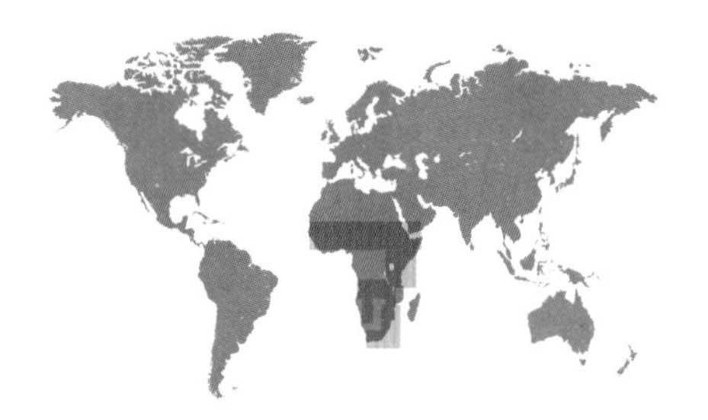

ORDER	Piciformes
FAMILY	Indicatoridae
BREEDING RANGE	Sub-Saharan Africa
BREEDING HABITAT	Dry, open woodlands, mixed savanna
NEST TYPE AND PLACEMENT	Like all honeyguides, this is an obligate brood parasite, laying its eggs in nests of cavity-dwelling hosts
CONSERVATION STATUS	Least concern
COLLECTION NO.	WFVZ 138027

ADULT BIRD SIZE
7–8 in (18–20 cm)

INCUBATION
15–17 days

CLUTCH SIZE
1 egg per host nest, up to 20 eggs per season

INDICATOR INDICATOR

GREATER HONEYGUIDE

PICIFORMES

Clutch

The Greater Honeyguide is a specialist on honeybee and wasp nests, to which it gains access by leading Honeybadgers or people to them. These agents tear the nest apart to feed themselves and the bird feeds on the remains. The Greater Honeyguide is also an obligate brood parasite. It never builds its own nest, but instead taps into the parental care of other species in its habitat, including woodpeckers, barbets, kingfishers, woodhoopoes, and bee-eaters.

To reduce competition with host nest mates, the female honeyguide punctures host eggs at the time of laying the parasitic egg. The rest of the competition is eliminated by the hatchling honeyguide itself: possessing a sharp bill-hook at the time of hatching, it bites, punctures, and kills all nest mates within days, to become the sole occupant of the host nest. The Greater Honeyguide egg hatches two to five days earlier than the host eggs, providing a size advantage for the parasitic chick when it attempts to fatally bite other hatchlings.

Actual size

The egg of the Greater Honeyguide is white and clear of spotting, and ⅞ x ⅔ in (22 x 16 mm) in dimensions. The photograph of the clutch shows the honeyguide's egg (the one on the left), together with the egg of a typical host species, the Little Bee-eater.

ORDER	Piciformes
FAMILY	Picidae
BREEDING RANGE	North Africa and temperate Eurasia
BREEDING HABITAT	Open woodland, forests with clearings and low undergrowth, orchard stands
NEST TYPE AND PLACEMENT	Tree cavity; uses holes carved by woodpeckers
CONSERVATION STATUS	Least concern
COLLECTION NO.	FMNH 2491

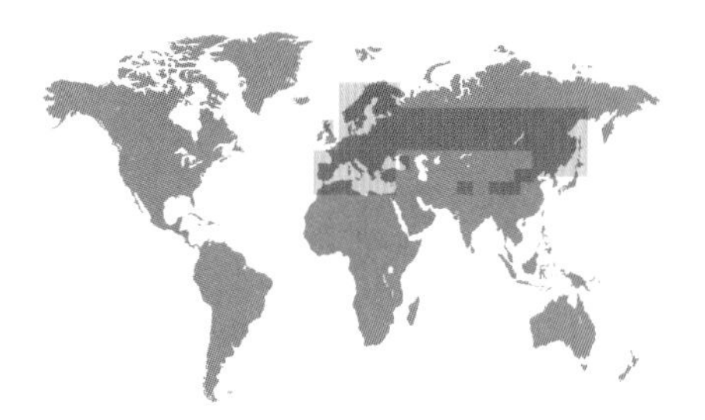

ADULT BIRD SIZE
6½–7 in (16–18 cm)

INCUBATION
12–14 days

CLUTCH SIZE
7–10 eggs

JYNX TORQUILLA

EURASIAN WRYNECK

PICIFORMES

Like its closest relatives, the woodpeckers, the Eurasian Wryneck has a large head, a long tongue, and a set of four toes, two of which point forward and two backward. However, unlike true woodpeckers, it lacks a strong bill and stiff tail feathers, so that it cannot drill holes into tree trunks in search of larvae or to construct a primary nest cavity. Instead, it is a specialist on ants, often seen foraging in decaying wood, on the forest floor, or in sandy areas where traveling ants are easily spotted.

For breeding, the relatively small and defenseless Wryneck relies on its cryptic plumage to approach its nest cavity without being seen by a predator. Once inside, its white eggs are rarely visible in the low light levels of the cavity, and this is especially true when the adult is incubating. The chicks are brooded at night for the first week, but afterwards their own thermoregulation keeps them warm, and the parents spend the nights outside the nest.

Clutch

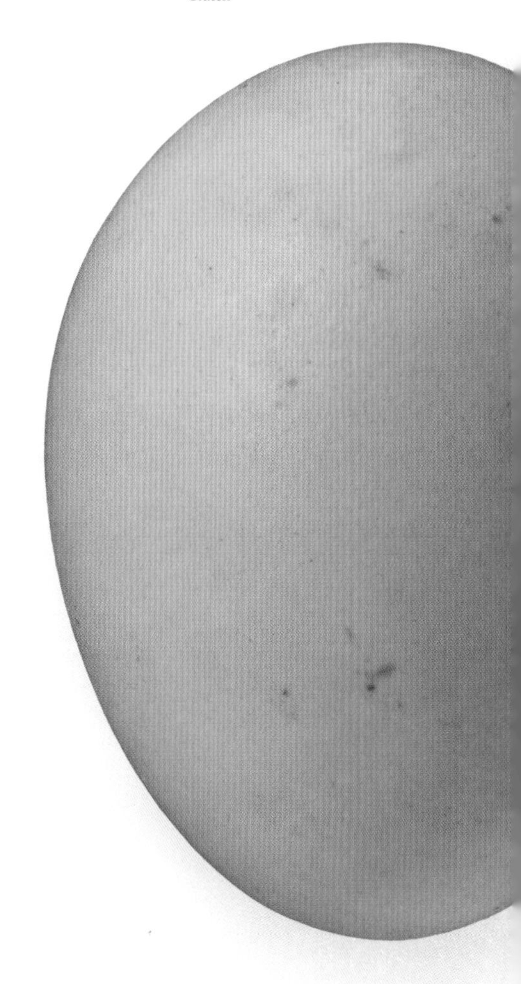

Actual size

The egg of the Eurasian Wryneck is white and immaculate in appearance, and measures ¾ x ⅝ in (20 x 15 mm) in dimensions. The incubating adult protects its eggs from inside the nest cavity by twisting its head nearly 180 degrees, and hissing like a snake.

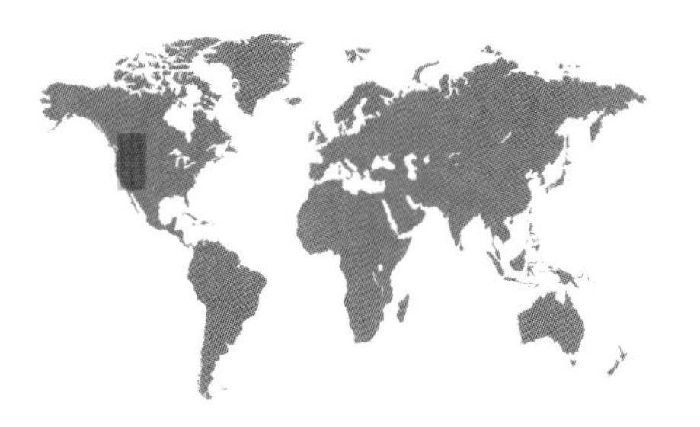

ORDER	Piciformes
FAMILY	Picidae
BREEDING RANGE	Western North America
BREEDING HABITAT	Open woodlands, grassy tree stands
NEST TYPE AND PLACEMENT	Cavity in live trees or snags; but typically reuses existing holes
CONSERVATION STATUS	Least concern
COLLECTION NO.	FMNH 16549

ADULT BIRD SIZE
10–11 in (26–28 cm)

INCUBATION
13–16 days

CLUTCH SIZE
5–7 eggs

MELANERPES LEWIS

LEWIS'S WOODPECKER

PICIFORMES

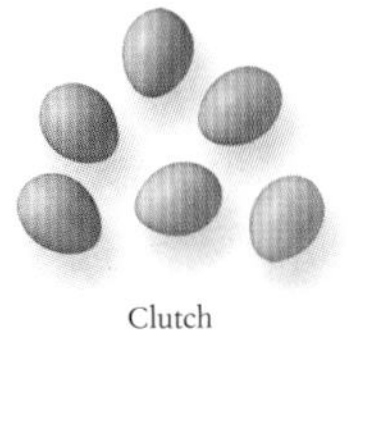

Clutch

This colorful and unique woodpecker lacks the typical black-and-white patterning of most North American woodpeckers. In preparation for breeding, males and females across the range arrive at the same time; pairs are likely philopatric and return to both the same feeding territory and the same nest cavity across multiple years. Parental investment is shared by the pair members, although not equitably; the males invest more time in incubating, as well as brooding the eggs and chicks.

Unlike other North American woodpeckers, instead of excavating tree bark for insect larvae, the Lewis's Woodpecker forages much like a flycatcher, perching at a prominent observation spot, scanning the area for flying insects, and sallying to them for capture. It also differs from many woodpeckers in that it often reuses cavities whose primary excavators were other woodpeckers and flickers.

Actual size

The egg of the Lewis's Woodpecker—like that of all woodpeckers—is immaculate white. It is oval in shape and measures 1 x ¾ in (26 x 20 mm) in size. Although the parents return to the same cavity to nest across years, the young disperse to unknown distances from their natal tree hole.

ORDER	Piciformes
FAMILY	Picidae
BREEDING RANGE	Eastern half of North America
BREEDING HABITAT	Deciduous forests, swampy or riparian woodlands, orchards
NEST TYPE AND PLACEMENT	Cavity nest, drilled into dead tree trunks
CONSERVATION STATUS	Near threatened
COLLECTION NO.	FMNH 16546

ADULT BIRD SIZE
7½–9½ in (19–24 cm)

INCUBATION
12–14 days

CLUTCH SIZE
4–7 eggs

MELANERPES ERYTHROCEPHALUS

RED-HEADED WOODPECKER

PICIFORMES

The Red-headed Woodpecker used to be one of the most commonly seen woodpeckers in North America, occurring in large forest tracts and orchards. For a while, it benefited from major wood-borne diseases devastating the eastern forests, yielding dead trees and stumps in which the bird could forage and excavate nesting holes. But today, this species' population has been declining dramatically, leading to its current heightened conservation status.

Both during the breeding season and beyond, these birds maintain exclusive territories against other Red-headed Woodpeckers. Pair members both work to drill out the nesting cavity, but typically the male selects the hole entrance, and if the female approves, she joins him in excavating. The same hole is typically used by the pair for their two annual nesting attempts, and they may also return to the same spot during consecutive years of breeding.

Clutch

The egg of the Red-headed Woodpecker is pure white in background color, has no markings, and measures 1 x ¾ in (25 x 19 mm) in dimensions. By being laid in a smooth stump, with all bark removed, the eggs are safer from nest predators, such as snakes, which cannot easily climb smooth surfaces.

Actual size

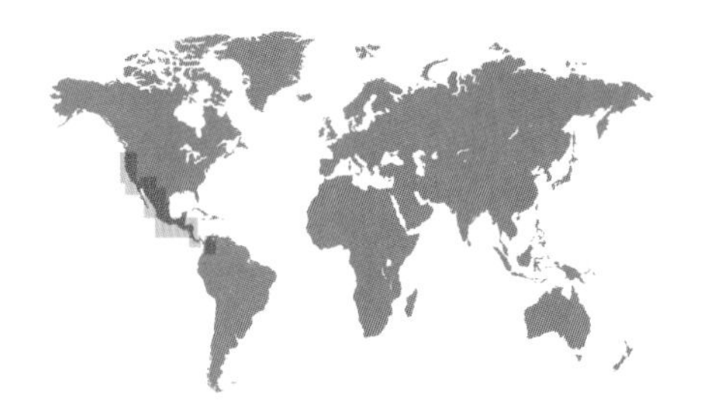

ORDER	Piciformes
FAMILY	Picidae
BREEDING RANGE	Southwestern North America, Central, and northernmost South America
BREEDING HABITAT	Open woodlands
NEST TYPE AND PLACEMENT	Cavity nest, dug into live or dead trees, reused across years
CONSERVATION STATUS	Least concern
COLLECTION NO.	FMNH 7449

ADULT BIRD SIZE
7½–9 in (19–23 cm)

INCUBATION
11–12 days

CLUTCH SIZE
3–6 eggs

MELANERPES FORMICIVORUS

ACORN WOODPECKER

PICIFORMES

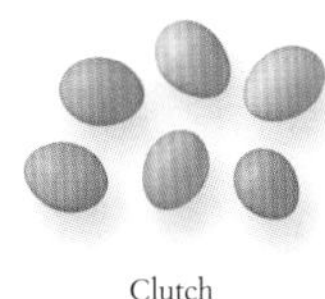

Clutch

The egg of the Acorn Woodpecker is white in background color with no maculation, and measures 1 x ¾ in (25 x 19 mm) in dimensions. In nests where more than one female lays, a breeding female may grasp and toss out all eggs that were already present before she herself begins laying.

There are few species of birds with a more complex family structure than Acorn Woodpeckers. Group living is the norm in this species, with small groups of woodpeckers protecting their group territory that includes not only the oak trees that produce acorns—their main source of food—but also the trunks of the trees, called "granaries." Here the group stores hundreds of acorns, with each acorn having its own hole in the granary. Acorn Woodpecker groups can be composed of one or more monogamous breeding pairs, all laying, incubating, and provisioning young in the same nest.

The breeders are often assisted by adult helpers, who do not themselves breed, but still contribute to incubation and feeding the chicks, as well as defending the territory. DNA studies have been used repeatedly to determine which individuals sire which young in the communal nest, and how much each breeding pair actually contributes to the genetic make-up of the next generation. The results, however, have been unclear, because group members are so closely related to each other that the task of telling daughters and sons apart from nieces and nephews has been hard to disentangle.

Actual size

ORDER	Piciformes
FAMILY	Picidae
BREEDING RANGE	Eastern North America
BREEDING HABITAT	Forests, open woodlands, wooded suburbs
NEST TYPE AND PLACEMENT	Cavity nest in dead trees or power poles; often reuses the same tree, but builds a new hole each year
CONSERVATION STATUS	Least concern
COLLECTION NO.	FMNH 1342

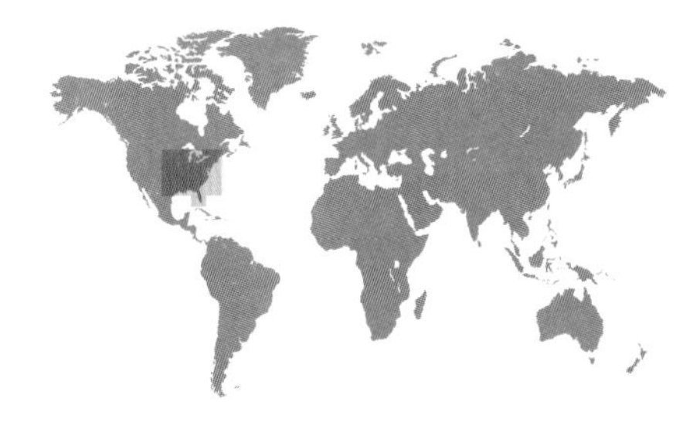

ADULT BIRD SIZE
9–10½ in (23–27 cm)

INCUBATION
12–14 days

CLUTCH SIZE
2–6 eggs

MELANERPES CAROLINUS

RED-BELLIED WOODPECKER

PICIFORMES

This is a common and widespread woodpecker that ranges from open woodlands to suburbia. Wherever it occurs, however, it performs the critical ecosystem task of excavating brand new cavities. These woodpeckers do not reuse their cavities, which thus become available for other hole-dwelling bird and mammal species to nest in. This species has adapted to urban sprawl and will nest in wooden fenceposts and power poles when large trees are not available.

For nesting, the male selects and begins to excavate in a suitable dead trunk or snag, although it is up to the female to make the final decision, and complete the cavity together with her mate. The pair is monogamous, and both parents incubate the eggs and feed the young, until they are ready to fledge, displaying their adultlike, but slightly duller, plumage patterns.

Clutch

Actual size

The egg of the Red-bellied Woodpecker is smooth and white in appearance, immaculate, and measures 1 x ¾ in (25 x 19 mm). Despite the bird's aggressive nature and formidable beak, in some areas nearly half of Red-bellied Woodpecker nests are usurped and destroyed by invasive Common Starlings (see page 515).

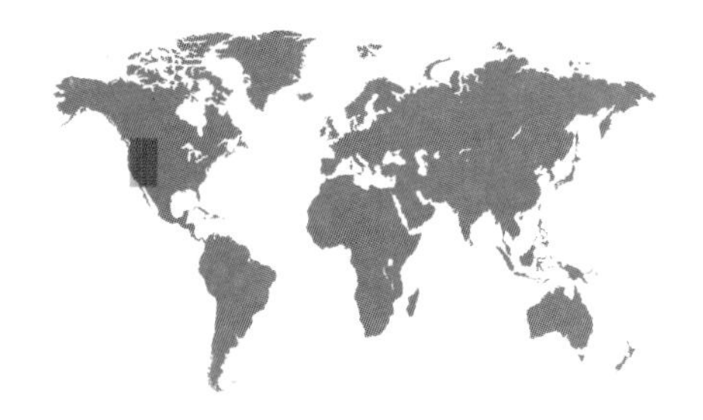

ORDER	Piciformes
FAMILY	Picidae
BREEDING RANGE	Western North America
BREEDING HABITAT	Coniferous and mixed montane forests
NEST TYPE AND PLACEMENT	Cavity nest, excavated anew in conifer stands
CONSERVATION STATUS	Least concern
COLLECTION NO.	FMNH 15361

ADULT BIRD SIZE
8–9 in (20–23 cm)

INCUBATION
12–14 days

CLUTCH SIZE
4–6 eggs

SPHYRAPICUS THYROIDEUS

WILLIAMSON'S SAPSUCKER

PICIFORMES

Clutch

Most black-and-white plumaged woodpeckers show slight sexual dichromatism, with the male having a little more red coloration around his head. But the sexes of this sapsucker are so dramatically different that early naturalists originally thought they were separate species; the male is predominantly black, white, and yellow while the female is more cryptically striped with earth tones. Pair bonds are based on a single female–male coalition and both parents provide parental duties.

For nesting, this sapsucker requires trees with cores that are softened by fungal infections; such a soft core facilitates nest excavation, which occurs anew each year. Old cavities are not reused. To feed the young, this species adds ants to its predominantly sap- and phloem-based diet to provide a better source of protein for the developing nestlings.

Actual size

The egg of the Williamson's Sapsucker is white, glossy, and clear in color, and $\frac{7}{8}$ x $\frac{2}{3}$ in (24 x 17 mm) in size. Both parents incubate the eggs, and develop an incubation patch to transfer heat more efficiently to the embryo through the eggshell.

ORDER	Piciformes
FAMILY	Picidae
BREEDING RANGE	Subarctic and north temperate North America
BREEDING HABITAT	Deciduous and mixed forests, orchards
NEST TYPE AND PLACEMENT	Cavity in a deciduous tree, excavated anew or reused from previous year
CONSERVATION STATUS	Least concern
COLLECTION NO.	FMNH 1340

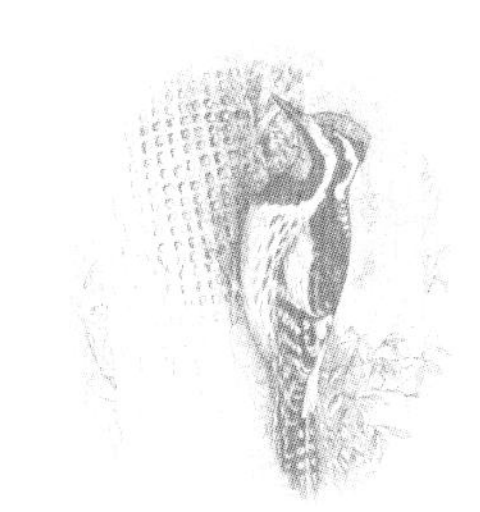

ADULT BIRD SIZE
8–8½ in (20–22 cm)

INCUBATION
10–13 days

CLUTCH SIZE
4–5 eggs

SPHYRAPICUS VARIUS

YELLOW-BELLIED SAPSUCKER

PICIFORMES

At the start of the breeding season, the Yellow-bellied Sapsucker drums loudly on trees with its stout bill. Males produce the sound to attract females, or to indicate to the returning mate that the previous year's breeding site is still active and available. The female arrives on the breeding grounds about one week after the male and he secures or excavates the nest hole. If the first nesting cavity of the season is lost, often due to competition with other hole-nesting species, then the female joins the male to drill the next nest.

The parents feed mostly insects to the hatchlings, but they might coat the prey in tree sap prior to handing it over to the young. Observations suggest that the hungriest chick vocalizes alone while the parents are away; it then jockeys for the position nearest the cavity entrance, and despite the other chicks also begging and calling when the parent arrives, it manages to obtain the food brought by the adult.

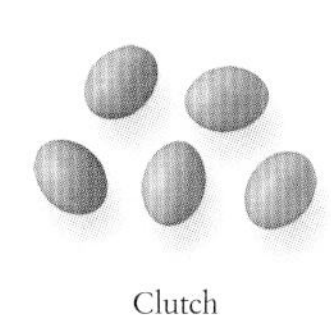

Clutch

The egg of the Yellow-bellied Sapsucker is white in background color, spotless, and measures ⅞ x ⅔ in (24 x 17 mm) in size. The eggs are rarely left unattended as both sexes share incubation duties throughout the day, with the male typically taking the night shift.

Actual size

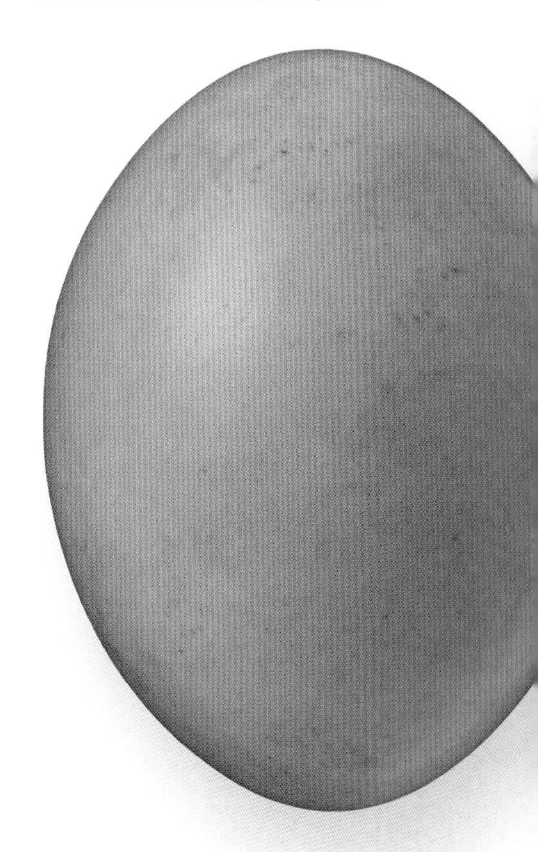

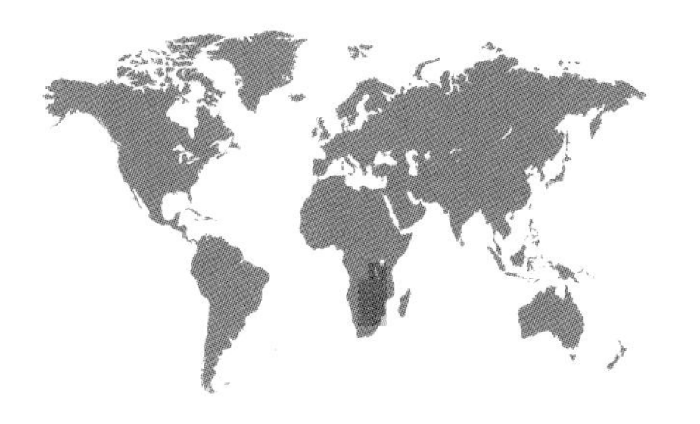

ORDER	Piciformes
FAMILY	Picidae
BREEDING RANGE	Southern Africa
BREEDING HABITAT	Open woodlands and parklands, typically with sandy soils, and often near water
NEST TYPE AND PLACEMENT	Excavated hole nest in tree
CONSERVATION STATUS	Least concern
COLLECTION NO.	FMNH 2541

ADULT BIRD SIZE
9–9½ in (23–24 cm)

INCUBATION
15–18 days

CLUTCH SIZE
2–6 eggs

CAMPETHERA BENNETTII

BENNETT'S WOODPECKER

PICIFORMES

Clutch

This is a common and widespread woodpecker that inhabits open woodlands and riverine parklands. Its foraging strategies include both drilling through bark to reach insects, and foraging on the ground for terrestrial arthropods; it can also sally between trees to catch prey on the wing. The genus *Campethera* is comprised of 13 species found only in Africa. Where members of the genus co-occur, they typically exploit different niches for feeding and breeding.

Bennett's Woodpeckers are considered to be resident in the areas where they occur, but few details are known about the breeding biology of the species. They are usually seen in pairs, implying a long-term pair bond throughout the year. The two members of the pair are thought to cooperate closely throughout breeding to share all aspects of parental care for the eggs and the young.

Actual size

The egg of the Bennett's Woodpecker is immaculate white and ⅞ x ¾ in (23 x 19 mm) in size. Prior to egg laying, both parents work together to excavate a new nest, or they may simply reuse the previous year's nesting hole.

ORDER	Piciformes
FAMILY	Picidae
BREEDING RANGE	North America
BREEDING HABITAT	Woodlands and parks, deciduous forests
NEST TYPE AND PLACEMENT	Excavated cavity in dead limbs of live trees or dead stumps
CONSERVATION STATUS	Least concern
COLLECTION NO.	FMNH 7352

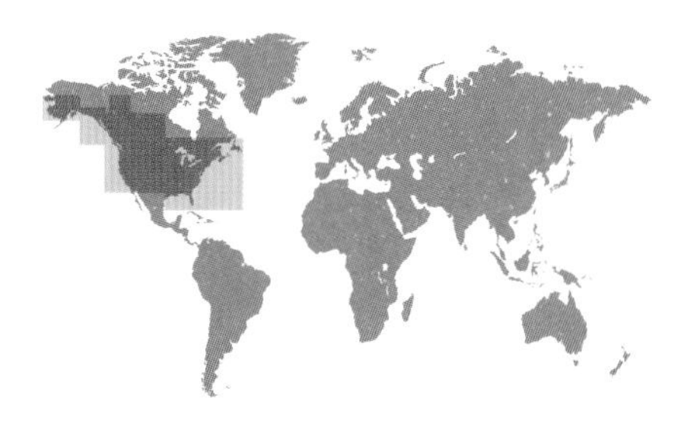

ADULT BIRD SIZE
5½–6½ in (14–17 cm)

INCUBATION
12 days

CLUTCH SIZE
3–8 eggs

PICOIDES PUBESCENS

DOWNY WOODPECKER

PICIFORMES

This is the smallest woodpecker in North America, not only in overall measurements, but also in relative beak length. It is mostly a year-round resident, with only the northernmost breeders migrating south. It looks and acts very much like the larger Hairy Woodpecker (see page 372), but the two species are not closely related, and so they provide an example of convergent plumage and behavioral evolution among woodpeckers.

Clutch

This woodpecker excavates its own nest each year, selecting trees or limbs with heart rot or other fungal diseases. Typically the female chooses a site, alerts the male to it by tapping at the bark, and if he accepts, he taps there too. Nest building can take up to two weeks, and both sexes must agree on the location, before excavation starts. The nest cavity becomes soiled at the end of the nestling period, when the parents stop taking fecal sacs out of it; after the chicks fledge, the cavity is abandoned and not reused for the next season.

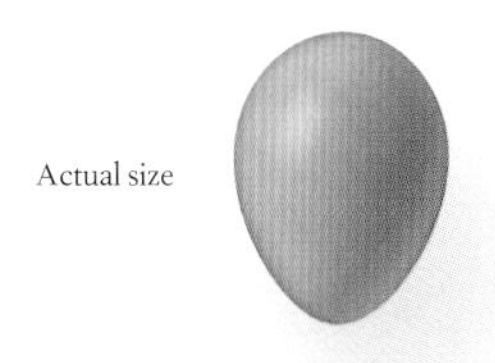

Actual size

The egg of the Downy Woodpecker is clear and white in color, and measures ¾ x ½ in (19 x 14 mm) in dimensions. Incubation typically begins on the day when the clutch is completed, less often on the day when the penultimate egg is laid.

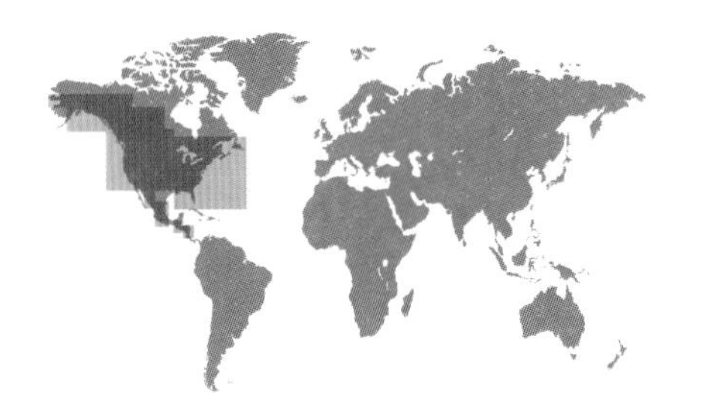

ORDER	Piciformes
FAMILY	Picidae
BREEDING RANGE	North America and montane Central America
BREEDING HABITAT	Mature forests, both coniferous and deciduous, also orchards, parklands, and suburban backyards
NEST TYPE AND PLACEMENT	Cavity drilled in dead parts of live trees, dead stumps, or live branches with advanced heart rot
CONSERVATION STATUS	Least concern
COLLECTION NO.	FMNH 1336

ADULT BIRD SIZE
7–10 in (18–26 cm)

INCUBATION
11–12 days

CLUTCH SIZE
3–6 eggs

PICOIDES VILLOSUS

HAIRY WOODPECKER

PICIFORMES

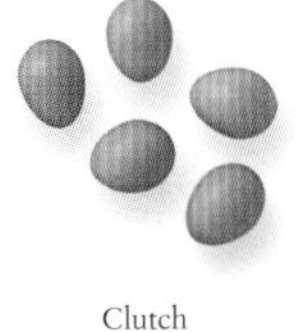

Clutch

The larger of the two most widely distributed North American woodpeckers, the Hairy Woodpecker is distinguished from the Downy Woodpecker (see page 371) by its larger size, relatively longer beak, and fully white and stiff tail feathers which it uses to balance itself while drilling the tree bark for food and nesting. This species is also known to sample sap from "sapsucker wells," shallow holes dug by sapsuckers in tree trunks.

The female typically selects and completes most of the cavity drilling; this could take one to two weeks. To avoid competition from sapsuckers and flying squirrels trying to take over the nest hole, the cavity entrance is often positioned on the hard-to-access underside of a sloping limb or trunk. Once breeding begins, some parents of this species sit tightly on the eggs and will not flush out of the nesting cavity, even when there is significant disturbance around the nest hole.

Actual size

The egg of the Hairy Woodpecker is pure white, and ⅞ x ¾ in (24 x 19 mm). Throughout the laying period, the male roosts in the nesting cavity at night, but does not start incubating until the last egg is laid. Even so, the earlier laid eggs hatch one or more days before the last-laid egg.

ORDER	Piciformes
FAMILY	Picidae
BREEDING RANGE	Southeastern United States
BREEDING HABITAT	Mature southern pine forests
NEST TYPE AND PLACEMENT	Cavity nest drilled into live trees
CONSERVATION STATUS	Vulnerable
COLLECTION NO.	FMNH 7381

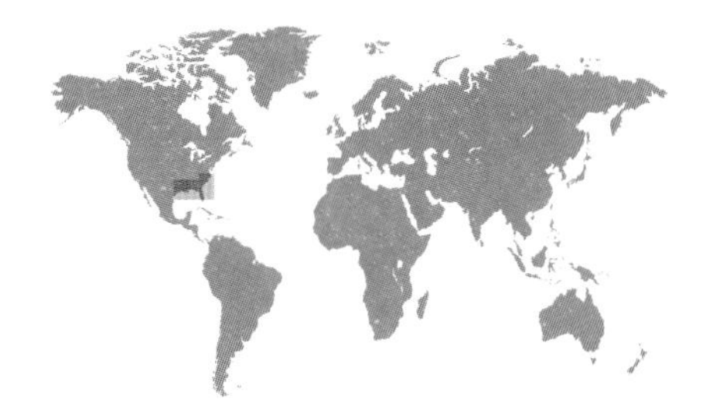

ADULT BIRD SIZE
8–9 in (20–23 cm)

INCUBATION
10–13 days

CLUTCH SIZE
2–5 eggs

PICOIDES BOREALIS

RED-COCKADED WOODPECKER

PICIFORMES

This endemic woodpecker is a specialist of mature pine forests, preferring long-leaf pine stands in the southeastern United States. To build their nests, they require long-leaf pine trees of a certain age and a certain degree of rot. With the clearing, cutting, and management of these pine forests, including the planting of different pine species for commercial harvesting, the native habitat of this woodpecker has declined so dramatically that the species now receives intensive protection efforts.

This bird's complex and specialized breeding biology depends on sociality: clans of two to five or more individuals carve nesting and roosting cavities in live trees, with soft cores, over a period of months to years. Clans include a monogamous breeding pair and several juvenile and adult male helpers. The breeding male roosts in the most recently excavated hole; this also becomes the nest site, with the male incubating the eggs at night.

Clutch

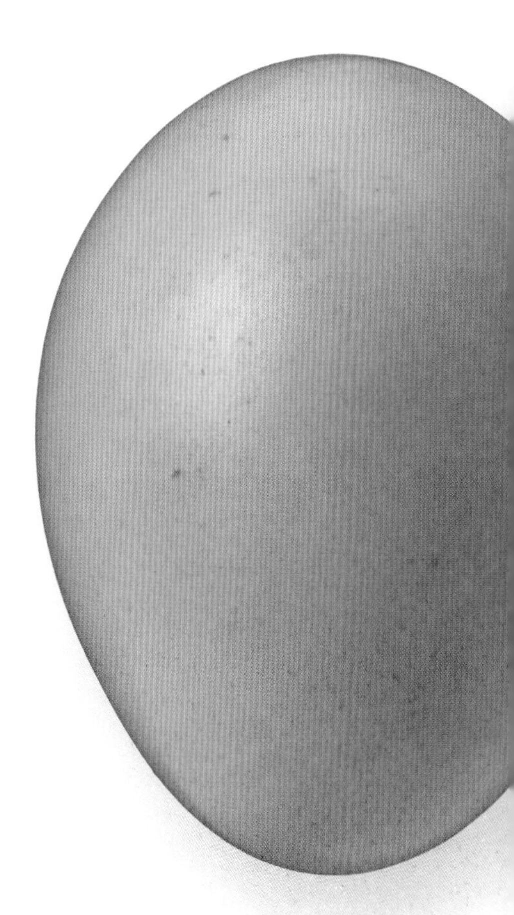

Actual size

The egg of the Red-cockaded Woodpecker is clear and shiny white in color, and $\frac{7}{8}$ x $\frac{2}{3}$ in (24 x 18 mm) in size. Both the breeding male and female have a large brood patch, while the helpers have smaller patches; all assist in incubation.

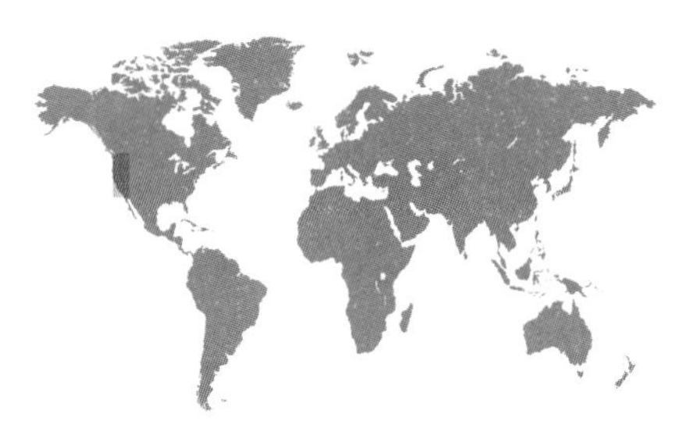

ORDER	Piciformes
FAMILY	Picidae
BREEDING RANGE	Western montane North America
BREEDING HABITAT	Montane mixed and pine forests, including arid areas and post-fire habitats
NEST TYPE AND PLACEMENT	Tree hole, drilled into dead trees, limbs, and fallen logs
CONSERVATION STATUS	Least concern
COLLECTION NO.	FMNH 15383

ADULT BIRD SIZE
8½–9 in (21–23 cm)

INCUBATION
14 days

CLUTCH SIZE
4–5 eggs

PICOIDES ALBOLARVATUS

WHITE-HEADED WOODPECKER

PICIFORMES

Clutch

With its white head and dark body, this woodpecker has a unique plumage pattern in North America. It lives in a narrow range of conifer forests in the mountains of the western United States, where it requires mature, but ailing or dead, trees in which to excavate a nesting hole.

Both sexes are actively involved in drilling the nesting cavity; nonetheless there are many abandoned or failed attempts, resulting in cavity entrances without completed nesting chambers throughout the breeding territories. Incubation also is conducted by both parents, with the male in charge of the nighttime shift, and both sexes brood and feed the chicks. Only about half the nesting attempts succeed in this species each year, and only about two-thirds of the adults survive to excavate a new cavity and try to breed again.

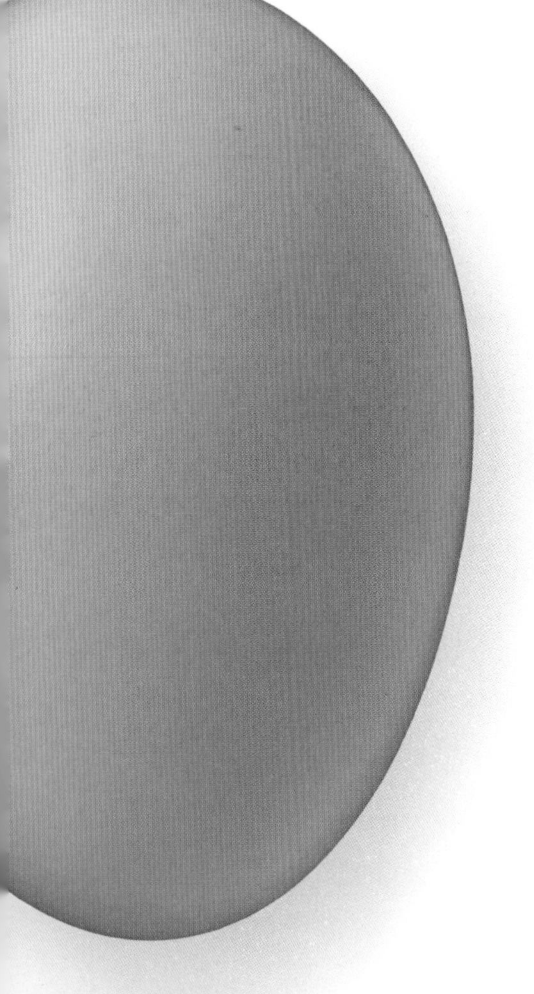

Actual size

The egg of the White-headed Woodpecker is white in background color, clear of spotting, and ⅞ x ⅔ in (24 x 18 mm) in size. Lying on the unlined bottom of the tree cavity, the eggs often become stained or spotted as the incubation proceeds.

ORDER	Piciformes
FAMILY	Picidae
BREEDING RANGE	North America, Central America
BREEDING HABITAT	Open woodlands, forests, parks, and grasslands with distant tree stands
NEST TYPE AND PLACEMENT	Tree hole, excavated anew; bottom is lined with a layer of woodchips, probably a side effect of drilling
CONSERVATION STATUS	Least concern
COLLECTION NO.	FMNH 7497

ADULT BIRD SIZE
11–12 in (28–30 cm)

INCUBATION
11–13 days

CLUTCH SIZE
5–7 eggs

COLAPTES AURATUS

NORTHERN FLICKER

PICIFORMES

Though this is an iconic backyard bird species, recent discoveries using individually marked birds and genetic analyses have revealed that the family life of the Northern Flicker is anything but simple and drama-free. Although most pairs are socially monogamous, in as many as one-in-six nests, some of the eggs may be laid parasitically, often by a neighboring flicker female, even though at the time she too typically has a full clutch of eggs. In around one-in-twenty territories, the female has two separate pair bonds with different males, and thus breeds polyandrously.

Gathering resources to lay extra eggs takes both time and energy, but female flickers can apparently afford it, because parental duties involve both sexes. In fact, the male is so attentive, incubating the eggs and feeding the chicks, that at most of the nests where the female disappears (either due to predation or because she joins another male) he alone can raise the brood. But at nests where the male disappears, the female typically fails to fledge the brood.

Clutch

The egg of the Northern Flicker is white in background color, and is highly variable in size, with average measurements of 1⅛ x ⅞ in (28 x 22 mm) in length and width. Incubation is done mostly by the male; he also keeps the eggs warm at night.

Actual size

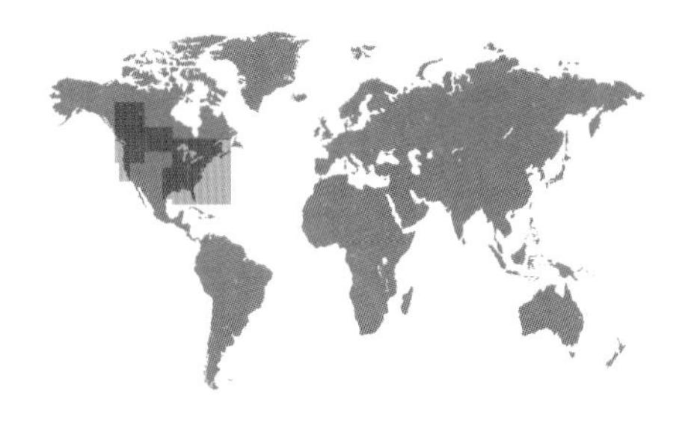

ORDER	Piciformes
FAMILY	Picidae
BREEDING RANGE	Temperate North America
BREEDING HABITAT	Deciduous and coniferous forests, wooded parks
NEST TYPE AND PLACEMENT	Large cavity, carved into dead trees and logs; unlined except for woodchips
CONSERVATION STATUS	Least concern
COLLECTION NO.	FMNH 7428

ADULT BIRD SIZE
15½–19½ in (40–49 cm)

INCUBATION
15–18 days

CLUTCH SIZE
3–5 eggs

DRYOCOPUS PILEATUS

PILEATED WOODPECKER

PICIFORMES

Clutch

This is the largest North American woodpecker, assuming that the Ivory-billed Woodpecker is extinct. It maintains its pair bond throughout the year, with the mates occupying a year-round feeding and nesting territory. It prefers to live in mature wood lots, but will inhabit smaller and younger stands, as long as several large trees are still standing for feeding and nesting. The male selects and begins to carve out the nesting chamber, and the female contributes if she accepts the location.

Once a nesting chamber is complete, the female begins laying her eggs on a daily basis. She spends many of the daylight hours in the nest during the laying period, while the male roosts in that cavity at night. During the incubation period, the pair is very quiet and secretive and the eggs are covered by one of the adults 99 percent of the time, with the male solely taking charge of nocturnal shifts.

Actual size

The egg of the Pileated Woodpecker is white and spotless, and 1⅓ x 1 in (33 x 25 mm) in dimensions. The nest cavity is used for a single clutch of eggs, and the birds excavate a brand new hole in the following year.

ORDER	Piciformes
FAMILY	Picidae
BREEDING RANGE	North Africa
BREEDING HABITAT	Dry montane forests, up to the tree line
NEST TYPE AND PLACEMENT	Cavity nest in a tree, lined with woodchips
CONSERVATION STATUS	Least concern
COLLECTION NO.	FMNH 20616

ADULT BIRD SIZE
12–13 in (30–33 cm)

INCUBATION
14–17 days

CLUTCH SIZE
4–8 eggs

PICUS VAILLANTII

LEVAILLANT'S WOODPECKER

PICIFORMES

This green woodpecker lives and breeds in montane and foothill forest stands, typically far from human settlements. For feeding, it normally lands on the ground in clearings and fields, and collects ants and other insects by flicking out its long sticky tongue to capture them. Although behaviorally similar to the *Colaptes* flickers of North America, the two genera, *Colaptes* and *Picus*, are not closely related.

Pairs begin nest building by selecting a dead, rotten, or occasionally a still-living part of the tree trunk, and they excavate a deep and spacious hole in the softer, often fungus-infected core of the tree. The couple remain in the vicinity of their nesting and foraging range all year long, except for the highest-elevation populations. These descend to spend the non-breeding season in warmer, but adjacent, lowlands, in a pattern termed "altitudinal migration."

Clutch

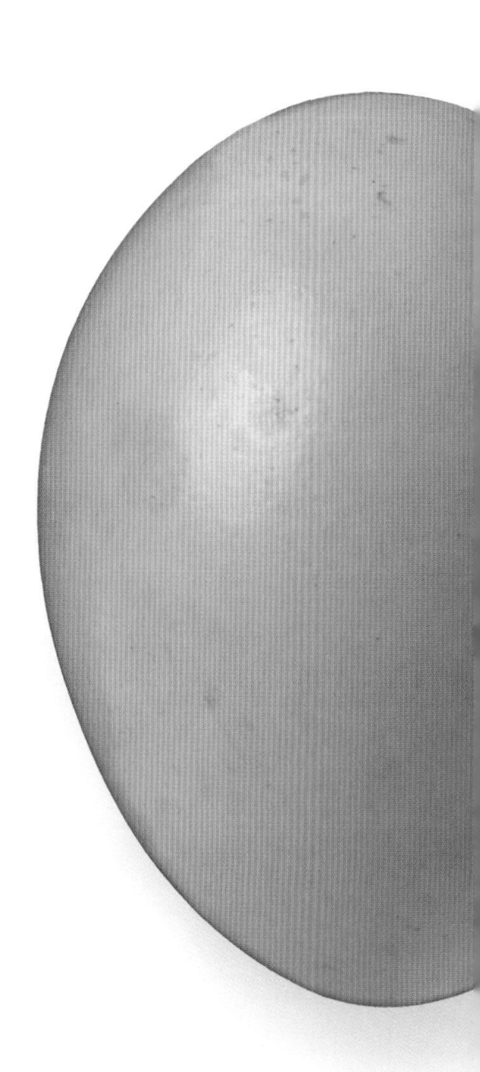

Actual size

The egg of the Levaillant's Woodpecker is glossy white and immaculate in color, and measures 1 1/16 x 3/4 in (27 x 20 mm) in dimensions. Both parents contribute to excavating the nest, taking turns to incubate the eggs and provision the young.

PASSERINES

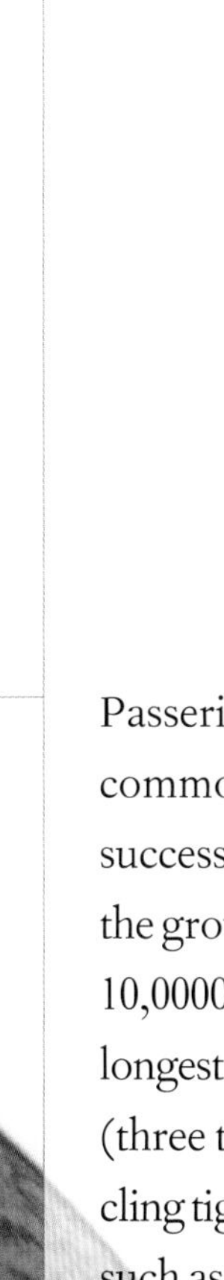

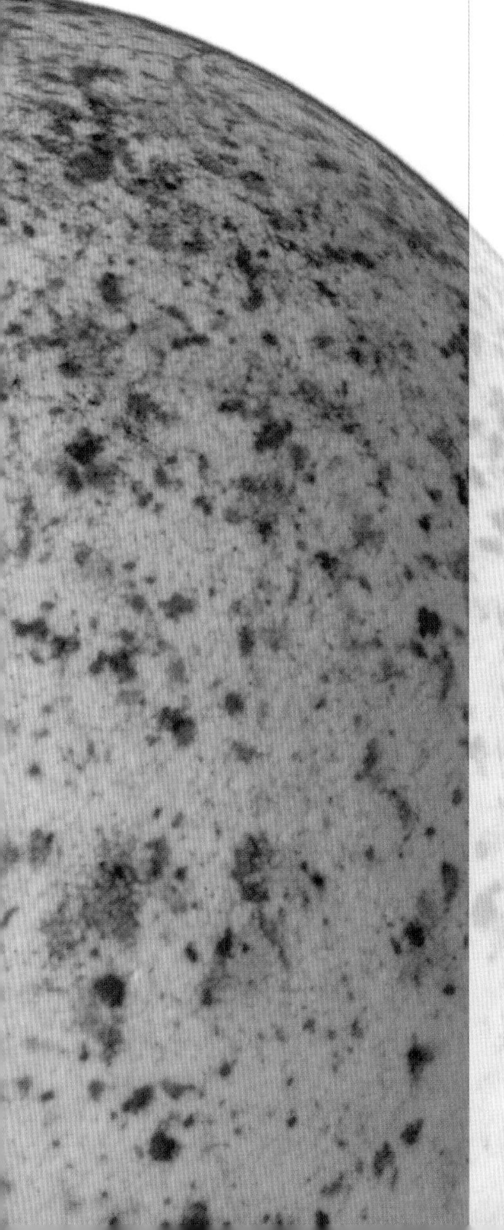

Passerines, members of the order Passeriformes, are commonly called perching birds. The evolutionary success of the Passerines is illustrated by the fact that the group contains more than 50 percent of the world's 10,0000 bird species. This final chapter, therefore, is the longest in the book. Passerines all share "perching feet" (three toes front and one back), which enable them to cling tightly to small branches and even vertical surfaces such as cliffs. One of the two main Passerine lineages, the Oscines or song birds, have evolved specialized vocal organs to produce elaborate calls and songs that are learned as early as embryos in the egg, but also as chicks, fledglings, and young adults, settling near an older neighbor. In contrast, Suboscines do not need a tutor to learn and practice a song. Passerines vary in size and occur in all habitats, on all continents except Antarctica, and have reached even the most remote oceanic islands. Their nests take all manner of forms, and their eggshell colors and patterns show a similarly broad variety. The chicks are always altricial, hatching naked and helpless to be cared for by one parent, both, or a cooperating group of helpers at the nest.

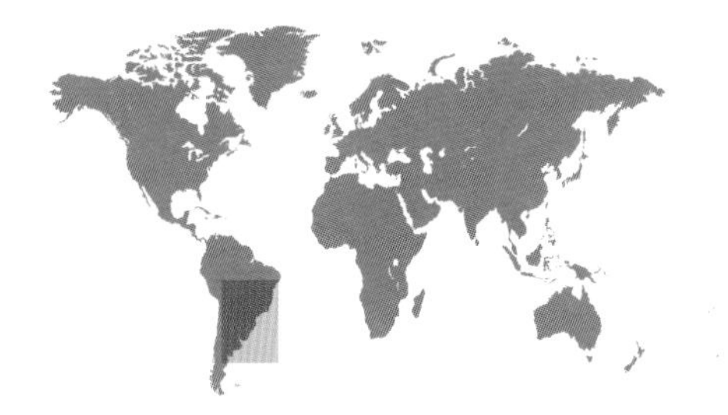

ORDER	Passeriformes
FAMILY	Furnariidae
BREEDING RANGE	Central and eastern South America
BREEDING HABITAT	Open forests and pastures
NEST TYPE AND PLACEMENT	Globular mud nests, placed on thick tree branches or fenceposts
CONSERVATION STATUS	Least concern
COLLECTION NO.	WFVZ 53777

ADULT BIRD SIZE
7–8 in (18–20 cm)

INCUBATION
16–17 days

CLUTCH SIZE
2–4 eggs

FURNARIUS RUFUS

RUFOUS HORNERO

PASSERIFORMES

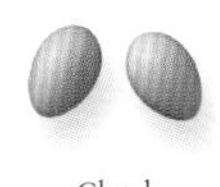

Clutch

The egg of the Rufous Hornero is white in background color, and measures $1\frac{1}{16}$ x $\frac{3}{4}$ in (27 x 21 mm) in size. If a hornero nest is parasitized by a Shiny Cowbird (*Molothrus bonariensis*), in the dark nesting chamber, the host uses egg size, rather than coloration, to detect the smaller, foreign egg and remove it.

This hornero is also known as the Red Ovenbird in English, but it is not to be confused with the North American ovenbird species, a type of warbler (see page 524). Horneros belong to a large, diverse family of South American birds, the Furnariidae. Nests of the Rufous Hornero are a common sight in the open pastures and pampas of temperate regions across most of South America, south of the Amazonian forests. The national bird of Argentina, it builds a pizza-oven-shaped nest, with a 180-degree turn leading to a pitch-dark nesting chamber. Nests are often placed on top of fence or electric poles; this adaptability has resulted in an expansion of the species' range to include not only all open country, but many urban areas as well.

The breeding pair is monogamous, remains on the territory all year, and can quickly build a new nest in just one week, or refurbish an old nest in the course of a few days. Old and unused nests are sometimes found next to, or on top of one another. Other bird species have evolved to usurp abandoned ovenbird nests, instead of building their own nests.

Actual size

ORDER	Passeriformes
FAMILY	Furnariidae
BREEDING RANGE	Subtropical lowlands of southern South America
BREEDING HABITAT	Seasonally dry grassland, pastures, degraded forests
NEST TYPE AND PLACEMENT	Large collection of sticks and branches, in the crutch of a tree, cactus, or on an electric wire pole
CONSERVATION STATUS	Least concern
COLLECTION NO.	WFVZ 160675

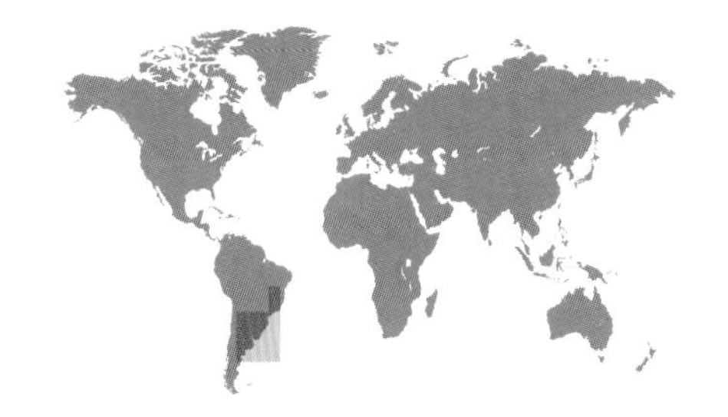

ANUMBIUS ANNUMBI

FIREWOOD-GATHERER

PASSERIFORMES

ADULT BIRD SIZE
7–8½ in (18–22 cm)

INCUBATION
12 days

CLUTCH SIZE
2–4 eggs

The Firewood-gatherer gets its name because it is frequently seen carrying large sticks. The bird, which is about the size of a blackbird, uses the sticks to build its massive nest, which can be as large as 6½ ft (2 m) across. Constructed somewhat like a giant house of cards, when scientists have attempted to take it apart, it most often simply fell to pieces. Careful examination of the nests reveals a single entrance, from which a spiral passageway leads to a central nesting chamber; the passageway provides an arduous and nearly impenetrable route for a predator to reach the eggs and the nestlings. Some of the Firewood-gatherer's successfully reared young opt to remain at home and help their parents to raise the next brood, instead of dispersing and starting to breed on their own.

The bulky nests, assembled with the twigs of heavily thorny trees and shrubs, are a valuable resource, eagerly sought after by other species that do not build their own. The Bay-winged Cowbird, for example, an icterid blackbird, does not construct a nest, but aggressively usurps the Firewood-gatherer's fortress.

Clutch

The egg of the Firewood-gatherer is white in background color, immaculate, and 1 x ⅔ in (25 x 17 mm) in size. The eggs are hidden and well protected deep inside the complex and bulky stick nest.

Actual size

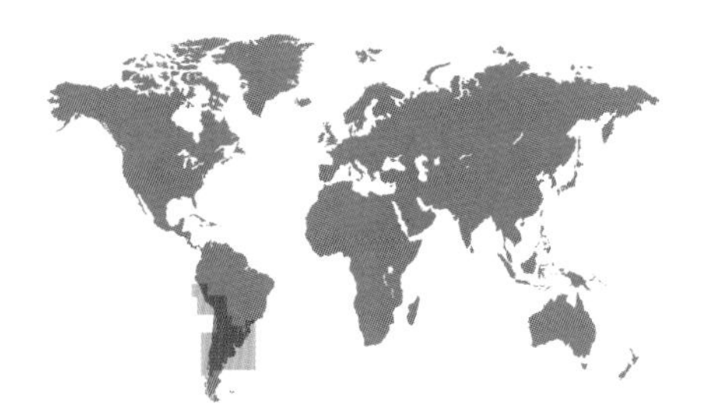

ORDER	Passeriformes
FAMILY	Furnariidae
BREEDING RANGE	Western and southern South America
BREEDING HABITAT	Reed beds, swamps; both fresh and coastal marshes
NEST TYPE AND PLACEMENT	Spherical nest of leaves and grasses, attached to reed stems, with a side entrance
CONSERVATION STATUS	Least concern
COLLECTION NO.	FMNH 3484

ADULT BIRD SIZE
5–6 in (13–15 cm)

INCUBATION
15–22 days

CLUTCH SIZE
2–3 eggs

PHLEOCRYPTES MELANOPS

WREN-LIKE RUSHBIRD

PASSERIFORMES

Clutch

This species is a specialist inhabitant of wetlands, using its small body and narrow bill to navigate through the thickets of reed stems and other swamp vegetation. In its appearance, foraging, and nesting behavior, it is highly reminiscent of the North American Long-billed Marsh Wren, but the two evolved from completely different parts of the perching bird order. The rushbird is a suboscine, and does not learn its song from tutors, whereas the wren is an oscine, and learns to sing by imitating others.

The breeding behavior of the rushbird is influenced by a need to find nesting sites that are safe from both flooding and terrestrial predators. They do not always succeed, because although most nests are built directly over water, some of the young are still reached and preyed upon by rodents, whereas many others perish due to fluctuating water levels. On average, less than half of rushbird nests produce fledglings successfully.

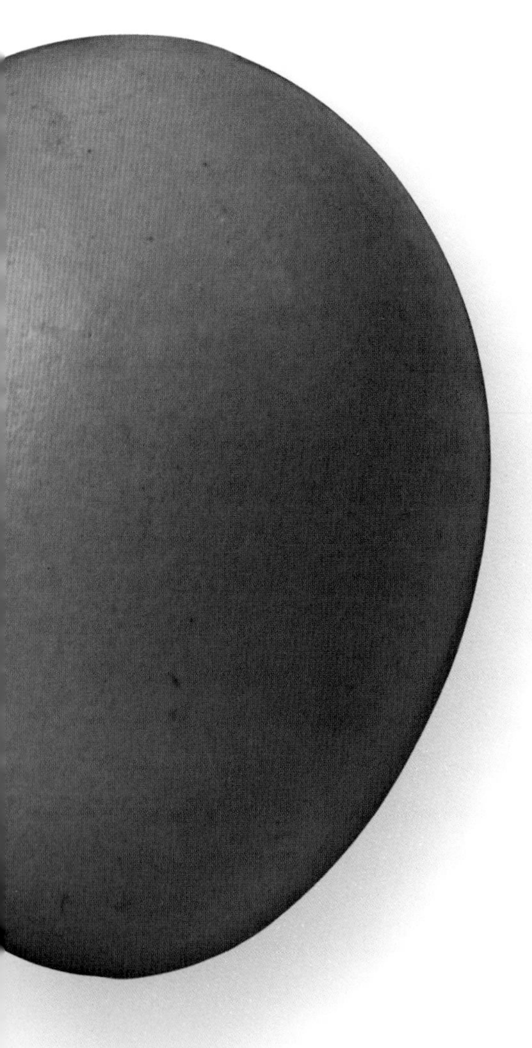

Actual size

The egg of the Wren-like Rushbird is blue green and immaculate in appearance, and ¾ x ⅝ in (20 x 15 mm) in size. Both parents incubate the eggs and share the duties of feeding the nestlings.

ORDER	Passeriformes
FAMILY	Furnariidae
BREEDING RANGE	Coastal Mexico and northern Central America
BREEDING HABITAT	Dry or moist lowland and montane woodlands; secondary growth forests.
NEST TYPE AND PLACEMENT	Cavity nest in a natural tree hole
CONSERVATION STATUS	Least concern
COLLECTION NO.	WFVZ 141279

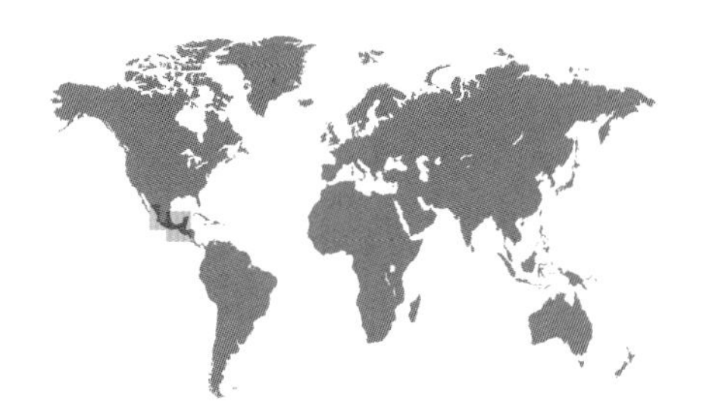

ADULT BIRD SIZE
8½–10½ in (22–27 cm)

INCUBATION
14–20 days

CLUTCH SIZE
2–3 eggs

XIPHORHYNCHUS FLAVIGASTER

IVORY-BILLED WOODCREEPER

PASSERIFORMES

The common name of this species is not to be confused with the extinct Ivory-billed Woodpecker. This woodcreeper is a prominent inhabitant of Central American forests, where it uses its long, curved bill to probe and pull arthropod prey from hiding-places in leaf litter and under tree bark. Typically a solitary feeder, Ivory-billed Woodcreepers are known to join mixed-species flocks of forest birds that forage together presumably to improve foraging opportunities and to benefit from multiple individuals on the look-out for potential predators.

Throughout the breeding and non-breeding season, this woodcreeper remains faithful to its territory, staying put within the same range irrespective of the stage of its breeding cycle. Though published data are sparse, one study suggests that females of this species are in sole charge of building the nest, incubating the eggs, and feeding the nestlings.

Clutch

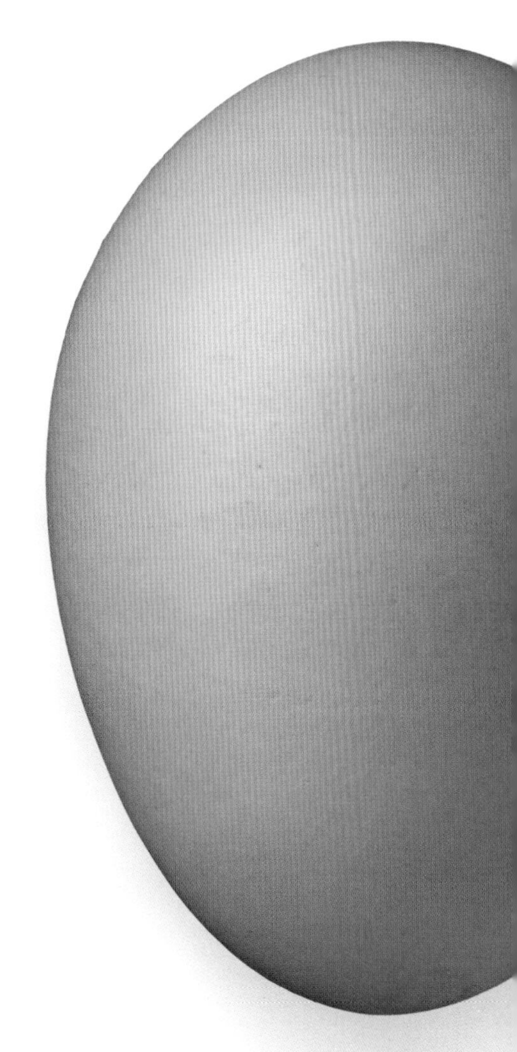

Actual size

The egg of the Ivory-billed Woodcreeper is white in background color, has no spotting, and measures 1⅛ x ⅞ in (29 x 21 mm) in dimensions. Little else is known about the breeding biology of this species or of other woodcreepers.

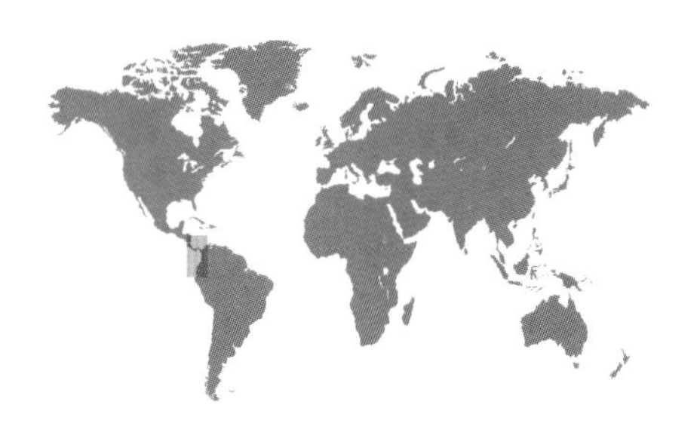

ORDER	Passeriformes
FAMILY	Thamnophilidae
BREEDING RANGE	Central America and western South America
BREEDING HABITAT	Tropical humid forest, edge and secondary growth
NEST TYPE AND PLACEMENT	Cup nest placed low in tree, made of vines, leaves, plant fibers
CONSERVATION STATUS	Least concern
COLLECTION NO.	WFVZ 64488

ADULT BIRD SIZE
5½–6½ in (14–16 cm)

INCUBATION
16 days

CLUTCH SIZE
1–2 eggs

MYRMECIZA EXSUL

CHESTNUT-BACKED ANTBIRD

PASSERIFORMES

Clutch

More is known about this antbird than about most species in this large and diverse family of tropical birds. This species maintains a year-long (probably longer) pair bond, with the mates remaining in close physical or acoustic contact in the dense forest undergrowth. They are territorial against other antbirds of the same species, with both the male and the female calling and singing, while also puffing their feathers, dropping their wings, and flicking their tails in response to an intruder. At the same time, they are tolerant of other species, especially when flocks of mixed composition move through their territory, following army ants and feeding on the insect prey flushed along these ants' march.

The nest of the Chestnut-backed Antbird is placed near the ground in dense foliage, and hidden from visual predators by a layer of moss on the outside, and the cryptic cup and egg colors on the inside. To protect the nestlings from smell-driven predators, the parents keep the nest clean from odorous cues by consuming or removing fecal sacs produced by the chicks after each feeding bout.

The egg of the Chestnut-backed Antbird is pinkish red in background color, with darker purplish chestnut maculation, and is ⅞ x ⅔ in (23 x 17 mm) in dimensions. Both parents incubate the eggs and provision the nestlings.

Actual size

ORDER	Passeriformes
FAMILY	Formicariidae
BREEDING RANGE	Southern Central America and northwestern South America
BREEDING HABITAT	Humid lowland and montane forests
NEST TYPE AND PLACEMENT	Cup nest built inside a tree cavity, lined with leaves and plant fibers
CONSERVATION STATUS	Least concern
COLLECTION NO.	WFVZ 154766

ADULT BIRD SIZE
8–8½ in (20–22 cm)

INCUBATION
20 days

CLUTCH SIZE
2 eggs

FORMICARIUS NIGRICAPILLUS

BLACK-HEADED ANTTHRUSH

PASSERIFORMES

Although similar in their miniature, rail-like appearance to other antthrushes in the same genus, a defining characteristic of this species is its calls, which, including its mate attraction song, are dramatically different from those of close relatives. This vocal signature provides error-free communication for species recognition by individuals seeking mates, and presumably minimizes territorial disputes between battling neighbors. The males are typically faithful to their breeding site all year round, while the females may stay with the same male across years or float between territories and pair with a new male when the local female disappears.

This antthrush adopts, rather than excavates, existing tree cavities for nesting; in its tropical range, such cavities in damaged or decaying trees are not as limited as in more temperate regions. The result is that even in bird-rich habitats, such as the humid mid-elevation forests of Costa Rica, competition for nesting sites inside natural cavities is not severe, and there is a lower risk of a cavity-nester being usurped in the middle of its nesting attempt.

Clutch

The egg of the Black-headed Antthrush is matte and white in coloration, not speckled, and 1⅓ x 1$^{1}/_{16}$ (33 x 27 mm) in dimensions. Once the eggs hatch, the chicks leave the nest several days before they are able to fly, relying on both parents for shelter and food.

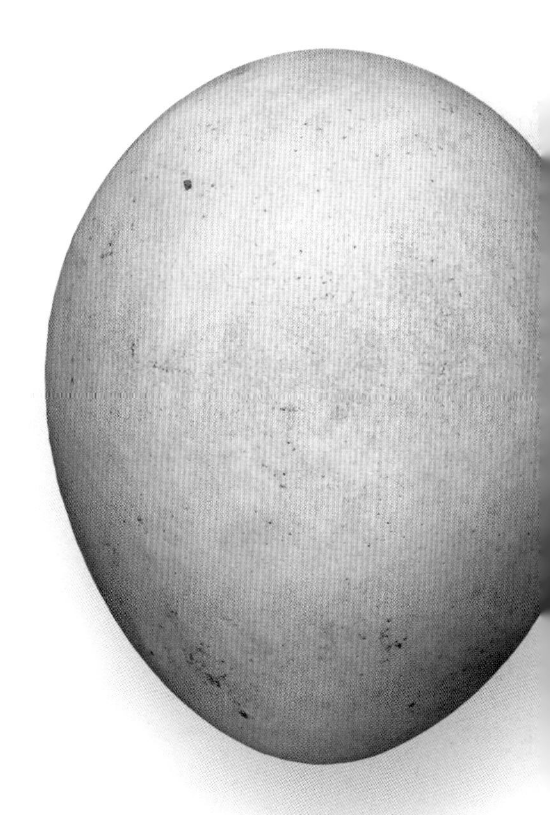

Actual size

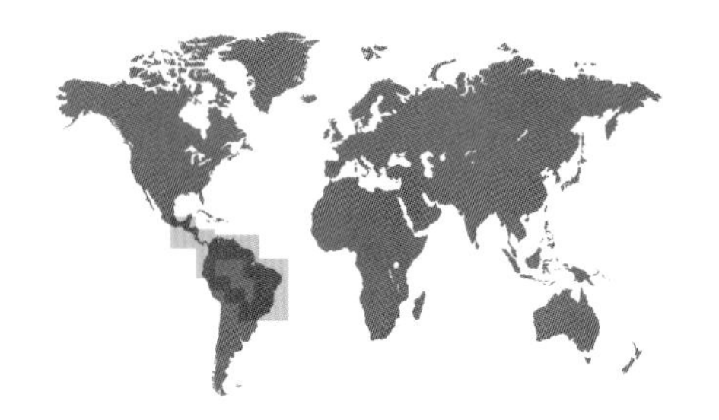

ORDER	Passeriformes
FAMILY	Tyrannidae
BREEDING RANGE	Central America and South America
BREEDING HABITAT	Humid second growth forests, open woodlands and clearings, plantations, and gardens
NEST TYPE AND PLACEMENT	Enclosed nest made of grasses and plant fibers, suspended from a thin branch or vine
CONSERVATION STATUS	Least concern
COLLECTION NO.	FMNH 2489

ADULT BIRD SIZE
3½–4 in (9–10 cm)

INCUBATION
15–16 days

CLUTCH SIZE
2 eggs

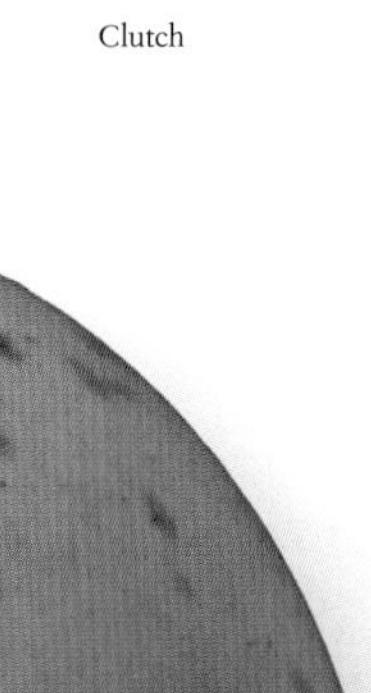

TODIROSTRUM CINEREUM

COMMON TODY-FLYCATCHER

PASSERIFORMES

Clutch

This is a small, common flycatcher that inhabits parks and backyard gardens throughout its tropical distribution. This bird is so small that it is occasionally entangled in the webs of large orb-weaving spiders and eaten. The species maintains a year-round residence; the pairs form long-lasting bonds, and mates can be seen moving and foraging near one other throughout the seasons.

The nest is a meticulously constructed, lightweight, globular structure attached to a long thin branch, with grasses and plant fibers woven around a case of dry leaves. It has a side entrance that allows the incubating female to peek outside while she is sitting on the eggs. Among birds, larger species tend to have longer incubation periods, but this Tody-flycatcher is an exception to this rule; for its small body size, it has a relatively long incubation period.

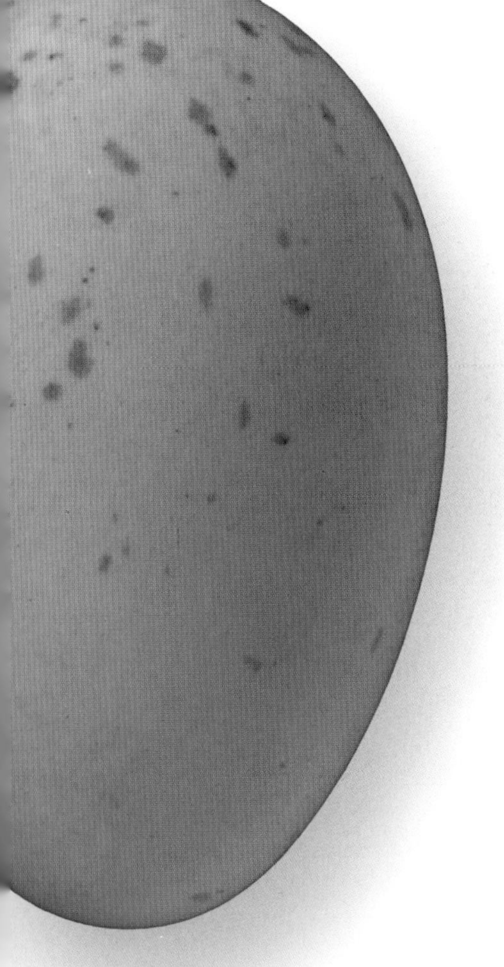

Actual size

The egg of the Common Tody-Flycatcher is immaculate white, and ⅝ x ½ in (15 x 12 mm) in size. Both parents build the nest, but only the female incubates the eggs; the male joins her again to provision the young together.

ORDER	Passeriformes
FAMILY	Tyrannidae
BREEDING RANGE	North America
BREEDING HABITAT	Mid- to high-elevation and coniferous forests, often near clearings and other edges
NEST TYPE AND PLACEMENT	Open cup of twigs, roots, and lichens, placed along a thin branch of a pine or spruce, in a fork or on top of a cluster of fallen needles
CONSERVATION STATUS	Near threatened
COLLECTION NO.	FMNH 16673

ADULT BIRD SIZE
7–8 in (18–20 cm)

INCUBATION
13–16 days

CLUTCH SIZE
3–4 eggs

CONTOPUS COOPERI

OLIVE-SIDED FLYCATCHER

PASSERIFORMES

This flycatcher really lives up to its name by hunting and capturing flying prey in the air after spotting it from an exposed perch in the forest or forest edge. The female and the male defend their feeding and breeding territory aggressively from other Olive-sided Flycatchers; relative to their small body size, they maintain a large exclusive area around the nest. Population numbers for this species have been steadily declining by 3 percent or more yearly over the past decade, resulting in its designated conservation status of "near threatened."

This flycatcher is a long-distance migrant, which complicates conservation efforts. It arrives on its breeding grounds very late in the spring, so it has limited time to establish the pair bond, initiate the nest, and fledge the young. Even so, if the first or second nesting attempt fails, the pair remains together and inspects and initiates a new nest at a new site, often repeatedly. However, later breeding attempts fail more frequently than earlier attempts because as summer turns into fall, there is little time to complete a full breeding cycle.

Clutch

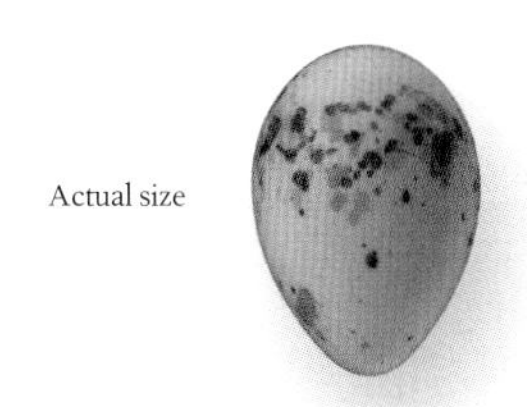

Actual size

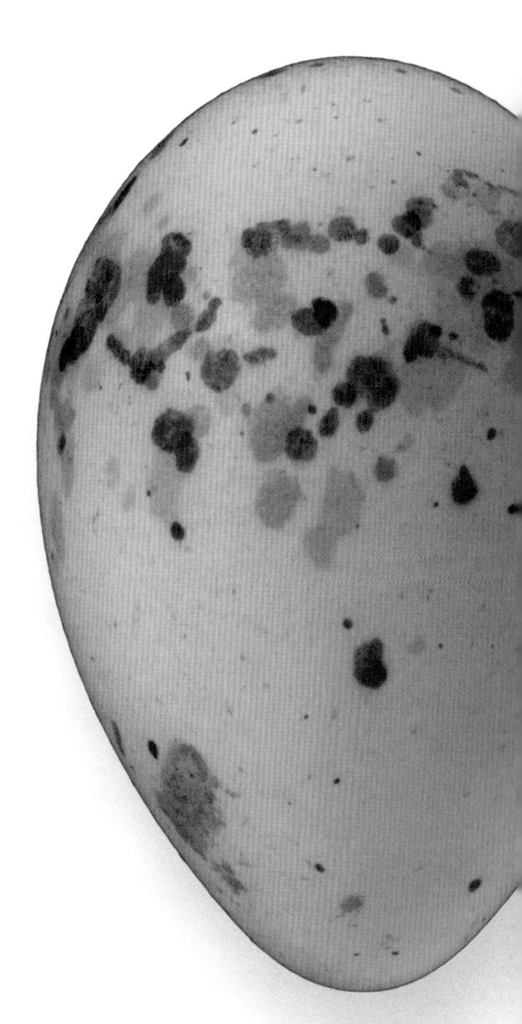

The egg of the Olive-sided Flycatcher is creamy white or buff in color, spotted with a ring of brown blotches, and measures ⅞ x ⅔ in (23 x 16 mm). The female builds the nest and incubates the eggs alone, with the male feeding her at the nest during the early incubation stage.

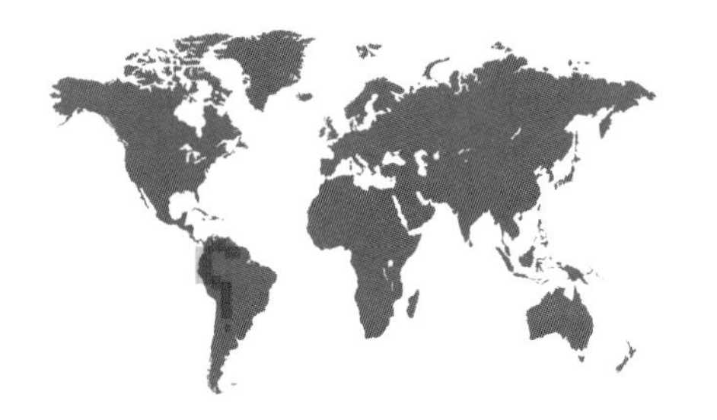

ORDER	Passeriformes
FAMILY	Tyrannidae
BREEDING RANGE	Western and northern South America
BREEDING HABITAT	Highland forests, open woodlands, secondary growth
NEST TYPE AND PLACEMENT	Shallow open cup, made of moss and covered with lichen, built over a horizontal branch
CONSERVATION STATUS	Least concern
COLLECTION NO.	FMNH 19082

ADULT BIRD SIZE
6½–6¾in (16–17 cm)

INCUBATION
16 days

CLUTCH SIZE
2 eggs

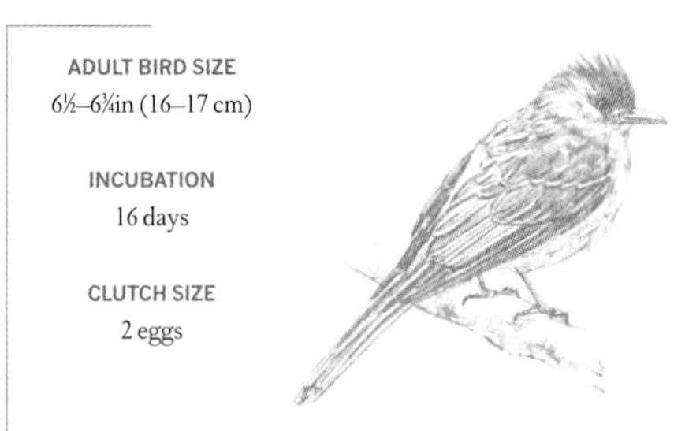

CONTOPUS FUMIGATUS

SMOKE-COLORED PEWEE

PASSERIFORMES

Clutch

This drab but loud flycatcher is an inhabitant of upper elevations, with its range in the Andes Mountains starting at around 3,300 ft (1,000 m) and reaching to the tree line at 8,500 ft/ 2,600 m). Like many members of this extremely large and mainly neotropical family, relatively little is known of its breeding biology. Flycatcher sexes are usually identical in plumage, and unless each member of a pair is captured and color-banded to distinguish them, no sex-specific parental roles can be observed.

Nest construction and incubation each take about two weeks, with the nest attended by a parent for most, but not all, of the day. This is feasible in the tropics (even in the mountains) where external temperatures are high enough to support embryonic growth, even when incubation is interrupted for long periods of time. During the nestling period, both parents attend the nest and provision the young with insects gleaned and caught in midair.

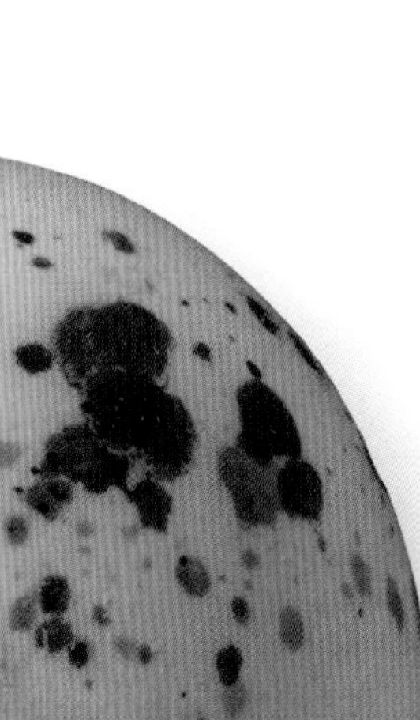

Actual size

The egg of the Smoke-colored Pewee is white in coloration, has sparse lavender spotting, and measures ¾ x ⅝ in (19 x 15 mm) in dimensions. The nest-building and egg-laying period coincides with the drier months of its local breeding range.

ORDER	Passeriformes
FAMILY	Tyrannidae
BREEDING RANGE	Eastern North America
BREEDING HABITAT	Deciduous, mixed, or coniferous forests
NEST TYPE AND PLACEMENT	Cup of woven grass with lichens on the outside, and lined with hair and plant fibers; placed on a thick horizontal limb of a tree
CONSERVATION STATUS	Least concern
COLLECTION NO.	FMNH 9297

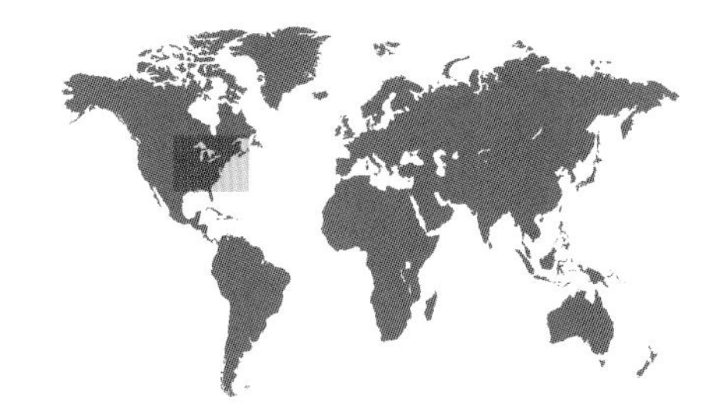

ADULT BIRD SIZE
6–6¾ in (15–17 cm)

INCUBATION
12–14 days

CLUTCH SIZE
2–4 eggs

CONTOPUS VIRENS

EASTERN WOOD-PEWEE

PASSERIFORMES

This dull-green-colored species is commonly heard calling "Pewee" as it sallies between trees capturing insect prey. However, stomach-contents analyses show that most prey items are likely taken from the foliage and even from the ground, with crickets, grasshoppers, and spiders dominating the prey base; these are also the foods provided for the chicks. Pairs maintain a common feeding and nesting territory in the summer, but the species is solitary outside of the breeding season.

The eggs are laid one day at a time, which is typical of most small birds. Prior to incubation, the female develops a brood patch, a featherless area of skin on her belly; this allows efficient transfer of heat from the parent to the eggs. The male does not develop a brood patch, but he feeds the female at the nest, and stays nearby to stand guard while the female departs briefly between incubation bouts.

Clutch

Actual size

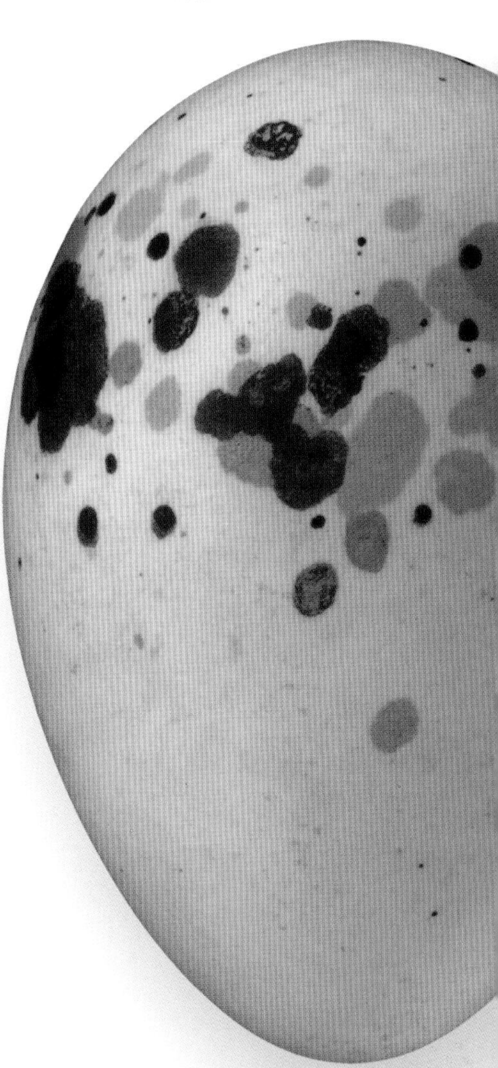

The egg of the Eastern Wood-Pewee is milky white in background color with purplish maculation around the blunt end, and is ⅔ x ⅝ in (18 x 14 mm) in size. The eggs in the nest, high up in a tree and with its lichen camouflage, are nearly impossible for predators or researchers to find.

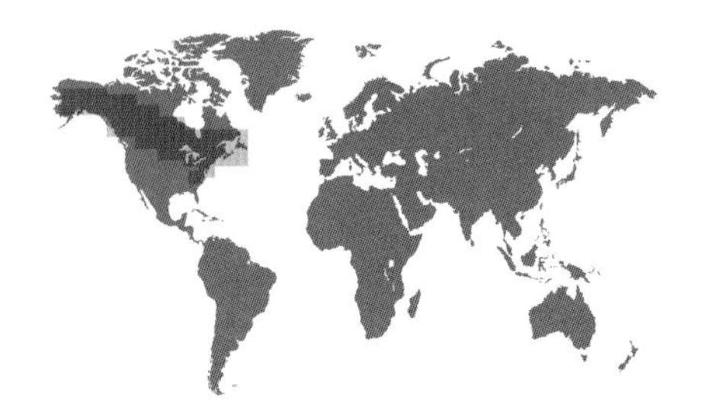

ORDER	Passeriformes
FAMILY	Tyrannidae
BREEDING RANGE	Northern and Southeastern North America
BREEDING HABITAT	Wet brush and shrubland, forest edges, early successional woods
NEST TYPE AND PLACEMENT	Coarse, loose cup of woven twigs and grasses, lined with finer grasses but no feathers; low in a tree fork or shrub
CONSERVATION STATUS	Least concern
COLLECTION NO.	FMNH 18969

ADULT BIRD SIZE
5–6¾ in (13–17 cm)

INCUBATION
11–14 days

CLUTCH SIZE
3–4 eggs

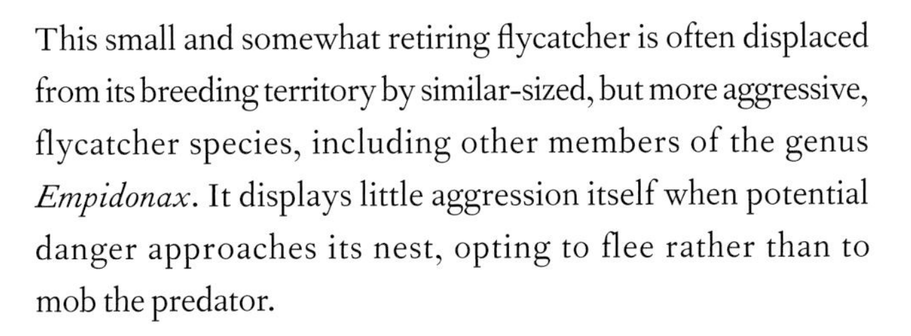

EMPIDONAX ALNORUM

ALDER FLYCATCHER

PASSERIFORMES

Clutch

The egg of the Alder Flycatcher is creamy white or buff in background color, has no or only sparse dark maculation, and measures ¾ x ⅝ in (19 x 14 mm). This species is one of the top 15 most frequently parasitized by the Brown-headed Cowbird (see page 616).

This small and somewhat retiring flycatcher is often displaced from its breeding territory by similar-sized, but more aggressive, flycatcher species, including other members of the genus *Empidonax*. It displays little aggression itself when potential danger approaches its nest, opting to flee rather than to mob the predator.

As a long-distance migrant (wintering in Central and South America), time is of the essence, and breeding begins soon after both members of the pair arrive and reestablish their bond. The female builds the nest alone, in as little as a day and a half (and it shows!), and then she lays her eggs one day apart. Both sexes incubate the eggs, with the male investing more time in the afternoon, before the longer nighttime shift taken usually by the female. Despite its occurrence near many densely populated regions in the northeastern United States and in Canada, surprisingly little is known about most aspects of the breeding biology of this species.

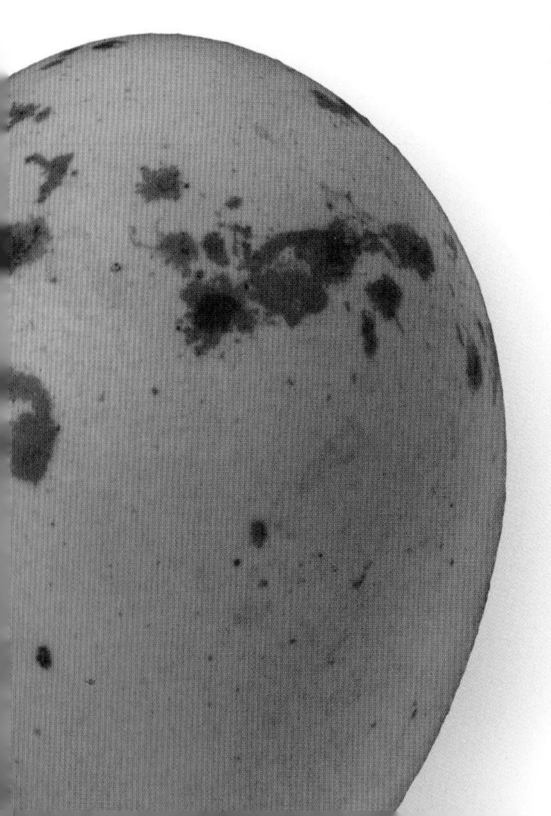

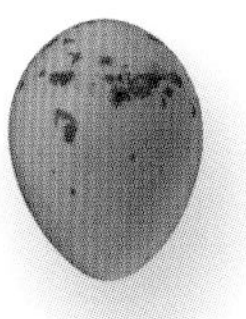

Actual size

ORDER	Passeriformes
FAMILY	Tyrannidae
BREEDING RANGE	Subarctic and eastern North America
BREEDING HABITAT	Mixed deciduous and conifer forests, swamps and bogs, orchards, and scrubby clearings
NEST TYPE AND PLACEMENT	Open cup of woven bark strips, grass, leaves and feathers, placed in a fork or crutch of a low tree
CONSERVATION STATUS	Least concern
COLLECTION NO.	FMNH 20954

ADULT BIRD SIZE
5–5½ in (13–14 cm)

INCUBATION
13–14 days

CLUTCH SIZE
3–5 eggs

EMPIDONAX MINIMUS

LEAST FLYCATCHER

PASSERIFORMES

This flycatcher is a long-distance migrant, and its breeding behaviors are influenced by timing; males return first, followed by the females a couple of days later. Males establish and defend territories near each other, forming loose clumps of two to 20 territories that border one another. Females prefer to settle and mate with males that are part of such a cluster, whereas solitary males typically remain unmated.

Both members of the pair are involved in selecting a nest site. The female generally leads and presses her chest into forks of branches, moving about as if to measure the strength or the dimension of the possible nesting site; the male follows her closely, but she alone is thought to construct the nest. During the egg-laying period, the female often forays into nearby territories, and mates with extra-pair males, resulting in mixed genetic paternity of her brood.

Clutch

Actual size

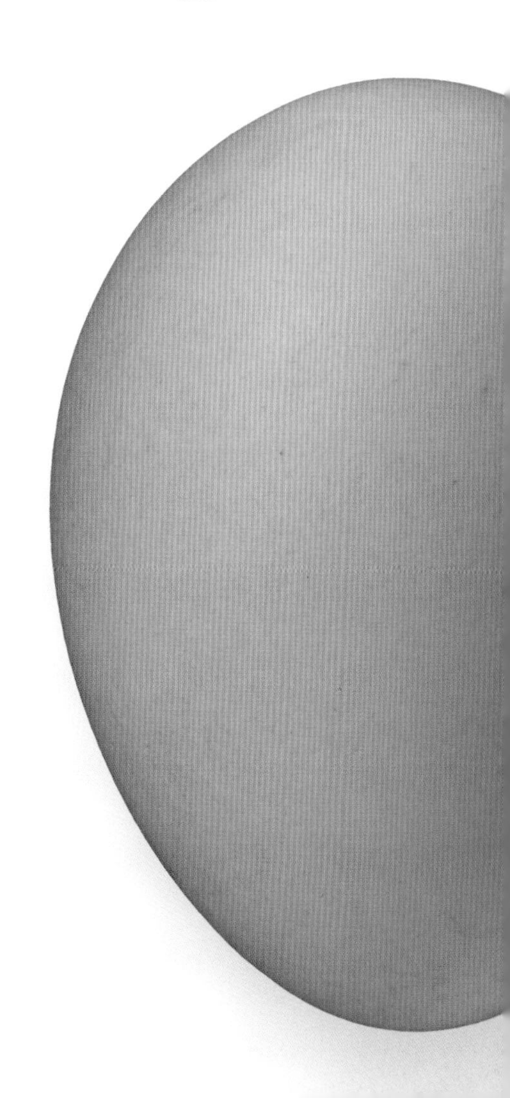

The egg of the Least Flycatcher is yellow or creamy white in background color, immaculate, and measures ⅔ x ½ in (17 x 13 mm) in size. The female alone incubates the eggs, often before the last egg is laid.

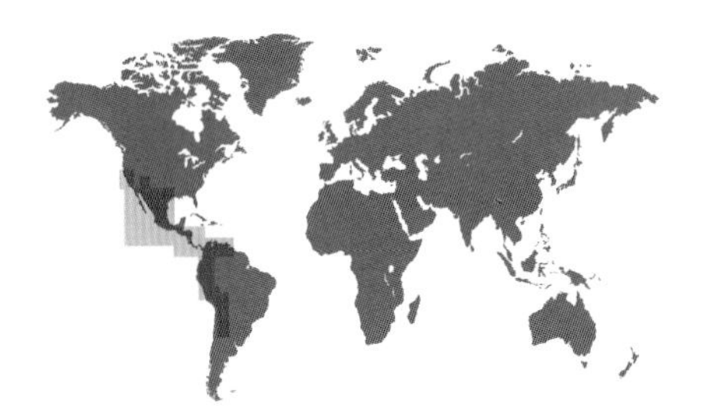

ORDER	Passeriformes
FAMILY	Tyrannidae
BREEDING RANGE	Western North America, Central America, and western South America
BREEDING HABITAT	Coastal and riparian habitats, open woods, clearings, suburban and city parks
NEST TYPE AND PLACEMENT	Mud cup nest, lined with grasses, and plastered on a vertical cliff or wall, or under a natural overhang or the eaves of a building
CONSERVATION STATUS	Least concern
COLLECTION NO.	FMNH 9273

ADULT BIRD SIZE
6½ in (16 cm)

INCUBATION
15–18 days

CLUTCH SIZE
3–5 eggs

SAYORNIS NIGRICANS

BLACK PHOEBE

PASSERIFORMES

Clutch

This phoebe has a vast distribution from the Pacific Northwest of the United States to Argentina. Throughout its range, the availability of breeding sites is patchy, as the species requires cliffs or buildings and proximity to water with muddy shores for nest construction. In most places, these birds are year-round residents. During the non-breeding season, members of a breeding pair will split up and defend individual but neighboring territories. In the spring, they re-form their pair bond, reestablish their joint breeding territory, and begin nesting weeks before any newly formed pairs.

The male often calls the female to several suitable nesting sites, but it is she alone who builds the nest, or refurbishes an existing nest, and incubates the eggs. If the female dies or disappears, the male abandons the eggs. However, if the eggs hatch, both parents feed the young, and they often succeed in raising two broods per season.

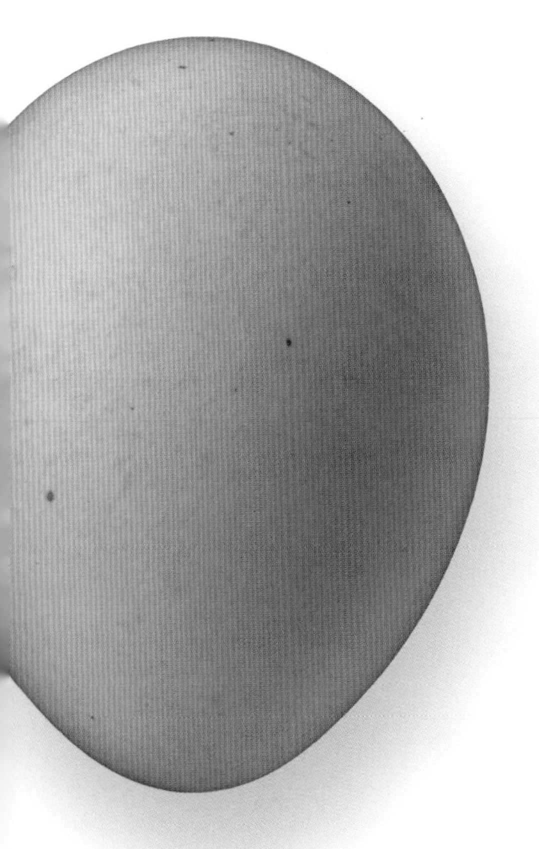

Actual size

The egg of the Black Phoebe is pure white and glossy in appearance, occasionally lightly speckled, and measures ¾ x ⅝ in (19 x 15 mm) in size. Unlike the Eastern Phoebe, shown on the facing page, this species is not parasitized by cowbirds, even though the two phoebe species have a similar nest placement. The reason for this is unknown.

ORDER	Passeriformes
FAMILY	Tyrannidae
BREEDING RANGE	Eastern and central North America
BREEDING HABITAT	Forests with clearings, edge habitats, suburban and rural settlements
NEST TYPE AND PLACEMENT	Mud nest, just inside a cave, on a cliff wall, or, commonly, under a wooden deck or a bridge; open cup, lined with grasses and cattle or horse hair, plastered on the outside with green moss
CONSERVATION STATUS	Least concern
COLLECTION NO.	FMNH 9242

SAYORNIS PHOEBE

EASTERN PHOEBE

PASSERIFORMES

ADULT BIRD SIZE
5½–6½ in (14–17 cm)

INCUBATION
15–16 days

CLUTCH SIZE
4–5 eggs

The Eastern Phoebe is best known for its placement and reuse of mud nests in the same location, sometimes across generations. Many people have reported that a phoebe nest was active under their porch or inside their barn in rural areas for decades, even though a single phoebe typically lives at most seven to ten years. Experiments have shown that refurbishing last year's nest, whether it's their own or another bird's nest, provides for a five-day jump start to lay the first clutch in the early spring, and allows the pair time to breed for a second or third time in the same season.

Eastern Phoebes are also common hosts of the brood parasitic Brown-headed Cowbird (see page 616), and although the parasite chick hatches first in the nest, one or two phoebes can also survive. Having nest mates, surprisingly, is beneficial for the young cowbird, because more open beaks bring in more parental provisions, which are then monopolized by the large and aggressively begging parasite chick, which grows even faster.

Clutch

The egg of the Eastern Phoebe is white in background color, with no or sparse reddish brown maculation, and measures ¾ x ⅝ in (19 x 15 mm) in dimensions. Clutch size typically decreases, and hatching asynchrony increases, between first and second nesting attempts in the same year.

Actual size

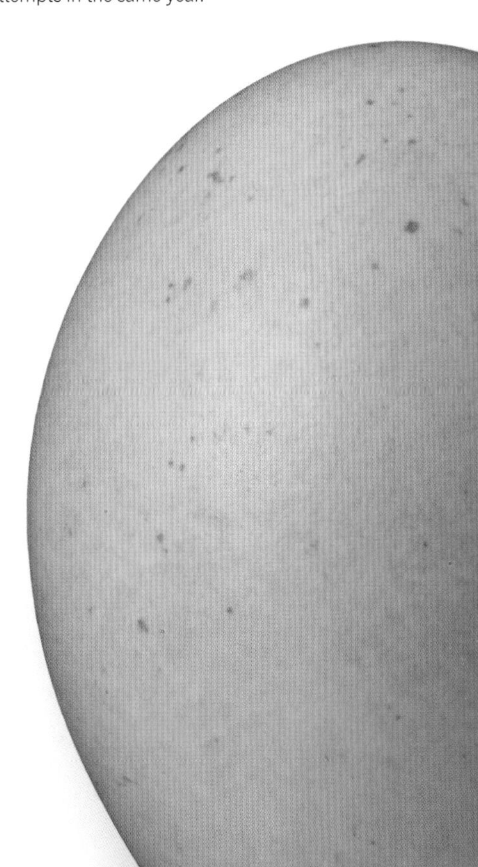

ORDER	Passeriformes
FAMILY	Tyrannidae
BREEDING RANGE	Western North America
BREEDING HABITAT	Arid open country, prairie, semi-deserts, often near open water
NEST TYPE AND PLACEMENT	Open cup nest, lined with hair and fine grass, placed on a ledge with cover, such as in a cave, cliff, or on a building or under a bridge; occasionally usurps mud-built swallow or robin nests
CONSERVATION STATUS	Least concern
COLLECTION NO.	FMNH 9264

ADULT BIRD SIZE
6½–7½ in (16–19 cm)

INCUBATION
12–17 days

CLUTCH SIZE
4–5 eggs

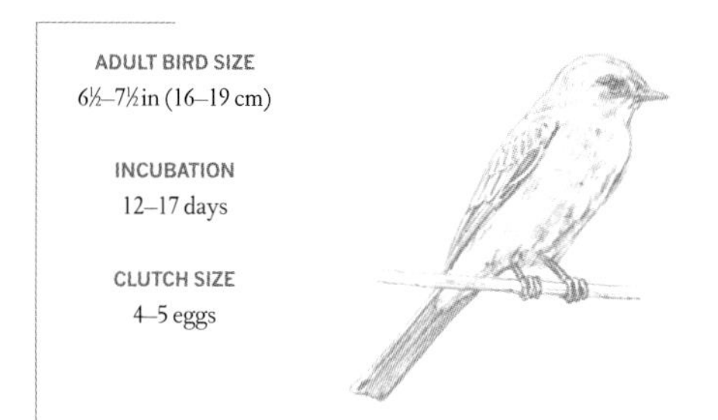

SAYORNIS SAYA

SAY'S PHOEBE

PASSERIFORMES

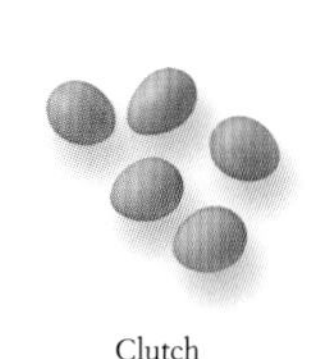

Clutch

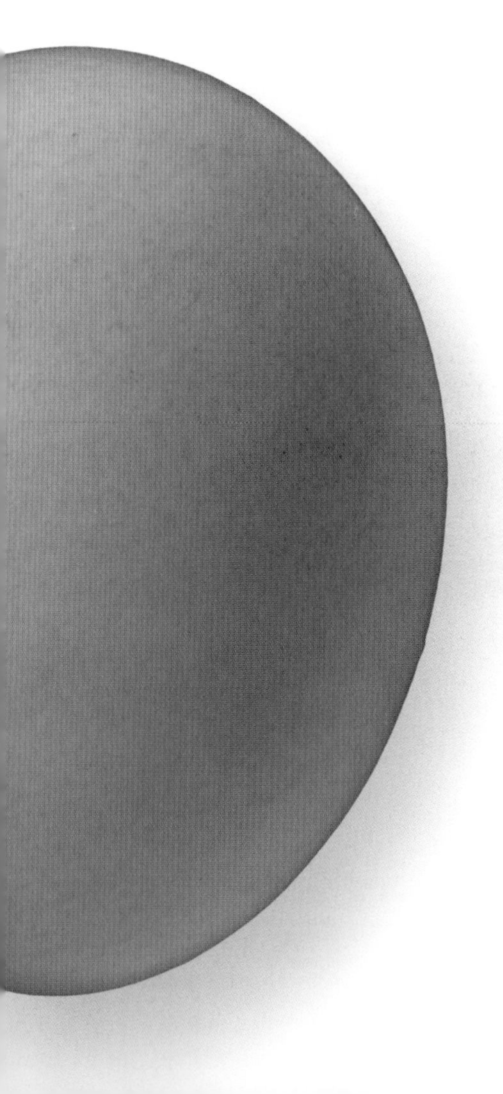

This widely distributed open country species behaves similarly to the other phoebe species; it is often seen sallying for flying insects in midair and over open water. It also builds or reuses mud nests of its own or other species, presumably to conserve time and energy for egg laying and incubation that would otherwise have been needed for nest construction. Nest building requires this bird to make hundreds of flights from a puddle to the nest, carrying a pellet of mud in its beak on each trip.

The male arrives on the breeding ground first, and when the female arrives, the nest building and laying proceed within days. When the eggs hatch, one of the parents stays with the young chicks to keep them warm, but both parents eventually become full-time providers of food for the brood. After the first breeding cycle is complete, this species is quick to re-nest and start a new clutch to complete that too before the end of the breeding season.

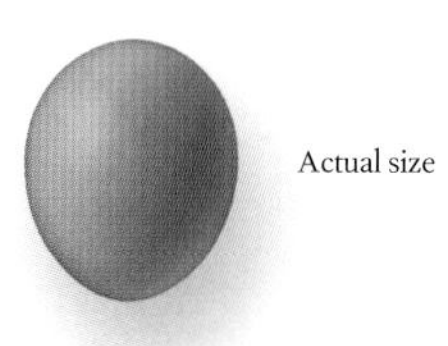

Actual size

The egg of the Say's Phoebe is white in background color with no maculation, and ¾ x ⅝ in (20 x 16 mm) in size. Despite this species' frequent association with human settlements, little is known about its breeding behavior; for example, it is not known if the female alone builds the nest, as other phoebes do.

ORDER	Passeriformes
FAMILY	Tyrannidae
BREEDING RANGE	Southwestern North America, Central America, South America, and the Galápagos
BREEDING HABITAT	Riparian woodlands, open scrubland, fields with sparse tree stands; arid areas, but near open water
NEST TYPE AND PLACEMENT	Open, loose cup of twigs and stems, lined with grasses, feather, and hair, in a fork in a horizontal tree branch
CONSERVATION STATUS	Least concern
COLLECTION NO.	FMNH 9426

ADULT BIRD SIZE	5–5½ in (13–14 cm)
INCUBATION	12–14 days
CLUTCH SIZE	2–4 eggs

PYROCEPHALUS RUBINUS

VERMILION FLYCATCHER

PASSERIFORMES

As if its brilliant crimson color and handsome black feather patches were not enough, the male of this flycatcher performs a spectacular mate-attraction flight, bouncing on its wings across the sky, calling all the way. Once the pair bond is established, the male also introduces the female to feasible nesting sites by pressing his chest into suitable forks of tree branches while uttering a conspicuous chatter call. Whether the female pays attention to the male is unclear, because she eventually settles on a nesting site of her own choice.

In contrast to the brightly colored male, the only reddish hues on the female's plumage are her rosy stomach feathers. These are conveniently hidden from the sight of any potential predators while she warms the eggs, displaying only her cryptic brown-and-grey plumage colors when viewed from above.

Clutch

Actual size

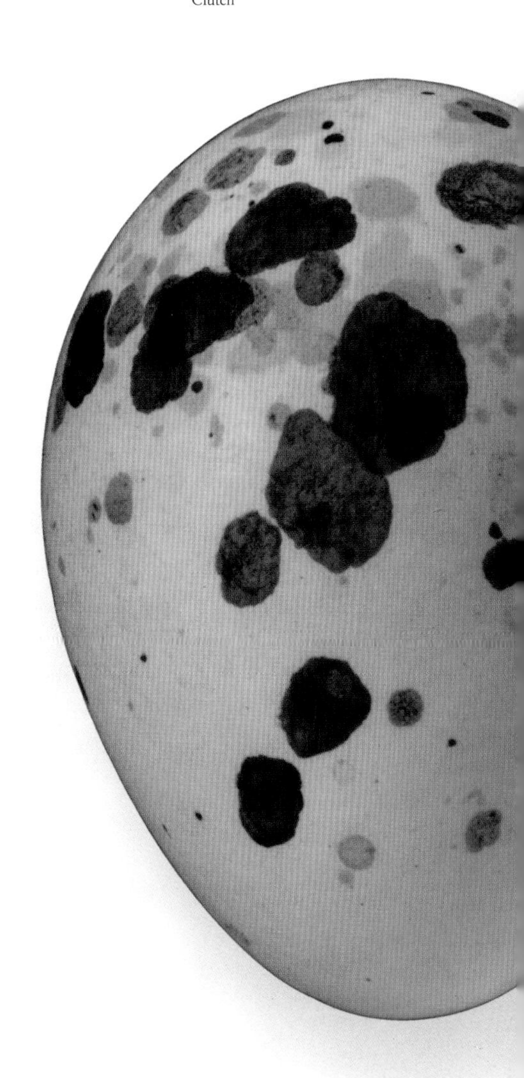

The egg of the Vermilion Flycatcher is creamy white in background color, has bold, darker maculation, and measures ⅔ x ½ in (17 x 13 mm) in size. The female alone builds the nest and the male feeds her while she incubates the eggs.

ORDER	Passeriformes
FAMILY	Tyrannidae
BREEDING RANGE	Southwestern North America, Central America, and South America
BREEDING HABITAT	Oak and pine forests, riparian woodlands
NEST TYPE AND PLACEMENT	Cavity nest of feathers, twigs, and finer grasses and hair, placed inside natural or woodpecker-carved holes; will also breed in nestboxes
CONSERVATION STATUS	Least concern
COLLECTION NO.	FMNH 16513

ADULT BIRD SIZE
6½–7½ in (16–19 cm)

INCUBATION
13–14 days

CLUTCH SIZE
3–5 eggs

MYIARCHUS TUBERCULIFER

DUSKY-CAPPED FLYCATCHER

PASSERIFORMES

Clutch

A non-social forager and breeder, this flycatcher remains on its territory year-round in most of its range; it is typically seen singly or in pairs. The smallest member of its genus, it defends its territory from other Dusky-capped Flycatchers, but is not aggressive to similar-looking species in the same genus. This species may be mimicking the coloration of the larger more aggressive flycatchers in order to gain some protection to avoid being attacked itself.

The egg of the Dusky-capped Flycatcher is buff in background color, with brown swirls and blotching, and 1 x ⅝ in (23 x 16 mm) in dimensions. Its cavity nest is difficult to find, and some egg collectors have used nestboxes to attract the nesting pair, only to confiscate the eggs to sell.

Relatively little is known about the breeding biology of this species; with no modern studies (as of yet) capitalizing on the opportunity provided by its willingness to use artificial nestboxes. It seems that the female alone gathers the nesting material, and she alone also incubates the eggs. Both parents provision the chicks with insects, carried in the bill by the feeding adult. As with other aspects of parental chores, the female appears to spend more time feeding the chicks than does the male.

Actual size

ORDER	Passeriformes
FAMILY	Tyrannidae
BREEDING RANGE	Western North America, including northern Mexico
BREEDING HABITAT	Desert scrub, deciduous and mixed forests, riparian corridors
NEST TYPE AND PLACEMENT	Nests in cavities, including a broad range of natural or artificial holes. Inside, the nest cup is made of dry grass, stems, manure, and leaves; lined with hair, feathers, and plant fibers
CONSERVATION STATUS	Least concern
COLLECTION NO.	FMNH 3019

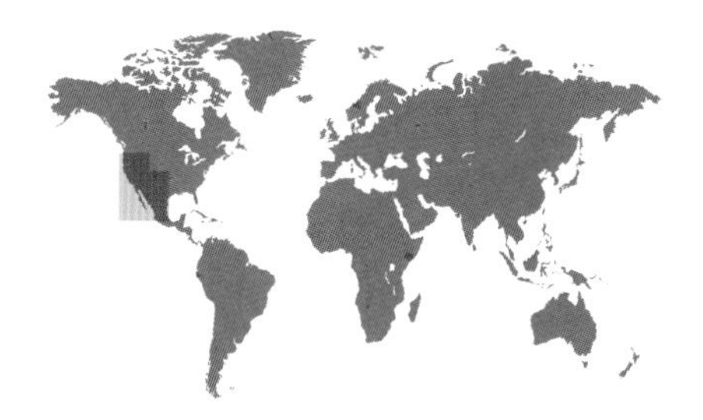

ADULT BIRD SIZE
7½–8½ in (19–21 cm)

INCUBATION
15 days

CLUTCH SIZE
3–5 eggs

MYIARCHUS CINERASCENS

ASH-THROATED FLYCATCHER

PASSERIFORMES

Though this species looks and acts like a flycatcher, sallying from perch to perch and observing its surroundings for potential prey and danger, it catches most of its prey by gleaning insects off foliage and branches, rather than taking them in midair. It is very similar in plumage coloration and stature to other flycatchers in the same genus, but the different species can be told apart by their distinctive and species-specific vocalizations. As with all suboscine birds, the chicks do not learn their calls, but inherit them from their parents.

Across most of its range, this species is migratory, which places a definite time constraint on its breeding schedule: further north, there is only time for a single breeding attempt per summer. Even more restrictive is the availability of natural cavities for nesting. Many hole-breeding species are resident or local migrants, giving them a head-start to claiming a cavity over these late-arriving, long-distance migrant flycatchers.

Clutch

Actual size

The egg of the Ash-throated Flycatcher is creamy white or buff in color, has lines of darker maculation, and measures ⅞ x ⅔ in (22 x 17 mm) in size. The use of human-made cavities, including nestboxes and open-ended metal poles, has allowed this species to expand its nesting range into previously unoccupied regions and habitats.

ORDER	Passeriformes
FAMILY	Tyrannidae
BREEDING RANGE	Eastern and middle North America
BREEDING HABITAT	Mixed or deciduous woodlands, orchards, swamp forests
NEST TYPE AND PLACEMENT	Nest in natural cavities, woodpecker holes, or nestboxes
CONSERVATION STATUS	Least concern
COLLECTION NO.	FMNH 16606

ADULT BIRD SIZE
6½–8½ in (17–21 cm)

INCUBATION
13–15 days

CLUTCH SIZE
4–8 eggs

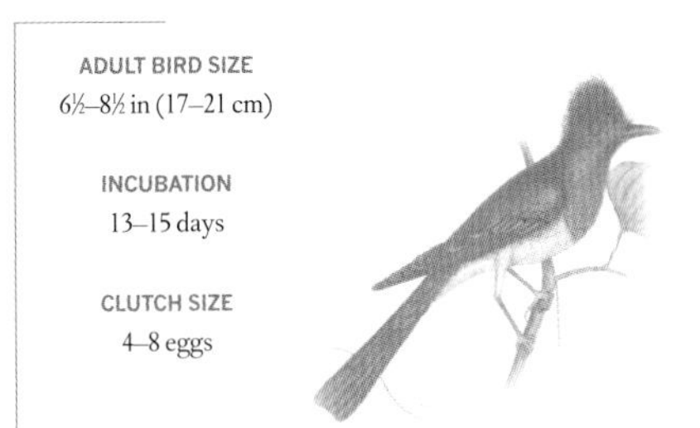

MYIARCHUS CRINITUS

GREAT CRESTED FLYCATCHER

PASSERIFORMES

Clutch

This large species is the only cavity-nesting flycatcher in eastern North America. It often sets up a territory near human settlements, including farms, orchards, and parks. Pairs form quickly once the birds return from spring migration, and previously used nest cavities are often refurbished and reused in subsequent years.

Both sexes are involved in inspecting potential nesting cavities. They prefer natural tree holes, and only switch to woodpecker cavities or artificial spaces where mature trees are not available. Deep cavities are typically filled with branches, twigs, leaves, and trash, and then the cup nest is placed on top at the line of the opening. Great Crested Flycatchers often collect shed snake skins (or plastic wrappers near human settlements) for lining the nest underneath the eggs. Only the female incubates the eggs, and she also delivers the majority of provisions to the young, while the male is in charge of territory defense.

Actual size

The egg of the Great Crested Flycatcher is creamy white to pinkish in background color, with darker lines and maculation, and is ⅞ x ⅔ in (23 x 17 mm) in dimensions. Once the chicks hatch, the parents quickly remove the broken eggshells and dispose of them by dropping them on the ground while flying away from the nest entrance.

ORDER	Passeriformes
FAMILY	Tyrannidae
BREEDING RANGE	Southern North America, Central America, South America
BREEDING HABITAT	Mature woodlands, riparian corridors, second growths, cactus forests
NEST TYPE AND PLACEMENT	Cup nest, made of grasses, leaves, feathers, and wool, inside an abandoned woodpecker cavity in a tree or cactus, in a natural tree hole, or in an artificial space
CONSERVATION STATUS	Least concern
COLLECTION NO.	FMNH 16610

ADULT BIRD SIZE
8–9½ in (20–24 cm)

INCUBATION
13–15 days

CLUTCH SIZE
2–7 eggs

MYIARCHUS TYRANNULUS

BROWN-CRESTED FLYCATCHER

PASSERIFORMES

This large flycatcher prefers to nest in bulky cavities excavated by larger species of woodpeckers, rather than in natural holes or artificial chambers. As a result, its local breeding range and distribution depend more on an ability to find abandoned woodpecker cavities, than on a need for a specific type of habitat. It is unknown what the role of each sex is in selecting the nest site, constructing the nest structure inside the cavity, or incubating the eggs. Both parents, however, are known to feed the nestlings.

The Brown-crested Flycatcher is ubiquitous throughout the American tropics and subtropics, as it sallies after flying prey amid the forest canopy. It is a drab and relatively quiet species, and with its cavity-dwelling breeding habits, it remains mostly unseen by, and unknown to, observers, including scientists.

Clutch

Actual size

The egg of the Brown-crested Flycatcher is creamy white in background color, has purple and lavender blotching, and measures $^{15}/_{16}$ x ⅔ in (24 x 18 mm). The parent is so dedicated to incubation that it may be hesitant to flush from the nest when disturbed.

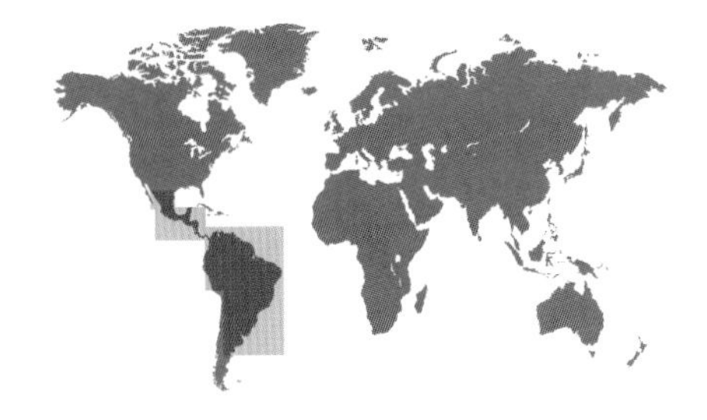

ORDER	Passeriformes
FAMILY	Tyrannidae
BREEDING RANGE	Southern North America, Central America, and South America
BREEDING HABITAT	Open woodlands, cultivated areas with tree stands, gardens and parks
NEST TYPE AND PLACEMENT	Built in a tree or on a telephone pole, the nest is an enclosed dome made from twigs and grasses, and also cotton, ribbons, and other trash; accessed through a side entrance
CONSERVATION STATUS	Least concern
COLLECTION NO.	FMNH 3131

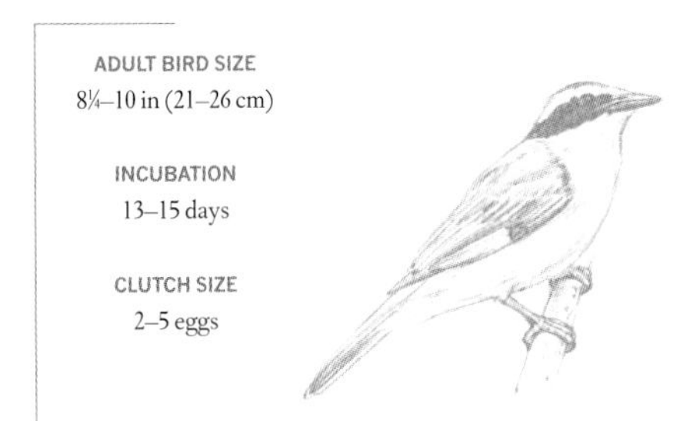

ADULT BIRD SIZE
8¼–10 in (21–26 cm)

INCUBATION
13–15 days

CLUTCH SIZE
2–5 eggs

PITANGUS SULPHURATUS

GREAT KISKADEE

PASSERIFORMES

Clutch

This brightly and boldly patterned bird is one of the more common sights in tropical regions of the Americas. It has expanded its breeding range even into urban settlements, using telephone poles instead of nesting trees, and stealing food from unattended plates and bowls instead of hunting for insects. To protect itself and its conspicuous, bulky nest, the kiskadee relies on loud calls, its large beak, and brave mobbing flights to chase cats on the ground and hawks in the air. Predators quickly learn to avoid the Great Kiskadee's yellow-white-and-black plumage patterns. It is likely that a number of smaller flycatcher species that share the same coloration patterns as the Kiskadee are actually mimics hoping to fool other animals into thinking they are Kiskadees.

Despite the species being a close associate of the fast-paced urban expansion in the neotropics, there is little scientific information about detailed breeding behaviors. Because only the female has a brood patch, the male is not thought to contribute to incubation duties; this is consistent with a lack of sightings of two parents changing guard at the nest during the incubation period.

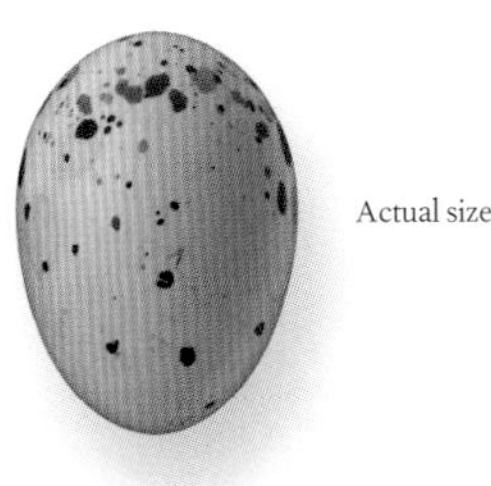

Actual size

The egg of the Great Kiskadee is buff to cream in color, marked with light reddish brown blotches, and measures 1⅛ x ⅞ in (29 x 22 mm) in size. Both parents construct the enclosed nest, and they defend it vigorously by mobbing approaching potential predators, including people.

ORDER	Passeriformes
FAMILY	Tyrannidae
BREEDING RANGE	Southwestern North America and Central America
BREEDING HABITAT	Woodlands and riparian forest corridors in canyons, in high-elevation mountain areas
NEST TYPE AND PLACEMENT	Cup nest, built inside a natural tree cavity or abandoned woodpecker hole; made of twigs, rootlets, and grasses
CONSERVATION STATUS	Least concern
COLLECTION NO.	FMNH 16602

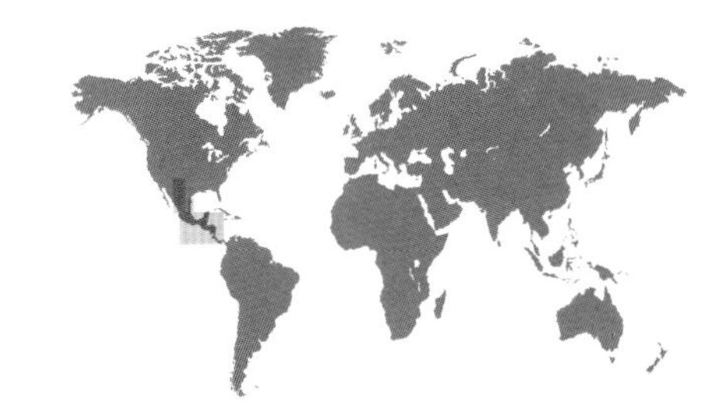

ADULT BIRD SIZE
7–8 in (18–20 cm)

INCUBATION
15–16 days

CLUTCH SIZE
3–4 eggs

MYIODYNASTES LUTEIVENTRIS

SULPHUR-BELLIED FLYCATCHER

PASSERIFORMES

This is a distinctively patterned, conspicuously striped species. It relies on natural cavities for nesting, and is absent in regions where woodpeckers or mature trees are lacking in sufficient density to provide potential nesting holes. In areas where there are few nest cavities, nests and eggs of this species have been found on top of the nest and eggs of other flycatchers and woodpeckers; this is consistent with severe interspecific competition for suitable nesting cavities.

Clutch

To initiate breeding, the female and male perform a coordinated sequence of movements and calls, closely following each other as they slowly cross their territory. Once the pair bond is formed, the female begins to construct the nest, while the male follows her closely or stands guard nearby when she is busy inside the nest hole.

Actual size

The egg of the Sulphur-bellied Flycatcher is white to buff in background color, has reddish brown and lavender maculation, and measures 1 x ¾ in (26 x 19 mm) in size. To provide a clear line of sight for the incubating female, deeper nesting chambers are frequently filled with branches and twigs before the nest cup is formed at the same height as the cavity entrance.

ORDER	Passeriformes
FAMILY	Tyrannidae
BREEDING RANGE	Southern North America, Central America
BREEDING HABITAT	Semi-open country, shrub, arid fields with tree stands, gardens
NEST TYPE AND PLACEMENT	Loose open cup of vines and rootlets, lined sparsely with animal fur; placed in high crotch of isolated tree
CONSERVATION STATUS	Least concern
COLLECTION NO.	FMNH 907

ADULT BIRD SIZE
7–9 in (18–23 cm)

INCUBATION
15–16 days

CLUTCH SIZE
2–4 eggs

TYRANNUS COUCHII

COUCH'S KINGBIRD

PASSERIFORMES

Clutch

The egg of the Couch's Kingbird is whitish or pale pink in background color, with dark spotting, and ⅞ x ⅔ in (23 x 18 mm) in size. The female alone incubates the eggs and broods the young.

This kingbird was long believed to be part of the more widespread Tropical Kingbird (*Tyrannus melancholicus*), before being recognized as a separate species. Associated closely with human settlements, it occurs in many open or deforested landscapes scattered with lone trees and criss-crossed with power lines, which provide excellent perching and hunting stake-out sites for these birds. To defend their breeding territory, and especially the vicinity of the nest, the kingbird repeatedly engages with intruders much larger than itself, including harmless frigatebirds and toucans.

This species forms socially monogamous pairs, with the male providing much of the guarding and defense behaviors of the territory while the female builds the nest. Both parents feed the chicks with insects and berries, and occasionally they misdirect their parental care by raising the obligate parasitic Bronzed Cowbird (see page 615).

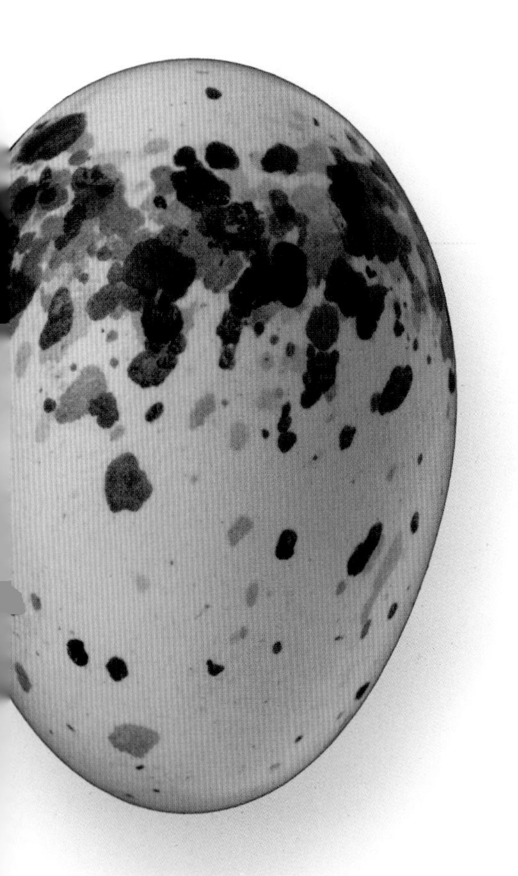

Actual size

ORDER	Passeriformes
FAMILY	Tyrannidae
BREEDING RANGE	Southern North America
BREEDING HABITAT	Open woodlands, riparian corridors, edges of conifer forests, desert scrub
NEST TYPE AND PLACEMENT	Cup nest, built from twigs and bark, lined with rootlets and grasses, occasionally feathers; saddled over a horizontal branch
CONSERVATION STATUS	Least concern
COLLECTION NO.	FMNH 16576

ADULT BIRD SIZE
8½–9 in (22–23 cm)

INCUBATION
18–19 days

CLUTCH SIZE
3–4 eggs

TYRANNUS VOCIFERANS

CASSIN'S KINGBIRD

PASSERIFORMES

This kingbird shares much of its range with the closely related Western Kingbird (see page 404). Yet, despite their similarities in size and prey-base, there appears to be little aggression between the two species, and their nests can even be found on the same tree or shrub. The Cassin's Kingbird's aggressive behavior and loud warning calls (which accounts for its Latin species name, *vociferans*) are typically directed at potential nest-predators, and intruding conspecifics.

In the spring, when the pairs have formed, the female takes charge of nesting: she collects plant matter to construct the nest, while the male follows closely or stands guard nearby. Such mate-guarding behavior may be to protect the female from harassment by unpaired male kingbirds, and also an early warning system concerning approaching danger while she is busy with her parental duties. The male benefits by siring most or all of the chicks.

Clutch

The egg of the Cassin's Kingbird is creamy white in background color, has brown and lavender maculation, and is 15⁄16 x ⅔ in (24 x 18 mm) in size. In addition to building the nest, the female is in charge of incubating the eggs, and spends almost twice as much time feeding the young as the male.

Actual size

ORDER	Passeriformes
FAMILY	Tyrannidae
BREEDING RANGE	Western North America
BREEDING HABITAT	Open areas with sparse tree stands, pastures, cultivated fields, grasslands, desert shrub
NEST TYPE AND PLACEMENT	Cup nest, built in the crotch of a tree or shrub, also on fence poles
CONSERVATION STATUS	Least concern
COLLECTION NO.	FMNH 9163

ADULT BIRD SIZE
8–9½ in (20–24 cm)

INCUBATION
12–19 days

CLUTCH SIZE
2–7 eggs

TYRANNUS VERTICALIS

WESTERN KINGBIRD

PASSERIFORMES

Clutch

This is a bird of open grasslands. But it still requires trees or poles as high vantage points from which to survey the habitat for prey, including flying or ground insects, and it needs a high place for its nest. At the start of the nesting season, the Western Kingbird's territory includes much of the large open space around the trees and bushes where the nest is eventually built. As the season progresses, the defended area diminishes to focus around just the pair's nest-tree.

To attract a mate, the male performs a flight that involves heading upward, followed by a prominent twitch at the peak of the flight, and ending in a tumbled fall accompanied by bursts of calling. Once the pair bond is formed, the male performs a nest-showing display, during which he calls and leads the female to a potential nesting site, where he shakes and wiggles his body. Eventually, the female alone constructs the nest.

Actual size

The egg of the Western Kingbird is white, creamy, or pinkish in background color, with heavy dark brown maculation, and measures 15⁄16 x ⅔ in (24 x 18 mm) in size. Only the female incubates the eggs and broods the chicks, but both parents feed the young.

ORDER	Passeriformes
FAMILY	Tyrannidae
BREEDING RANGE	Eastern and central North America
BREEDING HABITAT	Fields scattered with trees, orchards, forest edges
NEST TYPE AND PLACEMENT	Open cup nest, made of wiry twigs and glasses, placed in a shrub, tree, or on top of a stump or a pole
CONSERVATION STATUS	Least concern
COLLECTION NO.	FMNH 9114

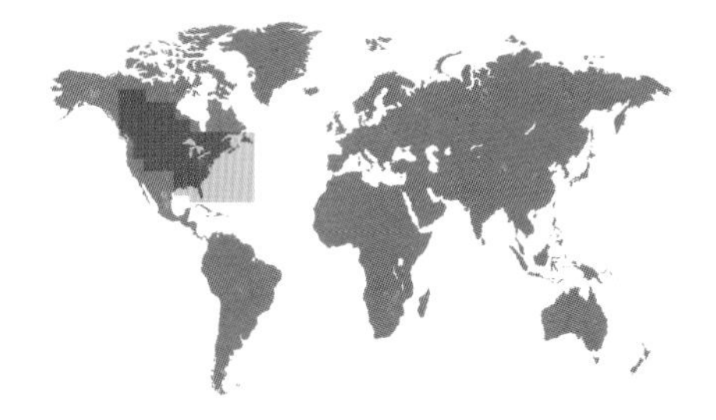

ADULT BIRD SIZE
8–9 in (20–23 cm)

INCUBATION
14–17 days

CLUTCH SIZE
2–5 eggs

TYRANNUS TYRANNUS

EASTERN KINGBIRD

PASSERIFORMES

This species is a long-distance migrant, moving between temperate North America and Amazonian South America. Once the birds return to their breeding site, the male typically settles and begins to defend his territory, often the same general area as the year before. If the female is his mate from the prior year, the breeding can begin in a week after arrival; otherwise it takes two to three weeks for a new pair to start nesting. Although the female alone builds the nest and incubates the eggs, the male can also be seen visiting feasible nesting spots, and the actual nest structure is often rebuilt in the same spot where it was in the previous year(s). Often, the exposed nest is lost or damaged due to the weather, and the female quickly builds another nest for another clutch of eggs.

This kingbird is well known for its aggressive territorial defense, which includes protecting the often conspicuously placed nest from larger potential predators, including jays, hawks, and squirrels. Clearly, camouflage for the nest is not required when it is under the protection of a true 'tyrant' flycatcher, as implied by the Latin genus name *Tyrannus* for this species and its close relatives.

Clutch

The egg of the Eastern Kingbird is pale white in background color, maculated with a ring of reddish spots, and measures 15⁄16 x 2⁄3 in (24 x 18 mm) in size. There is no second breeding attempt if the first nest successfully fledged the young.

Actual size

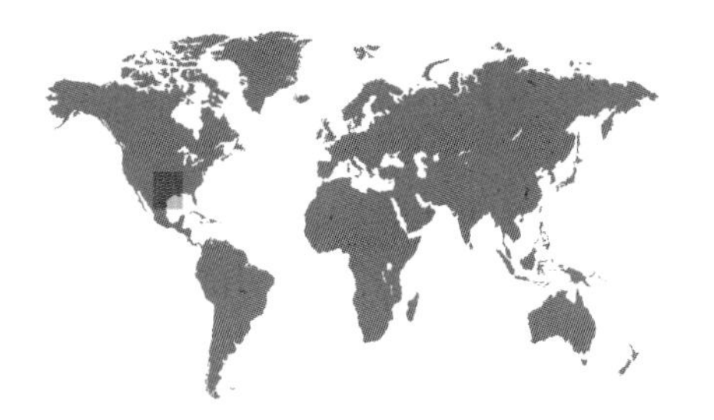

ORDER	Passeriformes
FAMILY	Tyrannidae
BREEDING RANGE	Southern-central North America
BREEDING HABITAT	Grasslands, savanna
NEST TYPE AND PLACEMENT	Cup nest of twigs and grass blades, lined with hair and other fibers; in an isolated shrub or tree
CONSERVATION STATUS	Least concern
COLLECTION NO.	FMNH 9093

ADULT BIRD SIZE
8½–15½ in (22–40 cm)

INCUBATION
13–23 days

CLUTCH SIZE
4–5 eggs

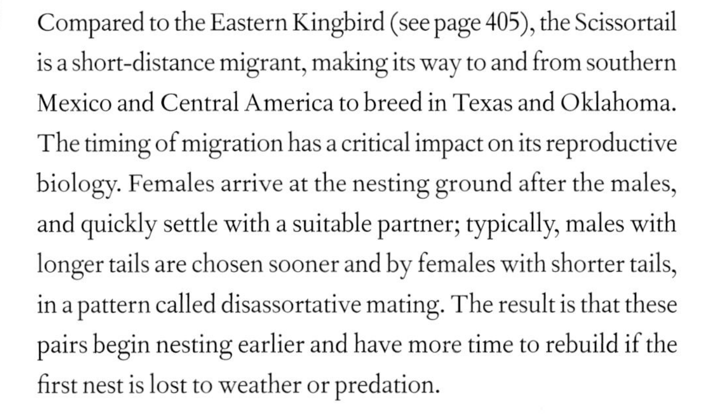

TYRANNUS FORFICATUS

SCISSOR-TAILED FLYCATCHER

PASSERIFORMES

Clutch

Compared to the Eastern Kingbird (see page 405), the Scissortail is a short-distance migrant, making its way to and from southern Mexico and Central America to breed in Texas and Oklahoma. The timing of migration has a critical impact on its reproductive biology. Females arrive at the nesting ground after the males, and quickly settle with a suitable partner; typically, males with longer tails are chosen sooner and by females with shorter tails, in a pattern called disassortative mating. The result is that these pairs begin nesting earlier and have more time to rebuild if the first nest is lost to weather or predation.

The nest site is selected by the female and the male, with both birds pressing their breasts into forks of tree branches, as if to "size up" the potential sites. Once construction begins, the female collects and weaves the nesting materials together, often while the male follows closely. Incubation is carried out fully by the mother, but both parents contribute to feeding the nestlings.

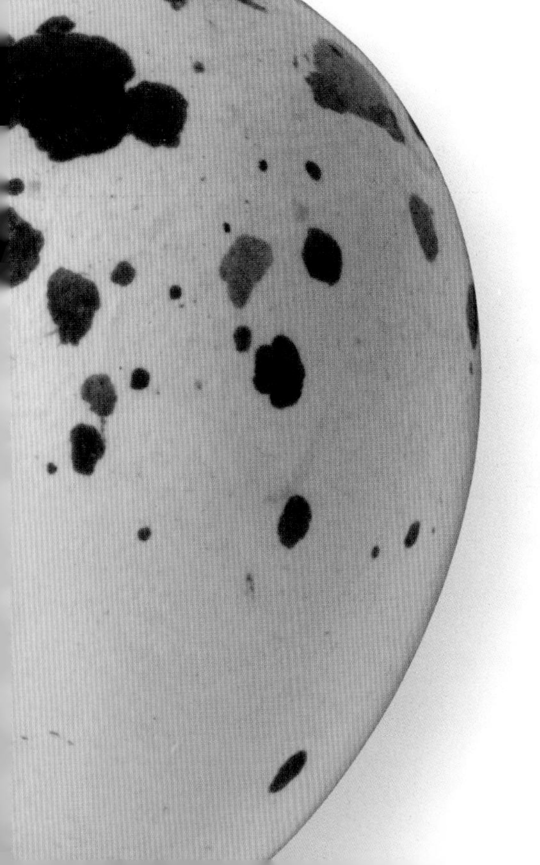

Actual size

The egg of the Scissor-tailed Flycatcher is white or cream in background color, has distinctive reddish brown spots and blotches, and measures ⅞ x ⅔ in (23 x 17 mm). With the nest placed in an isolated tree, the eggs and the chicks are especially vulnerable to damage or total loss during severe summer storms in the tornado-belt region.

ORDER	Passeriformes
FAMILY	Pipridae
BREEDING RANGE	Central America
BREEDING HABITAT	Dry or moist, seasonal tropical forests, typically with scrubby undergrowth
NEST TYPE AND PLACEMENT	Shallow cup nest, suspended from a tree branch, woven from leaves, bark, and spider silk
CONSERVATION STATUS	Least concern
COLLECTION NO.	WFVZ 146784

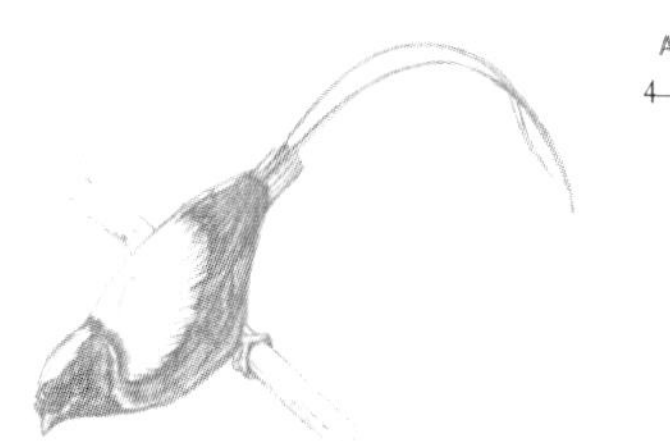

ADULT BIRD SIZE
4–4½ in (10–11 cm)

INCUBATION
18 days

CLUTCH SIZE
1–2 eggs

CHIROXIPHIA LINEARIS

LONG-TAILED MANAKIN

PASSERIFORMES

The Long-tailed Manakin's breeding behaviors range from solitary parental duties performed by the female only, to highly social group-based mating performances by the males. The sexes are highly dimorphic, with the female being dull green, an effective camouflage in the lush tropical foliage, whereas mature males possess a striking red-black-and-blue plumage with an elongated pair of central tail feathers.

To attract the female, groups of four to ten males band together in "leks." The two most dominant males then perform a stereotyped dance-show, which involves calling loudly and conducting aerial acrobatics along a long branch that has been cleared of all leaves and other obstructions. Once a female shows interest in a pair of males, the alpha of the group sends his beta away, performs the dance on his own, and soon he, and only he, copulates with the female. Betas typically do not sire any young on their own until years later, when the alpha dies and they ascend to the top rank.

Clutch

The egg of the Long-tailed Manakin is buff to tan in color, with darker maculation, and is 13⁄16 x 9⁄16 in (21 x 15 mm) in dimensions. While the father continues to dance and attract other females with his male partner, the female alone builds the nest, incubates the egg(s), and feeds the young.

Actual size

ORDER	Passeriformes
FAMILY	Tityridae
BREEDING RANGE	Southwestern North America, Central America
BREEDING HABITAT	Dry forests, both deciduous and mixed coniferous woods, riparian galleries, and mangroves
NEST TYPE AND PLACEMENT	Large enclosed nest, suspended from a branch
CONSERVATION STATUS	Least concern
COLLECTION NO.	FMNH 151

ADULT BIRD SIZE
6½–7 in (17–18 cm)

INCUBATION
15–17 days

CLUTCH SIZE
2–6 eggs

PACHYRAMPHUS AGLAIAE

ROSE-THROATED BECARD

PASSERIFORMES

Clutch

This small insectivore displays attributes of behavior and appearance that are like those of the tyrant flycatchers, but recent genetic work has placed it and other becards with the Tityras in the Tityridae. Its foraging strategies involve gleaning insects off foliage, and also sallying to capture them in flight. During the breeding season, the male calls and displays his typically hidden white shoulder patches to attract the female.

The female alone can build a nest, but the male may also assist; the nest is a covered structure, an untidy woven mass of grasses, leaves, Spanish moss, and twigs, suspended from the outer reaches of a long, thin tree limb for extra safety. The entrance to the nest is at the bottom, which allows quick escape for the incubating bird.

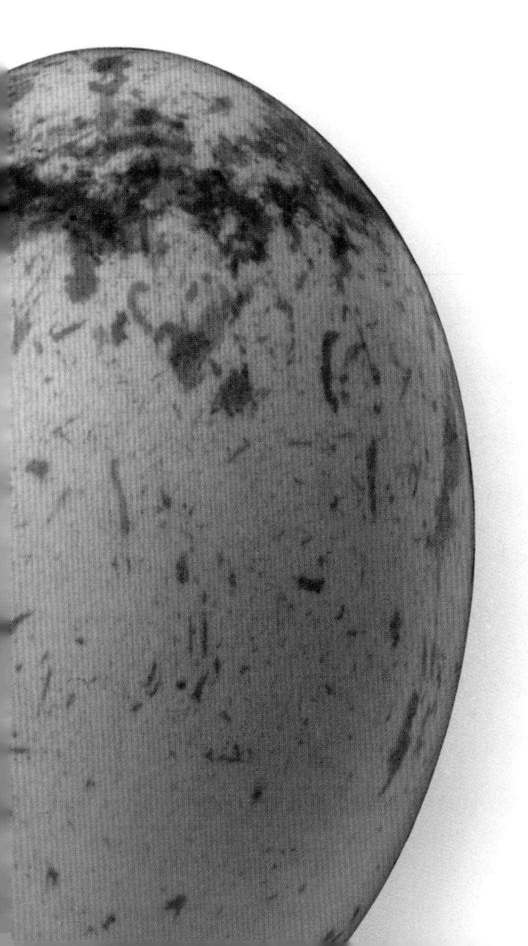

Actual size

The egg of the Rose-throated Becard is white in background color, has brown maculation, and is 1$\frac{1}{16}$ x ¾ in (27 x 19 mm) in dimensions. The female alone incubates the eggs, but both of the parents take turns to feed the young.

ORDER	Passeriformes
FAMILY	Ptilonorhynchidae
BREEDING RANGE	Northern Australia
BREEDING HABITAT	Dry forests, rainforest edges, riverine wood lots, suburban parks and gardens
NEST TYPE AND PLACEMENT	Cup nest constructed from twigs and lined with leaves, placed in a tree or bush
CONSERVATION STATUS	Least concern
COLLECTION NO.	WFVZ 143858

ADULT BIRD SIZE
13–15 in (33–38 cm)

INCUBATION
21 days

CLUTCH SIZE
1–2 eggs

CHLAMYDERA NUCHALIS

GREAT BOWERBIRD

PASSERIFORMES

Bowerbirds are some of the bird world's most original architects and decorators. Yet a bower is not a home but an aphrodisiac, built by males to attract females. Males build a bower consisting of a long avenue, lined along both edges with a wall of vertical sticks and twigs pushed into the ground. The entrance and the exit of the bower are lined with pebbles and shells, and occasionally berries and flower petals. In contrast to his fanciful display site, the male's own plumage is relatively drab, although he does have brilliant, fuchsia-colored crown feathers.

When a female is attracted to a bower, she enters the avenue and observes the male as he dances and calls for her attention. Once they mate, the male continues to attract other females, while the female builds her nest, incubates the eggs, and feeds the young, all on her own. In the next year, she typically returns to the same bower and the same male if he is still around and has not lost his bower; such mate fidelity reduces her costs of time and travel to try out different males.

Clutch

The egg of the Great Bowerbird is creamy with a green or gray tint in background color, has deep chestnut scrawling maculation, and is 1⅔ x 1⅛ in (42 x 29 mm) in size. To protect her eggs, the female attacks intruders and mimics calls of predators to divert attention from the nest.

Actual size

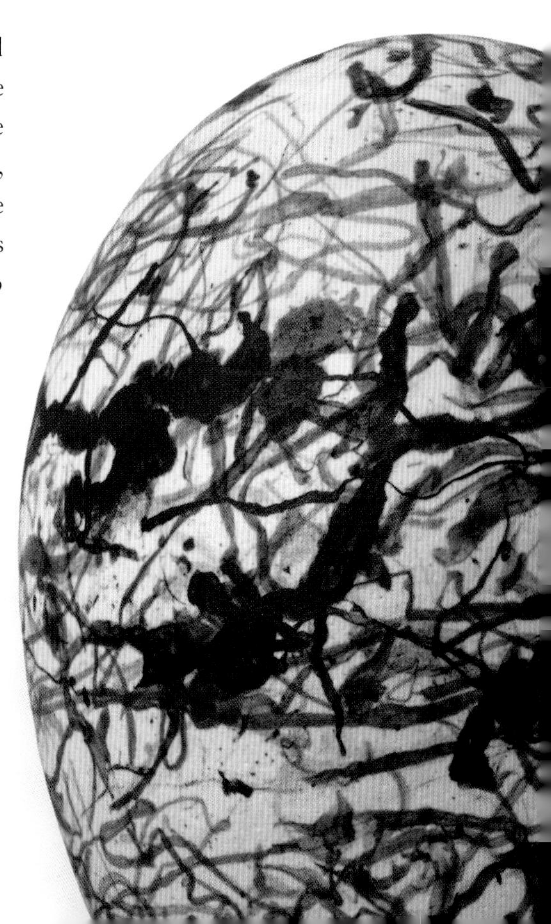

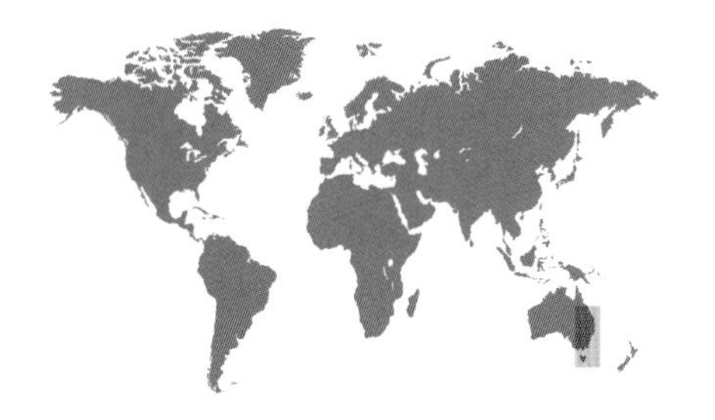

ORDER	Passeriformes
FAMILY	Maluridae
BREEDING RANGE	Southeastern Australia, Tasmania
BREEDING HABITAT	Open woodlands, mixed with dense undergrowth, also city parks and gardens
NEST TYPE AND PLACEMENT	Enclosed nest, made of grasses, leaves, and cobwebs; hidden in thick and thorny vegetation, near the ground
CONSERVATION STATUS	Least concern
COLLECTION NO.	WFVZ 75778

ADULT BIRD SIZE
5½–6¼ in (14–16 cm)

INCUBATION
14–15 days

CLUTCH SIZE
2–4 eggs

MALURUS CYANEUS

SUPERB FAIRYWREN

PASSERIFORMES

Clutch

The egg of the Superb Fairywren is dull white in background color, marked with reddish brown spots, and measures ⅝ x ½ in (16 x 12 mm). Occasionally, the Fairywren's nest is taken over by just a single inhabitant: a parasitic Bronze-Cuckoo chick. Just before hatching, the mother calls to her embryos inside the eggs and teaches them a password which the parents use to distinguish between their own young and parasitic cuckoo chicks.

The small breeding male of the Superb Fairywren is iconic throughout Australia, in part because this bright blue and black bird has made its home in urban areas throughout the most densely populated states and cities. The life of fairywrens revolves around the family, with the breeding female and male raising a brood of chicks assisted by reproductively mature but non-breeding helpers, typically sons from a prior nesting attempt. The presence of helpers allows the female to invest more heavily into producing the yolk of her eggs; she can then recover from this output of energy while the helpers provide many of the feedings to the nestlings.

When it comes to mate fidelity, however, female fairywrens are anything but family-centric; in some nests, most or all chicks may be sired by fathers from neighboring territories, instead of the resident blue male. Female fairywrens often mate with neighboring territorial males, possibly to increase the genetic diversity of their young. According to one hypothesis, they do this to bring in good genes which are involved in immune function and disease-resistance.

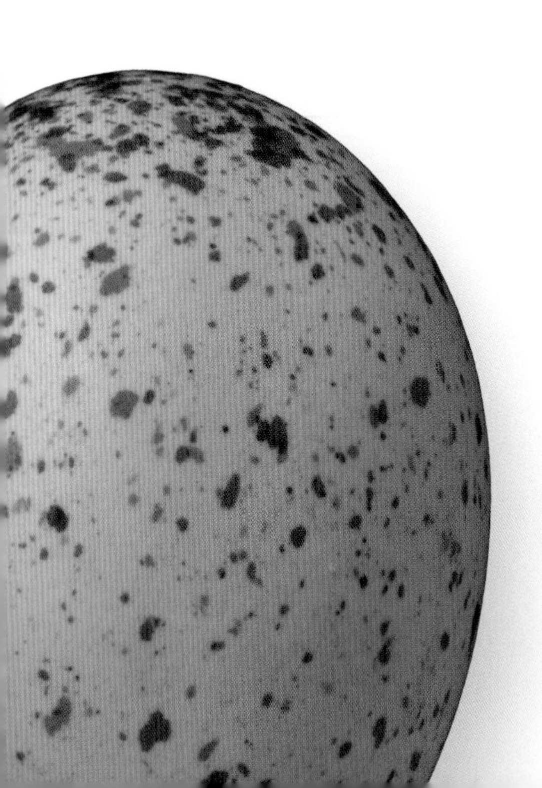

Actual size

ORDER	Passeriformes
FAMILY	Vangidae
BREEDING RANGE	Madagascar
BREEDING HABITAT	Dry and wet tropical forests
NEST TYPE AND PLACEMENT	Cup nest made from woven plant fibers and twigs, placed in the fork of a tree
CONSERVATION STATUS	Least concern
COLLECTION NO.	WFVZ 74247

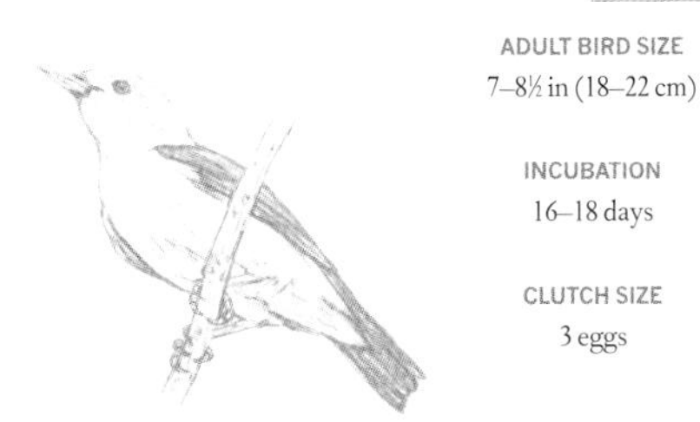

ADULT BIRD SIZE
7–8½ in (18–22 cm)

INCUBATION
16–18 days

CLUTCH SIZE
3 eggs

ARTAMELLA VIRIDIS

WHITE-HEADED VANGA

PASSERIFORMES

Vangas are a diverse group of birds endemic to Madagascar. They have many differences in their bill shapes that allow them to forage in a variety of ways. For example, the White-headed Vanga has a big stout bill for catching mid-size insects, which differentiates it from other vangas that share its range. In recent years, habitat destruction has pushed many of Madagascar's endemic lineages toward increasingly threatened status, and even extinction.

Based on a single scientific report, following individually banded birds at a handful of nests, we know that White-headed Vanga males maintain a year-round territory, where they are joined by the female prior to the onset of nesting. Both sexes participate in nest building, incubating, and provisioning the young, and some pairs are assisted by immature males during the breeding period. These younger males do not feed the young, but allopreen (exchange preening sessions) with the parents and help to defend the territory from intruders.

Clutch

The egg of the White-headed Vanga is creamy white in background color, has reddish brown speckling, and measures 1 x ⅔ in (25 x 18 mm) in dimensions. Females have been seen to mate only with males on whose territories they reside, but genetic confirmation of such monogamy is still lacking.

Actual size

ORDER	Passeriformes
FAMILY	Laniidae
BREEDING RANGE	Arctic and temperate North America and Eurasia
BREEDING HABITAT	Open grassland interspersed with trees, large clearings near boreal forests, pastures with fence lines
NEST TYPE AND PLACEMENT	Large, bulky cup of twigs and roots, woven together with feathers and hair; placed on tree branches
CONSERVATION STATUS	Least concern
COLLECTION NO.	FMNH 20819

ADULT BIRD SIZE
9–11 in (23–28 cm)

INCUBATION
15–16 days

CLUTCH SIZE
4–9 eggs

LANIUS EXCUBITOR

NORTHERN SHRIKE

PASSERIFORMES

Clutch

This large and multi-continental shrike species is a strong and aggressive predator. It obtains nearly half of its daily food by preying on small rodents, but also eats lizards, frogs, small birds, and large invertebrates. If the prey is too large to swallow whole, the shrike impales it on a large tree thorn to tear it apart, and devours it in several bites. This species is territorial, and defends its breeding site from intruder shrikes and larger birds to reduce competition for food and to protect its nest.

A pair bonds monogamously, but the bond only lasts a single breeding season. In the winter, the pair breaks up and the next year often a new bond is forged. The male attracts the female to potential nesting sites with song and food; he also begins to collect and place twigs as the base of a nest. If the female accepts, the male provides the rest of the nesting materials and the female constructs the nest. Some males maintain large enough territories to attract two nesting females; these males, however, pay a cost of more time spent hunting for prey for their larger families and less time spent on preening and resting.

The egg of the Northern Shrike is white with a gray or green tint in background color, speckled with brown spots, and 1 1/16 x 3/4 in (27 x 20 mm) in dimensions. In Europe, to defend its nest from Carrion Crows and other predators, this shrike often nests near a loud thrush species called a Fieldfare; birds of both species mob approaching danger together.

Actual size

ORDER	Passeriformes
FAMILY	Laniidae
BREEDING RANGE	Temperate and tropical North America
BREEDING HABITAT	Open fields, orchards, pastures with sparse bushes and trees
NEST TYPE AND PLACEMENT	Cup nest in tree, constructed of twigs, lined with fine grasses, feathers, hair, and other filaments
CONSERVATION STATUS	Least concern, but declining in some regions
COLLECTION NO.	FMNH 12401

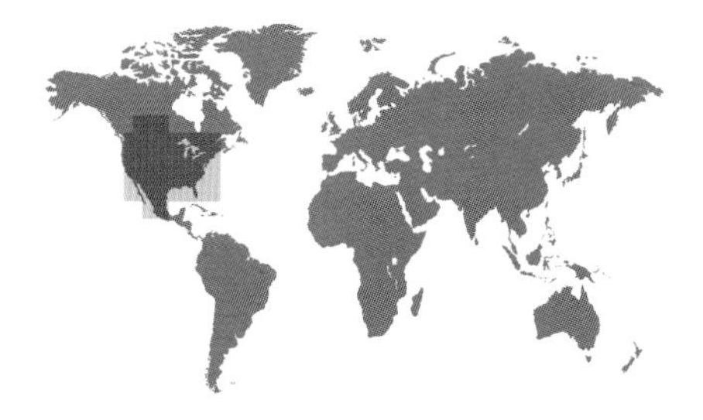

ADULT BIRD SIZE
8–9 in (20–23 cm)

INCUBATION
16–17 days

CLUTCH SIZE
5–6 eggs

LANIUS LUDOVICIANUS

LOGGERHEAD SHRIKE

PASSERIFORMES

This shrike is the only endemic North American species in a large genus and family that is distributed mostly in the Old World. Despite its small stature, it is a bird of prey that surveys the habitat from a high vantage point and uses its thick, hooked bill to capture and crush prey, from arthropods to lizards and mice.

Clutch

Breeding begins early in the spring, and in southern regions, where this species is resident all year long. Pairs may stay together through the winter, and come into reproductive condition synchronously and more quickly than newly formed pairs. Nesting materials are gathered by both sexes, but the female constructs the nest with little help. She is also in charge of incubating the eggs and brooding the chicks, while the male feeds her at the nest and provides food for the young family during the first days after hatching. Later, both sexes feed the growing young and defend the nest.

Actual size

The egg of the Loggerhead Shrike is grayish buff in background color, has brown maculation, and measures 1 x ¾ in (25 x 20 mm). The female regularly rotates the eggs during the day; on hotter days, she turns them more often and she also stands above them to provide shade.

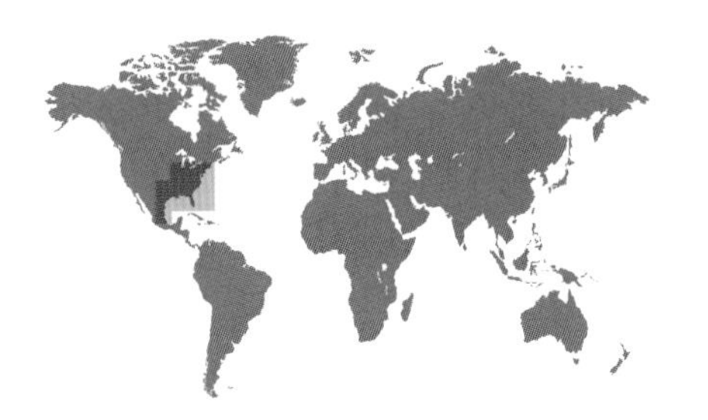

ORDER	Passeriformes
FAMILY	Vireonidae
BREEDING RANGE	Eastern and southern North America, the Bahamas, and Bermuda
BREEDING HABITAT	Secondary-growth forests, scrubland, pastures with tree stands
NEST TYPE AND PLACEMENT	Open, penduline cup, suspended with spider silk from the fork of a tree limb, typically hidden in dense foliage
CONSERVATION STATUS	Least concern
COLLECTION NO.	FMNH 12499

ADULT BIRD SIZE
4½–5 in (11–13 cm)

INCUBATION
13–15 days

CLUTCH SIZE
3–5 eggs

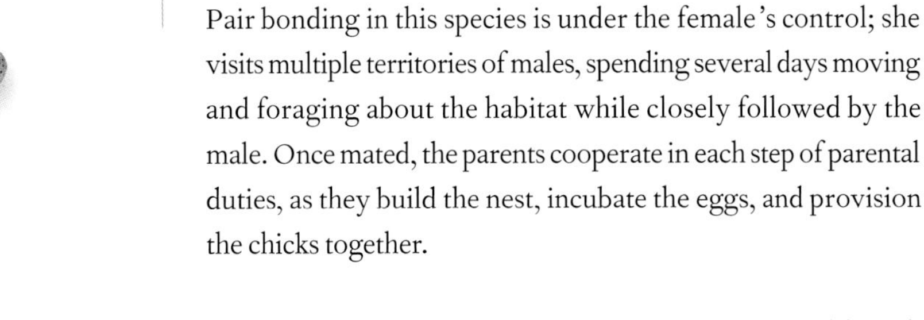

VIREO GRISEUS

WHITE-EYED VIREO

PASSERIFORMES

Clutch

Pair bonding in this species is under the female's control; she visits multiple territories of males, spending several days moving and foraging about the habitat while closely followed by the male. Once mated, the parents cooperate in each step of parental duties, as they build the nest, incubate the eggs, and provision the chicks together.

Male and female both develop incubation patches, although the female's patch is more heavily vascularized with blood vessels for more efficient heat transfer from the body to the eggs. The female spends all night on the nest and continues to stay there to brood the young after they hatch. In turn, the male provides most of the food delivered to the nest. He passes food directly to the female who then turns around and feeds it to the nestlings.

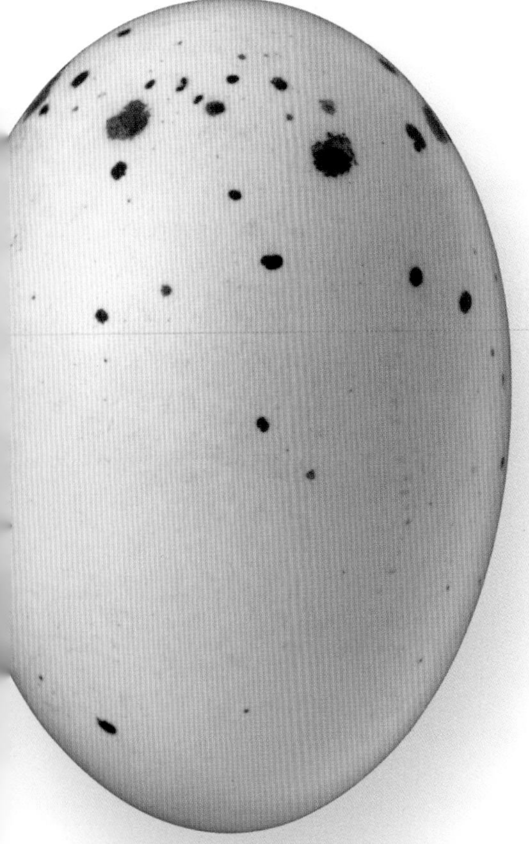

Actual size

The egg of the White-eyed Vireo is white in background color, maculated with dark spots, and ⅔ x 9⁄16 in (18 x 14 mm) in dimensions. In some regions, nearly half of nests are parasitized with eggs of the Brown-headed Cowbird (see page 616), severely reducing the breeding success of this otherwise common species.

ORDER	Passeriformes
FAMILY	Vireonidae
BREEDING RANGE	Central and southwestern North America
BREEDING HABITAT	Shrubs, low trees, riparian brush; also parkland
NEST TYPE AND PLACEMENT	Basketlike open cup, hanging from forked branches, lined with fine grasses
CONSERVATION STATUS	Near threatened
COLLECTION NO.	FMNH 12438

VIREO BELLII

BELL'S VIREO

PASSERIFORMES

ADULT BIRD SIZE
4½–4¾ in (11–12 cm)

INCUBATION
14–15 days

CLUTCH SIZE
3–5 eggs

This species, though widespread throughout much of North America, has been the subject of intense conservation efforts, as its westernmost subspecies, in California, has been reduced in range and number to a small fraction of its prior size. The primary factors leading to decline are heavy brood parasitism by Brown-headed Cowbirds (see page 616), which also prefer open and shrubby habitats; and land modification that turns riparian and open shrub land into suburban developments. Today, conservation efforts, including habitat protection and restoration, and intensive trapping and removal of parasitic cowbirds, have improved the species' numbers.

Finding a Bell's Vireo nest is relatively easy for both parasites and researchers alike, because of a ritualized "changing of the guard" process: the male sings on several trees as he approaches, eventually landing on the nest tree to sing again. The female also calls from the nest repeatedly, and eventually leaves so that the male can inspect the eggs and settle on the nest for his bout of incubation.

Clutch

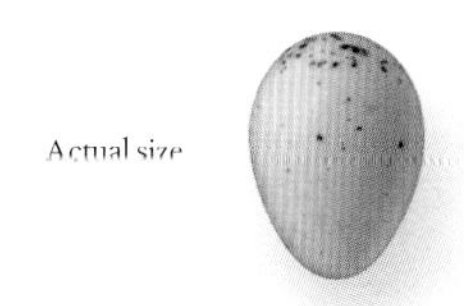

Actual size

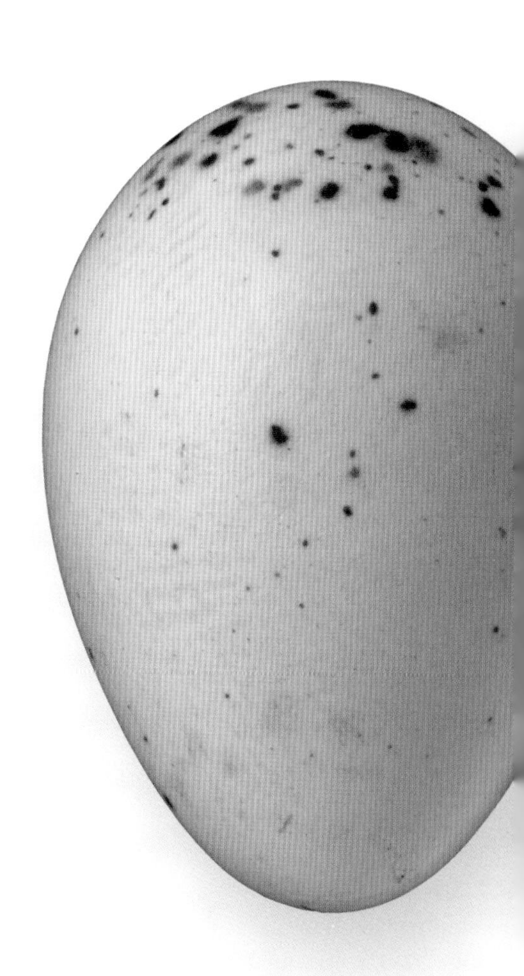

The egg of the Bell's Vireo is white in background color, with sparse brown maculation, and is ⅔ x ½ in (18 x 13 mm) in size. Building the nest takes four to five days when it is constructed by both sexes, but the female alone can also fully build a nest.

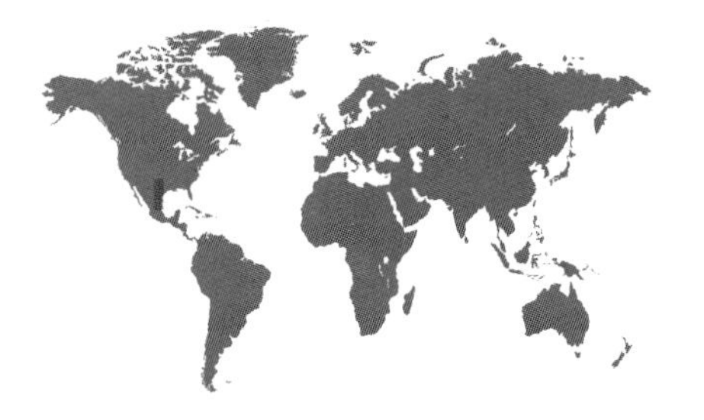

ORDER	Passeriformes
FAMILY	Vireonidae
BREEDING RANGE	Southern North America
BREEDING HABITAT	Disturbed, regenerating scrub and low forest
NEST TYPE AND PLACEMENT	Open cup, hanging from a branch fork, made of leaves, grasses, and silk, lined with fine grass
CONSERVATION STATUS	Vulnerable; endangered in the United States
COLLECTION NO.	FMNH 12424

ADULT BIRD SIZE
4½–5 in (11–13 cm)

INCUBATION
14–17 days

CLUTCH SIZE
3–4 eggs

VIREO ATRICAPILLUS

BLACK-CAPPED VIREO

PASSERIFORMES

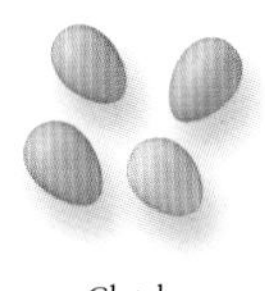

Clutch

The habitat requirements of this species are so specialized that even before human-made changes to the landscape, it was patchily distributed. Today it only occurs in a handful of populations, where the preferred disturbed scrub and low-forest habitat meets its narrow foraging and nesting criteria. One of these surprisingly productive sites is based at an active bombing practice range used by the military; there, instead of recurrent wild fires, explosives and heavy motorized equipment provide the repeated disturbance to the habitat preferred by this bird.

Nest building and incubation duties are shared by the pair, whereas the female broods the chicks after hatching and the male provides them with food. The fledging period is only 10 to 12 days, and then the female is quickly off to build a new nest and lay a new clutch, while the male feeds the fledglings until independence.

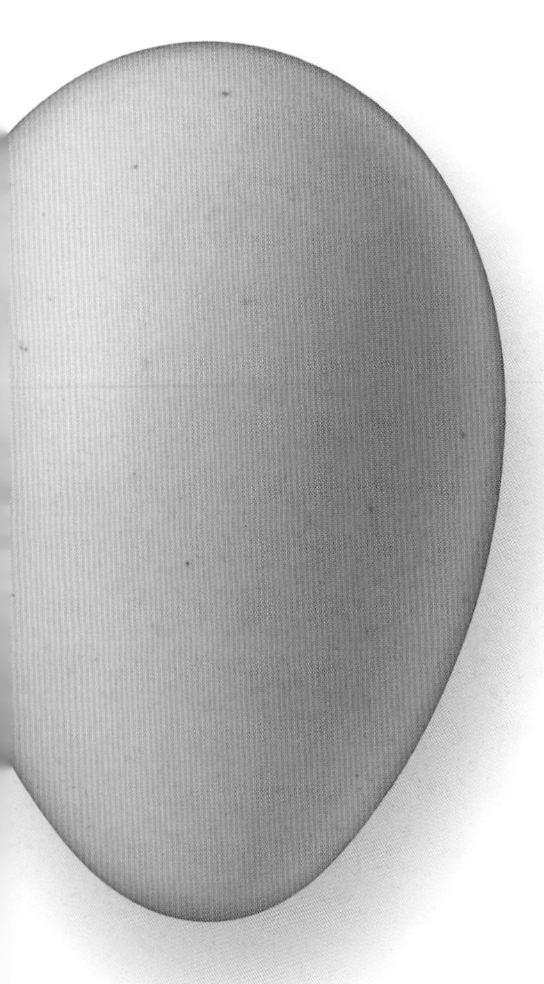

Actual size

The egg of the Black-capped Vireo is white in background color, immaculate, and measures ⅔ x 9⁄16 in (18 x 14 mm). The male's puzzling habit of singing while sitting in the nest makes the eggs vulnerable to both predators and brood parasites.

ORDER	Passeriformes
FAMILY	Vireonidae
BREEDING RANGE	Southwestern North America
BREEDING HABITAT	Juniper and oak scrub, chaparral
NEST TYPE AND PLACEMENT	Free-hanging cup on a horizontal fork of a branch, woven from plant fibers, grasses, and animal silks, lined with soft plant fibers
CONSERVATION STATUS	Least concern
COLLECTION NO.	FMNH 16931

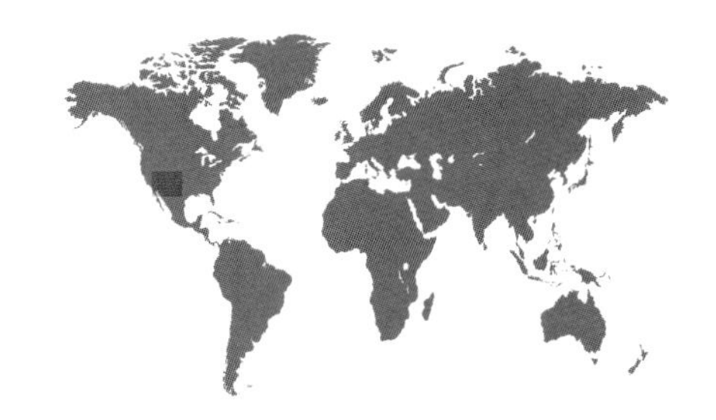

VIREO VICINIOR

GRAY VIREO

PASSERIFORMES

ADULT BIRD SIZE
5–6 in (13–15 cm)

INCUBATION
13–14 days

CLUTCH SIZE
2–4 eggs

This vireo is a habitat specialist, occurring in arid scrublands and open juniper forests, where its simple grayish plumage easily blends into the background colors of the landscape. It is a short-distance migrant; once they arrive at the breeding ground, the males swiftly establish their territories, and the returning females quickly pair up—within days, if not hours.

Clutch

Nest site selection is spearheaded by the female, as she moves meticulously through bushes, looking and pressing into forks in the branches, while the male follows, singing along the way. At some sites, the female leaves a few blades of grass, and the pair may return there to continue nest construction together. At the same time, the males may build loosely woven "bachelor" nests, perhaps as a practice prior to constructing the real thing; these remain empty and are never used for laying the eggs.

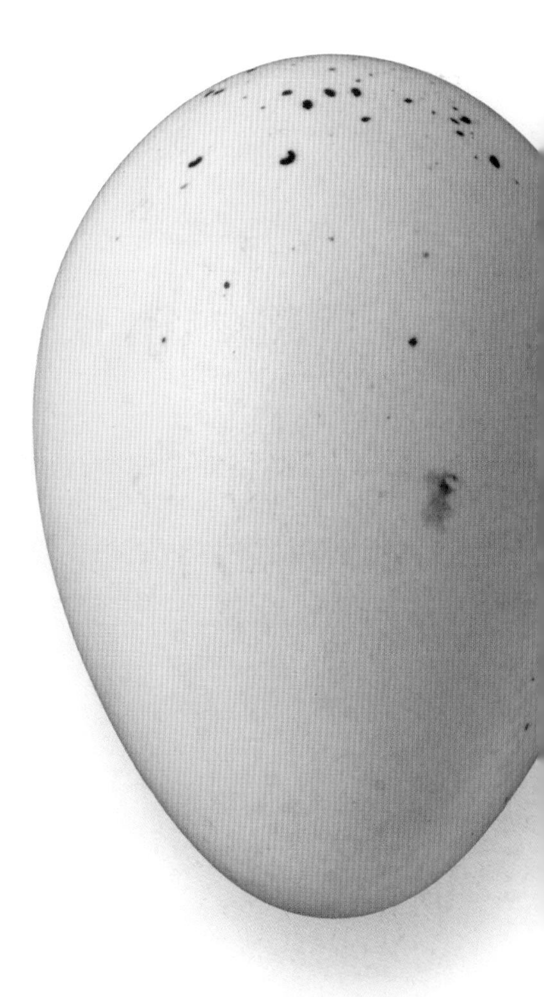

Actual size

The egg of the Gray Vireo is white in background color, with sparse, dark maculation, and measures ¾ x 9⁄16 in (19 x 14 mm) in size. Both the female and the male call and sing briefly from the nest, while incubating the eggs or changing guard between the pair.

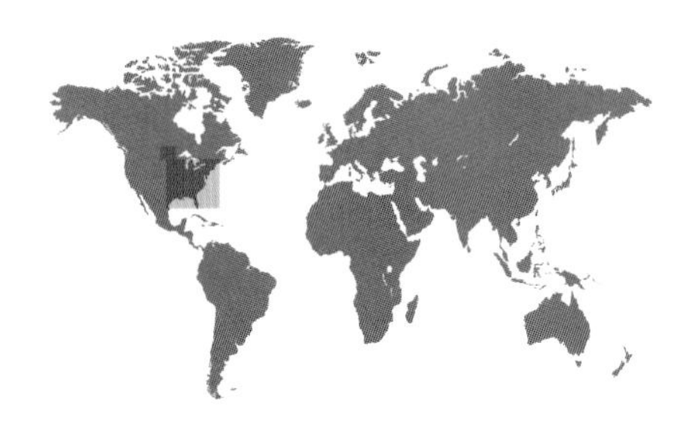

ORDER	Passeriformes
FAMILY	Vireonidae
BREEDING RANGE	Eastern and central North America
BREEDING HABITAT	Edge habitats of mature deciduous and mixed forests, including woodlands bordering rivers, swamps, roads, and parks
NEST TYPE AND PLACEMENT	Open, suspended cup in a tree, with its rim attached to a horizontal fork of a small branch
CONSERVATION STATUS	Least concern
COLLECTION NO.	FMNH 12462

ADULT BIRD SIZE
5–6 in (13–15 cm)

INCUBATION
14 days

CLUTCH SIZE
3–5 eggs

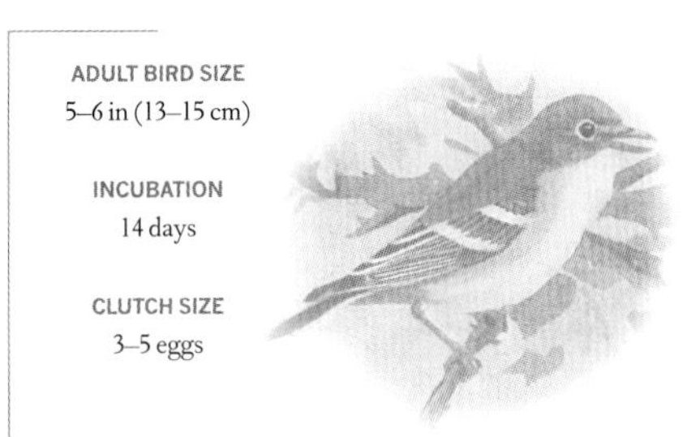

VIREO FLAVIFRONS

YELLOW-THROATED VIREO

PASSERIFORMES

Clutch

The egg of the Yellow-throated Vireo is creamy white in color, has some brown maculation, and is ⅞ x ⅝ in (21 x 15 mm) in size. While the clutch is being laid, one parent stays with the eggs but does not sit on them, so incubation does not begin until the clutch is complete.

This vireo used to inhabit large urban parks, including in New York City and Boston, but during the era of intensive spraying of insecticides, the species disappeared from most of these sites. Nonetheless, the overall population size of this naturally uncommon species has been steadily increasing over the last few decades in rural regions, perhaps associated with the maturation of many previously clear-cut Eastern deciduous forests, whose edges provide the most suitable breeding habitat for the Yellow-throated Vireo.

The males return from migration just days before the females do, and potential mates assess each other during courtship displays, which include the male leading the female to several feasible nesting sites. Once the mates have settled, the male continues building the nest at one of these sites; the female soon takes over construction duties and she alone completes the nest. The female also spends more time incubating the eggs, as she sits on them all night, and changes guard with the male only during daytime.

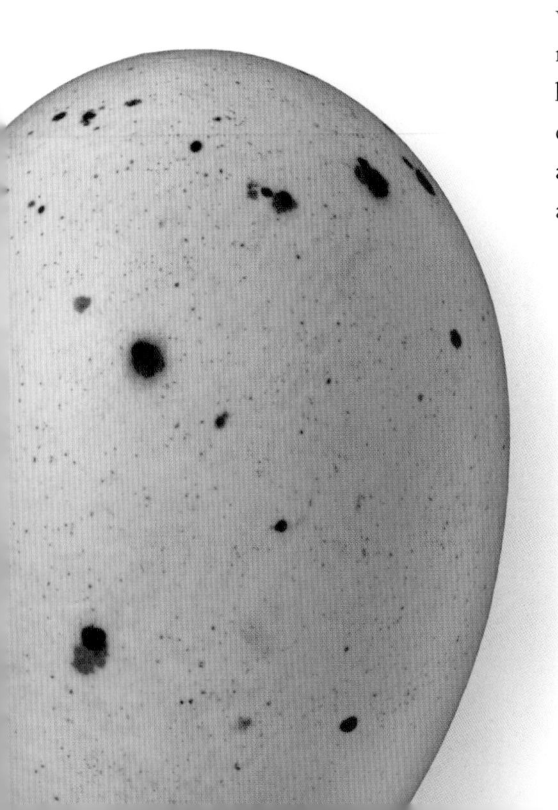

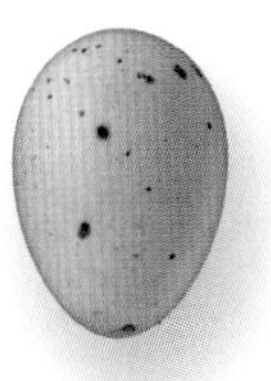

Actual size

ORDER	Passeriformes
FAMILY	Vireonidae
BREEDING RANGE	Eastern and sub-boreal North America
BREEDING HABITAT	Coniferous and mixed forests, with dense deciduous undergrowth
NEST TYPE AND PLACEMENT	Open cup, suspended with its rim attached to a branch of a shrub or sapling below the canopy
CONSERVATION STATUS	Least concern
COLLECTION NO.	FMNH 12571

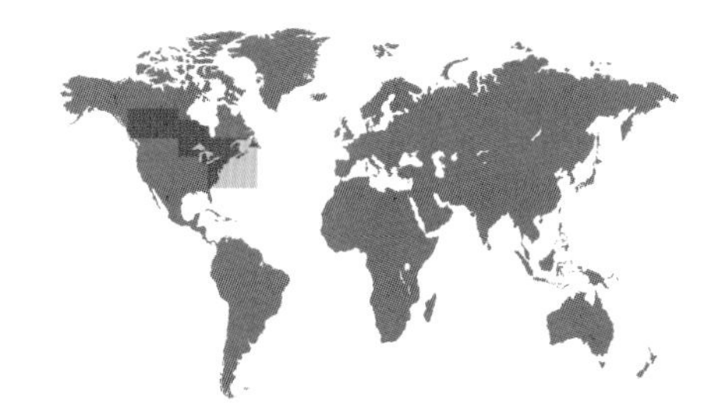

ADULT BIRD SIZE
5–6 in (13–15 cm)

INCUBATION
13–15 days

CLUTCH SIZE
3–4 eggs

VIREO SOLITARIUS

BLUE-HEADED VIREO

PASSERIFORMES

Genetic studies have only recently led scientists to recognize the Blue-headed Vireo as a separate species. Previously, it was considered the eastern population of a larger species complex known as the Solitary Vireo. A migrant to and from wintering grounds in the southern United States and the Caribbean, the Blue-headed Vireo returns quickly in spring to its preferred mature forest habitats for breeding. There the male first roams widely, then establishes a territory, following up with quick courtship displays for the later-arriving females. Nest building begins soon after the pairs are established.

The eggs and the young are the focal points of social and foraging behaviors in this species; members of the pair typically spend most of their time within a hundred yards of the nest, which may be situated near a hawk nest. This does not endanger the vireo's own breeding attempt, and probably provides added safety from other predators, which are hesitant to approach the hawks.

Clutch

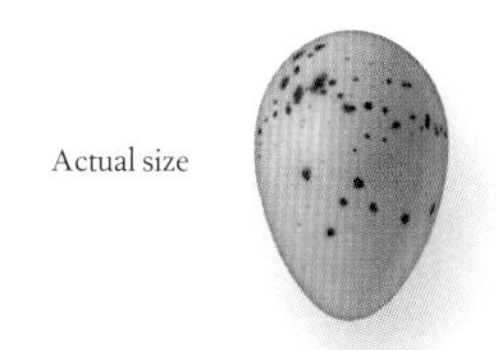

Actual size

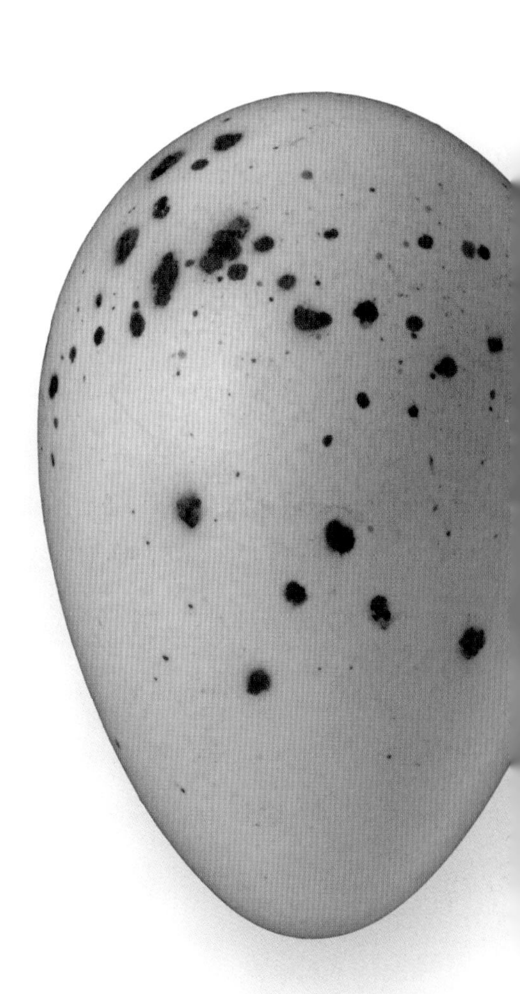

The egg of the Blue-headed Vireo is creamy white in background color, has irregular spotting, and its dimensions are ¾ x ⅝ in (20 x 15 mm). Both parents incubate the eggs, with the female developing a full, and the male a partial, incubation patch.

ORDER	Passeriformes
FAMILY	Vireonidae
BREEDING RANGE	Western North America and Central America
BREEDING HABITAT	Mixed forests, especially oak woodlands
NEST TYPE AND PLACEMENT	Globular cup, suspended by its rim from a fork on a terminal branch, made of lichens, grasses, leaves, and silk
CONSERVATION STATUS	Least concern
COLLECTION NO.	FMNH 12518

ADULT BIRD SIZE
4½–5 in (12–13 cm)

INCUBATION
14–16 days

CLUTCH SIZE
4 eggs

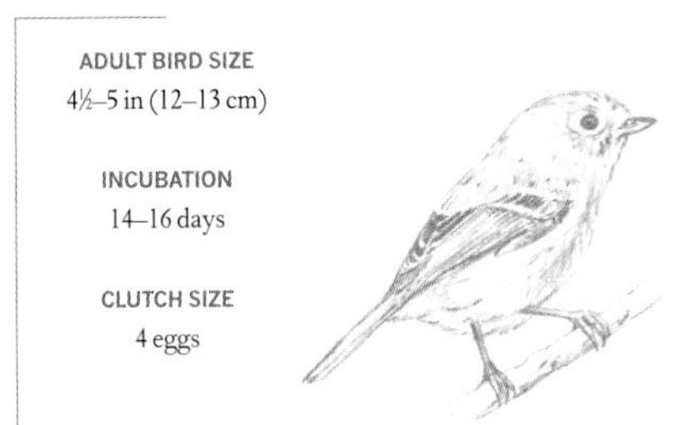

VIREO HUTTONI

HUTTON'S VIREO

PASSERIFORMES

Clutch

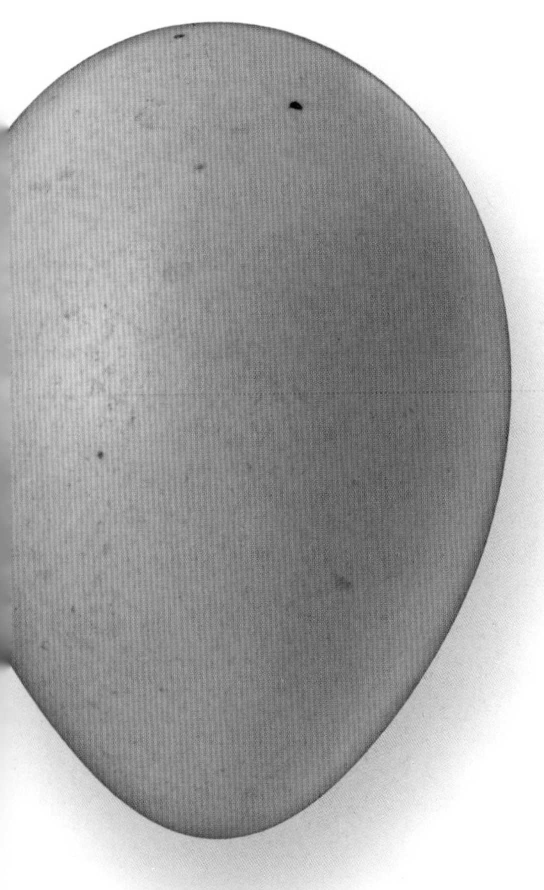

This is a small vireo species, and in most of its range it is resident or shows only short regional movement patterns between breeding and wintering areas. There may be several different species known as the "Hutton's Vireo" but more genetic work is required to delineate species limits between coastal and inland resident populations. Pair formation is not as hurried as it is in other migratory vireos, and the breeding season's onset is signaled when the males start singing at increasingly higher rates, often when the pair bond with the female is already established.

Each reproductive attempt begins with the construction of a brand new nest. Prior to egg laying, this vireo is skittish and often abandons the half-constructed nest to start building elsewhere, frequently reusing the first nest's materials. In turn, materials of old and successful vireo nests are recycled by goldfinches, which begin their own nesting much later in the season than the vireos.

Actual size

The egg of the Hutton's Vireo is white in background color, sometimes with fine brown speckling, and is ⅔ x 9/16 in (18 x 14 mm) in size. The parents sit on the eggs, staying with them from the day the first egg is laid. Full heat transfer, and thus incubation, is delayed until the clutch is nearly complete; accordingly, the eggs hatch within the course of just one day.

ORDER	Passeriformes
FAMILY	Vireonidae
BREEDING RANGE	Temperate North America
BREEDING HABITAT	Deciduous forests, riparian corridors
NEST TYPE AND PLACEMENT	Rough and messy hanging cup, suspended from the forks of horizontal branches; made of grasses, stems, and spider silk
CONSERVATION STATUS	Least concern
COLLECTION NO.	FMNH 12469

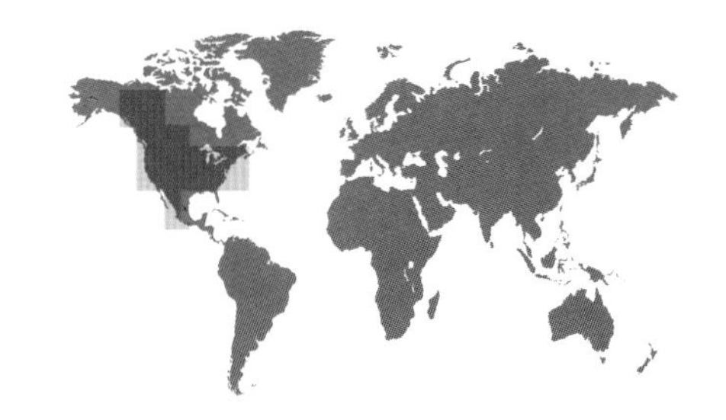

ADULT BIRD SIZE
4½–5 in (12–13 cm)

INCUBATION
12–14 days

CLUTCH SIZE
3–4 eggs

VIREO GILVUS

WARBLING VIREO

PASSERIFORMES

The Warbling Vireo has a near continent-wide distribution where it prefers open forest edge; it builds its nests often high up in the canopy. It is a migratory species, and the two sexes arrive together or within days of one another. Pairs form rapidly and breeding territories are defended from intruders by both members of the pair. Nest building is carried out mostly by the female, but both parents incubate the eggs, brood the hatchlings, and provision the nestlings.

Curiously, the different subspecies of this vireo have very different responses to parasitic eggs laid in the nest by Brown-headed Cowbirds (see page 616). In the Great Plains, females consistently puncture, pierce, and eject cowbird eggs (the males' attempts to do this typically fail). Further west, where the birds are smaller in size (and in beak strength), cowbird eggs are accepted and bring about major losses of reproductive success in parasitized nests.

Clutch

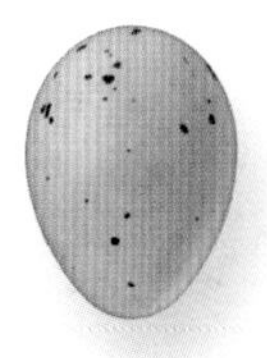

Actual size

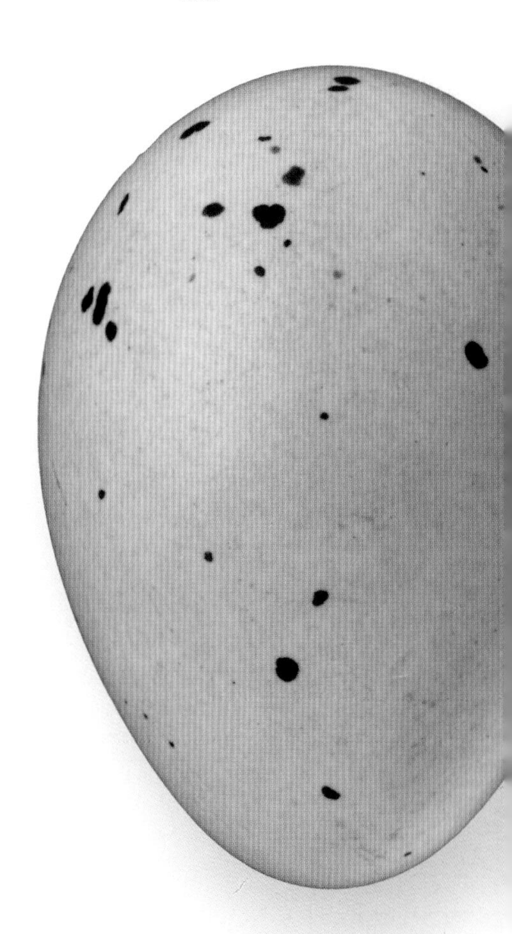

The egg of the Warbling Vireo is white in background color, has small reddish or dark brown maculation, and measures ¾ x 9/16 in (20 x 14 mm). During nest building and egg laying, the male follows and guards the female closely; they frequently copulate within sight of the nest.

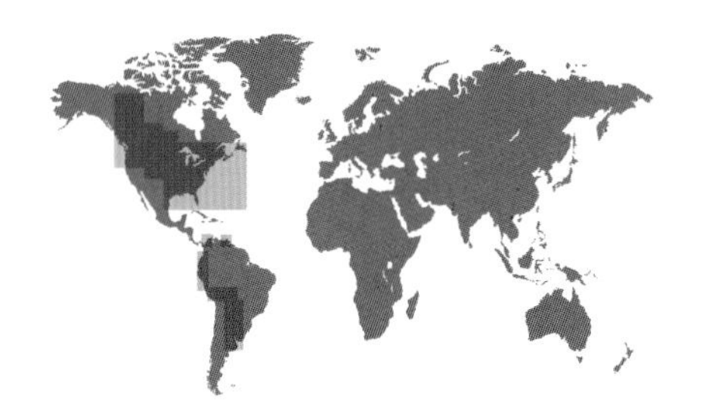

ORDER	Passeriformes
FAMILY	Vireonidae
BREEDING RANGE	North and South America
BREEDING HABITAT	Contiguous forests, mixed conifer or deciduous wood lots, riparian corridors, with brushy understory
NEST TYPE AND PLACEMENT	Open cup, suspended by its rim from a fork of twigs; consists of bark strips, grasses, pine needles, lichen, and spider silk
CONSERVATION STATUS	Least concern
COLLECTION NO.	FMNH 12546

ADULT BIRD SIZE
4½–5 in (12–13 cm)

INCUBATION
11–14 days

CLUTCH SIZE
2–5 eggs

VIREO OLIVACEUS

RED-EYED VIREO

PASSERIFORMES

Clutch

This species is one of the most common forest-nesting songbirds in North America; it can be present both in forest interiors and edge habitats. In the dense forest foliage, it is more often heard than seen, singing its monotonous but melodious song. The males use their song to establish and defend the territorial boundaries. Displays to attract the females are limited to mutual chases, feather-puffing, and courtship feeding. Throughout nest building, incubation, and even during the nestling period, the male continues to feed the female, but typically at some distance from the nest.

The female alone selects the nesting site and constructs the nest, while the male sings nearby and also helps to collect some of the materials. Throughout incubation and during the nestling stage, the female continues to add nesting materials to build up the nest. The result is a sturdy structure that can last up to two years, yet the nest is never actually reused for subsequent breeding attempts.

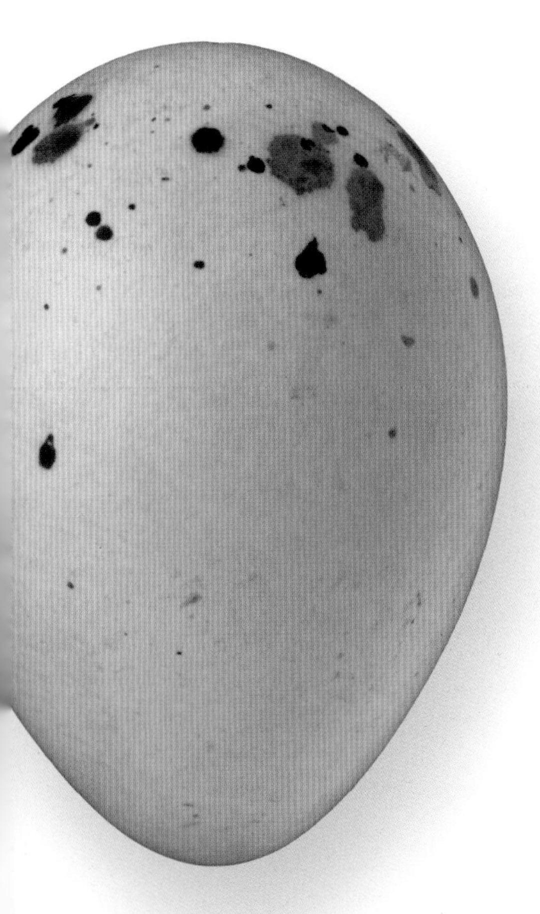

Actual size

The egg of the Red-eyed Vireo is dull white, with irregular reddish brown maculation, and is ¾ x ⅝ in (20 x 15 mm) in size. The female alone incubates the eggs, although the male may stand above them to provide cover while she is away foraging between incubation bouts.

ORDER	Passeriformes
FAMILY	Oriolidae
BREEDING RANGE	Temperate Europe, north Africa, western and central Asia
BREEDING HABITAT	Broadleaf and mixed forests, riparian wood lots, orchards, gardens
NEST TYPE AND PLACEMENT	Cup nest of grasses and filaments, with lichen and moss, placed in a tree fork
CONSERVATION STATUS	Least concern
COLLECTION NO.	FMNH 20732

ADULT BIRD SIZE
8–12 in (20–30 cm)

INCUBATION
14–15 days

CLUTCH SIZE
3–4 eggs

ORIOLUS ORIOLUS

EURASIAN GOLDEN-ORIOLE

PASSERIFORMES

This oriole is a long-distance migrant that winters in southern Africa. Upon returning to its breeding grounds, the male does not establish an exclusive territory for feeding and nesting, but instead feeds on various trees opportunistically while singing to attract a female. Once the pair is formed, the female alone builds the nest, but both parents incubate the eggs and provision the nestlings.

Not to be confused with New World orioles (Icterdiae), this species belongs to a family of Old World species that includes other brightly colored orioles and figbirds. Despite the bright yellow coloration of the male, both he and the drabber, greenish female are difficult to spot; their plumage colors blend well into the green-blue light bouncing off the dense foliage of the high canopy, where they feed on chunky caterpillars or fruits. They are more easily recognized by their territorial, screeching call, and by the male's unique, flutelike, melodious mating song.

Clutch

The egg of the Eurasian Golden-Oriole is white in background color, has black or brown maculation, and is 1 3⁄16 x 7⁄8 in (30 x 21 mm) in dimensions. The eggs and the nestlings are secured in an elaborately constructed nest, woven into and hung from a forking tree branch, like a hammock.

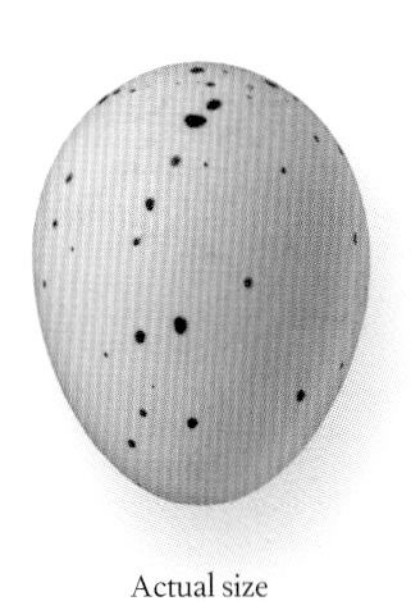

Actual size

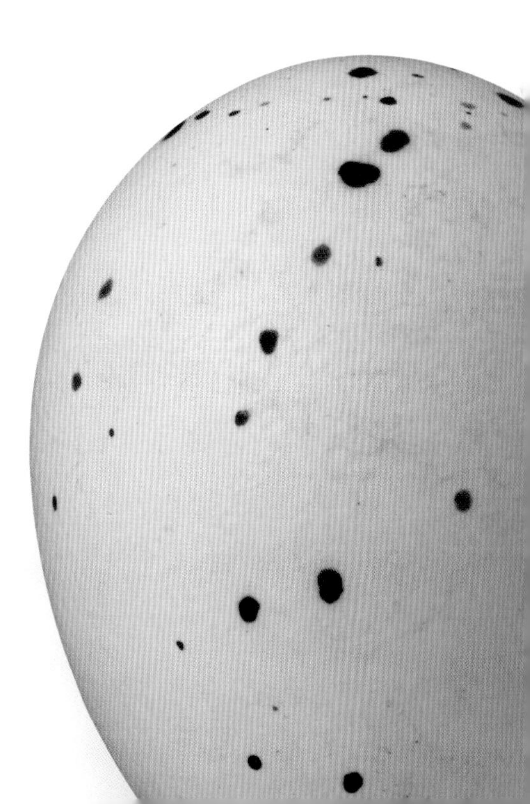

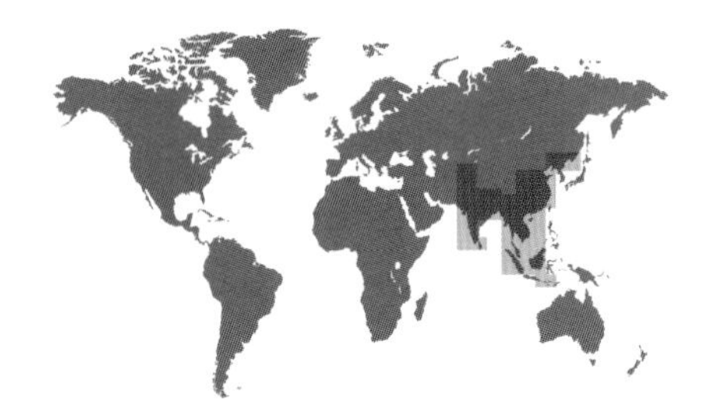

ORDER	Passeriformes
FAMILY	Monarchidae
BREEDING RANGE	East and south Asia
BREEDING HABITAT	Dense and mature tropical forests
NEST TYPE AND PLACEMENT	Tightly constructed cup nest, made with twigs, leaves, and spider webs
CONSERVATION STATUS	Least concern
COLLECTION NO.	FMNH 20842

ADULT BIRD SIZE
8–20 in (20–50 cm), with or without the 12 in (30 cm) tail

INCUBATION
12–16 days

CLUTCH SIZE
3–4 eggs

TERPSIPHONE PARADISI

ASIAN PARADISE FLYCATCHER

PASSERIFORMES

Clutch

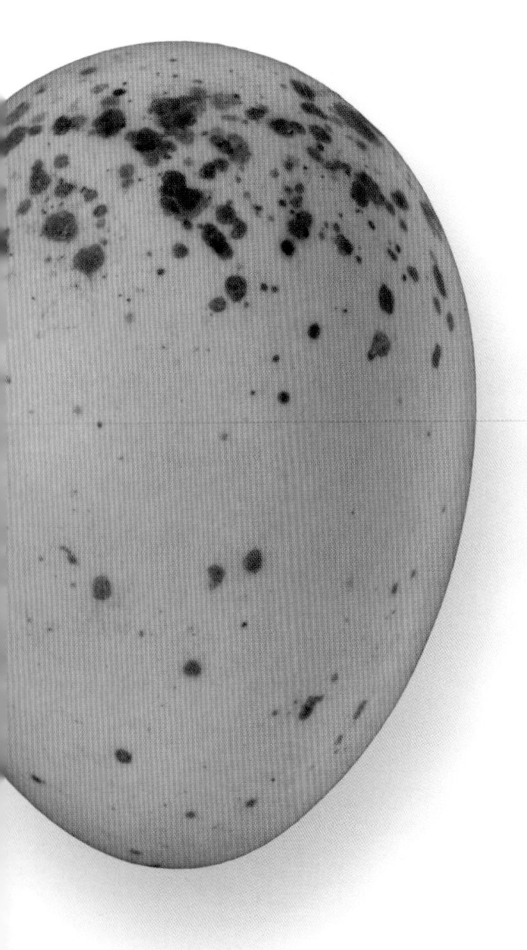

This flycatcher, along with several of its relatives, is perhaps best known not for its elongated and elaborate tail plumes, but for the male's plumage polymorphism which includes long-tailed white males, long-tailed rufous males, and short-tailed rufous males. All three of these male morphs are sexually mature, and attempt to attract females to breed in socially monogamous pairs. Those females who mate with long-tailed males begin to nest earlier in the season and lay more eggs in their nests than females mated with short-tailed males. However, researchers have found no apparent difference in reproductive success between the white and rufous long-tailed males.

Once the eggs hatch, the chicks solicit parental care by begging loudly and opening their bright yellow gape whenever an adult lands on the nest's rim. Nourished by both parents, who deliver insects, these chicks grow quickly, and are able to leave and flutter out of the nest on their own wings just 10 to 11 days after hatching.

Actual size

The egg of the Asian Paradise Flycatcher is dull white or buff in color, spotted with brown speckles, and measures ¾ x ⅝ in (20 x 15 mm). Nest building, incubation, and provisioning are shared by both members of the pair.

ORDER	Passeriformes
FAMILY	Corvidae
BREEDING RANGE	Northern and western North America
BREEDING HABITAT	Boreal and subalpine conifer forests
NEST TYPE AND PLACEMENT	Cup nest lined with feathers or fur, made of twigs and branches, placed high in a mature tree
CONSERVATION STATUS	Least concern
COLLECTION NO.	FMNH 101

ADULT BIRD SIZE
10½–12 in (27–30 cm)

INCUBATION
18–19 days

CLUTCH SIZE
2–5 eggs

PERISOREUS CANADENSIS

GRAY JAY

PASSERIFORMES

The Gray or Canada Jay is a year-round resident of pine forests at high elevation and in the far north. In these cold climates it is familiar to many skiers as the gray bird waiting at the top of the chair-lift for a hand-out of energy bars or nuts. Gray Jays rely on their resourcefulness, long-term spatial memory, and copious sticky saliva to collect seeds and other nonperishable foods during the summer. These they store in small crevices or glued under tree bark, retrieving them during the foodless winter months to feed themselves and their chicks.

The breeding pair lives on a permanent, all-purpose territory, where they tolerate few companions. For instance, after the nesting season, all but one of the year's fledglings are kicked off the natal territory, often by their most dominant sibling. The dispersing young then must seek out adoptive pairs who will tolerate their presence from the end of the summer and into the fall. Because these young help to collect the next winter's essential food stores, adopting strange chicks may be beneficial for pairs without a chick of their own.

Clutch

The egg of the Gray Jay is white with a green tint in background color, has dark olive to rusty maculation, and measures 1⅛ x ⅞ in (29 x 21 mm) in size. Nesting begins during the late winter or early spring, with the eggs often incubated in subzero temperatures. If a nest fails, a replacement clutch is not laid in the seemingly more hospitable spring months.

Actual size

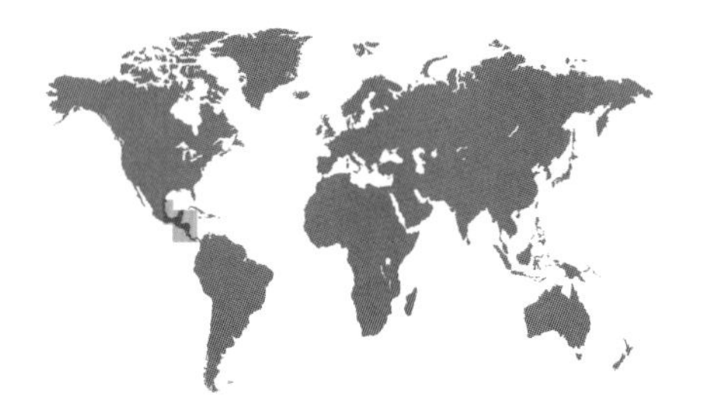

ORDER	Passeriformes
FAMILY	Corvidae
BREEDING RANGE	Southern North America, Central America
BREEDING HABITAT	Riparian corridors, open woodlands, forest edges, second growths, farmland
NEST TYPE AND PLACEMENT	Stick and twig platform, with a softer, lined, inner cup nest; built high up in a tree on a thin branch
CONSERVATION STATUS	Least concern
COLLECTION NO.	FMNH 20264

ADULT BIRD SIZE
15–17 in (38–44 cm)

INCUBATION
18–20 days

CLUTCH SIZE
3–6 eggs

PSILORHINUS MORIO

BROWN JAY

PASSERIFORMES

Clutch

Brown Jays are cooperative breeders; the mated pair are assisted by several other helpers of various ages. Helpers protect the nest from predators, and also provision the nestlings; the larger the group, the better the reproductive success. It appears that helping is also a learning opportunity for young jays: as the helpers gain experience with feeding the chicks, their efficiency as surrogate parents becomes greater, so that they bring more food at each feeding visit to the growing young.

This large and loud jay is one of the very few birds that has seemingly benefited from deforestation, habitat modification, and farm development. In its Central American range, the logging of extensive forests has generated vast new areas of suitable habitat for the Brown Jay. Groups of five to ten individuals are often seen moving about and foraging together. They defend their territory communally from intruders and predators.

Actual size

The egg of the Brown Jay is pale blue-green in coloration, has brown speckling, and its dimensions are 1⅓ x 1 in (35 x 25 mm). Among the Brown Jays studied in Monteverde, Costa Rica, two or more females may contribute to laying eggs in the nest, doubling the clutch size.

ORDER	Passeriformes
FAMILY	Corvidae
BREEDING RANGE	Southern North America, Central America, and northern and western South America
BREEDING HABITAT	Open woodlands and thicket, also humid tropical forests
NEST TYPE AND PLACEMENT	Flimsy cup of twigs, lined with fine roots, vine stems, moss, and dry grass; placed on thin branches in a tree or shrub
CONSERVATION STATUS	Least concern
COLLECTION NO.	FMNH 9669

ADULT BIRD SIZE
10½–11½ in (27–29 cm)

INCUBATION
17–18 days

CLUTCH SIZE
2–6 eggs

CYANOCORAX YNCAS

GREEN JAY

PASSERIFORMES

With its green body and bright blue head, the Green Jay is a lusciously adorned member of the corvid lineage. It occurs in distinct populations in the northern and southern subtropics and tropics of the Americas. These populations all show some level of sociality, where pairs are accompanied by young and subadults from previous years' broods who help protect the territory from intruders. However, the birds in the north and the south are different in their breeding behaviors: in northern populations, group members do not assist in the breeding attempt, but in the southern populations, helpers participate in defending the territory and feeding the nestlings.

Nest construction is shared between the nuclear pair, whereas only the mother incubates the eggs, and only the father delivers food to the hatchlings. Later, the sexes both feed the growing chicks equitably, but once the chicks leave the nest, it is the female who mostly provisions the fledglings.

Clutch

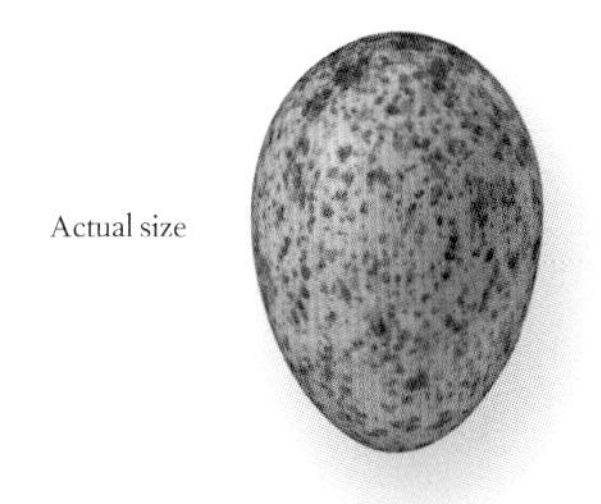

Actual size

The egg of the Green Jay is white with a pale green background color, with dark blotching, and is 1 1/16 x 13/16 in (27 x 21 mm) in dimensions. The nest platform is so thin that often the eggs are visible when viewed from directly below the nest.

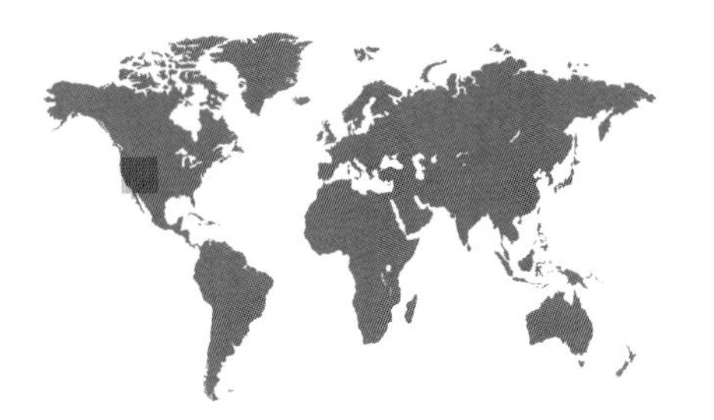

ORDER	Passeriformes
FAMILY	Corvidae
BREEDING RANGE	Western North America
BREEDING HABITAT	Pinyon pine forests, in foothills of mountain ranges
NEST TYPE AND PLACEMENT	In pine or juniper trees; a bulky cup made of twigs and bark, also of paper, plastic, and other trash near urban areas
CONSERVATION STATUS	Vulnerable
COLLECTION NO.	FMNH 20306

ADULT BIRD SIZE
10–11½ in (26–29 cm)

INCUBATION
16–17 days

CLUTCH SIZE
3–4 eggs

GYMNORHINUS CYANOCEPHALUS

PINYON JAY

PASSERIFORMES

Clutch

The Pinyon Jay forms some of the most permanent and sizable social systems in the bird world, with flocks of several hundred individuals foraging, roosting, mobbing, and breeding together all year round. Many fledglings never leave their natal flock, and those that do are typically females. This is unlike what mammals typically do, but it is a general pattern in birds: the males are more site-faithful than the females.

This jay is a specialist on pine nuts and usually feeds communally. The onset of breeding depends on the availability of vast amounts of this preferred food item. During incubation, only the female stays on the nest. The male joins other males in foraging flocks, and returns to his mate to feed her on a regular, hourly basis. Once the eggs hatch, both parents search for food to provision the young.

Actual size

The egg of the Pinyon Jay is pale blue in background coloration, has dark brown markings, and measures 1⅛ x ⅞ in (29 x 22 mm). Egg laying is synchronized across the colony, through communal courtship "parties," so that most, if not all, of the chicks fledge within just one week of each other.

ORDER	Passeriformes
FAMILY	Corvidae
BREEDING RANGE	Eastern and central North America
BREEDING HABITAT	Deciduous and coniferous forests, parks, suburban developments, and wooded urban areas
NEST TYPE AND PLACEMENT	Open cup, made of twigs, sticks, and mud, lined with grass and rootlets; placed in trees
CONSERVATION STATUS	Least concern
COLLECTION NO.	FMNH 618

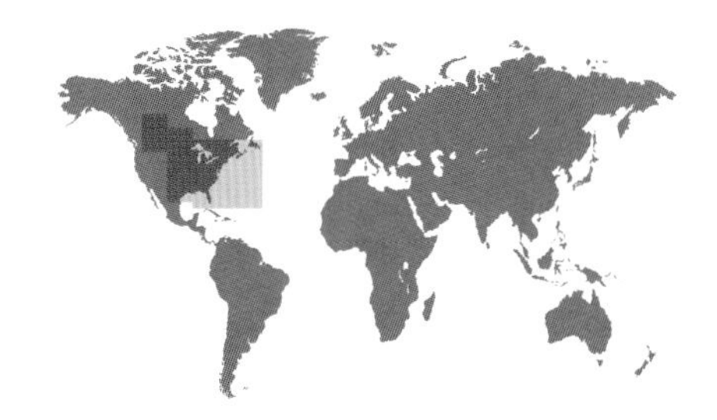

ADULT BIRD SIZE
10–12 in (25–30 cm)

INCUBATION
17–18 days

CLUTCH SIZE
2–7 eggs

CYANOCITTA CRISTATA

BLUE JAY

PASSERIFORMES

The Blue Jay is a common, bright, and loud inhabitant of forests and wood lots as well as cities, towns, and farms throughout its range. Most people can count this native North American bird species in their personal list of familiar birds, in addition to introduced sparrows, starlings, and pigeons. The Blue Jay occurs even in the most densely settled of streets, including New York City (I can hear one now as I am writing these lines), as long as a few bushes and trees line and grow on the quieter side streets, where these birds can build their nests in the high canopy.

To keep the nest clean and hygienic, Blue Jay parents consume broken eggshells soon after hatching and swallow the fecal sacs produced by the chicks after each feeding bout. This also serves to eliminate any visual and olfactory cues that predators might use to find the nest. Only the female incubates the eggs and she also broods the young chicks, while the male provides her and the whole young family with all the food they need during the first few days after hatching.

Clutch

The egg of the Blue Jay is bluish or light brown in color, has brownish maculation throughout, and measures $1\frac{1}{8} \times \frac{3}{4}$ in (28 x 20 mm). There are only a handful of cases of natural cowbird parasitism in Blue Jays; when parasitic eggs were experimentally introduced to nests, the jays easily removed them.

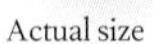

Actual size

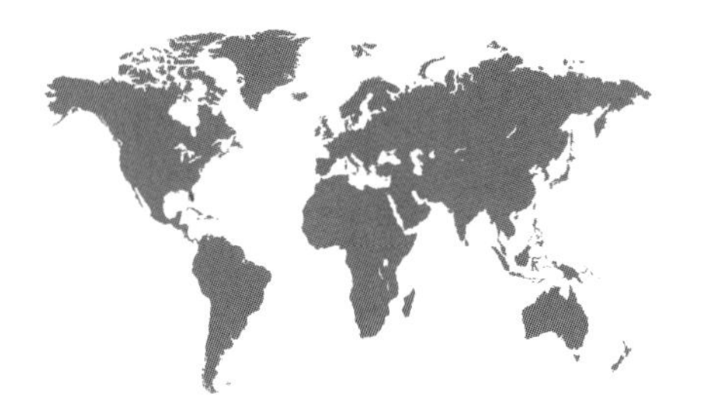

ORDER	Passeriformes
FAMILY	Corvidae
BREEDING RANGE	Florida
BREEDING HABITAT	Low-growth oak and palmetto scrub, on well-drained, sandy ridges
NEST TYPE AND PLACEMENT	Open cup of twigs and rootlets, lined with palmetto fibers, placed in low, dense shrub
CONSERVATION STATUS	Vulnerable
COLLECTION NO.	FMNH 1409

ADULT BIRD SIZE
9–11 in (23–28 cm)

INCUBATION
17–18 days

CLUTCH SIZE
2–5 eggs

APHELOCOMA COERULESCENS

FLORIDA SCRUB-JAY

PASSERIFORMES

Clutch

The egg of the Florida Scrub-Jay is cream with a green tint as its background color, has brown maculation, and measures 1 1/16 x ¾ in (27 x 20 mm). The female alone incubates the eggs, but once they hatch, the male and older male helpers feed them frequently, while younger male and female helpers only visit the chicks occasionally.

Breeding for the Florida Scrub-Jay is a highly cooperative affair; a monogamous pair is accompanied and assisted by up to six helpers at all stages of the nesting cycle, and also in territorial defense during and outside the breeding season. These helpers are offspring from prior broods, both male and female, that do not leave the natal territory. They benefit individually because they help their own parents to raise more siblings, with whom they share just as many genes as they would share with their own offspring.

This is the only species of bird that lives only in Florida. It occurs on both coasts, in remnant patches of low-growing oak and palmetto thickets, and also suburban parks and yards. Most of its historic habitat has been converted to citrus groves and suburban development. Wherever this species lives near human settlements, roadkills and feral cats reduce its reproductive success and lifespan dramatically compared to what these are in the remaining, protected, natural habitats.

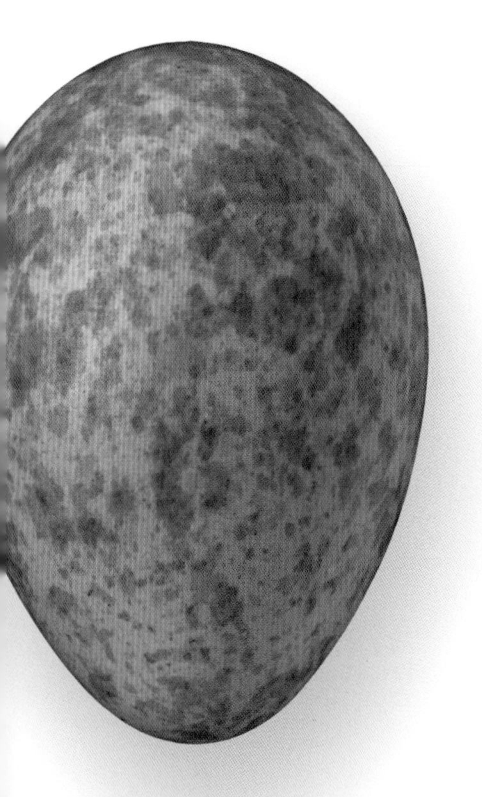

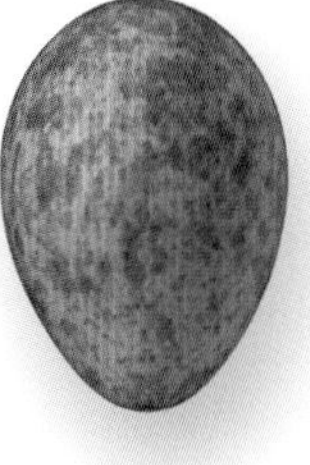

Actual size

ORDER	Passeriformes
FAMILY	Corvidae
BREEDING RANGE	Europe, north Africa, and temperate and tropical Asia
BREEDING HABITAT	Mixed and deciduous woodlands, especially oak forests
NEST TYPE AND PLACEMENT	Cup nests in forking branches of a tree or large shrub; made of twigs, stems, and roots, and lined with plant fibers and rootlets
CONSERVATION STATUS	Least concern
COLLECTION NO.	FMNH 1405

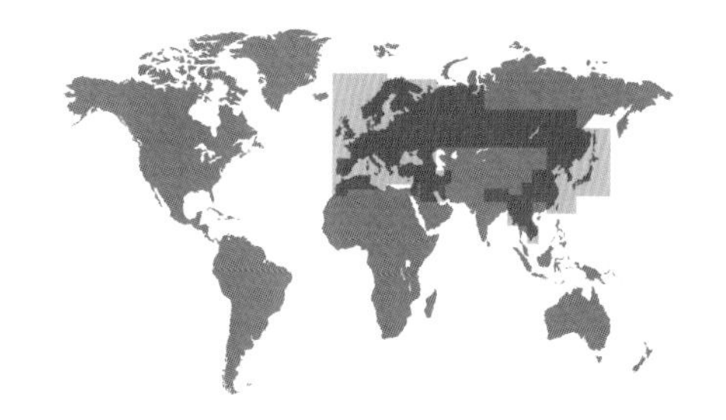

ADULT BIRD SIZE
12½–14 in (32–36 cm)

INCUBATION
16–19 days

CLUTCH SIZE
4–6 eggs

GARRULUS GLANDARIUS

EURASIAN JAY

PASSERIFORMES

The pair bond is at the core of family life and breeding in the Eurasian Jay. To impress his mate, and to eventually secure as many matings with her as possible, the male frequently "courtship feeds" the female. Jays and crows are well known for their intelligence and the male crow apparently does more than simply pass available food items: in captivity, at least, he closely monitors what type of food items the female has been consuming, and attempts to offer her a different, unusual type of food whenever he gets a chance.

The male usually selects the nest site and both sexes build the nest, each incorporating materials that they themselves collected. Later, only the female incubates the eggs, but the male feeds the brooding mother and the hatchlings. Feedings become more evenly shared by the pair as the chicks continue to grow.

Clutch

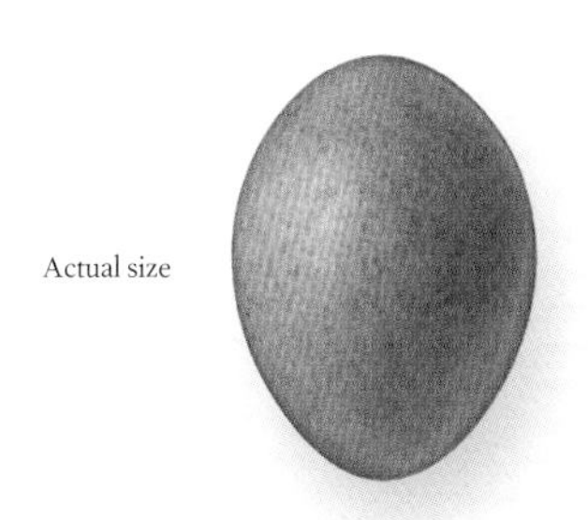

Actual size

The egg of the Eurasian Jay is beige, pale blue-green to olive in coloration, has buff maculation, and measures 1¼ x ⅞ in (32 x 22 mm). The female removes the sharp-edged eggshell fragments from the nest immediately after hatching; the shell edges can harm the chicks and their bright white interiors can attract predators.

ORDER	Passeriformes
FAMILY	Corvidae
BREEDING RANGE	Western North America
BREEDING HABITAT	Open fields with isolated stands of trees, forest edges near grassland or pastures; suburban developments
NEST TYPE AND PLACEMENT	Domed nest, made of branches and sticks, cemented with soil, and placed in a large tree or shrub
CONSERVATION STATUS	Least concern
COLLECTION NO.	FMNH 16720

ADULT BIRD SIZE
17½–18 in (44–46 cm)

INCUBATION
16–21 days

CLUTCH SIZE
5–8 eggs

PICA HUDSONIA

BLACK-BILLED MAGPIE

PASSERIFORMES

Clutch

The egg of the Black-billed Magpie is buff with an olive sheen in background color, marked with brown to gray maculation, and is 1¼ x ⅞ in (32 x 22 mm) in dimensions. The young will stay in the nest and continue to be fed by the adults for up to a month after hatching.

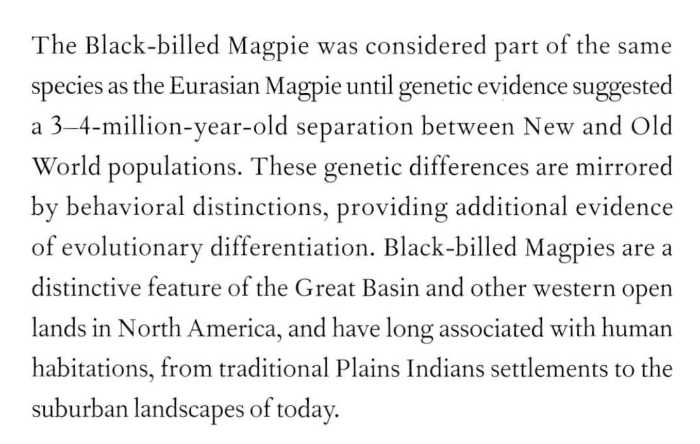

The Black-billed Magpie was considered part of the same species as the Eurasian Magpie until genetic evidence suggested a 3–4-million-year-old separation between New and Old World populations. These genetic differences are mirrored by behavioral distinctions, providing additional evidence of evolutionary differentiation. Black-billed Magpies are a distinctive feature of the Great Basin and other western open lands in North America, and have long associated with human habitations, from traditional Plains Indians settlements to the suburban landscapes of today.

Magpie pairs mate for life and stay together year-round. Males do most of the work to build the domed nest structure, while the female brings in mud that will form the base of the internal nest cup. The female alone incubates the eggs while the male brings her food. In the next breeding season, they may repair or reuse the old nest or build on top of it. Often-used nests can grow to be 4 ft (1.2 m) high and 3 ft (90 cm) wide.

Actual size

ORDER	Passeriformes
FAMILY	Corvidae
BREEDING RANGE	California
BREEDING HABITAT	Central Valley oak savanna, and nearby mountain woody scrub
NEST TYPE AND PLACEMENT	High up in trees, often in clumps of mistletoe; dome made of sticks and anchored with mud, lined with hair, grass, bark, or rootlets
CONSERVATION STATUS	Least concern
COLLECTION NO.	FMNH 16723

PICA NUTTALLI

YELLOW-BILLED MAGPIE

PASSERIFORMES

ADULT BIRD SIZE
17–21½ in (43–54 cm)

INCUBATION
16–18 days

CLUTCH SIZE
4–7 eggs

The Yellow-billed Magpie is a California endemic: it resides and breeds nowhere else in the world. A specialist of open oak savanna, it has suffered from destruction of its habitat, especially in coastal areas. Although the species is not shy of human neighbors, population trends have shown a sharper decline near urbanized regions.

Nest construction is a laborious process and can take up to two months. Both sexes contribute, although the final touch of forming a mud bowl inside the dome of sticks is typically done by the female. She then spends most of her time in the nest, while laying and incubating, and the male feeds her at the nest following a regular schedule. She also broods the chicks during the early nestling stage, and then joins the male to find and provision food for the developing chicks.

Clutch

Actual size

The egg of the Yellow-billed Magpie is greenish blue or olive in background color with dark maculation, and measures 1¼ x ⅞ in (32 x 22 mm) in size. The female incubates alone starting soon after the first egg is laid; the eggs therefore hatch asynchronously.

ORDER	Passeriformes
FAMILY	Corvidae
BREEDING RANGE	Western North America
BREEDING HABITAT	Montane pine forests
NEST TYPE AND PLACEMENT	Cup nest, made of fine fibers and other insulating material; built in a tree fork, on a platform of sticks and branches
CONSERVATION STATUS	Least concern
COLLECTION NO.	FMNH 9831

ADULT BIRD SIZE
10½–12 in (27–30 cm)

INCUBATION
18 days

CLUTCH SIZE
2–4 eggs

NUCIFRAGA COLUMBIANA

CLARK'S NUTCRACKER

PASSERIFORMES

Clutch

The Clark's Nutcracker is a specialist on pine seeds: it feeds on and stores tens of thousands of unripe and ripe seeds during its life. It can recall for months the places where it has stashed some of its essential food stores for the cold winter months, even when there is snow cover. Forgotten, or simply unrecovered, seeds often germinate when stored in the soil, providing a self-sustaining habitat and establishing a mutualism between plant and bird.

The parents work side by side to gather materials and complete the nest in about a week. Nesting is timed early in the spring, when no pine cones are available and so the parents feed the chicks from the previous year's stored crop of seeds. In turn, when the chicks fledge, they spend several months with adults, perhaps to learn the tricks of feeding on pine cones and caching, remembering, and harvesting stored seeds.

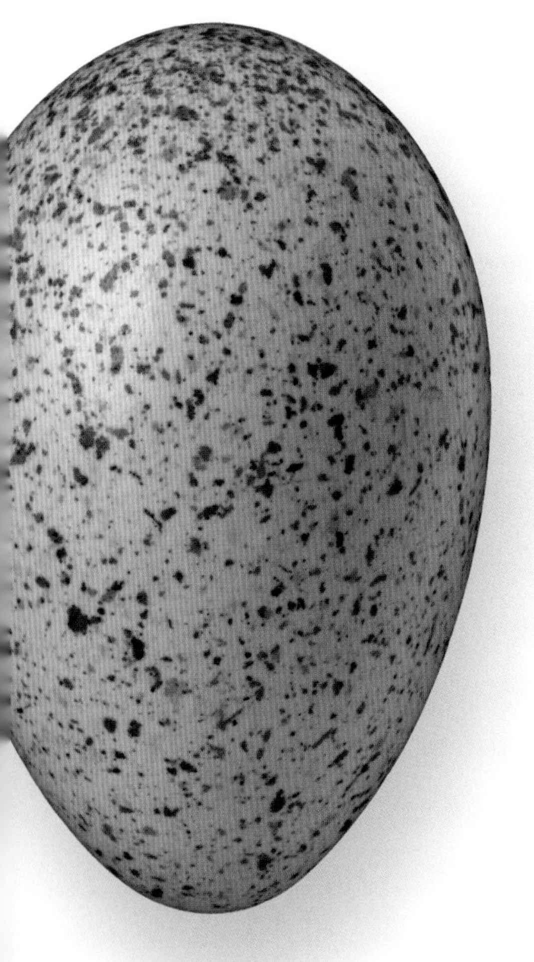

Actual size

The egg of the Clark's Nutcracker is pale green in background color; it has fine, brown, olive, or gray spotting, and dimensions of 1⅓ x ⅞ in (33 x 23 mm). Both parents incubate the eggs, and both sexes develop an incubation patch to transfer heat efficiently to the eggs in the cold springtime.

ORDER	Passeriformes
FAMILY	Corvidae
BREEDING RANGE	North and east Africa, southern and northern Europe, central and western Asia
BREEDING HABITAT	Rocky high mountains, coastal cliffs, open grasslands near the tree line
NEST TYPE AND PLACEMENT	Cup nest made of sticks and roots, lined with wool or hair; placed in the crevice of a cliff face
CONSERVATION STATUS	Least concern
COLLECTION NO.	FMNH 3477

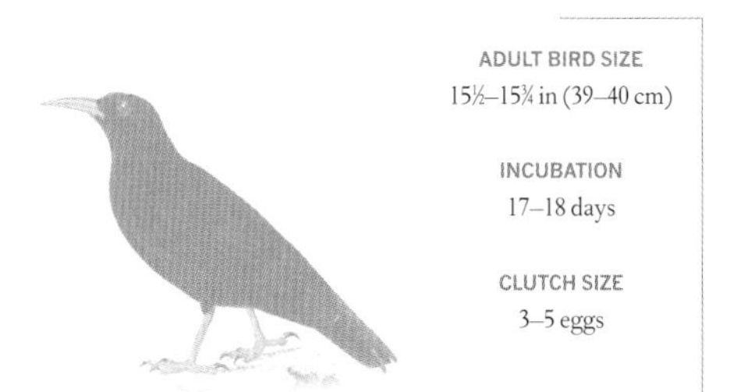

ADULT BIRD SIZE
15½–15¾ in (39–40 cm)

INCUBATION
17–18 days

CLUTCH SIZE
3–5 eggs

PYRRHOCORAX PYRRHOCORAX

RED-BILLED CHOUGH

PASSERIFORMES

This species is a distinctive-looking corvid with a bright red bill and feet. Most populations show long-term "philopatry," that is, they remain in their birthplace with few exchanges in pair bond membership or population affiliation. For example, two island-dwelling populations in Scotland occur only 6 miles (10 km) apart, but fewer than ten banded birds have been seen to transfer between the islands in over 20 years of study of hundreds of color-banded birds.

These banding studies also showed that female choughs age reproductively much as humans and elephants do: initially, reproductive effort (as measured by clutch size) increases with age, and then it declines with advancing years. There is no such relationship between breeding and age for males. Researchers also found that chough mothers that lay more eggs when they are young die earlier, while those that lay fewer eggs while they are young tend to have longer lives.

Clutch

The egg of the Red-billed Chough is cream to buff in background color, with brown to gray maculation, and measures 1½ x 1⅛ in (39 x 28 mm) in size. The probability of the eggs hatching successfully varies between years, being lower when the preceding winter was colder and had more rainfall.

Actual size

ORDER	Passeriformes
FAMILY	Corvidae
BREEDING RANGE	North America
BREEDING HABITAT	Diverse habitats including woodland, forest edges, grasslands with scattered trees, parks, orchards, and urban developments
NEST TYPE AND PLACEMENT	Bulky stick nest, with an inner cup lined with pine needles, weeds, bark, or hair; placed high in trees
CONSERVATION STATUS	Least concern
COLLECTION NO.	FMNH 9749

ADULT BIRD SIZE
17–21 in (43–53 cm)

INCUBATION
16–18 days

CLUTCH SIZE
3–7 eggs

CORVUS BRACHYRHYNCHOS

AMERICAN CROW

PASSERIFORMES

Clutch

Sociality rules the life of the American Crow. Outside of the breeding season, it forms large flocks that move and settle in communal roosts; these roosting sites remain stable and can be reused year after year. During the breeding season, the flock breaks into cooperative family units, which consist of a central monogamous pair, and their adult offspring from prior seasons.

The egg of the American Crow is pale blue-green to olive-green in coloration, with dense brown and gray blotching, and is 1⅝ x 1⅛ in (41 x 29 mm) in size. Despite the large and conspicuous perimeter of this crow's nest, easily spotted from the ground, the inner cup holding the eggs is surprisingly small and tight, and is lined with soft materials by the female.

Both sexes construct the nest, but the female is in charge of lining the inner cup with soft materials before the eggs are laid. The mother alone incubates the eggs, typically starting before they are all laid. The father and helpers assist during incubation by feeding the female directly at the nest, so that the incubation bouts are longer and less interrupted. Once the eggs hatch, the oldest and the youngest chick can vary by as much as three days in age, which generates a natural size- and behavioral-dominance hierarchy in the nest.

Actual size

ORDER	Passeriformes
FAMILY	Corvidae
BREEDING RANGE	Temperate and Arctic northern hemisphere, including montane subtropical regions
BREEDING HABITAT	Forested areas, with open fields or coastal areas nearby
NEST TYPE AND PLACEMENT	Large platform of sticks, branches, and twigs, lined with mud, grasses, lichens, and plant and animal fibers; on cliff ledges, in trees, and on power lines
CONSERVATION STATUS	Least concern
	FMNH 16726

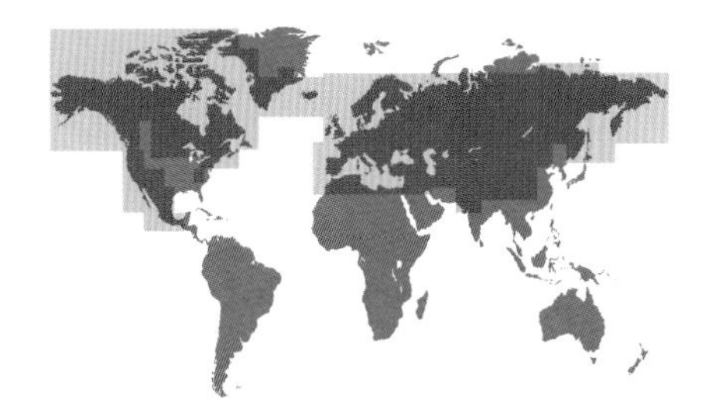

ADULT BIRD SIZE
22–27 in (56–69 cm)

INCUBATION
20–25 days

CLUTCH SIZE
3–7 eggs

CORVUS CORAX

COMMON RAVEN

PASSERIFORMES

The Common Raven is one of the most widespread perching birds in the world. Ravens are well known for their high intelligence, including the ability to solve problems that require insight and foresight. Their powers include a detailed long-term memory, along with fine-tuned vocal signaling that allows for sophisticated communication between individuals.

Ravens mate for life, but unpaired males occasionally visit a female when her mate is out of sight. Once the pair bond is formed, and a breeding territory is established, nest building can begin. The male assists with providing the larger sticks and branches for the platform of the nest, and then the female completes the nest and lines the inner cup with soft wool and hair for the eggs. Few of the nests survive in usable condition for the next year, but many pairs still re-nest or repair what little is left at the previous year's location.

Clutch

The egg of the Common Raven is white or beige, with a green, olive, or blue tint in background color, has dark green or purplish brown maculation, and measures 1⅞ x 1⅓ in (48 x 34 mm). Incubation does not begin until the last egg is laid, but the female spends the night on the nest as soon as she starts laying. She also remains on the nest during severe weather, such as early spring snowstorms.

Actual size

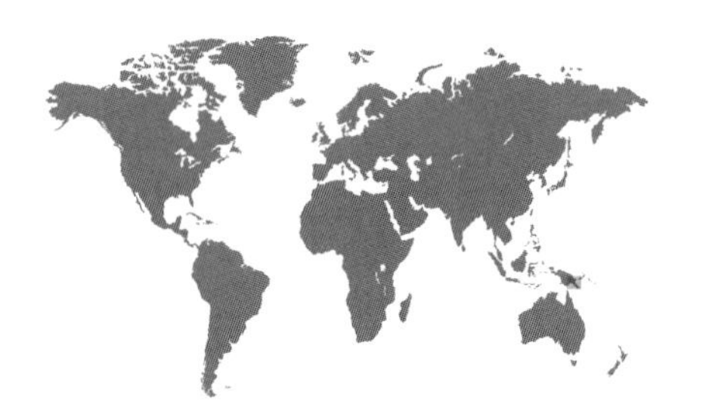

ORDER	Passeriformes
FAMILY	Paradisaeidae
BREEDING RANGE	Eastern New Guinea
BREEDING HABITAT	Montane, moist forests
NEST TYPE AND PLACEMENT	Placed in a tree fork in the canopy; a cup made of ferns, vines, and leaves
CONSERVATION STATUS	Vulnerable
COLLECTION NO.	FMNH 2959

ADULT BIRD SIZE
12–13 in (30–33 cm)

INCUBATION
16–22 days

CLUTCH SIZE
1 egg

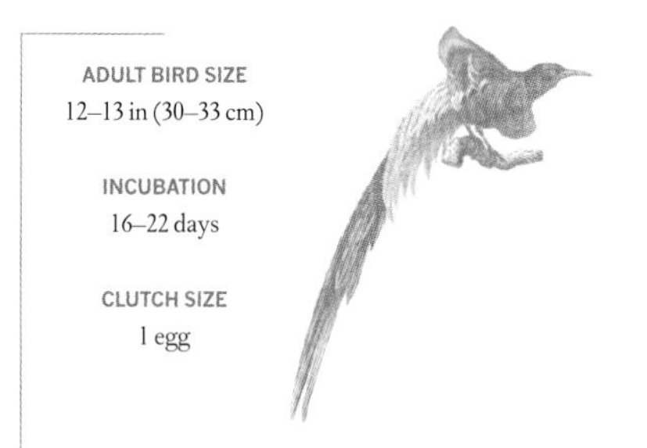

PARADISAEA RUDOLPHI

BLUE BIRD-OF-PARADISE

PASSERIFORMES

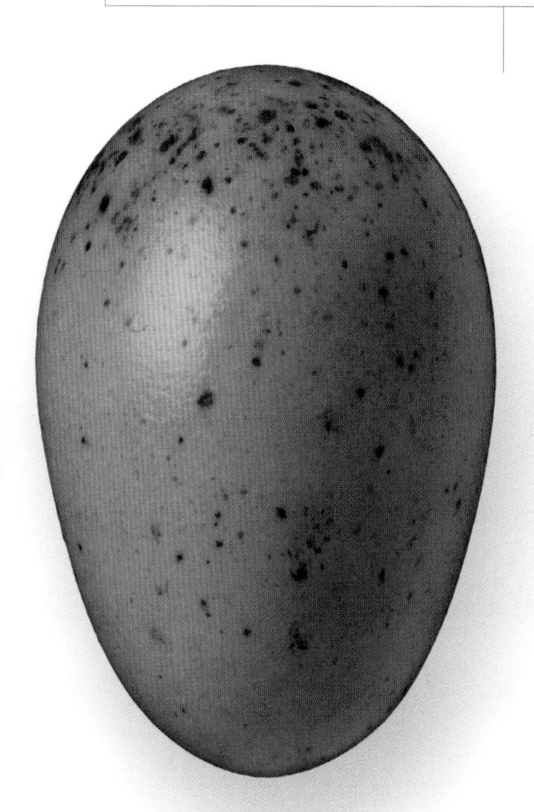

There are few greater pleasures for an ornithologist than seeing the Blue Bird-of-Paradise; many enthusiasts and scientists agree that this may be the most beautiful bird species of all. The male is adorned with brilliant colors and plumes with a diversity of shape, length, composition, and consistency that defy the basic structure of typical bird feathers. Unfortunately, this bird-of-paradise's plumes also attract hunters, who typically do not use them for traditional, local tribal ceremonies, but instead sell the feathers to Western collectors. Today, all birds-of-paradise are protected in Papua New Guinea and Australia.

The male lives for a single purpose: to hang upside down on a branch in the rainforest, where he vibrates his ornamental body and tail plumes and calls loudly, in order to attract and mate with females. The female makes the critical choice about which male she mates with, and she takes away nothing more than sperm from the chosen one.

The egg of the Blue Bird-of-Paradise is salmon or cinnamon in coloration, has reddish and tawny spotting, and measures 1½ x 1 in (39 x 25 mm) in dimensions. Aided by her green, cryptic plumage, the female remains inconspicuous throughout the breeding cycle, and parental duties are carried out solely by her.

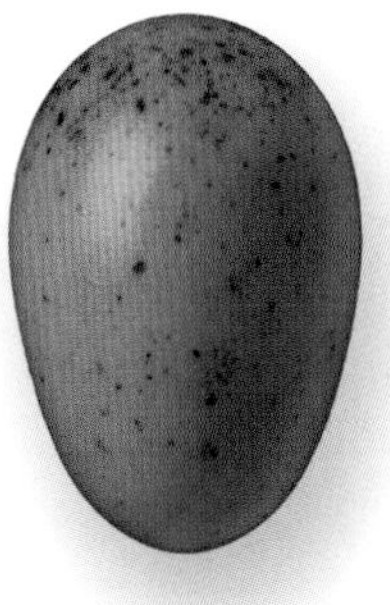

Actual size

ORDER	Passeriformes
FAMILY	Panuridae
BREEDING RANGE	Temperate Europe and Asia
BREEDING HABITAT	Wetlands with extensive reed beds
NEST TYPE AND PLACEMENT	Cup nest woven around reed stems with cobwebs, made of grass and strips of reed leaves, and lined with plant and animal fibers
CONSERVATION STATUS	Least concern
COLLECTION NO.	FMNH 20707

ADULT BIRD SIZE
5½–6½ in (14–16 cm)

INCUBATION
10–14 days

CLUTCH SIZE
4–8 eggs

PANURUS BIARMICUS

BEARDED REEDLING

PASSERIFORMES

This species is a specialist inhabitant of large, freshwater reed beds; blending in to the background of reed stems and flowers, it uses its long tail and flexible feet to balance. It preys on aphids during the summer, and alters its digestive physiology to feed on reed seeds during the winter, as it remains a resident in its home marsh or swamp year-round. In fact, juveniles often disperse just outside the immediate vicinity of their natal territory, and DNA fingerprinting has revealed that inbreeding in some of these small, closed populations occurs quite frequently.

This species adopts two different tactics for breeding. Pairs may settle on their own, or nest as part of a loose colony. Females are larger and in better condition when breeding in groups, and more often engage in extra-pair matings; they are less faithful to the pair bond when neighboring males have longer beard and tail plumes relative to their own social partner.

Clutch

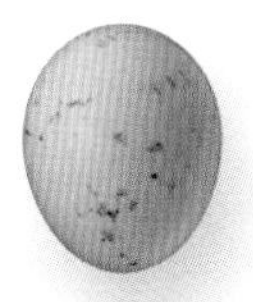

Actual size

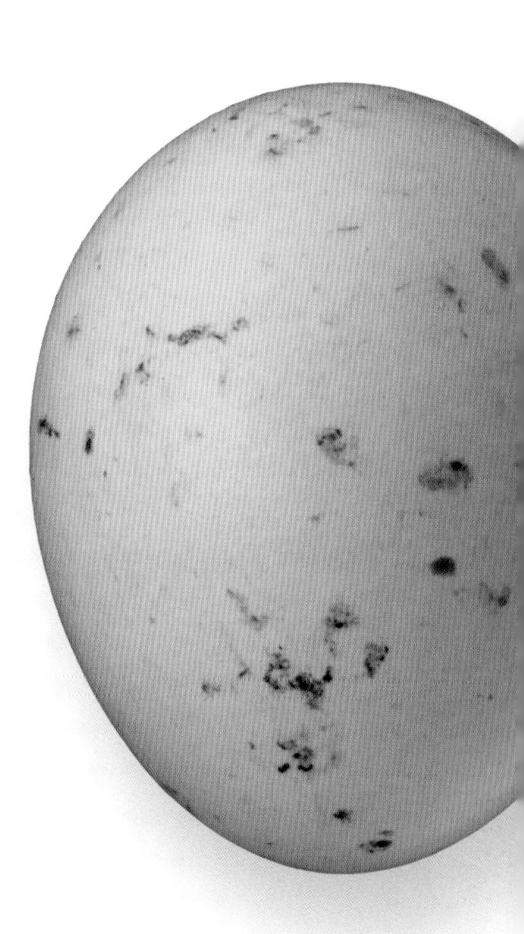

The egg of the Bearded Reedling is round in shape, white in background color, reddish in maculation, and ⅔ x 9/16 in (17 x 14 mm) in size. Colonially nesting females also seek out extra-pair fertilizations, whereas solitary nesting females are faithful to their mates.

ORDER	Passeriformes
FAMILY	Alaudidae
BREEDING RANGE	Saharan Africa and western Asia
BREEDING HABITAT	Lowland stone desert, arid rocky areas
NEST TYPE AND PLACEMENT	Ground nest, cup made of grasses, in a crevice or among rocks and pebbles
CONSERVATION STATUS	Least concern
COLLECTION NO.	FMNH 20836

ADULT BIRD SIZE
6–6½ in (15–17 cm)

INCUBATION
10–11 days

CLUTCH SIZE
3–4 eggs

AMMOMANES DESERTI

DESERT LARK

PASSERIFORMES

Clutch

The egg of the Desert Lark is grayish white in coloration, has brownish maculation, and measures ⅞ x ⅝ in (22 x 15 mm) in dimensions. The female builds the nest and she also is in charge of incubating the eggs, leaving them uncovered briefly while she feeds herself.

This species occurs in some of the most inhospitable habitats throughout the dry desert regions of Africa and the Middle East. To attract the female, the male sings on the wing, as do most lark species. Temperatures in the desert fluctuate widely, from chilly nights to scorching days, and the female chooses the location and placement of the nest to reduce the impact of these daily temperature changes on the eggs. She often places the nest under a rock ledge or dry bush for shade during the hottest part of the day, but orients it so that it will catch the morning sun and also receive breezes in the afternoon.

Feeding on seeds and insects, the Desert Lark remains a resident throughout the year, and there is little movement between its distant populations across its broad distribution. The result is that this species is divided into two-dozen subspecies, which are morphologically, and likely genetically, distinct.

Actual size

ORDER	Passeriformes
FAMILY	Alaudidae
BREEDING RANGE	Southern Europe, northern Africa and western Asia
BREEDING HABITAT	Cultivated fields, pastures, open steppe
NEST TYPE AND PLACEMENT	On the ground; hay and straw cup lined with fine grasses on the inside
CONSERVATION STATUS	Least concern
COLLECTION NO.	FMNH 20634

ADULT BIRD SIZE
7–8 in (18–20 cm)

INCUBATION
16 days

CLUTCH SIZE
4–5 eggs

MELANOCORYPHA CALANDRA

CALANDRA LARK

PASSERIFORMES

This chunky lark species vigorously defends its breeding territory by posturing, or even attacking intruders, whether other larks or potential predators, and mobbing them in flight. It prefers to forage and nest in large, continuous open spaces, making it less common in regions where agricultural practices result in small plots of land being used for diverse crops.

The male's song is varied and melodious, and this, together with its ability to mimic other birds' songs, has led to this species being kept as a popular cage bird throughout the history of Mediterranean civilizations. The male uses the song in his long circular ascending and descending display flight to attract the female, who typically listens and watches while foraging on the ground. He also makes himself more noticeable to her by displaying bright white spots on the underside of the wings, which are only visible from underneath when he is flying.

Clutch

The egg of the Calandra Lark is white with a faded green tint, blotched with dark brown to gray maculation, and is $^{15}/_{16}$ x $^{2}/_{3}$ in (24 x 18 mm) in size. The female alone builds the nest and incubates the eggs, but both parents feed the young.

Actual size

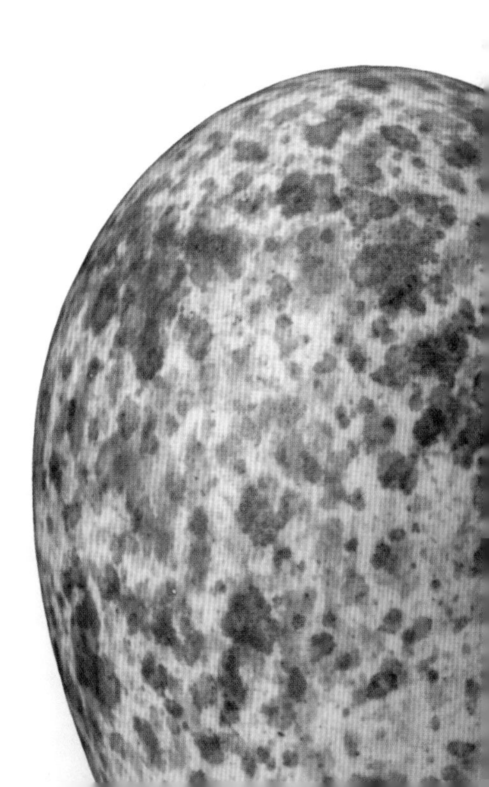

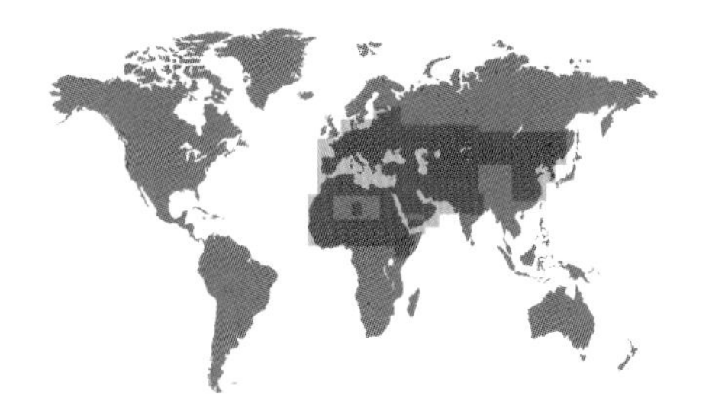

ORDER	Passeriformes
FAMILY	Alaudidae
BREEDING RANGE	Temperate Eurasia, northern Africa
BREEDING HABITAT	Dry and flat grasslands, pastures, roadsides
NEST TYPE AND PLACEMENT	Shallow cup nest on the ground, or on flat building roofs
CONSERVATION STATUS	Least concern
COLLECTION NO.	FMNH 20678

ADULT BIRD SIZE
6½ in (17 cm)

INCUBATION
11–13 days

CLUTCH SIZE
4–5 eggs

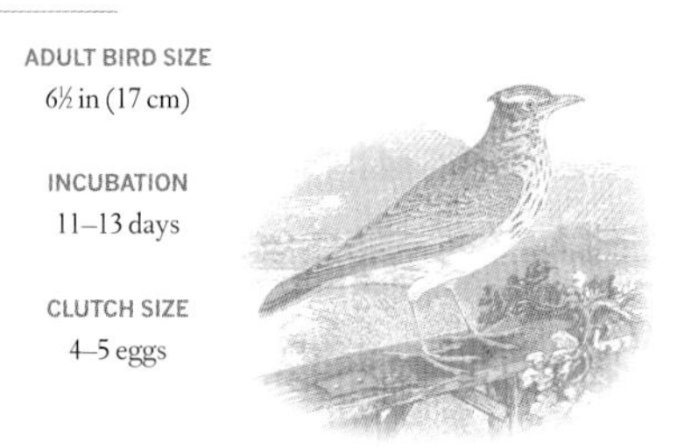

GALERIDA CRISTATA

CRESTED LARK

PASSERIFORMES

Clutch

This lark used to be familiar throughout urban and suburban areas; it easily established itself in developed lands with parks, lawns, and flat-roofed buildings, much like the Killdeer (see page 140) has done in North America. Yet urban populations of this lark have severely declined throughout Europe because of shrinking parklands, unpredictable weather, and the recent invasion of the cities by magpies, crows, and numerous other nest predators. Due to these factors, up to 40 percent of nesting attempts can be lost in any given year before the chicks fledge.

In a classic case of an "ecological trap," these larks are attracted to urban areas and can survive there; but their ability to raise young is severely hampered compared to natural areas. Research shows that nests in natural areas (at the foot of earth banks or in tufts of grass) are much more successful than those in urban parklands.

Actual size

The egg of the Crested Lark is grayish to olive in background color, with gray and brown maculation, and measures ⅞ x ⅔ in (22 x 17 mm) in size. The female alone incubates the eggs, but both parents provision the young with an insect-based diet.

ORDER	Passeriformes
FAMILY	Alaudidae
BREEDING RANGE	Europe, north Africa, and Asia
BREEDING HABITAT	Open habitats, with few or no trees, including pastures, fallow fields, managed grasslands, including airport grounds
NEST TYPE AND PLACEMENT	Cup nest on the ground; a depression in the soil, filled with coarse grasses, rootlets, and flower stems, lined with fine grasses
CONSERVATION STATUS	Least concern
COLLECTION NO.	FMNH 9433

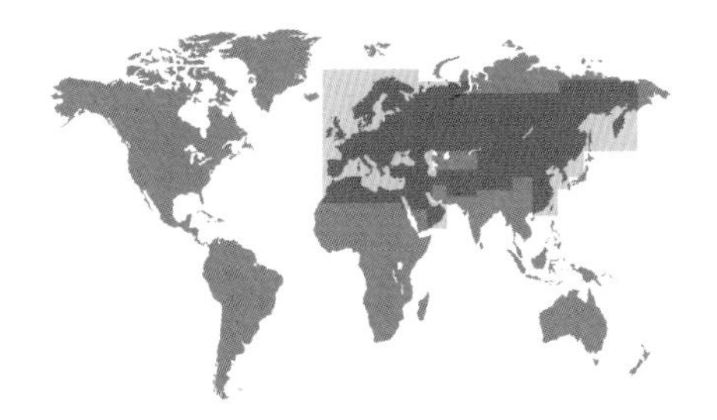

ALAUDA ARVENSIS

SKY LARK

PASSERIFORMES

ADULT BIRD SIZE
6–8 in (15–20 cm)

INCUBATION
11 days

CLUTCH SIZE
3–5 eggs

The complex and trilling songs of the Sky Lark seem to fill the air as breeding males slowly ascend in steep spirals up to 330 ft (100 m), circle repeatedly while continuing to sing, then descend slowly back to the ground. A single display can last for five minutes. With this display song, the male both attracts females and defends its breeding territory. Once a male has established a territory and a pair bond, the female selects her nest site, and loosens the soil into a depression before building a grass cup, in an exposed, grassy area.

The Sky Lark is common in agricultural and grazing fields, as well as city parks of Europe. So beloved is the Sky Lark's aerial display song that Europeans settling on other continents repeatedly introduced the bird to their new homes, including Australia, New Zealand, Hawaii, and British Columbia, with variable degrees of success.

Clutch

Actual size

The egg of the Sky Lark is cream in coloration, has heavy dark brown maculation, and measures ⅞ x ⅔ in (22 x 17 mm) in dimensions. Only the female incubates, and all the chicks hatch synchronously, within 8–10 hours of each other.

ORDER	Passeriformes
FAMILY	Alaudidae
BREEDING RANGE	North Africa, Europe, western Asia
BREEDING HABITAT	Open heaths, scattered woodlands, often with pines in sandy soils, also farmland
NEST TYPE AND PLACEMENT	Cup nest made of grass, on the ground, placed in an open field, or forest clearing
CONSERVATION STATUS	Least concern
COLLECTION NO.	FMNH 20672

ADULT BIRD SIZE
5½–6 in (14–15 cm)

INCUBATION
13–14 days

CLUTCH SIZE
4–6 eggs

LULLULA ARBOREA

WOOD LARK

PASSERIFORMES

Clutch

As a year-round resident in much of its range, nesting begins early in the spring, and after a successful first brood, there is often sufficient time for a second and even a third clutch of eggs. The timing of the onset of breeding, and the effort invested into laying eggs, are closely influenced by the climatic conditions during the previous winter; larger clutches are usually preceded by warmer and drier winters.

However, as in most bird species, nesting success is most dramatically influenced by predation, rather than by weather conditions directly. In some years, predation reduces productivity by as much as 50 percent. While Wood Larks tend to avoid disturbed areas (in contrast to the Crested Lark, see page 442, which is drawn to them), their nesting attempts are equally successful whether the area is disturbed or nondisturbed. So for these larks, local land use, predator population cycles, and climate all have some influence on their decisions about where to settle and where to breed.

Actual size

The egg of the Wood Lark is grayish or buff, has heavy dark brown maculation, with a denser ring around the blunt pole in some cases, and its dimensions are ¾ x 9/16 in (19 x 14 mm). Both sexes build the nest, but only the female incubates the eggs; both parents provision the chicks.

ORDER	Passeriformes
FAMILY	Alaudidae
BREEDING RANGE	Most of the northern hemisphere, except for central and southeastern Asia and sub-Saharan Africa
BREEDING HABITAT	Low vegetation, open country, semi-deserts, arid montane meadows, agricultural fields and pastures
NEST TYPE AND PLACEMENT	In a cavity or deep depression in the ground, woven basket of fine stems and grasses, lined with softer grasses and fibers
CONSERVATION STATUS	Least concern
COLLECTION NO.	FMNH 19090

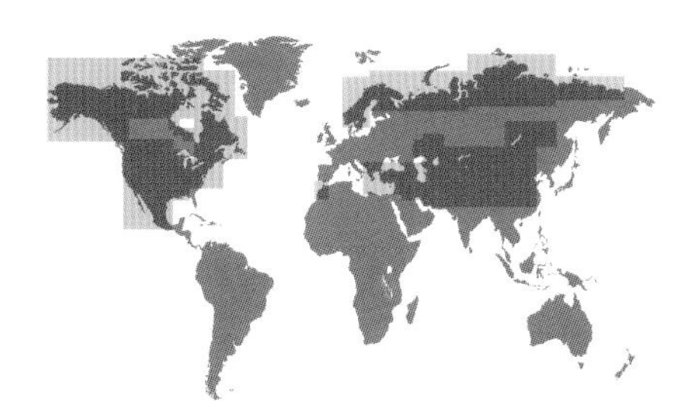

ADULT BIRD SIZE
6½–8 in (16–20 cm)

INCUBATION
10–12 days

CLUTCH SIZE
2–5 eggs

EREMOPHILA ALPESTRIS

HORNED LARK

PASSERIFORMES

The Horned Lark, known as the Shore Lark in Europe, is a circumpolar species and the only true lark native to the continental New World. As with most other larks, it sings on the wing, as the males demarcate their territory and attract females. Following pair bonding, the female swiftly seeks out a suitable nest site, typically in the open grass, under tufts of stems and stalks of plants. There, the female uses her body to loosen the soil, and digs a nesting cup in the ground.

Clutch

Throughout much of the year, the species consumes seeds and shoots of grasses and other plants. But once the eggs hatch, the adults move to a diet of insects and other arthropods, and feed these to the nestlings. Both parents provision the young, although the female also spends much of her day brooding the hatchlings, and single mothers have been known to raise full broods to fledging.

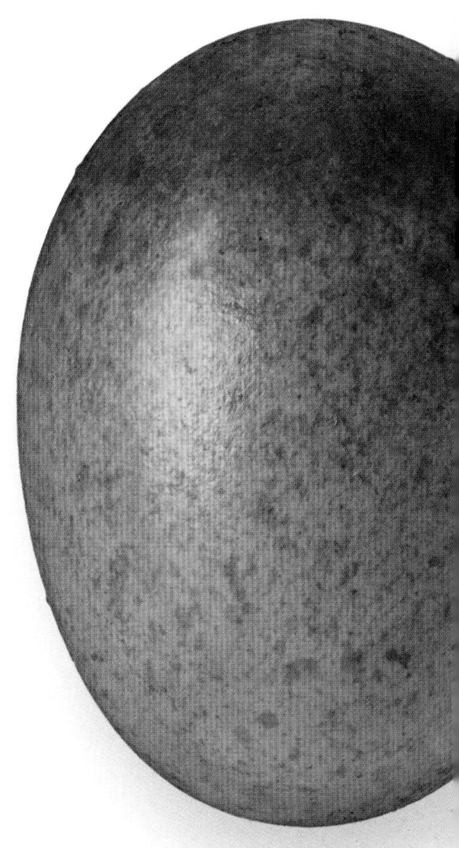

Actual size

The egg of the Horned Lark is pale gray in color, with very fine rusty maculation, and is ⅞ x ⅔ in (22 x 17 mm) in size. Once the female begins to incubate, she is occasionally fed by the male, but mostly she leaves the nest briefly to feed on her own.

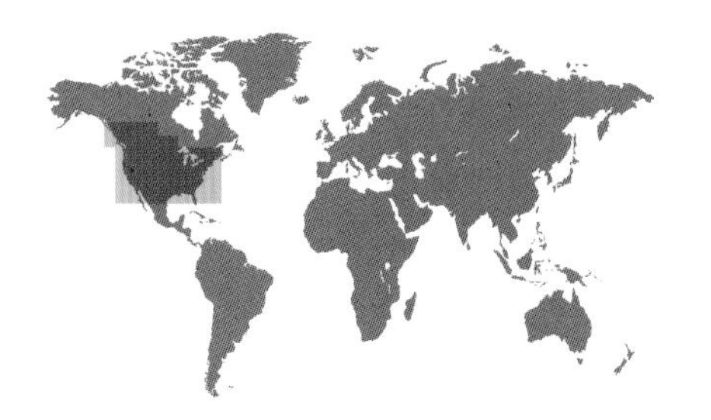

ORDER	Passeriformes
FAMILY	Hirundinidae
BREEDING RANGE	Temperate North America
BREEDING HABITAT	Open fields, valleys, and hillsides near rivers
NEST TYPE AND PLACEMENT	Cavity breeder, nests in burrow dug into soil banks or in artificial tubing; nesting chamber lined with stems, rootlets, and grasses
CONSERVATION STATUS	Least concern
COLLECTION NO.	FMNH 13697

ADULT BIRD SIZE
5–6 in (13–15 cm)

INCUBATION
16–18 days

CLUTCH SIZE
3–6 eggs

STELGIDOPTERYX SERRIPENNIS

NORTHERN ROUGH-WINGED SWALLOW

PASSERIFORMES

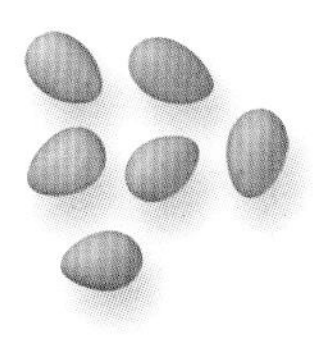

Clutch

A drab and inconspicuous aerial hunter, this swallow is not colonial; hence it is less well known than many of the other North American swallows that typically appear in large and conspicuous flocks with the arrival of springtime. Like other swallows, however, the rough-winged swallow also has adapted to nesting in and around harbors and other human development along waterways where it can find holes and crevasses in human-built structures in which to nest.

There is debate about whether this species digs new cavities or just usurps existing ones; what is known is that a new nest cup is built each time, often on top of a preexisting nest structure in the end chamber of the burrow. Only the female is in charge of incubation, but both parents take turns in provisioning the young; the male often begins to visit the nest several days after the chicks hatch.

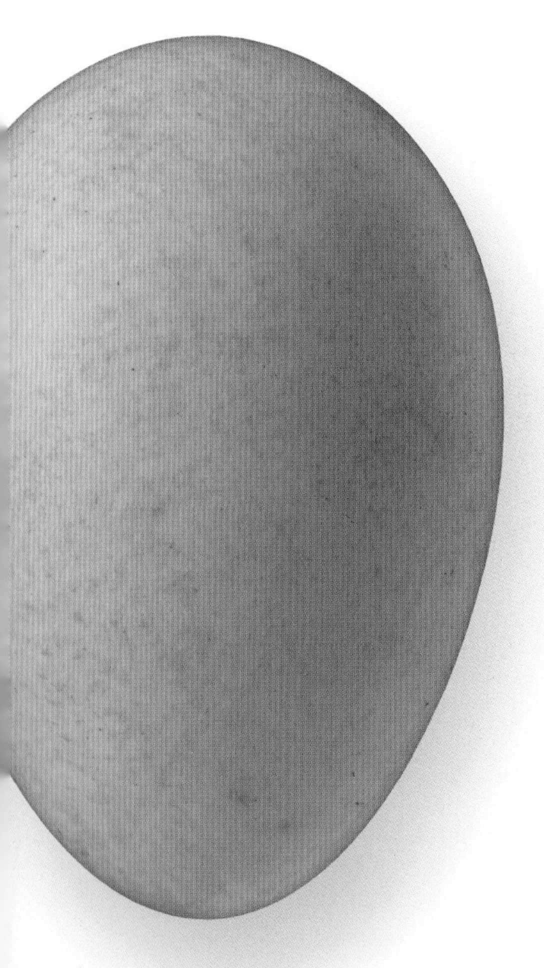

Actual size

The egg of the Northern Rough-winged Swallow is glossy white in background, clear of maculation, and ¾ x 9⁄16 in (20 x 14 mm) in dimensions. The female spends the night on the nest incubating from the time the first egg is laid; this yields a hatching sequence of eggs across several hours to days.

ORDER	Passeriformes
FAMILY	Hirundinidae
BREEDING RANGE	Temperate and subtropical North America
BREEDING HABITAT	Forest edges, riparian woodlands, city parks and gardens
NEST TYPE AND PLACEMENT	Nest platform made of twigs, plant stems, and mud, lined with grass, bark, and leaves; placed inside a cavity, in cliffs and trees, old holes excavated by large woodpeckers, or inside bird houses
CONSERVATION STATUS	Least concern
COLLECTION NO.	FMNH 13670

ADULT BIRD SIZE
7–8⅔ in (18–22 cm)

INCUBATION
15–18 days

CLUTCH SIZE
3–6 eggs

PROGNE SUBIS

PURPLE MARTIN

PASSERIFORMES

Originally a solitary nesting species, most Purple Martins today breed in colonial structures; this has been encouraged by the construction of artificial martin houses, which are erected in backyards and front lawns across North America. Unlike boxes for Eastern Bluebirds (see page 494) and Tree Swallows (see page 448), Purple Martin houses must have entrances that are big enough for the somewhat larger martins. But these larger holes also allow Common Starlings (see page 515) to enter the cavity; this introduced species is a fierce nest competitor and will even kill the martin adults, as well as destroy eggs or chicks, which stand in their way for a cavity take-over.

Despite the costs of group-nesting (increased aggression from neighbors, greater conspicuousness to predators, and competition for food) nesting in groups has been embraced by these martins. It may have increased extra-pair mating, as females who are socially mated to younger males often go outside the pair bond and obtain extra-pair fertilizations from older colony males.

Clutch

The egg of the Purple Martin is pure white and $\frac{15}{16}$ x $\frac{2}{3}$ in (24 x 17 mm) in dimensions. The female is in charge of constructing most of the nest, but the male also brings in fresh, green leaves before the eggs are laid.

Actual size

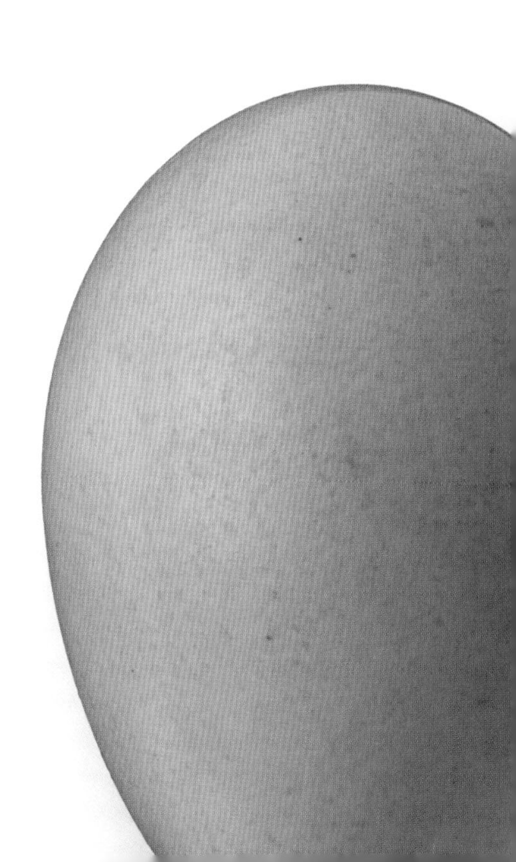

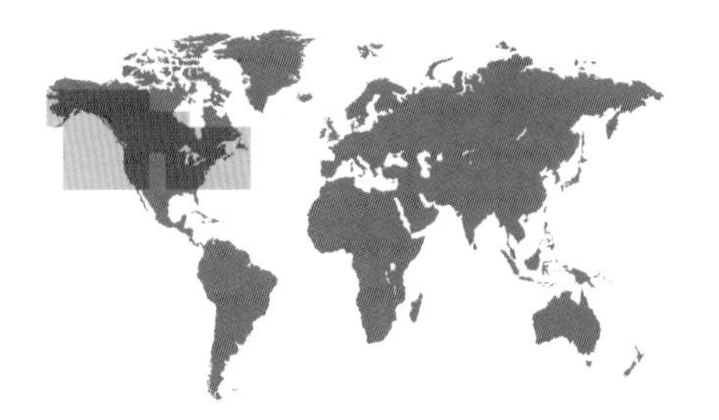

ORDER	Passeriformes
FAMILY	Hirundinidae
BREEDING RANGE	Temperate North America
BREEDING HABITAT	Open fields, low-vegetation wetlands, pastures; typically near water
NEST TYPE AND PLACEMENT	Natural cavities or old woodpecker holes in standing dead trees, also nestboxes; grass nest bowl lined with many large feathers
CONSERVATION STATUS	Least concern
COLLECTION NO.	FMNH 13610

ADULT BIRD SIZE
4½–6 in (12–15 cm)

INCUBATION
11–20 days

CLUTCH SIZE
4–7 eggs

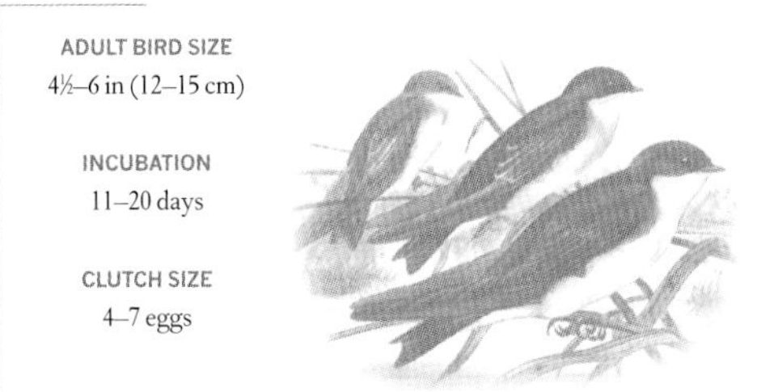

TACHYCINETA BICOLOR

TREE SWALLOW

PASSERIFORMES

Clutch

This species is a common backyard bird throughout much of North America. Its nesting habitats have been vastly extended by people placing nestboxes to attract bluebirds; these boxes are also suitable for the swallows. Established pairs tend to return to the site where they nested before to breed, whereas young birds may be the ones who discover new nestboxes in habitats lacking natural stands of mature trees with cavities.

Unlike many bird species, where the males take several years to acquire the adult plumage, sexually mature male Tree Swallows already display their shiny, green-blue iridescent plumes. Females in their second year of life remain brownish, and only acquire the shiny green plumage in their third year. Irrespective of her plumage, female Tree Swallows are sexually adventurous and most clutches contain eggs sired by males different from the mother's social mate.

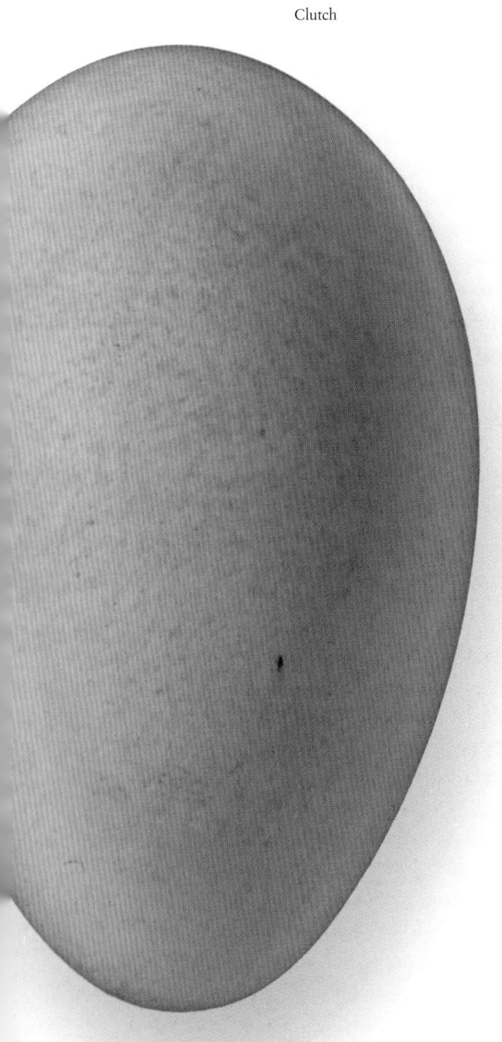

Actual size

The egg of the Tree Swallow is bright, pure white in coloration, and can be translucent at laying; it is immaculate and measures ¾ x 9⁄16 in (19 x 14 mm) in size. The eggs of this swallow have been used extensively in environmental toxicology research: swallows breeding near more acidic lakes have more mercury in their bodies and lay smaller eggs.

ORDER	Passeriformes
FAMILY	Hirundinidae
BREEDING RANGE	North Africa, Europe, northern and Central Asia, North America
BREEDING HABITAT	Lowland coastal areas, open fields along rivers, lakes, and reservoirs
NEST TYPE AND PLACEMENT	Excavated tunnel ending in a chamber, with straw and feathers as lining; in vertical buffs, quarries, or sand or gravel river banks
CONSERVATION STATUS	Least concern
COLLECTION NO.	FMNH 13690

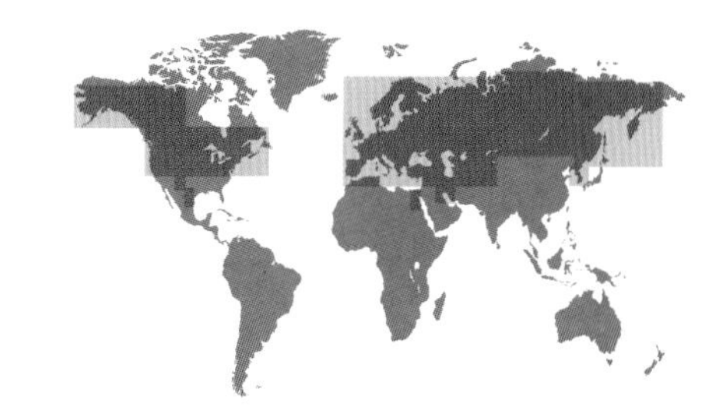

ADULT BIRD SIZE
4½–5¼in (12–13 cm)

INCUBATION
12–16 days

CLUTCH SIZE
4–5 eggs

RIPARIA RIPARIA

BANK SWALLOW

PASSERIFORMES

Bank Swallows, as they are known in North America, or Sand Martins in Europe, are some of the most gregariously breeding songbirds; colonies in natural sand banks or gravel quarries range from five to 3,000 pairs. The sheer density of tunnels dug into a vertical soil face often results in a collapse and erosion of the nesting site between years, or even during a single breeding season. The birds benefit from prospecting for other suitable colonies where they can restart the breeding process if their original nesting site has collapsed or flooded.

The male establishes a small territory around a tunnel opening on the sand bank, and flies around it to attract a female. When one is interested, she inspects the burrow by hovering in front of it. Once the pair bond is formed, the male completes the tunnel, the female builds the nest, and both sexes incubate the eggs and feed the young, although it is typically the female who spends the night on the eggs. The parents continue feeding the young after they become mobile and leave the natal nest. To avoid wasting food, the parents rely on the chicks' vocal signatures to recognize them among the other fledglings in the colony.

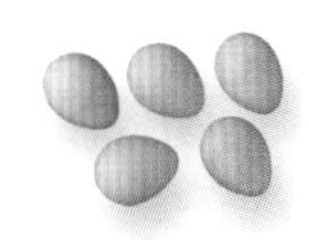

Clutch

The egg of the Bank Swallow is glossy white in background color, with no maculation, and is ⅔ x ½ in (18 x 13 mm) in size. Typically, new burrows are excavated each spring; if an old tunnel is reused, the birds remove the old nest, and put in a new lining before the eggs are laid.

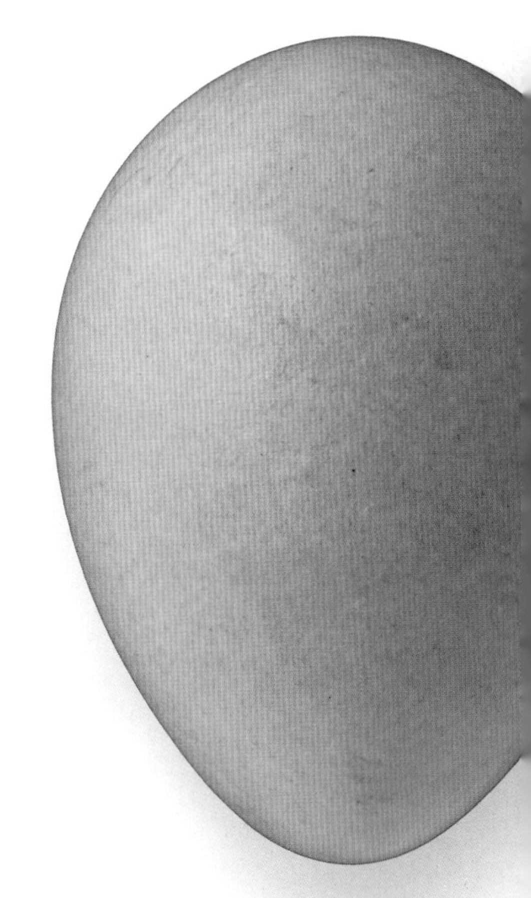

Actual size

ORDER	Passeriformes
FAMILY	Hirundinidae
BREEDING RANGE	Temperate North America, Buenos Aires, Eurasia, north Africa
BREEDING HABITAT	Open fields, farmland, pastures, punctuated by buildings, barns, bridges, and road culverts
NEST TYPE AND PLACEMENT	Mud cup built on a ledge of a natural cliff, most commonly under the eaves of a porch and bridge overhangs, often reused across years; inside lined with grasses and feathers
CONSERVATION STATUS	Least concern
COLLECTION NO.	FMNH 1464

ADULT BIRD SIZE
6–7½ in (15–19 cm)

INCUBATION
12–17 days

CLUTCH SIZE
3–7 eggs

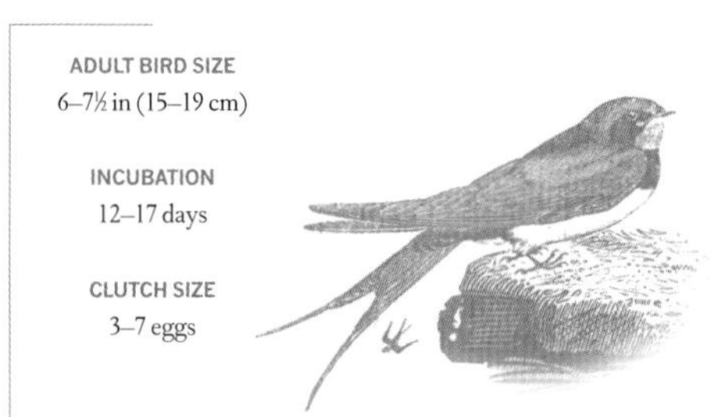

HIRUNDO RUSTICA

BARN SWALLOW

PASSERIFORMES

Clutch

The Barn Swallow occurs in all inhabited continents of the globe, although it is mostly a migratory bird in the southern hemisphere. The exception is a small population that established itself near Buenos Aires, Argentina, from North American stock, and has fully adapted to breeding in the southern hemisphere's spring time.

Across their vast range, Barn Swallows show extensive variability in their plumage patterns, from white to rusty chest feathers and from shorter to longer tail streamers. In Europe, males with long tail streamers do not often incubate the eggs, but they valiantly defend them from nearby males. By contrast, males in North America have shorter tails and spend more time sitting on the eggs. Females in turn pay close attention to these traits, and appear to use different combinations of these variable cues to choose their mates.

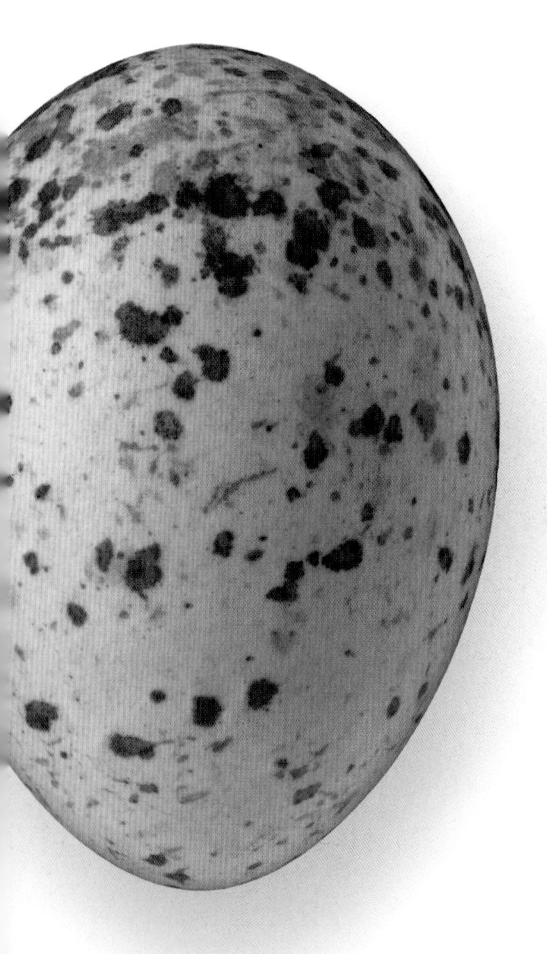

Actual size

The egg of the Barn Swallow is white in background, has sparse reddish, brown, or purple maculation, and measures ¾ x 9⁄16 in (19 x 14 mm). Experienced breeders return to lay eggs at the same site where they bred before; young birds in turn use the presence of old swallow nests to make their decisions to settle and breed for the first time.

ORDER	Passeriformes
FAMILY	Hirundinidae
BREEDING RANGE	Temperate and boreal North America
BREEDING HABITAT	Open canyons, foothills, open areas near water
NEST TYPE AND PLACEMENT	Covered bowl, with a side entrance tunnel, attached to cliffs, bridge overhangs, culverts, and other artificial structures; made of mud pellets, lined with grass
CONSERVATION STATUS	Least concern
COLLECTION NO.	FMNH 3159

ADULT BIRD SIZE
5–6 in (13–15 cm)

INCUBATION
13–17 days

CLUTCH SIZE
3–6 eggs

PETROCHELIDON PYRRHONOTA

CLIFF SWALLOW

PASSERIFORMES

Cliff Swallows are known for their colonial breeding, both because of the sheer number of their mud nests pasted against each other underneath midwestern highway overpasses, and because the most important decision prior to mating in this species is to gauge the distribution and suitability of different-sized colonies in the spring. Early experience matters critically for these birds: those individuals that were hatched in small or large colonies tend to prefer to settle in small or large colonies, respectively.

Once the colony site is chosen, the males quickly claim ownership of an old nest, or start building a new nest, even if they have not yet been chosen by a female. The females tend to spend a few days prospecting between suitable colonies, then quickly settle and find a mate. Both sexes help to complete the nest, incubate the eggs, and feed the young.

Clutch

Actual size

The egg of the Cliff Swallow is creamy white in coloration, spotted throughout with light and dark brown speckles, and measures ¾ x 9⁄16 in (20 x 14 mm). Female Cliff Swallows are known to occasionally carry an egg in their bill to move it from their own nest into another female's nest as a form of intraspecific brood parasitism.

ORDER	Passeriformes
FAMILY	Paridae
BREEDING RANGE	Southeastern and central North America
BREEDING HABITAT	Hardwood forests, riparian, wetland and coastal woodlands
NEST TYPE AND PLACEMENT	Excavated new cavity in rotting wood, or repurposed existing cavities; cup nest made of grasses, stems, and lined with hair, fur, and soft fibers
CONSERVATION STATUS	Least concern
COLLECTION NO.	FMNH 19174

ADULT BIRD SIZE
4–5 in (10–13 cm)

INCUBATION
12–15 days

CLUTCH SIZE
3–10 eggs

POECILE CAROLINENSIS

CAROLINA CHICKADEE

PASSERIFORMES

Clutch

A common species of southeastern North America, many people confuse this bird with the similar Black-capped Chickadee (see facing page). Despite shared behaviors and plumage traits, their distinct mating songs are still a solid characteristic to tell them apart in regions where they both occur. Sometimes, however, even the birds appear to get it wrong, because they are known to hybridize.

The Carolina Chickadee spends its winters in flocks formed by nearby breeding pairs and their fledged young. There is thus a continued association of pairs from the previous breeding season, and so, unless predation or weather causes mortality, surviving mates often return together to the breeding territory to nest in subsequent years. During the breeding season, the male is both aggressive against intruding males, and follows the female closely from nest site selection to the onset of incubation. The male occasionally feeds the female in the nest, and both parents provision the young.

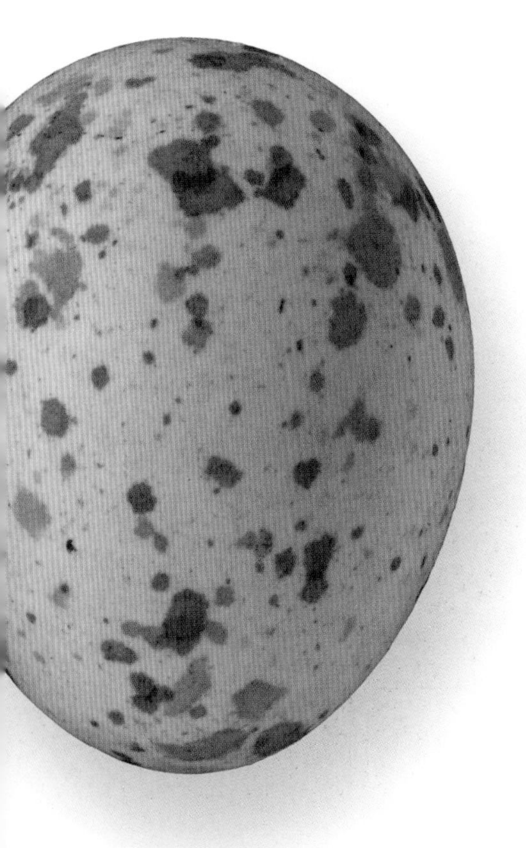

Actual size

The egg of the Carolina Chickadee is white in background color, has reddish brown maculation, and measures ⅝ x ½ in (15 x 12 mm) in size. The female is in charge of most of the nest excavation, all of the nest cup construction, and also of the incubation.

ORDER	Passeriformes
FAMILY	Paridae
BREEDING RANGE	Northern and boreal North America
BREEDING HABITAT	Deciduous and mixed woodlands, open forests and edges, parks, and gardens
NEST TYPE AND PLACEMENT	Small natural cavities, or excavated into soft wood, reused woodpecker cavities; cup nest made of grasses, leaves, and rootlets, and lined with moss, cobwebs, and fibers
CONSERVATION STATUS	Least concern
COLLECTION NO.	FMNH 19140

ADULT BIRD SIZE
4½–6 in (12–15 cm)

INCUBATION
12–13 days

CLUTCH SIZE
6–8 eggs

POECILE ATRICAPILLUS

BLACK-CAPPED CHICKADEE

PASSERIFORMES

Black-capped Chickadees are familiar backyard birds, and one of the more common visitors to suet cakes and feeders stocked with sunflower seeds. Wherever there is a temporary abundance of food, especially in the fall, chickadees store the surplus in crevices, cracks, and holes in the soil; then, in the winter, they rely on their enlarged hippocampus, the brain area responsible for spatial memory, to remember and recover their hoarded food stocks.

Singing contests between males are carefully monitored by females; when a female is mated to a male whose song is often challenged, she seeks out nearby males to fertilize many of her eggs. In contrast, when she is mated to a male with a dominant song, she remains genetically monogamous during the breeding season. When researchers experimentally challenged the dominance of a male, by using interactive playbacks, the females treated these males as if they had lost their dominance and sought other males to sire their young.

Clutch

Actual size

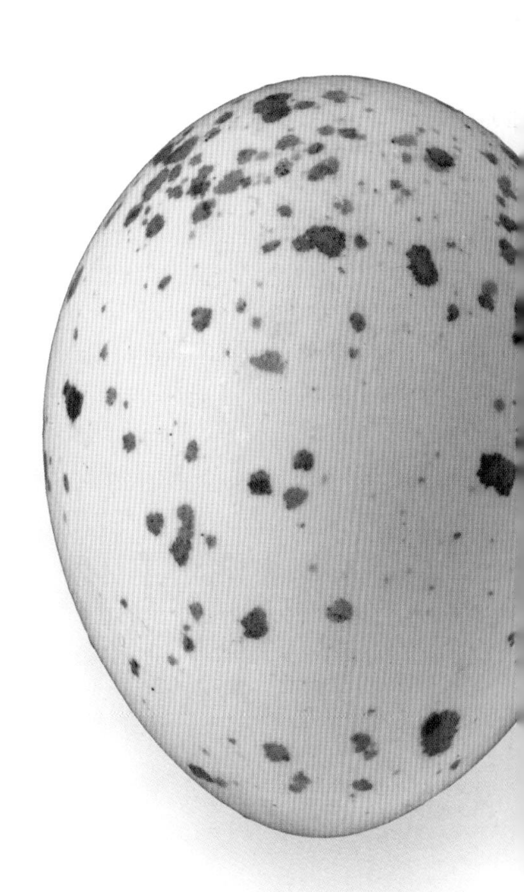

The egg of the Black-capped Chickadee is white with reddish brown speckling, and is ⅝ x ½ in (15 x 12 mm) in dimensions. Eggs in cavities that are excavated in rotting tree stumps are often lost to predators tearing the nest wall open; nests in cavities excavated originally by woodpeckers have thicker walls and are safer.

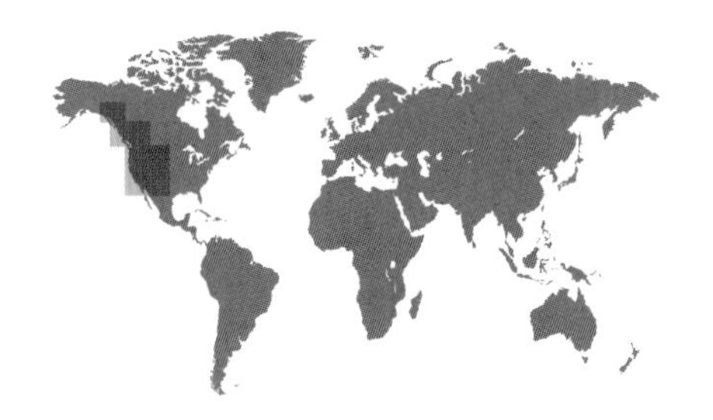

ORDER	Passeriformes
FAMILY	Paridae
BREEDING RANGE	Western North America
BREEDING HABITAT	Montane, coniferous forests
NEST TYPE AND PLACEMENT	Cavity nest, in existing holes of trees, or rarely in the ground and between roots; cup nest of wet bark, lichen, mosses, with a lining of soft animal hair and fur
CONSERVATION STATUS	Least concern
COLLECTION NO.	FMNH 11395

ADULT BIRD SIZE
4½–5½ in (12–14 cm)

INCUBATION
12–15 days

CLUTCH SIZE
5–7 eggs

POECILE GAMBELI

MOUNTAIN CHICKADEE

PASSERIFORMES

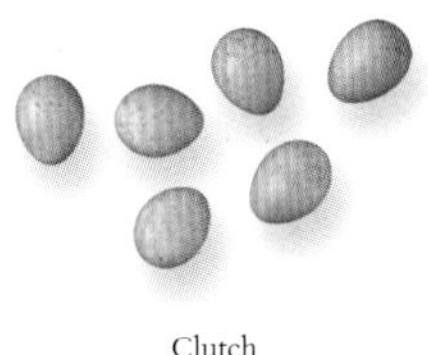

Clutch

This species is a year-round resident in its high-elevation, evergreen habitat; in the cold and snowy winter, it relies on stored food stashes for nourishment. Because many of the seeds are stored in relatively accessible locations, a fine-scale memory is not always needed to recover these stocks, and so, instead, it becomes critical to defend the territory from trespassers and intruders. Winter group territoriality also allows mates to remain with each other, or new pairs to form in late fall and soon after. Once the weather warms up, and springtime is near, the established pairs break away and start defending a breeding territory around their preferred nesting cavity.

Nest construction is shared between the parents, as is provisioning the young. However, most of the nesting material is gathered by the female, both before and after incubation begins, whereas the males appear to bring nesting materials more often after the eggs have hatched.

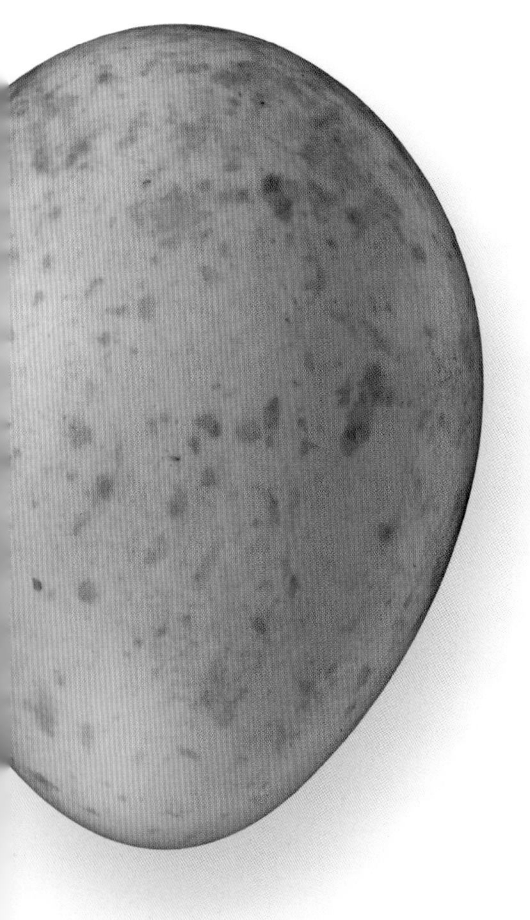

Actual size

The egg of the Mountain Chickadee is white in background color, has small red spotting, and measures ⅔ x ½ in (16 x 12 mm). If the first nest fails, re-nesting attempts and second clutches are initiated in different cavities from the first; but, generally, nesting holes are reused from year to year.

ORDER	Passeriformes
FAMILY	Paridae
BREEDING RANGE	Southwestern North America
BREEDING HABITAT	Oak and oak-pine forests, riparian woodlands in montane areas
NEST TYPE AND PLACEMENT	Cavities in trees, also in nestboxes; inside, a cup of grass, cottonwood down, flowers, fur, and cocoons, lined with soft fibers
CONSERVATION STATUS	Least concern
COLLECTION NO.	FMNH 11335

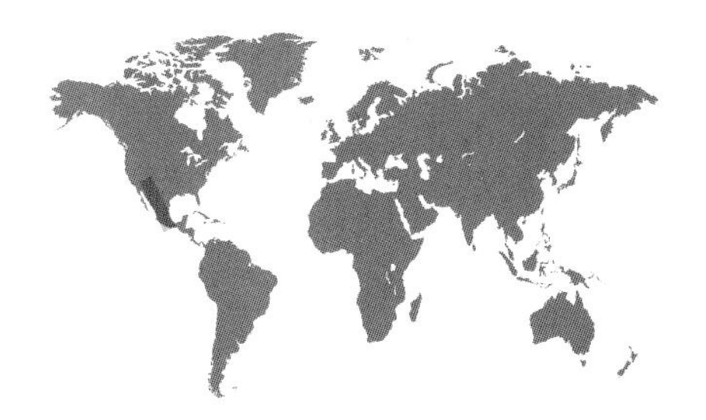

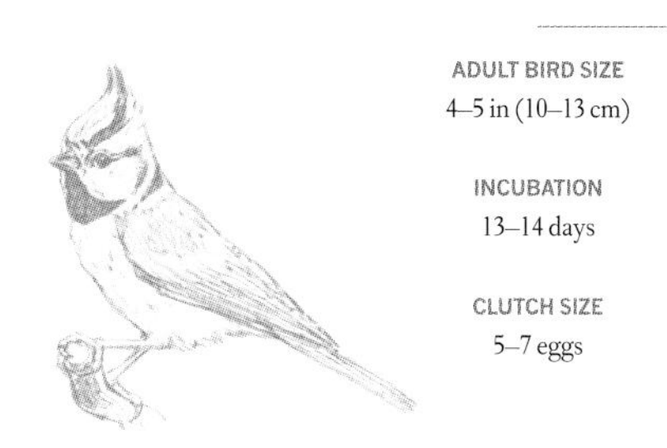

ADULT BIRD SIZE
4–5 in (10–13 cm)

INCUBATION
13–14 days

CLUTCH SIZE
5–7 eggs

BAEOLOPHUS WOLLWEBERI

BRIDLED TITMOUSE

PASSERIFORMES

Like other titmice and chickadees, this species shares a social structure of winter flocking and summer pair bonding. During the non-breeding season, in its montane forest habitats, other species join the winter flocks of bridled titmice, and these mixed-species groups move and forage together, while keeping watch for potential predators. Once the winter is over, the flock breaks apart, and pairs establish and defend exclusive feeding and nesting territories.

Clutch

Titmice do not excavate cavities, but instead select and clear out debris, bark, and old nesting material from existing holes before laying their own eggs. Females typically stay with their pair-bonded mate from the previous year; the pair is occasionally assisted in feeding the young by an adult helper. The helper also participates in keeping the nest clean by removing fecal sacs produced by the nestlings, and keeps the territory safer by mobbing predators.

Actual size

The egg of the Bridled Titmouse is pure white in color, immaculate, and ⅔ x ½ in (16 x 12 mm) in size. Typically, only the female sits on the eggs, but some males also develop a partial brood patch, and may be able to contribute to incubation.

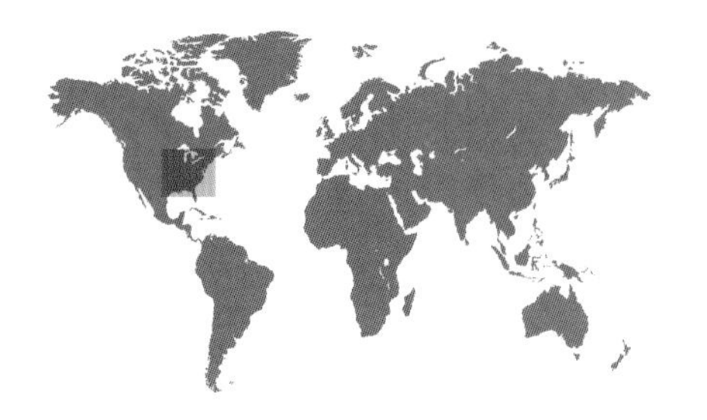

ORDER	Passeriformes
FAMILY	Paridae
BREEDING RANGE	Eastern and southeastern North America
BREEDING HABITAT	Mature deciduous and mixed coniferous forests, parkland, suburban backyards
NEST TYPE AND PLACEMENT	Inside a natural or an abandoned woodpecker cavity; cup-shaped, built from damp leaves, moss, grasses, and bark strips; lined with soft hair and plant fibers
CONSERVATION STATUS	Least concern
COLLECTION NO.	FMNH 11312

ADULT BIRD SIZE
5½–6½ in (14–16 cm)

INCUBATION
12–14 days

CLUTCH SIZE
5–6 eggs

BAEOLOPHUS BICOLOR

TUFTED TITMOUSE

PASSERIFORMES

Clutch

Tufted Titmice are common visitors to backyard bird feeders throughout their range, and their recent expansion toward more northerly breeding areas may be a direct result of reliable, people-provided food sources available throughout the winter. As year-long residents, these birds live in a fusion-fission society, where pairs and their young join winter flocks, only to disband during the early spring when the dominant pair in the group establishes its territory in the middle of the winter home range.

To breed, these birds rely on natural cavities or holes carved and abandoned by woodpeckers. There the pair builds the cup nest together, but then only the female incubates the eggs. The male feeds the female at and away from the nest while she sits or takes short breaks from incubating to find water and to preen and relieve herself.

Actual size

The egg of the Tufted Titmouse is white to cream in coloration, with chestnut-red, brown, lavender, or purple maculation, and is ¾ x 9⁄16 in (19 x 14 mm) in dimensions. The eggs are laid on a soft cup of hair and wool, often plucked by the birds from grazing mammals.

ORDER	Passeriformes
FAMILY	Remizidae
BREEDING RANGE	Southwestern North America
BREEDING HABITAT	Desert scrub, arid open woodland
NEST TYPE AND PLACEMENT	Large sphere of sticks with a hole usually located near the bottom, inside lined with leaves and smaller twigs; placed in a shrub
CONSERVATION STATUS	Least concern
COLLECTION NO.	FMNH 11292

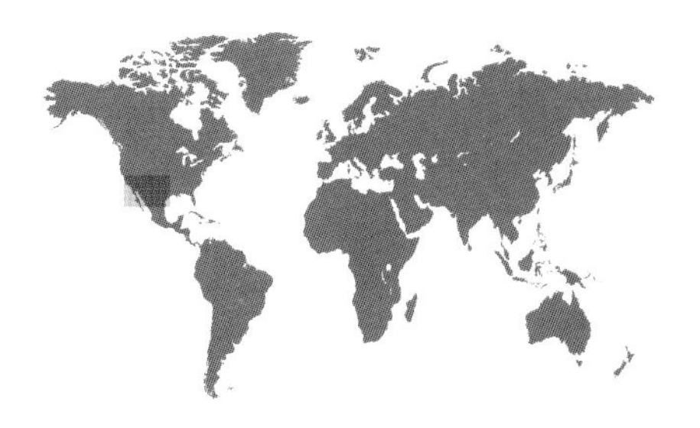

AURIPARUS FLAVICEPS

VERDIN

PASSERIFORMES

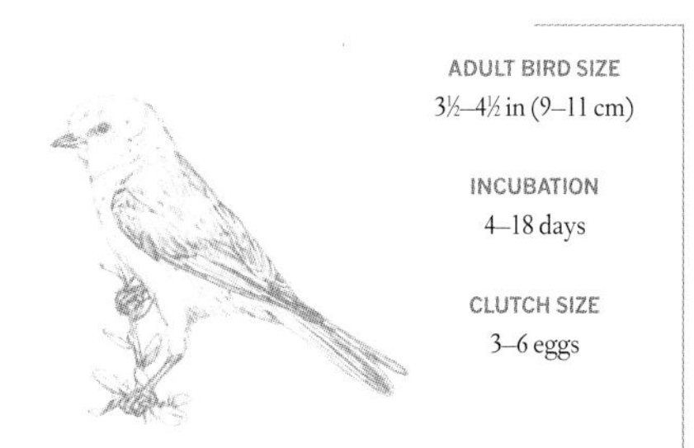

ADULT BIRD SIZE
3½–4½ in (9–11 cm)

INCUBATION
4–18 days

CLUTCH SIZE
3–6 eggs

Solitary during most of the year, these birds pair up for the breeding season after the males start showing increasing aggression toward conspecific males, and begin to defend and maintain a territory against neighboring males. The male may also build one or more display nests, one of which is then completed by both members of the pair. Nests are prominently placed on the edge of a shrub, and not hidden in dense foliage. The eggs are safely tucked inside the soft lining of the enclosed breeding nest. These birds also build smaller roosting nests where they keep warm during the night throughout the year.

This small, desert-dwelling songbird is the sole North American representative of a unique Old World family that includes the penduline tits. Verdins share their delicate size and shape, and their complex nest construction behaviors, with these Old World relatives.

Clutch

Actual size

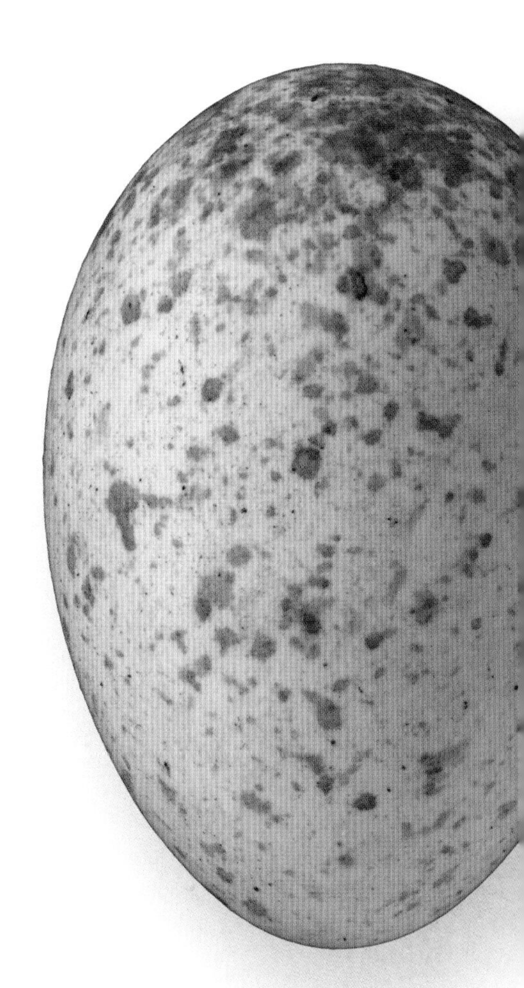

The egg of the Verdin is light greenish in background color, has irregular dark reddish spotting, and measures ⅔ x 7⁄16 in (16 x 11 mm). If a first clutch is lost to predators or weather, the materials from roosting nests and other pairs' main nests are often borrowed to quickly build replacement nests in which to lay a new set of eggs.

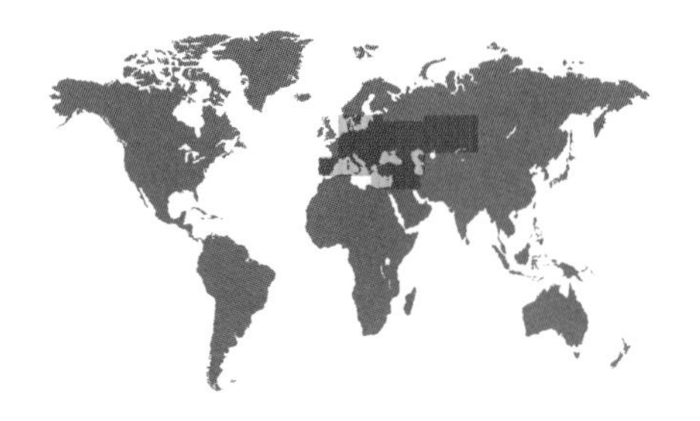

ORDER	Passeriformes
FAMILY	Remizidae
BREEDING RANGE	Eurasia, North Africa
BREEDING HABITAT	Open country, near lakes, slow-moving rivers and deltas, scattered with trees, bushes, and reeds
NEST TYPE AND PLACEMENT	Woven basket of spiderwebs, wool, and other animal hair, soft plant fibers, suspended from the ends of twigs and branches in trees, such as weeping willows
CONSERVATION STATUS	Least Concern
COLLECTION NO.	FMNH 20700

ADULT BIRD SIZE
3–4½ in (8–11 cm)

INCUBATION
13–14 days

CLUTCH SIZE
6–10 eggs

REMIZ PENDULINUS

EURASIAN PENDULINE TIT

PASSERIFORMES

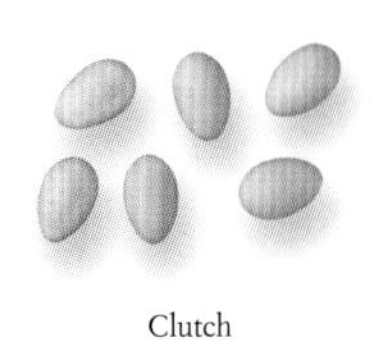

Clutch

This species is apparently a typical, widespread small songbird, seen and heard by many in its habitat around lakes and rivers. Yet scientists following individually banded birds have discovered that it has an extraordinary family life. Pair formation starts with males establishing a territory and building an elaborate nest. This is then completed and lined by the female, until she begins to lay her eggs. However, one member of the pair often deserts the nest and seeks another mate soon after laying, so that either the female or the male can end up incubating the eggs, and raising the young, alone.

There is an inherent risk in a breeding strategy where both sexes may desert the pair bond and parental duties. In up to a quarter of cases, both parents desert the nest. Deserted nests represent a total failure of both the male's investment into nest building and the female's energy expenditures into laying eggs.

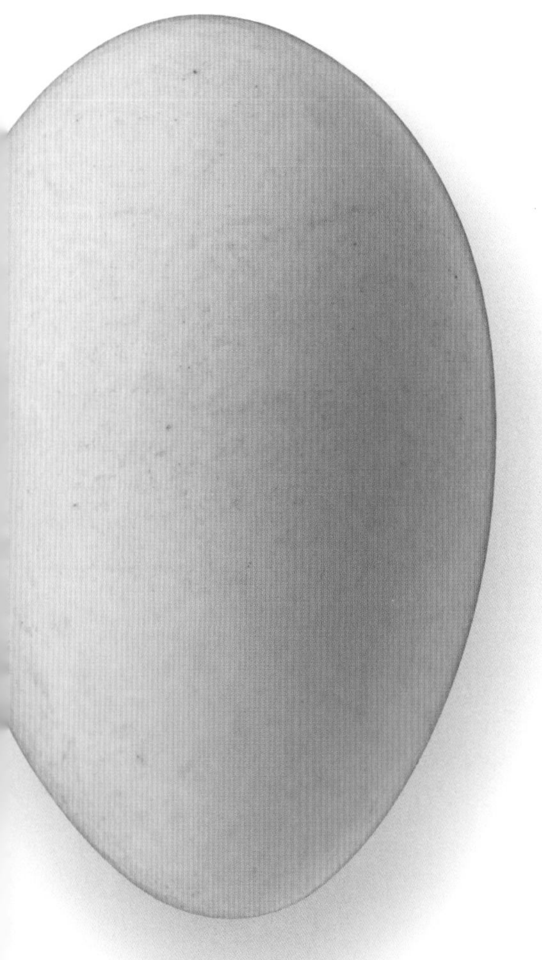

Actual size

The egg of the Eurasian Penduline Tit is white, cream, or beige in background color, immaculate, and ⅔ x ⅜ in (16 x 10 mm) in size. Females prefer to lay eggs in larger nests, where the energetic costs of keeping eggs warm during incubation are lower.

ORDER	Passeriformes
FAMILY	Aegithalidae
BREEDING RANGE	Europe, central Asia
BREEDING HABITAT	Deciduous or mixed woodlands, with dense shrub understory
NEST TYPE AND PLACEMENT	An enclosed sac of soft, flexible plant fibers, silk, and animal hair, camouflaged with lichen; suspended in a dense bush or in the forks of tree branches
CONSERVATION STATUS	Least concern
COLLECTION NO.	FMNH 1603

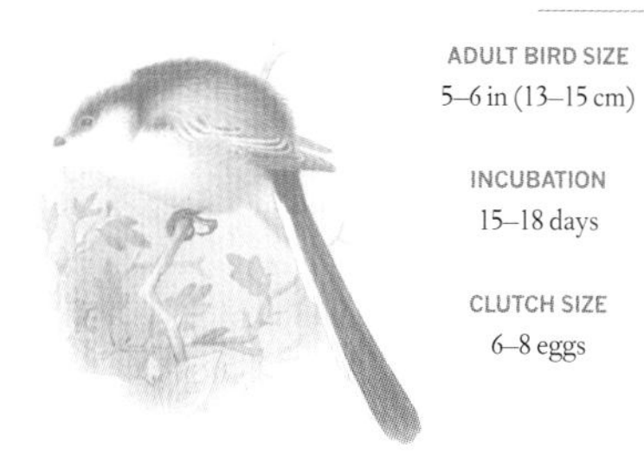

ADULT BIRD SIZE
5–6 in (13–15 cm)

INCUBATION
15–18 days

CLUTCH SIZE
6–8 eggs

AEGITHALOS CAUDATUS

LONG-TAILED TIT

PASSERIFORMES

The Long-tailed Tit has a small body, as most of its length is the tail. But despite the fact that smaller birds tend to lose body heat quicker than larger birds, this species remains resident in cooler climates throughout the year. To keep warm on winter nights, they roost together, body pressed against body, in groups of up to 30 individuals. Early in the spring, these flocks disband, and pairs establish breeding territories, where they build their elaborate and well-camouflaged nests for roosting and for egg laying. However, no camouflage is perfect, and many nests are torn apart by predators.

If the breeding attempt fails, pair members face a decision to abandon all breeding attempts for the year, or, as many do, to join a nearby territory and help to raise another pair's brood. The female and the male often split and, using calls that they learned from their own parents and helpers, locate other relatives in the habitat. Helping at a relative's nest is still a means to raise a brood of genetically related young.

Clutch

Actual size

The egg of the Long-tailed Tit is white in background color, has fine, sparse, reddish spotting, and its measurements are $\frac{9}{16}$ x $\frac{7}{16}$ in (14 x 11 mm). To better hide the nest, its camouflage includes up to 3,000 bits of lichen pasted on the outside surface.

ORDER	Passeriformes
FAMILY	Aegithalidae
BREEDING RANGE	Western North America, Central America
BREEDING HABITAT	Mixed open woodlands, typically oak with dense understory; parks and gardens
NEST TYPE AND PLACEMENT	Hanging nest with small entrance at top, suspended from branches or mistletoe; made of cobwebs and other filaments, covered in leaves and bark, and lined with soft hair and feathers
CONSERVATION STATUS	Least concern
COLLECTION NO.	FMNH 19115

ADULT BIRD SIZE
4½–5 in (11–13 cm)

INCUBATION
12–13 days

CLUTCH SIZE
4–10 eggs

PSALTRIPARUS MINIMUS

BUSHTIT

PASSERIFORMES

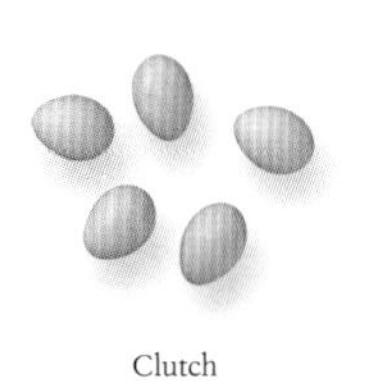

Clutch

An inhabitant of montane areas, this small species relies on flocks for feeding, moving, and roosting in the non-breeding season. Once early spring arrives, pairs leave the flock to search together for a suitable nesting site, and to begin constructing the intricate nest. However, if temperatures drop, the breeders rejoin the flocks and suspend the breeding attempt until it gets warmer. The nest is often exposed to the direct sun, which, together with the thick insulation, helps with incubation efficiency and hatching success.

The great American ornithologist, Alexander Skutch, first coined the term "helpers at the nest" in 1935 with regards to the Bushtit which he was studying in the mountains of Guatemala. Indeed, in the southern distribution, adult males, females, and juveniles often help provision the young at a nest of the breeding pair. Further north, pair-based reproduction becomes the norm for the local populations.

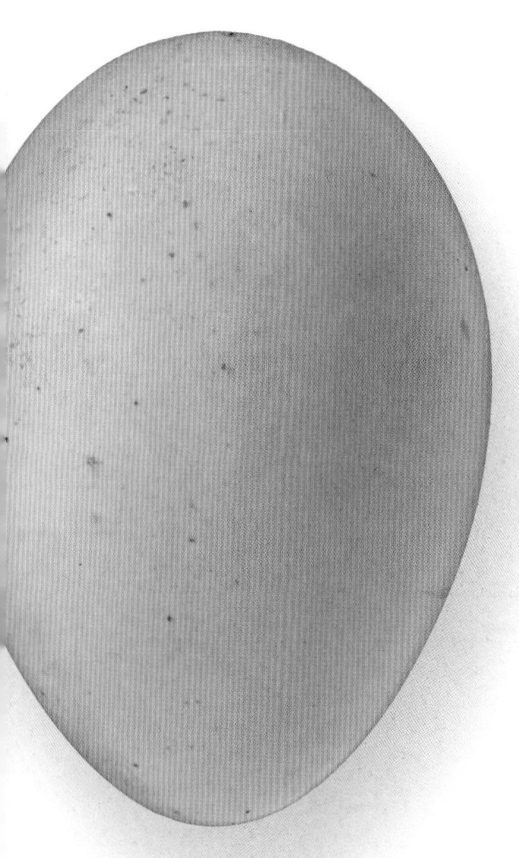

Actual size

The egg of the Bushtit is smooth white and immaculate in color, and 9⁄16 x 3⁄8 in (14 x 10 mm) in dimensions. Despite the presence of several adult male helpers at many nests, DNA analysis indicates that all the eggs are sired by the social mate of the laying female.

ORDER	Passeriformes
FAMILY	Sittidae
BREEDING RANGE	Temperate North America
BREEDING HABITAT	Mature deciduous woodlands, mixed forests, city parks, and orchards
NEST TYPE AND PLACEMENT	Cavity inside a natural or woodpecker-excavated hole; lined with bark, lumps of dirt, and fur, the cup is made of fine grass, feathers, and soft fibers
CONSERVATION STATUS	Least concern
COLLECTION NO.	FMNH 11456

ADULT BIRD SIZE
5–5½ in (13–14 cm)

INCUBATION
13–14 days

CLUTCH SIZE
5–9 eggs

SITTA CAROLINENSIS

WHITE-BREASTED NUTHATCH

PASSERIFORMES

This nuthatch has a unique body shape and foraging strategy shared only with other nuthatches around the world; they use their sturdy beak to pry under tree bark in search of insects, while moving head up or head down along the trunk of a mature tree. The White-breasted Nuthatch uses some of its captured prey to smear against the nest cavity, where the smell may deter mammalian predators from entering the hole.

The female is in charge of most parental behaviors, from nest building to egg hatching, while the male joins her in feeding the nestlings and fledglings until independence. Once ready to feed on their own, the juveniles are chased away from the natal territory. They move away, often long distances, in search of suitable, and empty, territories elsewhere. Some juveniles dispersing from home may have been responsible for these nuthatches being spotted, ranging far off the continental shoreline, over the open Atlantic Ocean.

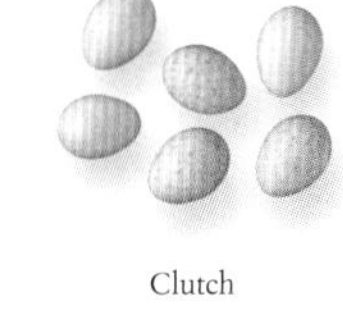

Clutch

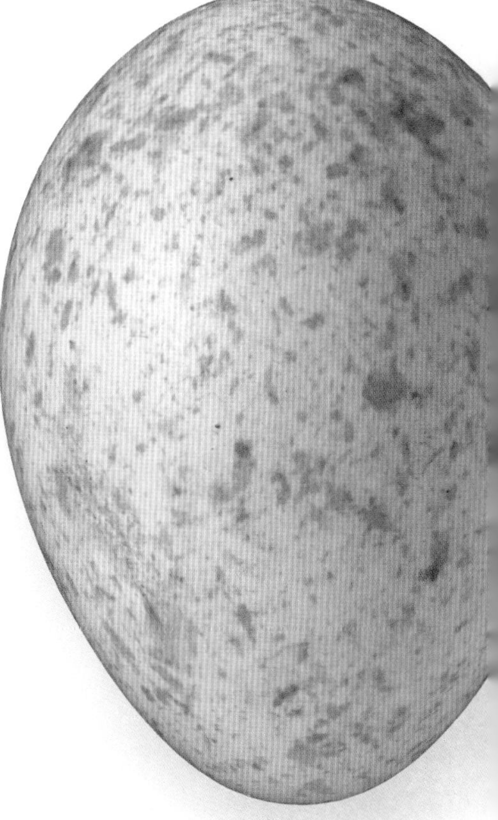

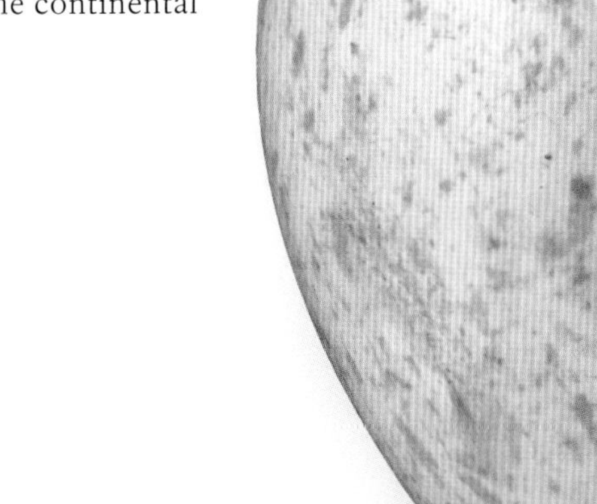

Actual size

The egg of the White-breasted Nuthatch is creamy white to pinkish white in coloration, has reddish brown or gray spotting, and measures ¾ x 9/16 in (19 x 14 mm in size). The female alone incubates the eggs and broods the young, and the male only enters the nest to feed his mate.

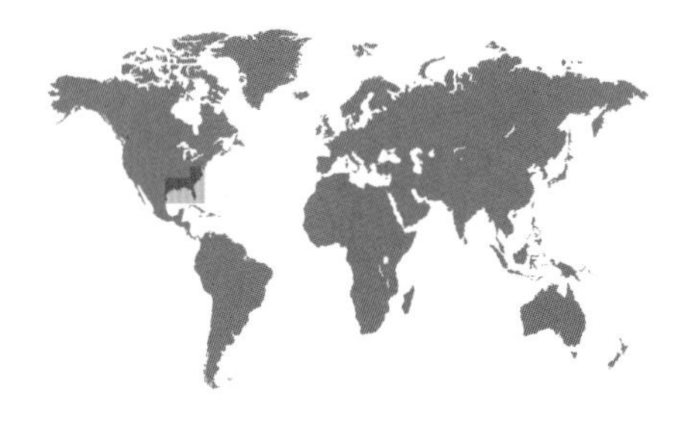

ORDER	Passeriformes
FAMILY	Sittidae
BREEDING RANGE	Southeastern North America
BREEDING HABITAT	Native mature, and cultivated, southern pine forests
NEST TYPE AND PLACEMENT	Cavity nest, typically newly excavated, but sometimes reusing old woodpecker holes or adopting nestboxes
CONSERVATION STATUS	Least concern
COLLECTION NO.	FMNH 11470

ADULT BIRD SIZE
4–4½ in (10–11 cm)

INCUBATION
13–15 days

CLUTCH SIZE
3–7 eggs

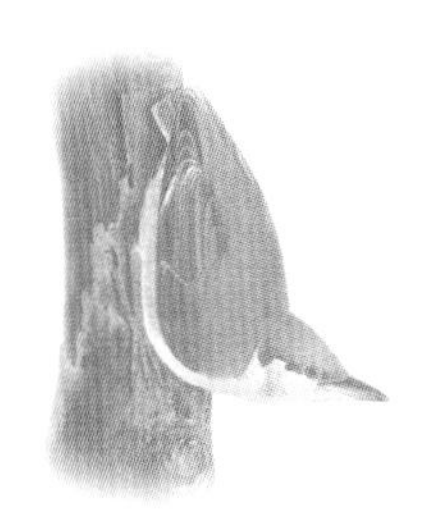

SITTA PUSILLA

BROWN-HEADED NUTHATCH

PASSERIFORMES

Clutch

On breeding territories the Brown-headed Nuthatch has a social structure which is based around a monogamous pair bond, assisted at many nests by one to three helpers. These helpers are younger but sexually mature males who forgo direct breeding, and probably assist their father, older brother, or some other relative across the reproductive cycle. Specifically, they contribute toward excavating the nest, lining the nest hole, feeding the chicks, removing excrement, and defending the territory. Nests with at least one helper have more than 50 percent greater breeding success than nests without helpers.

This small nuthatch species occurs today only in the southeastern United States, and its strong preference to breed in mature pine stands makes it an ecological indicator species for habitat health and quality. Its narrow niche requirements likely led to extinction of its Bahamas populations following the clear-cutting of much of the larger islands' native pine forests.

Actual size

The egg of the Brown-headed Nuthatch is white to buff in background color, with reddish brown blotching, and measures ⅔ x ½ in (17 x 12 mm) in size. The female alone warms the eggs but is provisioned by the male during incubation—he calls to her to come out of the cavity or feeds her directly inside the nest.

ORDER	Passeriformes
FAMILY	Sittidae
BREEDING RANGE	Temperate and boreal North America, including western mountain ranges
BREEDING HABITAT	Conifer or mixed forests with fir or spruce stands
NEST TYPE AND PLACEMENT	Excavated cavity nest in dead trees or rotting trunks; the cup nest inside consists of a bed of grass, bark strips, and pine needles, lined with fur, feathers, and grasses
CONSERVATION STATUS	Least concern
COLLECTION NO.	FMNH 11448

ADULT BIRD SIZE
4½–5 in (11–12 cm)

INCUBATION
12–13 days

CLUTCH SIZE
5–8 eggs

SITTA CANADENSIS

RED-BREASTED NUTHATCH

PASSERIFORMES

A small and widespread nuthatch, this species coexists with some of the other nuthatch species in North America. But its plumage patterns, as well as its preference to excavate its own nest in conifer forests, are reminiscent of nuthatches distributed in Eurasia, thus placing it apart from its American relatives.

Prior to egg laying, lone males establish and defend a breeding territory, and may even begin to excavate the nest burrow. But once successfully mated, the female typically completes the cavity. She alone incubates, although the male may join her in the nest to deliver food and to roost at night; the parents also work together to feed the nestlings and the fledglings. To keep the nest clean, the parents remove waste materials produced by the chicks; carrying the fecal sacs, they fly some distance directly away from the nest and drop them.

Clutch

Actual size

The egg of the Red-breasted Nuthatch is white, creamy, or pink, has reddish rusty blotching, and measures ⅔ x ½ in (17 x 12 mm). To discourage predators and competitors for the nest hole, the parents smear pungent and sticky pine resin around the cavity opening.

ORDER	Passeriformes
FAMILY	Certhiidae
BREEDING RANGE	Temperate Eurasia
BREEDING HABITAT	Deciduous or mixed coniferous forests
NEST TYPE AND PLACEMENT	In tree crevices or between bark flakes, also in wall holes and rock crevices; the platform of the nest is built from twigs and bark, and the cup is lined with finer grasses and mosses
CONSERVATION STATUS	Least concern
COLLECTION NO.	FMNH 2397

ADULT BIRD SIZE
4¾–6 in (12–15 cm)

INCUBATION
12–17 days

CLUTCH SIZE
3–5 eggs

CERTHIA FAMILIARIS

EURASIAN TREECREEPER

PASSERIFORMES

Clutch

The egg of the Eurasian Treecreeper is white to cream in background color, with pink or red speckling throughout, and is ⅔ x ½ in (16 x 12 mm) in size. The female alone incubates the eggs and broods the young chicks, but both parents feed them.

These tiny birds are a common and widespread inhabitant of European forests and other wooded areas, where they spend most of their time clinging on and climbing upward along bark, using their curved bills to search for hidden insect larvae and eggs. Within a mated pair, the female forages higher up on the tree trunk and branches compared to her mate.

A favorite nesting tree of this species is the introduced Giant Sequoia (brought to European gardens from California), whose soft wood is exploited by many other birds to carve a nest hole, which can then be usurped by treecreepers. Occasionally, this species will also settle in nestboxes placed for Great and Blue Tits in backyards. Territorial all year round, they sometimes are in direct competition for foraging and nesting sites with other treecreeper species; in such sites, the Eurasian Treecreeper typically moves into more coniferous forests. In contrast, treecreepers, as secondary cavity nesters, are not in direct competition with woodpeckers, which excavate their own tree holes. However, woodpeckers occasionally prey on treecreeper eggs and chicks, despite the loud alarm calls uttered by the parents.

Actual size

ORDER	Passeriformes
FAMILY	Troglodytidae
BREEDING RANGE	Southwestern North America
BREEDING HABITAT	Desert scrub, cactus stands
NEST TYPE AND PLACEMENT	Large messy dome with a tunnel entrance; made of coarse grass or plant fibers, lined with feathers; placed among spines on a cactus or other thorny tree
CONSERVATION STATUS	Least concern
COLLECTION NO.	FMNH 11572

CAMPYLORHYNCHUS BRUNNEICAPILLUS

CACTUS WREN

PASSERIFORMES

ADULT BIRD SIZE
7–8½ in (18–22 cm)

INCUBATION
16 days

CLUTCH SIZE
3–5 eggs

This species is the largest wren in North America, and it calls loudly and conspicuously from prominent perching sites atop thorny bushes. These birds are resident all year round, with the pair bond and the territory remaining the same across consecutive years and breeding bouts.

Clutch

The breeding season begins when the female initiates the nest construction, or starts to repair an already existing nest (perhaps one used for roosting outside of the breeding season). Both members of the pair help to complete the nest, but only the female incubates the eggs. She begins incubating soon after the first egg is laid, which results in a staggered hatching order, often over a period of two to three days. The brood, therefore, consists of differently aged and sized chicks, which are provisioned by both parents taking turns to deliver food to the nest.

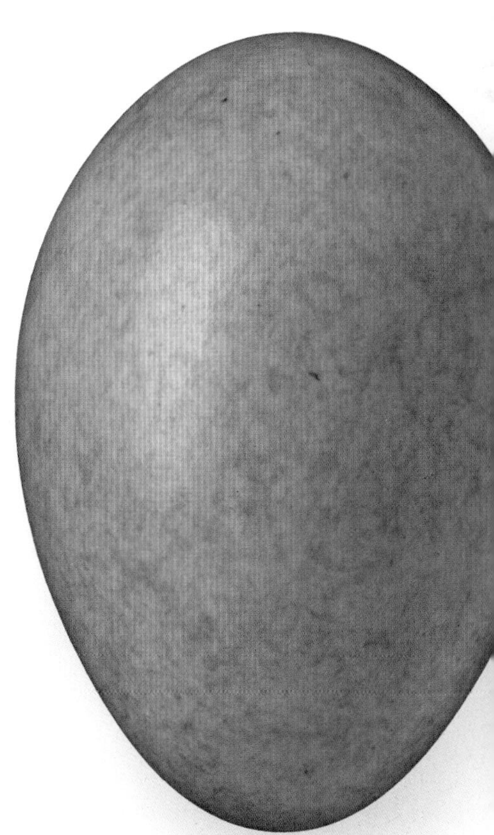

Actual size

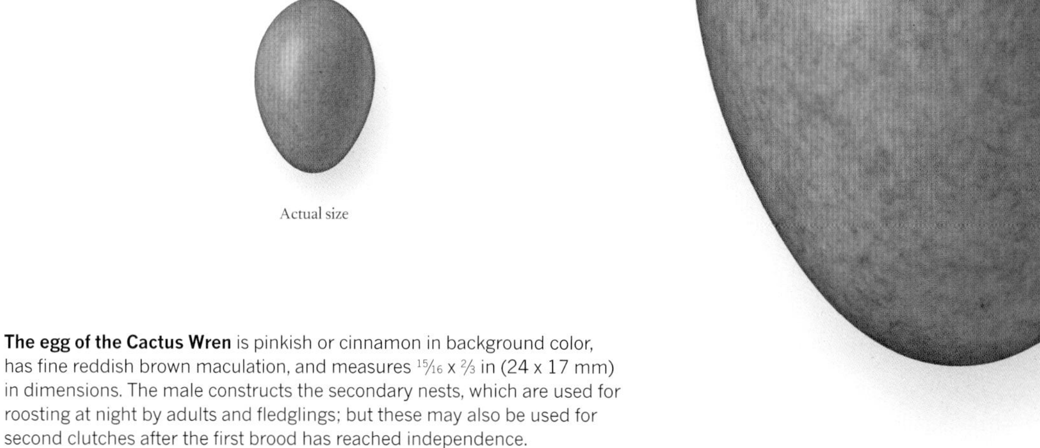

The egg of the Cactus Wren is pinkish or cinnamon in background color, has fine reddish brown maculation, and measures $^{15}/_{16}$ x $^{2}/_{3}$ in (24 x 17 mm) in dimensions. The male constructs the secondary nests, which are used for roosting at night by adults and fledglings; but these may also be used for second clutches after the first brood has reached independence.

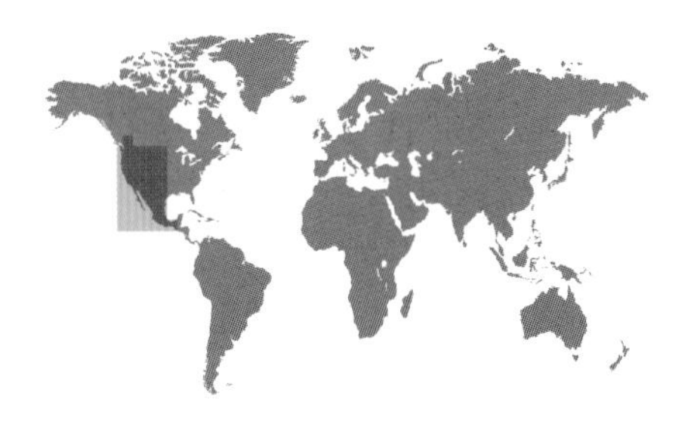

ORDER	Passeriformes
FAMILY	Troglodytidae
BREEDING RANGE	Western North America
BREEDING HABITAT	Arid, rocky outcroppings, and valleys
NEST TYPE AND PLACEMENT	Cavity or crevice in a rock face or cliff; cup made of twigs and rootlets, lined with lichens, soft plant fibers, wool, cobwebs, and feathers
CONSERVATION STATUS	Least concern
COLLECTION NO.	FMNH 11522

ADULT BIRD SIZE
5½–6 in (14–15 cm)

INCUBATION
12–18 days

CLUTCH SIZE
4–6 eggs

CATHERPES MEXICANUS

CANYON WREN

PASSERIFORMES

Clutch

This small wren species is a specialist of rocky hillsides, cliff faces, and arid terrains with exposed boulders. It remains resident on its territory throughout the year, including the colder winter months, which are also likely the time when pair formation occurs, although this has not yet been studied in detail. Once a pair bond is established, the mates often remain together throughout the year, as well as across multiple nesting seasons.

Perhaps because the nests are often in hidden or inaccessible crevices, little is known about the nest-building behaviors of this species. The male, however, is known to provide food for both the incubating female and, together with her, he also feeds the nestlings. To keep the nest sanitary, the adults remove, and carry in their beaks, all nestling excrement in fully wrapped up packages, called fecal sacs, produced by the chicks at the end of each feeding bout.

Actual size

The egg of the Canyon Wren is white in coloration, with gray, reddish, or rusty speckling, and is ⅔ x 9⁄16 in (17 x 14 mm) in size. Only the female develops a brood patch, incubates the eggs, and broods the chicks, while the male provisions her at the nest.

ORDER	Passeriformes
FAMILY	Troglodytidae
BREEDING RANGE	Eastern and southern North America
BREEDING HABITAT	Open woodlands, parks, and gardens, with some dense understory
NEST TYPE AND PLACEMENT	In a low cavity in trees, overhangs, and stumps, also in metal boxes or tilted flower pots; builds a domed nest of twigs, rootlets, and grasses, with a small opening toward the top
CONSERVATION STATUS	Least concern
COLLECTION NO.	FMNH 11703

ADULT BIRD SIZE
4½–5½ in (12–14 cm)

INCUBATION
12–14 days

CLUTCH SIZE
4–6 eggs

THRYOTHORUS LUDOVICIANUS

CAROLINA WREN

PASSERIFORMES

The Carolina Wren is a nesting opportunist; it relies on cavities or other enclosed spaces for breeding, but the species is highly flexible in choosing such sites. As long as there are a few feasible nesting cavities within the territory, at the base of broken limbs or in old woodpecker holes, the monogamous pairs remain settled year-round. Where natural cavities are scarce, these birds can occupy just about any closed space with an opening wide enough for the parents to squeeze through: holes in walls, open pipes, even unused mailboxes.

Clutch

As a nonmigratory, insectivorous species, cold and wet winters often devastate the most northerly populations. However, numbers quickly recover with the production of two or three annual broods and the species has been expanding its distribution northward. Clutch size often decreases between first and subsequent clutches, but a typical wren nest always seems full of tightly packed eggs and nestlings pressing against each other.

Actual size

The egg of the Carolina Wren is white, cream, or pinkish white in background color with fine rusty brown maculation, and measures ¾ x ⅝ in (19 x 15 mm) in size. Only the female incubates the eggs, but the male often visits, brings food to her, and sings softly nearby.

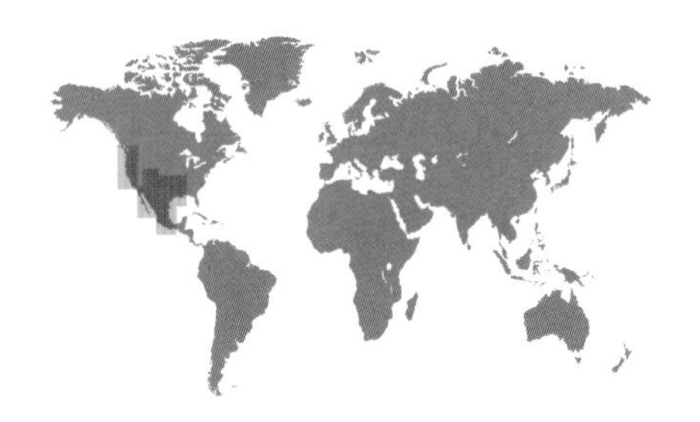

ORDER	Passeriformes
FAMILY	Troglodytidae
BREEDING RANGE	Western, central, and southern North America
BREEDING HABITAT	Brushy areas, open woods, thickets, and hedgerows, often near water
NEST TYPE AND PLACEMENT	In a cavity or other covered space; builds an open-cup nest of sticks, roots, mosses, and leaves
CONSERVATION STATUS	Least concern
COLLECTION NO.	FMNH 11679

ADULT BIRD SIZE
5–5½ in (13–14 cm)

INCUBATION
14–16 days

CLUTCH SIZE
5–7 eggs

THRYOMANES BEWICKII

BEWICK'S WREN

PASSERIFORMES

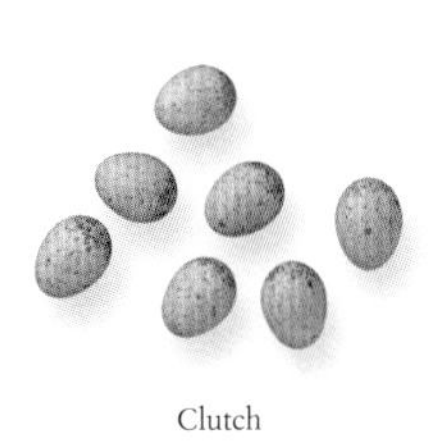

Clutch

The egg of the Bewick's Wren is white in background coloration, has reddish brown or purplish spotting, and measures ⅔ x ½ in (17 x 13 mm). After hatching, the female broods the chicks, and also promptly removes the empty eggshells by carrying them in her beak as she flies away from the nest.

This species was first described to science by John J. Audubon, the father of North American natural history illustration, who named the bird "Bewick" after an artist friend. Once widespread across most of the southern United States, today this species is still common in the southwest, but in the southeast two of its subspecies have already become extinct.

There are both resident and migrant populations. Among migrants, pair bonding occurs soon after the sexes settle on the breeding territory, with the males arriving first from the wintering grounds. In resident populations, the pairs form by the end of the first year of life, and remain stable across multiple years. The male selects the nesting site, and both members of the pair construct the nest in a hollow log, or other cavity. Only the female incubates, but the male feeds her while she is on the eggs, and closely follows her while she is off the nest.

Actual size

ORDER	Passeriformes
FAMILY	Troglodytidae
BREEDING RANGE	Temperate North America, northern Central America
BREEDING HABITAT	Forests and edges, open woodlands, parks, and gardens
NEST TYPE AND PLACEMENT	Cavity nests: old woodpecker holes, natural crevices, nestboxes; fills cavity with twigs and branches, and builds soft cup nest with grasses and cobwebs with a narrow passageway through the twigs and sticks
CONSERVATION STATUS	Least concern
COLLECTION NO.	FMNH 1567

ADULT BIRD SIZE
4½–5 in (11–13 cm)

INCUBATION
9–16 days

CLUTCH SIZE
4–8 eggs

TROGLODYTES AEDON

NORTHERN HOUSE WREN

PASSERIFORMES

Together with the Red-winged Blackbird (see page 607), the Northern House Wren is one of the most thoroughly and broadly studied songbird species in North America. This is not surprising, since it has a ubiquitous distribution and readily shares human habitats, including an eagerness to occupy nestboxes, often before birds that are long-distance migratory competitors have a chance to do so. Even if a nest hole is not used, the male often fills it up with twigs and sticks, so as to discourage other birds from settling there, or perhaps to confuse, misguide, and delay nest predators and brood parasites.

Male Northern House Wrens are often polygynous, singing their melodious songs to attract several females which end up nesting in the same male's territory. Polygyny seems to benefit both females and males: females preferentially choose males who have several mates; and males with more mates sire more young in each nest, and across all nests combined. Males are mated to just a single female, in social monogamy; it is the female that becomes promiscuous and often seeks paternity for her eggs outside the pair bond.

Clutch

The egg of the House Wren is white, pink-white, or grayish in coloration, with reddish brown maculation, and is ⅔ x ½ in (17 x 13 mm) in dimensions. This species is highly aggressive, and will peck, pierce, and remove the eggs of other cavity-nesting bird species in its territory in order to usurp the cavity or reduce the competition for food.

Actual size

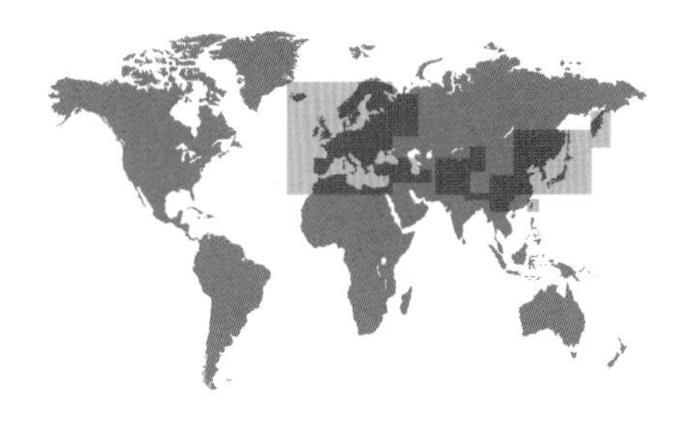

ORDER	Passeriformes
FAMILY	Troglodytidae
BREEDING RANGE	Temperate Europe, Asia
BREEDING HABITAT	Mixed and coniferous forests; also parklands, farms, and boulders on ocean beaches
NEST TYPE AND PLACEMENT	Domed nest of mosses, lichens, and grasses; built inside a hole in a wall, tree trunk, or rock crack, but also in bushes, piles of branches, or firewood stacks
CONSERVATION STATUS	Least concern
COLLECTION NO.	FMNH 20708

ADULT BIRD SIZE
3½–4½ in (9–11 cm)

INCUBATION
12–16 days

CLUTCH SIZE
5–8 eggs

TROGLODYTES TROGLODYTES

EURASIAN WREN

PASSERIFORMES

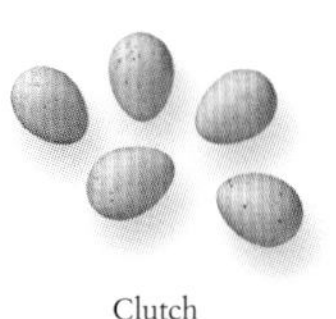
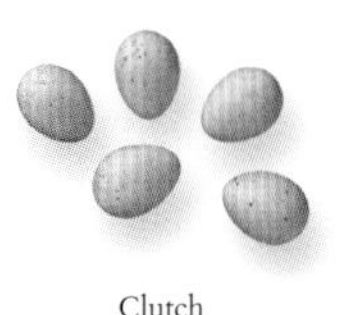

Clutch

This is the only member of the wren family Troglodytidae to occur in the Old World, and the first wren to be described by scientists; ironically, the family is much more diverse in the New World. This wren's distribution spans the entire landmass of Eurasia, from several insular subspecies off the Atlantic coast of Scotland to the birds of the Japanese coastline of the Pacific. The family name Troglodytidae means "cave-dwelling" in reference to the covered and domed nest this species builds.

A small and mousey species, in most of its range it occurs year-round, surviving the cold winter nights by roosting together with other wrens in tree holes to share body heat. In the spring, the males become aggressive and noisy; despite its small size, the male produces a very loud and melodious, tonal song, used to attract one or (typically) more females to settle and nest on his breeding territory.

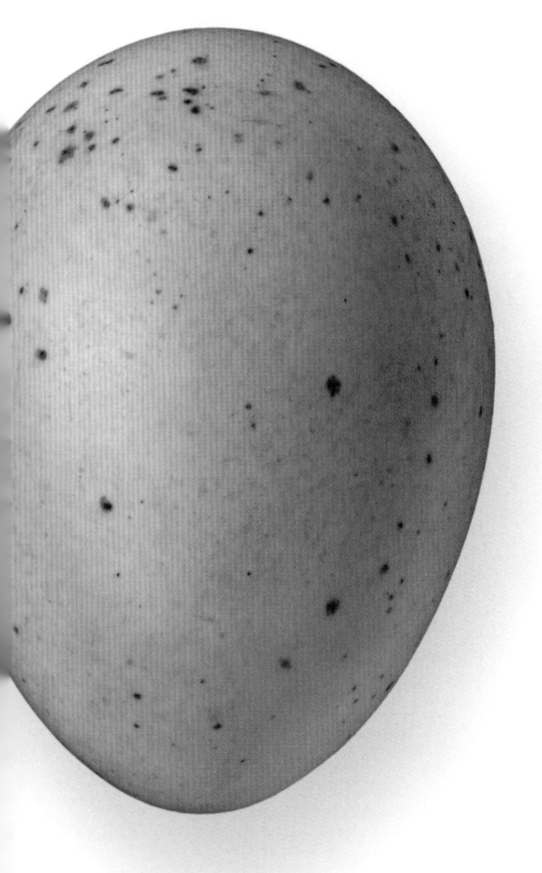

Actual size

The egg of the Eurasian Wren is mostly white, with sparse and small reddish spotting, and is ⅝ x ½ in (16 x 13 mm) in size. This bird's adaptability and resourcefulness are well known, as the species can make use of any available cavities, including old coat pockets left hanging outside in the spring, to build its nest.

ORDER	Passeriformes
FAMILY	Troglodytidae
BREEDING RANGE	Temperate North America, including both coastlines, and western mountains
BREEDING HABITAT	Marshland, swamps with large standing beds of vegetation
NEST TYPE AND PLACEMENT	Domed nest, woven from leaves of grasses and reeds, with entrance on side, lined with soft fine grasses; attached to stems of standing marsh vegetation
CONSERVATION STATUS	Least concern
COLLECTION NO.	FMNH 20937

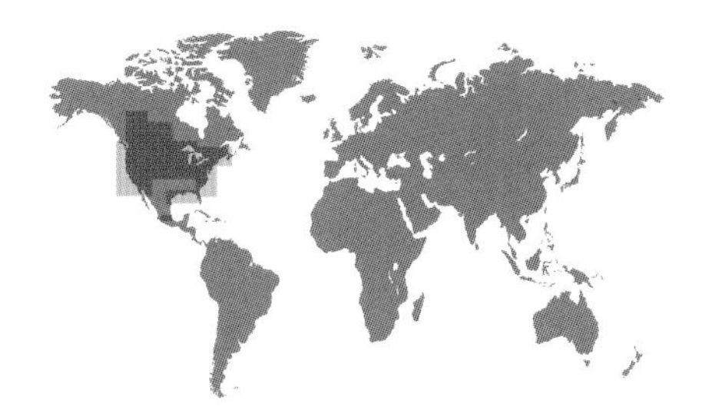

ADULT BIRD SIZE
7–9 in (18–23 cm)

INCUBATION
12–16 days

CLUTCH SIZE
4–6 eggs

CISTOTHORUS PALUSTRIS

MARSH WREN

PASSERIFORMES

Singing is what males of this species do: an individual Marsh Wren can produce anywhere from 50 to 200 different song types. Males spend most of their days engaged in vocal duets with neighboring males; dominance is assessed as one individual switches between song types and the other individual follows to match to the new type, but only if he knows it. Females pay attention to these songs when selecting mates; they prefer males with larger repertoires; these males also have more complex brain architecture involved in the learning and production of varied songs.

As part of pair formation, the male escorts a female throughout his territory, showing her several mostly completed nest shells that he has built as potential breeding sites. The female chooses one of these "dummy nests," and completes the shell and its lining before starting to lay her eggs. Males with no dummy nests fail to attract mates. Unused nests become temporary shelters and overnight roosting sites for fledglings, and their parents, later in the season.

Clutch

The egg of the Marsh Wren is buff to light brown in background coloration, has heavy dark brown maculation, and its dimensions are ⅝ x ½ in (16 x 13 mm). The female begins to incubate the eggs one to two days before the clutch is complete, and so the chicks hatch asynchronously, across several days.

Actual size

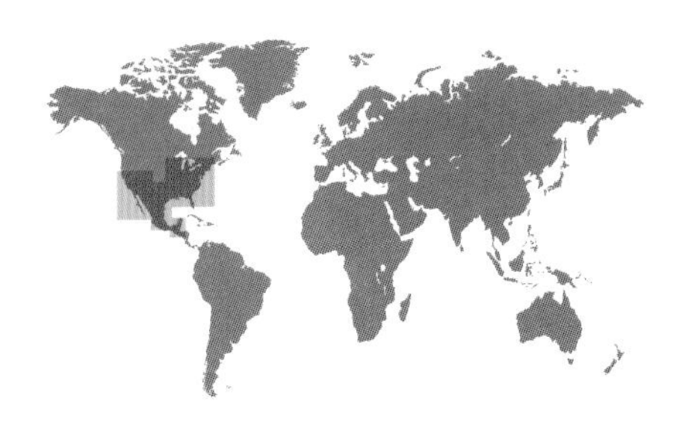

ORDER	Passeriformes
FAMILY	Polioptilidae
BREEDING RANGE	Temperate North America, Central America
BREEDING HABITAT	Deciduous woodlands, from scrub to mature forest, often near water or other habitat edges
NEST TYPE AND PLACEMENT	Deep, open cup, built from grasses, cobwebs, and lichens; attached to top of branches and limbs of shrubs and trees with dense canopy
CONSERVATION STATUS	Least concern
COLLECTION NO.	FMNH 12244

ADULT BIRD SIZE
4–4½ in (10–11 cm)

INCUBATION
11–15 days

CLUTCH SIZE
3–6 eggs

POLIOPTILA CAERULEA

BLUE-GRAY GNATCATCHER

PASSERIFORMES

Clutch

This is a small and mostly migratory songbird, and the northernmost representative of a typically tropical, New World lineage. Once it returns from the wintering grounds, pairs of females and males can form within just one day. The two sexes contribute equally to all aspects of parental investment, from the construction of the nest, through incubation and the feeding of the nestlings. In warmer regions, where there is time for second broods, the male constructs the bulk of the second nest, while the female completes the first breeding bout.

Although the geographic range of this species has been expanding northward in recent years, so has exposure to the negative effects of brood parasitism by Brown-headed Cowbirds (see page 616). With the large and intensively begging cowbird chick in the nest of the gnatcatcher, the host's own chicks invariably starve to death.

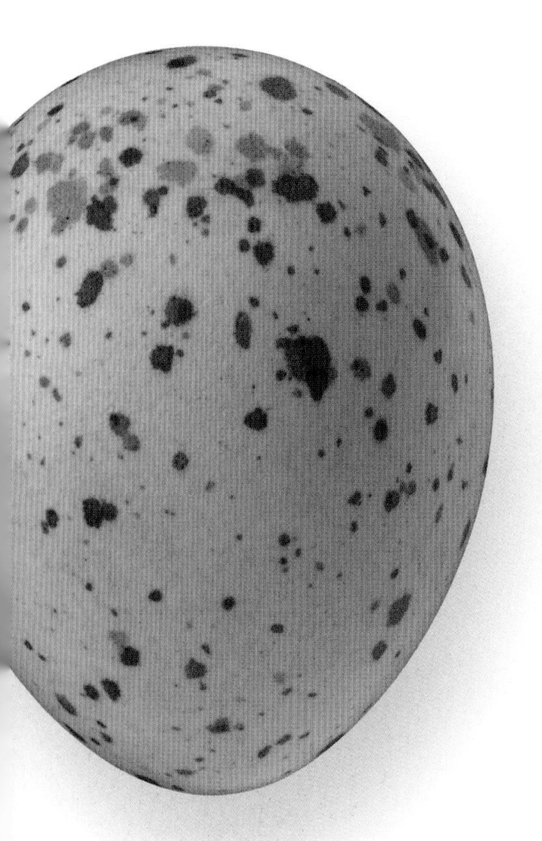

Actual size

The egg of the Blue-gray Gnatcatcher is white to pale blue in coloration, with darker and sharp spotting, and is $\frac{5}{8}$ x $\frac{7}{16}$ in (15 x 11 mm) in size. If the eggs perish, these birds re-nest elsewhere, with up to six failed re-nesting attempts recorded in one season for one particularly unlucky pair.

ORDER	Passeriformes
FAMILY	Cinclidae
BREEDING RANGE	Western North America
BREEDING HABITAT	Montane or coastal streams in undisturbed areas, including permanent desert streams
NEST TYPE AND PLACEMENT	Bowl-shaped nest, made of mosses and rootlets; placed on cliff ledges, behind waterfalls, under boulders, and overhanging dirt banks
CONSERVATION STATUS	Least concern
COLLECTION NO.	FMNH 11505

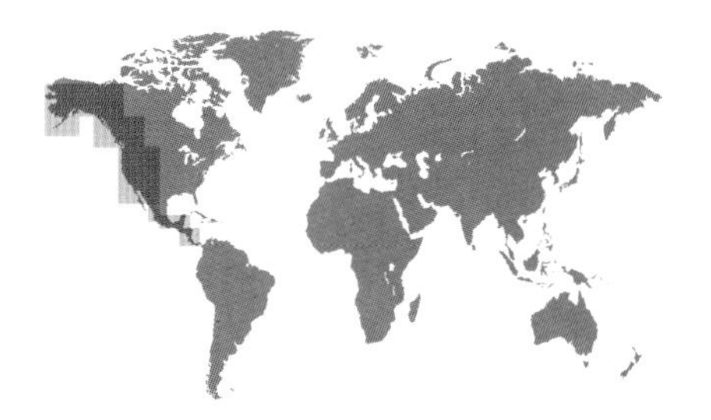

CINCLUS MEXICANUS

AMERICAN DIPPER

PASSERIFORMES

ADULT BIRD SIZE
5½–8 in (14–20 cm)

INCUBATION
14–17 days

CLUTCH SIZE
4–5 eggs

Dippers, as a genus, have a worldwide distribution, inhabiting every continent except for Australia and Antarctica. The related species share a characteristic compact body shape and short-tailed stature, as well as a unique ability to run and flap underwater against the flow of fast-moving streams, while hunting for aquatic insect larvae. This author's most cherished childhood experience was watching a dipper move under his feet, while he played ice-hockey on an (almost) fully frozen forest stream. Dippers have specialized structures to keep the water out of their eyes and nostrils while they forage under the surface.

The density of the American Dipper's populations depends on the suitability of nesting sites and the abundance of prey items. The territories are strictly linear, as all feeding and breeding activities run strictly in and along stream shores. The pair shares the chores of building the nest, with the male often contributing nesting materials while the female weaves these together. The chicks are also provisioned together, with each parent taking turns.

Clutch

The egg of the American Dipper is white in background color, free of spotting, and 1 x ⅔ in (25 x 17 mm) in dimensions. Only the female develops a brood patch, and she alone incubates the eggs; the male provides her with some food at the nest.

Actual size

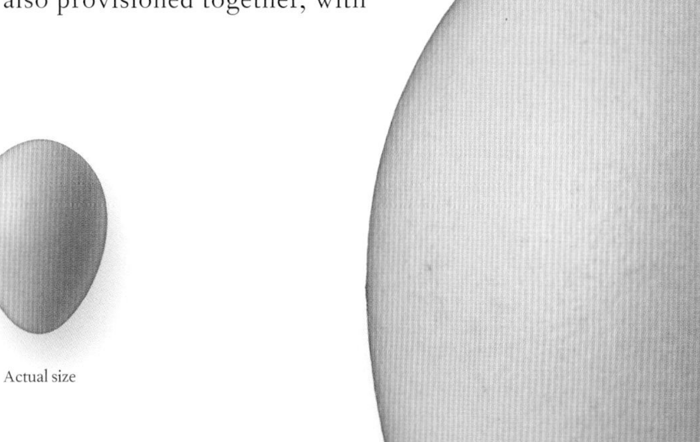

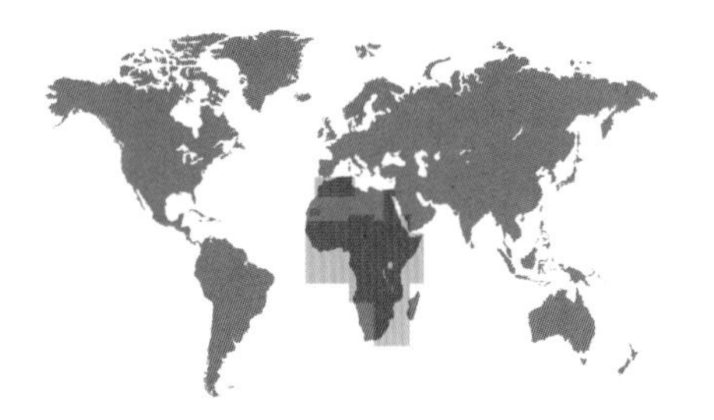

ORDER	Passeriformes
FAMILY	Pycnonotidae
BREEDING RANGE	Northern and sub-Saharan Africa
BREEDING HABITAT	Forest edges, open woodlands, riparian and coastal galleries; also city parks and gardens
NEST TYPE AND PLACEMENT	Rigid, thick-walled cup made from grasses and twigs, lined with hair, moss, and rootlets; placed in branch forks inside the canopy
CONSERVATION STATUS	Least concern
COLLECTION NO.	FMNH 20726

ADULT BIRD SIZE
7–7½ in (18–19 cm)

INCUBATION
12–14 days

CLUTCH SIZE
2–3 eggs

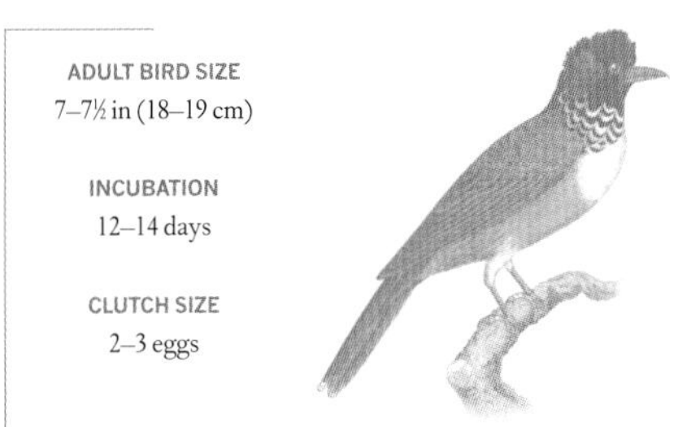

PYCNONOTUS BARBATUS

COMMON BULBUL

PASSERIFORMES

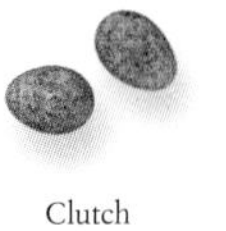

Clutch

Bulbul is the ancient Arabic word for "small bird." Today the name refers to over 100 species of closely related songbirds, each with quite impressive displays of songs and vocal mimicry. As omnivores, many of these species, including the Common Bulbul, have quickly adapted to life near human settlements. The monogamous pair remains resident all year round, and individuals can become so well adjusted that when house windows are left open, they steal left-over food from plates.

In sub-Saharan Africa, this and other bulbuls are often parasitized by the Jacobin Cuckoo. Surprisingly, the bulbul does not eject the parasitic egg, even though the cuckoo's egg looks quite different. The reason for the lack of antiparasitic response may be that cuckoo eggs are so large that ejection is physically impossible. In addition, cuckoos are just as likely to parasitize replacement nests, so nest abandonment of parasitized first clutches is also not a feasible strategy.

Actual size

The egg of the Common Bulbul is cinnamon or pale mauve in background, has darker blotching all over, and measures 13/16 x 5/8 in (21 x 16 mm). Only the female builds the nest and incubates the eggs, but the male often visits and feeds her.

ORDER	Passeriformes
FAMILY	Regulidae
BREEDING RANGE	Boreal and mountainous North America
BREEDING HABITAT	Coniferous or mixed forests
NEST TYPE AND PLACEMENT	Globular cup nest, made of mosses, lichens, cobwebs, and fine grasses; suspended between small twigs and branches, often high up in the canopy
CONSERVATION STATUS	Least concern
COLLECTION NO.	FMNH 16913

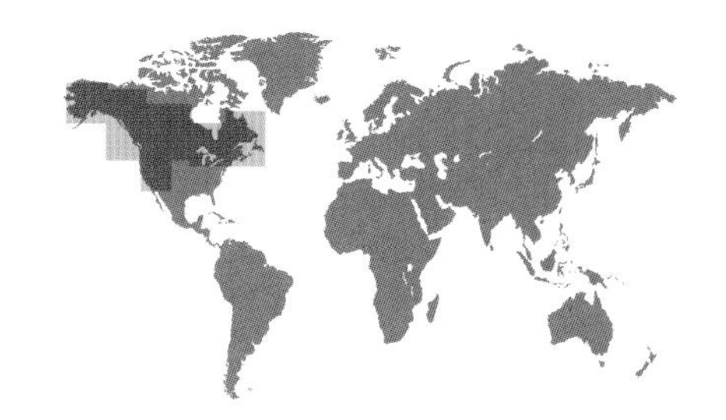

ADULT BIRD SIZE
3½–4½ in (9–11 cm)

INCUBATION
13–16 days

CLUTCH SIZE
4–9 eggs

REGULUS CALENDULA

RUBY-CROWNED KINGLET

PASSERIFORMES

This is one of the smallest songbirds in North America. It crosses the temperate regions of the continent between its southern wintering grounds and subarctic breeding sites. The male and the female are nearly identical in size and plumage, except for the male's red crown, which only becomes visible when displaying to other kinglets or mobbing predators.

Clutch

The reproductive period starts swiftly, with females gathering nesting materials as soon as they arrive on the breeding territory. Nest construction and incubation are both performed by females, while the male feeds the female on the nest, especially during bouts of cold weather. Once the brood is fledged, insufficient time remains to start a second clutch. Unlike most migratory species, juvenile kinglets get a head start on migration, and move south before the adults leave the breeding grounds.

Actual size

The egg of the Ruby-crowned Kinglet is drab white in coloration, with slight, reddish brown maculation, and is $\frac{9}{16}$ x $\frac{7}{16}$ in (14 x 11 mm) in dimensions. Some nests contain 12 eggs, representing one of the largest clutch sizes, relative to body size, among all birds.

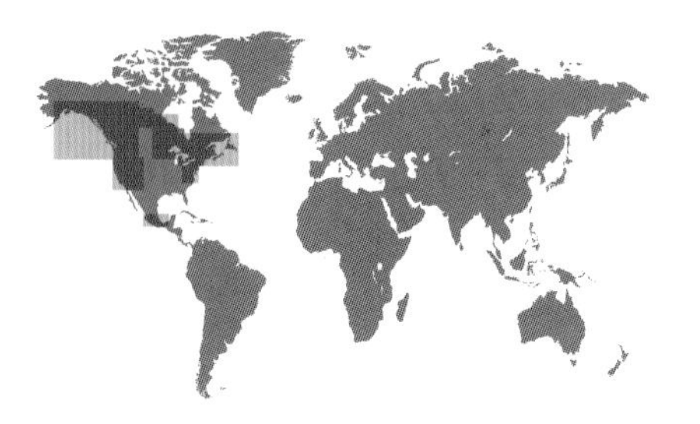

ORDER	Passeriformes
FAMILY	Regulidae
BREEDING RANGE	Northern and montane North America
BREEDING HABITAT	Old growth, boreal spruce and pine forests; expanding into pine plantations and more temperate mixed forests
NEST TYPE AND PLACEMENT	Open, deep cup of moss, lichen, spider silk, and bark, lined with feathers, fine grasses, plant down, lichens, and fur; hung from a forking branch, near the tips
CONSERVATION STATUS	Least concern
COLLECTION NO.	FMNH 16910

ADULT BIRD SIZE
3–4½ in (8–11 cm)

INCUBATION
14–15 days

CLUTCH SIZE
5–11 eggs

REGULUS SATRAPA

GOLDEN-CROWNED KINGLET

PASSERIFORMES

Clutch

Because of this species' northerly distribution, small size, and high-canopy habits, there are still important gaps in our knowledge of its reproductive biology and behavioral interactions. For example, it is unclear how and when pairs are formed, as most breeding studies of this kinglet have only noticed the presence of already formed pair bonds. It is also unclear which sex selects the nest site, but both parents are known to collect and move nesting materials to the chosen branch fork from which the woven nest cup is suspended.

Once the nest is complete, the female begins to lay the eggs, spending each night on the nest. She alone develops an incubation patch, but the male provisions her while sitting on the nest. When the chicks hatch, she broods to keep them warm, otherwise she joins the male and the parents take turns to provision the nestlings.

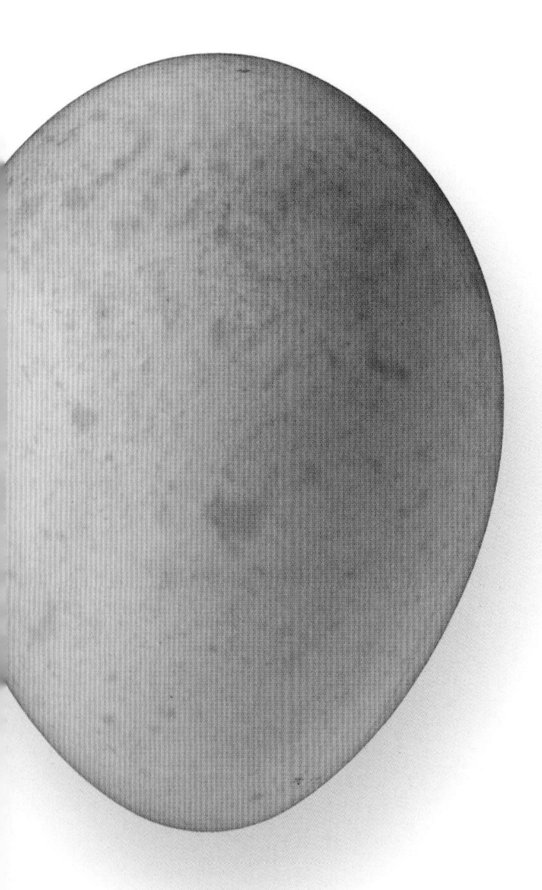

Actual size

The egg of the Golden-crowned Kinglet is drab white in background color, with light brown maculation, and measures ½ x ⅜ in (13 x 10 mm) in size. To assure sufficient nest depth for the eggs, the female uses her chest and rotates her body around to smooth the cup's inner walls.

ORDER	Passeriformes
FAMILY	Cettiidae
BREEDING RANGE	Western Asia, north Africa
BREEDING HABITAT	Arid grasslands with scattered bushes, coastal scrub
NEST TYPE AND PLACEMENT	Spherical, with a lateral entrance, made of twigs, grass, and plant fibers, lined inside with down, wool, and other hair; placed low to the ground, in thorny bushes
CONSERVATION STATUS	Least concern
COLLECTION NO.	FMNH 20694

ADULT BIRD SIZE
4–4½ in (10–11 cm)

INCUBATION
13–15 days

CLUTCH SIZE
2–4 eggs

SCOTOCERCA INQUIETA

STREAKED SCRUB-WARBLER

PASSERIFORMES

An inhabitant of arid semi-deserts and coastal scrub, this species relies on its cryptic plumage, and a nest placement in thorny bushes, to protect the incubating adult and the eggs and chicks from predators. It is a resident species throughout most of its distribution, with movements only during periods of extreme droughts. Otherwise, it survives in the absence of open water by consuming soft and moist body parts of its insect and other arthropod prey.

Clutch

Breeding takes place typically twice per year, depending on the timing of the onset of local rains, and begins with both members of the pair gathering nesting materials for its spherical, dome nest. The pair attends the construction site together, and shares other parental duties; after fledging, family groups remain together, with both parents provisioning the young until independence.

Actual size

The egg of the Streaked Scrub-Warbler is white with a pinkish tint, with sparse bright red maculation, and measures ⅔ x ⅝ in (17 x 15 mm). In some cases the nest includes a second entrance, perhaps to escape predators at the nest more easily.

ORDER	Passeriformes
FAMILY	Cettiidae
BREEDING RANGE	Southern and central Europe, north Africa, southwest Asia
BREEDING HABITAT	Moist areas, near water, including reed beds near rivers, ponds, and swamps
NEST TYPE AND PLACEMENT	Loose cup, made of stems and dry leaves, lined with fine grass, hair, flowers, and feathers; placed low in dense shrubs or marsh reeds
CONSERVATION STATUS	Least concern
COLLECTION NO.	FMNH 19143

ADULT BIRD SIZE
5–5½ in (13–14 cm)

INCUBATION
16–17 days

CLUTCH SIZE
3–6 eggs

CETTIA CETTI

CETTI'S WARBLER

PASSERIFORMES

Clutch

The egg of the Cetti's Warbler is glossy red or deep brick-red, immaculate, and ⅔ x 9/16 in (17 x 14 mm) in dimensions. It is the only temperate-zone passerine which lays allover red eggs; it remains unknown which pigments are responsible for this unique shell coloration, but scientists are working on it.

This species is one of the few songbirds in Europe whose populations have been increasing and whose range has been widening in recent decades. It is easy to notice the appearance of the Cetti's Warbler in a newly inhabited area, because of the conspicuous songs and loud calls broadcast by territorial males. These territories are large and exclusive, and typically follow a narrow strip along the shores of rivers, canals, and lakes.

The male does not stop singing after a female settles on his territory, as he often mates polygynously with up to three different females. Females appear to use male body size, not his territory, as a cue for mate choice, and larger males have more mates, and they lay larger clutches in each nest. The females alone incubate the eggs. Monogamous males assist their mates more often to feed the chicks; yet growth rates of nestlings are similar across territories, suggesting that females mated with polygynous males are able to make up for any lost paternal provisioning.

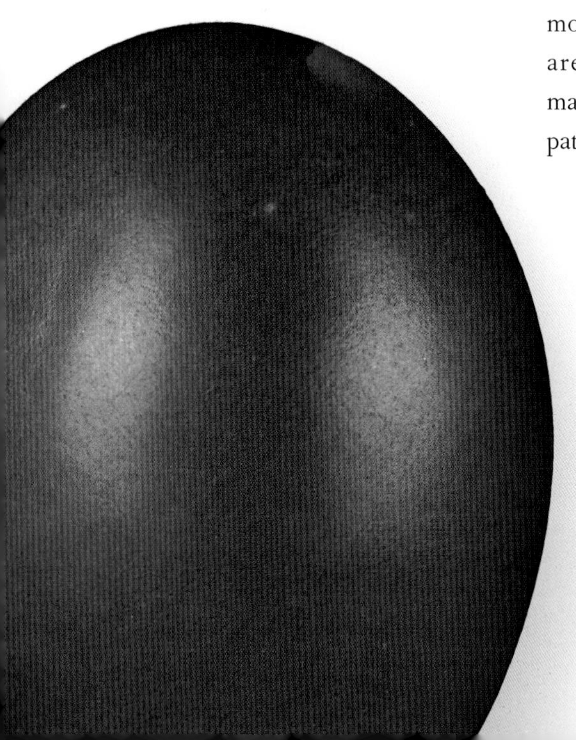

Actual size

ORDER	Passeriformes
FAMILY	Phylloscopidae
BREEDING RANGE	Boreal and temperate Eurasia
BREEDING HABITAT	Open woodlands, second-growth forests, younger tree plantations
NEST TYPE AND PLACEMENT	Dome-shaped nest usually placed on the ground, well concealed among grass or at the base of shrubs or trees; composed of grass, moss, rotten bark, and rootlets
CONSERVATION STATUS	Least concern
COLLECTION NO.	FMNH 15352

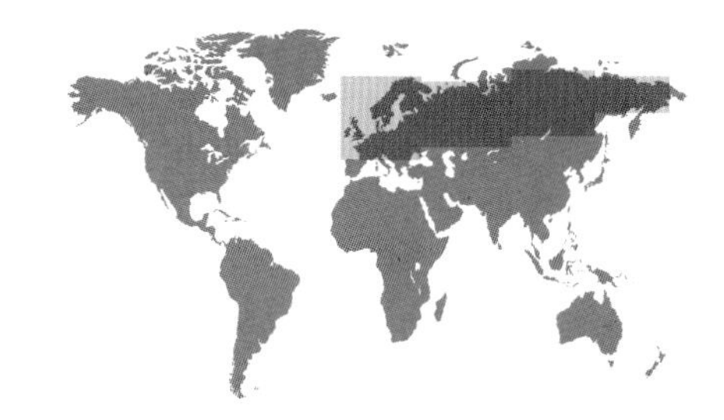

ADULT BIRD SIZE
4½–5½ in (11–14 cm)

INCUBATION
10–16 days

CLUTCH SIZE
4–8 eggs

PHYLLOSCOPUS TROCHILUS

WILLOW WARBLER

PASSERIFORMES

The Willow Warbler is a northern breeding songbird that engages in distinct routes of migratory flights to and from its tropical wintering grounds in Africa. The males return in the spring one to two weeks ahead of the females, and begin song displays to claim their territory. Males that sing more frequently often attract the first females to arrive, but they also allocate less time to feeding so they must be in top condition. Their reward is an earlier onset of breeding in a region where summers are short.

Clutch

Once a pair bond has been established, Willow Warblers remain monogamous, and nesting begins. Females that are widowed (or have their mates experimentally removed) are able to complete the incubation and nesting stage on their own. However, lone females pay a cost for losing paternal care, as they take longer to fledge the young, and their chicks grow more slowly.

Actual size

The egg of the Willow Warbler is white in coloration, has reddish brown maculation, and measures ⅔ x 9⁄16 in (17 x 14 mm). Most nesting attempts fail not during the incubation period, but while the parents are provisioning the chicks, implying that predators may locate nests by cluing in on the parents' frequent feeding visits.

ORDER	Passeriformes
FAMILY	Acrocephalidae
BREEDING RANGE	Temperate Europe and Asia
BREEDING HABITAT	Wetlands, canals, and lakes with emergent vegetation
NEST TYPE AND PLACEMENT	Basket-shaped nest, suspended from reed stems, typically over water; reed leaves and grasses woven together, lined with softer, finer grasses
CONSERVATION STATUS	Least concern
COLLECTION NO.	FMNH 18763

ADULT BIRD SIZE
6½–8 in (16–20 cm)

INCUBATION
14–15 days

CLUTCH SIZE
4–6 eggs

ACROCEPHALUS ARUNDINACEUS

GREAT REED-WARBLER

PASSERIFORMES

Clutch

The egg of the Great Reed-Warbler is beige, pale green, or bluish in background color, has variable blue-gray and dark brown maculation, and measures ⅞ x ⅔ in (23 x 17 mm) in size. To combat egg mimicry by cuckoos, these hosts remember what their own eggs look like; when scientists replaced all but one egg with foreign eggs, the birds ejected all the new eggs and left only their own egg in the nest.

Great Reed-Warblers are specialists of large marshes. Males advertize their breeding territory by singing loudly atop long reed stems, often within meters of where the female is building her nest. The mating system is opportunistically polygynous, and males with larger song repertoires are more likely to attract two or more females to nest in their territories. Nest building and incubation are carried out by the female only, and the male may assist her with feeding the young in one or more nests.

This species is one of the European birds most commonly parasitized by Common Cuckoos. To protect the nests and its contents, these hosts attack approaching cuckoos, and also look for foreign eggs to eject from the nest. But the cuckoo's eggs are excellent mimics: they resemble this host's eggs in size, color, and maculation, although not in shape. The cuckoos try to avoid being spotted near the nest by laying their eggs in the afternoon when the female warbler is busy feeding before she lays her next egg, the following morning.

Actual size

ORDER	Passeriformes
FAMILY	Locustellidae
BREEDING RANGE	Central and eastern Europe and western Asia
BREEDING HABITAT	Dense, deciduous forests, bogs, wooded swamps, riparian corridors
NEST TYPE AND PLACEMENT	Cup nest at the base of a tree or low in a shrub; made of grasses, stems, and leaves, lined with finer grass and hair
CONSERVATION STATUS	Least concern
COLLECTION NO.	FMNH 18983

ADULT BIRD SIZE
5½–6½ in (14–16 cm)

INCUBATION
11–12 days

CLUTCH SIZE
5–7 eggs

LOCUSTELLA FLUVIATILIS

EURASIAN RIVER WARBLER

PASSERIFORMES

This species and its relatives produce one of the most monotone songs among songbirds. A simple trill repeated over and over, it resembles the call of a hefty locust or cricket (as reflected in the bird's Latin name, *Locustella*). The song is produced from a prominent perch, but at other times this drab bird remains cryptic and hard to spot. The song functions to advertize and maintain the territorial boundaries defended by each male, after they return from spring migration days ahead of the females.

A common inhabitant of riverine and boggy forests and wooded swampland, this species relies on hiding in the dense understory for the success of its nesting attempts. It is extremely sensitive to disturbance during the laying period, so that many nests are abandoned, even with a partial clutch already laid. The nestlings disperse from the vulnerable ground nest before they are fully flighted (volant), and wait in low shrubs to be fed until they reach full independence.

Clutch

Actual size

The egg of the Eurasian River Warbler is olive gray in coloration, has dense, dark brown maculation, and its dimensions are ¾ x 9⁄16 in (19 x 14 mm). During incubation, only the female attends the nest.

ORDER	Passeriformes
FAMILY	Cisticolidae
BREEDING RANGE	Northeast Africa, southwestern Asia
BREEDING HABITAT	Understory in forests, open woodlands, and subtropical scrub and gardens
NEST TYPE AND PLACEMENT	Grass nest cup, placed in low bush or in tussocks of grasses
CONSERVATION STATUS	Least concern
COLLECTION NO.	FMNH 19022

ADULT BIRD SIZE
4–4½ in (10–11 cm)

INCUBATION
11–13 days

CLUTCH SIZE
3–5 eggs

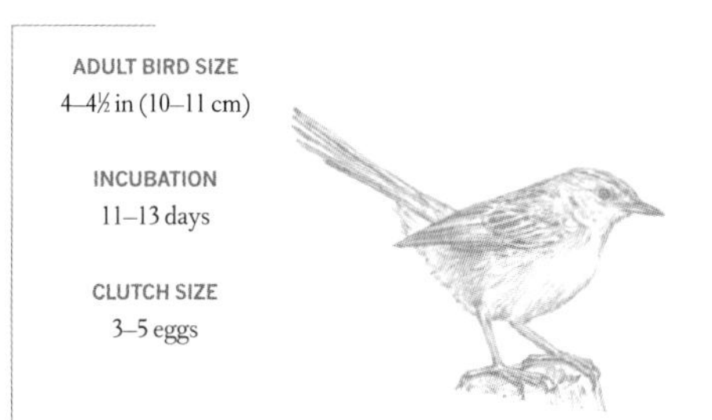

PRINIA GRACILIS

GRACEFUL PRINIA

PASSERIFORMES

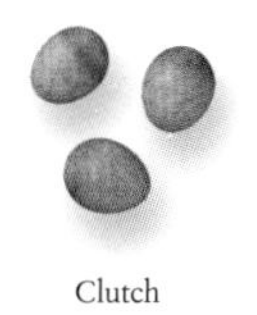

Clutch

Prinias are a widespread group of long-tailed, warblerlike birds, occurring in the tropical and subtropical ranges of Africa and Asia. The Graceful Prinia inhabits thickets and brush, but it also readily takes to living in plantations and parks, as long as there is plenty of cover for feeding and breeding.

The onset of the reproductive period is marked by the aggressive territorial displays, threats, and songs of the male, which function to repel intruders and to attract females. Once he is mated, the male also initiates construction of the nest, although both members of the pair complete it, and incubate the eggs. The nestlings are also provisioned by both parents, but the female may begin a second brood soon after fledging, while the male continues to feed the mobile, but dependent young.

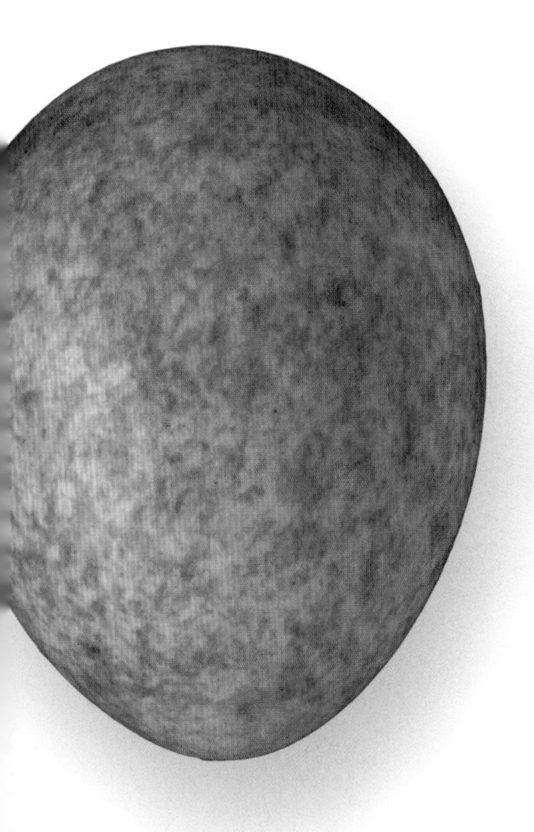

Actual size

The egg of the Graceful Prinia is pale green in background coloration, has reddish maculation, and measures ⅔ x ½ in (17 x 12 mm). In much of this species' range, daytime temperatures are so high that the eggs are left uncovered for most of the incubation period.

ORDER	Passeriformes
FAMILY	Sylviidae
BREEDING RANGE	Temperate Europe and western Asia
BREEDING HABITAT	Mature forests with a dense understory, also hedgerows, gardens and parks with bushes
NEST TYPE AND PLACEMENT	Cup made of rootlets and grasses, lined with hairs and other fibers, in a low bramble or scrub
CONSERVATION STATUS	Least concern
COLLECTION NO.	FMNH 2439

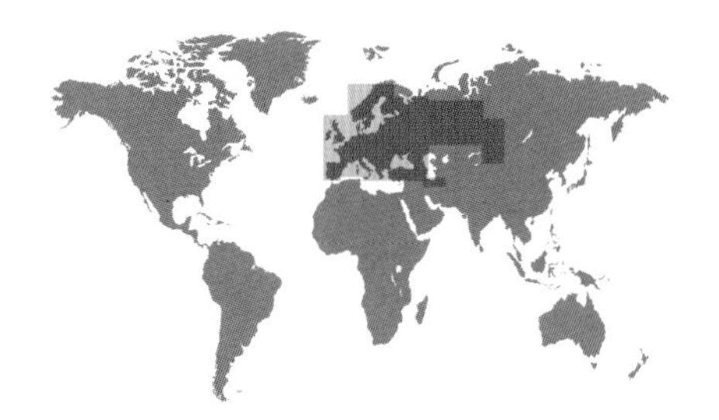

SYLVIA ATRICAPILLA

BLACKCAP

PASSERIFORMES

ADULT BIRD SIZE
4½–5 in (11–13 cm)

INCUBATION
11 days

CLUTCH SIZE
4–6 eggs

The Blackcap is a common and familiar warbler that often sings from tall trees at forest edges and in parks across Europe. In most populations, it is migratory to and from its African wintering grounds, with individuals from the same population following the same migratory routes. Cross-breeding experiments have shown that the direction of migratory flight is under strong genetic control, and hybrid offspring of parents from different populations with different migratory paths show intermediate flight directions.

On the breeding grounds, to defend their nest from predators and parasites, Blackcaps mob the attacker with loud calls and flight approaches. These mobbing calls attract many other birds, mostly other species, to the vicinity of the nest. However, these birds do not join the mobbing, indicating that they are curious, but unhelpful, bystanders while the Blackcap fights on its own for the safety and survival of the eggs or nestlings.

Clutch

The egg of the Blackcap is buff in background coloration, with gray and brown blotching, and is ¾ x ⅝ in (20 x 15 mm) in size. Both parents incubate the eggs, and are very efficient in detecting brood parasitism. Experiments show they use the brightness and the color of the blunt pole of eggs to identify parasitic cuckoo eggs for ejection from the nest.

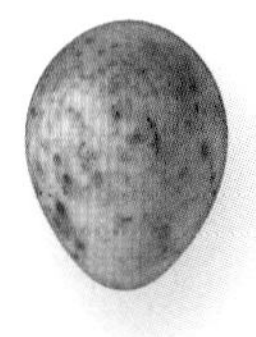

Actual size

ORDER	Passeriformes
FAMILY	Sylviidae
BREEDING RANGE	West coast of North America
BREEDING HABITAT	Sage scrub, chaparral in coastal plateaus and valleys
NEST TYPE AND PLACEMENT	Tight, open cup made out of bark and silk, lined with grasses; placed in branch forks and crotches of trees and shrubs
CONSERVATION STATUS	Least concern
COLLECTION NO.	FMNH 19129

ADULT BIRD SIZE
5½–6 in (14–15 cm)

INCUBATION
14 days

CLUTCH SIZE
3–4 eggs

CHAMAEA FASCIATA

WRENTIT

PASSERIFORMES

Clutch

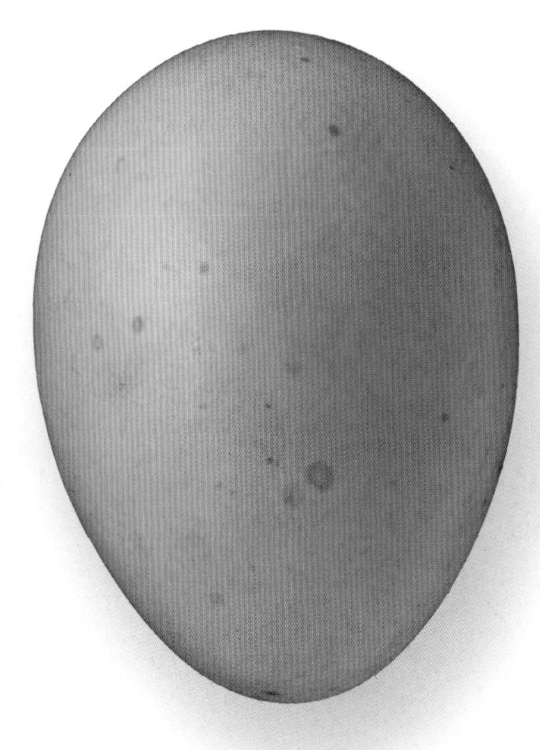

This is a long-lived and nonmigratory species; the breeding territories are defended year-round, and the pair bond remains in place across years until one of the mates dies or disappears. Parental care is shared equitably, from nest building, through incubation, and from provisioning the nestlings to feeding the fledglings. The female and male maintain close physical, visual, or acoustic contact at all times, both during the breeding season and outside.

There are a few species of birds whose biology is mostly and best known from the dedicated work of just a handful of scientists. The Wrentit is just such a species, with much of its life history described from a four-year study conducted as a Ph.D. thesis in the arid hills above the University of California, Berkeley, by a pioneering ornithologist, Mary Erickson. By color-banding individuals, she was able to make observations on the social, seasonal, and breeding behaviors of this bird.

Actual size

The egg of the Wrentit is white or pale greenish blue in background color, clear of spots (but may become stained during incubation) and its dimensions are ⅔ x 9⁄16 in (17 x 14 mm). Wrentits always build a brand new nest to house each clutch of eggs, but they may reuse a previous nest's construction material to save some time and energy.

ORDER	Passeriformes
FAMILY	Leiothrichidae
BREEDING RANGE	Southeast and east Asia
BREEDING HABITAT	Moist hillside forests, with dense understory
NEST TYPE AND PLACEMENT	Hanging, open cup, composed of dry leaves, grasses, mosses, and lined with fibers and hairs; suspended from the fork of a tree or a brush branch
CONSERVATION STATUS	Least concern
COLLECTION NO.	FMNH 20779

ADULT BIRD SIZE
6–6½ in (15–16 cm)

INCUBATION
11–14 days

CLUTCH SIZE
2–4 eggs

LEIOTHRIX LUTEA

RED-BILLED LEIOTHRIX

PASSERIFORMES

Outside the breeding season, young and adult birds flock and move together through the forest, gleaning insects off leaves and branches. When the breeding season arrives, pairs re-form on their own, and in captivity this species mates for life. Both members of the pair contribute to the construction of the elaborate nest, and to the parental duties to raise the next generation. To defend the eggs, when an incubating parent is flushed from the nest, it flies off quietly, but then it puts on a loud and conspicuous wing-flutter display to entice the predators away from the nest.

The bright plumage colors, and pleasantly melodious song, make this species a popular cage bird in its native Asian distribution, and around the globe. Its popularity may have also contributed to intentional releases of captive birds in Hawaii, Australia, Tahiti, and even England. While some introduced populations fail to establish, the Hawaiian birds now represent a stable but exotic member of the local avifauna.

Clutch

Actual size

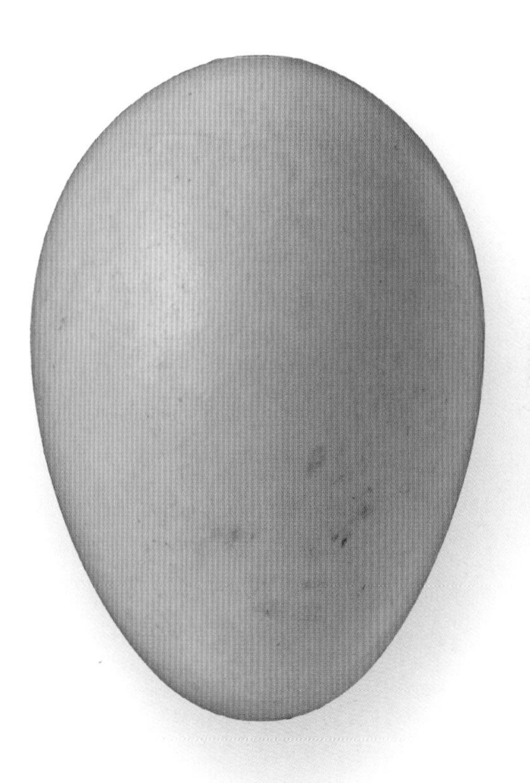

The egg of the Red-billed Leiothrix is pale blue in background, with reddish brown maculation, and is ¾ x ½ in (19 x 14 mm) in size. Both parents incubate the eggs, but only the female develops an incubation patch and she also remains on the eggs at night.

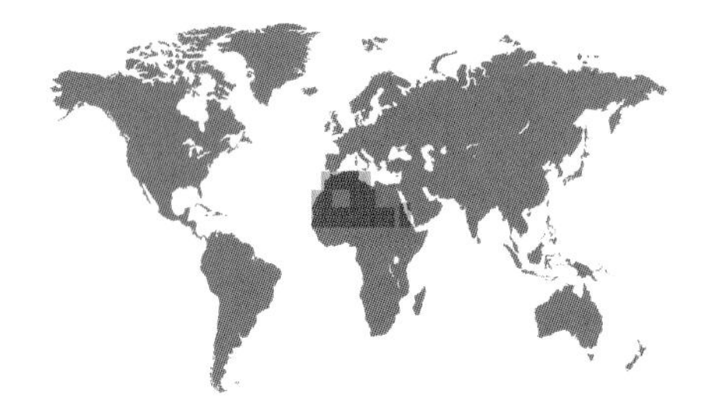

ORDER	Passeriformes
FAMILY	Leiothrichidae
BREEDING RANGE	Northern Africa
BREEDING HABITAT	Desert scrub, arid grassland with thickets, dry river beds
NEST TYPE AND PLACEMENT	Loose, deep cup, made of thin twigs and grasses; placed in dense canopy in palm crown or thicket
CONSERVATION STATUS	Least concern
COLLECTION NO.	FMNH 18784

ADULT BIRD SIZE
8½–10 in (22–25 cm)

INCUBATION
13–15 days

CLUTCH SIZE
4–5 eggs

TURDOIDES FULVA

FULVOUS CHATTERER

PASSERIFORMES

Clutch

The egg of the Fulvous Chatterer is light blue in coloration, immaculate, and measures 1 x ¾ in (25 x 19 mm). In territory defense and throughout breeding, the core pair is assisted by their older progeny who help at the nest.

The dozens of species in the genus *Turdoides*, often referred to as babblers, are widespread across subtropical and tropical regions in Africa and Asia, and have shown much affinity for human settlements. For example, the Fulvous Chatterer can become very tame in desert oases and arid backyards, especially when food is provided. This species is resident all year round, unless prolonged dry periods, or lack of permanent water sources, force it to migrate locally looking for regions that have received rain.

Perhaps one reason for this species to become tame near people is that the life history of babblers revolves around sociality, and repeated interactions with familiar individuals. For instance, groups of four to five birds are seen often, and these also represent the breeding unit: a core pair, one or more of that pair's mature progeny, and recently fledged juveniles, which have yet to decide to disperse or remain.

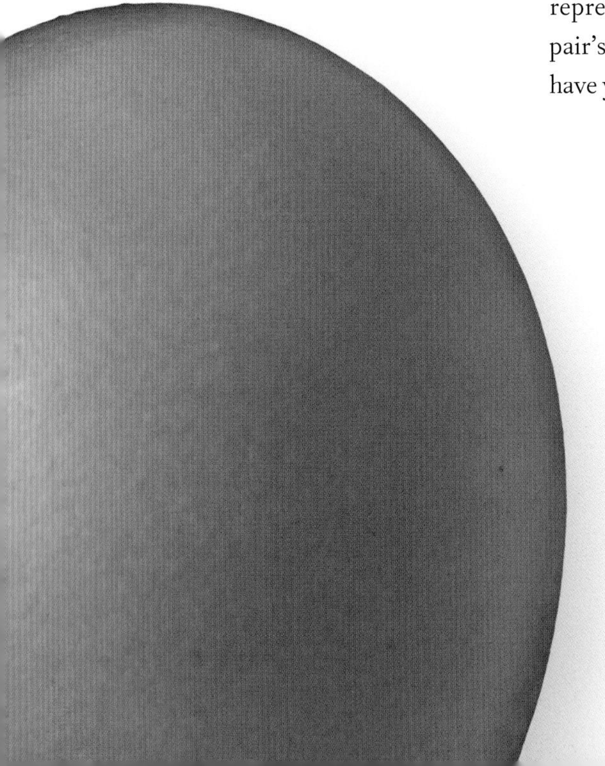

Actual size

ORDER	Passeriformes
FAMILY	Muscicapidae
BREEDING RANGE	Europe, western Asia, and north Africa
BREEDING HABITAT	Deciduous forests, near clearings and edges; also parks and orchards
NEST TYPE AND PLACEMENT	Cup nest made of grasses, twigs, and mosses, lined with finer grasses and rootlets; placed in open natural cavities, ledges, shallow recesses in walls, or open-faced nestboxes
CONSERVATION STATUS	Least concern
COLLECTION NO.	FMNH 1620

ADULT BIRD SIZE
4½–5½ in (12–14 cm)

INCUBATION
11–15 days

CLUTCH SIZE
4–6 eggs

MUSCICAPA STRIATA

SPOTTED FLYCATCHER

PASSERIFORMES

For this migratory species, establishing territories soon after arrival from the wintering grounds is paramount. The males settle as soon as they arrive and defend their territory by singing loudly and continuously from high perches in the canopy, separated by 650–1,000 ft (200–300 m) from neighboring males. Once the female arrives, a pair bond has been established, and the nest is completed, the territory size often shrinks and becomes limited to a foraging range of just 165–330 ft (50–100 m) from the nest site itself.

The relatively open nests of this species are vulnerable to predation by mammals and birds, especially by squirrels. But buildings, walls, and bird houses also provide suitable nesting sites, resulting in a pattern that, near human settlements, these flycatchers often are more successful in hatching the eggs and raising the young than they are in more natural habitats.

Clutch

The egg of the Spotted Flycatcher is beige, buff, or brown in background color, with darker maculation, and is ¾ x ½ in (18 x 14 mm) in size. The female alone incubates the eggs, but the chicks and the fledglings are provisioned by both parents.

Actual size

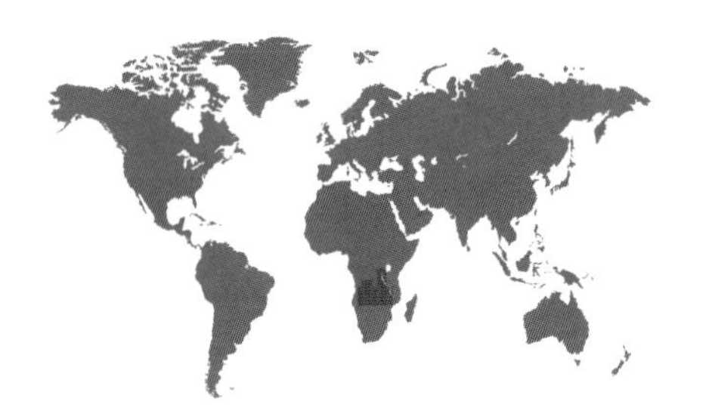

ORDER	Passeriformes
FAMILY	Muscicapidae
BREEDING RANGE	Southern tropical Africa
BREEDING HABITAT	Dry woodlands, open scrub
NEST TYPE AND PLACEMENT	Lined cup nest, in low bushes or on the ground
CONSERVATION STATUS	Least concern
COLLECTION NO.	FMNH 14773

ADULT BIRD SIZE
5½–6¾ in (14–17 cm)

INCUBATION
11–17 days

CLUTCH SIZE
3–8 eggs

CERCOTRICHAS BARBATA

MIOMBO SCRUB-ROBIN

PASSERIFORMES

Clutch

Miombo is a specific ecosystem in sub-Saharan African, comprised of mostly savanna-type, open woodlands, and trees in the genus *Brachystegia*, or "miombo" in Swahili. The several species of scrub-robins that occur in this habitat share similar plumage patterns and vocalizations; the taxonomic status of each species and their boundaries remain unclear and require further fieldwork and genetic analyses. For example, the Miombo Scrub-Robin quickly responds to playbacks of the songs of both its own species, and the similar Bearded Scrub-Robin (*Cercotrichas quadrivirgata*) with which it shares genetic and ecological affinities.

The reproductive strategies of the Miombo Scrub-Robin remain poorly studied; its relatives show a male-based territorial system, with the females in charge of nest building and incubation. These birds forage predominantly on the ground, seeking and capturing insects to feed themselves. Some females have been seen carrying insects in their beak, presumably to feed to their hungry young in the nest.

The egg of the Miombo Scrub-Robin is pale green in background color, has heavy chestnut blotching and spotting, and is ⅞ x ⅝ in (22 x 16 mm) in size. Little is known about the nesting biology of this species, illustrating how much we have to learn about many tropical birds.

Actual size

ORDER	Passeriformes
FAMILY	Muscicapidae
BREEDING RANGE	Central and eastern Europe, western Asia
BREEDING HABITAT	Forests, wet woodlands, and open parkland, with a dense understory
NEST TYPE AND PLACEMENT	Cup nest, lined with grass; placed on the ground or low in dense and thorny brush
CONSERVATION STATUS	Least concern
COLLECTION NO.	FMNH 18842

ADULT BIRD SIZE
4½–5½ in (12–14 cm)

INCUBATION
13–15 days

CLUTCH SIZE
4–6 eggs

ERITHACUS LUSCINIA

THRUSH NIGHTINGALE

PASSERIFORMES

Like its fabled relative the Common Nightingale (*Luscinia megarhynchos*), this species is well known for its melodious and varied song, which it also often performs during the quietest time in the forest—at night. The songs both attract mates and announce their territories to competitors and neighbors. Females seeking a potential mate, listening to their mate, or considering a quick foray into a neighboring male's territory, can easily hear the nocturnal songs more than half a mile away.

Clutch

The females arrive from migration seven to ten days after the males, and are immediately courted by a resident male. Once the monogamous pair is formed, the female begins nest building, followed by incubation, unassisted by the male. She also broods the hatchlings on her own, but once the nestlings demand more food, both parents contribute to provisioning the next generation.

Actual size

The egg of the Thrush Nightingale is dark, olive brown in coloration all over the eggshell, and ⅞ x ⅔ in (22 x 17 mm) in size. If the eggs are lost late in the incubation period, this species does not attempt a second brood, probably due to the brief breeding season in the far north.

ORDER	Passeriformes
FAMILY	Muscicapidae
BREEDING RANGE	Central and eastern Europe, temperate Asia
BREEDING HABITAT	Deciduous forests, open woodlands, parks
NEST TYPE AND PLACEMENT	In a tree hole, excavated by woodpeckers, also in nestboxes; builds an open cup of grasses, leaves, and rootlets
CONSERVATION STATUS	Least concern
COLLECTION NO.	FMNH 20849

ADULT BIRD SIZE
4½–5½ in (12–14 cm)

INCUBATION
13–15 days

CLUTCH SIZE
5–7 eggs

FICEDULA ALBICOLLIS

COLLARED FLYCATCHER

PASSERIFORMES

Clutch

The egg of the Collared Flycatcher is pale, light blue in background coloration, and measures ⅔ x ½ in (17 x 13 mm) in dimensions. Incubating females use less energy and so maintain a stable and high body mass; once the eggs hatch, they spend more time on the wing, in search of insects for the young to eat, and quickly lose body mass.

Collared Flycatchers are migratory woodland songbirds. The boldly patterned, black-and-white males arrive from the wintering grounds before the females, and establish breeding territories which invariably include one or more suitable cavities for nesting. The drabber females evaluate the quality of the male's territory and the size of his bright white forehead patch to choose a mate. Once the pair bond is formed, the male often goes off seeking extra-pair matings or even to attract another female, especially when his plumage is extra bright.

Females are in charge of incubating the eggs, but males may help feed the nestlings. However, males typically only help one of the females breeding on their territory, the one whose eggs hatch soonest. This complicates the mate-choice decisions for the females. The result is that females choose less bright males early in the season, to avoid losing him to extramarital forays; in turn, a late-nesting female tends to choose a brighter male, because although he is unlikely to help at all, at least his good genes will contribute to the genome of the female's own young.

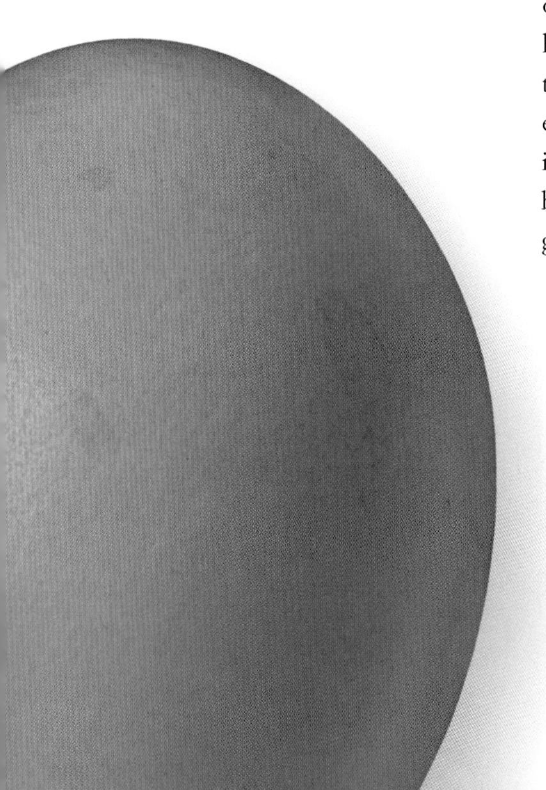

Actual size

ORDER	Passeriformes
FAMILY	Muscicapidae
BREEDING RANGE	North Africa, warm temperate Eurasia
BREEDING HABITAT	Open, grassy areas in arid mountains and hills
NEST TYPE AND PLACEMENT	In rock crevices, holes in cliffs or walls, ground holes under boulders; grass cup nest, lined with rootlets and moss
CONSERVATION STATUS	Least concern
COLLECTION NO.	FMNH 18843

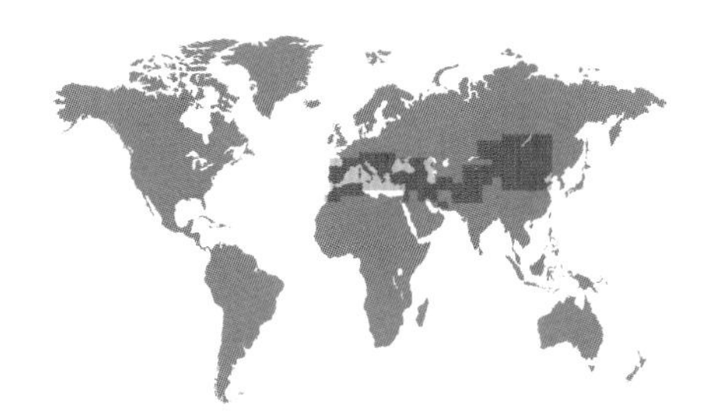

ADULT BIRD SIZE
6½–8 in (17–20 cm)

INCUBATION
14–15 days

CLUTCH SIZE
4–5 eggs

MONTICOLA SAXATILIS

RUFOUS-TAILED ROCK-THRUSH

PASSERIFORMES

This is a dramatically dichromatic species, with the males boasting a conspicuous blue-and-rusty plumage, while the female is drabber with gray and brown patterns. The male uses his colorful feathers, including a tail-spread display, and melodious, tonal song to attract the female. In captivity, the male will incorporate other birds' songs and calls into his own song, but it is not known whether such vocal mimicry also occurs in the wild.

Throughout much of its range, this species has declined because of the transformation of open habitats into agricultural fields. Its need to forage for insects and small lizards is still met, however, on steeper, arid montane meadows and fields. The parents feed their young a predominantly insect-based diet, which they capture on the ground, after dropping from a perch. On foraging forays to feed their young, they make short flights from perch to perch looking for prey.

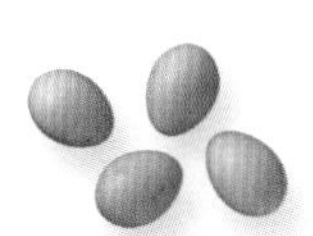

Clutch

The egg of the Rufous-tailed Rock-Thrush is greenish blue in coloration, and 1 x ¾ in (25 x 19 mm) in size. The female alone incubates the eggs, relying on her cryptic plumage to access the nest entrance without being spotted by falcons.

Actual size

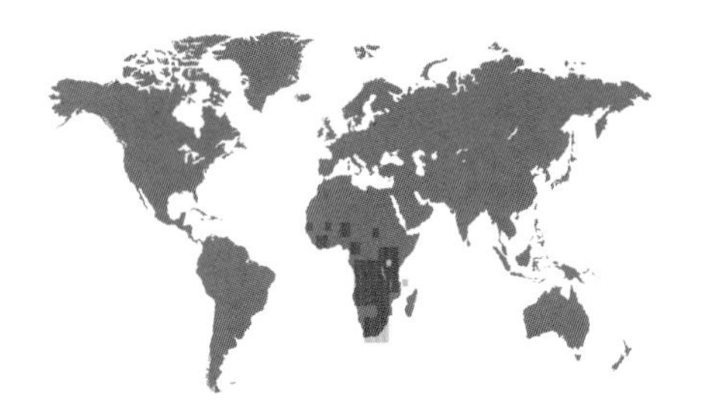

ORDER	Passeriformes
FAMILY	Muscicapidae
BREEDING RANGE	Eastern and southern sub-Saharan Africa
BREEDING HABITAT	Montane, open, shrubby grasslands, pastures
NEST TYPE AND PLACEMENT	Deep, open cup, made with leaves, stems, hay, and rootlets, lined with finer materials, including grasses, hair, wool, and feathers; hidden under a tuft or dry grasses
CONSERVATION STATUS	Least concern
COLLECTION NO.	FMNH 14775

ADULT BIRD SIZE
4½–5 in (12–13 cm)

INCUBATION
12–14 days

CLUTCH SIZE
2–4 eggs

SAXICOLA TORQUATUS

AFRICAN STONECHAT

PASSERIFORMES

Clutch

The female African Stonechat builds the nest, incubates and broods the young; the male assists with feeding the chicks. The chicks leave the nest around two weeks after hatching, but remain with their parents for up to four months, after which they disperse to look for their own mates and territories. Stonechats nesting at the equator complete just a single breeding cycle per season, whereas in captivity, they raise two to three broods in a year.

This species was recently split taxonomically from other stonechats that are widespread across Europe and Asia. African Stonechats live in fields in open areas, but their local populations are scattered across the continent, many of them in arid highland and semi-arid montane regions. While temperate zone stonechat species are migratory, most African Stonechats defend year-round territories.

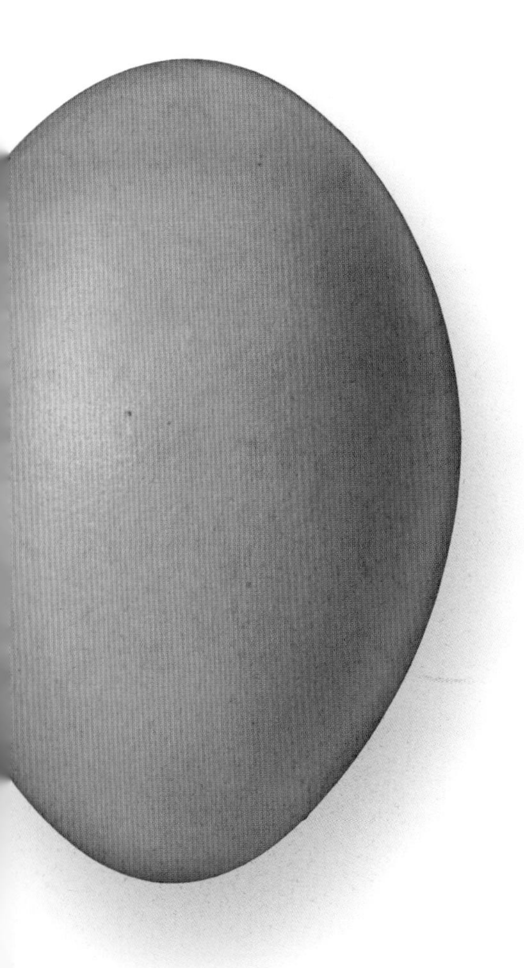

Actual size

The egg of the African Stonechat is bluish green in color, has light yellow or cinnamon blotching, and measures ¾ x 9⁄16 in (19 x 14 mm) in size. The clutch size is under strong genetic control; when bred in captivity, indoors or outdoors in the temperate zone, African Stonechats continue to lay three eggs per nest.

ORDER	Passeriformes
FAMILY	Muscicapidae
BREEDING RANGE	Western Asia
BREEDING HABITAT	Dry grassy, semidesert, and stony hillsides
NEST TYPE AND PLACEMENT	In rock crevices; built from dry grass, lined with finer grass, hair and feathers
CONSERVATION STATUS	Least concern
COLLECTION NO.	FMNH 20689

ADULT BIRD SIZE
5½–6½ in (14–16 cm)

INCUBATION
12–13 days

CLUTCH SIZE
4–6 eggs

OENANTHE FINSCHII

FINSCH'S WHEATEAR

PASSERIFORMES

This species inhabits dry and desert areas; it does not require permanent or seasonal access to water to begin breeding. After a relatively short-distance migration from the Middle East, the males establish breeding territories; they advertize by perching on high boulders in their distinctive black-and-white nuptial plumage, both to attract females and to keep intruder males out. On the wintering grounds, both sexes maintain feeding territories, with the males being dominant over the females in territorial disputes.

To build a nest, these birds seek out shaded and hidden crevices under rocks and on cliff faces, but also in old stone walls. Where boulders and walls are sparse, such as in remote and sandy semidesert regions, they may build a nest inside the tunnel of an abandoned rodent hole. The female's brown-and-gray plumage keeps her camouflaged from predators during the incubation period.

Clutch

The egg of the Finsch's Wheatear is pale blue in background color, with reddish brown maculation, and is ⅞ x ⅝ in (22 x 16 mm) in dimensions. The female alone incubates the eggs, and the male contributes to feeding the young, with both parents carrying caterpillars and other arthropods in their beak to the nest.

Actual size

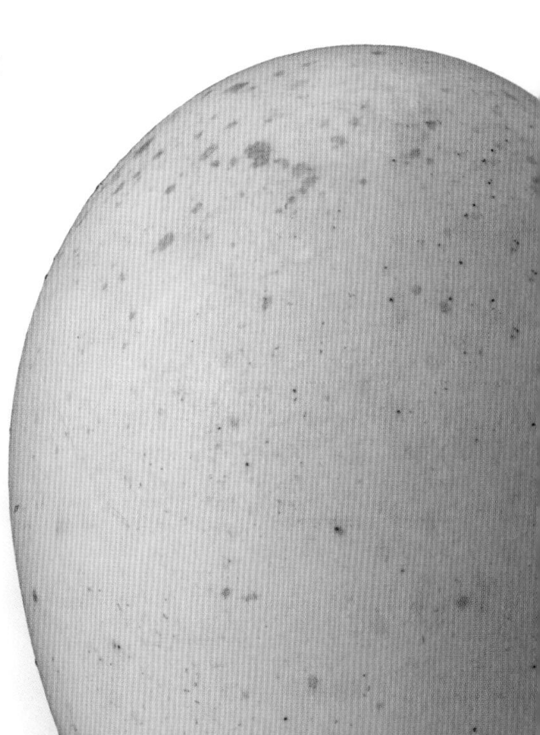

ORDER	Passeriformes
FAMILY	Turdidae
BREEDING RANGE	Eastern and central North America, and northern Central America
BREEDING HABITAT	Forest clearings, savanna, open woodlands, farms, yards, and parks
NEST TYPE AND PLACEMENT	Cavity nest, in natural holes or nestboxes; a cup built on a straw platform, lined with fine grass and hair
CONSERVATION STATUS	Least concern
COLLECTION NO.	FMNH 12147

ADULT BIRD SIZE
6½–8½ in (16–21 cm)

INCUBATION
11–19 days

CLUTCH SIZE
3–7 eggs

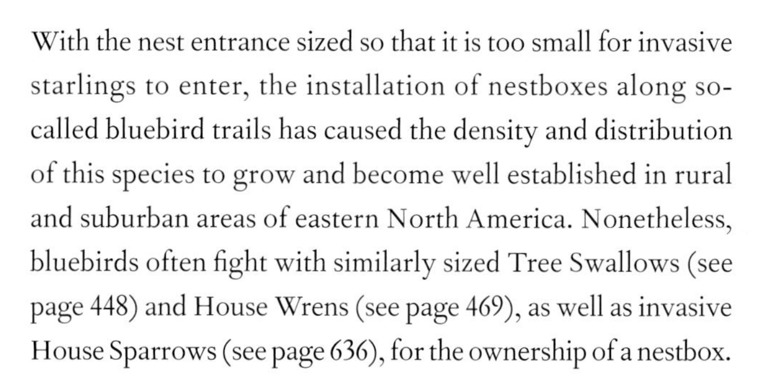

SIALIA SIALIS

EASTERN BLUEBIRD

PASSERIFORMES

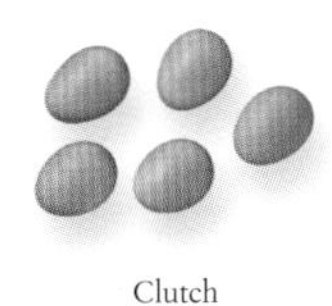

Clutch

The egg of the Eastern Bluebird is light blue in coloration, immaculate, and 13⁄16 x ⅔ in (21 x 17 mm) in dimensions. The clear blue egg is a trait shared among many thrush-like birds, consistent with the close evolutionary affinities between bluebirds and thrushes.

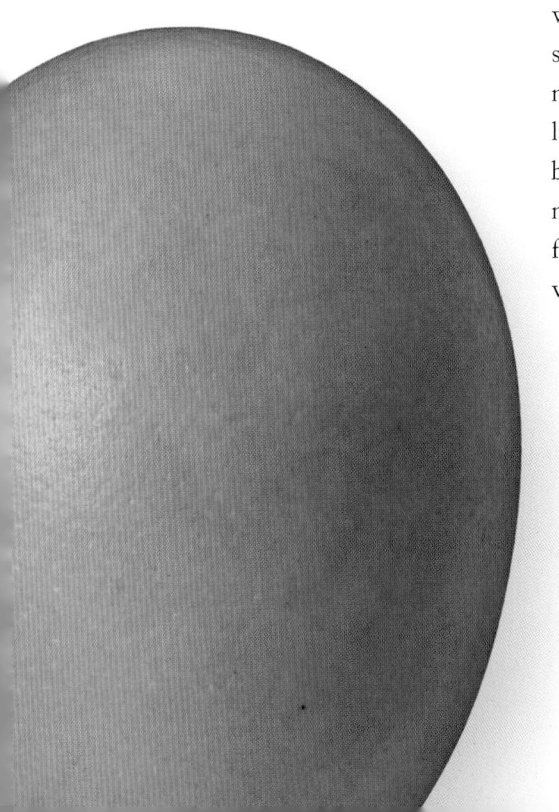

With the nest entrance sized so that it is too small for invasive starlings to enter, the installation of nestboxes along so-called bluebird trails has caused the density and distribution of this species to grow and become well established in rural and suburban areas of eastern North America. Nonetheless, bluebirds often fight with similarly sized Tree Swallows (see page 448) and House Wrens (see page 469), as well as invasive House Sparrows (see page 636), for the ownership of a nestbox.

The male attracts a female to a potential nest by carrying straw in his bill, but it is the female who is in full charge of building the nest and incubating the eggs; the male then helps equitably with feeding the chicks. When nesting cavities or boxes are scarce, younger birds continue to prospect for abandoned or new cavities throughout the spring and summer, and may start laying eggs very late in the season, instead of missing out on breeding for a full year altogether. Throughout this bird's range, nest cams have been placed to capture life in the bluebird nest, from building and laying through hatching and fledging; the video footage is streamed directly to the internet.

Actual size

ORDER	Passeriformes
FAMILY	Turdidae
BREEDING RANGE	Western, coastal North America
BREEDING HABITAT	Open woodlands, gallery forests, riparian areas
NEST TYPE AND PLACEMENT	In natural tree cavities, old woodpecker holes, or nestboxes; nest cup lined with dry grasses, rootlets, feathers, and hair
CONSERVATION STATUS	Least concern
COLLECTION NO.	FMNH 12120

ADULT BIRD SIZE
6–7 in (15–18 cm)

INCUBATION
12–17 days

CLUTCH SIZE
4–5 eggs

SIALIA MEXICANA

WESTERN BLUEBIRD

PASSERIFORMES

The bright blue feathers of Western Bluebirds are not only pleasant to look at against the dry grasses of its open woodland habitat, they also convey critical information about each individual. For example, the bluer the head, the older the male; and the larger his rufous breast patch, the healthier he is. Pairs are socially monogamous, with helpers—often mature sons of the resident pair—contributing to territory defense and feeding the young.

Western Bluebirds can plan for the future. In the fall, before food becomes scarce, the dominant male makes a decision about whether to tolerate independent young (who may later become helpers) or to actively chase them away from the family territory. He bases this decision on the potential availability of a critical winter food resource: mistletoe berries. Mistletoes produce berries only later in the year, so he apparently assesses the presence of mistletoe bunches alone: when scientists removed mistletoe bunches from the bluebirds' territories in the summer, the young were less likely to remain in the territory through the winter.

Clutch

The egg of the Western Bluebird is pale blue in background color, clear of spotting, and 13⁄16 x 5⁄8 in (21 x 16 mm) in size. Like many birds, while the female is laying her eggs, she spends little time at or around the nest until the clutch is complete and incubation begins in earnest.

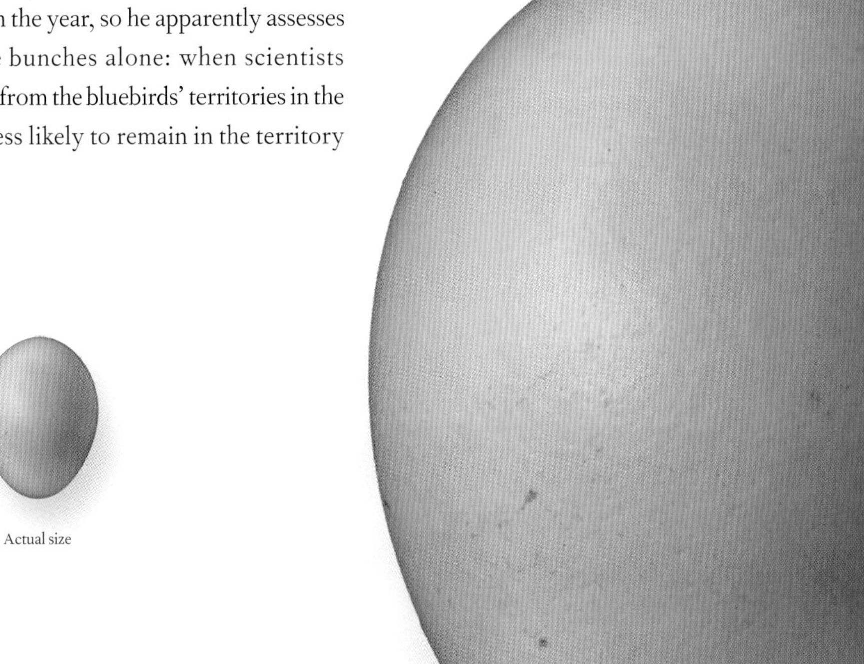

Actual size

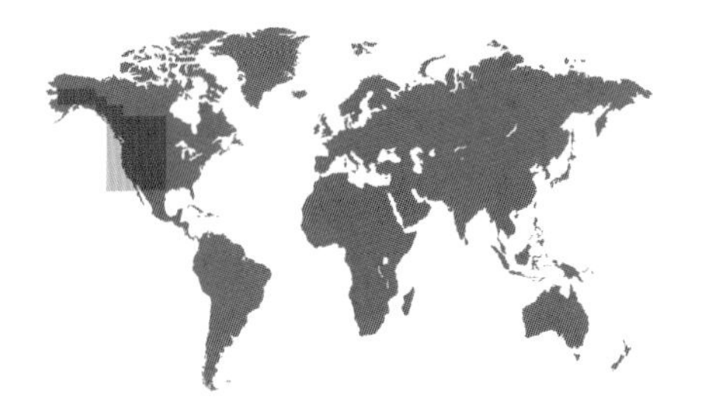

ORDER	Passeriformes
FAMILY	Turdidae
BREEDING RANGE	Western, montane and Arctic North America
BREEDING HABITAT	Open wooded fields, burnt slopes, forest clearings and clear-cuts, pastures with trees
NEST TYPE AND PLACEMENT	Preexisting cavities in trees or snags, also nestboxes; cup nest of woven grasses, lined with finer grass, shredded bark, hair, and feathers
CONSERVATION STATUS	Least concern
COLLECTION NO.	FMNH 16904

ADULT BIRD SIZE
6½–8 in (16–20 cm)

INCUBATION
13–14 days

CLUTCH SIZE
4–5 eggs

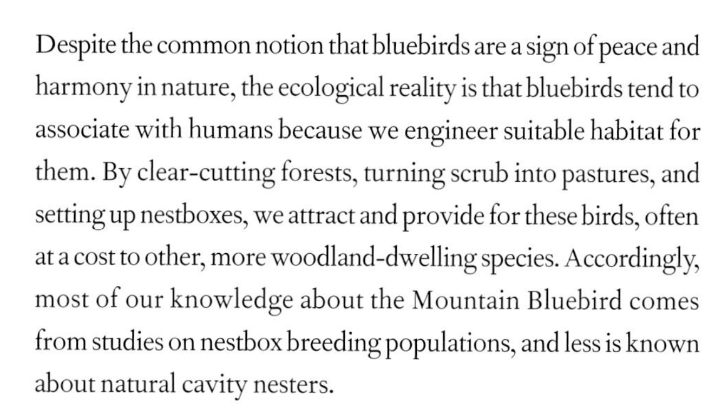

SIALIA CURRUCOIDES

MOUNTAIN BLUEBIRD

PASSERIFORMES

Clutch

The egg of the Mountain Bluebird is light blue, immaculate, and measures ⅞ x ⅔ in (22 x 17 mm) in size. The female bluebird sits very tightly on the eggs, and she may not flush even if a nestbox is approached and opened for inspection.

Despite the common notion that bluebirds are a sign of peace and harmony in nature, the ecological reality is that bluebirds tend to associate with humans because we engineer suitable habitat for them. By clear-cutting forests, turning scrub into pastures, and setting up nestboxes, we attract and provide for these birds, often at a cost to other, more woodland-dwelling species. Accordingly, most of our knowledge about the Mountain Bluebird comes from studies on nestbox breeding populations, and less is known about natural cavity nesters.

Although the male engages in nest-building movements and motions, he is only displaying to the female, and does not actually carry nesting materials in his bill. Instead, the nest is constructed only by the female, and she is also in charge of incubating the eggs and brooding the hatchlings. However, the male does respond to her begging displays and calls, feeding her on or near the nest throughout the process. The male also shares the feeding of the nestlings with the female.

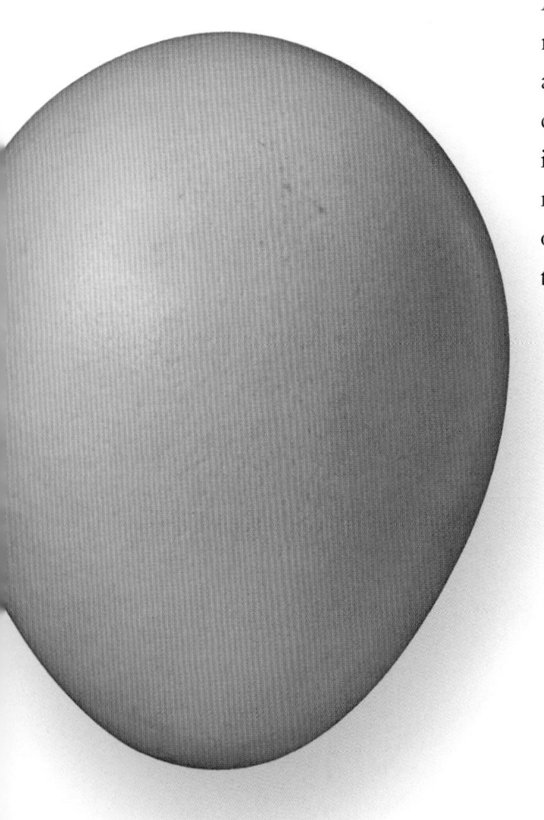

Actual size

ORDER	Passeriformes
FAMILY	Turdidae
BREEDING RANGE	Western North America
BREEDING HABITAT	High mountains, coniferous forests near the tree line
NEST TYPE AND PLACEMENT	Cup of pine needles, lined with grass stems or bark strips; placed on sloping ground or under cliff overhang, especially often along soil banks near dirt roads
CONSERVATION STATUS	Least concern
COLLECTION NO.	FMNH 2540

MYADESTES TOWNSENDI

TOWNSEND'S SOLITAIRE

PASSERIFORMES

ADULT BIRD SIZE
8–8½ in (20–22 cm)

INCUBATION
11 days

CLUTCH SIZE
2–5 eggs

This species breeds in montane forest habitats at high elevation, often reaching into areas just above tree level. It is unknown whether the pairs form locally at the breeding site, or earlier, during migration. Once the mates settle, nesting begins in a hurry. The pair selects a potential site together, visiting it repeatedly, until the female begins nest construction. The male occasionally picks up nesting materials, but he does not carry them. The female alone completes the nest: she first scrapes a horizontal depression into sloping dirt, then builds a platform of twigs, before weaving a grass cup on top, in which she will lay and incubate the eggs.

Despite the solitaire's high-elevation habitats and hidden nest locations, predators often find and eat their eggs and broods. To avoid a fully failed breeding season, these birds follow several strategies to secure reproductive success: they begin nesting early, quickly replace lost clutches, and continue breeding late into the season.

Clutch

The egg of the Townsend's Solitaire is dull white, pink, or greenish blue in background coloration, has darker maculation, and measures ⅞ x ⅔ in (23 x 17 mm). The male visits the nest and provisions the female while she incubates the eggs and broods the hatchlings.

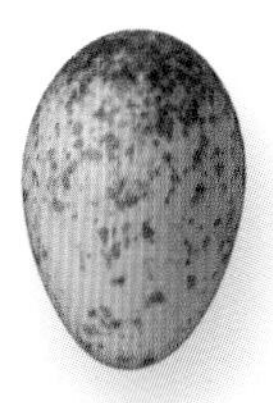

Actual size

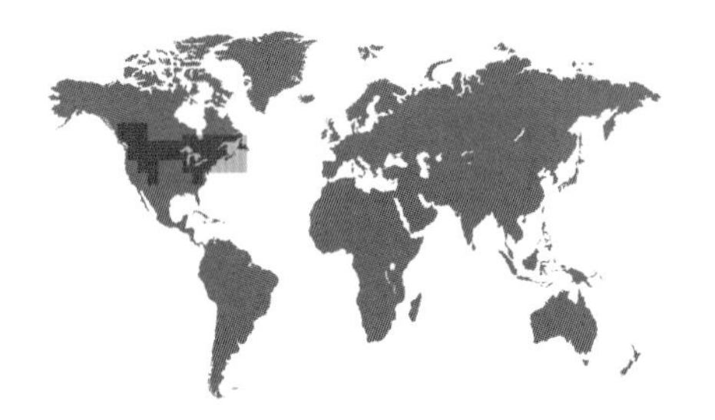

ORDER	Passeriformes
FAMILY	Turdidae
BREEDING RANGE	North temperate North America
BREEDING HABITAT	Moist, or swampy, deciduous forests, with dense understory
NEST TYPE AND PLACEMENT	Open cup of dead leaves, bark, stems, lined with rootlets and fibers; on ground, at base of or low in small trees and stumps
CONSERVATION STATUS	Least concern
COLLECTION NO.	FMNH 15149

ADULT BIRD SIZE
6½–7 in (17–18 cm)

INCUBATION
10–14 days

CLUTCH SIZE
3–5 eggs

CATHARUS FUSCESCENS

VEERY

PASSERIFORMES

Clutch

The egg of the Veery is pale greenish blue in color, has no speckles, and its dimensions are ⅞ x ⅔ in (22 x 17 mm). The female alone builds the nest and incubates the eggs, while the male patrols the territory. He only joins in with the parental duties to feed the chicks and fledglings.

The Veery is a long-distance migrant, making its way twice yearly across the Gulf of Mexico to and from South American wintering grounds. As part of the northwardly spring migration, males prepare for breeding by coming into reproductive condition, with enlarged testes, while en route. The females arrive a week later, and use the males' eerie descending song to guide their decision about pair bonding. Both the female and male sing in courtship duets at the early stages of the mating decision; this is quickly followed by the onset of nesting.

With forest fragmentation over the past several centuries, cowbirds now penetrate Veery habitat, and lay their eggs in these previously unexploited hosts' nests. Despite the dramatic difference in color, maculation, and shape between cowbird and Veery eggs, the Veery accepts the parasitic eggs, perhaps because of a lack of historical coexistence between edge-living cowbirds and forest-dwelling hosts.

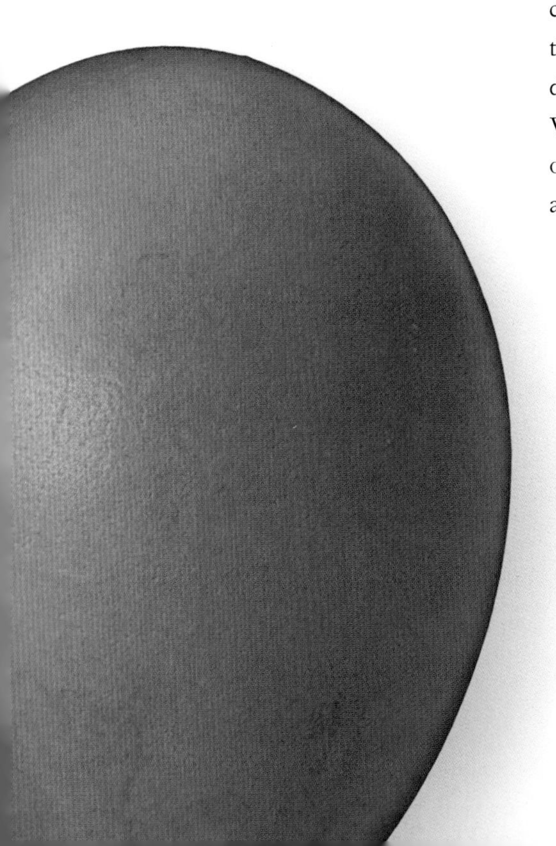

Actual size

ORDER	Passeriformes
FAMILY	Turdidae
BREEDING RANGE	Arctic North America, northeast Asia
BREEDING HABITAT	Boreal, coniferous forests
NEST TYPE AND PLACEMENT	Open cup, made of twigs and stems, and lined with rootlets, grass, and moss; placed in crotches of branches of low shrubs, or at the base of a tree, on the ground
CONSERVATION STATUS	Least concern
COLLECTION NO.	FMNH 11978

ADULT BIRD SIZE
6–6½ in (15–17 cm)

INCUBATION
13–14 days

CLUTCH SIZE
3–5 eggs

CATHARUS MINIMUS

GRAY-CHEEKED THRUSH

PASSERIFORMES

The Gray-cheeked Thrush is nearly identical in coloration and patterning to its cryptic, and rarer, Eastern sister species, the Bicknell's Thrush, but the two species are genetically distinct, slightly different in size, and sing different songs. Behaviorally, the Bicknell's Thrush is polyandrous in its mating system, whereas there is no reliable information on much of the breeding biology of the Gray-cheeked Thrush at all.

For instance, we do not know what the spacing and territorial behaviors of this species are, when and how pairs form on the breeding ground, what the male's role is during the female's nest-building and incubation periods, and how much he helps provision the chicks. Probably, the far-northern breeding distribution of this species, including nesting in dense taiga forests, is partially the cause for the lack of ornithological research.

Clutch

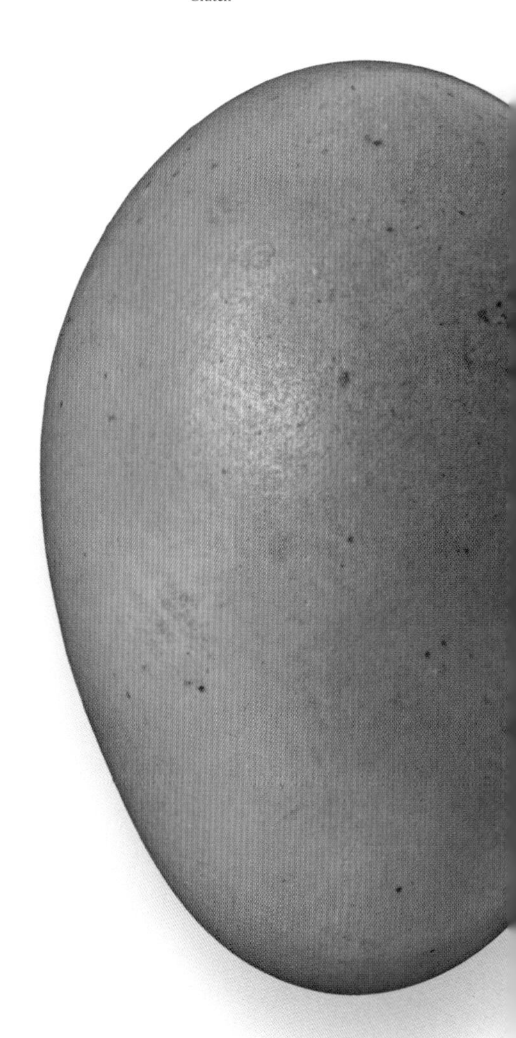

Actual size

The egg of the Gray-cheeked Thrush is light greenish blue in background, maculated with brown flecks, and ⅞ x ⅔ in (22 x 17 mm) in dimensions. Limited observations suggest that only the female builds the nest, develops a brood patch, and incubates the eggs.

ORDER	Passeriformes
FAMILY	Turdidae
BREEDING RANGE	Temperate and subarctic North America
BREEDING HABITAT	Coniferous forests with dense undergrowth, also deciduous woods on the west coast
NEST TYPE AND PLACEMENT	In crotch of a branch in understory shrubs; open cup of grasses, stems, and twigs, lined with dry leaves, rootlets, lichens, and moss
CONSERVATION STATUS	Least concern
COLLECTION NO.	FMNH 12085

ADULT BIRD SIZE
6½–7½ in (16–19 cm)

INCUBATION
12–14 days

CLUTCH SIZE
4–5 eggs

CATHARUS USTULATUS

SWAINSON'S THRUSH

PASSERIFORMES

Clutch

The Swainson's Thrush is a widespread breeder across the coniferous forests of much of North America. It is a long-distance migrant to tropical wintering grounds, with western populations traveling to Central America and eastern birds to South America. Males return earlier than the females to the breeding range; they quickly establish a breeding territory, which they defend through songs, alarm calls, and flight chases.

The pair bond is established through a series of territorial defense-like interactions, with the male initially appearing to be chasing the female away. If she persists in staying, his aggressive calls quickly turn into song displays, this time to keep the female from leaving. Eventually, she mates with the male, and begins to build the nest. She starts laying the eggs often on the same day nest construction is completed. The female incubates the eggs. She always approaches and leaves the nest secretively; after the eggs hatch, the male assists her in feeding the chicks.

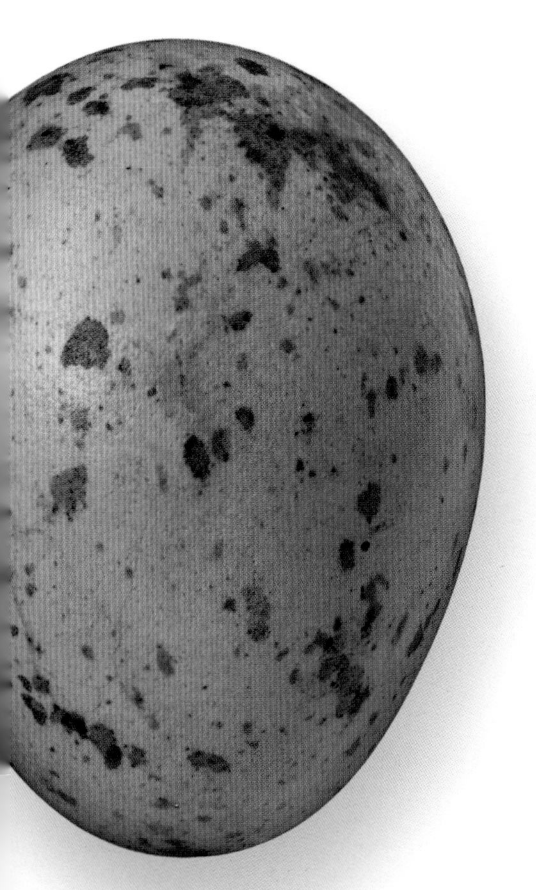

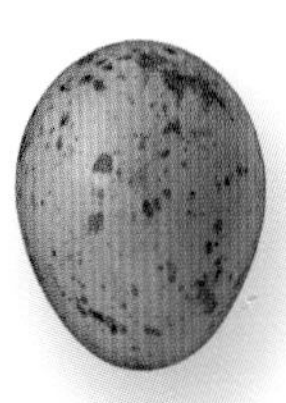

Actual size

The egg of the Swainson's Thrush is blue to greenish blue in coloration, with reddish or brown spotting, and is ⅞ x ⅔ in (22 x 17 mm) in size.

ORDER	Passeriformes
FAMILY	Turdidae
BREEDING RANGE	Temperate and subarctic North America
BREEDING HABITAT	Coniferous and mixed coniferous/deciduous forests
NEST TYPE AND PLACEMENT	Bulky, well-formed cup made of leaves, grasses, twigs, lichens, and mud, lined with fine plant materials; placed on ground, or low in small trees
CONSERVATION STATUS	Least concern
COLLECTION NO.	FMNH 15150

CATHARUS GUTTATUS

HERMIT THRUSH

PASSERIFORMES

ADULT BIRD SIZE
5½–7 in (14–18 cm)

INCUBATION
12–13 days

CLUTCH SIZE
3–6 eggs

Division of labor is the rule at the Hermit Thrush's nest. The male sets up the territory and defends it from intruders. He also gathers most of the food for the young. The female builds the nest and does all the incubation. As in most other birds, hatching chicks "pip" the egg: they break the shell with their bills, cutting it in two along the perimeter near its widest breadth. Females remove empty eggshells quickly from the nest; the whitish egg interiors would otherwise draw the attention of predators.

The Hermit Thrush is vulnerable to threats that plague many open-cup nesters, including nest parasitism by cowbirds and predation by squirrels, chipmunks, and feral cats. The helpless young in the nest are provisioned frequently by the parents; they grow quickly, and disperse from the nest as soon as they can, to hide in the undergrowth until they can fly. Despite these dangers, the Hermit Thrush can tolerate human proximity on the breeding grounds, and also often displaces the more secretive Swainson's Thrush (previous page) from mutually suitable, shared habitats.

Clutch

The egg of the Hermit Thrush is light blue, sometimes with minute brown flecks. It measures ⅞ x ⅔ in (22 x 17 mm). Typical of many thrush species, its blue eggs are richly colored by the shell pigment biliverdin.

Actual size

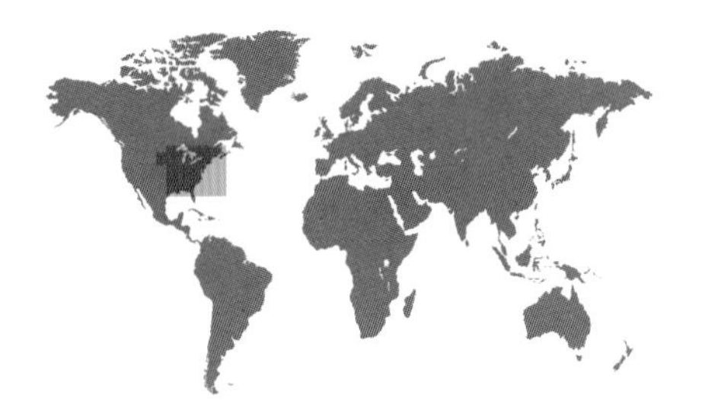

ORDER	Passeriformes
FAMILY	Turdidae
BREEDING RANGE	Temperate eastern and central North America
BREEDING HABITAT	Mature, deciduous and mixed forests, with a dense shrub layer
NEST TYPE AND PLACEMENT	Cup nest in a forked branch, or near the trunk; bulky and made of twigs, bark, and straw, lined with leaves and grasses
CONSERVATION STATUS	Least concern
COLLECTION NO.	FMNH 12035

ADULT BIRD SIZE
7–8½ in (18–22 cm)

INCUBATION
11–14 days

CLUTCH SIZE
3–4 eggs

HYLOCICHLA MUSTELINA

WOOD THRUSH

PASSERIFORMES

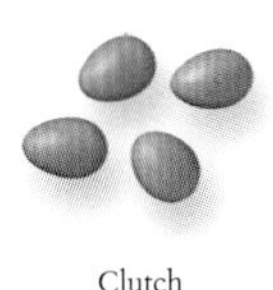

Clutch

Admired by many as the bird singing the most beautiful melodies in North America, the song of the Wood Thrush is a sure sign of spring and the onset of the arrival of migrant birds from the neotropics. The flute-like quality of this bird's vocalizations is produced by the syrinx, the unique apparatus that allows songbirds to generate two different voices, at two different pitches, that are heard as one combined sound by the listener.

Despite the current "least concern" conservation status of the Wood Thrush, it has become an indicator species of ecological and anthropogenic changes negatively affecting some deciduous forest birds; for example, on the breeding grounds, forest fragmentation has directly resulted in habitat loss and increased exposure to predation and parasitism on the nests. Pairs of Wood Thrush remain together on the nesting territory during the days leading up to egg laying, when the female is fertile; this means that both the female and the male have little chance to engage in extramarital affairs. Accordingly, in one study, more than 90 percent of the young were sired by both members of the resident pair, making this one of the most genetically faithful songbird species.

The egg of the Wood Thrush is sky blue in coloration, with no maculation, and is 1 x ¾ in (26 x 19 mm) in dimensions. When the nests are close to forest edges, this species is vulnerable to parasitism by Brown-headed Cowbirds (see page 616), with some nests having most of the blue thrush eggs replaced by beige and speckled cowbird eggs.

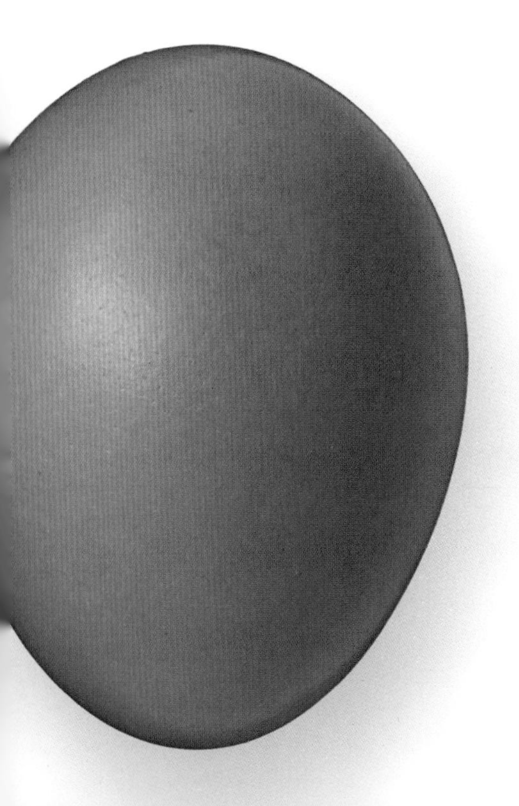

Actual size

ORDER	Passeriformes
FAMILY	Turdidae
BREEDING RANGE	Temperate and Arctic North America
BREEDING HABITAT	Forests, open woodlands, parks, backyards, city streets
NEST TYPE AND PLACEMENT	Open cup, made of grasses and hay, reinforced with mud into a sturdy, open cup; placed on horizontal branches, cliff ledges, and also bridge overhangs, eaves, barn interiors, and window sills
CONSERVATION STATUS	Least concern
COLLECTION NO.	FMNH 2320

ADULT BIRD SIZE
10–11 in (25–28 cm)

INCUBATION
12–14 days

CLUTCH SIZE
3–4 eggs

TURDUS MIGRATORIUS

AMERICAN ROBIN

PASSERIFORMES

Named after the robin of Europe, this species is not closely related to its smaller, Old World, red-breasted counterpart. One prominent example of the mix-up of the two species is the Disney movie, *Mary Poppins*, where a robotic American Robin, instead of a European Robin, sings a duet with the titular character, played by Julie Andrews, against a painted cityscape of London. In its native range, the melodious songs and bulky nests of the American Robin are familiar features of the spring and summer landscape both in rural and urban areas, including outside my own windows, in New York City.

Clutch

Late in the spring, broken robin-blue eggshells, or loud but flightless chicks on the sidewalk, frequently alarm pedestrians passing underneath a robin nest on a busy city street. Robin chicks, like many other thrush nestlings, often leave the nest before they are fully grown or able to fly, and unless they are in the middle of a busy path or roadway, it is best to leave the fledglings, or move them nearby into a low shrub, to let the parents continue to look after them. Robins often return to the same site year after year to breed, reusing the remains or building on top of the old nest structure to start the current year's reproductive bout.

The egg of the American Robin is light blue or blue green in coloration, clear of any spotting, and measures 1⅛ x ¾ in (28 x 21 mm) in size. Robins can recognize and eject cowbird eggs easily, because of the smaller size and distinctive coloration of the parasite's egg, compared to the robin's own eggs.

Actual size

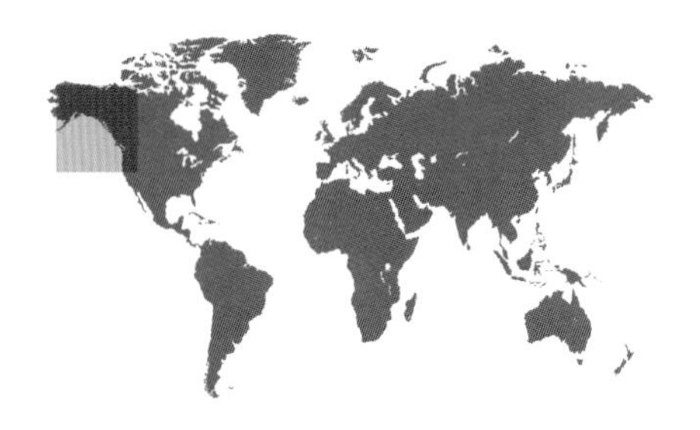

ORDER	Passeriformes
FAMILY	Turdidae
BREEDING RANGE	Northwestern North America
BREEDING HABITAT	Tall and moist evergreen or mixed forests, with a dense understory of mosses, ferns, and shrubs
NEST TYPE AND PLACEMENT	Open cup, made of twigs and branches, tree bark and wet leaves, mosses and fine grasses, soft leaves, and moss; placed near the trunk of a low tree, often near or on top of old nests
CONSERVATION STATUS	Least concern
COLLECTION NO.	FMNH 12097

ADULT BIRD SIZE
7½–10½ in (19–27 cm)

INCUBATION
12 days

CLUTCH SIZE
2–5 eggs

IXOREUS NAEVIUS

VARIED THRUSH

PASSERIFORMES

Clutch

The egg of the Varied Thrush is pale, light blue in background, and has large but sparse, distinctive dark brown maculation. Its measurements are 1³⁄₁₆ x ¾ in (30 x 21 mm). The eggs are kept in place by a network of woven and pasted soft and wet materials, which harden into a sturdy and deep nest cup.

Populations of this species can include year-round residents, short-distance migrants, and latitudinal migrants. They can form medium-sized flocks of up to 20 individuals in the non-breeding season, but maintain an aggressive demeanor toward many other species, including larger jays and smaller sparrows, at bird feeders and on open lawns. In the spring, the flocks break up, the males expand their combativeness toward other Varied Thrush males, and they can be seen chasing, attacking, and bill-locking with one another during territorial disputes.

Once nesting begins, the female is in charge of both gathering the nesting materials, and incubating the eggs. She often nests in the proximity of several previously used nests, occasionally by building the new structure on top of an old one. The nests are not at all disguised. Artificial nests made and positioned by researchers to look like natural Varied Thrush nests are more often depredated near forest edges than in the interior.

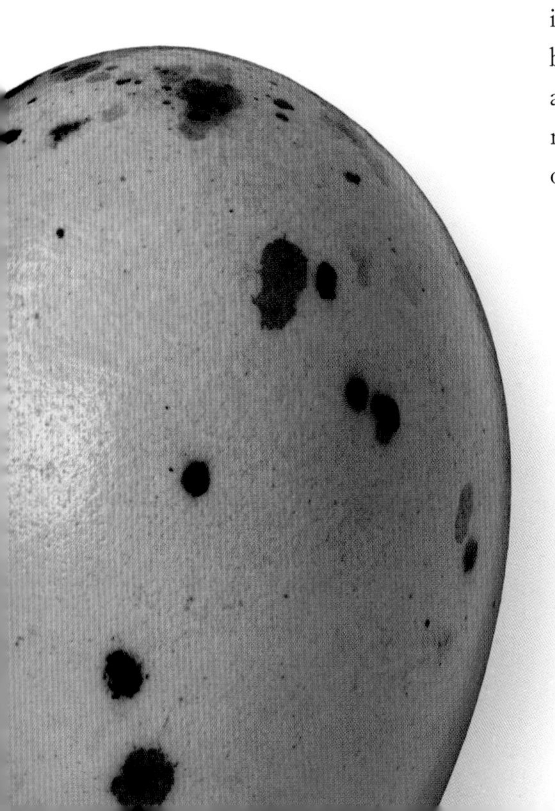

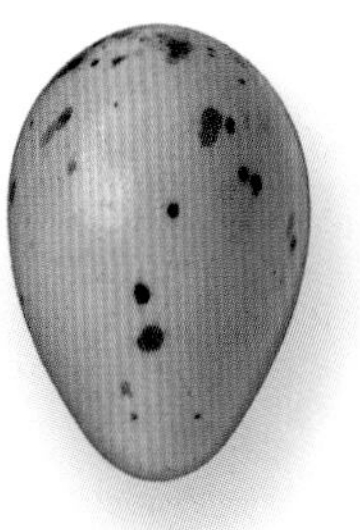

Actual size

ORDER	Passeriformes
FAMILY	Mimidae
BREEDING RANGE	Temperate eastern and central North America; introduced to Bermuda
BREEDING HABITAT	Dense shrubs, open areas with vines and thickets, city parks and suburban backyards
NEST TYPE AND PLACEMENT	Cup nest in the center of dense foliage in a brush, small trees, or in vines; made of sticks and twigs, lined with rootlets and grasses
CONSERVATION STATUS	Least concern
COLLECTION NO.	FMNH 16829

ADULT BIRD SIZE
8½–9½ in (21–24 cm)

INCUBATION
12–15 days

CLUTCH SIZE
3–5 eggs

DUMETELLA CAROLINENSIS

GRAY CATBIRD

PASSERIFORMES

The Gray Catbird lays some of the brightest and most intensely colored eggs among birds, with a uniform shiny green hue. But few people or predators ever see these eggs because the nest is well hidden deep in the dense foliage of low shrubs. The male stands guard over the nest while the incubating female is off to feed herself. Catbirds are also known to peck and remove eggs from active nests of other birds breeding near their own nest.

Together with mockingbirds and thrashers, the Gray Catbird belongs to the New World family Mimidae, which is known for its mimetic, variable, and continuously varied vocal displays. Gray Catbirds are among the handful of songbird species that are frequently parasitized by Brown-headed Cowbirds (see page 616); however, the catbirds have evolved to recognize and eliminate a foreign egg from the nest by piercing and grasping it, then tossing or flying away with it.

Clutch

The egg of the Gray Catbird is turquoise green in background coloration, immaculate, and 15⁄16 x ⅔ in (24 x 18 mm) in size. Catbirds remove cowbirds' eggs from their nests so quickly that, at first, many researchers thought they were not being parasitized.

Actual size

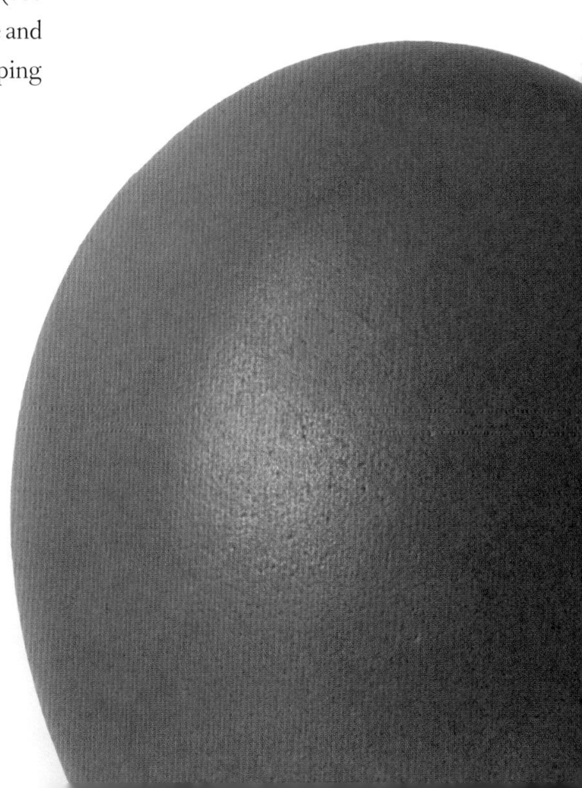

ORDER	Passeriformes
FAMILY	Mimidae
BREEDING RANGE	Temperate North America, Central America, the Caribbean
BREEDING HABITAT	Fields, forest edges, open scrub, backyards, and urban parks
NEST TYPE AND PLACEMENT	Open cup nest, in a low or mid-height bush or tree; made from twigs lined with grasses, rootlets, leaves, and occasionally trash filaments
CONSERVATION STATUS	Least concern
COLLECTION NO.	FMNH 11830

ADULT BIRD SIZE
8½–10 in (21–26 cm)

INCUBATION
11–14 days

CLUTCH SIZE
3–5 eggs

MIMUS POLYGLOTTOS

NORTHERN MOCKINGBIRD

PASSERIFORMES

Clutch

The egg of the Northern Mockingbird is pale blue or greenish white in coloration, has red or brown blotches, and measures 1 x ⅔ in (25 x 18 mm). The pair builds the nest together, but the female alone incubates the eggs; both parents provision the hatchlings.

This superficially drab-looking bird is anything but that in attitude: its song is long and varied, due to fine skills as a mimic of other birds' voices and artificial sounds. Its presence is easily noticed, because of its habit of perching dominantly on top of a tree, or sallying toward intruders, humans and mockingbirds alike, flashing its bright white wing patches. The nest of the mockingbird is quite bulky and often exposed; instead of camouflage, these birds rely on active mobbing to protect the eggs and the young.

In addition to investing heavily in the protection and success of each breeding attempt, with both parents making ever more frequent feeding deliveries to the growing young in the nest, mockingbirds also take advantage of the relatively warm and long spring and summer in most of their distribution. They are able to fit two to three full breeding attempts into each year's reproductive cycle. Wherever a nest is successful in fledging young, chances are these birds will return there to breed again.

Actual size

ORDER	Passeriformes
FAMILY	Mimidae
BREEDING RANGE	Western North America
BREEDING HABITAT	Shrub-steppe, vegetated by dense sagebrush
NEST TYPE AND PLACEMENT	Cup on a platform made of twigs and sticks, lined with rootlets, grass, and hair; placed in sagebrush, other shrub, or on ground
CONSERVATION STATUS	Least concern
COLLECTION NO.	FMNH 11836

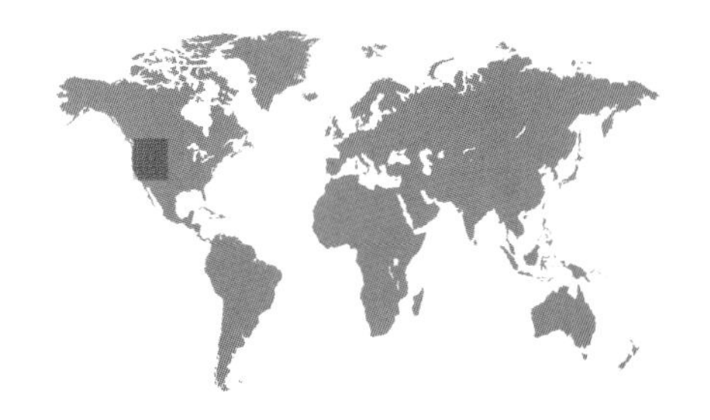

ADULT BIRD SIZE
8–9 in (20–23 cm)

INCUBATION
13–17 days

CLUTCH SIZE
4–5 eggs

OREOSCOPTES MONTANUS

SAGE THRASHER

PASSERIFORMES

This species is a specialist of sagebrush steppe; clearing and plowing of its preferred habitat have caused critical losses in local population size and geographic distribution. A short-distance migrant, males of this species move just hundreds of miles south in the winter, to escape the coldest, snow-covered months. Once they return, the male establishes a breeding territory, and advertizes its boundaries to other males and later-arriving females by singing a melodious song; the song can last uninterrupted for several minutes.

During the summer daytime, temperatures may rise so high that the eggs are at risk of overheating. To cool the nest, this thrasher relies on natural ventilation; it places the nest on a branch, where possible, instead of directly on the ground, and on a platform of loose twigs and sticks. In response to heavy rains, both parents take turns covering the eggs, and brooding the chicks, to keep them dry and healthy.

Clutch

The egg of the Sage Thrasher is blue or greenish blue in background, with brownish speckling, and is 1 x ⅔ in (25 x 17 mm) in size. The female has a full brood patch, and the male a partial one, but both sexes contribute to incubating the eggs.

Actual size

ORDER	Passeriformes
FAMILY	Mimidae
BREEDING RANGE	Eastern and central North America
BREEDING HABITAT	Forest edges, thickets, hedgerows, shrubby parkland
NEST TYPE AND PLACEMENT	Cup made of twigs, dead leaves, bark, and rootlets, lined with finer grasses; bulky and placed low in a tree or thorny bush
CONSERVATION STATUS	Least concern
COLLECTION NO.	FMNH 11910

ADULT BIRD SIZE
9½–12 in (24–31 cm)

INCUBATION
10–14 days

CLUTCH SIZE
2–6 eggs

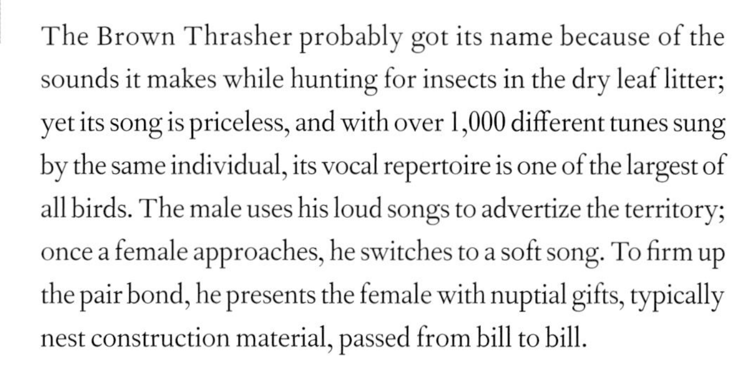

TOXOSTOMA RUFUM

BROWN THRASHER

PASSERIFORMES

Clutch

The egg of the Brown Thrasher is white or pale blue to greenish in coloration, has fine reddish brown maculation all over, and its dimensions are 1¹⁄₁₆ x ¾ in (27 x 19 mm). The breeding season is long enough in much of its breeding range for the laying of two or three clutches of eggs.

The Brown Thrasher probably got its name because of the sounds it makes while hunting for insects in the dry leaf litter; yet its song is priceless, and with over 1,000 different tunes sung by the same individual, its vocal repertoire is one of the largest of all birds. The male uses his loud songs to advertize the territory; once a female approaches, he switches to a soft song. To firm up the pair bond, he presents the female with nuptial gifts, typically nest construction material, passed from bill to bill.

Once the mates are ready to nest, both sexes construct the nest together in a well-hidden, thorny thicket. Both parents incubate the eggs, and stay tightly on the nest until danger approaches too closely; then they quietly slip off into the dense foliage. The male's contribution to parental duties increases from around 30 percent of time spent on the eggs to 50 percent while feeding the chicks.

Actual size

ORDER	Passeriformes
FAMILY	Mimidae
BREEDING RANGE	Southern North America
BREEDING HABITAT	Riparian woodlands, dense thickets and brushes
NEST TYPE AND PLACEMENT	Bulky cup, constructed from thorny twigs and sticks, lined with grass, hay, bark, and rootlets; placed in center of dense brush foliage, under large trees
CONSERVATION STATUS	Least concern
COLLECTION NO.	FMNH 16836

ADULT BIRD SIZE
10–11½ in (26–29 cm)

INCUBATION
13–14 days

CLUTCH SIZE
2–5 eggs

TOXOSTOMA LONGIROSTRE

LONG-BILLED THRASHER

PASSERIFORMES

As year-round residents, these birds occur alone or in pairs. The male perches conspicuously atop a bush or on a tree branch while he sings, and both sexes engage in light chases and calling when responding to intruders into the territory; they attack not only other Long-billed Thrashers but, where overlapping in their ranges, also Brown Thrashers (previous page). At other times, they remain quiet and secretive as they forage on the ground in dry undergrowth under thickets. In turn, they are often displaced by some other species, such as American Robins.

Both sexes seem to be involved in selecting the nest site and constructing it atop a platform made of thorny twigs. Once the chicks hatch, the parents spend little time brooding, because most nests are already in shaded spots. Instead, they deliver insect foods to the chicks, which grow fast and reach one-half of the adult weight in just one week after hatching.

Clutch

The egg of the Long-billed Thrasher is pale greenish or white in background, with reddish brown spotting, and is 1⅛ x ¾ in (28 x 20 mm) in size. Surprisingly little is known about the breeding biology of this species, but both parents are thought to incubate the eggs and provision the fast-growing nestlings.

Actual size

ORDER	Passeriformes
FAMILY	Mimidae
BREEDING RANGE	Southwestern North America
BREEDING HABITAT	Desert habitats from sea level
NEST TYPE AND PLACEMENT	Open cup of sticks, lined with soft plant and animal fibers; placed in bushes, cacti, or trees
CONSERVATION STATUS	Vulnerable
COLLECTION NO.	FMNH 1724

ADULT BIRD SIZE
9–10 in (23–25 cm)

INCUBATION
12–14 days

CLUTCH SIZE
3–4 eggs

TOXOSTOMA BENDIREI

BENDIRE'S THRASHER

PASSERIFORMES

Clutch

This is a secretive species. Pairs form soon after return from short-distance migrations. The few observations are of single birds or pairs during the breeding season, with small family groups appearing shortly after fledging. Successfully completed nesting attempts appear to be followed by a second, or even a third, clutch. To initiate new clutches, these birds may reuse an old nest, or build a brand new nest nearby.

This species was the last of its genus to be described from North America, and it was not an easy start: the first specimen collected was mistaken for the female of another thrasher species. Even today, the ecology and the behavior of this species is known mostly from anecdotal observations, instead of published research. Such cases stress the importance of all observers keeping a permanent record of all types of natural history details, observations, and data.

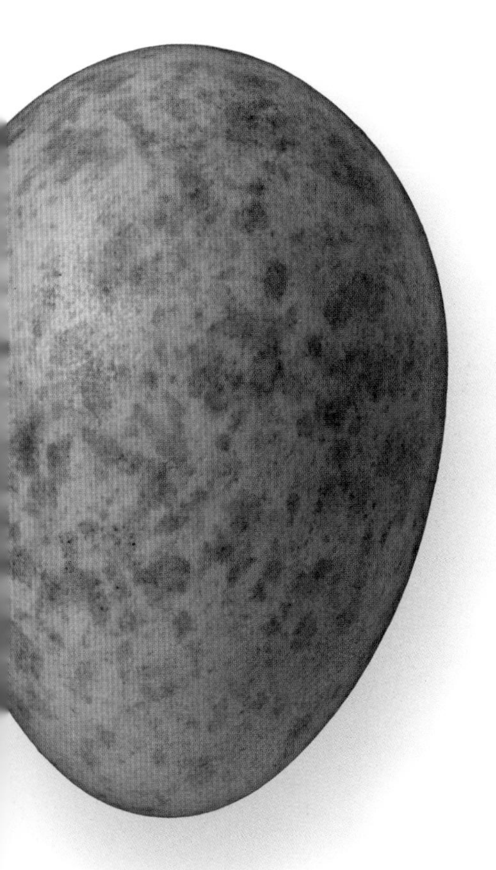

Actual size

The egg of the Bendire's Thrasher is white or pale blue-green, with darker maculation across the shell, and measures 1⅛ x ¾ in (28 x 19 mm). There is no information on incubation or brooding behavior, but both parents are known to provision the chicks.

ORDER	Passeriformes
FAMILY	Mimidae
BREEDING RANGE	Southwestern North America
BREEDING HABITAT	Open deserts and arid shrubland
NEST TYPE AND PLACEMENT	Twig platform, with a deep cup lined with grasses, rootlets, and fibers; placed in a cactus or spiny shrub
CONSERVATION STATUS	Least concern
COLLECTION NO.	FMNH 11866

TOXOSTOMA CURVIROSTRE

CURVE-BILLED THRASHER

PASSERIFORMES

ADULT BIRD SIZE
10–11 in (25–28 cm)

INCUBATION
12–15 days

CLUTCH SIZE
3–5 eggs

This is a resident species all year round, with pairs maintaining a permanent territory for feeding and nesting, as long as both parents are alive. Intruders and challengers for territory ownership also come in pairs, and this can lead to physical chases and fights. Nest construction, incubation, and parental provisioning of the chicks are shared between the mates, although females develop larger brood patches for incubation and provide more of the feedings.

This species practices selective parental feeding tactics. In years with plentiful food, all chicks are provisioned. In poorer years, only the larger and more vigorous chicks are fed, and the younger, smaller ones are left hungry and eventually perish. Successful first nesting attempts—those that fledge at least one young—are followed almost immediately by a new clutch. To save time, the pair builds a new nest during the last day of the nestling period.

Clutch

Actual size

The egg of the Curve-billed Thrasher is light bluish green in background color, has heavy reddish brown spotting, and measures 1⅛ x ¾ in (28 x 19 mm) in size. The male gathers fine grasses and rootlets, and the female forms them into a deep cup before she begins laying the eggs.

ORDER	Passeriformes
FAMILY	Mimidae
BREEDING RANGE	Southwestern North America
BREEDING HABITAT	Desert valleys, riparian shrub corridors, lower montane scrub
NEST TYPE AND PLACEMENT	Cup on a platform of twigs, lined with rootlets and finer vegetation; placed in a dense bush
CONSERVATION STATUS	Least concern
COLLECTION NO.	FMNH 11885

ADULT BIRD SIZE
10½–12½ in (27–32 cm)

INCUBATION
12–15 days

CLUTCH SIZE
2–3 eggs

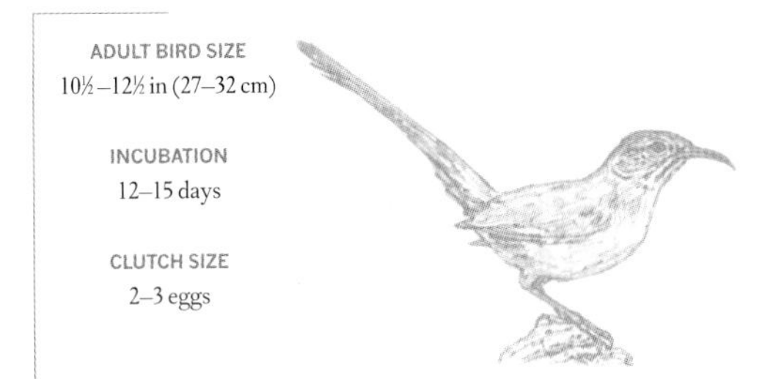

TOXOSTOMA CRISSALE

CRISSAL THRASHER

PASSERIFORMES

Clutch

The egg of the Crissal Thrasher is pale blue in coloration, immaculate, and measures 1⅛ x ¾ in (28 x 19 mm). Using an equation based on the adult's body size and the egg's weight, the incubation period of this species can be calculated to be 14 days, which closely matches observations in the wild.

During the breeding season, the Crissal Thrasher inhabits sparsely vegetated desert habitats, and can be easily spotted singing loudly perched prominently at the top of one of the few shrubs and thickets in its territory. These are aggressive birds, with the pairs defending their range all year round. The territory boundaries, however, become more fluid in the winter, and this is also a time when pair bonds may break and new pairs form.

Parental duties are shared in this species, with both sexes contributing to incubation, brooding the hatchlings, and making parental feeding visits. Unlike most other thrashers, but like catbirds and many other birds in the family Mimidae, this species lays clear, immaculate eggs. To remove the bright and conspicuous eggs from the nest, the parents carry away broken eggshells soon after hatching is complete.

Actual size

ORDER	Passeriformes
FAMILY	Mimidae
BREEDING RANGE	Southwestern North America
BREEDING HABITAT	Dry rocky dunes, shrubby desert flats
NEST TYPE AND PLACEMENT	Open cup, outer layer is made of branches and twigs, followed by a second layer of finer twigs and rootlets, lined with a compact layer of plant fuzz and fibers; on the branches in thorny shrub or cholla cactus
CONSERVATION STATUS	Least concern
COLLECTION NO.	FMNH 16861

TOXOSTOMA LECONTEI

LE CONTE'S THRASHER

PASSERIFORMES

ADULT BIRD SIZE
9½–11 in (24–28 cm)

INCUBATION
14–20 days

CLUTCH SIZE
2–5 eggs

This is a nonmigratory species, and it remains on its territory all year long. A pair can form during any season, when and where an opening in pair bonds occurs due to the death of a mate. The nest site is chosen by the female, while the male follows her closely. The male contributes to nest building, but not consistently; he picks up his share of parental duties during incubation, and also by providing most of the food delivered to the chicks.

Clutch

Le Conte's Thrashers are a little like miniature roadrunners: they use their long bills to probe for insects beneath dry sticks and leaf litter on their arid, sparsely vegetated territories. They also run frequently, rather than fly, as they cross their territories. This species may see rain only for a few days each year, getting the water it needs to survive from its insect prey.

Actual size

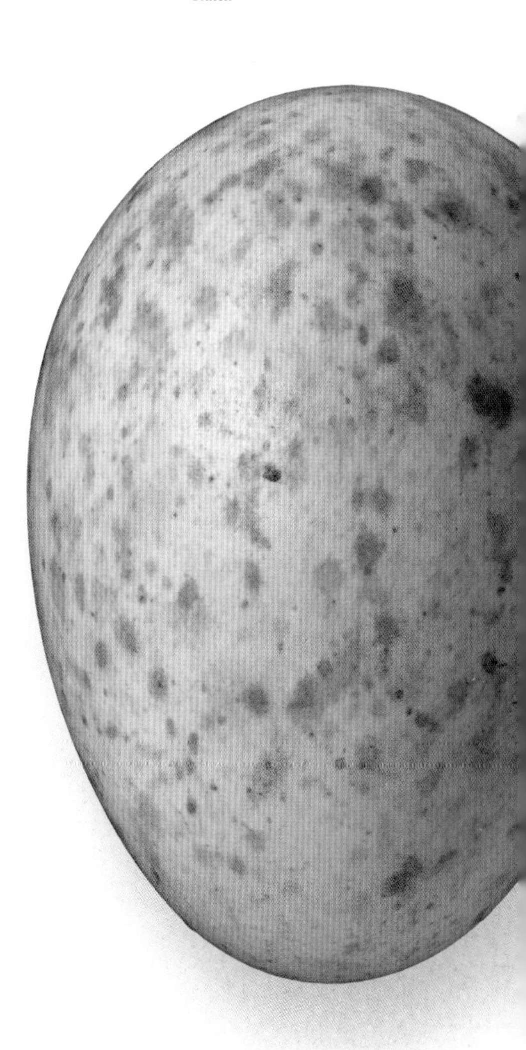

The egg of the Le Conte's Thrasher is white, beige, or pale blueish green in coloration, has darker blotching, and measures 1$\frac{1}{16}$ x ¾ in (27 x 19 mm) in dimensions. To protect the eggs from the desert heat and direct sunlight, the nest is placed under thick and dense branches for maximum shading.

ORDER	Passeriformes
FAMILY	Sturnidae
BREEDING RANGE	Eastern Europe, southern temperate Asia; occasionally eruptive into western Europe
BREEDING HABITAT	Open steppe, short-crop agricultural land, pastures
NEST TYPE AND PLACEMENT	Made of grass and twigs, with a lining of feathers and finer grasses; placed in tree holes, woodpecker cavities, rock crevices, quarries, or ground burrows
CONSERVATION STATUS	Least concern
COLLECTION NO.	FMNH 20805

ADULT BIRD SIZE
8½–10 in (22–26 cm)

INCUBATION
12–15 days

CLUTCH SIZE
3–8 eggs

PASTOR ROSEUS

ROSY STARLING

PASSERIFORMES

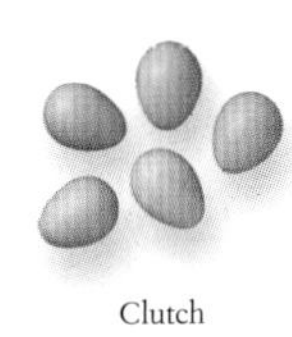

Clutch

The egg of the Rosy Starling is pale blue or whitish in coloration, free of spotting, and 1⅛ x ¾ in (29 x 21 mm) in measurements. Nesting chambers are highly variable in size and location, even within the same, densely populated breeding colony, perhaps because laying eggs at the same time as others in the colony is more important than taking time to find the ideal nesting space.

In the breeding season this relative of the Common Starling (facing page) settles in large colonies, and during the wintering period it forms enormous flocks that often outnumber local starlings, including mynahs, on the Indian subcontinent. Its reproductive strategies follow closely its predominant prey, the locust. In years when locusts breed in vast numbers, the Rosy Starling follows them closely, and can range and breed as far west as the Atlantic coastline of Europe.

Many birds incorporate green foliage into their nests. Rosy Starlings, for example, often use pungent wild fennel branches, whose function is still up for much debate and research. Birds do have a keen sense of smell and it is thought that the more odorous leaves in bird nests may function as a natural pesticide to help control nest mites and other ectoparasite levels, which can be especially prevalent in colonially nesting birds, such as this starling.

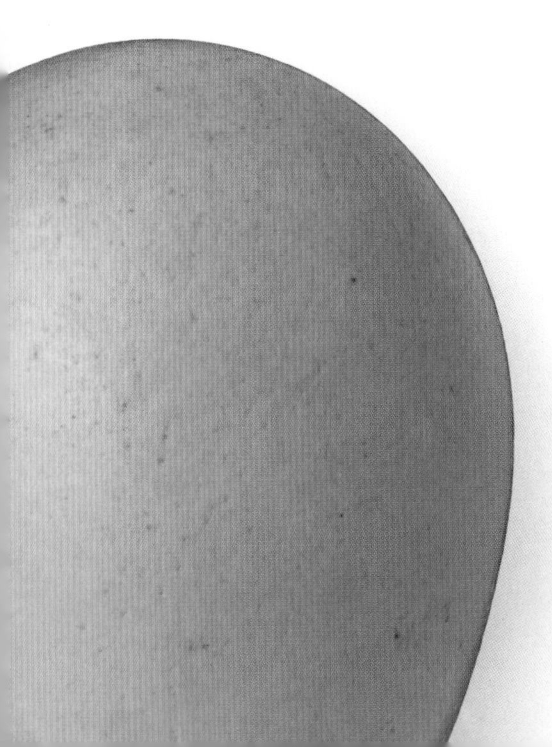

Actual size

ORDER	Passeriformes
FAMILY	Sturnidae
BREEDING RANGE	Europe and northwestern Asia; introduced and broadly established on other continents
BREEDING HABITAT	Open forests, parklands, orchards, city parks
NEST TYPE AND PLACEMENT	Untidy bowl or cup nest, in a natural or artificial cavity, box, tractor engine, or other enclosed space; made from straw, grasses, and twigs, lined with feathers, wool, and soft leaves
CONSERVATION STATUS	Least concern
COLLECTION NO.	FMNH 9841

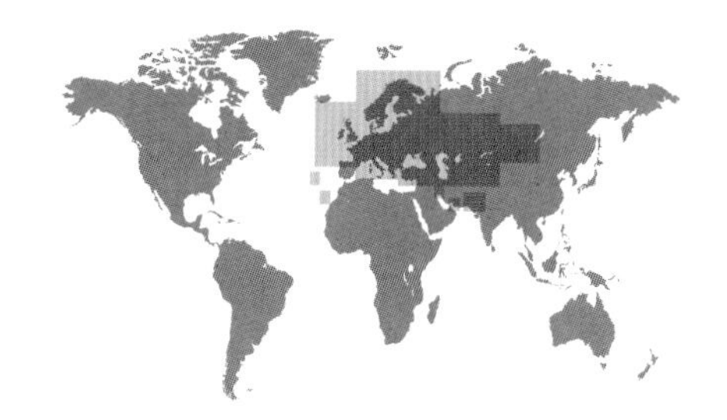

ADULT BIRD SIZE
7½–9 in (19–23 cm)

INCUBATION
12–14 days

CLUTCH SIZE
4–5 eggs

STURNUS VULGARIS

COMMON STARLING

PASSERIFORMES

The Common, or European, Starling is a nearly ubiquitous songbird. Its raspy but melodious song can be heard on each inhabited continent, either because it is a native species or because introduced populations have established themselves. The starling as an invasive species represents a severe danger for many other birds, as they are aggressive and successfully destroy or evict other native birds' nesting attempts from highly contested natural cavities in the forest.

Once a pair of starlings is settled to breed, the sexes share incubation and feeding duties equitably. However, if a female settles and mates with a male who is already mated, this secondary female will only receive paternal assistance if the first female's nesting attempt fails. Female starlings also often lay eggs parasitically in other starlings' nests, especially when the parasites themselves have lost their own nest to predation during the laying period.

Clutch

The egg of the Common Starling is glossy, light blue in background color, immaculate or, rarely, spotted, and is 1³⁄₁₆ x ¹³⁄₁₆ in (30 x 21 mm) in size. The number of eggs in the nest is under genetic control; an artificial selection experiment, which eliminated females that lay small clutches, and lasted a decade in New Zealand where the starling has been introduced, successfully increased the average clutch size from four to five eggs.

Actual size

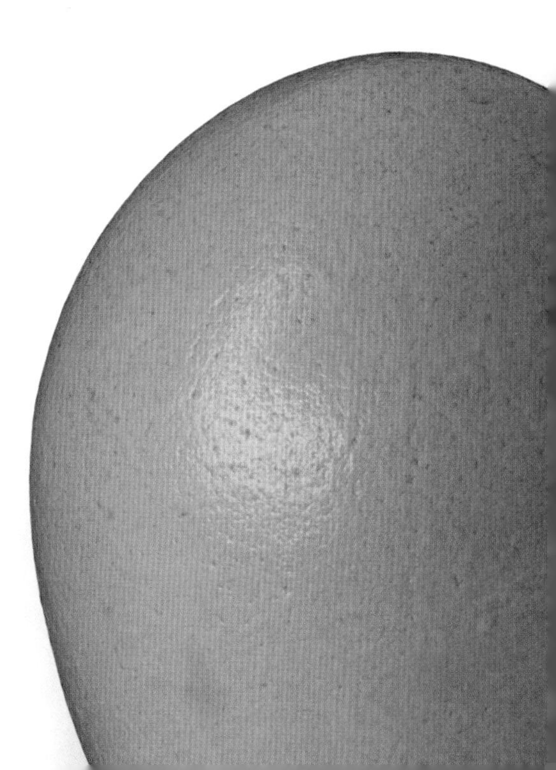

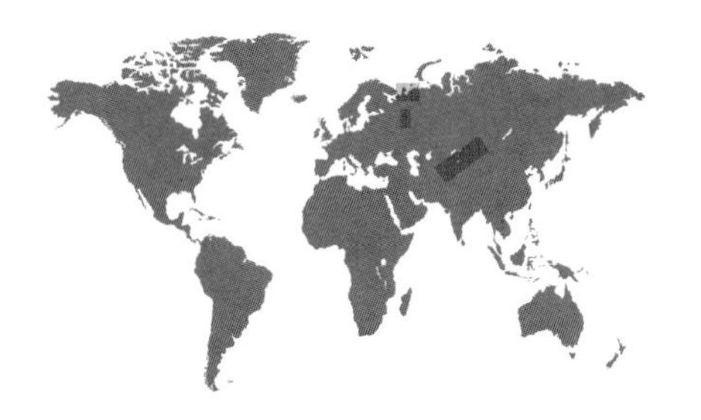

ORDER	Passeriformes
FAMILY	Prunellidae
BREEDING RANGE	North Eurasia
BREEDING HABITAT	Coniferous woodland, stunted alpine spruce forests
NEST TYPE AND PLACEMENT	Built on a branch, low in spruce thickets; open cup of twigs and mosses, lined with fine grasses
CONSERVATION STATUS	Least concern
COLLECTION NO.	FMNH 20719

ADULT BIRD SIZE
5½–6¼ in (14–16 cm)

INCUBATION
11–14 days

CLUTCH SIZE
3–5 eggs

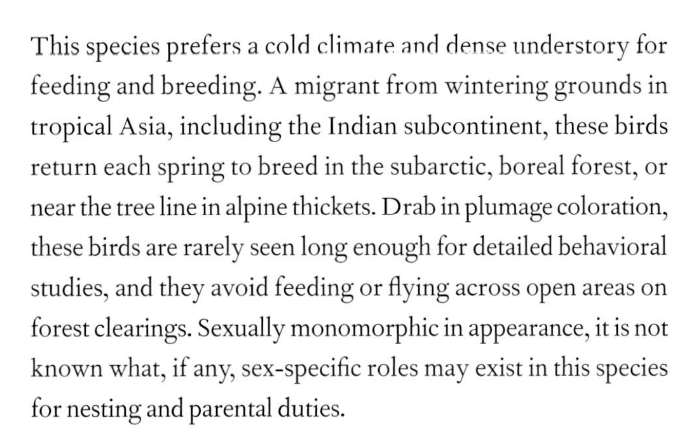

PRUNELLA ATROGULARIS

BLACK-THROATED ACCENTOR

PASSERIFORMES

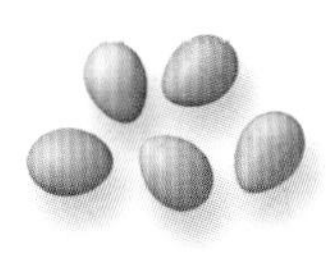

Clutch

The egg of the Black-throated Accentor is light blue to deep green in background color, unspotted, and ¾ x ⅔ in (19 x 17 mm) in dimensions. To keep the eggs safe and the nest location hidden from predators, these birds approach the nest quietly through dense thickets.

This species prefers a cold climate and dense understory for feeding and breeding. A migrant from wintering grounds in tropical Asia, including the Indian subcontinent, these birds return each spring to breed in the subarctic, boreal forest, or near the tree line in alpine thickets. Drab in plumage coloration, these birds are rarely seen long enough for detailed behavioral studies, and they avoid feeding or flying across open areas on forest clearings. Sexually monomorphic in appearance, it is not known what, if any, sex-specific roles may exist in this species for nesting and parental duties.

The Black-throated Accentor is well known to ornithologists, both because of its migratory habits and because of its broad distribution, yet, for the reasons described above, very little systematic information has been collected about its breeding biology. In Europe, there is a small breeding population, numbering in the thousands, but because Europe represents only 5 percent of the total area occupied by these birds, it has been assumed that this species' population sizes and temporal trends remain large, viable, and stable, and of little conservation concern.

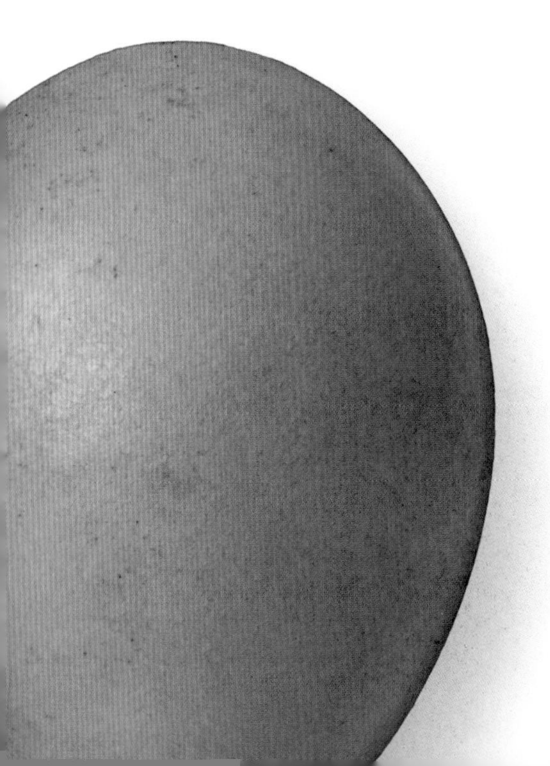

Actual size

ORDER	Passeriformes
FAMILY	Motacillidae
BREEDING RANGE	Northwestern Africa, temperate and subarctic Eurasia
BREEDING HABITAT	Open, flat grasslands, riverine corridors, coastal plateaus, subalpine clearings, parklands
NEST TYPE AND PLACEMENT	In a cavity, crevice, hole in the wall, or under an overhang in a soil bank; bulky nest of leaves, stems, and rootlets, lined with plant fluff, hair, and fur
CONSERVATION STATUS	Least concern
COLLECTION NO.	FMNH 12286

ADULT BIRD SIZE
6½–7½ in (17–19 cm)

INCUBATION
11–16 days

CLUTCH SIZE
4–7 eggs

MOTACILLA ALBA

WHITE WAGTAIL

PASSERIFORMES

The White Wagtail is a charismatic species, whose characteristic pied plumage and distinctive tail motions are familiar to many people, especially because these birds often migrate through city parks alongside rivers and streams. This wagtail maintains territories both summer and winter. In the winter, a satellite juvenile or an adult female is allowed to join the male's territory, whereas in the summer, the breeding pair together maintains an all-purpose range, used for both feeding and nesting.

The pair chooses the nesting site together; the male may lead the female to suitable crevices, often carrying a piece of nesting material, while the female inspects each site closely. Once a site is accepted, the nest is built by both members of the pair, and incubation is also shared. After the eggs hatch, both parents begin their frequently delivery of arthropod prey to the nestlings, but the mother may stay behind after each feeding to brood the chicks to keep them warm.

Clutch

The egg of the White Wagtail is cream in coloration, with a blueish green tint and heavy reddish brown maculation, and is 13⁄16 x ⅝ in (21 x 15 mm) in size. Perhaps the most unique place where this bird's eggs have been found is in a nest built inside an empty walrus skull.

Actual size

ORDER	Passeriformes
FAMILY	Motacillidae
BREEDING RANGE	Canary and Madeira Islands
BREEDING HABITAT	Coastal scrub, arid grass fields
NEST TYPE AND PLACEMENT	Deep cup nest on the ground and hidden under sedges, grasses, or other low plants, also in rock crevices; made from dry stems, grasses, and rootlets
CONSERVATION STATUS	Least concern
COLLECTION NO.	FMNH 21352

ADULT BIRD SIZE
5–6 in (13–15 cm)

INCUBATION
12–13 days

CLUTCH SIZE
3–5 eggs

ANTHUS BERTHELOTII

BERTHELOT'S PIPIT

PASSERIFORMES

Clutch

The breeding season begins with the onset of nesting activities within a pair's territory; in years with more frequent winter precipitation, nesting starts earlier for most of these birds. Early nestings are advantageous because they leave sufficient time for second clutches to be completed. The hatchlings are covered in filamentous gray fluff, display bright yellow-orange gapes, and depend on the parents for brooding and feeding until ready to leave the nest. Breeding usually ceases with the long dry season typical of these Atlantic islands, whose weather is influenced by wind patterns over the Sahara Desert.

This species is endemic to the Canary and Madeira Islands, archipelagos in the eastern Atlantic off the coast of northwest Africa. Adapted to life on the ground, this bird rarely perches or roosts in shrubs and trees. It spends most of the day foraging for insects, including larvae and eggs, but also seeds. With the presence of humans on their native islands, these birds have also moved into villages and towns, making use of open areas, grassy roadsides, and parklands.

The egg of the Berthelot's Pipit is buff with a yellow tinge in background color, it has pale purplish gray maculation, and measures ¾ x 9⁄16 in (19 x 14 mm).

Actual size

ORDER	Passeriformes
FAMILY	Bombycillidae
BREEDING RANGE	Temperate and subarctic North America
BREEDING HABITAT	Open and edge woodlands, fields with scattered trees, riparian galleries, parks, and golf courses
NEST TYPE AND PLACEMENT	Open cup of grasses and twigs, lined with fine roots, grasses, and pine needles; placed at a fork on a horizontal tree branch
CONSERVATION STATUS	Least concern
COLLECTION NO.	FMNH 12309

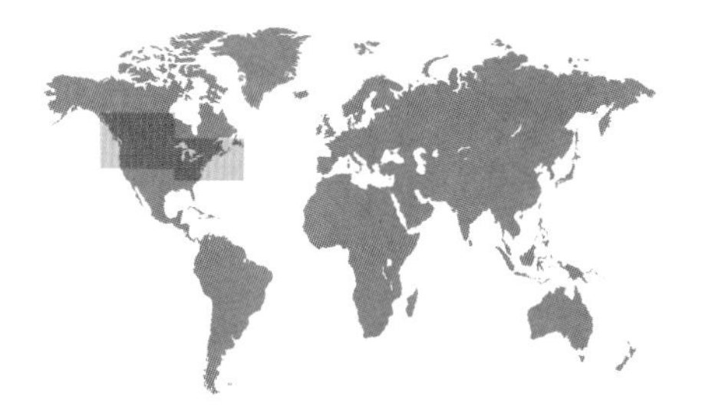

ADULT BIRD SIZE
6–7 in (15–18 cm)

INCUBATION
11–13 days

CLUTCH SIZE
4–6 eggs

BOMBYCILLA CEDRORUM

CEDAR WAXWING

PASSERIFORMES

The name "waxwing" comes from the unique bright red waxy cover of the wing feathers of this species. A specialist that eats mostly fruit, the Cedar Waxwing has become an increasingly common sight in highly populated areas, as it has undergone a recent expansion of both breeding range and population sizes. This may be because the preferred breeding and feeding habitat includes forest edges, wooded areas interspersed with open lands, and suburban parks and gardens with ornamental trees and shrubs bearing bright red, and sugary, berries.

These birds follow seasonal and erratic food sources in both their nesting behaviors and flocking tendencies. Based on the availability of food, for example, individuals, pairs, and large flocks are likely to move to new breeding sites across different years, and sometimes even within a single year. Egg laying often begins late in the season, so that the feeding period of the nestlings coincides with the ripening of sweet berries and other fruits.

Clutch

The egg of the Cedar Waxwing is pale blue or blueish gray in coloration, with black or gray spotting, and is ⅞ x ⅝ in (22 x 16 mm) in dimensions. The pair bond is very tight in this species, as both sexes take turns to construct the nest, incubate the eggs, and feed the chicks.

Actual size

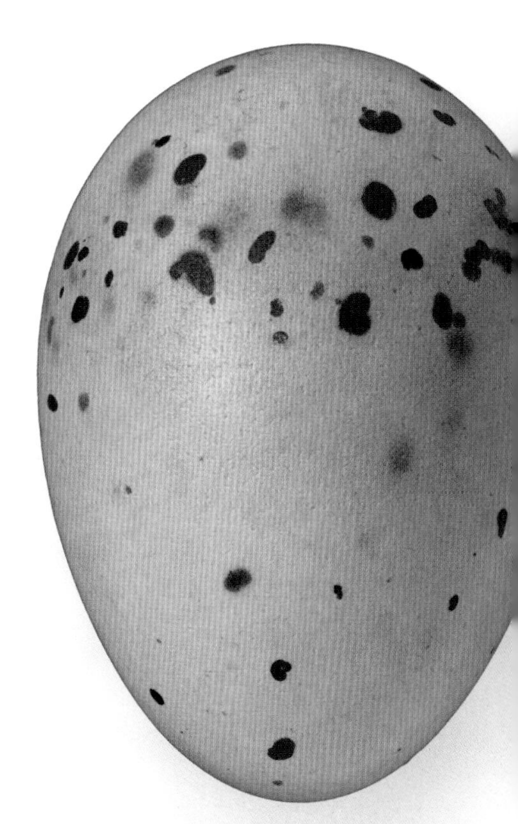

ORDER	Passeriformes
FAMILY	Bombycillidae
BREEDING RANGE	Subarctic northern Eurasia, northwestern North America
BREEDING HABITAT	Coniferous forests
NEST TYPE AND PLACEMENT	Bulky open cup, made of twigs, grasses, and moss; placed on tree branch near trunk
CONSERVATION STATUS	Least concern
COLLECTION NO.	FMNH 15162

ADULT BIRD SIZE
7–8½ in (18–21 cm)

INCUBATION
13–15 days

CLUTCH SIZE
4–6 eggs

BOMBYCILLA GARRULUS

BOHEMIAN WAXWING

PASSERIFORMES

Clutch

The Bohemian Waxwing is a food specialist, making use of an ephemeral and highly unpredictable source of nutrition: sugary berries and fruits in the far north. Typically, each fruiting tree and bush bears many fruits, but only for a short period of time, so that protecting a territory large enough to provide nutrition for a full breeding season becomes challenging. In response, these waxwings have given up aggression and territoriality and, instead, move and breed in flocks, using each other as sources of information to locate the nearest fruiting shrub and tree.

A pair builds their nest together, but only the female incubates the eggs and broods the hatchlings. The male, however, provides her and the chicks with regurgitated fruits, berries, and insects. The young grow fast, and within days the female joins the male to provision the nestlings by taking turns in food deliveries.

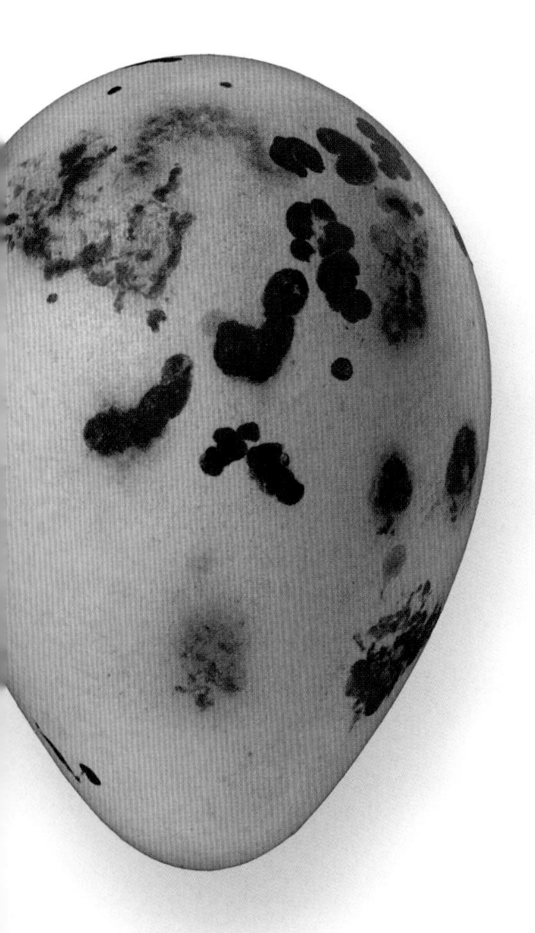

Actual size

The egg of the Bohemian Waxwing is pale blueish gray in background color, with sparse distinct black blotching, and is 1 x ⅔ in (25 x 17 mm) in size. To protect the eggs and the chicks, the nest is camouflaged with an outer covering of mosses and lichens.

ORDER	Passeriformes
FAMILY	Ptilogonatidae
BREEDING RANGE	Southwestern North America
BREEDING HABITAT	Desert and semidesert shrubland; also, arid woodlands and forested canyons
NEST TYPE AND PLACEMENT	Platform nest with a shallow cup, made of twigs and fibers, lined with plant fluff and animal hair; placed on a tree limb or fork
CONSERVATION STATUS	Least concern
COLLECTION NO.	FMNH 12341

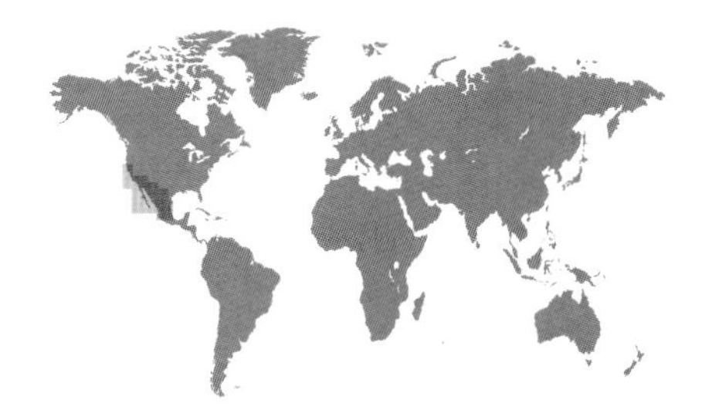

ADULT BIRD SIZE
7–8½ in (18–21 cm)

INCUBATION
14–15 days

CLUTCH SIZE
2–4 eggs

PHAINOPEPLA NITENS

PHAINOPEPLA

PASSERIFORMES

Phainopeplas are nomadic birds that breed in two distinct regions and habitats, within the same season but with some temporal delay. In the early spring, the Sonoran desert-dwelling populations establish breeding territories, and often build a nest in the same tree as the desert mistletoe bunch that will provide much of the nutrition during chick rearing. Later, the breeding season moves to the riparian forests in coastal California and Mexico, but there the birds nest in loose colonies, and rely on scattered fruiting trees to feed themselves and the chicks. It remains a question whether the same birds breed in both habitats in the same year.

Parental roles vary with the stage of breeding: the male is in charge of most of the nest construction, and both sexes contribute to incubating the eggs. Once the chicks hatch, the male may or may not share the feeding duties fairly; while each member of the pair does equal duty in riparian sites, in the desert habitat the male plays less of a parental role as the chicks grow larger.

Clutch

The egg of the Phainopepla is light grayish, has black maculation throughout, and measures $^{13}/_{16}$ x $^{5}/_{8}$ in (21 x 16 mm). The female typically has several options before she lays her eggs; as part of courtship, the male builds several nests for her, and she makes her choice by completing the nest rim and the lining.

Actual size

ORDER	Passeriformes
FAMILY	Calcariidae
BREEDING RANGE	Arctic North America and Eurasia
BREEDING HABITAT	Shrubby and swampy tundra, treeless mountain tops
NEST TYPE AND PLACEMENT	Shallow depression carved into dense vegetation, under tufts of grass, or in soil banks; open cup on a foundation of mosses, grasses, and stems, lined warmly with feathers and hair
CONSERVATION STATUS	Least concern
COLLECTION NO.	FMNH 13884

ADULT BIRD SIZE
6–6½ in (15–16 cm)

INCUBATION
10–14 days

CLUTCH SIZE
4–6 eggs

CALCARIUS LAPPONICUS

LAPLAND LONGSPUR

PASSERIFORMES

Clutch

The egg of the Lapland Longspur is pale greenish white to greenish gray in background color, with dense brown and black spots and dark lines, and is ¹³⁄₁₆ x ⅝ in (21 x 15 mm) in dimensions. The female alone builds the nest and incubates the eggs; the male is often absent from the territory during incubation, soliciting extra-pair matings from other, often distant females.

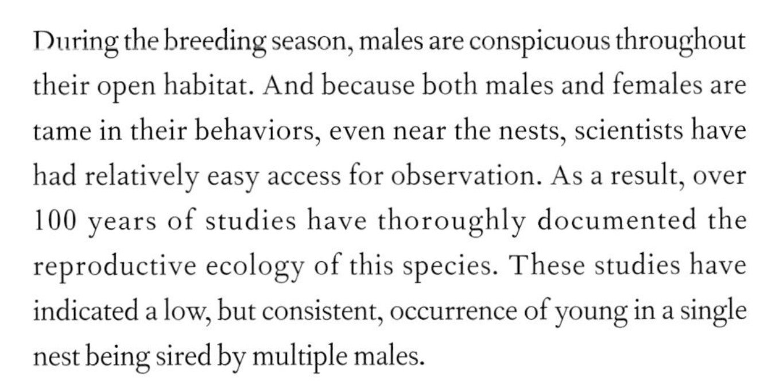

During the breeding season, males are conspicuous throughout their open habitat. And because both males and females are tame in their behaviors, even near the nests, scientists have had relatively easy access for observation. As a result, over 100 years of studies have thoroughly documented the reproductive ecology of this species. These studies have indicated a low, but consistent, occurrence of young in a single nest being sired by multiple males.

A gregarious species on the temperate wintering grounds, the males make their way to the nesting grounds and establish territories several days before the females arrive. Pair formation includes a courtship display in which the male sings in flight over the prospective female. Male coloration, especially a black bib, influences female choice. But while a male with a prominent black bib may be favored by females, he is less likely than drabber males to provide parental provisions to the chicks during the nestling stage.

Actual size

ORDER	Passeriformes
FAMILY	Calcariidae
BREEDING RANGE	Arctic North America and Eurasia
BREEDING HABITAT	Treeless tundra, rocky fields, bare mountain tops
NEST TYPE AND PLACEMENT	Deep inside a rock crevice or fissure; open cup of moss and grass, lined with fine grasses, rootlets, hair, and feathers
CONSERVATION STATUS	Least concern
COLLECTION NO.	FMNH 21508

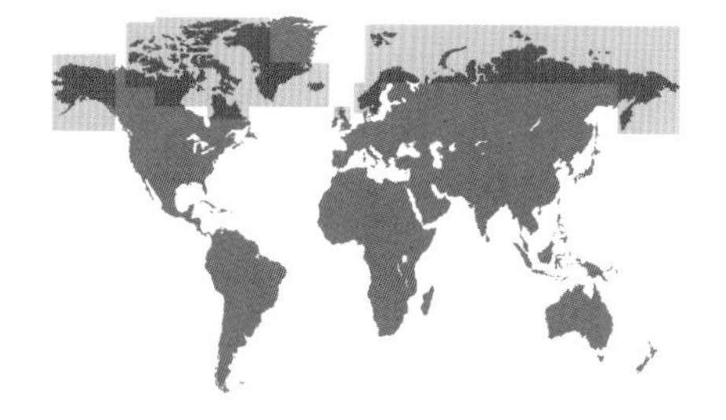

ADULT BIRD SIZE
6–7 in (15–18 cm)

INCUBATION
10–15 days

CLUTCH SIZE
2–7 eggs

PLECTROPHENAX NIVALIS

SNOW BUNTING

PASSERIFORMES

The Snow Bunting's white plumage in the winter serves both for camouflage in the snow and to reflect body heat back into the bird. Together with the much larger, and black, Common Raven, this bunting is the most northerly wintering songbird. Wintering far north saves time for the males, who arrive several weeks before the female to set up and defend breeding territories in the high Arctic. The value of each territory depends on the availability of rock crevices for nesting. In turn, cavity nesting protects the buntings from discovery by predators that use vision to hunt in the open, treeless Arctic breeding grounds.

When the females arrive, showing off the feasible nest sites to the potential mate is a part of the males' courtship displays. Once a pair bond is formed, nest building can begin. If there is an old nest in the crevice, the female may reuse it as a base to build her new nest. The female alone incubates the eggs and broods the hatchlings, but the male joins in to provision the growing young.

Clutch

The egg of the Snow Bunting is creamy white in coloration, has brown maculation, and its dimensions are ⅞ x ⅔ in (23 x 17 mm) in size. To minimize the female's time off the eggs, and the resulting loss of her body heat, the male feeds her directly on the nest.

Actual size

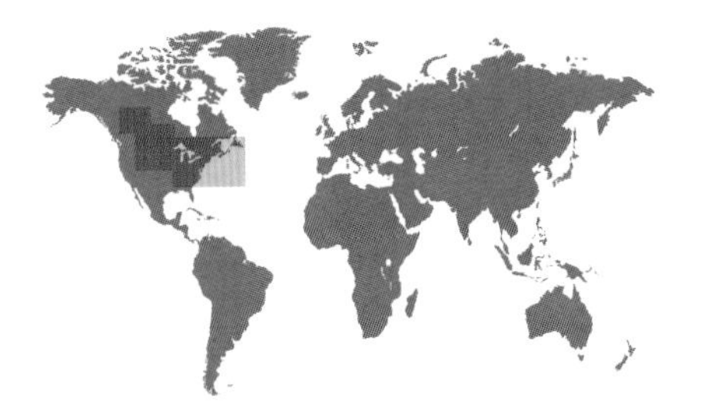

ORDER	Passeriformes
FAMILY	Parulidae
BREEDING RANGE	Eastern and central temperate North America
BREEDING HABITAT	Mature deciduous and mixed forests, with little undergrowth
NEST TYPE AND PLACEMENT	Woven dome, made of dead leaves and plant stems, lined with hair, with an entrance on the side; placed on the ground
CONSERVATION STATUS	Least concern
COLLECTION NO.	FMNH 13218

ADULT BIRD SIZE
4½–6¼ in (11–16 cm)

INCUBATION
11–13 days

CLUTCH SIZE
3–6 eggs

SEIURUS AUROCAPILLA

OVENBIRD

PASSERIFORMES

Clutch

This is North America's oscine ovenbird, building an oven-shaped grass and leaf nest on the forest floor, as opposed to the solid mud nest built on top of a fence pole by the hornero (see page 380), the suboscine ovenbird of South America. A migratory species from Central America and the Caribbean, the males arrive on the breeding grounds ahead of the females, divide up the available forest floor for nesting and feeding, and begin calling to attract a female, with their loud songs of "tea-cher, tea-cher, tea-cher."

Once the pair is formed, the female begins nest construction, and she alone incubates the eggs and broods the hatchlings. The male then joins her in provisioning the chicks. To avoid revealing the location of the nest, both parents approach it on the ground, moving carefully and quietly. Even so, many Ovenbird nests fail due to predation by chipmunks, and to brood parasitism by Brown-headed Cowbirds (see page 616). Parasitism is especially common when the breeding territory is near the forest edge, including along roads, clear-cuts, and power line strips.

The egg of the Ovenbird is white in background color, has dark brown speckling, and measures ¾ x ⅝ in (19 x 16 mm). To provide the best possible camouflage for the eggs, the nest is made entirely of materials collected from immediately around the nest site.

Actual size

ORDER	Passeriformes
FAMILY	Parulidae
BREEDING RANGE	Southeastern North America
BREEDING HABITAT	Deciduous, mature forests
NEST TYPE AND PLACEMENT	Open cup, placed on the ground under leaves; made of skeletonized foliage, lined with rootlets, grasses, hair, and mostly reddish moss stems
CONSERVATION STATUS	Least concern
COLLECTION NO.	FMNH 12067

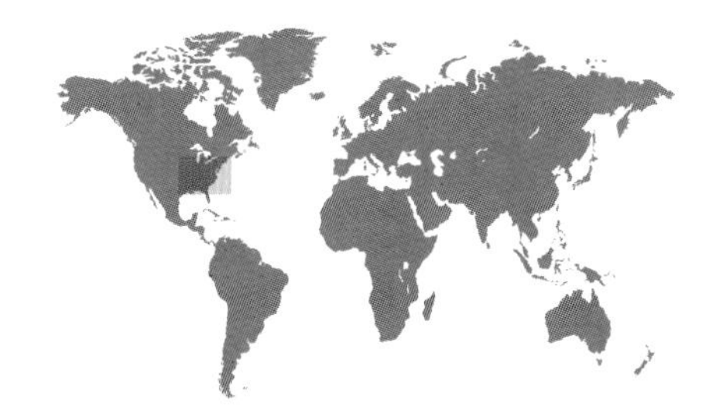

ADULT BIRD SIZE
4½–5 in (11–13 cm)

INCUBATION
12–13 days

CLUTCH SIZE
4–6 eggs

HELMITHEROS VERMIVORUS

WORM-EATING WARBLER

PASSERIFORMES

The Worm-eating Warbler occurs in mature forests of eastern mountain and foothill regions in the United States. An inhabitant of leafy forest floors, it feeds on caterpillars and gathers nest materials from the dry leaf litter. The female alone constructs the nest but the male often accompanies her while she searches for a feasible site. To build the base of the nest, the female first collects skeletonized leaves. Then, wetting her chest feathers to moisten the leaves, she shapes them into a bowl. Once the leaves are dried into a cup shape, she lines the nest with soft mosses, hair, and grasses, before egg laying begins.

Clutch

This warbler is an atypical member of the large New World warbler family Parulidae, placed in its own genus based on extensive recent molecular studies on the evolutionary relationships of these small New World insectivores. Its distribution and productivity have been declining; the causes include deforestation as well as reproductive failure due to brood parasitism.

Actual size

The egg of the Worm-eating Warbler is white to pink in coloration, with brown maculation, and measures ¾ x 9⁄16 in (19 x 14 mm) in size. If a nest is successful, there is no second brood; however, if a first nest fails, the female will lay another clutch of eggs as a replacement.

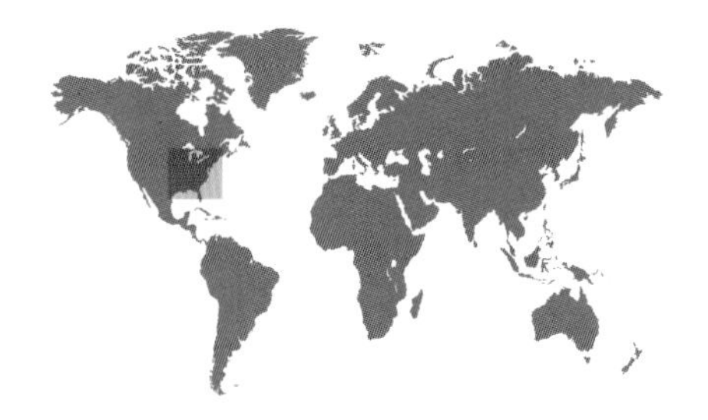

ORDER	Passeriformes
FAMILY	Parulidae
BREEDING RANGE	Eastern and middle North America
BREEDING HABITAT	Along clear, permanent streams, flowing through medium- or high-gradient forest slopes
NEST TYPE AND PLACEMENT	Open cup of mud, leaves, plant stems, and needles, built on a platform of wet leaves; typically in ground cavities or hollows with overhangs in stream banks or between roots of fallen trees
CONSERVATION STATUS	Least concern
COLLECTION NO.	FMNH 13318

ADULT BIRD SIZE
5½–6¾ in (14–16 cm)

INCUBATION
14–16 days

CLUTCH SIZE
2–6 eggs

PARKESIA MOTACILLA

LOUISIANA WATERTHRUSH

PASSERIFORMES

Clutch

This species specializes in breeding near permanent flowing streams and brooks, where it can find a higher diversity and greater biomass of benthic invertebrates to feed on. The females arrive on the breeding ground later than the males; older females typically settle in the same territory they occupied in previous years. If the resident male there is already mated, the experienced female may drive out the other female. To select their nesting site, the mates walk along their territory, inspecting potential cavities, assessing sturdiness, and pulling leaves and other materials into the cavity.

The Louisiana Waterthrush's song is loud and ringing to carry across long distances despite the inherently noisy waterside habitat. The bird draws even more attention to itself by frequently wagging its tail, justifying the Latin species name "motacilla," which is the same as the genus name for the

Actual size

The egg of the Louisiana Waterthrush is creamy white in background coloration, has reddish brown maculation, and its measurements are ¾ x ⅝ in (19 x 16 mm). Unlike many wood warblers, both sexes participate in nest-site selection, material collection, and construction; once the eggs are laid, only the female incubates.

ORDER	Passeriformes
FAMILY	Parulidae
BREEDING RANGE	Boreal North America
BREEDING HABITAT	Forested swamps, wetlands, river and lake shores
NEST TYPE AND PLACEMENT	Open cup of moss, bark, and leaves, lined with fine plant stems, rootlets, and hair; in a small hollow along a soil bank, or a cavity under a fallen log or uprooted tree
CONSERVATION STATUS	Least concern
COLLECTION NO.	FMNH 15185

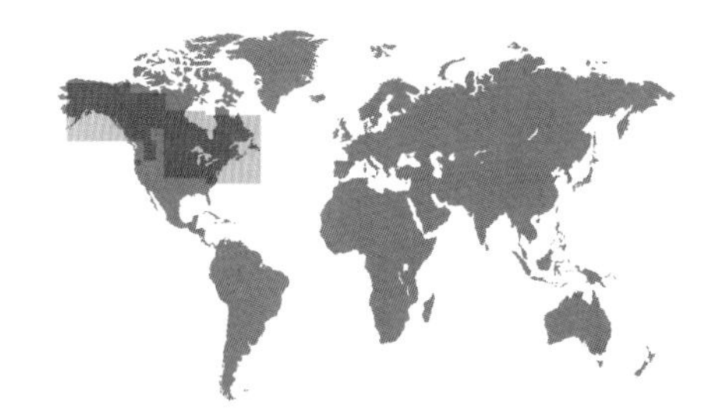

ADULT BIRD SIZE
4½–5½ in (12–14 cm)

INCUBATION
13 days

CLUTCH SIZE
4–5 eggs

PARKESIA NOVEBORACENSIS

NORTHERN WATERTHRUSH

PASSERIFORMES

The Northern Waterthrush uses many town parks, wetlands, and stream banks during its migration in the spring and the fall, but most of its breeding range is far from human settlements and even roads. The male arrives ahead of the female, and pair bonding often takes place on the day when the female is first seen in a territory. The newly formed pair moves about the habitat in search of suitable nesting cavities, one of which is selected by the female for construction.

While the female builds the nest, the male moves his regular singing perch to within 30–60 ft (10–20 m) of the site. He alternates his loud songs with more subtle and quiet calls, and continues vocalizing even after the female begins to incubate the eggs in the nest. Once the chicks hatch, both parents provision them, until about four weeks of age, at which point the fledglings can fly and feed independently.

Clutch

Actual size

The egg of the Northern Waterthrush is white, with dark brown spotting, and is ¾ x ⅝ in (19 x 16 mm) in size. To speed up the timing and efficiency of hatching, the female's brood patch may start to develop just before she leaves the wintering grounds in Mexico and Central America.

ORDER	Passeriformes
FAMILY	Parulidae
BREEDING RANGE	Southeastern North America
BREEDING HABITAT	Low, wet forested areas, wooded swamps
NEST TYPE AND PLACEMENT	Deep, open cup nest, made of leaves, mosses, and grasses, and lined with lichen and moss; placed near the ground or at a low height, in dense undergrowth
CONSERVATION STATUS	Likely extinct
COLLECTION NO.	FMNH 16949

ADULT BIRD SIZE
4–4½ in (10–11 cm)

INCUBATION
Unknown

CLUTCH SIZE
3–5 eggs

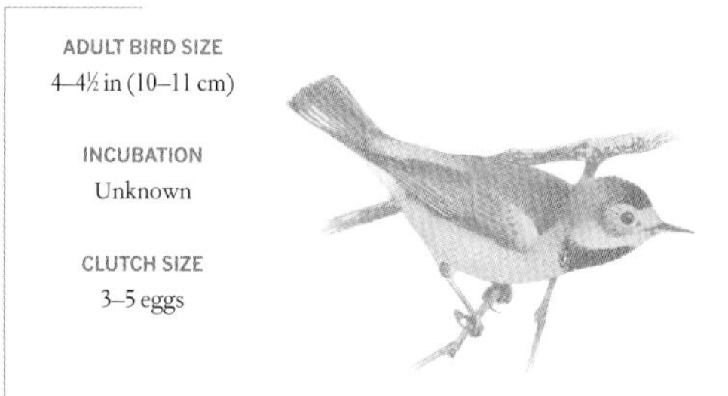

VERMIVORA BACHMANII

BACHMAN'S WARBLER

PASSERIFORMES

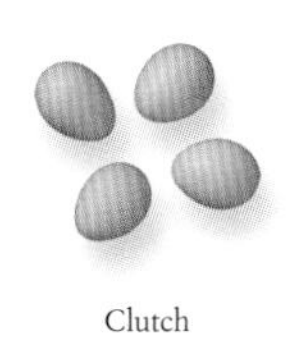

Clutch

The last confirmed sighting of a Bachman's Warbler was in 1988. Loss and modification of its canebrake habitat across the southeastern United States and the clearing of dry, semi-deciduous forests on its Cuban wintering grounds are likely the causes for the terminal decline of this species. Throughout its breeding range, this species relied on dense understories for foraging and for foliage cover over the ground nest.

The male Bachman's Warbler defended a small territory, with dense undergrowth, where the nests would be sited. The latest records of active nests are from 1937. Incubation duration was never firmly established for this species, and only the female was ever observed sitting on the eggs. In turn, both parents were seen carrying food to the nestlings, but the lengths of the nestling period and post-fledgling parental care were again never documented.

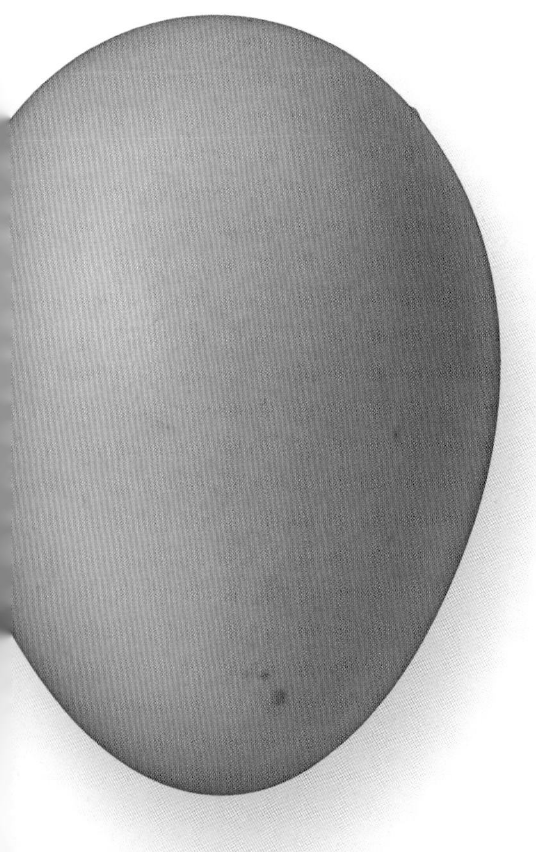

Actual size

Most of the surviving eggs of the Bachman's Warbler in collections are white in background color; some eggs have dark brown spotting at the blunt end; the measurements are ⅝ x ½ in (16 x 12 mm). There are no records of nest construction, but only females were ever observed carrying nesting materials.

ORDER	Passeriformes
FAMILY	Parulidae
BREEDING RANGE	Southeastern North America
BREEDING HABITAT	Overgrown fields, shrubby pastures, forest edges
NEST TYPE AND PLACEMENT	Open cup, made of grasses, bark strips, and often under leaves; on or near the ground, often near a tall tree
CONSERVATION STATUS	Least concern
COLLECTION NO.	FMNH 16947

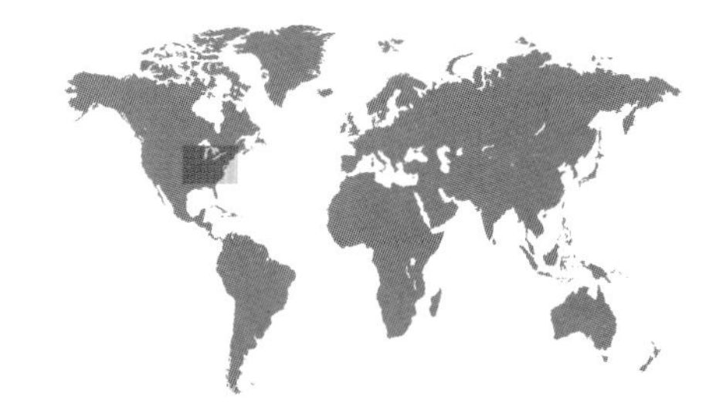

ADULT BIRD SIZE
4½–4¾ in (11–12 cm)

INCUBATION
10–11 days

CLUTCH SIZE
4–5 eggs

VERMIVORA CYANOPTERA

BLUE-WINGED WARBLER

PASSERIFORMES

This small wood warbler favors regenerating wooded habitats. It has benefited from hundreds of years of human habitat modification in North America, which included opening up dense forests and generating more edges and shrubby clearings. At the same time, this preferred habitat also favored large numbers of parasitic Brown-headed Cowbirds (see page 616), which have expanded into the east across the last centuries, and which specialize on ground-nesting birds; the result is that in some populations, nearly half of all warbler nests have cowbird eggs.

The Blue-winged Warbler is also known for its propensity to hybridize with its close relative, the Golden-winged Warbler (see page 530). Depending on the relative genetic make-up of the parents, the hybrid offspring are distinctively patterned, such that they were initially described as two separate, full species (Brewster's and Lawrence's Warblers). More often than not, though, the genome of the Blue-winged Warbler comes to dominate, and eventually causes the genetic extinction of the local Golden-winged Warbler population.

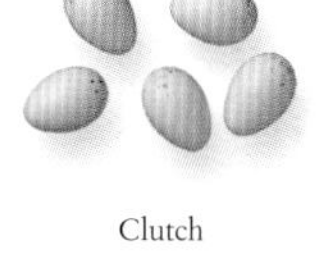

Clutch

The egg of the Blue-winged Warbler is white in coloration, with brown speckling, and is ⅝ x ½ in (16 x 12 mm) in dimensions. If individual eggs go missing from the nest, the female continues to lay and incubate; if all eggs disappear, she abandons it and starts nesting elsewhere.

Actual size

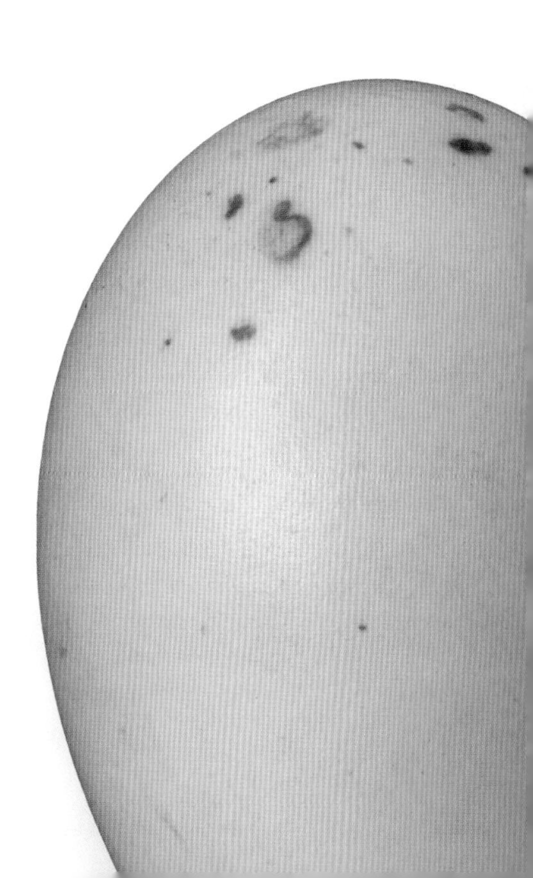

ORDER	Passeriformes
FAMILY	Parulidae
BREEDING RANGE	Eastern North America
BREEDING HABITAT	Near moist, forested habitats, in scrubby natural or artificial openings, swamp forests with beaver glades, edges, and regenerating clear-cuttings
NEST TYPE AND PLACEMENT	Ground nest often placed at the base of a plant, a shrub, or plant stem; open cup made from long leaves and bark strips
CONSERVATION STATUS	Near threatened
COLLECTION NO.	FMNH 16946

ADULT BIRD SIZE
4½–5 in (12–13 cm)

INCUBATION
10–12 days

CLUTCH SIZE
4–6 eggs

VERMIVORA CHRYSOPTERA

GOLDEN-WINGED WARBLER

PASSERIFORMES

Clutch

The egg of the Golden-winged Warbler is pale pink or pale cream in background color, has fine darker markings, and measures ⅝ x ½ in (16 x 12 mm). Prior to laying, the male accompanies the female while she searches for a nest, but all nest-building and incubation duties are carried out by the mother.

Together with his brightly patterned plumage, the male's song is his main means of attracting females to the territory. When the female accepts him, his vocal output diminishes, only to pick up again if she leaves, the nesting attempt fails, or incubation has begun fully. Clearly, the male's song can also be a signal to females other than his first mate; accordingly, some polygamous males are known to feed chicks at more than one nest. In addition, at more than 50 percent of nests where blood samples were taken for genotype analyses, males other than the territorial father were known to have sired the young.

The Golden-winged Warbler benefited from forest fragmentation and the spread of artificial, edgelike habitats in the form of regenerating forest stands and abandoned farmland. At many such sites, however, it also co-occurs with the genetically more dominant Blue-winged Warbler (see page 529), and the costly brood parasite, the Brown-headed Cowbird (see page 616). The result is a continually declining population size, yielding a conservation warning for this species.

Actual size

ORDER	Passeriformes
FAMILY	Parulidae
BREEDING RANGE	Temperate and southeastern North America
BREEDING HABITAT	Mature or advanced second-growth deciduous and mixed forests; increasingly spruce forests, too
NEST TYPE AND PLACEMENT	Open cup of dry leaves, bark strips, grass, and pine needles, lined with finer grasses, hair, and moss; placed on the ground next to a tree trunk
CONSERVATION STATUS	Least concern
COLLECTION NO.	FMNH 13039

ADULT BIRD SIZE
4½–5 in (11–13 cm)

INCUBATION
10–12 days

CLUTCH SIZE
4–5 eggs

MNIOTILTA VARIA

BLACK-AND-WHITE WARBLER

PASSERIFORMES

This distinctively patterned wood warbler species behaves like a nuthatch or treecreeper. Unique among parulids, the Black-and-white Warbler often forages on bark, moving along branches and up and down tree trunks in search of hiding insects or dormant pupae. Perhaps because of its bark-based feeding strategy, this species is one of the earliest spring warblers to arrive on the breeding grounds, often before the threat of snow has fully receded.

These warblers also begin breeding early, sometimes before the foliage has developed. When a pair is established, nest building begins within days; despite being a well-known species, we do not know much about the nest building, except that a female was seen collecting bark for a nest. Once the eggs are laid, the female sits so tightly on the nest that she can sometimes be touched before she flushes. The male may feed her on the nest, and both parents collect food to provision the hatchlings.

Clutch

Actual size

The egg of the Black-and-white Warbler is white in color, with brown and lavender maculation, and is ⅔ x ½ in (17 x 12 mm) in size. There is a consistent variation in the size of the eggs within the clutch: the earlier-laid eggs are larger than the later-laid ones.

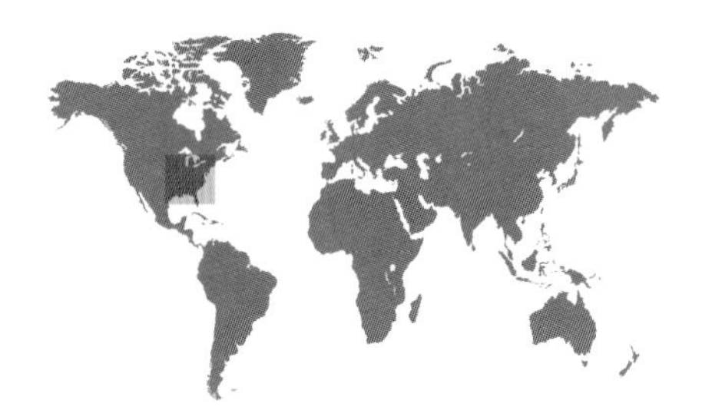

ORDER	Passeriformes
FAMILY	Parulidae
BREEDING RANGE	Central and southeastern North America
BREEDING HABITAT	Lowland swamp forests, riparian woods
NEST TYPE AND PLACEMENT	In cavities made by woodpeckers, natural tree holes, or nestboxes; cup made of rootlets, bark, and plant down, lined with grasses, sedges, and petioles, occasionally fishing line
CONSERVATION STATUS	Least concern
COLLECTION NO.	FMNH 13110

ADULT BIRD SIZE
4¾–5½ in (12–14 cm)

INCUBATION
12–14 days

CLUTCH SIZE
3–7 eggs

PROTONOTARIA CITREA

PROTHONOTARY WARBLER

PASSERIFORMES

Clutch

Unlike most other parulid warblers, this species breeds in cavities, and its nesting biology has been studied extensively in populations provided with artificial nestboxes in the form of cut-out milk cartons. Each year, by following hundreds of breeding pairs marked with unique combinations of color bands on their feet, scientists have amassed large records of reproductive details within and between years about the same individuals. Importantly, the reproductive success in natural or artificial cavities is very similar in this species. Male warblers defend several potential cavities, whereas females are in charge of choosing the site, building the nest, and incubating the eggs; once the chicks hatch, both parents provision the young.

Female Prothonotary Warblers exhibit what appear to be strong family ties in their breeding strategies. Daughters settle in nearby nesting sites whose vegetation resembles that of their mother: thus, the density of the forest around their nestboxes, and the likelihood of being parasitized by Brown-headed Cowbirds (see page 616), appear to be shared across the generations.

Actual size

The egg of the Prothonotary Warbler is white in background coloration, with lavender to rust-brown spotting, and measures ¾ x ⅝ in (19 x 16 mm) in size. If the eggs are lost to predation, the nest cavity is not used again; however, if the nest is parasitized, the parents will not abandon the nest site, continuing to use it in successive breeding seasons.

ORDER	Passeriformes
FAMILY	Parulidae
BREEDING RANGE	Southeastern North America
BREEDING HABITAT	Bottomland and riverine forests with dense understory vegetation, also pine plantations and wooded, shrubby swamps
NEST TYPE AND PLACEMENT	Open, bulky cup of dry leaves, sticks, and vines, lined with pine needles, hair, fine grasses, and Spanish moss; placed in dense foliage of shrub or vines
CONSERVATION STATUS	Least concern
COLLECTION NO.	FMNH 13030

ADULT BIRD SIZE
5–5½ in (13–14 cm)

INCUBATION
13–15 days

CLUTCH SIZE
3–4 eggs

LIMNOTHLYPIS SWAINSONII

SWAINSON'S WARBLER

PASSERIFORMES

This warbler is a relatively rare and patchily distributed species that maintains large breeding territories with low nesting densities. For a long time, most knowledge about this species was derived from one dedicated long-term study; today, more information is being gathered. This is because Swainson's Warblers are considered both indicators for habitat quality and a potential conservation concern due to their comparatively low population sizes.

Breeding begins with the later-arriving female's choosing a male; to entice her, he initially gently chases, then approaches, and eventually sings to her, while she inspects his territory. If she accepts him, then nest building begins with the placement of a large clump of dry leaves in dense vine foliage. Despite this camouflage, parasitic Brown-headed Cowbirds (see page 616) frequently locate and lay in the nest; and even though the cowbird does not evict host nest mates, the hatchling parasite chick grows faster, begs more intensively, and is fed by the warbler foster-parents nearly twice as often as their own young.

Clutch

The egg of the Swainson's Warbler is white, sometimes with faint reddish brown maculation, and measures ¾ x ⅝ in (19 x 16 mm) in size. The female remains tightly on the nest during incubation, and only leaves the bright white eggs uncovered when in imminent danger; she then drops to the ground and performs a broken-wing display to lead predators away from the nest.

Actual size

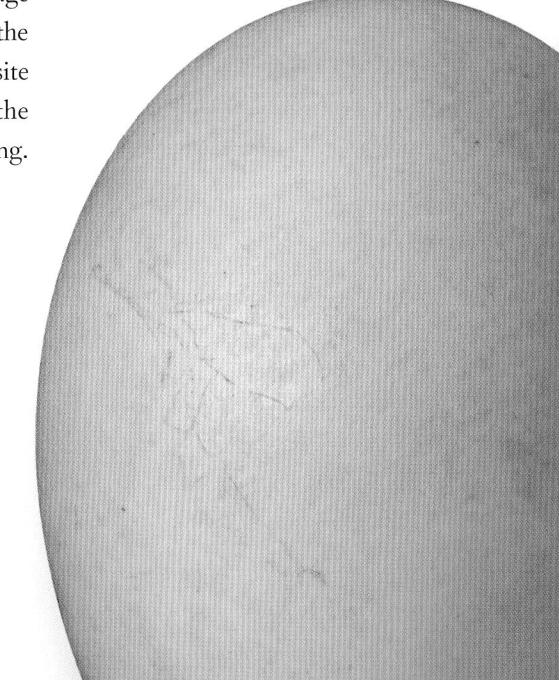

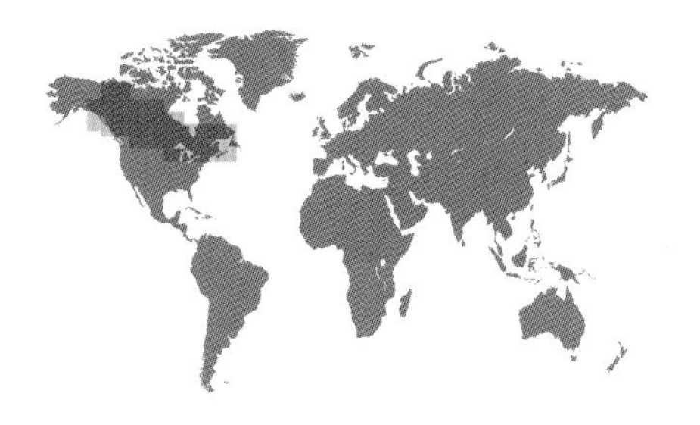

ORDER	Passeriformes
FAMILY	Parulidae
BREEDING RANGE	Boreal and subarctic North America
BREEDING HABITAT	Boreal coniferous or mixed forests with clearings and other open areas containing grasses, dense shrubs, and growing deciduous trees
NEST TYPE AND PLACEMENT	Open cup of dry grass, weed stems, leaves, and bark strips, lined with fine grass, moss, rootlets, and hair; placed on the ground, hidden in a tuft of vegetation, moss, or upturned roots
CONSERVATION STATUS	Least concern
COLLECTION NO.	FMNH 15180

ADULT BIRD SIZE
4–5 in (10–13 cm)

INCUBATION
11–14 days

CLUTCH SIZE
3–8 eggs

OREOTHLYPIS PEREGRINA

TENNESSEE WARBLER

PASSERIFORMES

Clutch

This warbler's misleading common name honors the region where it was first collected. But the species only passes through Tennessee as it migrates between its Central American wintering grounds and its breeding range in boreal Canada and Alaska. The Tennessee Warbler's population size, breeding range, and productivity follow closely the outbreaks of spruce budworm caterpillars, on which this bird is a specialist feeder. In the winter, it switches to feed mostly on nectar and sugary fruits.

On the breeding grounds, the nest building frenzy is quickly followed by the laying of the eggs, with one egg produced per day, as is typical of most smaller bird species. The female apparently starts incubation at around the time when the last egg is laid, because all chicks hatch synchronously, within just one day of each other. The parents feed the chicks by taking turns, and both provision the fledglings until independence.

Actual size

The egg of the Tennessee Warbler is white in background coloration, with reddish brown maculation, and measures ⅝ x ½ in (16 x 12 mm) in size. The female alone incubates the eggs, sitting very tightly on them unless in imminent danger; the male occasionally feeds her while on the nest.

ORDER	Passeriformes
FAMILY	Parulidae
BREEDING RANGE	Northern and western North America
BREEDING HABITAT	Woodlands with shrubby understory, chaparral, oak galleries, scrub, and thickets
NEST TYPE AND PLACEMENT	Open cup, on or near the ground in dense shrub foliage; made of leaves and fine twigs, bark strips, rootlets, plant down, or wool, lined with finer grasses, moss, or animal hair
CONSERVATION STATUS	Least concern
COLLECTION NO.	FMNH 13390

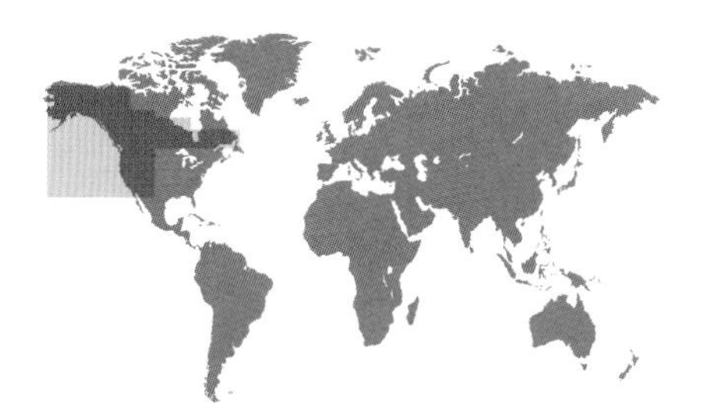

ADULT BIRD SIZE
4½–5½ in (11–14 cm)

INCUBATION
12–14 days

CLUTCH SIZE
4–6 eggs

OREOTHLYPIS CELATA

ORANGE-CROWNED WARBLER

PASSERIFORMES

The Orange-crowned Warbler has a vast distribution, breeding and feeding in diverse habitats. The males announce and defend their territory by their insistent songs, but once a female has settled in the territory, they often become silent, for a while at least. Territories with silent males indicate that nest building has begun, but it is never easy to find the nest. The females collect nesting materials from both the ground and above, in the shrub and tree canopy, and construct it in a well-hidden spot. The males do not participate in nest building and do not feed the females on the nest.

Once the chicks hatch, both parents contribute to parental duties by feeding the nestlings throughout the day, typically with equal frequency. If the nest is approached closely by a potential predator, the female quickly drops to the ground, moving some distance away on her feet, before beginning a broken-wing display; the male may also join in with an alarm call aimed at the predator to let it know that its attack is no longer a surprise.

Clutch

The egg of the Orange-crowned Warbler is white in background color, with fine reddish brown maculation, and measures ⅔ x ½ in (17 x 12 mm). To keep the eggs safe, the nest is always well camouflaged, typically under vegetation, woody stems, and even in shallow depressions, in crevices, and under rocks.

Actual size

ORDER	Passeriformes
FAMILY	Parulidae
BREEDING RANGE	Southwestern North America
BREEDING HABITAT	Dense mesquite shrub and riparian forest galleries
NEST TYPE AND PLACEMENT	In a cavity, excavated by woodpeckers or other natural holes in trees and cacti, also abandoned Verdin nests; small, woven cup of grasses, on top of a pile of sticks and debris
CONSERVATION STATUS	Least concern
COLLECTION NO.	FMNH 289

ADULT BIRD SIZE
3½–4½ in (9–11 cm)

INCUBATION
12 days

CLUTCH SIZE
3–7 eggs

OREOTHLYPIS LUCIAE

LUCY'S WARBLER

PASSERIFORMES

Clutch

The Lucy's Warbler is the smallest of the warblers breeding in North America; it is one of just a handful of parulid species that nests in cavities. However, unlike the Prothonotary Warbler (see page 536), Lucy's Warblers will not readily use artificial nestboxes, and consequently relatively little is known about their breeding biology. Despite the secluded location of the nest, many clutches are parasitized by Brown-headed Cowbirds (see page 616) and this, together with habitat loss, have likely contributed to the declining population trend.

This warbler nests in desert woodlands along water courses; suitable habitats are often packed very densely with territorial pairs, almost as if this solitary nesting species occurred colonially. In fact, breeding density is a side effect of the density of available nesting cavities. To avoid competition with larger cavity-dwelling species, the nest hole is often filled up with debris to the rim, and so the nest cup is placed at the level of the cavity entrance itself.

The egg of the Lucy's Warbler is white in color, has fine reddish maculation, and measures ⅝ x ½ in (16 x 12 mm). The female builds the nest and incubates the eggs; the chicks may be fed by both parents.

Actual size

ORDER	Passeriformes
FAMILY	Parulidae
BREEDING RANGE	Disjunct regions of temperate North America
BREEDING HABITAT	Secondary and re-growing mixed-species forests, scrubby bogs
NEST TYPE AND PLACEMENT	Open cup of moss, bark, sedge leaves, and grasses, lined with fine grass, pine needles, and hair; on the ground, under low brushy vegetation
CONSERVATION STATUS	Least concern
COLLECTION NO.	FMNH 13416

ADULT BIRD SIZE
4½–4¾ in (11–12 cm)

INCUBATION
11–12 days

CLUTCH SIZE
4–5 eggs

OREOTHLYPIS RUFICAPILLA

NASHVILLE WARBLER

PASSERIFORMES

The breeding behaviors of this species follow typical parulid patterns, with males arriving before the females to set up territories. Nest construction takes place right after pair formation is complete, with the female building the nest and incubating the eggs. The male sings and accompanies the female during nest construction, and he may also feed her at the nest; he then fully joins her to provision the nestlings, although single females have also been known to raise the young.

First seen and identified in western Tennessee in 1811, this bird got its name from a site where it stopped to rest on its migratory flight toward breeding grounds in eastern Canada. Its western subspecies, breeding in the mountainous west, was not discovered until half a century later, and was first described as a separate species. Today, these distinctive and geographically separated populations are unified under one, very unusual, common species name.

Clutch

The egg of the Nashville Warbler is white in background color, with brown maculation, and ⅝ x ½ in (16 x 12 mm) in size. In both the eastern and western populations, the eggs remain well hidden in the nests, placed under leaves, grass clumps, dense thickets, or at the foot of tree trunks.

Actual size

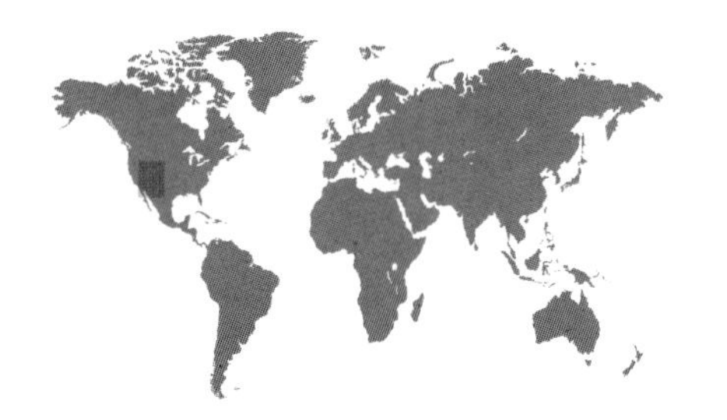

ORDER	Passeriformes
FAMILY	Parulidae
BREEDING RANGE	Southwestern montane North America
BREEDING HABITAT	Oak and pinyon forests, scrubby hillsides at high elevations
NEST TYPE AND PLACEMENT	Open cup on the ground, hidden in leaf litter and grass clumps, at the base of a shrub; constructed from grass, strips of bark, moss, and rootlets
CONSERVATION STATUS	Least concern
COLLECTION NO.	FMNH 267

ADULT BIRD SIZE
4–4½ in (10–11 cm)

INCUBATION
11–12 days

CLUTCH SIZE
3–5 eggs

OREOTHLYPIS VIRGINIAE

VIRGINIA'S WARBLER

PASSERIFORMES

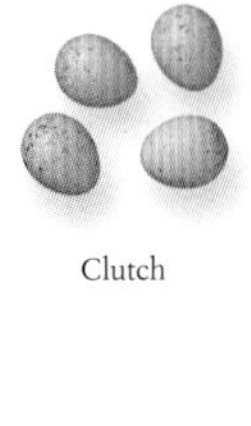

Clutch

Named after the wife of the army officer who discovered this species, 150 years later the Virginia's Warbler remains an enigmatic species. The breeding habits of this bird are best known from broad comparative studies of northern and southern hemisphere bird species breeding at the same latitude. However, basic data about the species' distribution and migratory behaviors remain to be documented.

The breeding season begins with the males' early arrival and the division of the suitable habitat into territories. The males are at their most vocal and aggressive by the time the females move in; the females quickly make decisions about pair bonding. Nest building and incubation are both conducted by the female, and she alone broods the chicks. Feeding the chicks is shared by both sexes, with nest hygiene maintained by removal of broken eggshells and fecal sacs by the parents.

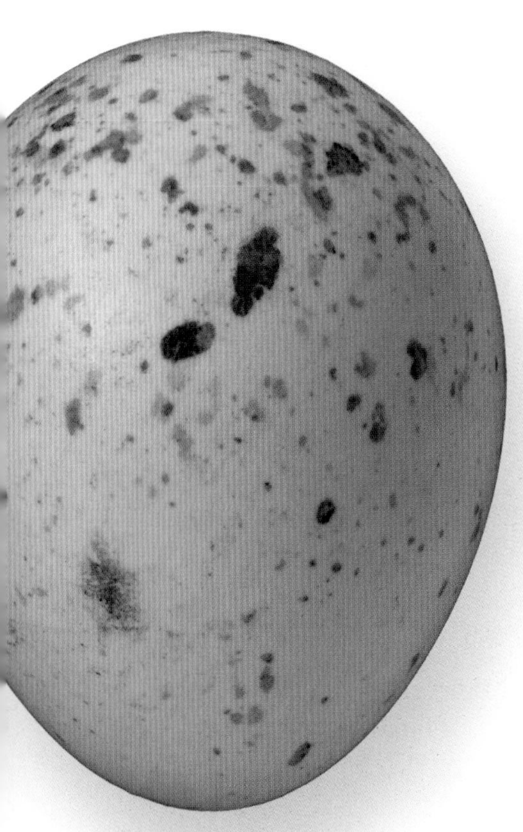

Actual size

The egg of the Virginia's Warbler is white in background coloration, has brown maculation, and its dimensions are ⅝ x ½ in (16 x 12 mm). The female keeps her nest clean and free of debris, deftly removing fallen branches, feces, ants, and even egg-shaped scientific data loggers, from crowding the real eggs in the nest cup.

ORDER	Passeriformes
FAMILY	Parulidae
BREEDING RANGE	North temperate North America
BREEDING HABITAT	Wooded bogs, mature coniferous or mixed forests, often near water
NEST TYPE AND PLACEMENT	Open cup of fine, dry grasses, leaves, stalks, stems, and rootlets; placed on the ground, hidden in moss or a clump of grass
CONSERVATION STATUS	Least concern
COLLECTION NO.	FMNH 2573

ADULT BIRD SIZE
5–6 in (13–15 cm)

INCUBATION
11–12 days

CLUTCH SIZE
3–5 eggs

GEOTHLYPIS AGILIS

CONNECTICUT WARBLER

PASSERIFORMES

With yet another misleading name, this warbler is predominately a Canadian breeding species. Perhaps due its nesting habitat in remote northern moist woodlands, far from many human settlements, few extensive studies have been conducted on its nesting biology. For example, we assume, but do not know for certain, that only the female is responsible for selecting the nest site and constructing the nest cup.

What we do know is that only the female develops an incubation patch and sits on the eggs until hatching. To reduce the chance of being detected by predators, she lands 30–40 ft (10–13 m) away from the nest, and sneaks through dense understory on foot to reach the eggs. Once the chicks hatch, both sexes provision them with caterpillars, moths, other insects, and small berries; even then, the parents are secretive, and approach the nest cautiously on the ground from a distance.

Clutch

Actual size

The egg of the Connecticut Warbler is creamy white in background, with darker maculation, and ¾ x ⅝ in (19 x 16 mm) in size. To distract and draw predators away from the nest, both parents call harshly, and can perform a broken-wing display if needed.

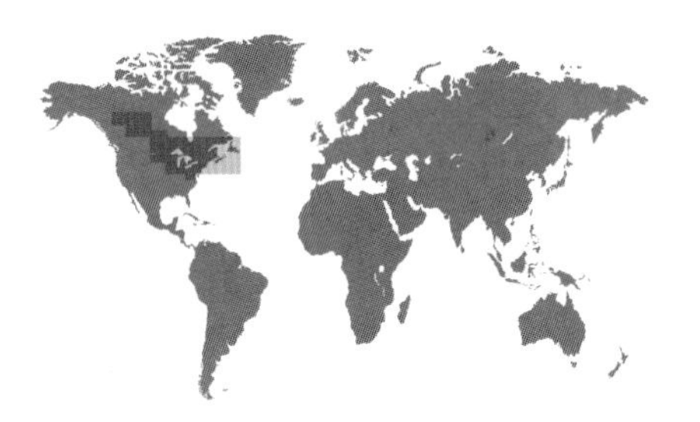

ORDER	Passeriformes
FAMILY	Parulidae
BREEDING RANGE	Boreal North America
BREEDING HABITAT	Second-growth coniferous forests, scrubby natural or artificial clearings
NEST TYPE AND PLACEMENT	Open cup near or on the ground, at the base of tufts of grasses and sedges; made of stalks, weed leaves, and bark, lined with roots, fine grasses, and fur
CONSERVATION STATUS	Least concern
COLLECTION NO.	FMNH 13082

ADULT BIRD SIZE
4–6 in (10–15 cm)

INCUBATION
12–13 days

CLUTCH SIZE
2–5 eggs

GEOTHLYPIS PHILADELPHIA

MOURNING WARBLER

PASSERIFORMES

Clutch

The egg of the Mourning Warbler is white in coloration, has reddish brown maculation, and measures ¾ x 9/16 in (19 x 14 mm). This species lays its eggs in the darkness, up to three hours before sunrise.

Actual size

This species is called an ecological "fugitive" because of its preference for early successional, regenerating forests, which it rapidly colonizes after a natural fire or a beaver-caused clearing event, or after clear-cutting associated with modern forestry practices. However, seven to ten years into the regeneration cycle, the canopy starts to fill in and the habitat becomes unsuitable for this warbler, resulting in local loss or extinction, until the next fire sweeps through or a storm uproots the mature woodland and the process can start again.

This species breeds in the north, and typically only has enough time to complete one full nesting cycle. Once the nest is built, the female alone incubates while the male feeds her on the nest; she is also the only parent to brood the hatchlings. By contrast, both parents feed the fast-growing chicks, who are ready to leave the nest in as little as nine days after hatching.

ORDER	Passeriformes
FAMILY	Parulidae
BREEDING RANGE	Temperate and southern North America
BREEDING HABITAT	Shrubs and thickets, often in or near wetlands
NEST TYPE AND PLACEMENT	Open cup on the ground, built on top of sedges or cattails; made of stems, grass leaves, and rootlets, lined with fine grasses, fibers, and hair
CONSERVATION STATUS	Least concern
COLLECTION NO.	FMNH 12839

GEOTHLYPIS TRICHAS

COMMON YELLOWTHROAT

PASSERIFORMES

ADULT BIRD SIZE
4½–5 in (11–13 cm)

INCUBATION
12 days

CLUTCH SIZE
3–5 eggs

This is a widespread and loud-singing bird of marshes and wet grasslands; however, its conspicuous black and yellow facial patterns can be hard to see, because of its preference for dense thickets and understory for nesting and feeding. The males in the migratory populations arrive about a week before the females, set up their territories, and sing throughout much of the day to keep the neighboring males away and to attract a nearby female.

The chicks are attended by both parents, who deliver arthropod prey in their bills, and feed each chick in response to the sound of their begging calls and sight of the open gapes. Once a nestling is fed, the parent remains at the rim of the nest, and waits for a fecal sac to be produced; this is an enclosed sac of excreta produced by the chicks, which is either consumed or carried away and dropped at a distance while flying from the nest. Nest sanitation becomes relaxed during the last two days prior to fledging and the nest is not reused later.

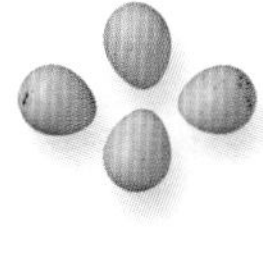

Clutch

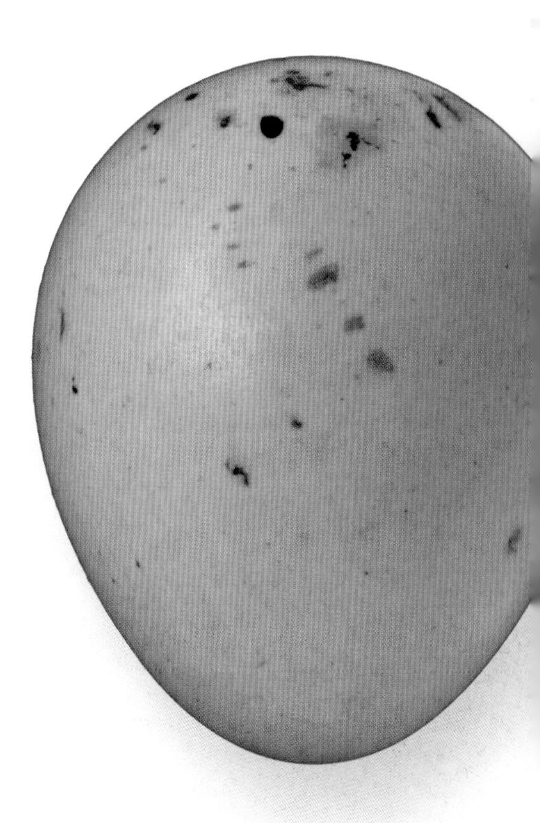

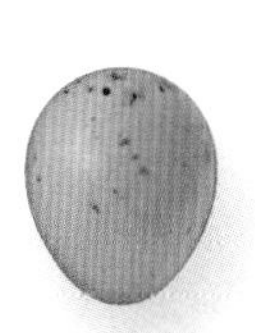

Actual size

The egg of the Common Yellowthroat is white in coloration, with gray, lilac, reddish brown, or black maculation, and is ⅔ x ½ in (17 x 12 mm) in size. The female alone builds the nest, and begins incubation on the day when the last egg is laid.

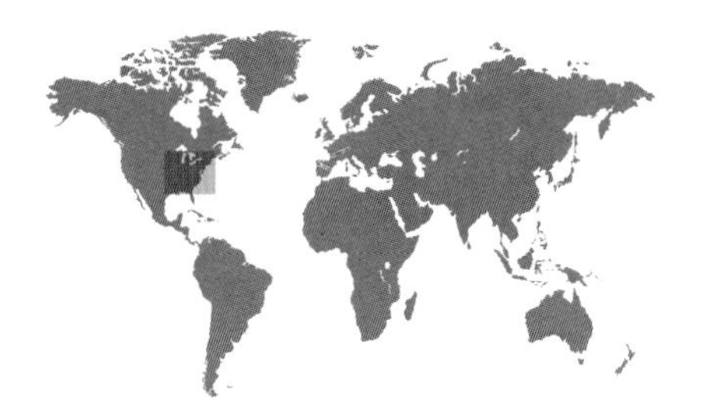

ORDER	Passeriformes
FAMILY	Parulidae
BREEDING RANGE	Southeastern and central North America
BREEDING HABITAT	Mixed hardwood forests, wooded swamps, including fragmented and remnant patches
NEST TYPE AND PLACEMENT	Cup nest, made of dead leaves, soft bark strips, sedges, and grasses, lined with rootlets and fur; placed low in the crotch of a leafy shrub
CONSERVATION STATUS	Least concern
COLLECTION NO.	FMNH 13455

ADULT BIRD SIZE
4½–5½ in (11–14 cm)

INCUBATION
12 days

CLUTCH SIZE
3–5 eggs

SETOPHAGA CITRINA

HOODED WARBLER

PASSERIFORMES

Clutch

The egg of the Hooded Warbler is creamy white in background color, with brown maculation, and ¾ x 9⁄16 in (19 x 14 mm) in size. Despite pressure from cowbird parasitism, most warbler eggs that fail to hatch are in fact lost to nest predators, rather than parasitism.

This species is highly vulnerable to brood parasitism by Brown-headed Cowbirds (see page 616), especially when breeding in fragmented forest patches, near edges facing open fields and cattle pastures. In some populations, over 50 percent of the nests have one or more cowbird eggs. The female warbler makes a chipping sound during nest construction, which may help female cowbirds to locate the warbler's nest and then return to parasitize it during the egg-laying period.

Like the birds they parasitize, cowbirds also lay their eggs just before sunrise; and so, because female Hooded Warblers do not typically roost in the nest prior to incubation, they are less likely to encounter, and defend their clutch from, the parasitic cowbirds. Curiously, when a female warbler encounters a cowbird near the nest, first-time breeders are just as aggressive as older, more experienced mothers, suggesting that prior experience is not critical for the recognition of the threat posed by the parasitic intruder.

Actual size

ORDER	Passeriformes
FAMILY	Parulidae
BREEDING RANGE	Temperate North America, Appalachian ranges
BREEDING HABITAT	Open woodlands, scrubby forest edges, wooded wetlands and shores
NEST TYPE AND PLACEMENT	Tightly woven, open cup made of grasses, bark, leaves, and hair, glued together with cobwebs; in forking branches in trees or shrub, from low to very high elevation
CONSERVATION STATUS	Least concern
COLLECTION NO.	FMNH 2604

ADULT BIRD SIZE
4½–5 in (11–13 cm)

INCUBATION
12 days

CLUTCH SIZE
3–5 eggs

SETOPHAGA RUTICILLA

AMERICAN REDSTART

PASSERIFORMES

A well-studied species, the conspicuous black-and-red plumage of the male is a critical chromatic indicator of his investment in mating: older males have darker plumages, and males with access to more carotenoid pigments in their food grow brighter red feathers. Females make decisions about males based on their plumages; a female mated to a redder male will be less likely to seek extra-pair fertilizations, and two or more females are more likely to settle with the same brighter male, even if he is already mated, leading to a facultatively polygynous mating system.

There is, however, also an advantage for the male in retaining a drabber juvenile plumage, at least in his first year of life: the phenomenon is called "delayed plumage maturation." Reproductively mature males with a female-like plumage can still obtain mates and fertilize eggs, but they are also less likely to be chased by neighboring males in territorial disputes, thereby benefiting from a relatively less confrontational breeding season.

Clutch

The egg of the American Redstart is creamy white in color, has darker maculation especially around the blunt pole, and measures ⅝ x ½ in (16 x 12 mm). Some females who are mated to the same male may not have direct evidence of his bigamy, because this species also practices polyterritorial polygyny, where males maintain two non-neighboring territories, paired with a single female in each of them.

Actual size

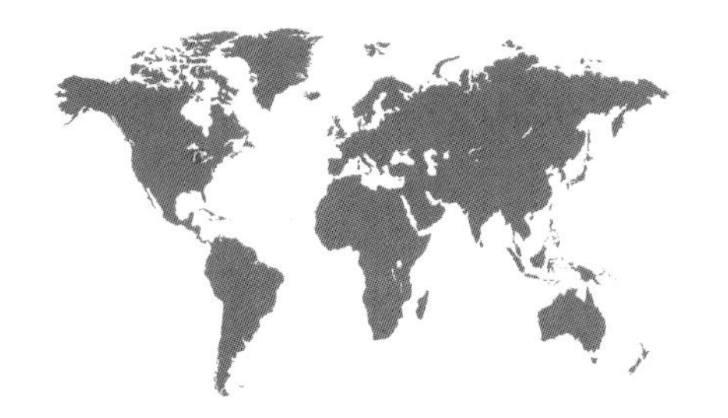

ORDER	Passeriformes
FAMILY	Parulidae
BREEDING RANGE	Northern Michigan
BREEDING HABITAT	Regenerating Jack Pine forests on sandy soils
NEST TYPE AND PLACEMENT	Open cup on ground, under tufts of grass or shrub leaves; made of grasses, sedges, pine needles, and leaves, lined with rootlets, fibers, and hair
CONSERVATION STATUS	Near threatened
COLLECTION NO.	FMNH 15179

ADULT BIRD SIZE
5½–6 in (14–15 cm)

INCUBATION
11–14 days

CLUTCH SIZE
3–6 eggs

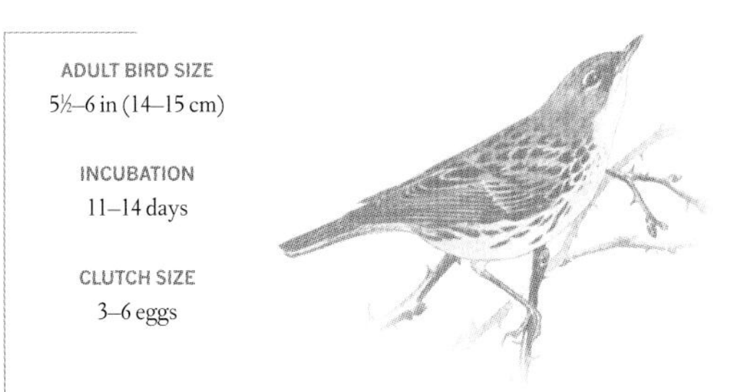

SETOPHAGA KIRTLANDII

KIRTLAND'S WARBLER

PASSERIFORMES

Clutch

The egg of the Kirtland's Warbler is white to buff with brown maculation, and ¾ x 9⁄16 in (19 x 14 mm) in size. Those of the Brown-headed Cowbird (the two larger eggs shown above with the smaller warbler egg below, all taken from the same nest) are typically similar in color and maculation to the eggs of many hosts' species, but the parasitic egg's shell is thicker so that puncturing and ejecting it is no easy task for small birds.

This is one of the rarest parulid warblers, surviving in a small portion of what was never a broad breeding habitat in the upper Midwest. Likely always sparse in numbers, it is a "fugitive" species, and its habitat requirements ecologically are met by cycles of forest fires that make space for stands of Jack Pines. This warbler dwells and nests only in middle-aged stands of these trees. Fire suppression in the breeding grounds, and logging of the small stands of forests in the Bahamas' wintering grounds, likely contributed to a decline of this species in the 1950s and the 1970s.

To further hinder the reproductive success of this rare species, Kirtland's Warblers have been heavily parasitized by Brown-headed Cowbirds (see page 616); when parasitized, the nest abandonment rates increase threefold, and with even just one cowbird chick in the nest, the warbler's own young typically all die. For over 40 years now, cowbirds have been trapped and killed within the breeding range of Kirtland's Warblers; together with recent, accidental forest fires and the resulting opening of new breeding habitats, the Kirtland's Warbler is en route to a delicate recovery.

Actual size

ORDER	Passeriformes
FAMILY	Parulidae
BREEDING RANGE	Boreal North America
BREEDING HABITAT	Coniferous forests with unbroken canopy
NEST TYPE AND PLACEMENT	Bulky cup of mosses, twigs, spruce needles, and bark strips, lined with rootlets, plant fluff, hair, and feathers; placed high in a spruce or fir tree
CONSERVATION STATUS	Least concern
COLLECTION NO.	FMNH 16970

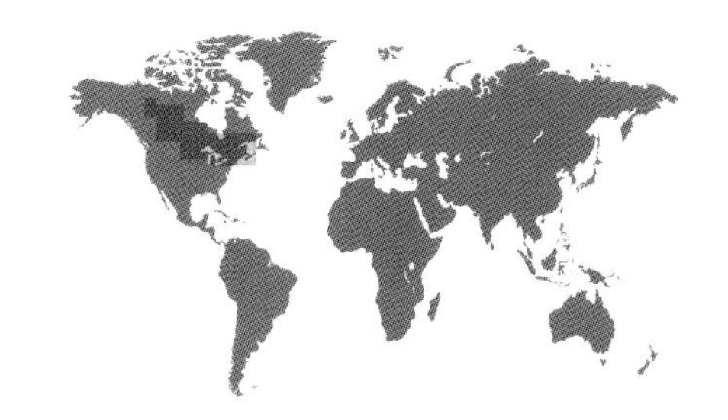

ADULT BIRD SIZE
4½–5½ in (12–14 cm)

INCUBATION
11–13 days

CLUTCH SIZE
4–9 eggs

SETOPHAGA TIGRINA

CAPE MAY WARBLER

PASSERIFORMES

To attract a female to the breeding territory, the male Cape May Warbler sings and hops around her with erect and rigid wings. He also follows her closely during nest construction, although he does not assist. The female stays on the nest tightly during incubation, and is reluctant to flush unless danger is imminent. The pair feeds the chicks a mostly larval diet through fledging. Perhaps because of the remote breeding habitat of this species, relatively little else is known about the breeding cycle and paternal care patterns.

This is another warbler species that was named after its first collecting site, where it was not seen again for another 100 years. A resident of the boreal coniferous forests of Canada, this species is a specialist feeder on caterpillars, which have characteristic eruption cycles, closely followed by the warbler's own population sizes.

Clutch

The egg of the Cape May Warbler is white in background color, has reddish brown maculation, and measures ⅔ x ½ in (17 x 12 mm) in dimensions. The variable, and higher than typical, clutch size of this warbler allows the pair to take advantage of raising a bigger brood in years with abundant spruceworm caterpillars as food for the chicks.

Actual size

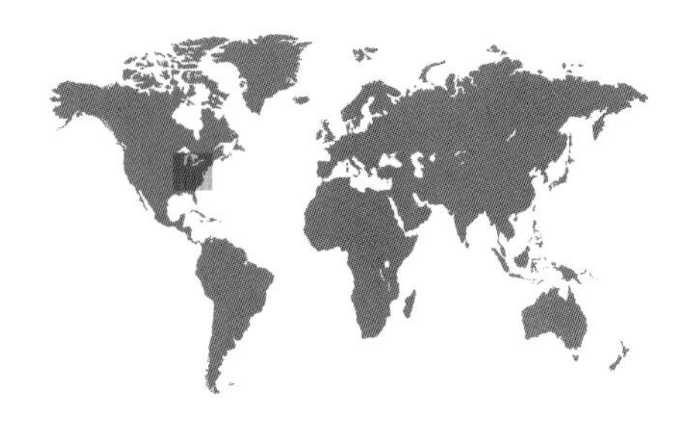

ORDER	Passeriformes
FAMILY	Parulidae
BREEDING RANGE	Eastern and central North America
BREEDING HABITAT	Mature, deciduous forests, both in riparian bottom lands, and drier, hillside woodlands
NEST TYPE AND PLACEMENT	Open cup of bark strips, grasses, stems, silk, and hair woven together; placed on a horizontal limb of a deciduous tree, typically high up, but under foliage cover
CONSERVATION STATUS	Vulnerable
COLLECTION NO.	FMNH 12706

ADULT BIRD SIZE
4½–4¾ in (11–12 cm)

INCUBATION
11–13 days

CLUTCH SIZE
3–4 eggs

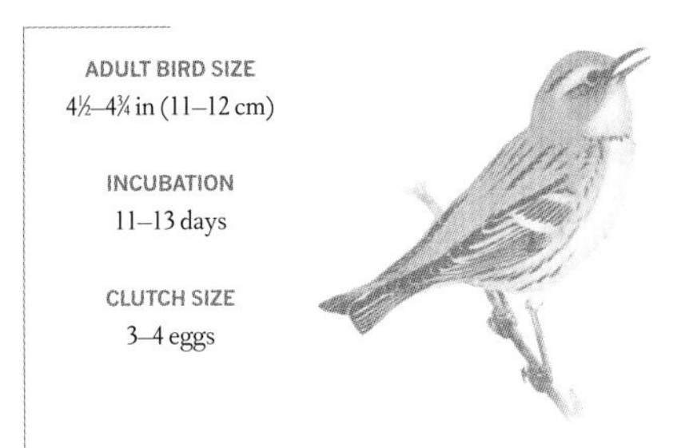

SETOPHAGA CERULEA

CERULEAN WARBLER

PASSERIFORMES

Clutch

The egg of the Cerulean Warbler is grayish to greenish white, speckled with brown, and ⅔ x ½ in (17 x 12 mm) in size. When a nest is lost to weather or predation, even if late in the season, this warbler will build a new nest and lay a replacement clutch of eggs.

During the breeding season, this warbler follows many of its relatives' breeding habits, with the males arriving first, and pair formation taking place just days after the females move in. There is much interaction between the mates during nest-site selection, and the male often contributes to nest construction by collecting spider silk to weave the outer cup materials together. The female alone incubates the eggs, and the parents feed the chicks equitably. The female, however, continues to repair and reupholster the outside of the nest by adding additional layers of spiderwebs, often until the day of fledging.

A once common forest-dwelling warbler, this blue-plumed, canopy insectivore has seen a consistent decline in its population sizes throughout the twentieth century, so much so that its conservation status has become a concern and it is listed as a vulnerable species. Habitat loss and modification have likely contributed to these patterns, but more troubling is the lack of knowledge about the exact causes of these negative trends.

Actual size

ORDER	Passeriformes
FAMILY	Parulidae
BREEDING RANGE	Eastern and central North America
BREEDING HABITAT	Dense, diverse, and tall mixed forests, wooded swamps
NEST TYPE AND PLACEMENT	Pendulum nest made inside hanging mass of mosses, lichens, or other epiphytes, typically high up in a tree; inner cup is lined with mosses, hair, fine grasses, or pine needles
CONSERVATION STATUS	Least concern
COLLECTION NO.	FMNH 19214

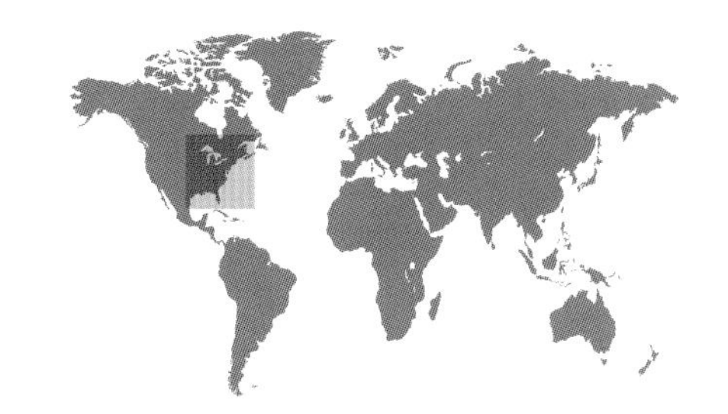

ADULT BIRD SIZE
4½–5 in (11–13 cm)

INCUBATION
12–14 days

CLUTCH SIZE
3–5 eggs

SETOPHAGA AMERICANA

NORTHERN PARULA

PASSERIFORMES

The Northern Parula builds a well-camouflaged, and hard-to-find nest within clumps of mosses and other epiphytes, often dozens of feet above ground; accordingly, information about the nesting biology of this species is sparse, at best. For example, it remains unclear to what extent the male contributes, if at all, to nest construction, incubation, and brooding, and feeding the chicks. What we do know is that the female takes the principal role in all these activities.

To keep the nest sanitary, during incubation the female leaves for short trips to feed and defecate, and both parents participate in removing the chicks' fecal sacs, produced right after each feeding bout. The result is that the nest remains clean and ready for reuse. In fact, unlike most other warblers, this species often lays eggs in its own previously occupied nest, both within and across seasons and years.

Clutch

The egg of the Northern Parula is white to creamy white in background coloration, with red, brown, purple, or gray maculation, and is ⅝ x ½ in (16 x 12 mm) in size. Despite the nest being well hidden, there is a low-level, but persistent, pattern of brood parasitism by Brown-headed Cowbirds (see page 616) throughout this warbler's range.

Actual size

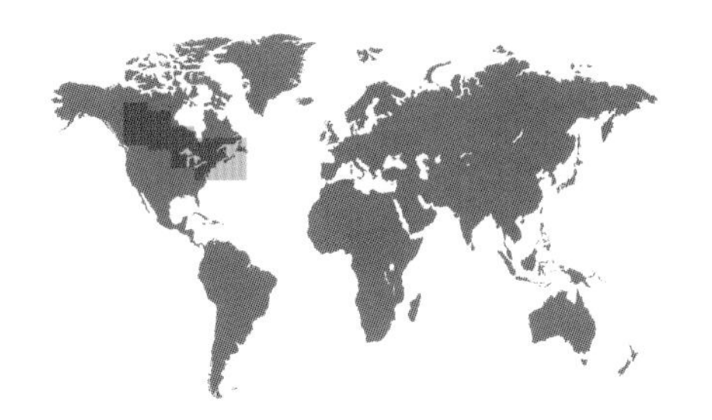

ORDER	Passeriformes
FAMILY	Parulidae
BREEDING RANGE	Subarctic and north temperate North America
BREEDING HABITAT	Young, regenerating conifer stands, or shrubby mixed forests
NEST TYPE AND PLACEMENT	Loose cup near the trunk of a small spruce or hemlock; built on a foundation of twigs, made of grasses, and lined with black rootlets
CONSERVATION STATUS	Least concern
COLLECTION NO.	FMNH 14716

ADULT BIRD SIZE
4½–5 in (11–13 cm)

INCUBATION
11–13 days

CLUTCH SIZE
3–5 eggs

SETOPHAGA MAGNOLIA

MAGNOLIA WARBLER

PASSERIFORMES

Clutch

This northern-breeding warbler establishes all-purpose territories that are used as exclusive feeding and breeding sites. In the non-breeding season, these aggressive birds continue to chase conspecifics, whether female or male. The breeding territories are essential for successful pair formation, and suitable habitats are likely saturated by territorial birds. We know this because, when adult males were removed as part of an experiment to examine these birds' roles in controlling spruce caterpillars, several other males showed up and attempted to claim the empty territory.

When a pair bond is formed, the female sets out to locate a suitable nest site, and begins construction. The male stays near her, singing throughout the day, but does not contribute to building the nest or incubating the eggs. He joins the female to provision the chicks, often taking an equitable share with the female in the rate of food delivery.

The egg of the Magnolia Warbler is white in background color, has variable brown spotting, and its measurements are ⅔ x ½ in (17 x 12 mm). During the laying period, the female (like many other birds) visits the nest only when she is ready to produce an egg, usually in the early mornings; she begins incubation once the clutch of eggs is completed.

ORDER	Passeriformes
FAMILY	Parulidae
BREEDING RANGE	Boreal eastern North America
BREEDING HABITAT	Spruce and mixed conifer forests, subalpine woodlands
NEST TYPE AND PLACEMENT	Open cup in dense foliage on a spruce branch; made of twigs, bark strips, lichen, spider silk, and plant down, lined with rootlets, pine needles, hair, moss, and fine grasses
CONSERVATION STATUS	Least concern
COLLECTION NO.	FMNH 2556

ADULT BIRD SIZE
5–5½ in (13–14 cm)

INCUBATION
12–13 days

CLUTCH SIZE
4–6 eggs

SETOPHAGA CASTANEA

BAY-BREASTED WARBLER

PASSERIFORMES

The Bay-breasted Warbler is a prominently sexually dichromatic species in the spring. During the non-breeding season, the males molt into drabber and more female- and juvenile-like plumage. This is also a highly aggressive and territorial species in the breeding season, with males maintaining exclusive ranges, and often excluding other sympatric warbler species too. The pair bond is a function of location and not necessarily of individual preferences, and pair bonds can shift quickly; for example, in years with outbreaks of caterpillars, the nest site selected by the female may fall outside a male's territory, leading to separation and pairing with a new male.

The female alone builds the nest and incubates the eggs, but the male may feed her on the nest and sing on top of a nearby tree, or follow her closely while she is off the nest to feed. The female also broods the hatchlings alone, while the male feeds her. Both parents then provision the growing chicks, although the female typically delivers twice as much food as does the male.

Clutch

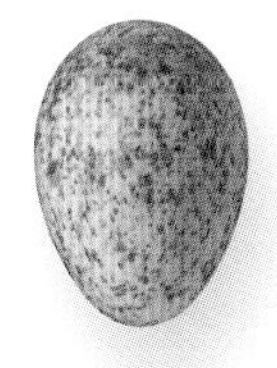

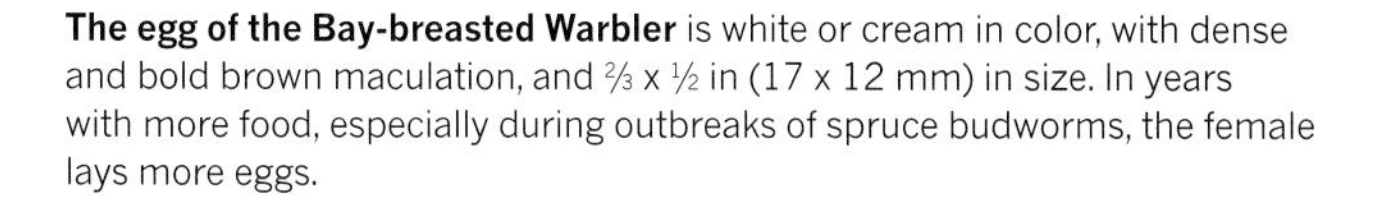

The egg of the Bay-breasted Warbler is white or cream in color, with dense and bold brown maculation, and ⅔ x ½ in (17 x 12 mm) in size. In years with more food, especially during outbreaks of spruce budworms, the female lays more eggs.

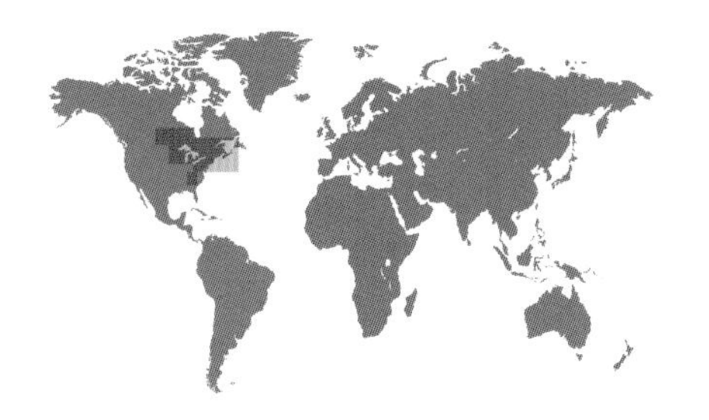

ORDER	Passeriformes
FAMILY	Parulidae
BREEDING RANGE	Northeastern North America
BREEDING HABITAT	Mature coniferous or mixed forests
NEST TYPE AND PLACEMENT	Open cup of twigs, bark, fibers, and rootlets, lined with lichens, moss, pine needles, and hair; attached to a horizontal branch with cobwebs, high up in a tree
CONSERVATION STATUS	Least concern
COLLECTION NO.	FMNH 2547

ADULT BIRD SIZE
4½–4¾ in (11–12 cm)

INCUBATION
12–14 days

CLUTCH SIZE
3–5 eggs

SETOPHAGA FUSCA

BLACKBURNIAN WARBLER

PASSERIFORMES

Clutch

A social, flocking bird during winters in the Amazon Basin, on the breeding grounds this warbler is solitary, only tolerating its mate within the territory. Immediately after pairing, the female and male move around in tight proximity to one another, until a nest site is chosen, and the female begins construction. During egg laying, when there is still time to fertilize some of the remaining eggs, the male again follows the female closely, presumably guarding her from the approaches of other suitors.

The female alone incubates the eggs and broods the chicks, but the male quickly joins her, as both parents provision the hatchlings with food. Feeding the young continues into the fledging period, with each parent taking part of the brood with them and drifting apart. They may never come back together again before the end of the breeding season, and the onset of fall dispersal and southwardly migration.

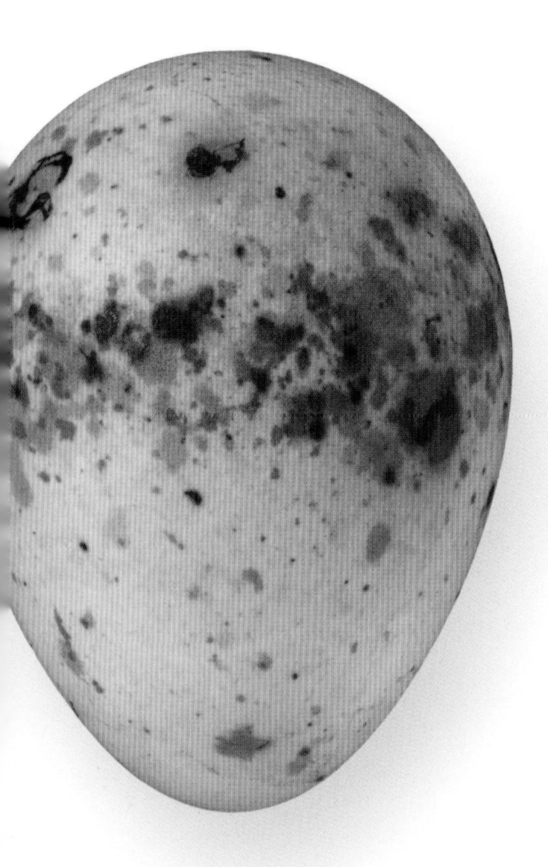

Actual size

The egg of the Blackburnian Warbler is white with an occasional green tint in background coloration, has brown maculation often forming a ring, and measures ⅔ x ½ in (17 x 12 mm). The female lays just one clutch of eggs per year, unless it is lost, in which case she may produce one or more replacement clutches.

ORDER	Passeriformes
FAMILY	Parulidae
BREEDING RANGE	North America, Caribbean islands, and Galapagos Islands
BREEDING HABITAT	Wooded or shrubby swamps, wet thickets, riparian corridors
NEST TYPE AND PLACEMENT	Small cup in vertical fork of a bush or small tree; made of grasses, bark, and stems, lined with deer hair, feathers, and fibers from cottonwood, dandelion, willow, and cattail seeds
CONSERVATION STATUS	Least concern
COLLECTION NO.	FMNH 15170

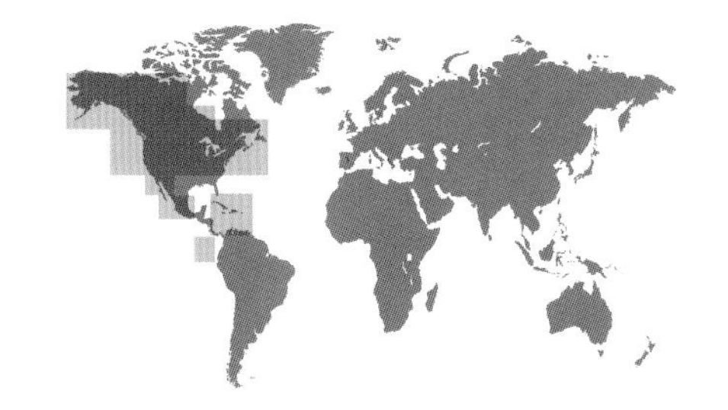

ADULT BIRD SIZE
4½–5 in (12–13 cm)

INCUBATION
10–13 days

CLUTCH SIZE
4–5 eggs

SETOPHAGA PETECHIA

YELLOW WARBLER

PASSERIFORMES

Yellow Warblers are conspicuous migrants in the early springtime, showing off their yellow plumes and singing in nearly leafless trees and shrubs, as they forage in diverse habitats. These include busy city streets, but also backyard thickets, and urban parks, which they use as refueling sites before taking off on another night's flight north. Once they arrive on the breeding grounds, the males fight and establish their territories, in anticipation of the female's arrival. Pair bonding takes just one day and then the female begins the nesting cycle.

Nest-site selection, construction, and incubation are all undertaken by the mothers alone, while the male often attempts to court and pair with a second female; in fact, bigamy is quite common, characterizing 10–50 percent of males in some populations. Older males have more orange-brown streaking on their chest feathers, and once the chicks hatch, they make a smaller contribution to provisioning than do younger males.

Clutch

Actual size

The egg of the Yellow Warbler is grayish or greenish white, with darker blotching, and ⅔ x ½ in (17 x 12 mm) in dimensions. The nest of the Yellow Warbler often contains an entombed Cowbird egg in the nest lining, which is not rotated during incubation, and will inevitably fail to hatch.

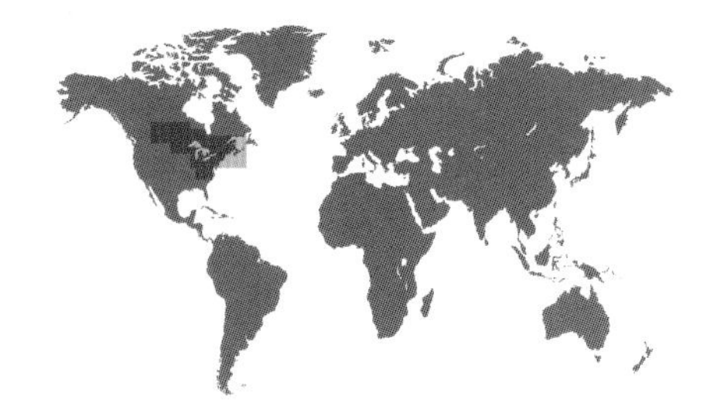

ORDER	Passeriformes
FAMILY	Parulidae
BREEDING RANGE	Northeastern and central North America
BREEDING HABITAT	Shrubby and second-growth deciduous forests, abandoned farmland and regenerating clear-cuts
NEST TYPE AND PLACEMENT	Open cup, made from bark, leaves, stems, grasses, and plant down, lined with grasses, hair, and rootlets; placed in small crotch of a low shrub
CONSERVATION STATUS	Least concern
COLLECTION NO.	FMNH 12752

ADULT BIRD SIZE
4–4½ in (10–11 cm)

INCUBATION
12–13 days

CLUTCH SIZE
3–5 eggs

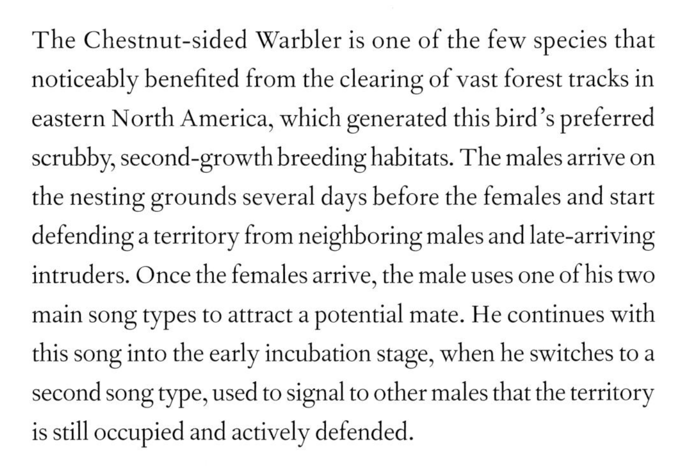

SETOPHAGA PENSYLVANICA

CHESTNUT-SIDED WARBLER

PASSERIFORMES

Clutch

The egg of the Chestnut-sided Warbler is creamy white to pale green in background coloration, has brown spotting, and measures ⅔ x ½ in (17 x 12 mm). The open, scrubby habitat of this species is shared with the brood parasitic Brown-headed Cowbird (see page 616), which often lays its eggs in this warbler's nests.

The Chestnut-sided Warbler is one of the few species that noticeably benefited from the clearing of vast forest tracks in eastern North America, which generated this bird's preferred scrubby, second-growth breeding habitats. The males arrive on the nesting grounds several days before the females and start defending a territory from neighboring males and late-arriving intruders. Once the females arrive, the male uses one of his two main song types to attract a potential mate. He continues with this song into the early incubation stage, when he switches to a second song type, used to signal to other males that the territory is still occupied and actively defended.

The female alone builds the nest cup, using her bill to weave filamentous nesting materials together to form the base and wall. She then uses her chest, while sitting inside, to form the final shape of the nest. The male often follows closely and sings nearby, but he does not contribute to nest building or incubation duties, and only joins the female when provisioning the nestlings.

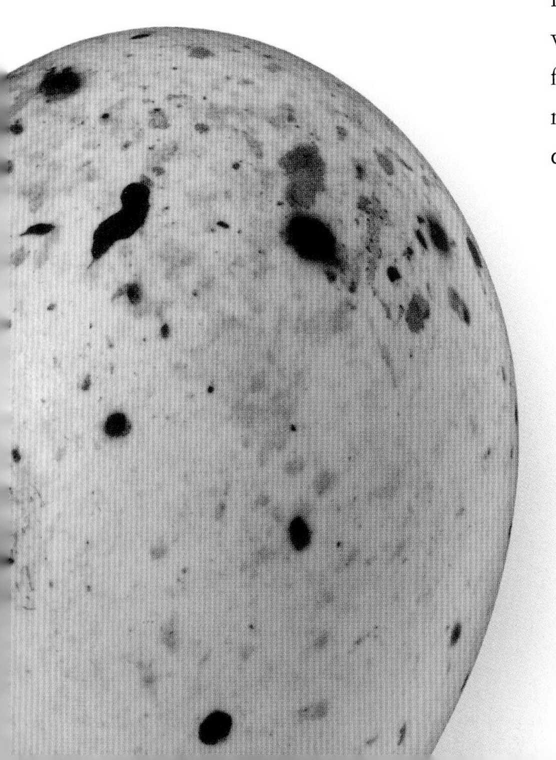

Actual size

ORDER	Passeriformes
FAMILY	Parulidae
BREEDING RANGE	Arctic and subarctic North America
BREEDING HABITAT	Montane woodland and shrubland, also boreal coniferous forests
NEST TYPE AND PLACEMENT	Open cup of twigs and lichens, lined with grasses, plant fibers, moose hair, and feathers; placed low against the trunk of a spruce or other tree
CONSERVATION STATUS	Least concern
COLLECTION NO.	FMNH 12812

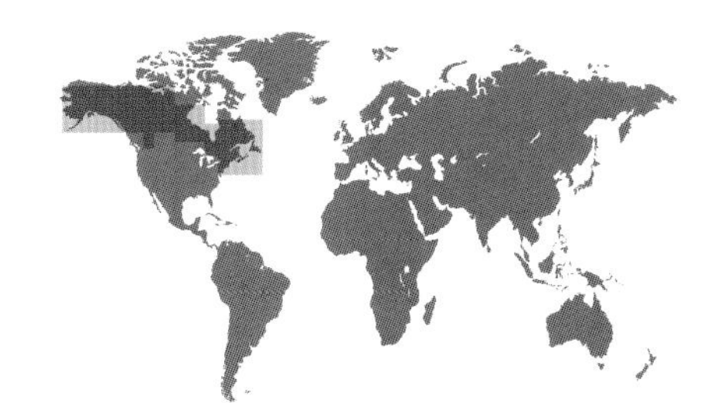

ADULT BIRD SIZE
5–6 in (13–15 cm)

INCUBATION
11–12 days

CLUTCH SIZE
4–5 eggs

SETOPHAGA STRIATA

BLACKPOLL WARBLER

PASSERIFORMES

A long-distance migrant to and from South America, with some individuals covering nearly 5,000 miles (8,000 km) from Alaska to Brazil, the Blackpoll breeds further north than its relatives. Despite the long migration and northerly breeding, in some populations most females attempt to start a second clutch after the first brood fledges; the female saves time by letting the male look after the fledglings while she starts building the second nest.

The female selects the nest site, builds the structure, and incubates the eggs, with the male singing nearby before egg laying is complete. There are only rare sightings of the male feeding the female on the nest; instead, he joins her to feed the hatchlings. Some males are polygynous, and in these cases the male provides an equitable share of feedings at one nest, but only about 20 percent of them at the second nest.

Clutch

The egg of the Blackpoll Warbler is white, buff, or pale green in coloration, with brown and purple markings, and is ⅔ x 9/16 in (17 x 14 mm) in size. During incubation, the female often pokes and pulls nest materials to adjust and reinforce the nest wall, and maintain the cup holding the eggs.

Actual size

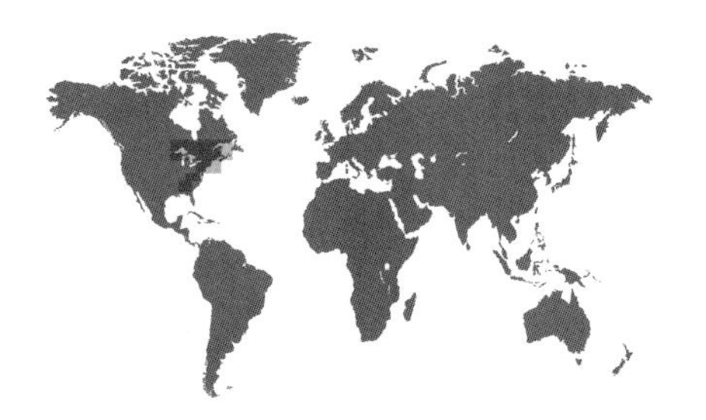

ORDER	Passeriformes
FAMILY	Parulidae
BREEDING RANGE	Eastern North America and the Appalachian Mountains
BREEDING HABITAT	Large, mature tracks of hardwood and mixed coniferous forests
NEST TYPE AND PLACEMENT	Open cup, made from strips of bark, held together with spider web and saliva; placed in the fork of a low shrub
CONSERVATION STATUS	Least concern
COLLECTION NO.	FMNH 12693

ADULT BIRD SIZE
4½–5 in (11–13 cm)

INCUBATION
11–12 days

CLUTCH SIZE
3–5 eggs

SETOPHAGA CAERULESCENS

BLACK-THROATED BLUE WARBLER

PASSERIFORMES

Clutch

The egg of the Black-throated Blue Warbler is creamy white in background color, with dark brown blotching, and is ⅔ x ½ in (17 x 12 mm) in size. After successfully fledging the first brood, about half of the females go on to lay a second clutch of eggs, more frequently when they are older, mated to an older male, and occupy territories with greater food availability.

The Black-throated Blue Warbler breeds in the temperate forests of eastern North America, and winters on islands of the Caribbean. While many studies have been conducted on the annual dynamics of this species' breeding populations in the north, as well as the wintering populations in the tropics, it has been difficult to understand its migratory pathways. However, since these birds molt into their spring plumage on the wintering grounds, sampling the feathers for stable isotopes of breeding birds can be informative to localize the site where those feathers were grown. Using this technology, scientists have now connected these warblers' northern breeding populations to Cuba and Jamaica, and the Appalachian populations to Hispaniola and Puerto Rico.

This species has conspicuous white wing spots which are visible from some distance in the forest, as the birds move from the high canopy to the low shrub layer between feeding and nesting. The female produces a characteristic chip sound while she collects nesting material, making it easier for the male to follow her, and for scientists more conveniently to locate nests. Despite mate-guarding by the male, many females mate with neighbors, who then sire some of the young in nests for which they do not provide paternal care.

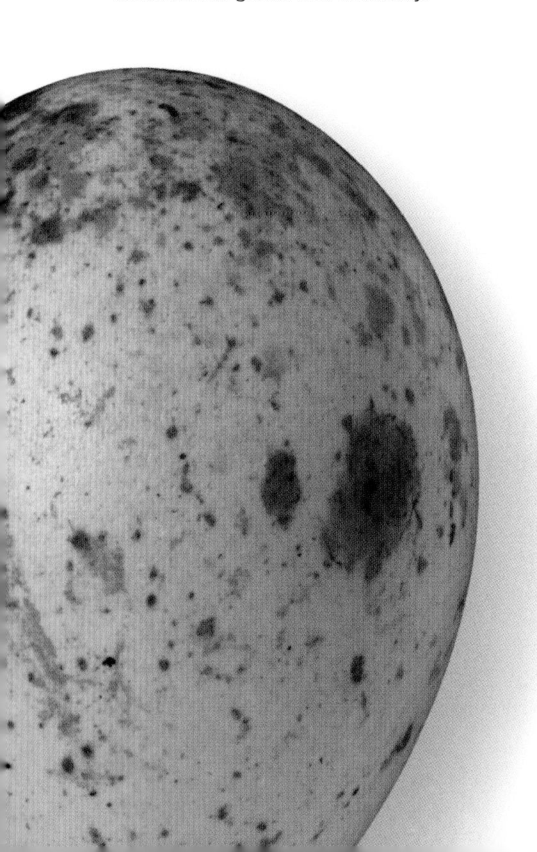

Actual size

ORDER	Passeriformes
FAMILY	Parulidae
BREEDING RANGE	Northern North America
BREEDING HABITAT	Subarctic bogs, scrubs, and open coniferous forests, with a dense understory
NEST TYPE AND PLACEMENT	Open cup, on the ground, near the stem of a pine tree or sapling; made of weed stalks, grasses, sedge leaves, bark strips, lined with fine grasses, rootlets, mosses, hair, and plant fluff
CONSERVATION STATUS	Least concern
COLLECTION NO.	FMNH 2560

ADULT BIRD SIZE
5–6 in (13–15 cm)

INCUBATION
12 days

CLUTCH SIZE
4–5 eggs

SETOPHAGA PALMARUM

PALM WARBLER

PASSERIFORMES

This species was first described from a specimen collected in the West Indian wintering range. During the breeding season it is an inhabitant of boreal wetlands and scrubby wood lots. It feeds and breeds on the ground, much more so than most wood warblers. Its remote distribution and well-hidden nests mean few detailed observations about its reproductive biology have been made. It is not known if the males arrive on the breeding grounds before the females do, as with most other migratory parulid warblers.

The female alone incubates the eggs, and the male may feed her while she sits on the nest. Only the female broods the chicks, but both parents are thought to provision the nestlings. However, detailed observations at just two nests revealed that the male's contribution may be minimal during the nestling stage, whereas he may join in feeding the young after they fledge from the nest.

Clutch

The egg of the Palm Warbler is whitish in coloration, has dark brown maculation, and measures ⅔ x ½ in (17 x 12 mm). Despite the short summers in this species' northern breeding range, females may attempt to lay a second clutch of eggs after the first brood has successfully fledged.

Actual size

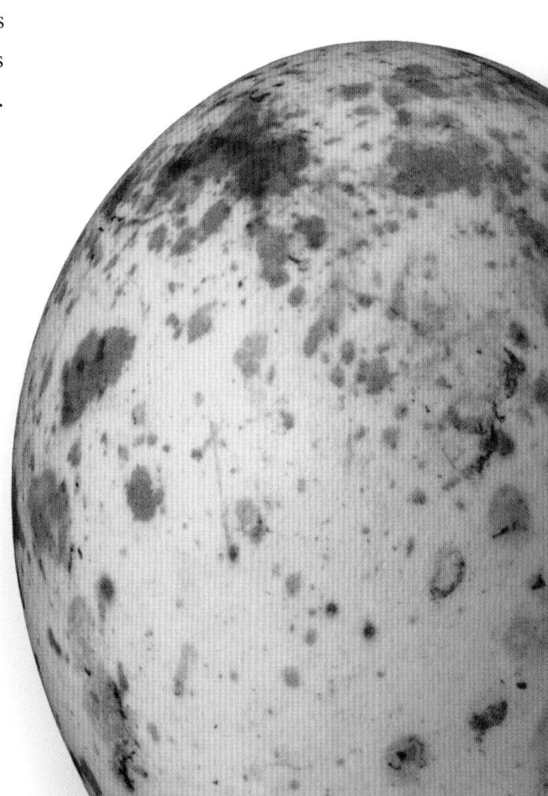

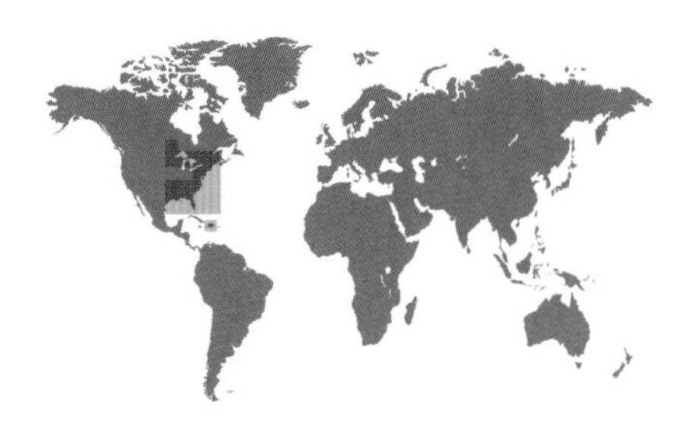

ORDER	Passeriformes
FAMILY	Parulidae
BREEDING RANGE	Eastern and central North America; also the Caribbean
BREEDING HABITAT	Coniferous and mixed, open forests, pine plantations
NEST TYPE AND PLACEMENT	Open cup made of grasses, twigs, and stems, bound together by cobwebs, lined with plant fibers and animal hair; placed high in a pine tree canopy, on a horizontal branch, among bunches of needles and cones
CONSERVATION STATUS	Least concern
COLLECTION NO.	FMNH 12804

ADULT BIRD SIZE
5–5½ in (13–14 cm)

INCUBATION
10–13 days

CLUTCH SIZE
3–5 eggs

SETOPHAGA PINUS

PINE WARBLER

PASSERIFORMES

Clutch

This is one of the earliest arriving and latest departing warblers in the United States. Its early arrival at the breeding grounds may be helped by its ability to feed in various ways. It probes for insects in bark and seeds from pine cones, and even eats from bird feeders. The Pine Warbler is also opportunistic in its strategies for collecting nesting materials; when the female locates a source of a specific nest component, she will return there repeatedly. Nesting materials may also be collected from active crow or old and abandoned tanager and warbler nests.

Timing is not of the essence for this species, as nest construction can take up to a full week, and another week is spent between completing it and starting to lay the first egg. As with many temperate-breeding bird species, the clutch size declines steadily with later nest initiation dates across the season.

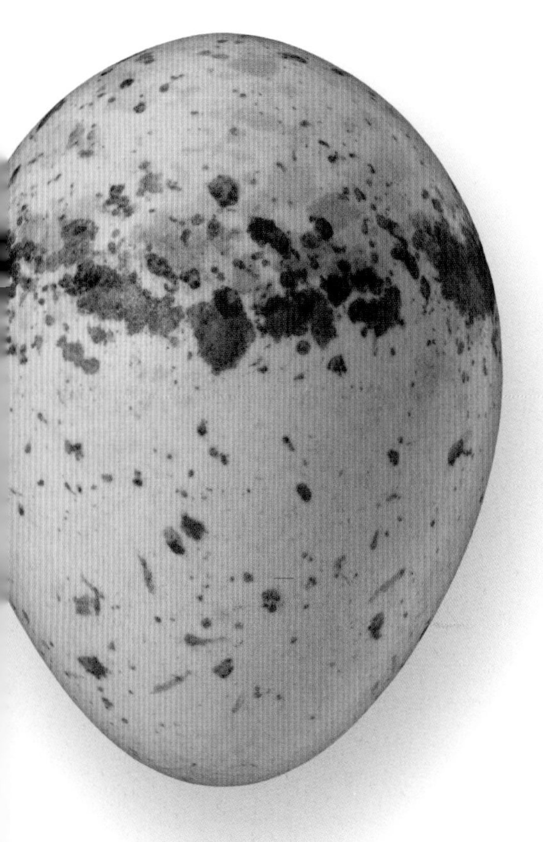

Actual size

The egg of the Pine Warbler is white, grayish, or greenish white in background color, with brown maculation, often forming a dense ring, and its measurements are ⅔ x 9⁄16 in (17 x 14 mm). The female incubates the eggs alone, and the male feeds her on the nest; both parents provision the chicks equitably.

ORDER	Passeriformes
FAMILY	Parulidae
BREEDING RANGE	Northern and western North America
BREEDING HABITAT	Mature coniferous and mixed forests
NEST TYPE AND PLACEMENT	Open cup of twigs, pine needles, grasses, and rootlets, woven together with silk and hair, lined with rootlets, hair and feathers; placed typically in a spruce, fir, or small pine tree, at mid-height
CONSERVATION STATUS	Least concern
COLLECTION NO.	FMNH 16972

ADULT BIRD SIZE
4½–6 in (11–15 cm)

INCUBATION
12–13 days

CLUTCH SIZE
4–5 eggs

SETOPHAGA CORONATA

YELLOW-RUMPED WARBLER

PASSERIFORMES

Formerly separated into two species (the Audubon's Warbler and the Myrtle's Warbler), this species is one of North America's ecologically most generalist species, feeding in diverse habitats using varied foraging techniques. During its nocturnal migration, thousands of individuals become trapped in the bright lights produced by cities, with perhaps the most famous example being the 9/11 Memorial Double Beam at Ground Zero, New York City. Volunteers lie at the base of these reflector lights and count the number of bird shadows, mostly Yellow-rumped Warblers, circling in the beam. If the numbers reach 1,000 per minute, the volunteers make a call and the City authorities turn off the lights for 20 minutes to allow the birds to pass through in the dark.

For breeding, this species is restricted by its habitat requirements. The males mostly establish all-purpose territories in predominantly mature pine forests. The female is in charge of constructing the nest and incubating the eggs, but the male may also gather nesting materials and feed the female at the nest. Both parents provision the young, often staying together at the nest after food delivery.

Clutch

The egg of the Yellow-rumped Warbler is white in color, with brown, reddish brown, gray, or purplish gray spotting, and is ⅔ x 9⁄16 in (17 x 14 mm) in size. The first egg is laid within a day after the nest is completed, with each additional egg added on a daily basis.

Actual size

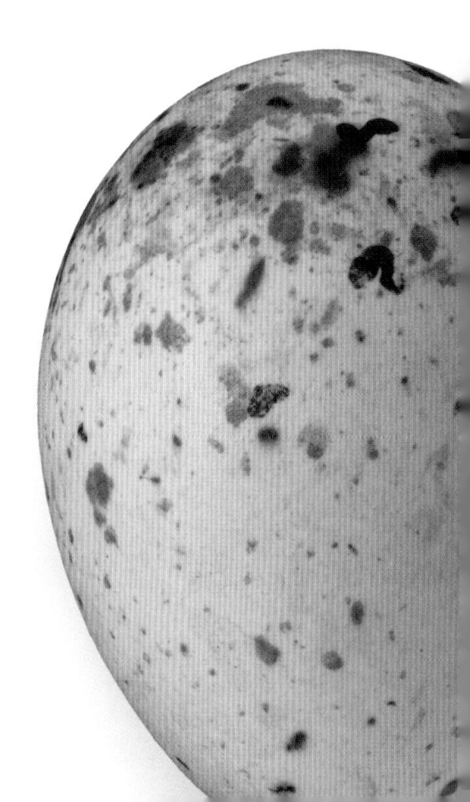

ORDER	Passeriformes
FAMILY	Parulidae
BREEDING RANGE	Southeastern and central North America
BREEDING HABITAT	Mixed and pine stands, mature riparian bottomland woods, swamp forests
NEST TYPE AND PLACEMENT	A pocket-shaped cup inside Spanish moss, lined with grasses, weeds, mosses, plant fluff, and feathers; also in open cups, made of rootlets and bark strips, placed on forks of limbs situated high up in the canopy
CONSERVATION STATUS	Least concern
COLLECTION NO.	FMNH 2548

ADULT BIRD SIZE
5–5½ in (13–14 cm)

INCUBATION
12–13 days

CLUTCH SIZE
3–5 eggs

SETOPHAGA DOMINICA

YELLOW-THROATED WARBLER

PASSERIFORMES

Clutch

This species is a short-distance migrant, wintering in northern Central America and the Caribbean and breeding in the southeastern continental United States. Until now, only the Brown-headed Cowbird (see page 616) has been known as a brood parasite of this warbler; however, because of its southern distribution, and the recent invasion of south Florida Gulf by the Shiny Cowbird (*Molothrus bonariensis*) from South America, a second parasitic species could represent a new threat to the reproductive success of the Yellow-throated Warbler.

Nest building is completed by the female, with few reports of the male also gathering and weaving together nesting materials. It takes nearly a week to complete the nest and to lay the first egg. Only the female develops an incubation patch, and once the nest is complete, she sits on the eggs alone. The male stays nearby, singing throughout the day, but does not assist in the task of incubation. The female also broods the hatchlings, but both parents feed the nestlings.

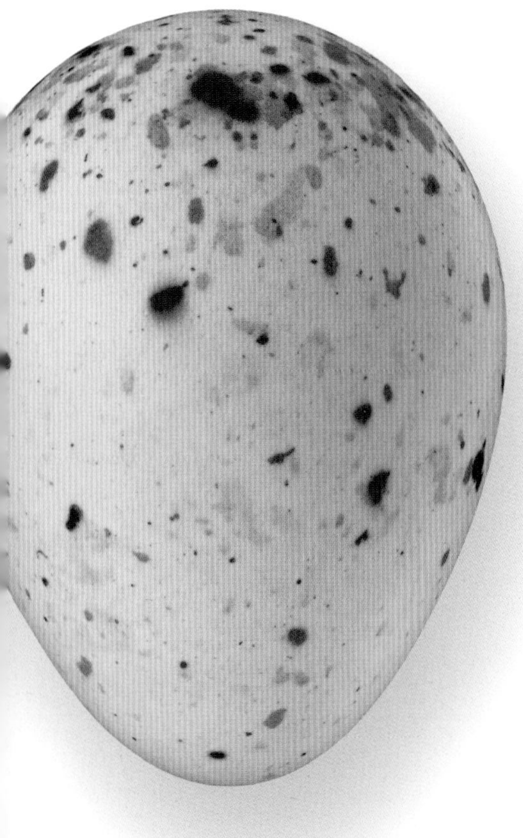

Actual size

The egg of the Yellow-throated Warbler is pale green in coloration, but in museum collections it often fades to white; it has sparse purple-brown blotching, and is ⅔ x ½ in (17 x 12 mm) in size. To keep the nest clean and safe, the parents remove broken eggshells soon after hatching.

ORDER	Passeriformes
FAMILY	Parulidae
BREEDING RANGE	Central and eastern North America
BREEDING HABITAT	Shrubby fields, forest edges
NEST TYPE AND PLACEMENT	Open cup of long plant fibers, sedge leaves, grasses, and stems, lined with fine grasses, mosses, and feathers; placed in an upright fork or on a horizontal branch, often near the ground
CONSERVATION STATUS	Least concern
COLLECTION NO.	FMNH 12725

ADULT BIRD SIZE
4½–5 in (11–13 cm)

INCUBATION
11–14 days

CLUTCH SIZE
3–5 eggs

SETOPHAGA DISCOLOR

PRAIRIE WARBLER

PASSERIFORMES

Based on detailed field research, there is much individual variation in the reproductive behaviors of this warbler. Individuals tend to be faithful to previously used breeding sites across years. Some males and females also show personal consistency in arriving from the wintering grounds either in the early or in the late spring, year after year. Young males typically maintain smaller territories than older males; also, territories are smaller when rectangular in shape and larger when elongated, such as edges and other habitat corridors.

In the context of territoriality and mating, the males use two different song types for communication. One song variant is employed to attract females before pair bonding, and, later, sung during the day near the female after nest building and laying has begun; the other song type is used for male–male interactions, between neighbors, and to keep intruders away.

Clutch

Actual size

The egg of the Prairie Warbler is pale brownish or gray in background, has brown maculation, and its dimensions are ⅝ x ½ in (16 x 12 mm). The female alone selects the nest site, builds it, and incubates the eggs; the male feeds her at the nest and assists in provisioning the chicks.

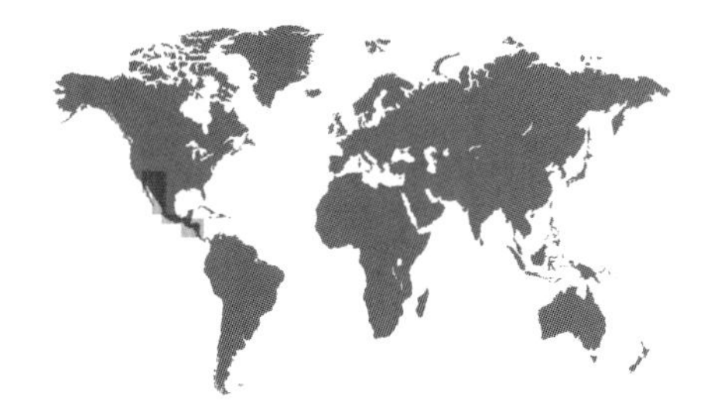

ORDER	Passeriformes
FAMILY	Parulidae
BREEDING RANGE	Southwestern North America, northern Central America
BREEDING HABITAT	Open woodlands, oak scrub, gallery forests
NEST TYPE AND PLACEMENT	Small open cup, made of plant fluff and fibers, lined with feathers and hair; placed high above ground on a pine tree branch
CONSERVATION STATUS	Least concern
COLLECTION NO.	FMNH 3080

ADULT BIRD SIZE
4½–5 in (11–13 cm)

INCUBATION
10–12 days

CLUTCH SIZE
3–5 eggs

SETOPHAGA GRACIAE

GRACE'S WARBLER

PASSERIFORMES

Clutch

This small, bright warbler lives in both temperate and tropical habitats where there are pine trees. Southern populations form year-round territories and pair bonds; northern populations are migratory, and establish an initial consortship soon after arrival at the breeding grounds. But courtship is prolonged for several days or weeks before the onset of nesting activities. The male does not contribute to nest building, and the female may venture outside the boundaries of the territory to gather nesting materials. She may situate the nest anywhere in the territory, including close to the edges.

During incubation, the female stays with the eggs and sleeps on the nest, while the male roosts in a tall tree, from the top of which he sings his dawn chorus song on the next day. The male may feed the female on the nest, and both parents feed the chicks, both during the nestling and the fledgling stages.

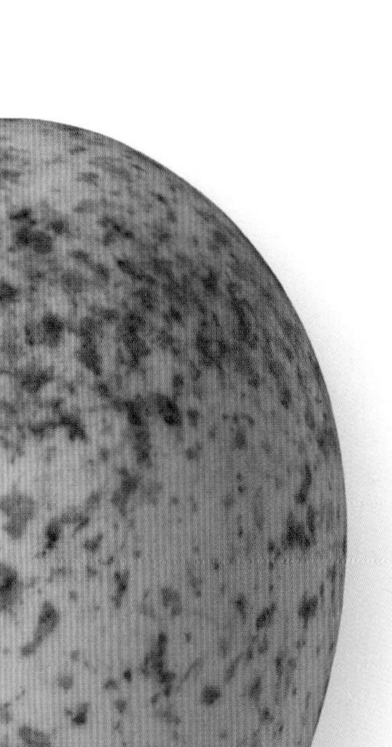

Actual size

The egg of the Grace's Warbler is cream or white in coloration, with brown maculation, and is ⅔ x ½ in (17 x 12 mm) in size. The eggs are rarely seen as the female spends most of her time on the nest, which is further protected by being placed in the middle of dense clumps of dead pine needles.

ORDER	Passeriformes
FAMILY	Parulidae
BREEDING RANGE	Western North America
BREEDING HABITAT	Open forests, coniferous and mixed woodlands with dense understory, scrub and chaparral
NEST TYPE AND PLACEMENT	Deep, open cup placed on a horizontal branch or limb, in a small tree or shrub; made with plant fibers, leaves, grasses, and stalks, lined with rootlets, feathers, and hair
CONSERVATION STATUS	Least concern
COLLECTION NO.	FMNH 12744

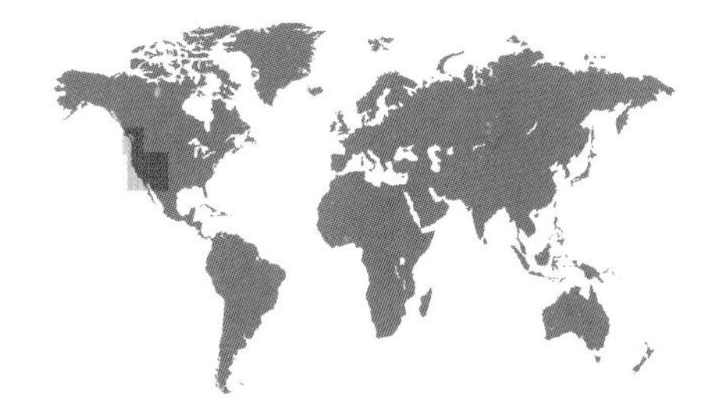

ADULT BIRD SIZE
4½–5 in (11–13 cm)

INCUBATION
12 days

CLUTCH SIZE
3–5 eggs

SETOPHAGA NIGRESCENS

BLACK-THROATED GRAY WARBLER

PASSERIFORMES

This common and widespread species has attracted surprisingly few detailed observations and systematic studies of its reproductive and social behaviors. We do know that it is a territorial species on the breeding ground, where individuals typically only consort with their mate. The female alone selects the nest site, while the male follows her closely and calls nearby. To assess a potential site, the female presses her body into crotches and forks of limbs and branches, as if measuring the available space for construction; she then returns to one of these spots and begins nest building.

This species' range overlaps with the western subspecies of the parasitic Brown-headed Cowbird (see page 616). When a parasitic egg is laid in the warbler's nest, the female may continue nest construction, but add an extra layer of lining, so as to entomb the cowbird egg to prevent it from being incubated and rotated, and thus doom its chances of hatching.

Clutch

The egg of the Black-throated Gray Warbler is creamy white in background coloration, with reddish brown maculation, and is ⅝ x ⅜ in (16 x 10 mm) in size. Relatively tame when approached at the nest by humans, the female responds with alarm calls and distress displays when approached by a jay, or other known nest predator.

Actual size

ORDER	Passeriformes
FAMILY	Parulidae
BREEDING RANGE	Northwestern North America
BREEDING HABITAT	Mature coniferous forests, from the taiga to montane pine and coastal fir forests
NEST TYPE AND PLACEMENT	Bulky open cup, made of bark, pine needles, twigs, grass, lichens, and cocoons, lined with finer grasses, moss, and elk and deer hair; placed on a main horizontal limb, halfway up in the foliage of a coniferous tree
CONSERVATION STATUS	Least concern
COLLECTION NO.	FMNH 16978

ADULT BIRD SIZE
4–4½ in (10–11 cm)

INCUBATION
11–14 days

CLUTCH SIZE
5–7 eggs

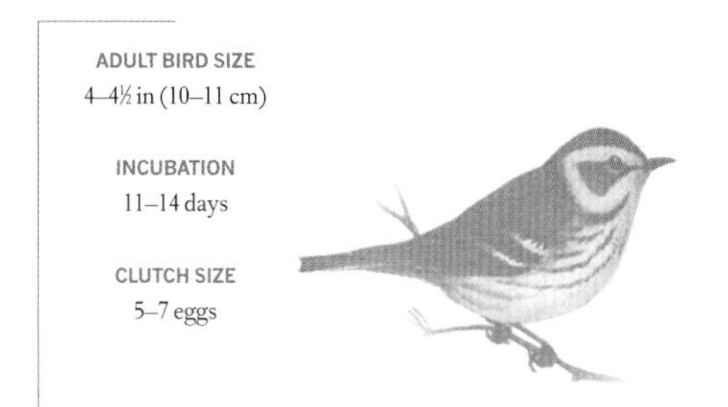

SETOPHAGA TOWNSENDI

TOWNSEND'S WARBLER

PASSERIFORMES

Clutch

The egg of the Townsend's Warbler is white in coloration, has brownish dark spotting, and measures ⅔ x ½ in (17 x 12 mm) in size. Once the nest is complete, egg laying begins within one to four days, and incubation begins on the day when the last egg is laid.

This small and colorful warbler works hard to establish breeding territories in the spring. Before the females arrive, the males use song exchanges, chase flights, and physical contests to secure a territory. However, perhaps because of the remote and mature woods in which this species breeds, relatively little specific information is known about its nesting biology.

The female selects the nest site and gathers the materials, followed closely by the unhelpful male. The female may build an incomplete nest first, only to decide to move elsewhere, and recycle the materials to complete a new structure. The female alone incubates the eggs, while the male often sings nearby. To distract predators at the nest, the female drops to the ground, flutters her wings, and moves away from the nest. Both parents feed the fast-growing young, and the chicks may fledge as quickly as ten days after hatching.

Actual size

ORDER	Passeriformes
FAMILY	Parulidae
BREEDING RANGE	Western North America
BREEDING HABITAT	Tall, mature forests of conifers
NEST TYPE AND PLACEMENT	Open cup, placed high in a tree, near the tip of a branch; made from fine twigs, rootlets, dry moss, bark strips, pine needles, and animal silk, lined with fine plant fibers and hair
CONSERVATION STATUS	Least concern
COLLECTION NO.	FMNH 2545

ADULT BIRD SIZE
4¾–5½ in (12–14 cm)

INCUBATION
12 days

CLUTCH SIZE
4–5 eggs

SETOPHAGA OCCIDENTALIS

HERMIT WARBLER

PASSERIFORMES

This secretive and solitary species has maintained stable population sizes despite much recent habitat modification across its breeding grounds. Living in mature coniferous forests, it rarely descends into the subcanopy and shrubby layers of the forest, as it feeds, sings, and breeds in the top foliage of the trees. To establish the boundaries of breeding territories, the males use song and physical aggression against intruders; in some regions, they also respond defensively to the songs of closely related Black-throated Gray Warblers (see page 561), and also Townsend's Warblers (see facing page), with which the species hybridizes.

The nesting cycle begins when the females arrive and pair bonds are established. Females select a nest site, and begin gathering nesting materials; no one has reported the sight of nest building itself for this species. The female is also in full charge of incubation, with the male following her closely when she is off the nest to feed and defecate. Both parents feed the chicks, with increasing frequency as they grow and fledging day approaches.

Clutch

The egg of the Hermit Warbler is creamy white in background coloration, has fine dark brown speckling, and measures ⅔ x ½ in (17 x 12 mm). To protect the eggs from overhead predators, the nest is typically placed under dense foliage, just below the top of the canopy.

Actual size

ORDER	Passeriformes
FAMILY	Parulidae
BREEDING RANGE	Central Texas
BREEDING HABITAT	Juniper and oak woodlands, in canyons and sloping hillsides
NEST TYPE AND PLACEMENT	Open cup, with bark strips woven together with arthropod silk, lined with finer grass, hair, and down; placed in a small tree
CONSERVATION STATUS	Endangered
COLLECTION NO.	FMNH 2543

ADULT BIRD SIZE
4½–5 in (11–13 cm)

INCUBATION
12–13 days

CLUTCH SIZE
3–4 eggs

SETOPHAGA CHRYSOPARIA

GOLDEN-CHEEKED WARBLER

PASSERIFORMES

Clutch

The egg of the Golden-cheeked Warbler is white in coloration, with reddish, dark brown to black maculation, and measures ¾ x ⅔ in (19 x 17 mm) in dimensions. This species is heavily impacted by Brown-headed Cowbirds (see page 616); at Ft. Hood, Texas, for example, the parasitism rates have dropped from over 50 percent to less than 10 percent after cowbird trapping and removal in the area was begun.

Unlike fugitive species, which are rare because they require constant cycles of habitat disturbance and regeneration, this warbler is rare because it is solely a mature (called "climax") habitat-dwelling species. Therefore, wherever its preferred oak and juniper forests are disturbed, due to natural causes such as fire or due to anthropogenic reasons such as housing construction and clearing for pastures, this warbler loses its breeding habitat; these forests will take many decades to regenerate. This species has lost much of its breeding habitat in recent years to expanding suburban developments surrounding growing Texan cities.

Most of the social and reproductive activities of this species are focused around the nest site; the female chooses a male often based on a suitable nesting site within his territory. Once the site is selected, she then calls to the male while she is constructing the nest, and they copulate on the nest platform. After egg laying, and during incubation, the female will defend the proximity of the nest from other females; the male is highly aggressive against other males throughout the entire breeding season. Older males are site faithful, and return to the same territory year after year, whereas younger males are less stationary, and often occupy poor-quality territories, atop plateaus, rather than preferred sites in ravines and canyons.

Actual size

ORDER	Passeriformes
FAMILY	Parulidae
BREEDING RANGE	Northern and eastern North America
BREEDING HABITAT	Coniferous and mixed forests, cypress swamps
NEST TYPE AND PLACEMENT	Open cup, placed near trunk of small spruce tree or magnolia bush; made of twigs, grass, bark strips, and spider silk, lined with moss, rootlets, hair, and feathers
CONSERVATION STATUS	Least concern
COLLECTION NO.	FMNH 16980

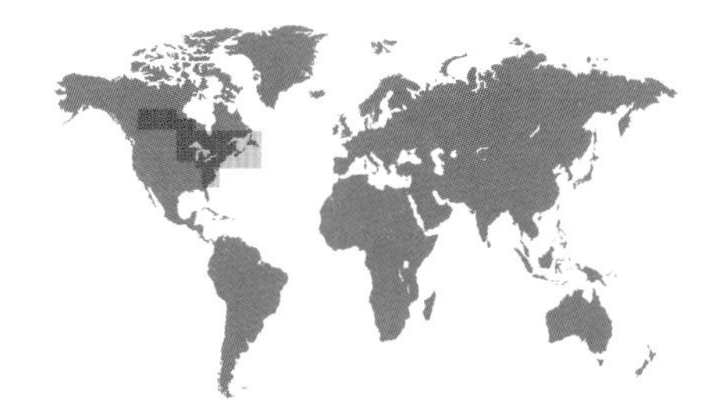

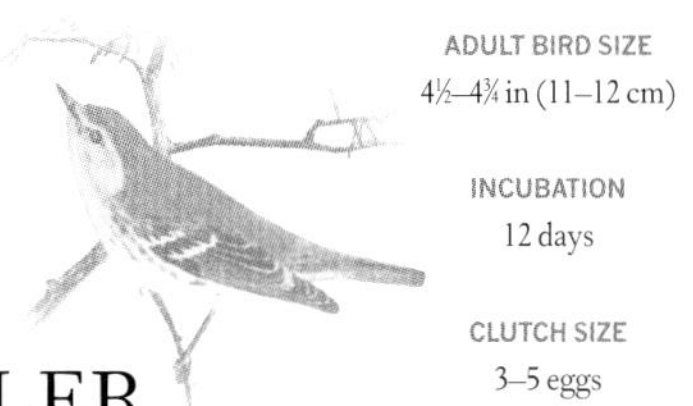

ADULT BIRD SIZE
4½–4¾ in (11–12 cm)

INCUBATION
12 days

CLUTCH SIZE
3–5 eggs

SETOPHAGA VIRENS

BLACK-THROATED GREEN WARBLER

PASSERIFORMES

This small warbler has a broad breeding distribution across northern and eastern North American coniferous forests. Its presence on the breeding grounds is easily detected, because of the male's bright colors and his high rate of song delivery, as many as ten songs per minute. Against other males, the territorial male maintains a strictly exclusive, all-purpose territory, suitable for both feeding and nesting; his female partner, too, is aggressive against intruder females of the same species.

To initiate breeding, the female and the male roam about the habitat, and select a suitable fork in a branch, typically close to the tree trunk, to begin nest construction. The male assists in gathering materials, but the female alone shapes the inner cup of the nest, before the eggs are laid. The female also alone incubates the eggs, and broods the hatchlings. She consumes the broken eggshells to keep the nest free of debris, and, together with the male, feeds the nestlings.

Clutch

Actual size

The egg of the Black-throated Green Warbler is whitish in color, with brown spotting, and is ⅔ x ½ in (17 x 12 mm) in size. Typically one clutch is laid per year, but if the first nest is destroyed, this species will build a new nest and lay a second clutch of eggs.

ORDER	Passeriformes
FAMILY	Parulidae
BREEDING RANGE	Northern and eastern North America
BREEDING HABITAT	Mixed coniferous and deciduous forests, with swampy or dense shrubby undergrowth
NEST TYPE AND PLACEMENT	Depression on the ground or pressed into mossy thickets, in or under low, dense shrub; cup made of grasses and bark strips, lined with deer or mouse hair
CONSERVATION STATUS	Least concern
COLLECTION NO.	FMNH 15190

ADULT BIRD SIZE
4½–6 in (11–15 cm)

INCUBATION
12 days

CLUTCH SIZE
3–5 eggs

CARDELLINA CANADENSIS

CANADA WARBLER

PASSERIFORMES

Clutch

Finally, a warbler species whose English and Latin names conform with its geographic distribution: this warbler breeds mostly throughout Canada, but also in the Appalachian mountain ranges in the eastern United States. Populations of this species have declined consistently during the last three decades, in parallel with forest fragmentation and the loss of many wetland forests, causing the species to be labelled as threatened by Canadian authorities.

Nest construction and incubation are both the tasks of the female, with the male accompanying her on trips to collect materials and to feed herself while off the eggs. Once the chicks hatch, both parents contribute to feeding, carrying insects and other prey in their beaks. They then wait for the nestlings to produce fecal sacs, which are carried and dropped off by the parents, while flying away from the nest.

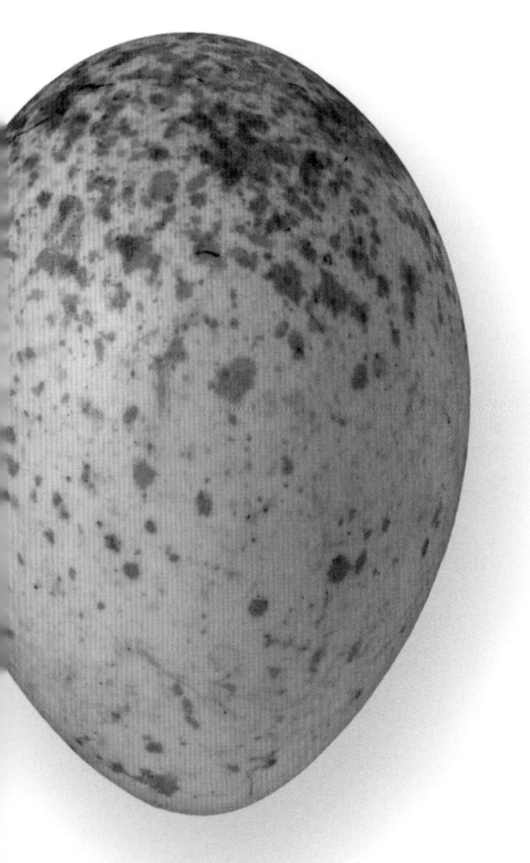

Actual size

The egg of the Canada Warbler is white to buff in background coloration, with brown, gray, or purple speckles, and is ⅔ x ½ in (17 x 12 mm) in size. Reusing an old nest after some essential repair to the lining, instead of building a new one, can save time for the female, so that she can start egg laying two to three days sooner.

ORDER	Passeriformes
FAMILY	Parulidae
BREEDING RANGE	Northern and western North America
BREEDING HABITAT	Coastal or mountain shrubland, forest edges, scrubby bogs
NEST TYPE AND PLACEMENT	Bowl of sedges, leaves, and grasses, lined with finer grasses and coarse hair; placed on the ground, at the base of a shrub, or at a low height, and under tufts of grass
CONSERVATION STATUS	Least concern
COLLECTION NO.	FMNH 2598

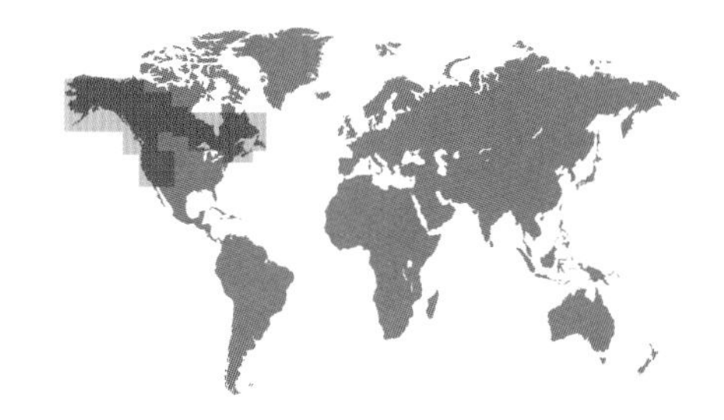

ADULT BIRD SIZE
4–4½ in (10–11 cm)

INCUBATION
11–13 days

CLUTCH SIZE
3–5 eggs

CARDELLINA PUSILLA

WILSON'S WARBLER

PASSERIFORMES

This is one of the five species of birds whose English name honors Alexander Wilson, the Scottish-born father of North American ornithology. A broadly distributed species, its breeding and foraging habitats vary extensively between the humid coastal forest of the Pacific coast, and the boreal woodlands of northern Canada. To begin breeding, the males arrive at suitable habitats one or two weeks before the females, and use song and visual displays to delineate the territory boundaries. Even then, intrusions by neighboring males, in search of females, are frequent throughout the breeding season.

Females typically settle, build nests, and lay eggs on territories held by males that are at least three years old. It is unclear whether territories of younger males are inferior in quality or are more likely to be lost to older males, or whether females prefer to settle with older males.

Clutch

Actual size

The egg of the Wilson's Warbler is cream or white in background color, with reddish maculation, and measures ⅝ x ½ in (16 x 12 mm) in size. Only the female incubates the eggs, and the male may be mostly, or totally, absent from the territory until the chicks hatch and receive his help in feeding.

ORDER	Passeriformes
FAMILY	Parulidae
BREEDING RANGE	Southwestern North America, northern Central America
BREEDING HABITAT	Open and dry pine and oak woodlands
NEST TYPE AND PLACEMENT	Open cup, made of coarse grasses and pine needles; placed on the ground, in rock walls, or on sloping terrain
CONSERVATION STATUS	Least concern
COLLECTION NO.	FMNH 270

ADULT BIRD SIZE
5–6 in (13–15 cm)

INCUBATION
12–14 days

CLUTCH SIZE
3–7 eggs

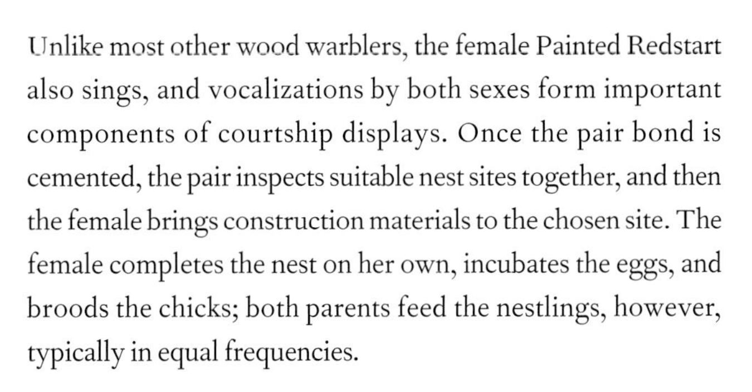

MYIOBORUS PICTUS

PAINTED REDSTART

PASSERIFORMES

Clutch

The egg of the Painted Redstart is white in color, has brown spotting, and measures ⅔ x ½ in (17 x 12 mm). A suitable nest site is often reused across years, even though a new structure is built each time, before the eggs are laid.

Unlike most other wood warblers, the female Painted Redstart also sings, and vocalizations by both sexes form important components of courtship displays. Once the pair bond is cemented, the pair inspects suitable nest sites together, and then the female brings construction materials to the chosen site. The female completes the nest on her own, incubates the eggs, and broods the chicks; both parents feed the nestlings, however, typically in equal frequencies.

This bright red, black, and white warbler and its neotropical relatives are not related to the redstarts of Eurasia, and are only distantly related to the parulid America Redstart; thus, sometimes they are called "whitestarts" to avoid confusion with the other lineages. Whitestart is a particularly appropriate name because the white patches against the black feathers of the tail and wings are used as flashy visual devices to scare and flush insects from hiding. When scientists colored these white patches black on redstart subjects that were being studied, their rate of insect capture decreased.

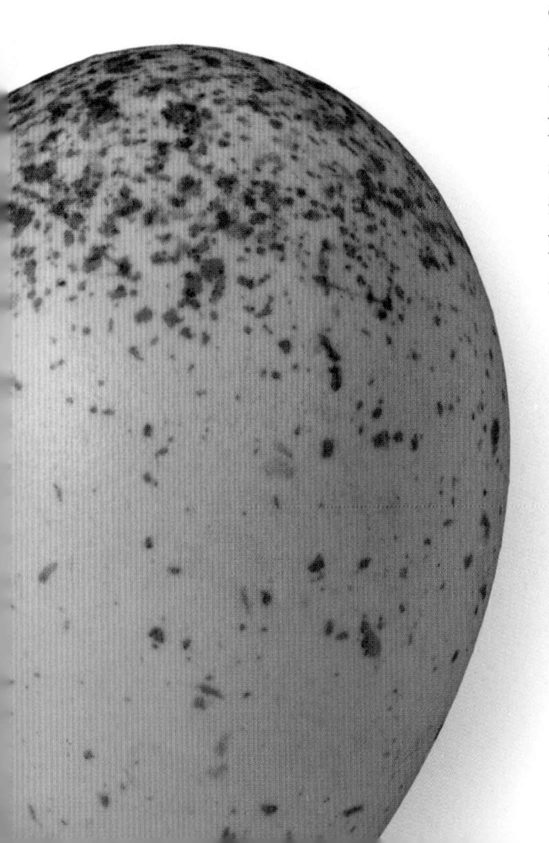

Actual size

ORDER	Passeriformes
FAMILY	Parulidae
BREEDING RANGE	Temperate North America
BREEDING HABITAT	Low, dense shrubby areas, regenerating clear-cuts, forest edges, riparian scrub
NEST TYPE AND PLACEMENT	Bulky cup of grasses, leaves, bark strips, artificial filaments, and stems, lined with finer grasses, wire rootlets, pine needles, and hair; placed in a low, dense shrub
CONSERVATION STATUS	Least concern
COLLECTION NO.	FMNH 12963

ADULT BIRD SIZE
6½–7½ in (17–19 cm)

INCUBATION
11–12 days

CLUTCH SIZE
3–5 eggs

ICTERIA VIRENS

YELLOW-BREASTED CHAT

PASSERIFORMES

This is the largest of the parulid warblers, and its unique gestalt, song behaviors, and genetic affiliations indicate no close relatives; it is placed in its own genus. Despite its size, this bird is difficult to spot because it likes dense, leafy shrubs for feeding and nesting. During the breeding season, however, its varied song repertoire makes it easier for researchers to census areas for occupied territories.

Clutch

Once the pair bond has been established, the female selects a nest site, and builds a bulky cup well hidden in dense foliage. She lays one egg a day, shortly after dawn, but spends the days during egg laying away from the nest. She begins incubation, on her own, the night before the last egg is laid, while the male often stays away from the nest during this period. Only the female broods the chicks, and the male may bring her food. Later, both parents take turns in provisioning the growing nestlings.

Actual size

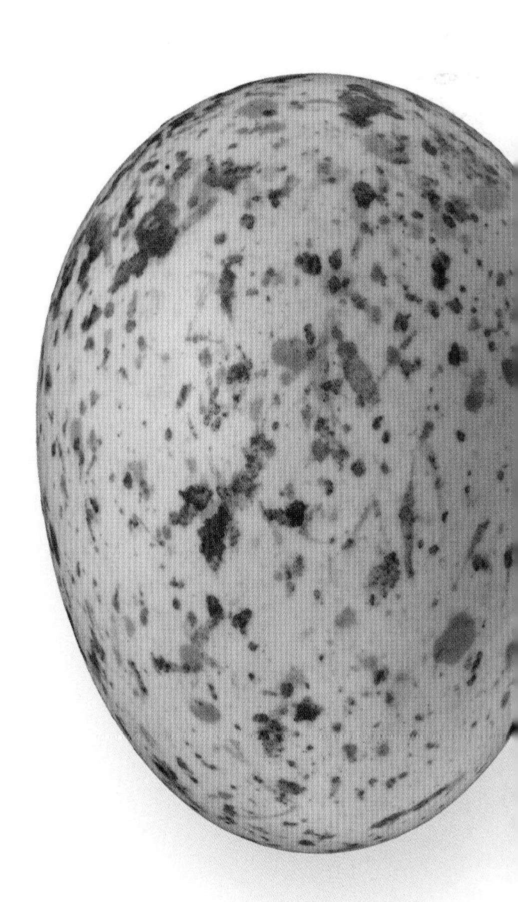

The egg of the Yellow-breasted Chat is white to off-white in coloration, with brownish or black maculation throughout, and is ⅞ x ⅔ in (22 x 17 mm) in size. Most nests perish during the winter between breeding seasons, but individual females often build the nest and lay their eggs in the exact same forking branch year after year.

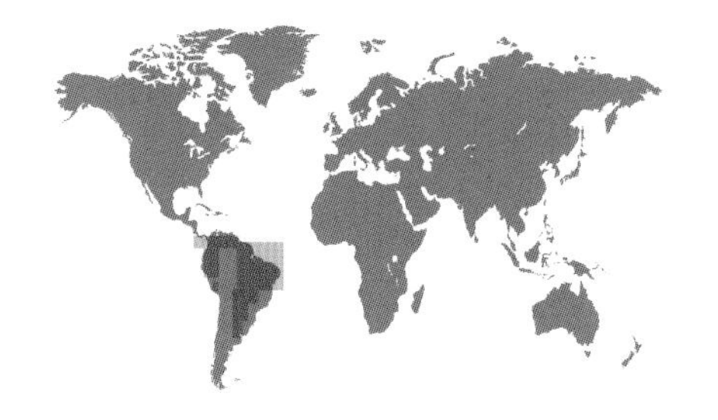

ORDER	Passeriformes
FAMILY	Thraupidae
BREEDING RANGE	Central America and tropical lowland South America
BREEDING HABITAT	Open forests, edges, gardens and parks
NEST TYPE AND PLACEMENT	Open, bulky cup, made out of grass and leaves; placed low in a tree or shrub
CONSERVATION STATUS	Least concern
COLLECTION NO.	FMNH 2499

ADULT BIRD SIZE
7–7½ in (18–19 cm)

INCUBATION
14–15 days

CLUTCH SIZE
2–4 eggs

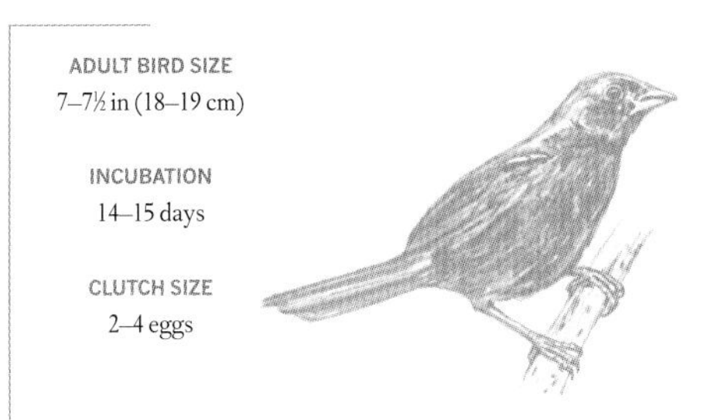

TACHYPHONUS RUFUS

WHITE-LINED TANAGER

PASSERIFORMES

Clutch

This common species is well known for its tame behaviors, pleasant songs, and year-round residence in and around human developments, including farms, plantations, parks, and gardens. There, the adults feed on nectar, fruit, and seeds, and also glean foliage and probe between bark for insects. By consuming and digesting the moist flesh of fruits, and defecating or regurgitating the seeds, this species plays an important ecological role in dispersing their food plants. In addition, seeds that pass through the digestive tract of these tanagers germinate faster than seeds that simply fall to the ground.

To establish exclusive access to food during and outside the breeding season, this species is highly territorial, using song and flight chases to keep intruders away. Despite its nonmigratory habits, these tanagers typically try to breed just once each season.

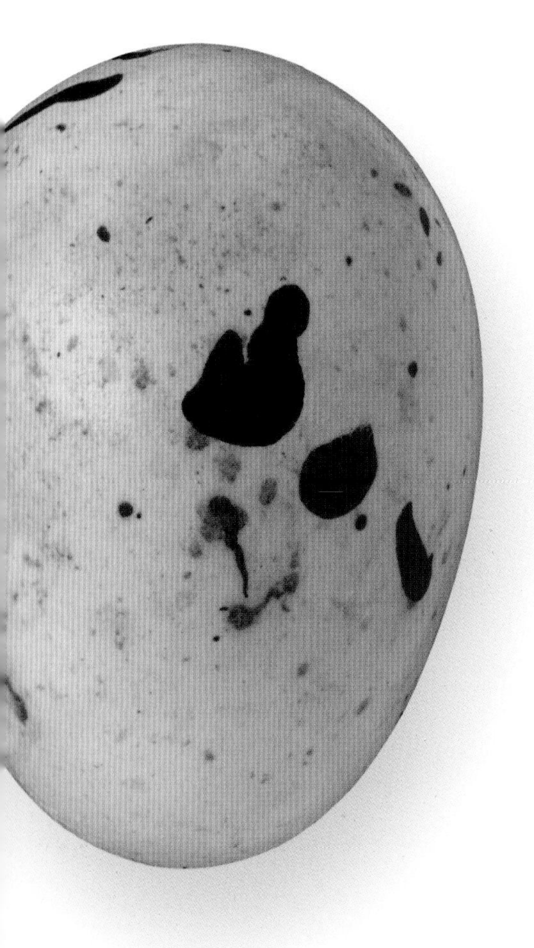

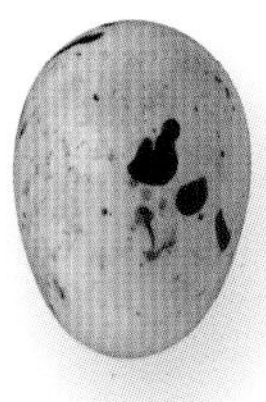

Actual size

The egg of the White-lined Tanager is cream in background coloration, with irregular large brown blotching, and measures 1 x ⅔ in (25 x 17 mm). Only the cryptically colored, brownish females incubate the eggs, and the nestlings and fledglings are also feathered in brown; adult males are glossy black.

ORDER	Passeriformes
FAMILY	Thraupidae
BREEDING RANGE	Central America, tropical South America
BREEDING HABITAT	Open forests, clearings, farms, parks and gardens
NEST TYPE AND PLACEMENT	Bulky and loose cup made of leaves, sticks, and needles; built in a palm, or under the eaves of buildings
CONSERVATION STATUS	Least concern
COLLECTION NO.	FMNH 14755

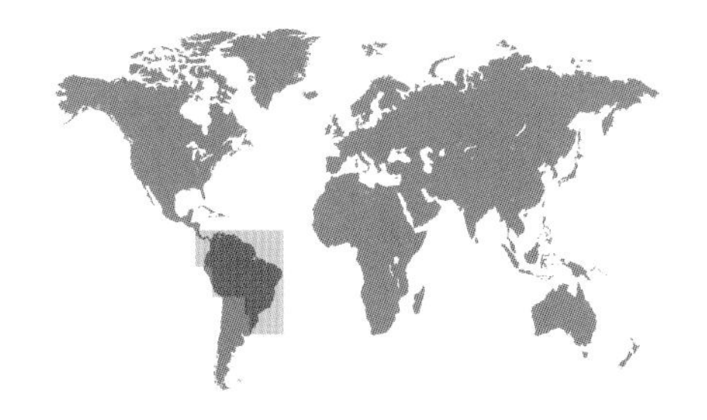

ADULT BIRD SIZE
6½–7½ in (17–19 cm)

INCUBATION
14 days

CLUTCH SIZE
2–3 eggs

THRAUPIS PALMARUM

PALM TANAGER

PASSERIFORMES

Loud and conspicuous, the Palm Tanager is a familiar bird to many people, especially when a pair decides to use a building ledge, or other flat surface under the eaves, to construct their nest platform. The female and the male collect materials and build the nest together. The parents also take turns feeding the nestlings and keeping them clean by removing fecal sacs. This species typically consumes fruits by swallowing them whole; to provision the young, the parents regurgitate fruit pulp to transfer directly into the hungry chicks' open gape.

The Palm Tanager is a broadly distributed tropical species, commonly heard and seen in gardens and parks, as it prefers to feed and breed at the interface between wooded and open areas, a habitat made more common by human settlement, throughout its distribution.

Clutch

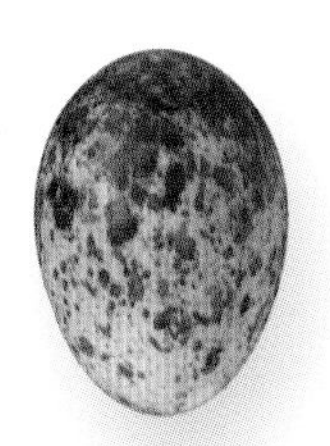

Actual size

The egg of the Palm Tanager is cream to rusty in background coloration, with heavy brown markings, and is 1 x ⅔ in (25 x 17 mm) in dimensions. The female alone incubates the eggs, and the nestlings require two and a half weeks in the nest before they are ready to fledge.

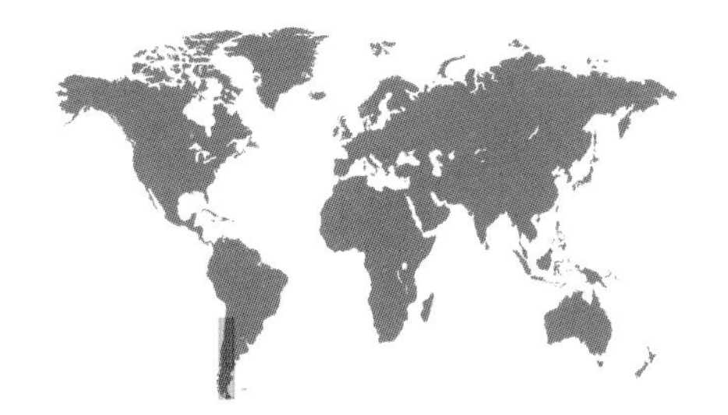

ORDER	Passeriformes
FAMILY	Thraupidae
BREEDING RANGE	Southern South America
BREEDING HABITAT	Dry grassland, shrubby steppe, forest edges
NEST TYPE AND PLACEMENT	Open cup on the ground, lined with fine grasses
CONSERVATION STATUS	Least concern
COLLECTION NO.	FMNH 2495

ADULT BIRD SIZE
6–6½ in (15–16 cm)

INCUBATION
14 days

CLUTCH SIZE
4 eggs

PHRYGILUS GAYI

GRAY-HOODED SIERRA-FINCH

PASSERIFORMES

Clutch

The egg of the Gray-hooded Sierra-Finch is whitish in color, with fine brownish maculation, and measures ¾ x ⅝ in (19 x 16 mm) in size. The female alone incubates the eggs, and both parents provision the chicks.

In plumage patterns, this species appears to be South America's yellowish equivalent of North America's juncos, but feather colors are often a poor indication of evolutionary affinities. For example, the Sierra-Finch genus, *Phrygilus*, is comprised of many species that were initially clumped based on body and plumage pattern similarities. But genetic studies have shown that *Phrygilus* is made up of at least four different evolutionary lineages, and convergent evolution in all those lines led to similar feather coloration.

Despite the advance of genetic tools, however, more natural history studies are required as these are critical in identifying the reproductive and ecological bases of species limits. For example, the open habitat-dwelling Gray-hooded Sierra-Finch and the forest-dwelling Patagonian Sierra-Finch have different habitat requirements, keeping the two species genetically distinct. However, since behavioral barriers to interbreeding have not (yet) evolved, where the two species meet, they form interspecific pairs in a narrow zone of hybridization.

Actual size

ORDER	Passeriformes
FAMILY	Thraupidae
BREEDING RANGE	Southern North America and Central America
BREEDING HABITAT	Successional scrub, shrubby grasslands, savanna
NEST TYPE AND PLACEMENT	Open cup, made from fibers, leaves, cobwebs, fine rootlets, and grass; placed near the center of small trees or dense shrubs
CONSERVATION STATUS	Least concern
COLLECTION NO.	FMNH 14545

ADULT BIRD SIZE
3¼–4½ in (9–11 cm)

INCUBATION
13 days

CLUTCH SIZE
2–4 eggs

SPOROPHILA TORQUEOLA

WHITE-COLLARED SEEDEATER

PASSERIFORMES

In many seedeater species, including the White-collared Seedeater, only the male plumages are bright combinations of black, white, and brown, whereas the female and the juvenile plumages are dull brown. Patterns of sex differences in plumage coloration, such as these, imply a strong role for male plumage in species recognition and female choice for mating, as well as strong selection for females to be cryptically colored, which provides camouflage when on the nest.

Clutch

In the non-breeding season, this seedeater is gregarious and moves about in large foraging flocks, roosting in dense grasses or cattails. In the spring, males separate from the flock and establish breeding territories, where they sing atop trees or bushes to attract the female. Once the pair bond is formed, the female constructs the nest, and incubates the eggs on her own. Both parents feed the nestlings, but it is unclear whether the male also contributes to provisioning the fledglings.

Actual size

The egg of the White-collared Seedeater is pale blue gray in background, with irregular dark brown blotching, and is ⅝ x ½ in (16 x 12 mm) in size. The female sleeps on the nest while incubating the eggs, while the male joins other males to roost together at night.

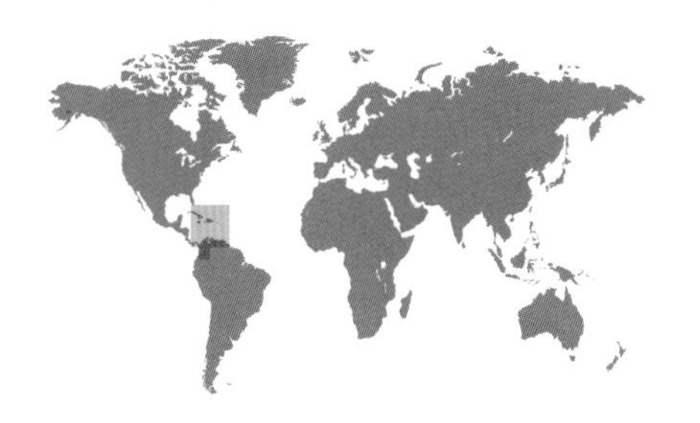

ORDER	Passeriformes
FAMILY	Thraupidae
BREEDING RANGE	Northern South America and the Caribbean, except Cuba
BREEDING HABITAT	Tall grassy fields, open scrub, rice patties, gardens, and roadsides
NEST TYPE AND PLACEMENT	Dome-shaped nest, made of grasses and long leaves, lined with finer grasses; placed in a soil bank, or in a dense shrub
CONSERVATION STATUS	Least concern
COLLECTION NO.	FMNH 15246

ADULT BIRD SIZE
4–4½ in (10–11 cm)

INCUBATION
9–12 days

CLUTCH SIZE
2–3 eggs

TIARIS BICOLOR

BLACK-FACED GRASSQUIT

PASSERIFORMES

Clutch

To attract females, the male flies above his breeding territory, and sings buzzing songs, while he vibrates his stiffened wings in a deliberate manner. If a female picks him, the pair seeks out a suitable nesting site together, collects the materials, builds the nest, and provides joint parental care all the way through feeding the nestlings and the fledglings. Today, this species often builds its nest in well-pruned ornamental bushes in gardens and parks.

A diverse and common group of neotropical birds, most recently the Tiaris grassquits have gained scientific and popular attention because of their strong phylogenetic linkage to the Darwin's finches of the Galapagos Islands. With hindsight, this seems a very feasible scenario, given the similarities between these two lineages of birds, including the males' dark blue or black plumage, the females' drab gray or brown coloration, and the building of enclosed, domed nests.

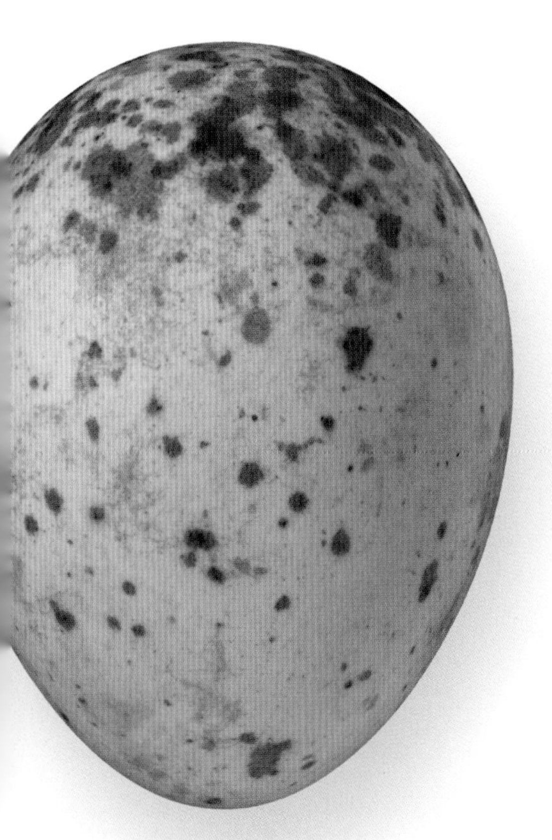

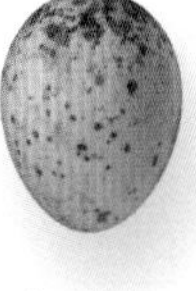

Actual size

The egg of the Black-faced Grassquit is white in background coloration, has red, brown, or chocolate spotting, and its dimensions are ⅔ x ½ in (17 x 12 mm).

ORDER	Passeriformes
FAMILY	Thraupidae
BREEDING RANGE	Galapagos Islands
BREEDING HABITAT	Coastal arid scrub, open forests
NEST TYPE AND PLACEMENT	Small, domed nest, made of grasses, rootlets, thorns, and stems, in a bush or cactus
CONSERVATION STATUS	Least concern
COLLECTION NO.	FMNH 2902

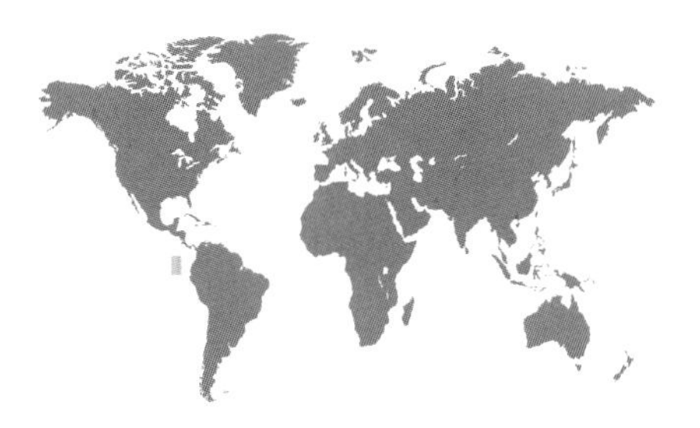

ADULT BIRD SIZE
4–4½ in (10–11 cm)

INCUBATION
12–13 days

CLUTCH SIZE
3 eggs

GEOSPIZA FULIGINOSA

SMALL GROUND-FINCH

PASSERIFORMES

The Small Ground-Finch is one of the better-known species of Darwin's finches, as it occupies most islands and habitats, even ones near urban developments. In wet years, this finch capitalizes on its ability to collect and process small grass seeds efficiently; however, in drier years, its beak size and strength often prove insufficient to crack larger seeds. On islands where habitats are naturally variable, from humid forested highlands to arid coastal scrub, birds of different morphs and beak sizes occur which can and do interbreed with one another.

As with with other types of ground-finch, breeding in the Small Ground-Finch is highly opportunistic, and depends on the extent and duration of the hot wet season on the islands. In many years, just a single clutch is completed, whereas in some years up to nine (!) broods may be initiated, and fledged successfully.

Clutch

Actual size

The egg of the Small Ground-Finch is buff to cream in coloration, with chocolate to chestnut maculation, and is ⅔ x ½ in (17 x 12 mm) in size. Egg laying is typically tied to the onset of local rains, and the clutch is incubated by the female only.

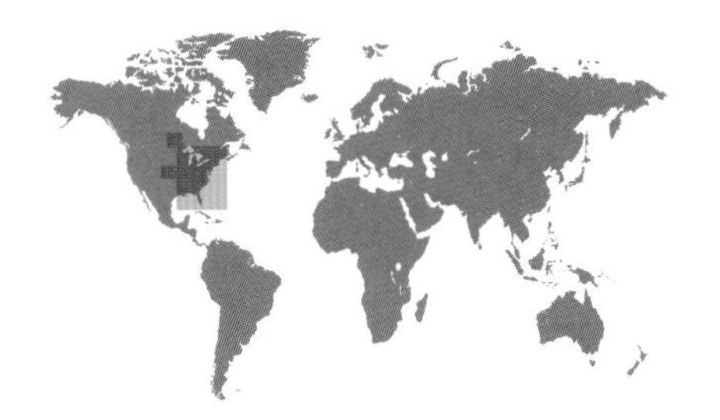

ORDER	Passeriformes
FAMILY	Emberizidae
BREEDING RANGE	Eastern and central North America
BREEDING HABITAT	Open woodlands, clearings, edges, scrubby fields
NEST TYPE AND PLACEMENT	Open cup, on the ground, embedded in leaf litter, or low in a shrub, always with dense vegetation cover; made of weeds, leaves, stems, and bark strips
CONSERVATION STATUS	Least concern
COLLECTION NO.	FMNH 14356

ADULT BIRD SIZE
6½–8½ in (17–21 cm)

INCUBATION
11–13 days

CLUTCH SIZE
3–5 eggs

PIPILO ERYTHROPHTHALMUS

EASTERN TOWHEE

PASSERIFORMES

Clutch

The egg of the Eastern Towhee is grayish or creamy white and occasionally greenish white, has chestnut brown and light gray maculation, and measures ⅞ x ⅔ in (22 x 17 mm) in size. While the female alone incubates the eggs, the male calls to her when danger approaches, and both sexes mob predators near the nest.

Eastern Towhees are colorfully patterned sparrows whose long tails and strong sexual dichromatism set them apart from other sparrows and make them amenable for detailed study and behavioral experimentation. Research using playbacks of familiar and unfamiliar songs of other males has shown that male towhees recognize their neighbors as individuals, even when the neighbor incorporates mimetic sounds and calls from other species' songs. Because adult males typically return to the same patch habitat year after year, remembering a neighbor's calls can save time by forgoing unnecessary territorial squabbles.

The breeding season of this towhee is prolonged, and can span nearly five months in the southern part of its range. The female selects the nest site, builds the nest, and incubates the eggs alone. We still do not know exactly what cues tell males to start bringing food to the nestlings: with this towhee, the male starts his food-delivery trips several days before the eggs even hatch.

Actual size

ORDER	Passeriformes
FAMILY	Emberizidae
BREEDING RANGE	Southwestern North America
BREEDING HABITAT	Open grasslands, scrub, chaparral, burnt clearings, oak galleries, and rocky canyons
NEST TYPE AND PLACEMENT	Bulky open cup, with a thick wall, made of dried grasses, leaves, and rootlets, with strips of bark and small twigs; placed on the ground or in a low bush, typically under overhanging rock or concealing vegetation
CONSERVATION STATUS	Least concern
COLLECTION NO.	FMNH 15229

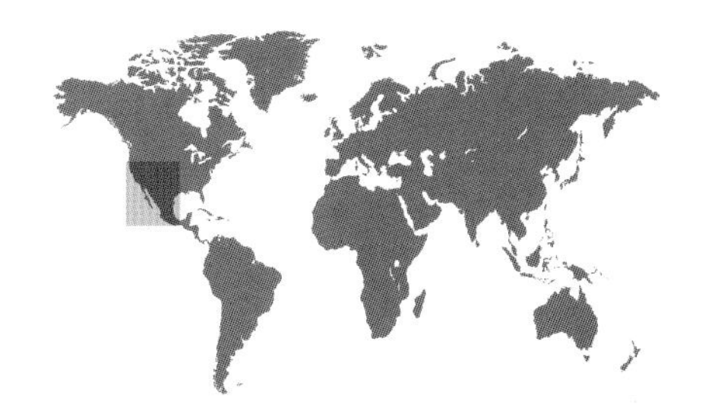

ADULT BIRD SIZE
5–6 in (13–15 cm)

INCUBATION
11–13 days

CLUTCH SIZE
3–4 eggs

AIMOPHILA RUFICEPS

RUFOUS-CROWNED SPARROW

PASSERIFORMES

This sparrow is an inconspicuous, territorial resident, forming distant and disconnected populations wherever its preferred arid habitats occur. Even within suitable areas, the territories are distributed in clumps, centered around the favored rocky hillsides. The males work hard at keeping intruders and neighbors out, and most territories are maintained year-round, with only the mated female and the juvenile offspring tolerated. But this tolerance may last only until winter, and the mated pair may not stay together through the next breeding season.

With the arrival of seasonal rains, the nest is built by the female only, typically under a dense shrub, a rocky overhang, or a cactus stem, or even inside ground holes and abandoned aluminum cans. She is also in charge of incubation, but both parents feed the nestlings, carrying insect prey in their beak, and shoving the food items down the open gape of the chicks begging in the nest.

Clutch

The egg of the Rufous-crowned Sparrow is pale blueish white in background, immaculate, and ¾ x ⅝ in (19 x 16 mm) in size. This bird is an exception to the rule that open-cup-nesting songbirds lay well-camouflaged, spotted, or deep blue eggs, in contrast to cavity nesters who produce clear and, typically, white eggs.

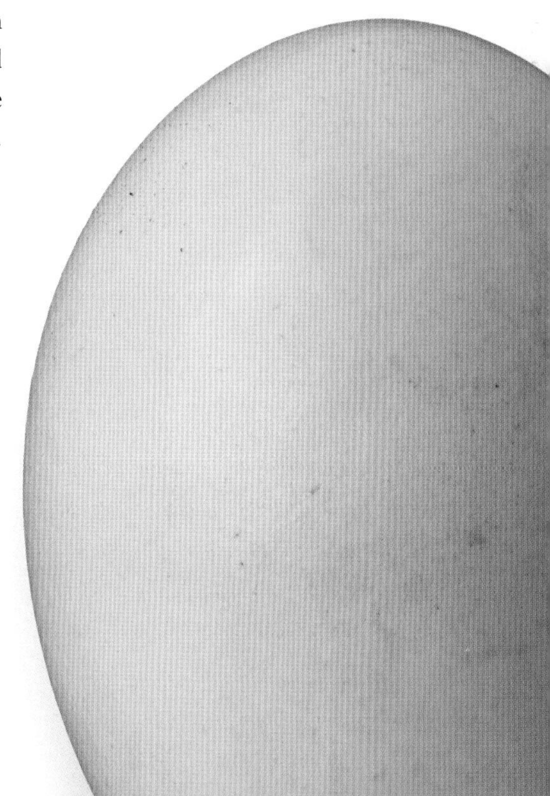

Actual size

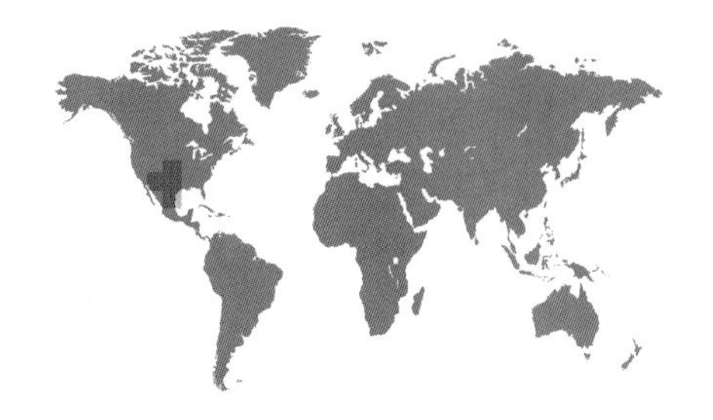

ORDER	Passeriformes
FAMILY	Emberizidae
BREEDING RANGE	Southwestern North America
BREEDING HABITAT	Open grasslands with occasional shrubs
NEST TYPE AND PLACEMENT	Open and deep cup nest, made of grasses and stems; on the ground, low in a tuft of grass, or in a shrub
CONSERVATION STATUS	Least concern
COLLECTION NO.	FMNH 13809

ADULT BIRD SIZE
5–6 in (13–15 cm)

INCUBATION
10–12 days

CLUTCH SIZE
3–5 eggs

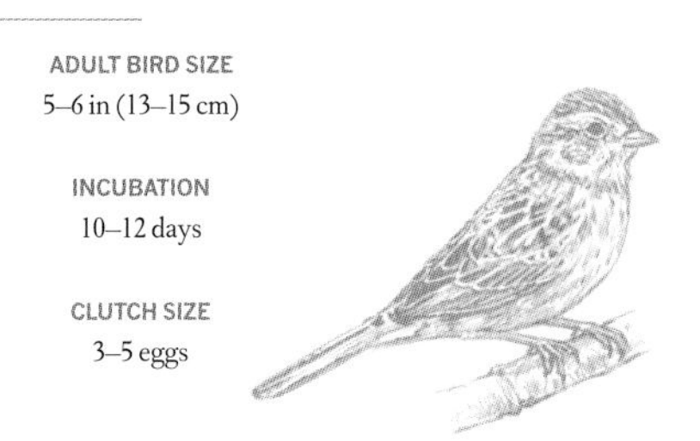

PEUCAEA CASSINII

CASSIN'S SPARROW

PASSERIFORMES

Clutch

The egg of the Cassin's Sparrow is white, with no patterning, and is ¾ x ⅝ in (19 x 16 mm) in size. This is an unusual songbird species: it sometimes nests on the ground, in an open cup, but lays immaculate eggs; the female begins incubation on the day of laying the penultimate egg.

The English term for males singing on the wing, "skylarking," is based on the behavior of Old World larks. Living in similar open grasslands of North America, the Cassin's Sparrow has evolved a parallel set of display behaviors; during the breeding season, the males sing a complex song while flying over their territories to attract females. The males also sing atop shrubs or dense grass, or even on the ground. Otherwise, this drab-colored bird remains inconspicuous, preferring to run through grass rather than fly.

Pair formation and breeding are tightly tied to the arrival of summer rains. Only the female is known to carry nesting materials, but the mechanics of nest construction have not been documented. The female alone develops a brood patch, but details of the incubation behaviors are also unknown. Both parents have been seen to provision the young and remove fecal sacs after feeding bouts.

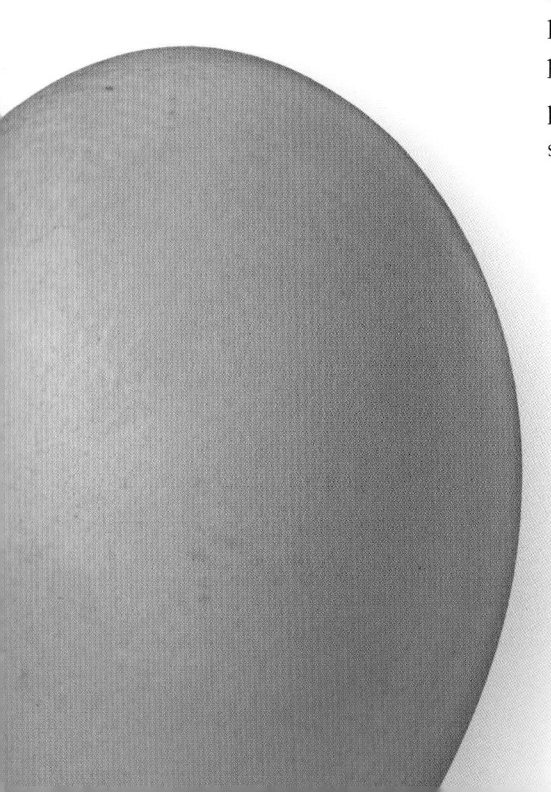

Actual size

ORDER	Passeriformes
FAMILY	Emberizidae
BREEDING RANGE	Southeastern United States
BREEDING HABITAT	Open, mature pine woodlands, with a dense undergrowth layer and open mid-canopy; also regrowth after forest fires
NEST TYPE AND PLACEMENT	Open or domed cup nest, made of grasses, stems, and pine needles; on the ground placed against a stem or clump of vegetation
CONSERVATION STATUS	Near threatened
COLLECTION NO.	FMNH 222

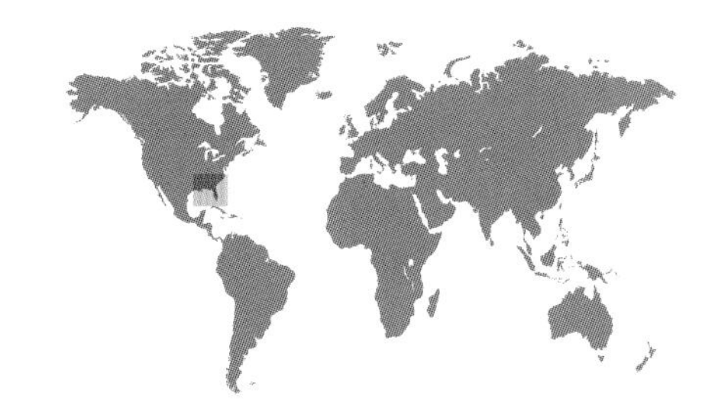

ADULT BIRD SIZE
4½–6 in (11–15 cm)

INCUBATION
13–14 days

CLUTCH SIZE
2–5 eggs

PEUCAEA AESTIVALIS

BACHMAN'S SPARROW

PASSERIFORMES

This enigmatic sparrow has become a conservation concern, because today it is rare or absent from many regions where it was common previously, as evidenced by springtime counts of singing males at the onset of the breeding season. However, this species' strong association with southern pine forests may perhaps be its saving grace: restored pine forests now managed for the endangered Red-cockaded Woodpecker (see page 373) are also favored by the Bachman's Sparrow.

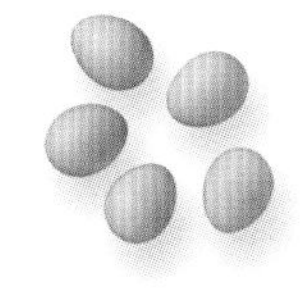

Clutch

Many of the adults return to their previously used breeding sites, and once the pair is formed, the females begin nest construction, with the singing male following closely behind. Only the mother develops a brood patch, and incubates the eggs. After hatching, both parents provision the chicks; they move quietly, landing near the nest and making their final approach on foot.

Actual size

The egg of the Bachman's Sparrow is white and immaculate in color, and ¾ x ⅝ in (19 x 16 mm) in dimensions. The breeding season of this southern sparrow is long, so that two full breeding attempts can be completed per year; if a nest fails, the female will also lay a replacement clutch.

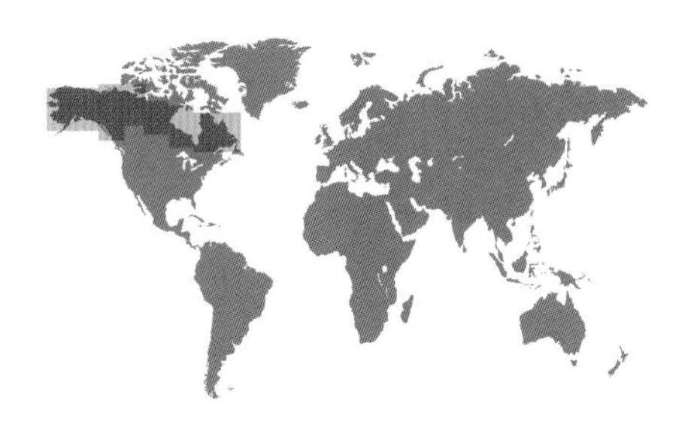

ORDER	Passeriformes
FAMILY	Emberizidae
BREEDING RANGE	Arctic and subarctic North America
BREEDING HABITAT	Open tundra, shrubby areas, boreal forest edges
NEST TYPE AND PLACEMENT	Open cup, made of mosses, grasses, bark shreds, and twigs, lined with finer grasses and downy feathers; placed on or near the ground, in a tussock of grass
CONSERVATION STATUS	Least concern
COLLECTION NO.	FMNH 2823

ADULT BIRD SIZE
5½–6¾ in (14–17 cm)

INCUBATION
10–14 days

CLUTCH SIZE
4–6 eggs

SPIZELLA ARBOREA

AMERICAN TREE SPARROW

PASSERIFORMES

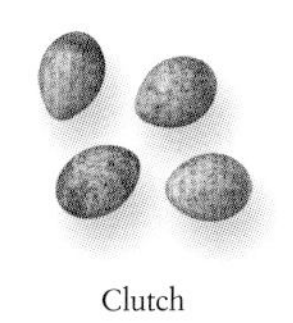

Clutch

This species is familiar throughout the North American temperate zone as a common winter-resident species, but its breeding range is far north, past the wide band of boreal pine forests. The males advertize their territory by loud song in the open tundra, and the females make decisions about pair bonding when they arrive on these breeding grounds. Females select the nest site, incubate the eggs, and brood the chicks, while the male often visits the nest, but he does not contribute until the chicks require regular feeding visits.

Lacking teeth, birds must soften foods like seeds before digestion; to do this they swallow and accumulate grit and small pebbles in their crop where the seeds are ground up. Nestling birds are fed grit by the parents; in American Tree Sparrows, grit first shows up in the crop at about three days after hatching.

Actual size

The egg of the American Tree Sparrow is pale blue in background coloration, has reddish speckling, and its dimensions are ¾ x ⅝ in (19 x 16 mm). After hatching, the female swallows the broken eggshells to keep the nest free of glowing whitish eggshell interiors.

ORDER	Passeriformes
FAMILY	Emberizidae
BREEDING RANGE	North and Central America
BREEDING HABITAT	Clearings and glades in coniferous forests, woodland edges, shrubby grass fields, and city parks
NEST TYPE AND PLACEMENT	Loose platform of rootlets and grasses, with an open, inner cup lined with animal hair and plant fibers; placed mid-height on an outer branch, in a conifer, or other trees and shrubs
CONSERVATION STATUS	Least concern
COLLECTION NO.	FMNH 2145

ADULT BIRD SIZE
5–6 in (13–15 cm)

INCUBATION
10–15 days

CLUTCH SIZE
2–7 eggs

SPIZELLA PASSERINA

CHIPPING SPARROW

PASSERIFORMES

This broadly distributed species occurs from boreal forest glades to Central American pine savanna, and the backyards of suburban developments. This bird's song is a simple, repetitive trill or buzz, and the male uses it both to attract the later-arriving female, and as a courtship display during pair formation. Nesting begins soon afterwards, but cold spells of weather may delay it by several days.

Clutch

Many nesting attempts are abandoned in favor of different sites, even during the egg-laying stage. The new nest, however, is often built with materials taken from the unfinished one. Once a site is chosen definitively, the female is particular to it. If a nest is lost, or removed, the replacement nest may be built in the same spot. From year to year, while the old nest disintegrates, the same site, or at least the same tree, is used for building a new nest.

Actual size

The egg of the Chipping Sparrow is pale blue to white in coloration, with black, brown, or purple maculation, and is ⅔ x ½ in (17 x 12 mm) in size. While the female incubates the eggs, the male ventures outside the territory, and solicits mating with other females; but he also may return to the nest and feed the female frequently.

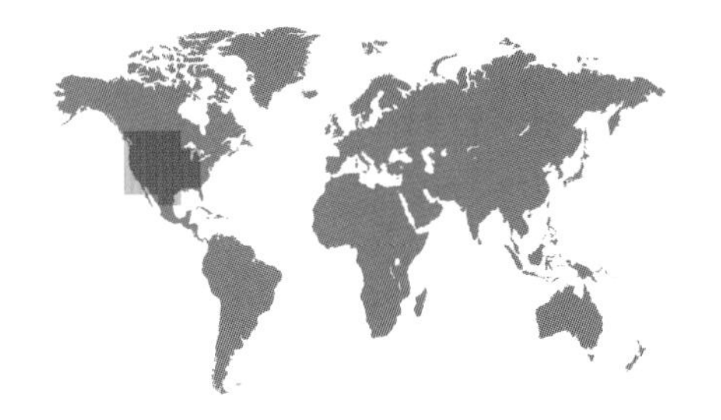

ORDER	Passeriformes
FAMILY	Emberizidae
BREEDING RANGE	Central and western North America
BREEDING HABITAT	Grasslands, edges with shrub or woodlands
NEST TYPE AND PLACEMENT	Open cup, made of grass, twigs, and stems, lined with finer grasses and horse hair; placed on the ground or low in a shrub or small tree
CONSERVATION STATUS	Least concern
COLLECTION NO.	FMNH 11002

ADULT BIRD SIZE
6–6½ in (15–17 cm)

INCUBATION
11–12 days

CLUTCH SIZE
3–6 eggs

CHONDESTES GRAMMACUS

LARK SPARROW

PASSERIFORMES

Clutch

The Lark Sparrow is a distinctive songbird of prairie edge habitats, where the males frequently fight beak and claw to keep their breeding territories free of intruders. Territorial defense, however, wanes quickly, and once the incubation period has begun, only the immediate vicinity of the nest is defended.

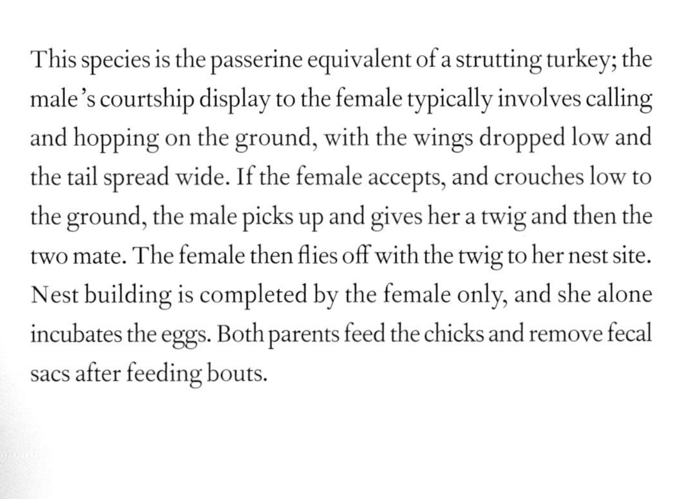

This species is the passerine equivalent of a strutting turkey; the male's courtship display to the female typically involves calling and hopping on the ground, with the wings dropped low and the tail spread wide. If the female accepts, and crouches low to the ground, the male picks up and gives her a twig and then the two mate. The female then flies off with the twig to her nest site. Nest building is completed by the female only, and she alone incubates the eggs. Both parents feed the chicks and remove fecal sacs after feeding bouts.

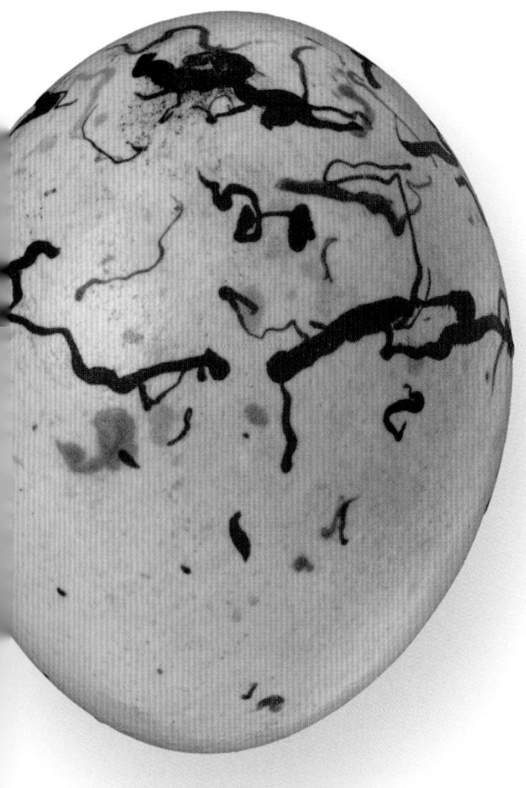

Actual size

The egg of the Lark Sparrow is creamy white in color, has intricate chestnut to chocolate spots and lines, and measures ¾ x ⅝ in (19 x 16 mm) in size. Occasionally, the female will reuse and lay eggs in a nest of a mockingbird or thrasher, but these attempts typically fail if the larger and more aggressive nest builder is still actively interested in using it.

ORDER	Passeriformes
FAMILY	Emberizidae
BREEDING RANGE	California
BREEDING HABITAT	Open thickets, typically sagebrush, but also other shrubs
NEST TYPE AND PLACEMENT	Open cup of twigs, rootlets, and grasses, lined with finer grasses and bark strips; placed low, in or under shrubs and tufts of grasses
CONSERVATION STATUS	Least concern
COLLECTION NO.	FMNH 13841

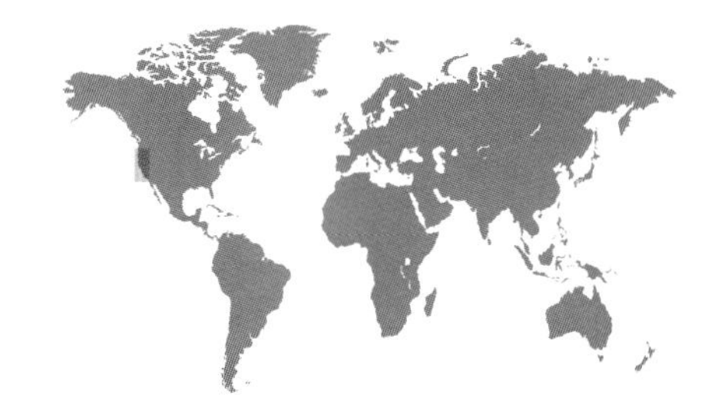

AMPHISPIZA BELLI

BELL'S SPARROW

PASSERIFORMES

ADULT BIRD SIZE
4½–6 in (11–15 cm)

INCUBATION
12–16 days

CLUTCH SIZE
2–6 eggs

In contrast to most passerines, pair bonds in migratory populations of Bell's Sparrows form prior to their arrival on the breeding grounds. In resident populations, pair bonds last throughout the year, although divorce and re-pairing may take place just before the next mating season. Throughout the early part of the breeding season, the male sings, follows and mate-guards the female closely. The female builds the nest and incubates, although some males may sit on the eggs while the female is briefly away. Both parents provision the young.

This sparrow is fairly common but relatively inflexible in its specialized habitat requirements, so much so that with any habitat modification, the population sizes typically drop. For example, on San Clemente Island off California, the endemic island population of Bell's Sparrows quickly dropped in numbers after the introduction of pigs, goats, and feral cats. It eventually received protection under the Endangered Species Act as a threatened subspecies.

Clutch

The egg of the Bell's Sparrow is pale blue in coloration, with brownish maculation, and measures ¾ x ⅝ in (19 x 16 mm) in size.

Actual size

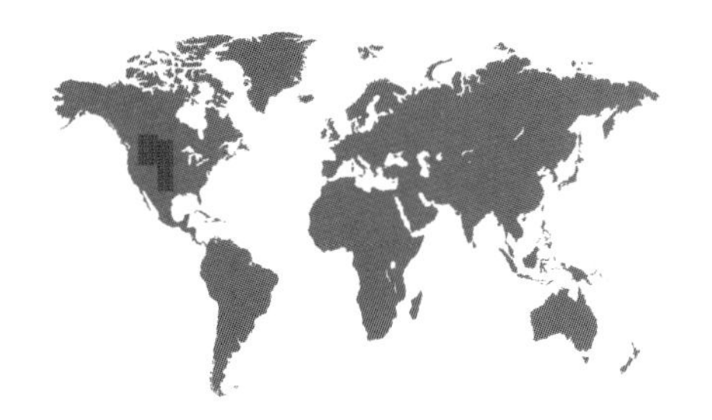

ORDER	Passeriformes
FAMILY	Emberizidae
BREEDING RANGE	Central and western North America
BREEDING HABITAT	Open prairie, grasslands
NEST TYPE AND PLACEMENT	Loose bowl of grasses, rootlets, and stems, lined with finer grasses and hair; placed in a scrape on the ground, usually under a shrub
CONSERVATION STATUS	Least concern
COLLECTION NO.	FMNH 13875

ADULT BIRD SIZE
5½–7 in (14–18 cm)

INCUBATION
11–12 days

CLUTCH SIZE
2–5 eggs

CALAMOSPIZA MELANOCORYS

LARK BUNTING

PASSERIFORMES

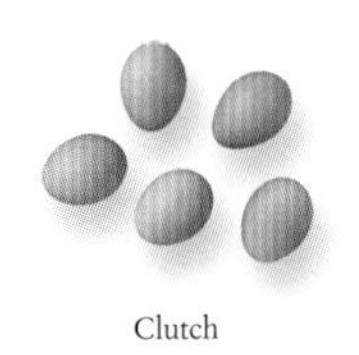

Clutch

The male Lark Bunting, with its black plumage and white wing bar, stands out against the green and yellow grass stalks of its prairie habitat. At the onset of the breeding season, the males return ahead of the females, and establish small territories, in patchy aggregations. They only defend these territories during the courtship stage, and once nesting begins, the breeding pattern is best described as loosely colonial.

Research has shown that females exhibit shifting preferences in what they prefer in potential male mates; in some years, it is males with large wing bars, and in other years, it is males with small white bars. Whatever the current preference, these choices seem to benefit the females, because the resulting couplings yield the highest reproductive success in that breeding season.

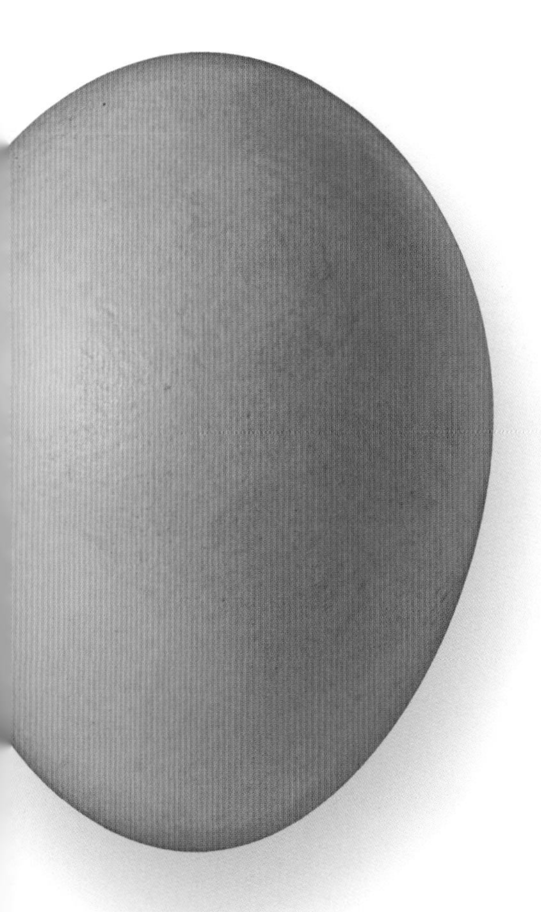

Actual size

The egg of the Lark Bunting is light blue in background color, immaculate, and ⅞ x ⅔ in (22 x 17 mm) in size. With decreasing density of the breeding populations, females may be mated to a polygynous male or have male helpers at the nest.

ORDER	Passeriformes
FAMILY	Emberizidae
BREEDING RANGE	North-central North America
BREEDING HABITAT	Tall grass prairie
NEST TYPE AND PLACEMENT	Open cup, made from coarse grasses and lined with finer grass and other plant fibers; placed in a depression on the ground or in a tuft of grass
CONSERVATION STATUS	Least concern
COLLECTION NO.	FMNH 10898

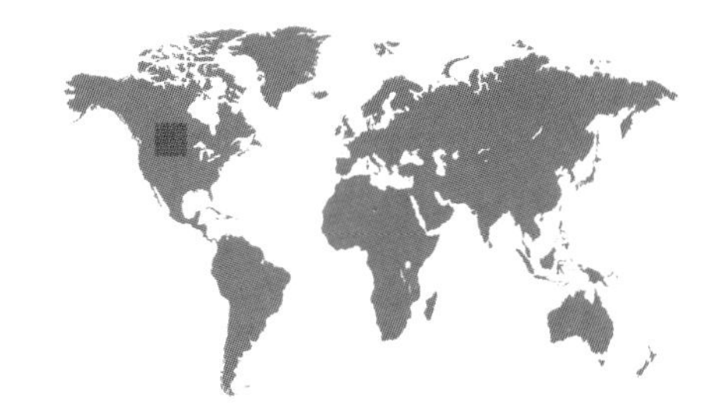

ADULT BIRD SIZE
4½–5½ in (12–14 cm)

INCUBATION
11–12 days

CLUTCH SIZE
4–5 eggs

AMMODRAMUS BAIRDII

BAIRD'S SPARROW

PASSERIFORMES

Male Baird's Sparrows arrive on the northerly breeding grounds about one week in advance of the females and establish display territories, which turn into all-purpose territories used for both breeding and feeding. In its homogenous grassy habitat, it is unclear what, if any, landmark or vegetation features are used to demarcate territory boundaries, but when these are crossed, vocal and physical fights quickly follow. With the appearance of later-arriving males in the spring, territory sites and shapes often shift.

Once the females arrive, and a pair bond is established, the female alone incubates the eggs, and the male does not provision her on the nest. Both parents, however, feed the chicks and maintain nest hygiene and safety by picking up fecal sacs and dropping them off while flying in different directions, away from the nest.

Clutch

Actual size

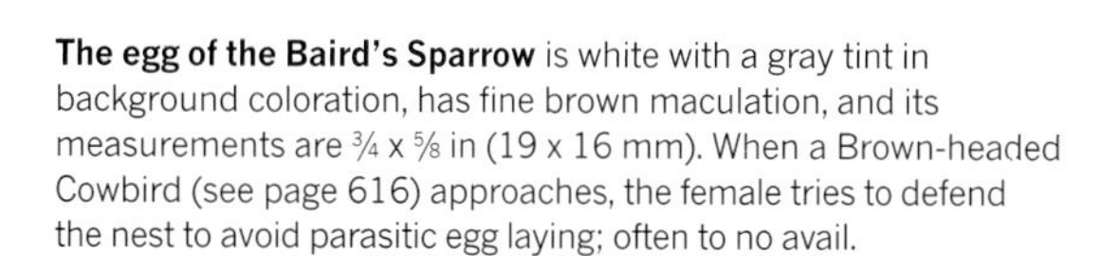

The egg of the Baird's Sparrow is white with a gray tint in background coloration, has fine brown maculation, and its measurements are ¾ x ⅝ in (19 x 16 mm). When a Brown-headed Cowbird (see page 616) approaches, the female tries to defend the nest to avoid parasitic egg laying; often to no avail.

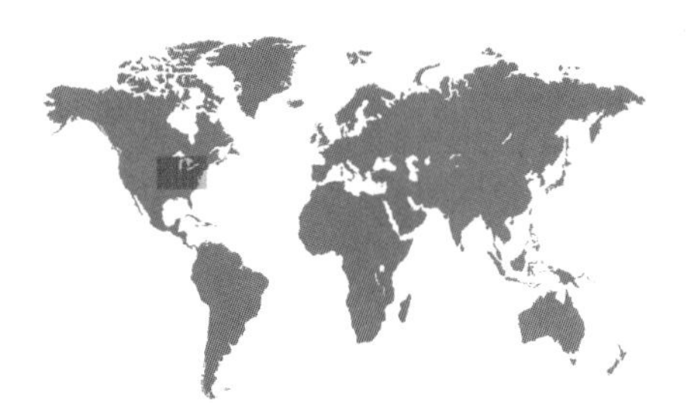

ORDER	Passeriformes
FAMILY	Emberizidae
BREEDING RANGE	Northeastern and central North America
BREEDING HABITAT	Tall-grass prairies, and moist fields with thickets
NEST TYPE AND PLACEMENT	Open bowl, with loosely woven dry grasses; placed in plant litter just off the soil, or in clumps of grass
CONSERVATION STATUS	Near threatened
COLLECTION NO.	FMNH 10922

ADULT BIRD SIZE
4½–5 in (11–13 cm)

INCUBATION
11 days

CLUTCH SIZE
2–5 eggs

AMMODRAMUS HENSLOWII

HENSLOW'S SPARROW

PASSERIFORMES

Clutch

The egg of the Henslow's Sparrow is glossy white in background coloration, with distinct black or brown spotting, and measures ¾ x 9⁄16 in (19 x 14 mm) in dimensions. If a nest is lost, the female can build a new nest and lay a replacement clutch of eggs in under a week.

The Henslow's Sparrow is a grassland-breeding species, whose habitat loss to agricultural crops has caused a consistent decline in population sizes across several decades, leading to its current conservation status (near threatened). The males arrive from wintering grounds in the southern United States to form exclusive breeding territories, where the boundaries and ownership are decided strictly on the basis of song displays given from favored perch sites throughout the territory. Males tend to settle and aggregate close to one another, and this spatial structure also translates into the loosely clumped distribution of nests.

The female alone builds the nest and incubates the eggs, spending about two-thirds of the daytime hours sitting on the nest. She is protective of the clutch, both during laying and after the onset of incubation, calling loudly to distract potential predators. Once the chicks hatch, both parents take an equal share in feeding the next generation, often collecting food outside the breeding territory's boundaries.

Actual size

ORDER	Passeriformes
FAMILY	Emberizidae
BREEDING RANGE	Eastern and southern coastline of North America
BREEDING HABITAT	Tidal marshes, muddy wetlands
NEST TYPE AND PLACEMENT	Open cup, made of grass stems and blades, lined with blades of finer grasses; placed on the ground or built-up mud
CONSERVATION STATUS	Least concern
COLLECTION NO.	FMNH 20257

ADULT BIRD SIZE
5–6 in (13–15 cm)

INCUBATION
12–13 days

CLUTCH SIZE
3–4 eggs

AMMODRAMUS MARITIMUS

SEASIDE SPARROW

PASSERIFORMES

This is a brackish and salt-marsh habitat specialist bird; it prefers a mix of grassy and open muddy wetland, where nesting and feeding are both suitable and profitable. It lives just within a narrow strip along the continental coastline, and although this species has a long distribution, several of its morphologically unique local subspecies and populations have become threatened, vulnerable, endangered, or even extinct. The presence of a breeding population is often taken as an indicator of the ecological health of the local salt-marsh community.

Storm-related floods of coastal wetlands pose a major threat to the survival of the eggs and chicks. This species is highly adapted to deal with these unpredictable events, and females can lay a new clutch of eggs in a new nest within a week of losing the previous breeding attempt. At other times, while the male continues to provision the nearly fledged young, the female initiates a new nest, and starts laying a second clutch of eggs.

Clutch

The egg of the Seaside Sparrow is blueish white to grayish white in coloration, with shades of dense brown maculation, and is ¾ x ⅝ in (19 x 16 mm) in size. If the female loses her mate during nest building, she will go on and complete the nest and lay a full clutch of eggs.

Actual size

ORDER	Passeriformes
FAMILY	Emberizidae
BREEDING RANGE	Northern and western North America
BREEDING HABITAT	Brushy woodland edges, riparian thickets, chaparral, and scrubland
NEST TYPE AND PLACEMENT	Open cup made of twigs, strips of bark, rotting wood, mosses, and lichens, lined with fine grass, rootlets, and fur and feathers; on the ground or in a low crotch of a bush or tree
CONSERVATION STATUS	Least concern
COLLECTION NO.	FMNH 15238

ADULT BIRD SIZE
6–7½ in (15–19 cm)

INCUBATION
12–14 days

CLUTCH SIZE
2–5 eggs

PASSERELLA ILIACA

FOX SPARROW

PASSERIFORMES

Clutch

Despite the extensive genetic knowledge amassed for this species, the life history traits of the Fox Sparrow, including breeding biology, remain to be explored further. Only the female has been seen gathering nesting materials, and developing a brood patch to incubate the eggs. The female is also in charge of brooding the hatchlings, and most, if not all, of feeding the chicks. She often consumes fecal sacs, rather than dropping them in midflight. The male may join the female in provisioning the young, either later in the nestling stage, or when they are fledglings.

Members of this species vary in size, plumage, and song across its distribution, which spans the entire northern reaches of the continent and most of the west. There are four major genetic lineages that correspond with significant morphologic patterns, such that many researchers feel there should be four species.

Actual size

The egg of the Fox Sparrow is pale blueish green in background color, is covered with cloudy reddish brown maculation, and its size is $^{15}/_{16}$ x ⅔ in (24 x 17 mm). In one rare record, a full nest was built anew in just a single day, before egg laying began.

ORDER	Passeriformes
FAMILY	Emberizidae
BREEDING RANGE	Temperate North America
BREEDING HABITAT	Forest edges, glades and clearings, shrubby, regenerating fields, riparian corridors; also brackish swamps
NEST TYPE AND PLACEMENT	Open cup made of grasses, weeds, and bark strip, and lined tightly with grasses, rootlets, and animal hair; on the ground or in a low dense bush
CONSERVATION STATUS	Least concern
COLLECTION NO.	FMNH 14125

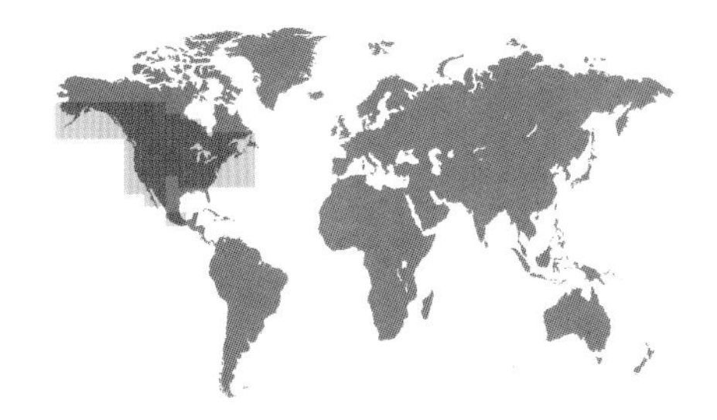

ADULT BIRD SIZE
4½–6½ in (11–17 cm)

INCUBATION
12–15 days

CLUTCH SIZE
3–5 eggs

MELOSPIZA MELODIA

SONG SPARROW

PASSERIFORMES

The Song Sparrow has long served as an exemplar subject of ornithological studies, ever since Margaret Nice published her detailed two-volume treatise of this species' behavioral ecology, based on her own original research culminating in the 1930s. Since then, morphologically and vocally extraordinarily diverse, both migratory (eastern) and stationary (western), as well as insular (Canadian), populations have been observed, marked, followed, and experimented with in detail. The aims of these studies included understanding the role of song learning, song matching, and song-based mate choice in social and genetic patterns of mating, as well as in avoidance of inbreeding.

Some Song Sparrow populations are heavily parasitized by Brown-headed Cowbirds (see page 616). Removing a cowbird egg from a parasitized nest does not necessarily help the sparrows, because of the higher predation rate of nests without cowbird eggs than of those with them. Later, once the cowbird chick hatches, female sparrow chicks are unable to compete, so the sex ratio of the surviving fledglings is heavily male-biased in parasitized broods.

Clutch

The egg of the Song Sparrow is blue, blue-green, or gray-green in background coloration, with brown, reddish brown, or lilac speckling, and is ¾ x ⅝ in (19 x 16 mm) in size. Video footage taken at nests has revealed the occurrence of intraspecific brood parasitism, where a female sparrow visited a nest that was not her own and removed one of the original eggs to lay one of her own in its place.

Actual size

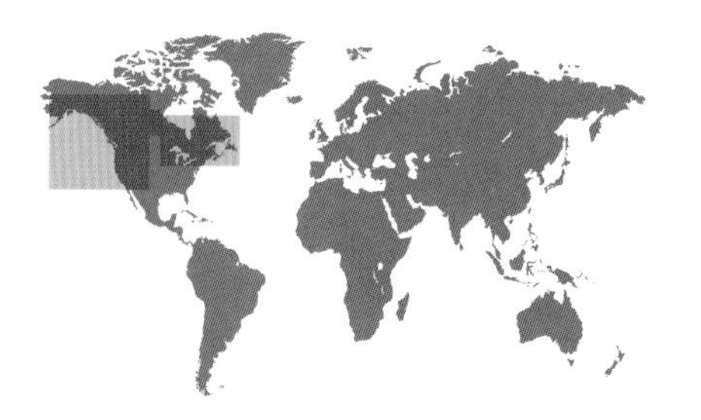

ORDER	Passeriformes
FAMILY	Emberizidae
BREEDING RANGE	Northern and western North America
BREEDING HABITAT	Subalpine bogs, riparian scrub, swampy forest clearings
NEST TYPE AND PLACEMENT	Open, bulky cup of sedges and grasses, with a lining of rootlets, finer grasses, and other soft plant matter; placed at the base of a scrub or tree, with dense foliage cover
CONSERVATION STATUS	Least concern
COLLECTION NO.	FMNH 14057

ADULT BIRD SIZE
5½–6 in (14–15 cm)

INCUBATION
10–13 days

CLUTCH SIZE
3–5 eggs

MELOSPIZA LINCOLNII

LINCOLN'S SPARROW

PASSERIFORMES

Clutch

Named after a friend of John James Audubon, well before the Lincoln presidency, this species breeds in scrubby montane and boreal bogs. Much of its behavioral ecology remains poorly known because of the inaccessibility of its breeding habitat, and because the sexes are similar, and so cannot be told apart unless the birds are banded. Even when the male sings his melodic tune, he often remains hidden from view in dense shrubby foliage.

The female selects the nest site and then she alone constructs it. The female is also in sole charge of incubating the eggs. Incubation begins on the day on which the penultimate egg is laid ; the result is hatching asynchrony, with the last chick emerging one day after the others. Both parents feed the chicks, but when fledging is also asynchronous, and the youngest chick remains in the nest, the parents are unlikely to return, and the chick most often perishes.

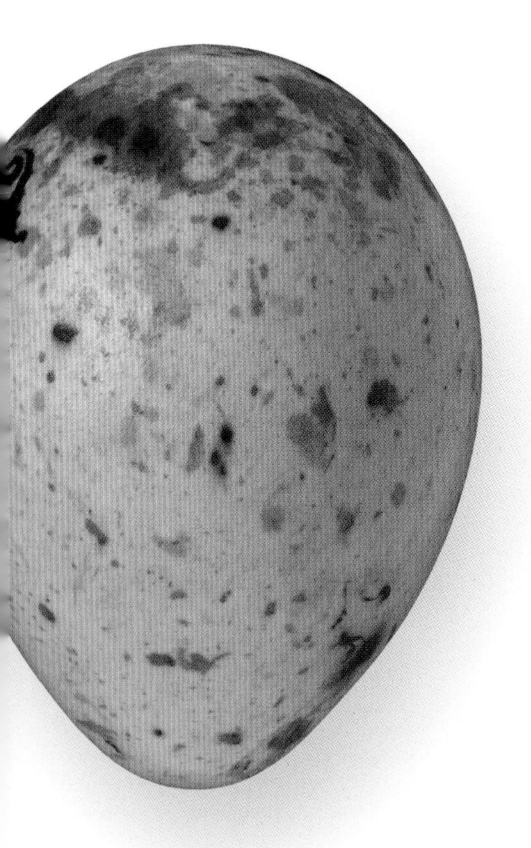

Actual size

The egg of the Lincoln's Sparrow is blue, green, pink, or white, has blotchy brown markings, and measures ¾ x ⅝ in (19 x 16 mm) in size. During egg laying, the female typically stays away from the nest, and does not alarm-call when the nest is approached.

ORDER	Passeriformes
FAMILY	Emberizidae
BREEDING RANGE	Eastern and northern North America
BREEDING HABITAT	Wetlands, marshes with thickets, brackish or freshwater swamps with standing vegetation
NEST TYPE AND PLACEMENT	Open cup of dry grasses, sedge leaves, and stalks, lined with fine grass, plant fibers, rootlets, and hair; placed in dense cattails or shrubs, some times on the ground under a tuft of grass
CONSERVATION STATUS	Least concern
COLLECTION NO.	FMNH 14027

MELOSPIZA GEORGIANA

SWAMP SPARROW

PASSERIFORMES

ADULT BIRD SIZE
5–6 in (13–15 cm)

INCUBATION
12–15 days

CLUTCH SIZE
3–6 eggs

To delineate and announce their territories, male Swamp Sparrows perform a monotonous trill-like song repeatedly throughout the day. The females pay close attention, and choose mates based on the trill rate and the frequency range of the song. Critically, the males' song performances are also influenced by the nutritional environment in which they grew up as nestlings: the brain area responsible for the production of complex and learned vocalizations in songbirds, the HVC, is smaller in males who grew up with limited food supplies. Perhaps, therefore, song also gives the female information about the male's ability to provision his own nestlings.

Swamp Sparrow territories rarely overlap with those of other wetland-dwelling songbirds, likely due to differences in preferred vegetation. For example, Marsh Wrens (see page 471) require stands of tall reeds for nesting, while Song Sparrows prefer the drier, grassy and shrubby upland edges of the wetlands. The end result is less competition for food and nesting sites between these species.

Clutch

The egg of the Swamp Sparrow is blueish green in coloration, with darker, cloudy speckling, and measures ¾ x ⅝ in (19 x 16 mm). Females whose nest was successful in the previous year build the new nest closer to that site than females with previously unsuccessful nests.

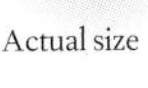

Actual size

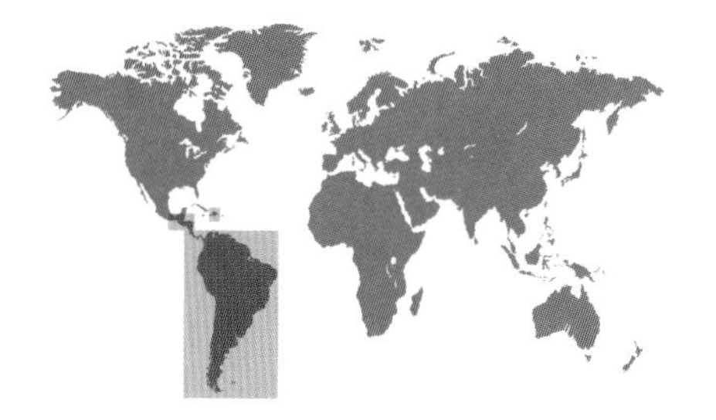

ORDER	Passeriformes
FAMILY	Emberizidae
BREEDING RANGE	Central and South America, Hispaniola
BREEDING HABITAT	Grasslands, open scrub, thickets, forest edges, suburban gardens, and parks
NEST TYPE AND PLACEMENT	Open cup, atop matted vegetation on the ground, low in a bush or in a crevice, lined with fine grasses
CONSERVATION STATUS	Least concern
COLLECTION NO.	FMNH 2498

ADULT BIRD SIZE
5½–6 in (14–15 cm)

INCUBATION
12–14 days

CLUTCH SIZE
2–3 eggs

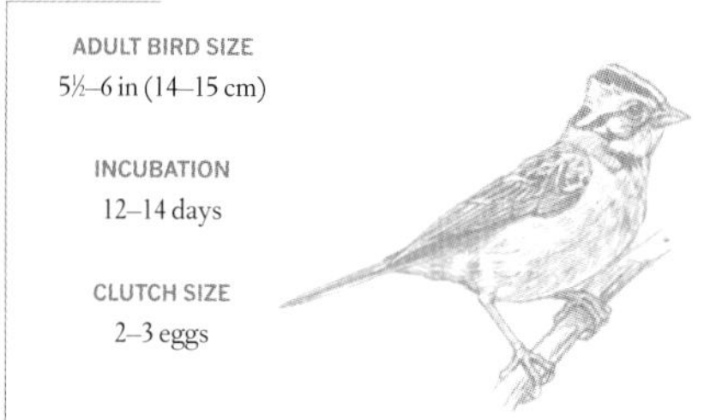

ZONOTRICHIA CAPENSIS

RUFOUS-COLLARED SPARROW

PASSERIFORMES

Clutch

This nonmigratory sparrow has a vast distribution and nearly 30 recognized subspecies; across its range from southern Mexico to Patagonia, it occupies diverse habitats, from open forests to arid grasslands, from lowlands to high in the mountains. To mark their territories and to attract mates, males in more open regions sing shorter and deeper trills, compared to longer and higher calls in shrubby and woodland habitats. These acoustic features ensure that neighboring males and interested females can hear and assess the quality of songs at greater distances in the more open habitats. Because the size of the male's body, and that of his vocal organ, the syrinx, does not correlate with song features, the implication is that males are flexible and learn to produce the song that gets the best results in a particular habitat.

The female alone incubates the eggs, and both parents provision the chicks and the fledglings. This sparrow serves as a common, or the main, host for the generalist brood parasite, the Shiny Cowbird, in much of their overlapping breeding ranges.

Actual size

The egg of the Rufous-collared Sparrow is pale greenish blue in background color, with heavy reddish brown maculation, and measures ¾ x ⅝ in (19 x 16 mm) in size. The development of the female's ovaries and subsequent egg laying are closely associated with the onset of the local rainy season.

ORDER	Passeriformes
FAMILY	Emberizidae
BREEDING RANGE	Northern and eastern North America
BREEDING HABITAT	Clearings and edges in forest stands, scrubby second growths, overgrown pastures
NEST TYPE AND PLACEMENT	Depression in the ground, built up with moss, grass, twigs, and bark, lined with rootlets, finer grasses, and hair; placed under a shrub or grass tussock
CONSERVATION STATUS	Least concern
COLLECTION NO.	FMNH 11091

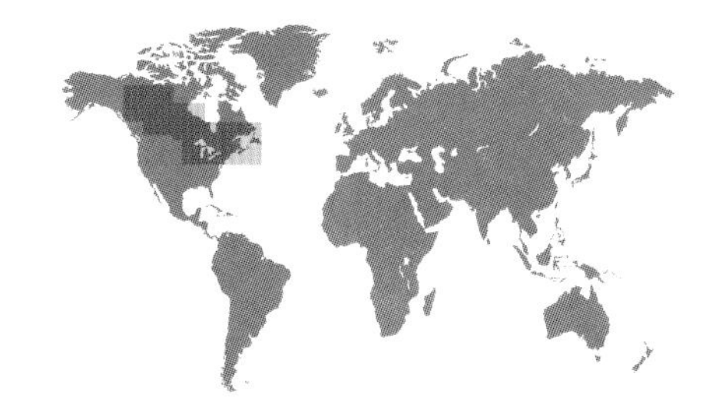

ADULT BIRD SIZE
6½–7 in (16–18 cm)

INCUBATION
11–14 days

CLUTCH SIZE
3–5 eggs

ZONOTRICHIA ALBICOLLIS

WHITE-THROATED SPARROW

PASSERIFORMES

The White-throated Sparrow is a common inhabitant, migrant, or winter resident throughout North America. It comes in two color morphs: white-striped or tan-striped males and females. These morphs represent what appears to be a genetic event from over two million years ago: in white-striped birds, the second pair of their 82 chromosomes is distinctly different from that of the tan-striped birds. These genetic differences, in turn, translate into differences in aggressiveness, breeding, and parental behaviors between the different color morphs, although they can still produce viable offspring.

Up to 95 percent of pairings are between a white-striped morph and a tan-striped morph; this phenomenon is called disassortative mating. Among these pairs, males and females with white stripes both sing and aggressively defend their territories; in contrast, males and females who are tan-striped provide more parental care to the nestlings.

Clutch

Actual size

The egg of the White-throated Sparrow is pale blue, greenish blue, or rusty green in background coloration, with cloudy chestnut and lilac maculation, and measures 13⁄16 x ⅝ in (21 x 15 mm) in size. In a given population, most females lay their first eggs synchronously; nest predation, however, leads to replacement clutches, and decreasing synchrony as the season progresses.

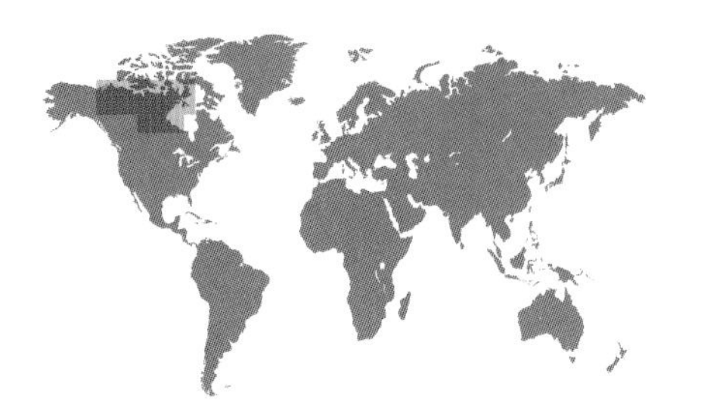

ORDER	Passeriformes
FAMILY	Emberizidae
BREEDING RANGE	North-central Canada
BREEDING HABITAT	Stunted spruce forests, boggy thickets, taiga-tundra edge
NEST TYPE AND PLACEMENT	Open cup, made of mosses, lichens, pine needles, and twigs, lined with dry grasses and caribou hair; placed on the ground, sunken into undergrowth, covered by shrubs and foliage
CONSERVATION STATUS	Least concern
COLLECTION NO.	FMNH 2833

ADULT BIRD SIZE
7–8 in (18–20 cm)

INCUBATION
13–14 days

CLUTCH SIZE
3–5 eggs

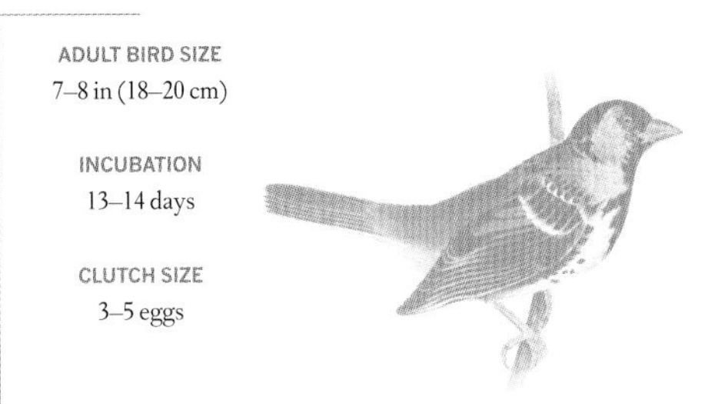

ZONOTRICHIA QUERULA

HARRIS'S SPARROW

PASSERIFORMES

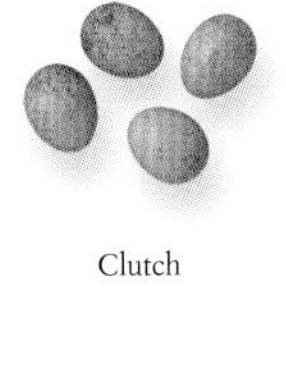

Clutch

This is the only bird species that breeds in Canada and nowhere else. Its large size, and distinctive black crown and throat patch would seem to make this sparrow conspicuous to observers. But its nesting sites, near the coastline of the Arctic Sea, are so far from most researchers' field sites or any other human settlements that little scientific data exist about its breeding habits.

Males migrate to the breeding grounds without having reached full reproductive condition. But their testis size and singing rates increase just days after their arrival, readying them for breeding. Males may gather in loose flocks in the evenings to vocalize together. Pair bonding takes place about a week after the females' arrival, followed closely by nest building, with only females ever having been seen to carry nesting materials. The mother is in full charge of incubation, and both parents feed the chicks. Older males feed from the day of hatching; other males may take up to two days to begin their provisioning trips to the nest.

Actual size

The egg of the Harris's Sparrow is pale green in color, has very fine spotting, and measures ⅞ x ⅔ in (22 x 17 mm). Probably because of its remote breeding grounds, the first nest with eggs was only discovered nearly a century after the species was first described to science.

ORDER	Passeriformes
FAMILY	Emberizidae
BREEDING RANGE	Northern and western North America
BREEDING HABITAT	Forest clearings and glades, alpine meadows and forest edges, shrubby fields, and grassy parks
NEST TYPE AND PLACEMENT	Open cup, made of twigs, coarse grasses, pine needles, moss, and dead leaves, lined with fine grasses and hair; placed on the ground or low in a shrub
CONSERVATION STATUS	Least concern
COLLECTION NO.	FMNH 11042

ZONOTRICHIA LEUCOPHRYS

WHITE-CROWNED SPARROW

PASSERIFORMES

ADULT BIRD SIZE
6–7 in (15–18 cm)

INCUBATION
11–13 days

CLUTCH SIZE
3–7 eggs

This widespread sparrow occurs in resident populations along the Pacific coast, where mated pairs remain affiliated and, typically, within the boundary of their breeding territory throughout the year. But the species is migratory elsewhere, with the males moving ahead of the females in springtime to establish territories in boreal scrub or high-elevation meadows. During migration, these birds reduce the time they sleep by two-thirds, with little apparent impact on cognitive performance. This could provide insights into how to increase work safety for some human professions, for example long-haul truck drivers.

The onset of breeding is marked with site selection and nest construction by the female; in southern parts of the breeding range, nests are typically built in low shrubs, whereas further north and at higher elevations, ground nests are more common. The female starts to feed the chicks soon after hatching, and her feeding trips are augmented by the male's provisioning efforts.

Clutch

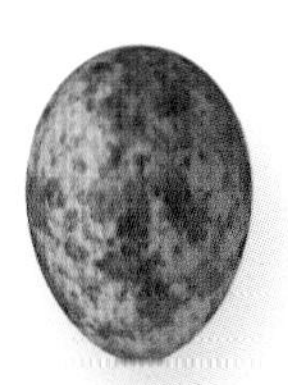

Actual size

The egg of the White-crowned Sparrow is greenish, greenish blue, or bluish in background color, with reddish brown spotting, and is ⅞ x ⅝ in (22 x 16 mm) in dimensions. Females in the northern populations have time to lay just one set of eggs; in resident western populations, they can complete as many as four nesting cycles in one year.

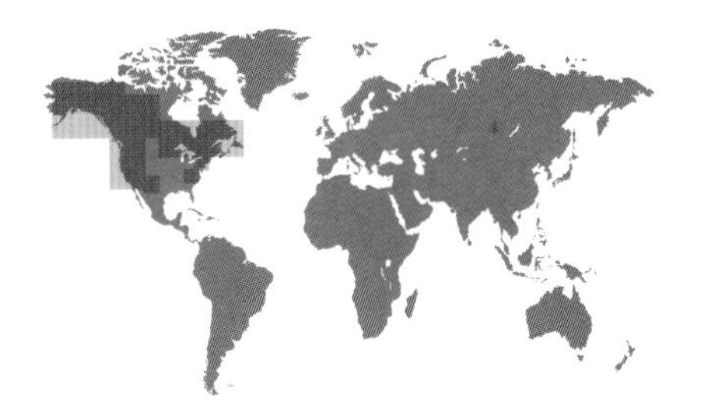

ORDER	Passeriformes
FAMILY	Emberizidae
BREEDING RANGE	Northern and western America, and Appalachian Mountains
BREEDING HABITAT	Coniferous and mixed forests, increasingly also urban parks
NEST TYPE AND PLACEMENT	Varies with placement; ground depression lined with fine grasses and hair, or open cup, low in a shrub or tree, on a platform of twigs and pine needles
CONSERVATION STATUS	Least concern
COLLECTION NO.	FMNH 13986

ADULT BIRD SIZE
6–6½ in (15–17 cm)

INCUBATION
12–13 days

CLUTCH SIZE
3–5 eggs

JUNCO HYEMALIS

DARK-EYED JUNCO

PASSERIFORMES

Clutch

The Dark-eyed Junco is a ubiquitous migrant, breeder, or winter resident throughout North America. Its formerly strictly mature-forest-dwelling populations have expanded into suburban developments, backyards, and city parks.

One particular contact with humans illustrates just how fast evolution can occur in some novel environments. About 30 years ago, a resident population became established on the University of California San Diego campus. In this location, the birds remain on their territories with their mates year-round, and instead of breeding during the warm spring and summer weeks, as in the mountains, they extend their nesting season to up to nine months of the year. Neighboring males come to know each other well, and so there is not much need for posturing and fighting with long-time neighbors. Typically, juncos use their white-spotted tail feathers in aggressive displays; in San Diego, where they do not need them, the birds have evolved darker tails in a handful of decades.

Actual size

The egg of the Dark-eyed Junco is grayish or pale bluish in coloration, with sparse spots overall and a ring-line blotching near the blunt end, and is ¾ x ⅝ in (19 x 16 mm) in dimensions. Where these birds are year-round residents, the same nest sites are often reused within and between years.

ORDER	Passeriformes
FAMILY	Emberizidae
BREEDING RANGE	Europe and Northwestern Asia
BREEDING HABITAT	Open scrub, shrubby grasslands, farms and pastures
NEST TYPE AND PLACEMENT	In hedges, scrub, and on the ground in sloping ditches; open cup, made of twigs and sticks, lined with woven grasses and rootlets
CONSERVATION STATUS	Least concern
COLLECTION NO.	FMNH 20676

ADULT BIRD SIZE
6–6½ in (15–17 cm)

INCUBATION
12–14 days

CLUTCH SIZE
3–6 eggs

EMBERIZA CITRINELLA

YELLOWHAMMER

PASSERIFORMES

Yellowhammers are colorful and melodious broadcasters of springtime, but recent changes in agricultural practices have led to consistent declines in many populations. For example, in Britain shrinking population sizes have been linked to losses in hedges and other boundary-type shrubby zones around large cultivated fields. These trends may be countered, however, by the growing number of organic farms, which are smaller in size and often surrounded by shrubby boundaries.

Surprisingly for a small insectivorous and open-cup nesting passerine, Yellowhammers are rarely parasitized by Common Cuckoos, despite their overlapping ranges. One explanation is that this potential host species is a very acute egg discriminator. Blue, non-mimetic eggs experimentally introduced to nests were nearly all rejected, and even a third of mimetic eggs, taken from other Yellowhammer nests, were rejected by the rightful nest owners. This implies that Yellowhammers might once have been cuckoo-hosts, but any parasite specializing on this species is now probably extinct.

Clutch

Actual size

The egg of the Yellowhammer is cream in background color, with ash-colored, scrawl-like maculation, and its measurements are ⅞ x ⅔ in (22 x 17 mm). Predation is the most common source of egg and nestling loss; over 60 percent of the nests fail to produce any fledglings.

ORDER	Passeriformes
FAMILY	Emberizidae
BREEDING RANGE	North Africa, and from southern Europe to central Asia
BREEDING HABITAT	Arid, montane, grassy and rocky, steep slopes and fields
NEST TYPE AND PLACEMENT	On the ground, in a rock crevice, or hole in a wall, also in a low bush; open cup made of grass, stalks, and roots, lined with finer grasses, rootlets, and some hair
CONSERVATION STATUS	Least concern
COLLECTION NO.	FMNH 20645

ADULT BIRD SIZE
6¼–6½ in (16–17 cm)

INCUBATION
10–13 days

CLUTCH SIZE
3–5 eggs

EMBERIZA CIA

ROCK BUNTING

PASSERIFORMES

Clutch

The Rock Bunting breeds mostly in high-elevation alpine habitats, where vegetation type, height, and coverage varies widely between different regions. There has been an ongoing decline in the population sizes of this species throughout Europe in recent decades, and scientists are attempting to uncover what, if any, modification of existing habitats may be needed to reverse these population losses. Such studies require a detailed understanding of all aspects of the breeding biology of the birds, in both the nestling and the fledgling phases.

For example, while many birds prefer to nest where there is cover, Rock Buntings prefer to breed on steeper slopes, with more rocky outcroppings, and less shrub cover. As a species which predominantly nests on the ground, it is thought that open, barren patches increase visibility to allow incubating buntings to detect predators better from a distance.

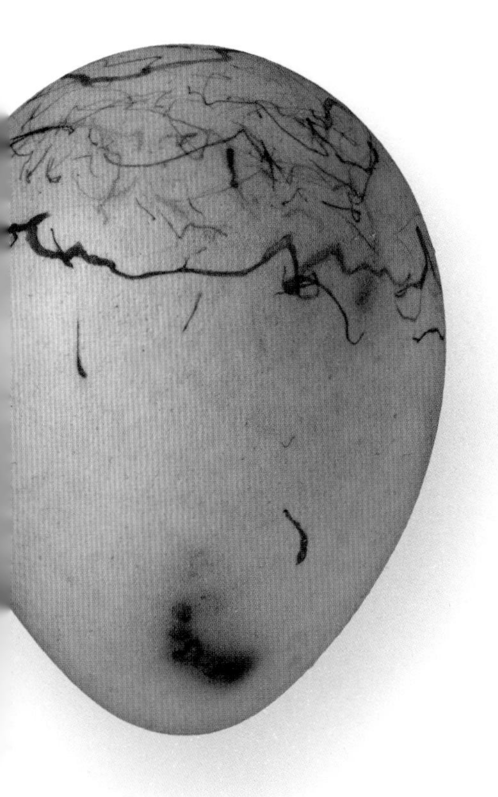

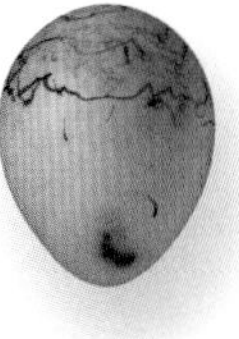

Actual size

The egg of the Rock Bunting is pale blueish white or rufous-buff in background color, with blackish brown and lavender lines and blotches, and is ¾ x 9⁄16 in (19 x 14 mm) in size. The female incubates the eggs alone, and in most regions there is time to complete two full nesting cycles.

ORDER	Passeriformes
FAMILY	Cardinalidae
BREEDING RANGE	Eastern and central North America
BREEDING HABITAT	Large tracts of mature deciduous and mixed forests
NEST TYPE AND PLACEMENT	Loosely woven, shallow pan of twigs, grasses, stalks, strips of bark, rootlets, and pine needles, lined with grasses, rootlets, vine tendrils, and plant fibers; placed high, near the terminus of a horizontal branch
CONSERVATION STATUS	Least concern
COLLECTION NO.	FMNH 13744

ADULT BIRD SIZE
6¼–6½ in (16–17 cm)

INCUBATION
12–14 days

CLUTCH SIZE
3–5 eggs

PIRANGA OLIVACEA

SCARLET TANAGER

PASSERIFORMES

Formerly associated with the neotropical tanager family, this flaming red and coal-black bird is now classified as a relative of cardinals and their allies. A long-distance migrant between South and North America, the males arrive before the females, and establish their breeding territories through song and flight. The females quickly select their mates, and nest building can be well on its way just one week after the females return.

The male occasionally follows the female while she gathers nesting materials, but she alone constructs the nest and incubates the eggs. Some males feed their mate on the nest, but others do not and the female will leave the nest periodically to feed herself. After the eggs hatch, the male begins food delivery to the nest, often while the female is still brooding the hatchlings to keep them warm before their feathers grow in and their own thermoregulatory systems become fully functional.

Clutch

Actual size

The egg of the Scarlet Tanager is greenish blue to light blue in appearance, with chestnut, purplish red, or lilac patterning, and is ⅞ x ⅔ in (22 x 17 mm) in size. Wherever forests are heavily fragmented, egg predation and nest parasitism increase, reducing reproductive success and sometimes leading to local extinction.

ORDER	Passeriformes
FAMILY	Cardinalidae
BREEDING RANGE	Southern temperate North America
BREEDING HABITAT	Open deciduous or mixed forests, near gaps and edges
NEST TYPE AND PLACEMENT	Open cup, made of dry grasses and stems; placed high in a tree, in a cluster of leaves or woven into a forking branch
CONSERVATION STATUS	Least concern
COLLECTION NO.	FMNH 1461

ADULT BIRD SIZE
6½–8 in (17–20 cm)

INCUBATION
11–13 days

CLUTCH SIZE
3–4 eggs

PIRANGA RUBRA

SUMMER TANAGER

PASSERIFORMES

Clutch

This is another conspicuously colored species of the spring forest; its red plumage is more extensive, although paler, than that of the Scarlet Tanager (see previous page). Across years, the otherwise yellowish female tanagers may grow increasingly more reddish feathers, perhaps indicative of an endocrine overproduction of steroids with advancing age.

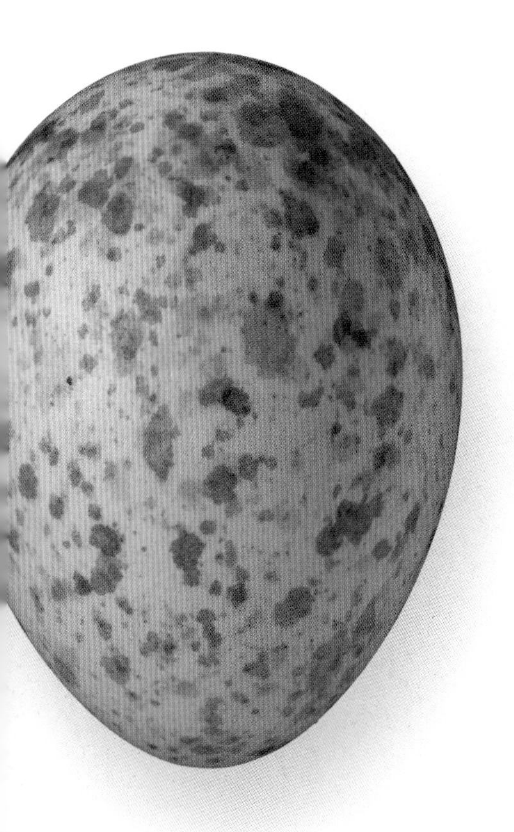

Settling in pairs soon after their return flight from South American wintering grounds, the female completes the nest, and she alone incubates the eggs. Her mate may feed her on the nest, calling her off the eggs and feeding her nearby, or he might forgo incubation-feeding entirely. Those males who do not feed the female during incubation still appear to be able to judge the timing of hatching, and bring food to the nest on that day. Afterwards, most males contribute equitably with the female to provision the chicks, either feeding the chicks alone, or transferring food to the beak of the female, who then feeds it to the young.

Actual size

The egg of the Summer Tanager is pale blue or pale green in background coloration, has reddish speckling, and its size is $^{15}/_{16}$ x ⅔ in (24 x 17 mm). To keep the nest sanitary and safe, the eggshells are removed after hatching, and fecal sacs produced by the chicks are swallowed or carried away from the nest and dropped.

ORDER	Passeriformes
FAMILY	Cardinalidae
BREEDING RANGE	Eastern, central, and southern North America
BREEDING HABITAT	Open woodlands, shrubby fields, gardens, city parks
NEST TYPE AND PLACEMENT	Open-cup nest, made of coarse twigs, dried leaves, and grapevine bark, lined with grasses, stems, rootlets, and pine needles; wedged into a fork of branches, low in a sapling, shrub, or vine tangle, hidden in dense foliage
CONSERVATION STATUS	Least concern
COLLECTION NO.	FMNH 14472

ADULT BIRD SIZE
8½–9½ in (21–24 cm)

INCUBATION
11–13 days

CLUTCH SIZE
2–4 eggs

CARDINALIS CARDINALIS

NORTHERN CARDINAL

PASSERIFORMES

This familiar and colorful bird is one of the brightest songsters in North American backyards and city parks. Unlike most northern hemisphere temperate zone birds, both sexes of this cardinal sing during springtime. Mutual singing may allow females and males to maintain individually recognizable contact with each other even in dense foliage. To the human ear, the songs of the different sexes sound similar, but the birds know the difference clearly; male cardinals respond to foreign male's songs by attacking, whereas females respond to foreign female's songs by counter-singing. Female cardinals learn their song nearly three times as fast as do males, but males produce a more stable and consistent song once they have mastered it.

To begin work on the nest, the female uses her powerful beak to chop and bend twigs and sticks to form its base. The male may provide some of the nesting materials, but she alone constructs the nest and incubates the eggs. The male feeds the incubating female at or near the nest, and continues to do so while she also broods the hatchlings. Within a couple of days, however, both parents begin to take turns to provision the young.

Clutch

The egg of the Northern Cardinal is grayish white, buffy white, or greenish white in color, with variable pale gray to brown maculation, and is 1 x ¾ in (25 x 19 mm) in dimensions. Due to the mild climate throughout the cardinal's breeding grounds, up to four cycles, from eggs to fledglings, may be completed in each year.

Actual size

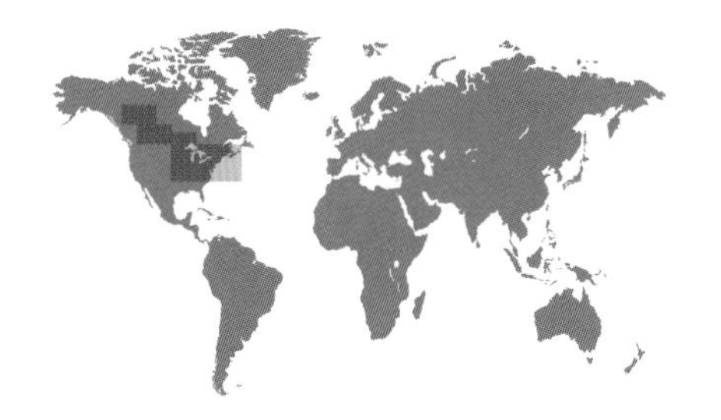

ORDER	Passeriformes
FAMILY	Cardinalidae
BREEDING RANGE	North and east temperate North America
BREEDING HABITAT	Deciduous woodlands, forest edges, secondary growths, thickets, parks, and gardens
NEST TYPE AND PLACEMENT	Loose, open cup, made of sticks, twigs, grasses, and stems, lined with fine twigs, rootlets, and hair; placed in a vertical fork or crotch of a sapling or tree, low to medium height
CONSERVATION STATUS	Least concern
COLLECTION NO.	FMNH 13909

ADULT BIRD SIZE
7–7½ in (18–19 cm)

INCUBATION
12–14 days

CLUTCH SIZE
3–5 eggs

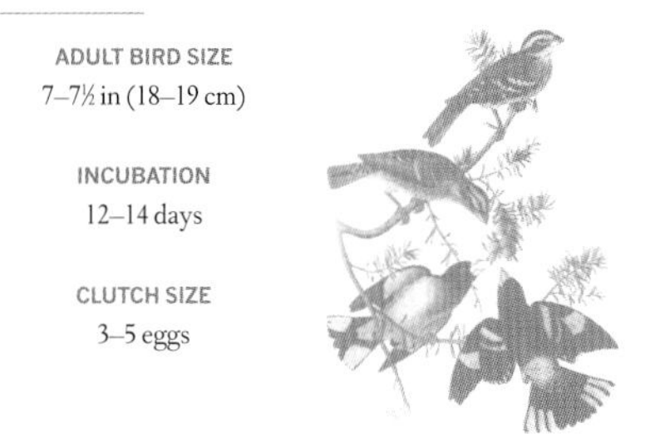

PHEUCTICUS LUDOVICIANUS

ROSE-BREASTED GROSBEAK

PASSERIFORMES

Clutch

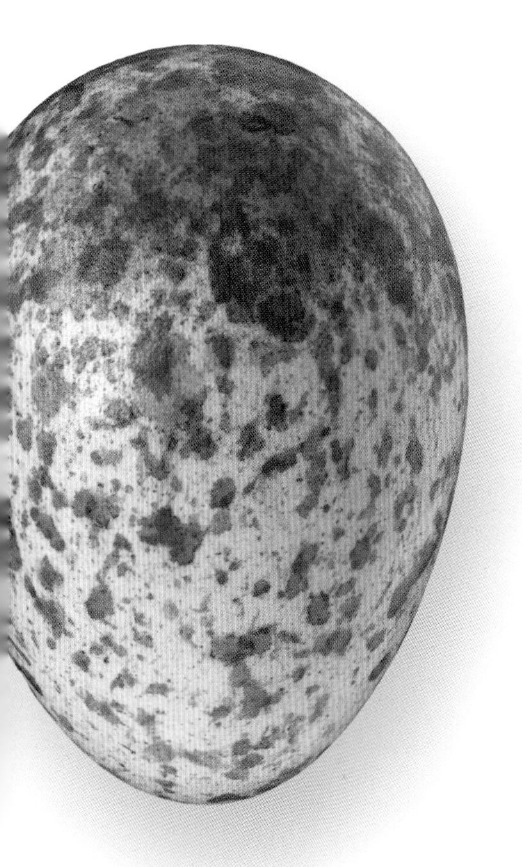

The Rose-breasted Grosbeak is a long-distance migrant that feeds on fruit, seeds, and flower buds in the winter, but switches to insects during the breeding season. Unlike most migratory passerines, this species is physiologically ready for mating while it is still en route from the wintering ground; male testes are fully functioning and produce sperm, even in birds passing through migratory stopover sites. The pair bond may already be established by the time these birds reach the breeding range, and nesting begins swiftly.

To select a suitable nest site, the pair stays together as both the female and the male inspect crotches and forks by pressing their chests against it as if measuring the space. Both mates also collect nesting materials and construct the nest together. The female develops a full, and the male a partial, brood patch, and both incubate the eggs. The nestlings are also cared for and provisioned by both parents.

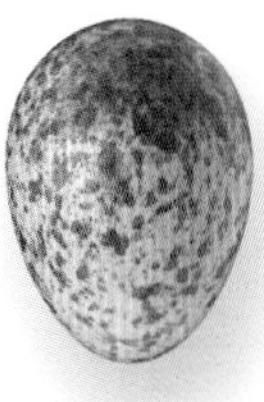

Actual size

The egg of the Rose-breasted Grosbeak is pale green to blue, with reddish brown or purplish speckling, and is 1 x ⅔ in (25 x 17 mm) in size. In the western distribution of this species, it hybridizes with the Black-headed Grosbeak, but hybrid females lay smaller clutches, and with less viable eggs, than non-hybrid females.

ORDER	Passeriformes
FAMILY	Cardinalidae
BREEDING RANGE	Southern North America, Central America
BREEDING HABITAT	Forest edges, fields with scattered trees, riparian corridors, thickets and hedgerows
NEST TYPE AND PLACEMENT	Open cup, made of twigs, bark, rootlets, grasses, and strips of plant matter; placed low in a shrub or small tree
CONSERVATION STATUS	Least concern
COLLECTION NO.	FMNH 13901

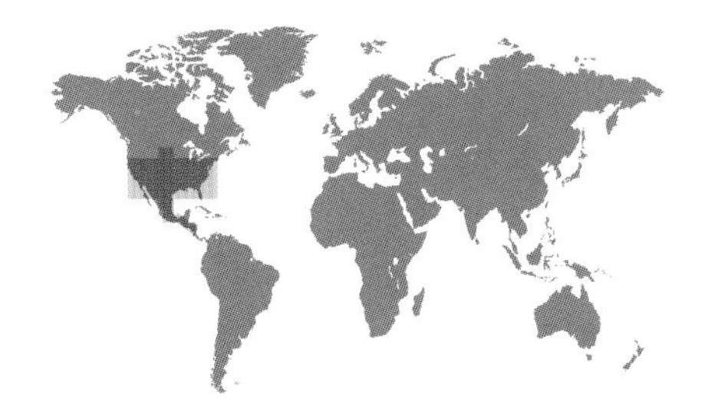

ADULT BIRD SIZE
6–6½ in (15–16 cm)

INCUBATION
11–12 days

CLUTCH SIZE
3–5 eggs

PASSERINA CAERULEA

BLUE GROSBEAK

PASSERIFORMES

The bright blue hue of the Blue Grosbeak's plumage is caused by the physical and optical properties of the feathers' microstructure, known as structural coloration. The visual system of other males and the females can fully appreciate the subtle variations of the grosbeak's deep blue plumage, although the color receptors of the human eye cannot. This is because the wavelengths at which these feathers reflect the most amount of light are in the violet and ultraviolet regions, which the birds, but not humans, can perceive.

Accordingly, a female grosbeak can instantaneously assess the quality of a potential mate by looking at his plumage. Specifically, the brightness of the male's ultraviolet and blue plumage is a true indicator of his qualities as a large and well-nourished individual, as a holder of a prey-rich and predator-safe territory, and as a father able to provision nestlings at high rates of food delivery.

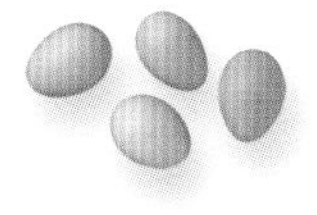

Clutch

Actual size

The egg of the Blue Grosbeak is pale blue or white in coloration, has no maculation, and measures ⅞ x ⅔ in (22 x 17 mm) in dimensions. The female alone builds the nest and incubates the clutch, with the male feeding her while she sits on the eggs; both parents provision the nestlings.

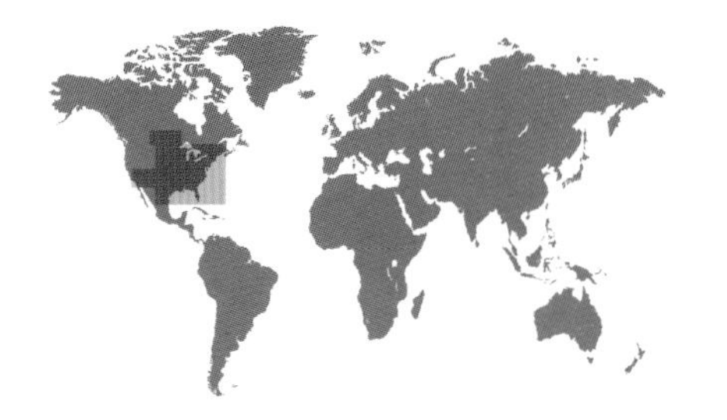

ORDER	Passeriformes
FAMILY	Cardinalidae
BREEDING RANGE	Eastern and southern North America
BREEDING HABITAT	Open thickets, forest edges, shrubby grasslands
NEST TYPE AND PLACEMENT	Open cup of woven leaves, grasses, stems, bark strips, and spider web, lined with thin grasses, rootlets, thistledown, and deer hair; placed in a crotch or fork in a low shrub or tree
CONSERVATION STATUS	Least concern
COLLECTION NO.	FMNH 701

ADULT BIRD SIZE
4½–5 in (12–13 cm)

INCUBATION
11–14 days

CLUTCH SIZE
3–4 eggs

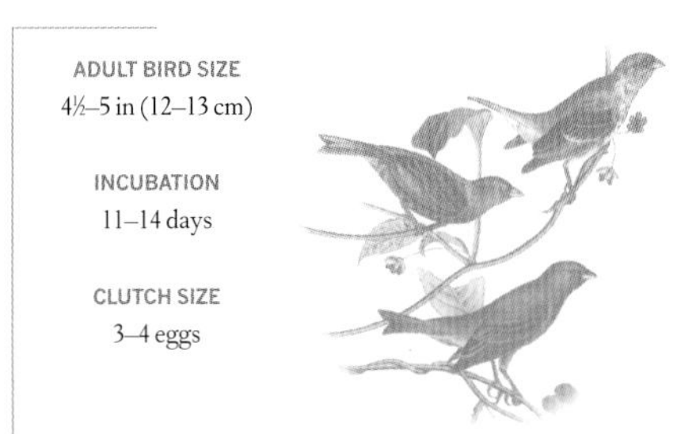

PASSERINA CYANEA

INDIGO BUNTING

PASSERIFORMES

Clutch

The male Indigo Bunting's bright blue spring color is an indication of advanced age; first-time breeding males often carry some brown-gray feathers, a phenomenon known as delayed plumage maturation. Apparently, young birds use plumage color to assess the suitability of tutors for song learning. Older males' songs are more likely to be learned by young neighbors; the songs of bluer young males' songs also are more often copied than those of browner young males.

The Indigo Bunting is heavily parasitized by Brown-headed Cowbirds (see page 616) in the edge habitats where both species prefer to breed. In the bunting's small nest cup, the fast-growing cowbird chick can marginalize, and even accidentally evict, some of the host young. Also, the cowbird's intensive begging attracts not only preferential provisioning from the foster parents over their own genetic young, but also risks the parasite's own survival chances by attracting predators cued in by its loud begging calls.

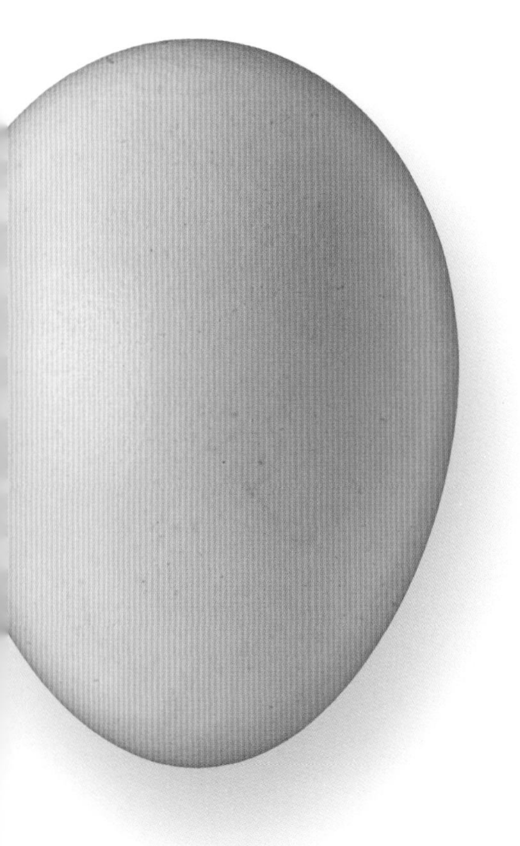

Actual size

The egg of the Indigo Bunting is mostly white, or very light blue, in background coloration, clear of markings, and $\frac{3}{4}$ x $\frac{9}{16}$ in (19 x 14 mm) in size. When Brown-headed Cowbirds parasitize a nest, those bunting chicks that survive to fledge are demonstrably weaker; they are less likely to return to breed in the following spring than are chicks from nonparasitized nests.

ORDER	Passeriformes
FAMILY	Cardinalidae
BREEDING RANGE	Southeastern and central North America
BREEDING HABITAT	Open grasslands, pastures, prairies
NEST TYPE AND PLACEMENT	Bulky cup, made from weeds and grasses, lined with finer grasses, rootlets, and hair; placed slightly above the ground in a dense patch of grasses or low in a sapling
CONSERVATION STATUS	Least concern
COLLECTION NO.	FMNH 14530

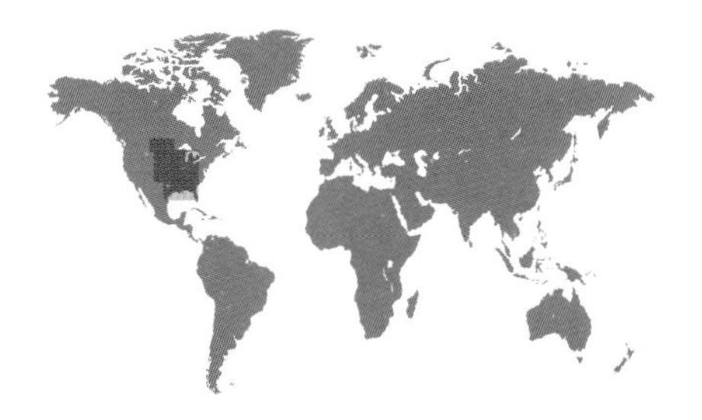

ADULT BIRD SIZE
5½–6½ in (14–16 cm)

INCUBATION
12–13 days

CLUTCH SIZE
3–5 eggs

SPIZA AMERICANA

DICKCISSEL

PASSERIFORMES

This broadly ranging specialist of grasslands and prairies has characteristics which make it stand apart from most other seed-eating songbirds. Dickcissels may engage in long-distance movements, both within and between seasons, to take advantage of patchy food sources. Even when settled on the breeding grounds, individual females choose mates based on the availability of food resources and suitable nest sites within a mate's territory. Males with more territorial resources get more mates.

As part of a female's settlement in the male's territory, he accompanies her to inspect feasible nesting sites. Nest building begins in the following few days, and the female goes on to take charge of the entire breeding cycle. The male does not assist in parental care, and instead continues to sing and display to defend his territory, and his harem of mates.

Clutch

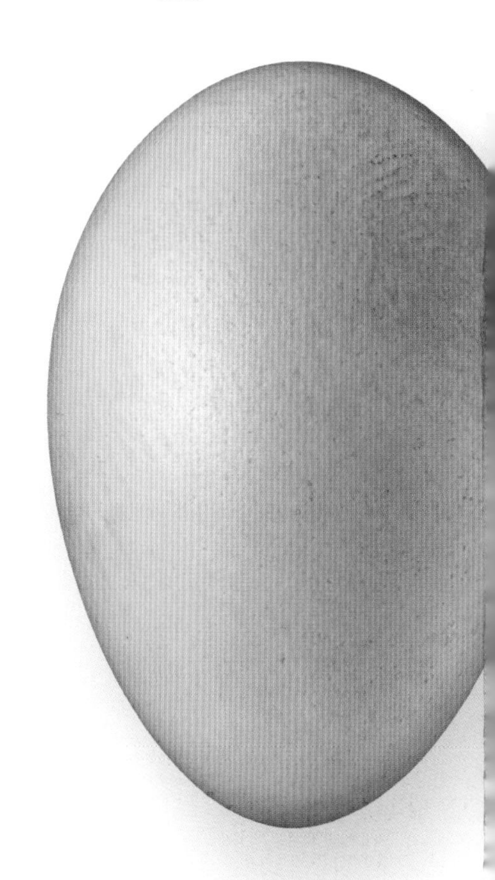

Actual size

The egg of the Dickcissel is light blue, immaculate, and 13⁄16 x 5⁄8 in (21 x 16 mm) in size. Typical nests are in dense grass patches, with overhead foliage for shading; to further protect the eggs from the hot sun, the female shades the eggs and the nestlings with her body.

ORDER	Passeriformes
FAMILY	Icteridae
BREEDING RANGE	Temperate North America
BREEDING HABITAT	Tall- or mixed-grass prairie, overgrown pastures, abandoned fields
NEST TYPE AND PLACEMENT	Thick walled cup of loose, dry grasses, with occasional overhang, lined with finer grass and sedges; placed on the ground, in sites with moist soil or poor drainage
CONSERVATION STATUS	Least concern
COLLECTION NO.	FMNH 9863

ADULT BIRD SIZE
6–8½ in (15–21 cm)

INCUBATION
11–13 days

CLUTCH SIZE
3–7 eggs

DOLICHONYX ORYZIVORUS

BOBOLINK

PASSERIFORMES

Clutch

The egg of the Bobolink is blueish gray or pale reddish brown in background coloration, with irregular darker maculation, and is ⅞ x ⅝ in (22 x 16 mm) in size. Older females arrive at the nesting grounds earlier, and begin nest construction and egg laying several days before the younger inexperienced females.

The Bobolink completes a nearly 12,500-mile (20,000-km) round trip each year to reach its South American wintering grounds. On the breeding grounds, males look like blond rockstars dressed for a black-tie event with their jacket on backwards. To mark their territory and attract females, males display their distinct plumages while performing low, helicopter-like flights over their territory. Many Bobolink males attract two or more nesting females.

Typically, the first arriving female is courted and followed intensely until she accepts the male and begins to copulate with him, initiating the nesting cycle. At that point, he resumes his mate-attraction displays and can successfully attract another female within a week. The female alone builds the nest and incubates the eggs, but the male starts to visit the nest late in the incubation period. He provisions the young differently at each nest, depending on the feeding rate of the mother and the hunger levels of the chicks.

Actual size

ORDER	Passeriformes
FAMILY	Icteridae
BREEDING RANGE	North America, Central America, western Caribbean
BREEDING HABITAT	Wetlands, marshes, thickets, fallow fields, roadside ditches, parklands
NEST TYPE AND PLACEMENT	Open cup, placed low among reeds and other marsh vegetation, but also in shrubs; made of leaves, wood strips, and mud, lined with fine, dry grasses
CONSERVATION STATUS	Least concern
COLLECTION NO.	FMNH 10228

ADULT BIRD SIZE
6½–7 in (17–18 cm)

INCUBATION
11–13 days

CLUTCH SIZE
2–4 eggs

AGELAIUS PHOENICEUS

RED-WINGED BLACKBIRD

PASSERIFORMES

The Red-winged Blackbird is likely the most numerous passerine bird in all of North America. The male's red and yellow shoulder patch, contrasting with his black plumage, is a characteristic sight along highways, rural roads, farms, salt marshes, city parks, and river walks; his call of "conk-a-reeee" is a typical sound of spring. In the fall, million-strong flocks of blackbirds, and other icterids, move, feed, and roost together.

To impress the female, the male blackbird sings and bows to her; he is able to project his voice directly at the female watching this courtship display. This blackbird is a territorial, colonial breeder, with males defending territories containing suitable nesting sites. The safest nesting sites tend to be in marshes where some males may have multiple females nesting on their territories. The females incubate the eggs alone, while the male patrols the territory, alarm-calling, and mobbing predators (including passing humans) to defend the nests.

Clutch

The egg of the Red-winged Blackbird is pale blueish green to gray in background coloration, maculated with brown and black blotches, and is 1 x ⅔ in (25 x 17 mm) in measurements. Living in large marsh-nesting colonies enables these blackbirds to mob and defend their clutch against brood parasitic Brown-headed Cowbirds (see page 616).

Actual size

ORDER	Passeriformes
FAMILY	Icteridae
BREEDING RANGE	Central California
BREEDING HABITAT	Wetlands and marshes, near large tracks of grasslands, pastures, and agricultural fields
NEST TYPE AND PLACEMENT	Deep cup, made from grass, leaves, and mud, lined with finer grass and stems; placed above water or ground on standing swamp vegetation or in thickets
CONSERVATION STATUS	Endangered
COLLECTION NO.	FMNH 10247

ADULT BIRD SIZE
7–9½ in (18–24 cm)

INCUBATION
11–13 days

CLUTCH SIZE
3–4 eggs

AGELAIUS TRICOLOR

TRICOLORED BLACKBIRD

PASSERIFORMES

Clutch

The egg of the Tricolored Blackbird is pale blue-green in background coloration, has fine darker lines and patterns of blotches, and measures $\frac{15}{16}$ x $\frac{2}{3}$ in (24 x 17 mm) in dimensions. Laying is highly synchronized within the nesting colony; the first egg of the clutch is often laid in tens of thousands of nests within the same week.

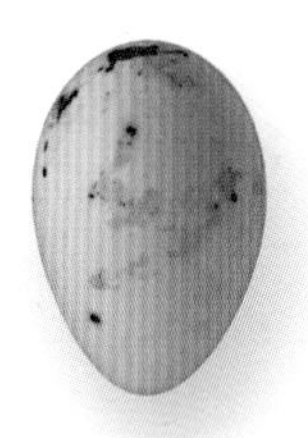

Actual size

This species only occurs in the Central Valley of California and nearby sites. To initiate breeding, the males settle in a large tract of swamp habitat, and display to defend a spot with suitable nest sites. The females may arrive simultaneously or within days of each other, and initiate nesting. Nest construction and incubation are carried out by the female, while the males actively stay away from the colony, making it appear quiet or abandoned. Once the eggs hatch, the males return, and both parents feed the chicks, provisioning the young up to three weeks after fledging.

Tricolored Blackbirds used to be one of the most colonial land birds in all of North America, with breeding aggregations of hundreds of thousands. Flocks were so large that they blocked out the sun when in flight. However, numbers declined steeply in the late twentieth century and are now causing grave concern. The population loss of these birds comes in the face of more breeding colonies in recent years, but these are much smaller and, due to fewer insects, less productive, than the historically vast colonies of the species.

ORDER	Passeriformes
FAMILY	Icteridae
BREEDING RANGE	Eastern and central North America, Central America, and northern South America
BREEDING HABITAT	Open grasslands and flood plains, sand dunes, prairies
NEST TYPE AND PLACEMENT	Cup nest, made from woven grasses, plant stems, and strips of bark; placed in small depression or hoof print, well concealed by dense vegetation
CONSERVATION STATUS	Least concern
COLLECTION NO.	FMNH 10273

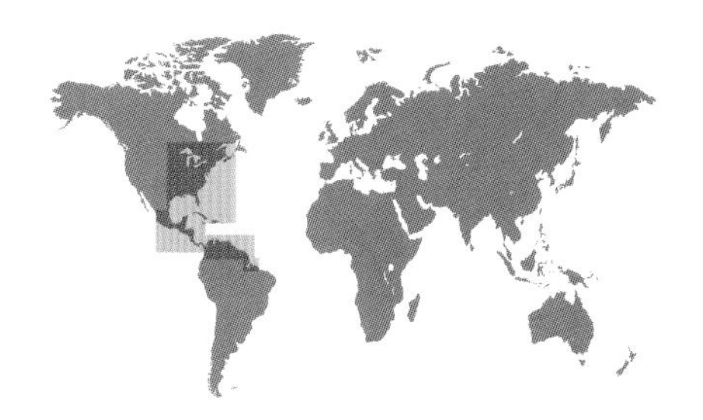

ADULT BIRD SIZE
7½–10 in (19–26 cm)

INCUBATION
13–16 days

CLUTCH SIZE
2–6 eggs

STURNELLA MAGNA

EASTERN MEADOWLARK

PASSERIFORMES

Male Eastern Meadowlarks settle on their all-purpose territory up to a month before the females; they sing to signal their territories, chasing out intruding males on the wing. Most males mate polygynously with two or even three females; the females initiate nest building in the same male's territory. The females undertake most of the nesting and parental chores, with the male only contributing sporadically to feed the growing nestlings.

The meadowlark is a close relative of blackbirds, grackles, and orioles. Although most songs are delivered while the bird is perching or on the ground, meadowlarks also have a flight song, but neither its duration, nor the height at which it is delivered, compares to the sky-high flight and songs of the true alaulid larks. These birds also produce a typical icterid chatter call, given by both sexes, often around the time of copulation.

Clutch

Actual size

The egg of the Eastern Meadowlark is white in coloration, with variable dark spotting across the shell, and measures 1⅛ x ¾ in (28 x 19 mm) in size. When nests are parasitized by Brown-headed Cowbirds (see page 616), about a third of females will eject the foreign egg from the nest.

ORDER	Passeriformes
FAMILY	Icteridae
BREEDING RANGE	Central and western North America
BREEDING HABITAT	Prairie, dry grasslands, pastures, roadsides, and abandoned fields
NEST TYPE AND PLACEMENT	Cup-shape depression in the ground; lined with soft, dry grasses and stems
CONSERVATION STATUS	Least concern
COLLECTION NO.	FMNH 10302

ADULT BIRD SIZE
7½–8½ in (19–22 cm)

INCUBATION
13–15 days

CLUTCH SIZE
5–6 eggs

STURNELLA NEGLECTA

WESTERN MEADOWLARK

PASSERIFORMES

Clutch

The egg of the Western Meadowlark is white in background coloration, with brown, rusty, and lavender blotches and spots, and is 1⅛ x 13⁄16 in (28 x 21 mm) in dimensions. When threatened, females abandon nests before egg laying begins, and start construction at a different site.

Pairs of the Western Meadowlark form quickly after returning from the wintering ground. The female selects a suitable nesting site, collects materials, and constructs the nest. She alone incubates the eggs; the male joins her once the chicks hatch, and both parents deliver an arthropod-based diet to the young. To keep the nest clean and free of cues for predators, the parents pick up fecal sacs, and fly long distances before they drop them in an open field.

The Latin name of this species, *neglecta*, reflects John James Audubon's opinion that it was overlooked because of its similarity to the Eastern Meadowlark. Noticeably distinct in their songs, even where Western Meadowlarks and Eastern Meadowlarks co-occur in the same habitat, mixed-species pairs are rare. A 12-year captive breeding study showed that when hybrids are sired, the resulting adult birds produce mostly unviable eggs, indicating that there is a genetic barrier to mixing between these two species.

Actual size

ORDER	Passeriformes
FAMILY	Icteridae
BREEDING RANGE	Central and western North America
BREEDING HABITAT	Cattail marshes, wetlands or lake shores with tall standing vegetation in prairie region
NEST TYPE AND PLACEMENT	Woven open cup, attached to reed stems, above standing water; made from long wet leaves, stems, and grasses, lined with fine, dry aquatic plants
CONSERVATION STATUS	Least concern
COLLECTION NO.	FMNH 659

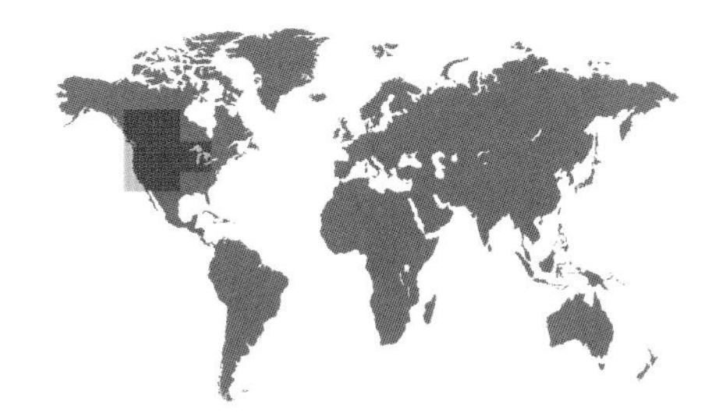

ADULT BIRD SIZE
8½–10½ in (22–27 cm)

INCUBATION
10–13 days

CLUTCH SIZE
3–5 eggs

XANTHOCEPHALUS XANTHOCEPHALUS

YELLOW-HEADED BLACKBIRD

PASSERIFORMES

The Yellow-headed Blackbird forms large and dense colonies in its breeding marshes, often mixed with Red-winged Blackbirds (see page 607). The male Yellow-headed Blackbird is territorial and displays his bright head plumes while making repeated songs as he perches prominently on emergent marsh vegetation. Several females will typically settle in the territory of the same male, establishing a polygynous, harem-based mating system, where all nest-building and incubation duties are conducted by the mother.

Female blackbirds appear to prefer territories with other active nests, so that males whose females lose broods to predators are less successful at recruiting additional females. Unlike most other harem-breeding species, whose males do not provision the chicks, male Yellow-headed Blackbirds do provide parental provisions to some of the nests. When a male disappears, a new male will take over the territory, but he will provision only his own young in later-built nests, ignoring unrelated offspring in nests fathered by the previous male.

Clutch

The egg of the Yellow-headed Blackbird is grayish to greenish white in coloration, with brown, rufous, and pearl gray speckling, and is 1 1⁄16 x ⅔ in (27 x 18 mm) in size. About half of all nesting attempts fail; the most important cause is the small and aggressive Marsh Wren (see page 471) which enters the blackbird nest and pecks holes in the eggs.

Actual size

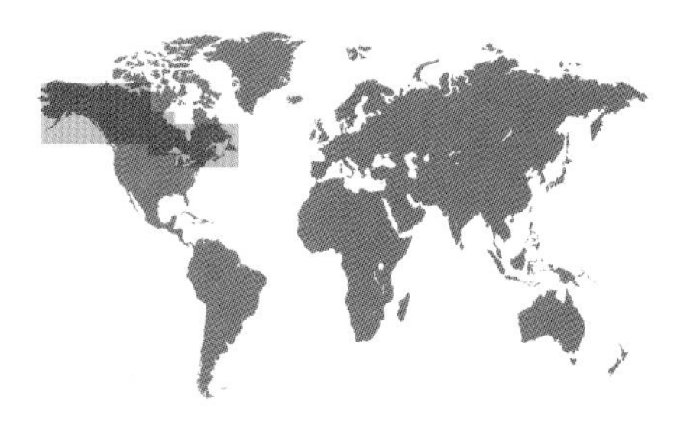

ORDER	Passeriformes
FAMILY	Icteridae
BREEDING RANGE	Boreal North America
BREEDING HABITAT	Coniferous and mixed forest, swampy woods, scrubby muskeg bogs
NEST TYPE AND PLACEMENT	Bulky bowl, made from twigs, grasses, and lichens, enforced with wet plant matter, which dries and hardens, lined with green leaves, fibers, and grasses; placed close to the trunk in trees and shrubs, near or over water
CONSERVATION STATUS	Vulnerable
COLLECTION NO.	FMNH 10415

ADULT BIRD SIZE
8½–10 in (21–25 cm)

INCUBATION
14 days

CLUTCH SIZE
4–6 eggs

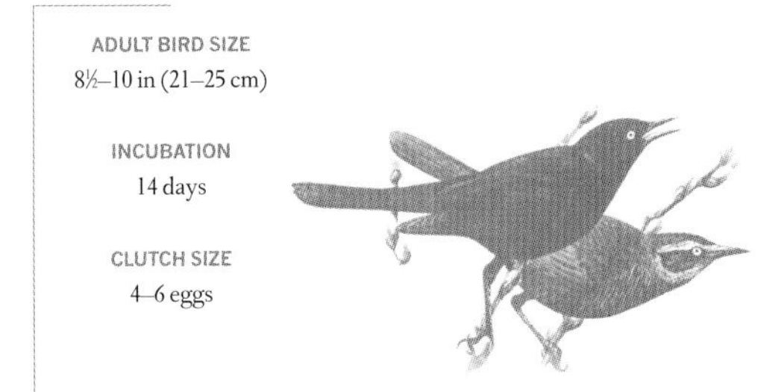

EUPHAGUS CAROLINUS

RUSTY BLACKBIRD

PASSERIFORMES

Clutch

To breed, the female Rusty Blackbird selects a suitable nesting site hidden by foliage from above, often near the remains of a nest surviving from the previous year. Only the female incubates the eggs, but the male assists by bringing food to the nest, often feeding the female on a nearby branch. Both parents provision the nestlings, bringing whole insects, including dragonflies, and putting them into the open gapes of the begging nestlings.

This species has become a conservation concern. In the winter, it joins large feeding and roosting flocks with grackles, blackbirds, and cowbirds, which may be poisoned or otherwise destroyed to protect crops. However, population declines may also be linked to habitat change on the breeding grounds, such as acid rain reducing the insect prey that is critical for this species' early spring arrival and breeding onset. Citizen science projects, including bird survey counts, have reported an alarming 90 percent decline in the numbers of this species, but more research is needed to understand why it is struggling.

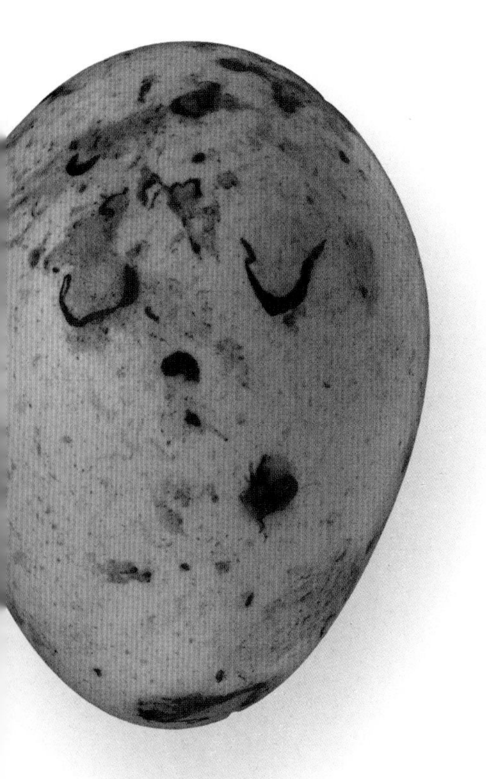

The egg of the Rusty Blackbird is blue-green to pale gray, with variable amounts of brown maculation, and is 1 x ¾ in (26 x 19 mm) in dimensions. The female begins to sit on the eggs after the first egg is laid, but only begins fully warming them after laying the last one.

Actual size

ORDER	Passeriformes
FAMILY	Icteridae
BREEDING RANGE	Southern and north temperate North America
BREEDING HABITAT	Open woodlands, edges, agricultural fields, pastures with thickets, city parks, and backyards
NEST TYPE AND PLACEMENT	Bulky cup, made from twig, leaves, grasses, strips of paper, and other materials, lined with mud, finer grasses, and horsehair; placed at medium height or high in a coniferous tree
CONSERVATION STATUS	Least concern
COLLECTION NO.	FMNH 10466

ADULT BIRD SIZE
11–13½ in (28–34 cm)

INCUBATION
11–15 days

CLUTCH SIZE
4–6 eggs

QUISCALUS QUISCULA

COMMON GRACKLE

PASSERIFORMES

The Common Grackle is a beneficiary of human-induced landscape changes: its preference to flock, feed, and court on the ground has allowed it to spread into many modified habitats, including agricultural areas, and the busiest of city parks. There, in areas of densely leafed bushes and trees, these birds readily set up nests, with the male occasionally helping to gather nesting materials for the female to construct the nest. During incubation, however, nearly half of the males desert their female, and seek new mates through displaying their shiny plumage and singing with a harsh, metallic tone.

The Common Grackle is sexually dimorphic: relative to the females' mostly drab brown feathers, the males' iridescent dark green and blue plumage is generated by a filmlike arrangement of melanin granules within the feather barbules, so that different wavelengths of light are reflected at different angles. Females, by contrast, are non-iridescent black and brown; in their plumage melanin granules are arranged in an unordered manner and reflect the same light irrespective of its angle.

Clutch

Actual size

The egg of the Common Grackle is light blue, pearl gray, white, or brownish in background coloration, with brown blotches, and is 1⅛ x 13/16 in (29 x 21 mm) in size. Brown-headed Cowbirds (see page 616) rarely parasitize grackle broods, not because grackles reject the foreign eggs, but because the large and fast-growing grackle chicks easily outcompete the cowbird chick.

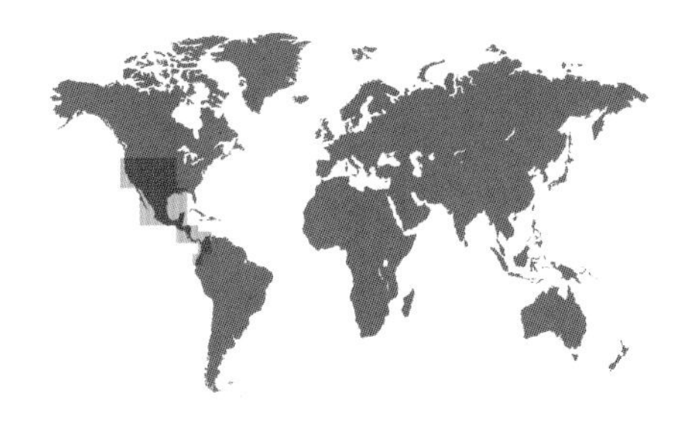

ORDER	Passeriformes
FAMILY	Icteridae
BREEDING RANGE	Southwestern North America, Central America, northern South America
BREEDING HABITAT	Open areas with scattered trees, coastal scrub and open forest, with standing water nearby; also parks and gardens
NEST TYPE AND PLACEMENT	Open cup, made with grasses, bark strips, weeds, and stems, lined with mud or dung, and pasted with fine grasses; in upward-forking branches, at medium or elevated height
CONSERVATION STATUS	Least concern
COLLECTION NO.	FMNH 10554

ADULT BIRD SIZE
15–18 in (38–46 cm)

INCUBATION
13–14 days

CLUTCH SIZE
1–5 eggs

QUISCALUS MEXICANUS

GREAT-TAILED GRACKLE

PASSERIFORMES

Clutch

The egg of the Great-tailed Grackle is bright blue to pale blueish gray in coloration, has dark brown to black lines and spots, and its measurements are 1¼ x ⅞ in (32 x 22 mm). Most eggs in a nest are sired by the territorial male, but a few may be sired by transient males that do not hold a territory.

In their open, grassy habitats, male Great-tailed Grackles maintain and defend nesting territories, which are centered around suitable trees and shrubs for nest construction. There, the males attempt to impress females by following them closely, puffing up their plumes, erecting their wings and tails, while calling with a raspy voice. Larger males with longer tails are able to successfully establish a territory, and they start doing so at about three years of age. By contrast, females start breeding in their second year.

With typically more than one female settling within a male's territory, the mating system is polygynous. However, females are also flexible in their mate choices, and may abandon a half-built nest, or move between different breeding attempts to another male's territory, especially if other females are already nesting nearby, thus acting polyandrously. Scientists term this complex social and genetic mating system "polygynandrous."

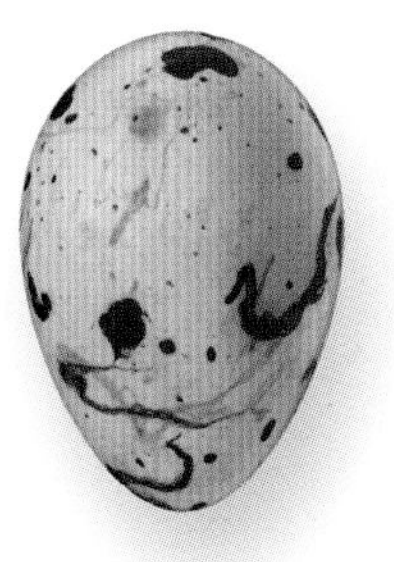

Actual size

ORDER	Passeriformes
FAMILY	Icteridae
BREEDING RANGE	Southern North America, Central America
BREEDING HABITAT	Open fields, scrubby areas, forest edges, golf courses, subtropical parks
NEST TYPE AND PLACEMENT	Obligate brood parasite: always lays eggs in other birds' nests, typically open-cup-nesting passerines
CONSERVATION STATUS	Least concern
COLLECTION NO.	FMNH 20428

ADULT BIRD SIZE
8–8½ in (20–22 cm)

INCUBATION
10–12 days

CLUTCH SIZE
1 egg laid per day

MOLOTHRUS AENEUS

BRONZED COWBIRD

PASSERIFORMES

Cowbirds never build a nest and do not incubate eggs or feed their own young. Instead, they are obligate brood parasites, always laying eggs in other birds' nests. Since the male offers nothing more than sperm to the female, she must assess his genetic contribution based on his courtship skills. The male Bronzed Cowbird does this well, as he turns himself into a miniature helicopter, and hovers 20 in (50 cm) directly over the female for ten seconds, after which he lands in front of her, sings, spreads his wings, bows, and takes off again.

Bronzed Cowbirds prefer to parasitize the nests of icterid orioles. Even though orioles are comparable in size and armed with a sharp beak, video recordings have revealed that female cowbirds are so determined to lay their egg that they will often do so while actively being struck at in the host nest. If the nest is occupied, the parasite will lie on top of a host female while she is sitting tightly on the nest, with the cowbird egg rolling off the back of the rightful nest owner.

Clutch

The egg of the Bronzed Cowbird is white to blueish green in background color, with no maculation, and 1 x ⅔ in (25 x 17 mm) in dimensions. The immaculate egg rarely matches the color and appearance of its commonly used host species, including those of the Hooded Oriole (see page 618) shown here, but many hosts continue to accept the foreign egg.

Actual size

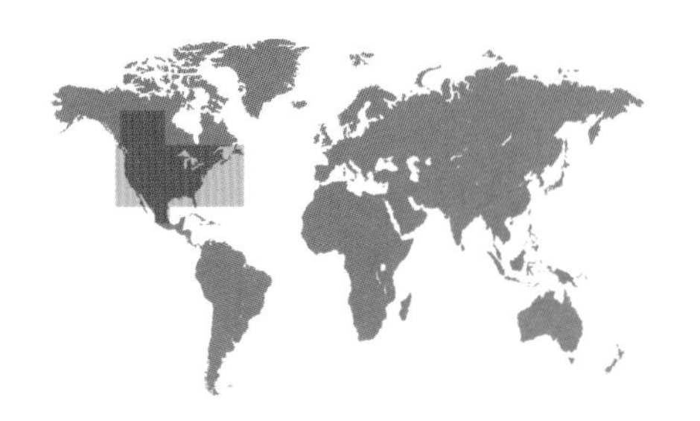

ORDER	Passeriformes
FAMILY	Icteridae
BREEDING RANGE	Temperate North America
BREEDING HABITAT	Forest edge, pastures, riparian corridors, and suburban parklands
NEST TYPE AND PLACEMENT	Obligate brood parasite: builds no nest but instead lays eggs in other species' nests; typically uses open-cup-nesting passerines as hosts, but also enters cavities, including bluebird and Prothonotary Warbler nestboxes
CONSERVATION STATUS	Least concern
COLLECTION NO.	FMNH 10012

ADULT BIRD SIZE
7–8½ in (18–21 cm)

INCUBATION
10–12 days

CLUTCH SIZE
1 egg laid per day

MOLOTHRUS ATER

BROWN-HEADED COWBIRD

PASSERIFORMES

Clutch

The egg of the Brown-headed Cowbird is beige in background coloration, with sparse to heavy brown spotting, and ¾ x ½ in (20 x 15 mm) in size. Unlike those of cuckoos and other egg-mimetic brood parasites, cowbird eggs generally stand out in color, pattern, shape, and size among the host eggs; their eggshell is relatively thicker compared to nonparasitic eggs. The parasite egg is shown here in the clutch of a Common Grackle host.

As obligate brood parasites, males spend most of their day displaying to each other and, in the process, attempting to attract females. Female Brown-headed Cowbirds defend a breeding territory to assure exclusive access to host nests; they lay throughout the season, sometimes over 40 eggs in a year. Cowbird eggs are unusually strong for their size, resisting attempts at puncture by foster parents who may be trying to remove the foreign egg.

Female Brown-headed Cowbirds are excellent nest searchers and the compartment of their brain associated with spatial memory is larger than in males. Females locate host nests suitable for parasitism days in advance; and lay and deposit their eggs before sunrise to avoid being detected and attacked by the rightful nest owners. Although the hatchling cowbird does not evict host nest mates, it begs more intensively than host chicks to commandeer most of the foster parents' provisions.

Actual size

ORDER	Passeriformes
FAMILY	Icteridae
BREEDING RANGE	Eastern and southern North America
BREEDING HABITAT	Open woodlands, riparian corridors, wooded marshes, flood plain forests
NEST TYPE AND PLACEMENT	Hanging pouch, made of woven grass, lined with finer grasses, plant down, wool, and feathers; hung from the fork of a branch, near the tip, on a small or medium-sized tree
CONSERVATION STATUS	Least concern
COLLECTION NO.	FMNH 10372

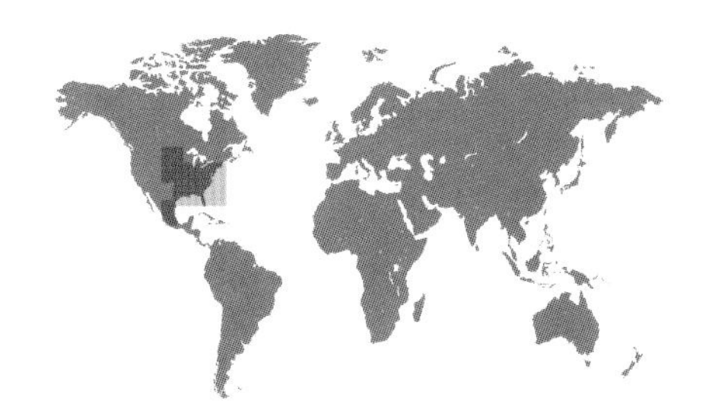

ADULT BIRD SIZE
6–7 in (15–18 cm)

INCUBATION
12–14 days

CLUTCH SIZE
4–6 eggs

ICTERUS SPURIUS

ORCHARD ORIOLE

PASSERIFORMES

The Orchard Oriole is a true neotropical bird that happens to breed in the north temperate zone; the breeding season lasts from May to July, but once the chicks have fledged, by mid-July, many adults quickly take off on their southwardly migration. Thus, nearly three-quarters of the annual cycle of this species is spent outside the breeding range.

There are at least three distinct plumage morphs of Orchard Orioles: females, second-year males, and older males. This poses the question why younger males should display their age so visibly: both types of males arrive at the breeding ground at the same time, and display to females, who generally choose an older male. Females move through territories of different males, assessing their song and plumage patterns, and take several days to settle into a pair bond. Nest building begins a few days later, and once it is under way, the female can complete the nest in under a week. She starts laying her clutch the next day.

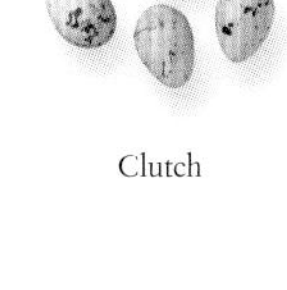

Clutch

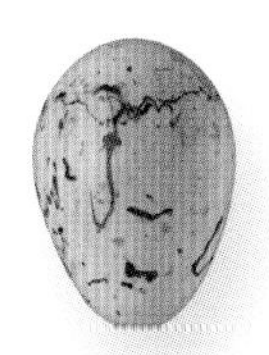

Actual size

The egg of the Orchard Oriole is light blueish in base color, with variably shaped black markings, and is 13⁄16 x 5⁄8 in (21 x 16 mm) in measurements. The female incubates the eggs alone, but the male may stand guard nearby and also regularly feeds her on the nest.

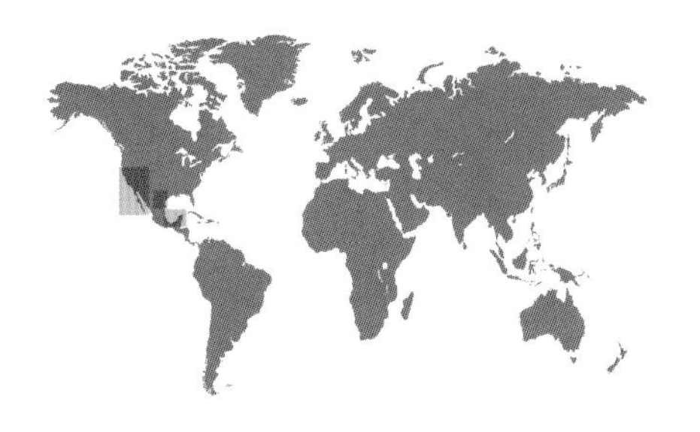

ORDER	Passeriformes
FAMILY	Icteridae
BREEDING RANGE	Southern and western North America, Central America
BREEDING HABITAT	Open woods, riparian corridors, parklands, oases, gardens and plantations
NEST TYPE AND PLACEMENT	Cup of woven plant fibers, suspended from leaf blades, hanging or attached by the rim; placed high in a tree or palm
CONSERVATION STATUS	Least concern
COLLECTION NO.	FMNH 266

ADULT BIRD SIZE
7–8 in (18–20 cm)

INCUBATION
12–14 days

CLUTCH SIZE
3–5 eggs

ICTERUS CUCULLATUS

HOODED ORIOLE

PASSERIFORMES

Clutch

Males arrive on the breeding grounds several days before females return; they establish and advertize their territory with songs and chatters that are softer than those of other orioles. The female alone builds her complex nest, but perhaps because of her frequent trips with nesting materials, these nests are easily discovered by brood parasites. Accordingly, in Texas, where both Bronzed and Brown-headed Cowbirds (see pages 615–16) co-occur with the Hooded Oriole, nests may contain eggs laid there by both of these parasitic species.

Originally an inhabitant of riparian forest galleries, this oriole has come to tolerate and occupy human-made habitats, including watered lawns, artificial oases, and tree plantations. It is particularly common among ornamental palm trees in the southern part of California, where the hanging nest structure, attached beneath broad palm leaf blades, remains well shaded and protected from the weather.

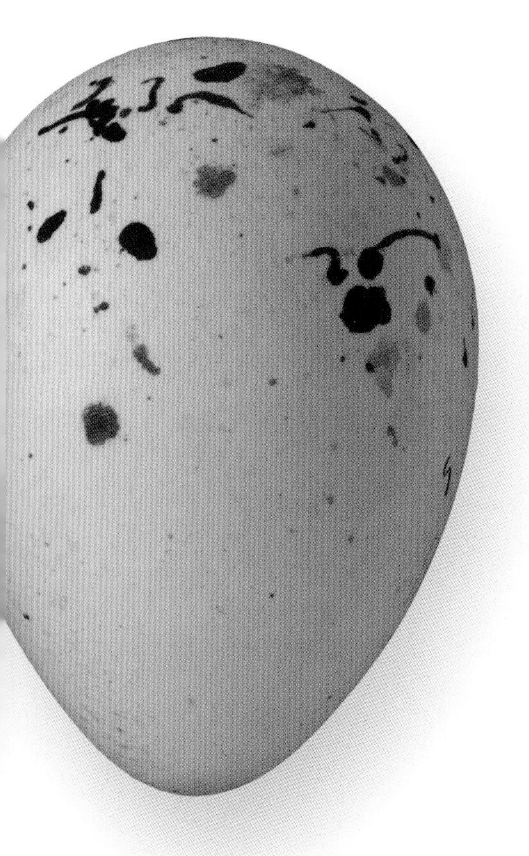

Actual size

The egg of the Hooded Oriole is white in coloration, with irregular and sparse brown blotching, and is ⅞ x ⅝ in (22 x 16 mm) in dimensions. This species is the preferred host of Bronzed Cowbirds (see page 615) throughout their overlapping ranges, with most nests containing one or more parasite eggs. To reduce competition for their own egg and chick, female cowbirds often peck holes into host eggs, further reducing the oriole's reproductive success.

ORDER	Passeriformes
FAMILY	Icteridae
BREEDING RANGE	Eastern and central North America
BREEDING HABITAT	Woodland edges, riparian forests, farmland, parks, and gardens
NEST TYPE AND PLACEMENT	Hanging pouch, woven from plant and animal fibers, including grass, bark, wool, hair, and even plastic fishing line; anchored high in a tree
CONSERVATION STATUS	Least concern
COLLECTION NO.	FMNH 2308

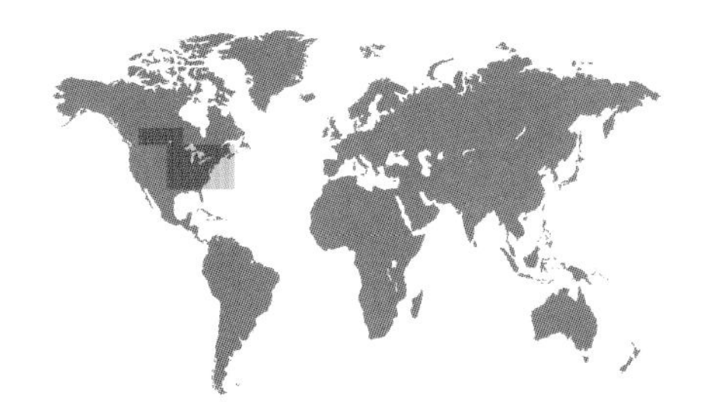

ADULT BIRD SIZE
6½–7½ in (16–19 cm)

INCUBATION
11–14 days

CLUTCH SIZE
3–7 eggs

ICTERUS GALBULA

BALTIMORE ORIOLE

PASSERIFORMES

This iconic species, famous for its orange-and-black coloration, is well known to birdwatchers and baseball fans alike. Pairs come together during the breeding season, after returning to the nesting grounds. The males take two years to mature into their familiar plumage pattern, with younger males resembling the females in feather coloration. Nonetheless, even younger males can be successful in attracting a mate, and fertilizing the eggs for a successful first breeding season. The female alone builds the nest and incubates the eggs, but the male may feed her on the nest, and shares the chore of feeding the chicks equitably.

The British term "oriole" is used for this species and its icterid relatives throughout the Americas, because of the color and pattern similarity with the arboreal orioles and figbirds in the African and Eurasian oriolid family. Genetically, however, New World orioles are most closely related to grackles, meadowlarks, blackbirds, and their brood parasitic nemeses, the cowbirds.

Clutch

The egg of the Baltimore Oriole is pale grayish or blueish white in background, with brown, black, or lavender lines and blotches, and measures ⅞ x ⅝ in (22 x 16 mm) in size.

Actual size

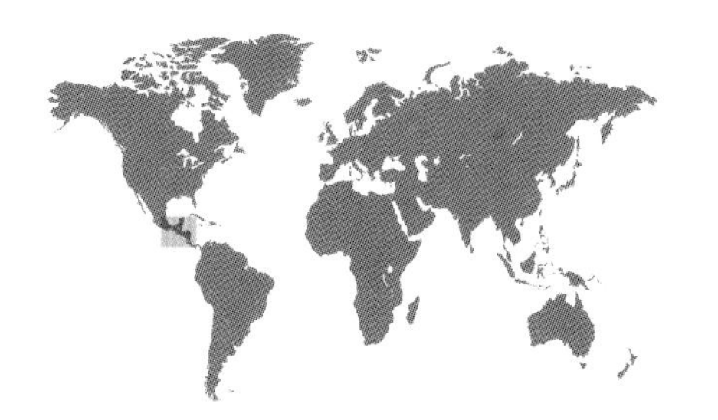

ORDER	Passeriformes
FAMILY	Icteridae
BREEDING RANGE	Central America
BREEDING HABITAT	Coastal lowland forests, woodland edges, and plantations
NEST TYPE AND PLACEMENT	Hanging woven nest, made of plant fibers and vines, in colonies of 30 or more nests; placed high in a large tree
CONSERVATION STATUS	Least concern
COLLECTION NO.	FMNH 2270

ADULT BIRD SIZE
15–19½ in (38–50 cm)

INCUBATION
15 days

CLUTCH SIZE
2 eggs

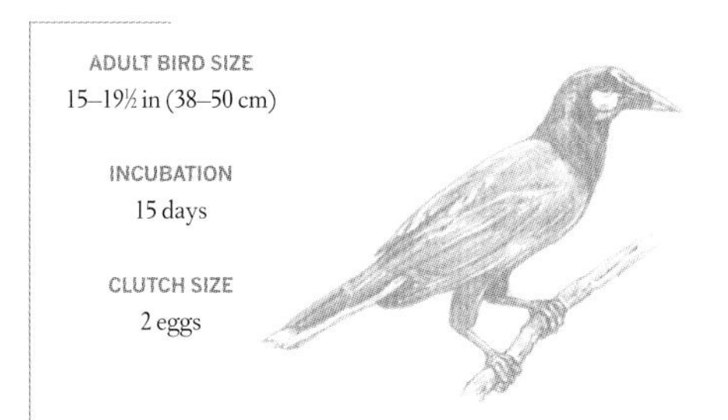

PSAROCOLIUS MONTEZUMA

MONTEZUMA OROPENDOLA

PASSERIFORMES

Clutch

The egg of the Montezuma Oropendola is white to buff in color, has brownish maculation, and its measurements are 1¼ x 1 in (33 x 25 mm). While the alpha male at the nesting tree sires the largest number of eggs, subordinate males, who mate with females inconspicuously and away from the colony, also father many of the young.

This oropendola is a common and conspicuous bird; it has a relatively limited distribution, but its affiliation with human-modified habitats, including parklands and farmlands with stands of tall trees, assures its stable population sizes. This species builds colonies of hard-to-reach nests hung from high branches; colony members collectively ward off predators and parasites. This bird is parasitized by a large brood parasite, the Giant Cowbird, whose females face severe physical attacks when attempting to sneak an egg into the nests of this host.

The mating system of this oropendola is called "female-defense harem polygyny." The nesting tree is attended by several males, with more males trying to gain access as more females complete their nests and come into their fertile period. Nevertheless, the dominant alpha male keeps most other males away, and mates with most of the females at the colony.

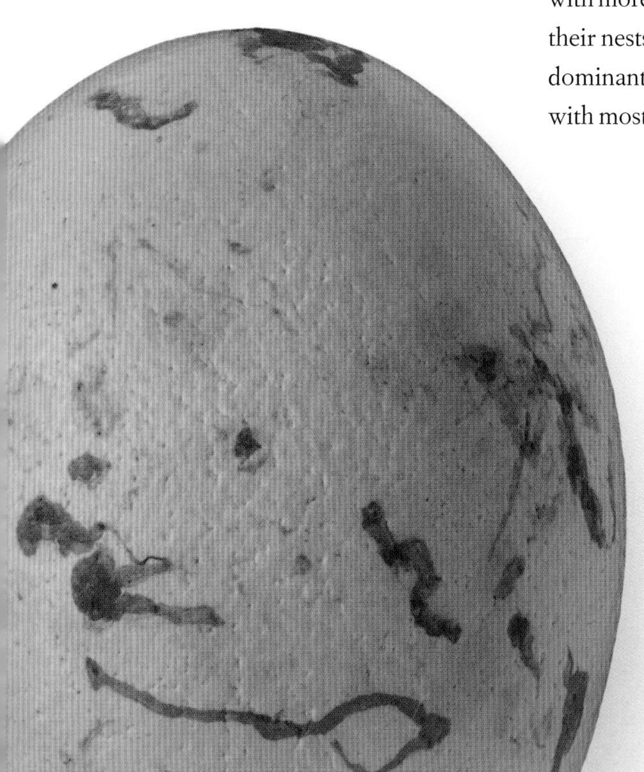

Actual size

ORDER	Passeriformes
FAMILY	Fringillidae
BREEDING RANGE	Western Asia, Europe, north Africa
BREEDING HABITAT	Open woodlands, forest edges, city parks, and backyards
NEST TYPE AND PLACEMENT	Open-cup nest in a tree fork; exterior covered with moss or lichen, interior lined with grasses and fibers, matted together with wool and cobwebs
CONSERVATION STATUS	Least concern
COLLECTION NO.	FMNH 20646

ADULT BIRD SIZE
5½–6½ in (14–16 cm)

INCUBATION
10–18 days

CLUTCH SIZE
4–6 eggs

FRINGILLA COELEBS

CHAFFINCH

PASSERIFORMES

In western Europe the Chaffinch is the most common finch. It sings a melodious tune to delineate its breeding territory and to attract the female. It is no surprise that English settlers arriving two centuries ago in places as far away as New Zealand insisted on bringing breeding pairs of this familiar songbird with them for release and propagation in the new colonies. In the absence of most of their typical mammalian and reptilian nest predators common in Eurasia, Chaffinches introduced to New Zealand have flourished and today they are some of the commonest and loudest urban and parkland birds.

Chaffinches possess regional song differences generated by the process of cultural evolution, whereby young birds learn to imitate closely, but not perfectly, some older tutors, and also invent new song variants on their own. These variants are then passed down to the next generation through imitative learning. Females prefer to mate with males whose songs end in a flourishing set of notes, and do not seem to distinguish between familiar or unfamiliar songs of comparable vocal complexity.

Clutch

The egg of the Chaffinch is light pink or gray in background color, with variable reddish brown markings, and is ¾ x ⅝ in (19 x 16 mm) in dimensions. Chaffinches in Europe are skilled at distinguishing their own eggs from the Common Cuckoo's egg; in fact, they readily eject even experimentally introduced eggs of other Chaffinches from the nest.

Actual size

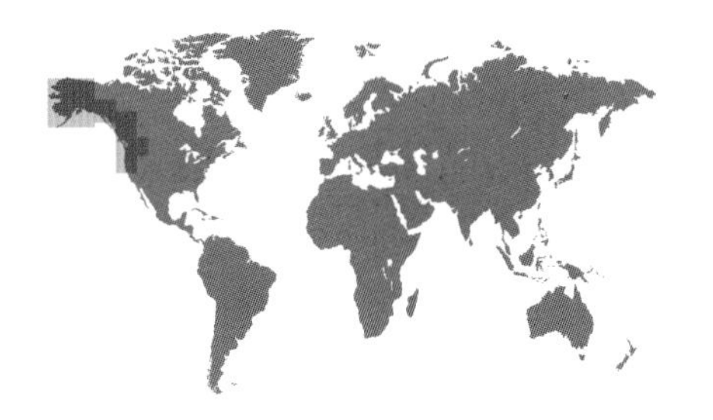

ORDER	Passeriformes
FAMILY	Fringillidae
BREEDING RANGE	Northwestern North America
BREEDING HABITAT	Open tundra, temperate steppe, and alpine grasslands
NEST TYPE AND PLACEMENT	Inside crevices in rocky outcrops and cliffs, also in holes in walls and buildings; open cup, made of twigs, mosses, and sedges, lined with fine grasses, hair, and feathers
CONSERVATION STATUS	Least concern
COLLECTION NO.	FMNH 2982

ADULT BIRD SIZE
5½–6½ in (14–16 cm)

INCUBATION
13–14 days

CLUTCH SIZE
3–5 eggs

LEUCOSTICTE TEPHROCOTI

GRAY-CROWNED ROSY-FINCH

PASSERIFORMES

Clutch

This species is socially monogamous, relying on speedy pair formation in the late winter and early spring to make the most of the warm months available for nesting. The female collects materials and builds the nest, while the male follows her closely. She alone incubates the eggs, but both parents feed the nestlings. Generally, two clutches of eggs are laid in northern and temperate populations, but there is only time for one full nesting cycle in high-elevation montane sites.

North American and Asian rosy-finch populations were only recently given fully separate species status. These birds breed and winter in northern or high-elevation habitats, and have served as subjects for scientific research on life in low-oxygen/high-altitude sites. For example, the birds in high-altitude populations have a more extensive network of capillaries carrying oxygen to the flight- and leg-muscle fibers compared to populations closer to sea level.

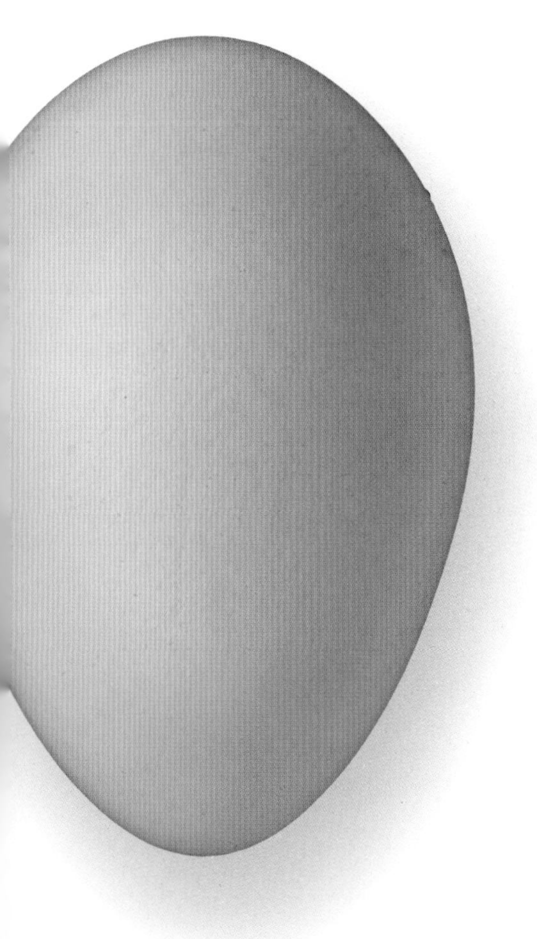

Actual size

The egg of the Gray-crowned Rosy-Finch is white with a pinkish tint, immaculate, and ¾ x ⅝ in (21 x 16 mm) in size. During incubation, the female frequently lifts her body and rolls the eggs; frequent rotation assures normal development of the embryo.

ORDER	Passeriformes
FAMILY	Fringillidae
BREEDING RANGE	North Africa, southern Europe, western Asia
BREEDING HABITAT	Dry cliff faces and boulder fields in rocky deserts
NEST TYPE AND PLACEMENT	Rocky crevice or underneath a shrub; untidy nest made of roots, twigs, leaves, grass, stalks, wool, and feathers
CONSERVATION STATUS	Least concern
COLLECTION NO.	FMNH 20660

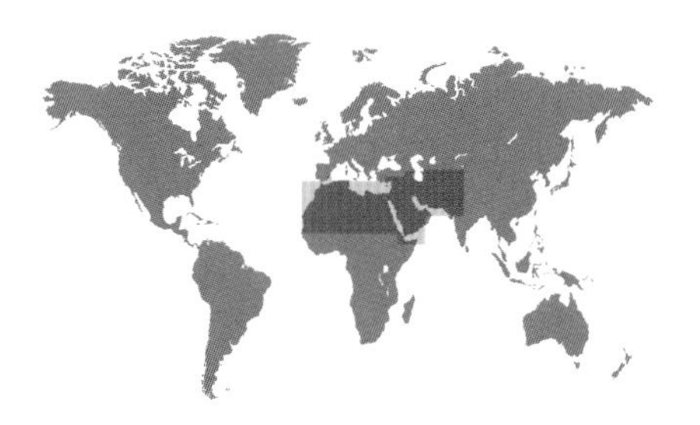

ADULT BIRD SIZE
4½–5 in (11–13 cm)

INCUBATION
11–14 days

CLUTCH SIZE
4–6 eggs

BUCANETES GITHAGINEUS

TRUMPETER FINCH

PASSERIFORMES

The Trumpeter Finch is a specialist resident of arid regions, with populations ranging as far west as the Canary Islands. In some years there are population eruptions, with large numbers of birds roaming into western Europe and other localities typically well outside the regular breeding area of this species.

As a habitat specialist of deserts and other arid fields, rain has long been thought to be a major influence on the onset and success of reproduction in this species. Still, as a grain-eater, this finch is more dependent on the availability of seeds from its preferred food plants, which in turn require higher temperatures for germination and growth. Accordingly, in years with lower than average temperatures, the onset of nesting can be delayed by as much as a month. However, once breeding begins, clutch sizes, incubation periods, and the age of fledging remain similar across the different years.

Clutch

Actual size

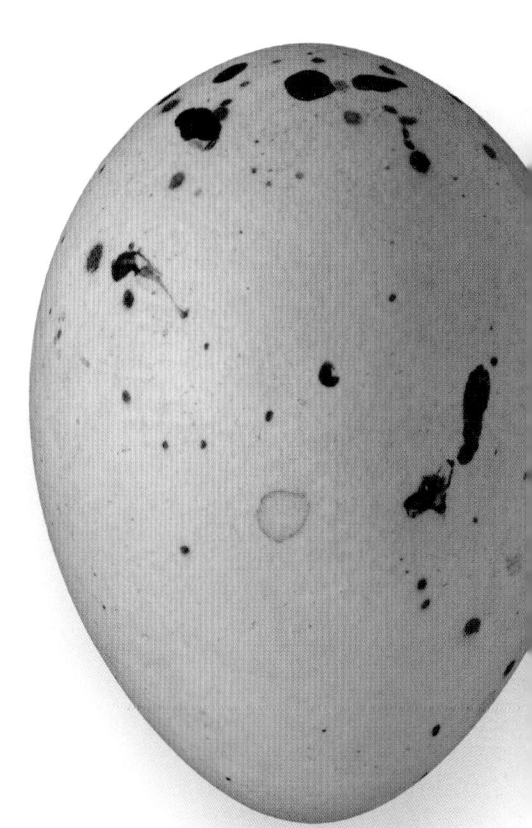

The egg of the Trumpeter Finch is buff to cream in background coloration, has sparse fine dark blotching, and its size is ⅔ x ⅝ in (18 x 16 mm). The preferred habitat around the nests is open cliffs; nests show higher rates of success from egg laying to fledging when surrounded by sparse vegetation.

ORDER	Passeriformes
FAMILY	Fringillidae
BREEDING RANGE	Northern and western North America, northern Eurasia
BREEDING HABITAT	Coniferous forests, near the tree line both in the north and on alpine slopes
NEST TYPE AND PLACEMENT	Open cup, made of twigs, pine needles, and rootlets, lined with grasses, lichen, and feathers; placed on a horizontal branch fork in a tree, near the trunk, under dense foliage
CONSERVATION STATUS	Least concern
COLLECTION NO.	FMNH 15197

ADULT BIRD SIZE
8–9½ in (20–25 cm)

INCUBATION
13–15 days

CLUTCH SIZE
3–4 eggs

PINICOLA ENUCLEATOR

PINE GROSBEAK

PASSERIFORMES

Clutch

Unlike several other grosbeaks in North America, this species is a relative of true fringillid finches, with a much greater proportion of its diet derived from seeds and buds, compared to the more insectivorous cardinalid buntings and grosbeaks. A plant-based diet is critical for the success of the Pine Grosbeak, as it remains resident in its northern range even during harsh winters, when insect prey becomes essentially unavailable.

This species is gregarious outside the reproductive season, and pairs may form in the winter flocks before settling on breeding territories. Both sexes assess the suitability of trees and branches for nest building, but the nest is built by the female only, and she alone incubates the eggs. The parents feed the chicks equitably, while also keeping the nest clean by removing fecal sacs after the feeding bouts.

Actual size

The egg of the Pine Grosbeak is pale blue or light blueish green, has brownish black maculation, and measures 1 x ⅔ in (26 x 18 mm) in dimensions. Incubation begins when the last egg is laid, and all eggs hatch within a 24-hour period.

ORDER	Passeriformes
FAMILY	Fringillidae
BREEDING RANGE	North temperate North America
BREEDING HABITAT	Coniferous and mixed forests, edges and clearings, parks
NEST TYPE AND PLACEMENT	Open cup, made from twigs, sticks, and roots, lined with fine grasses and hair; placed out on a limb of a tree
CONSERVATION STATUS	Least concern
COLLECTION NO.	FMNH 10580

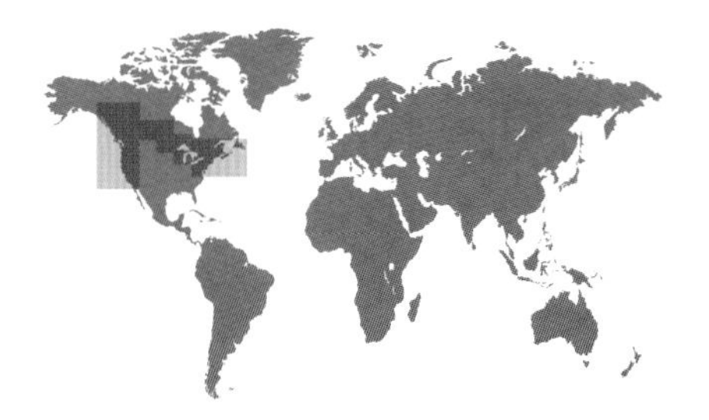

HAEMORHOUS PURPUREUS

PURPLE FINCH

PASSERIFORMES

ADULT BIRD SIZE
4½–6 in (12–15 cm)

INCUBATION
12–13 days

CLUTCH SIZE
3–5 eggs

The Purple Finch breeds in the northern temperate regions of North America, with year-round resident populations along the Pacific coast. With the expansion of House Finches (see next page) in the east, and the invasion of House Sparrows (see page 636) throughout the continent, the Purple Finch has seen consistent declines in its breeding population sizes and local productivity.

Clutch

Flocking during the non-breeding season, the Purple Finch male begins the breeding season by establishing and defending a nesting territory. He attempts to attract females by displaying his chest and wings, and showing off his singing voice. Once a pair is formed, the focal point of the territory shifts to the nest, with the male involved in selecting the site. Occasional records indicate that some males carry nesting materials during construction, and attend the eggs during incubation while the female is off the nest. Both parents provision the chicks with regurgitated seeds.

Actual size

The egg of the Purple Finch is pale grayish or greenish blue in background color, with some brown and black spotting, and is ¾ x ⅝ in (20 x 16 mm) in dimensions. Floaters in the population may also attempt to replace territorial males gone missing during breeding. By feeding the female on the nest and provisioning the chicks, the replacement male may father a second or replacement clutch of eggs.

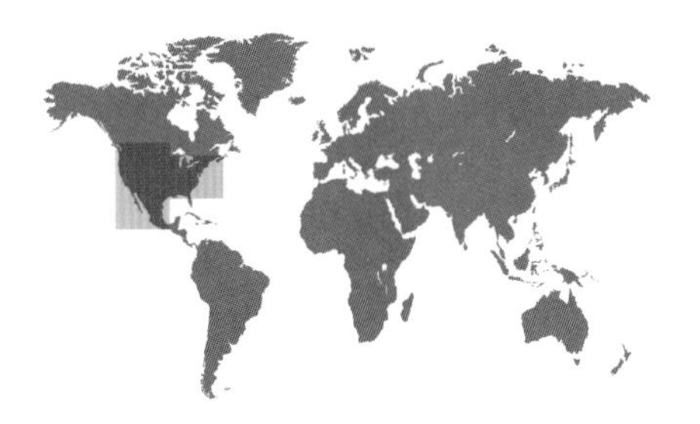

ORDER	Passeriformes
FAMILY	Fringillidae
BREEDING RANGE	North America
BREEDING HABITAT	Semi-deserts and arid scrubland; also farms, pastures, thickets, gardens, parks, and forest edges
NEST TYPE AND PLACEMENT	Cup nest, made of fine stems, leaves, rootlets, twigs, string, wool, and feathers; placed in deciduous and coniferous trees, on cliff or building ledges, street lamps, or hanging planters
CONSERVATION STATUS	Least concern
COLLECTION NO.	FMNH 2729

ADULT BIRD SIZE
5–5½ in (13–14 cm)

INCUBATION
13–14 days

CLUTCH SIZE
2–5 eggs

HAEMORHOUS MEXICANUS

HOUSE FINCH

PASSERIFORMES

Clutch

The egg of the House Finch is pale blue to white in coloration, with fine sparse black and pale purple maculation, and measures ¾ x 9⁄16 in (19 x 14 mm) in size.

A familiar bird to many people with active bird feeders, this species is both a native and an exotic in North America; the eastern populations represent an amalgamation of breeding populations of birds intentionally introduced in New York City and the leading edges of naturally expanding western populations. These eastern populations were also devastated by bacterial infections spread from bird to bird in this highly gregarious, flock-foraging species. However, the species seems to be holding its own. The males do not defend territories, and instead use their red plumage patches to establish dominance over other males, and to attract females for breeding. Occasionally, House Finches reuse old nests built by other birds; the male attends the nest to feed the female while she incubates the eggs and broods the hatchlings.

This species is commonly parasitized by Brown-headed Cowbirds (see page 616), especially in the early weeks of the spring when few other host bird species are nesting. This finch readily accepts nonmimetic foreign eggs, but the grain-based diet fed to the nestlings is unsuitable for the cowbird chicks. As a result, parasite hatchlings typically starve with a crop full of undigested seeds.

Actual size

ORDER	Passeriformes
FAMILY	Fringillidae
BREEDING RANGE	Temperate and subarctic Eurasia, northern and western North America, Central America
BREEDING HABITAT	Coniferous forests, both taiga and alpine spruce, pine, and fir stands
NEST TYPE AND PLACEMENT	Open cup, made from twigs and sticks, lined with grasses, rootlets, lichen, pine needles, bark shreds, hair, and feathers; placed in the dense canopy of a coniferous tree
CONSERVATION STATUS	Least concern
COLLECTION NO.	FMNH 2746

ADULT BIRD SIZE
5½–8 in (14–20 cm)

INCUBATION
12–18 days

CLUTCH SIZE
3–5 eggs

LOXIA CURVIROSTRA

RED CROSSBILL

PASSERIFORMES

The unusual cross-tipped beak of these finches is specifically adapted to pry open conifer cones to access seeds. In addition, the plumage colors of different populations vary distinctly (but not genetically) depending on the carotenoid pigment type and concentration derived from the seeds of various food trees. Different populations of the species produce different call and song types, and rarely mate with one another.

Clutch

Breeding begins with the discovery of a patch of forest with suitably large crops of pine cones to feed on. Several nesting attempts are initiated and completed, until food levels dip, at which point the flock moves on to search for another breeding site. Both parents provision the nestlings, feeding them with a blackened mixture of regurgitated seeds and saliva. To induce fledging, parents withhold feedings and call loudly to the large nestlings from a perch near the nest.

Actual size

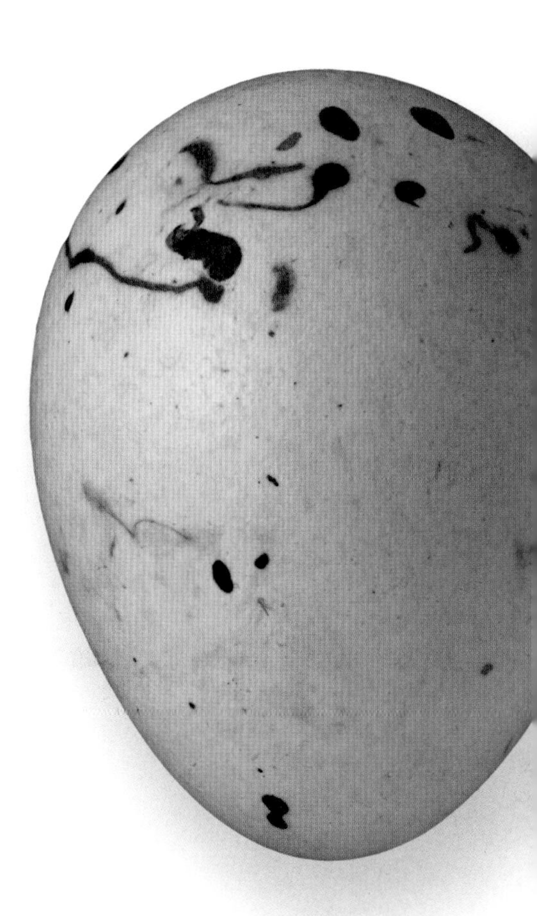

The egg of the Red Crossbill is whitish in background color, with variably shaped reddish markings, and is ¾ x ⅝ in (19 x 16 mm) in size. The female builds a bulkier structure to house the eggs in colder months. Both the eggs and the chicks are able to survive short periods of cold while the female is away from the nest.

ORDER	Passeriformes
FAMILY	Fringillidae
BREEDING RANGE	Western North America, Central America, northern South America
BREEDING HABITAT	Open forests, riparian galleries, fields with scattered trees and thickets, such as parks and gardens
NEST TYPE AND PLACEMENT	Open cup, made of plant fibers, bark, grasses, or wool, lined with seed fluff, hair, and fur; placed in the canopy of a cottonwood, willow, or other tree or shrub
CONSERVATION STATUS	Least concern
COLLECTION NO.	FMNH 10676

ADULT BIRD SIZE
3½–4½ in (9–11 cm)

INCUBATION
12–13 days

CLUTCH SIZE
3–6 eggs

SPINUS PSALTRIA

LESSER GOLDFINCH

PASSERIFORMES

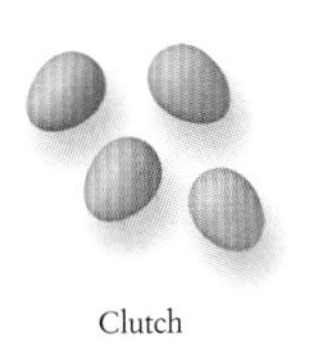

Clutch

The Lesser Goldfinch is a geographically restricted relative of the better-known American Goldfinch (see next page), with which it shares many reproductive traits. Both have a tendency to build tightly woven and compact cup nests using animal and plant fibers, and to begin the reproductive cycle weeks or months after other insectivorous migratory passerines start nesting. Even less is known about the reproductive biology of its Central and South American populations.

The small and compact, softly lined nest of the Lesser Goldfinch is constructed by the female, while the male follows his mate closely, singing and feeding nearby, but not assisting. The female gathers plant and animal fibers and hairs by holding materials in place under one foot, and using the beak to strip fibers and strips from it. The nest is then carefully woven into a cup after the rim is attached firmly to a fork or crotch of branches.

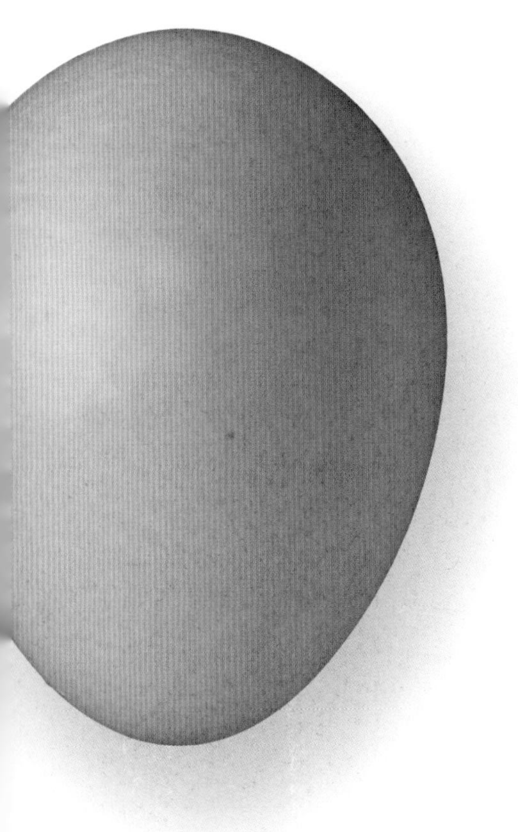

Actual size

The egg of the Lesser Goldfinch is pale blue or white in background coloration, free of markings, and ⅝ x 7⁄16 in (16 x 11 mm) in measurements. Like many other broadly distributed bird species, clutch sizes are smaller in the tropical populations (3–4 eggs), and larger in the temperate ones (4–6 eggs).

ORDER	Passeriformes
FAMILY	Fringillidae
BREEDING RANGE	Temperate North America
BREEDING HABITAT	Grassy fields with thickets, flood plains, shrubby lawns, gardens and parks, forest edges
NEST TYPE AND PLACEMENT	Open cup, made from rootlets, grasses, and plant fibers, lined with plant down; built high in a shrub and shaded by leaves or clusters of needles from above
CONSERVATION STATUS	Least concern
COLLECTION NO.	FMNH 10655

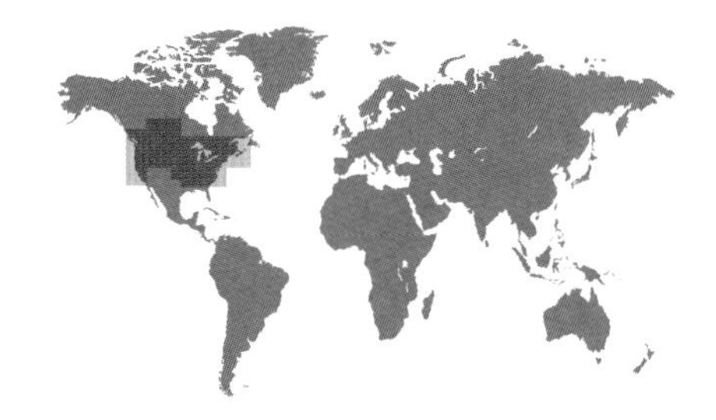

ADULT BIRD SIZE
4½–5 in (11–13 cm)

INCUBATION
12–14 days

CLUTCH SIZE
4–5 eggs

SPINUS TRISTIS

AMERICAN GOLDFINCH

PASSERIFORMES

The American Goldfinch is a widespread, familiar, and popular finch, whose presence at bird feeders is a welcome and colorful sight. The male's plumage and beak coloration reflect his recent history of health and exposure to food resources, and can be used by females to assess his potential value as a mate and co-parent.

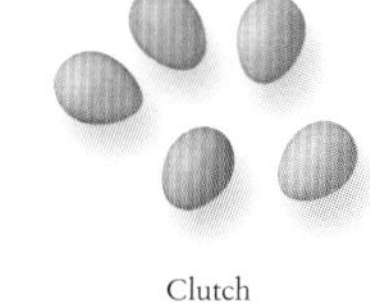

Clutch

American Goldfinches nest in loosely colonial aggregations. Males do not defend an all-purpose or feeding territory but only protect the vicinity of the nest from intruding males during the nest-building and egg-laying stages. Once incubation begins, male aggression drops, and the male turns his attention to feeding the female while she incubates the eggs. When the chicks hatch, the female continues to brood them, and the male continues to feed her; she then transfers the food to the hatchlings. After the first few days, the female joins the male to search for and deliver food directly to the nestlings.

Actual size

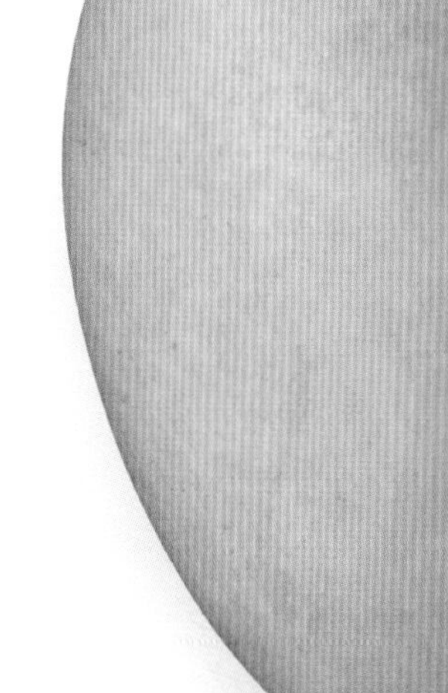

The egg of the American Goldfinch is pale blueish white in coloration, with faint brown maculation, and is ⅔ x ½ in (17 x 12 mm) in size. The species mostly escapes parasitism by Brown-headed Cowbirds (see page 616), probably because of the goldfinch's late onset of nesting, and also because its grain diet is unsuitable for cowbird chicks to digest.

ORDER	Passeriformes
FAMILY	Fringillidae
BREEDING RANGE	Northern Europe, western and northeastern Asia
BREEDING HABITAT	Coniferous and mixed forests, boggy woodland, pine plantations
NEST TYPE AND PLACEMENT	Cup nest in a tree; made from moss, lichens, rootlets, and twigs, lined with grasses and feather down
CONSERVATION STATUS	Least concern
COLLECTION NO.	FMNH 20844

ADULT BIRD SIZE
4½–5 in (11–13 cm)

INCUBATION
10–14 days

CLUTCH SIZE
2–6 eggs

SPINUS SPINUS

EURASIAN SISKIN

PASSERIFORMES

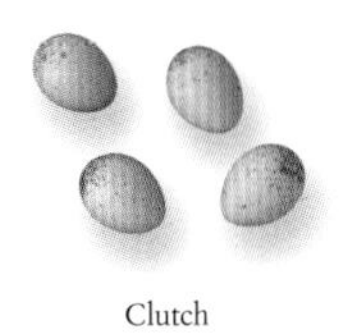

Clutch

The egg of the Eurasian Siskin is white or light gray or light blue, has fine reddish brown maculation, and measures ⅔ x ½ in (17 x 12 mm) in dimensions.

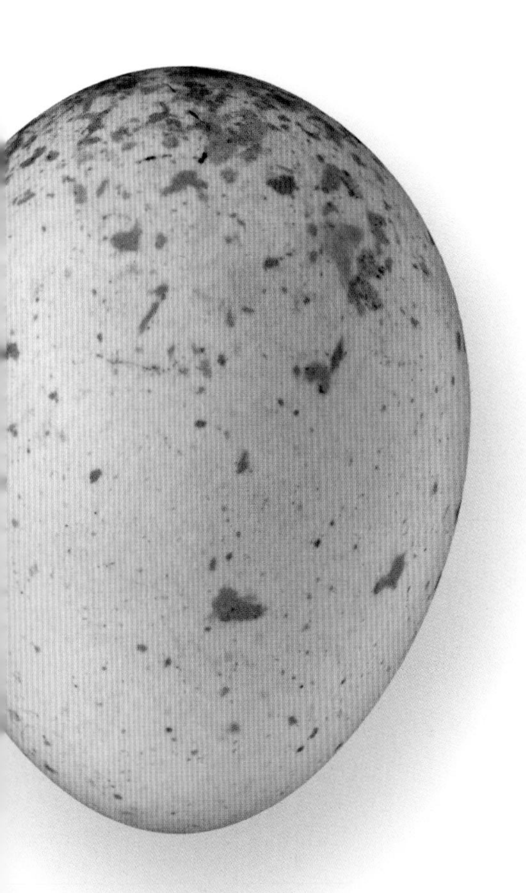

In the wild, this siskin is a far-ranging species; individuals rarely return to breed in the same site in consecutive years. Even during the winter, some populations wander to distant locations, while others migrate to the south to become resident at a single wintering site. Critically, the wing shape of transient birds is longer and more pointed compared to that of the resident birds, allowing for more efficient flights during long-distance movements. Siskin pairs nest near each other, in loose social groups of two to six pairs; the female incubates alone, but the male may make repeated visits while she sits on the eggs.

Eurasian Siskins, like other fringillid finches, are often kept in captivity for their songs. Siskins are easily hybridized with canaries, and the resulting genetic mix yields new colors, plumage patterns, and vocal performance capabilities. Easily adapted to life in a cage, captive siskins can live past ten years of age, whereas in the wild the estimated lifespan is two or three years for this species.

Actual size

ORDER	Passeriformes
FAMILY	Fringillidae
BREEDING RANGE	North Africa, temperate Europe, western Asia
BREEDING HABITAT	Open woodland, forest edges, scrubby fields, overgrown pastures, grassy parks, and gardens
NEST TYPE AND PLACEMENT	Open cup, built from grass, lichen, and moss, and lined with wool and plant down; placed in a tree near the terminus of a foliage-covered branch
CONSERVATION STATUS	Least concern
COLLECTION NO.	FMNH 14663

ADULT BIRD SIZE
4¾–5 in (12–13 cm)

INCUBATION
11–14 days

CLUTCH SIZE
4–6 eggs

CARDUELIS CARDUELIS

EUROPEAN GOLDFINCH

PASSERIFORMES

The European Goldfinch is a colorful songbird of open scrubland. Its melodious song has long attracted the attention of bird fanciers throughout Europe and beyond; released and escaped captive birds in Australia and New Zealand are firmly established as thriving exotics alongside the native avifauna.

Clutch

Nest building begins shortly after territory settlement and pair formation; the female alone constructs the nest. The female begins to lay her eggs within one or two days of completing the nest, and continues to lay one egg a day until the clutch is complete. She is also in full charge of incubating the eggs, spending over 90 percent of daylight hours, as well as the full night, on the nest. The male frequently visits and feeds her. Once the chicks hatch, the male continues to feed her regurgitated seeds on the nest, which she then feeds to the young. As the nestlings grow, both parents take turns to feed them directly.

Actual size

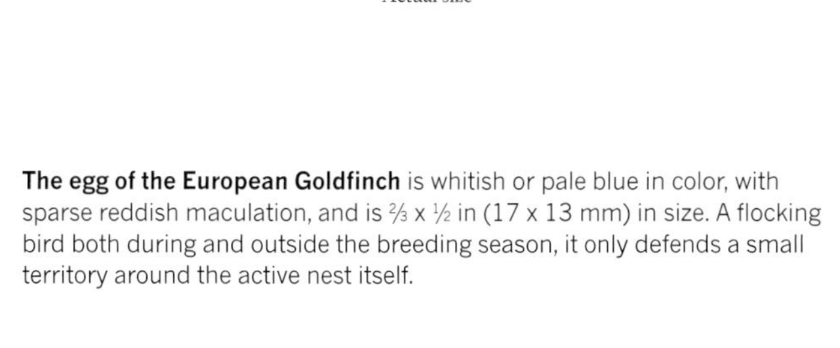

The egg of the European Goldfinch is whitish or pale blue in color, with sparse reddish maculation, and is ⅔ x ½ in (17 x 13 mm) in size. A flocking bird both during and outside the breeding season, it only defends a small territory around the active nest itself.

ORDER	Passeriformes
FAMILY	Fringillidae
BREEDING RANGE	Southwestern and central Europe
BREEDING HABITAT	Subalpine open coniferous forests, alpine meadows
NEST TYPE AND PLACEMENT	Open-cup nest, made of dry stalks, grasses, roots, and cobwebs, lined with hair, wool, and feathers; placed on a horizontal branch, near or halfway from the trunk of a spruce or other conifer tree or bush
CONSERVATION STATUS	Least concern
COLLECTION NO.	FMNH 20837

ADULT BIRD SIZE
4¾–5 in (12–13 cm)

INCUBATION
13–14 days

CLUTCH SIZE
3–5 eggs

SERINUS CITRINELLA

CITRIL FINCH

PASSERIFORMES

Clutch

The breeding success of the Citril Finch species appears to be tied in part to the placement of the nest within the nesting tree. Overall nesting success, from laying to fledging, is about 50 percent across populations, and higher-placed nests are more likely to be depredated by jays and other birds whereas lower nests are vulnerable to raids by mammals. In turn, nests near tree trunks are destroyed by ants that climb the trees and attack the helpless and naked nestlings.

This European species both breeds and winters within the confines of the continent; resident and insular populations on Corsica and nearby Mediterranean islands are now recognized as a separate species, based both on genetic isolation and on differences in habitat preferences. In its alpine, continental distribution, the Citril Finch is closely tied to open wooded areas, with a mix of meadows and forest edges.

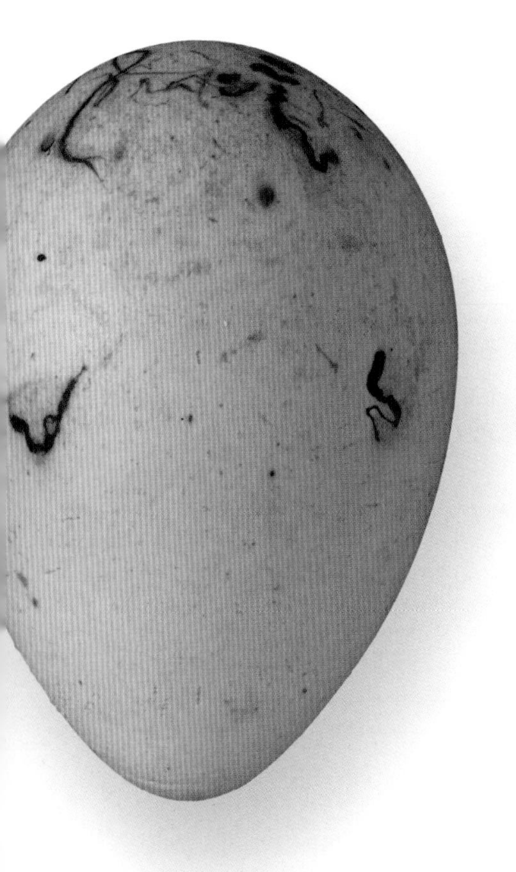

Actual size

The egg of the Citril Finch is white or pale greenish blue in background coloration, has sparse reddish brown markings, and its measurements are ⅔ x ½ in (17 x 12 mm). The female alone builds the nest and the male frequently feeds her while she is incubating the eggs.

ORDER	Passeriformes
FAMILY	Fringillidae
BREEDING RANGE	Temperate Eurasia
BREEDING HABITAT	Coniferous and mixed forests, woodland corridors, parks, and gardens
NEST TYPE AND PLACEMENT	Open-cup nest, made of twigs, mosses, and lichens, lined with rootlets and hair; built inside the canopy of a tree, or in dense foliage, or in a large bush
CONSERVATION STATUS	Least concern
COLLECTION NO.	FMNH 20679

ADULT BIRD SIZE
6–6½ in (15–17 cm)

INCUBATION
12–14 days

CLUTCH SIZE
4–6 eggs

PYRRHULA PYRRHULA

EURASIAN BULLFINCH

PASSERIFORMES

The male of this bullfinch is brightly colored, with a handsome blue-gray back, and a peach-pink chest. Its unique chest coloration is derived from the carotenoid pigmentation of the berries and fruits it consumes during the molt. In captivity, these yellow to rosy red pigments must be supplemented in the diet to prevent fading of these plumage patterns. Caged Eurasian Bullfinches will often learn to mimic their human captors' intonation and words.

Clutch

The male Eurasian Bullfinch is also unique among fringillid finches in that its gonads, the internal testes, are the smallest compared to other species, representing less than a third of 1 percent of its total body mass. Furthermore, the sperm produced by these small testes is highly variable in shape and speed. Taking these two factors together, scientists have predicted that the strength of sexual competition between males is fairly relaxed. Accordingly, instead of seeking reproductive opportunities elsewhere, males keep themselves busy assisting the female to incubate the eggs and feed the chicks.

Actual size

The egg of the Eurasian Bullfinch is pale blue, which often fades to white in museum collections, with brownish spotting, and measures it ¾ x ⅝ in (19 x 16 mm). Territoriality in this species is limited to the male's defense of the area immediately around the nest.

ORDER	Passeriformes
FAMILY	Fringillidae
BREEDING RANGE	North Africa, temperate Eurasia
BREEDING HABITAT	Deciduous and mixed forests, often near water, also orchards, parks, and gardens
NEST TYPE AND PLACEMENT	Open-cup nest, placed in a high tree or a shrub; built from sticks, bark, grass, and lichen, and lined with hair and rootlets
CONSERVATION STATUS	Least concern
COLLECTION NO.	FMNH 20843

ADULT BIRD SIZE
6½–7 in (17–18 cm)

INCUBATION
9–14 days

CLUTCH SIZE
3–7 eggs

COCCOTHRAUSTES COCCOTHRAUSTES

HAWFINCH

PASSERIFORMES

Clutch

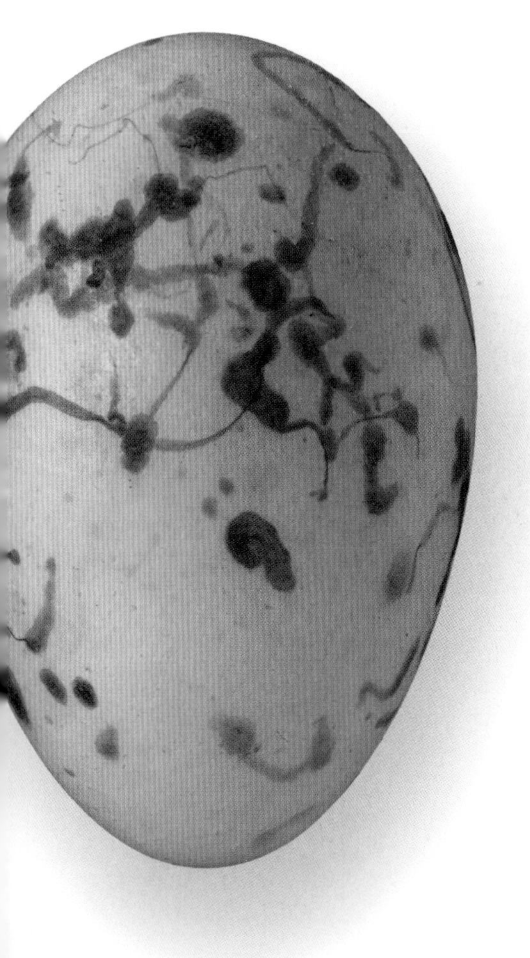

The Hawfinch has an impressive and massive beak, used to cut and crack through the hardest of seeds and fruit pits. Despite its powerful beak, this is a shy bird of large tracts of forest interiors, although it has also become more widespread in edge habitats, shrubby vegetations, and in cherry and sour cherry orchards, especially in eastern Europe. In Britain, this species went from a rare winter migrant in the nineteenth century to a widespread nesting species in the nineteenth to twentieth centuries, with a sharp and sudden decline of its breeding population sizes in recent decades.

In suitable habitats, this species is common and pairs maintain small nesting territories in loose aggregations. Even where this bird coexists with people, it remains sensitive during the breeding season, and nest abandonment is one of the main causes of reproductive failure. Lost nests are readily followed up by second attempts, making for a prolonged breeding season.

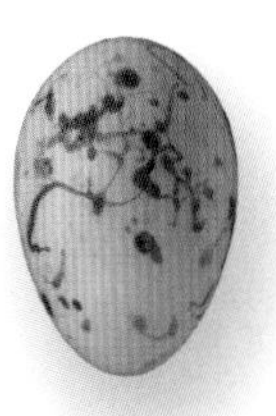

Actual size

The egg of the Hawfinch is light blue or gray green in color, with variably shaped, black markings, and is 15⁄16 x ⅔ in (24 x 17 mm) in dimensions. Once the eggs hatch, instead of seeds, the parents feed a mostly insect-based diet to the nestlings.

ORDER	Passeriformes
FAMILY	Fringillidae
BREEDING RANGE	Hawaiian Islands
BREEDING HABITAT	Higher-elevation, wet native forests on most of the larger Hawaiian Islands
NEST TYPE AND PLACEMENT	Cup nest, placed on a high, terminal branch of its food tree, the ohia-lehua, but also on tree ferns, lava tubes, or in cavities; made of twigs, grasses, mosses, and lichens
CONSERVATION STATUS	Least concern
COLLECTION NO.	FMNH 14773

ADULT BIRD SIZE
4–5 in (10–13 cm)

INCUBATION
13–14 days

CLUTCH SIZE
2–4 eggs

HIMATIONE SANGUINEA

APAPANE

PASSERIFORMES

The Apapane is a bright red member of the diverse Hawaiian honeycreeper lineages. Both parents of Apapane pairs contribute to the reproductive efforts, but each in different roles. The pair builds the nest together in just under a week; the female then lays and incubates the eggs alone, but the male stays and sings nearby. He feeds nectar to the female during the incubation and brooding stages. Both parents provision the nestlings, and keep the nest clean by removing fecal sacs produced by the chicks.

Descended from some far-reaching finches that landed on the newly formed volcanic islands several million years ago, most honeycreepers are now extinct, threatened, or endangered. The Apapane egg shown here is from an extinct population on Laysan Island, where the bird's habitat was destroyed by introduced rabbits. On the main Hawaiian Islands, Apapane are now restricted to forests above 3,300 ft (1,000 m), where mosquitoes that carry avian pox and malaria cannot survive.

Clutch

The egg of the Apapane is white in coloration, has sparse reddish spotting, and its size is 15⁄16 x 2⁄3 in (24 x 17 mm). The breeding season lasts long enough to allow two full egg-to-fledging cycles, and the parents typically broaden their diet from nectar to include caterpillars and spiders, to be fed to the nestlings.

Actual size

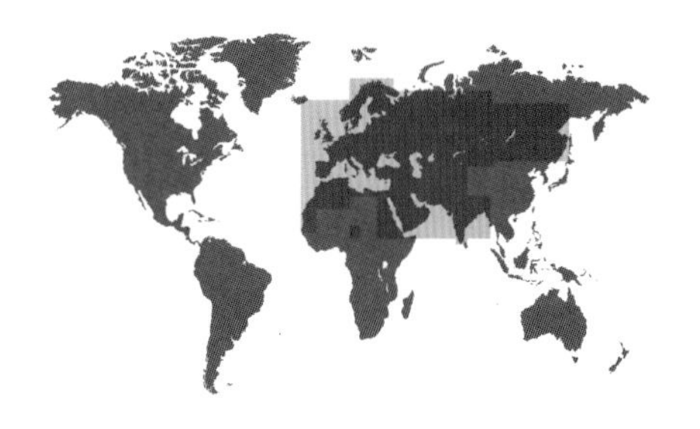

ORDER	Passeriformes
FAMILY	Passeridae
BREEDING RANGE	North Africa, Europe, northern, western, and south Asia
BREEDING HABITAT	Farms, rural settlements, city streets and parks
NEST TYPE AND PLACEMENT	Messy domed nest, stuffed into a natural cavity or accidental crevice; made of coarse dried vegetation, lined with finer grasses, hair, feathers, string, and paper
CONSERVATION STATUS	Least concern
COLLECTION NO.	FMNH 10709

ADULT BIRD SIZE
6–6½ in (15–17 cm)

INCUBATION
10–14 days

CLUTCH SIZE
4–5 eggs

PASSER DOMESTICUS

HOUSE SPARROW

PASSERIFORMES

Clutch

As its Latin name *domesticus* suggests, this species is the ultimate city bird. House Sparrows can successfully nest in any crevice with a large enough opening to allow entry: from woodpecker holes, to broken tree limbs, traffic light cases, wall crevices, nestboxes, and abandoned car engines. In areas where cavities are sparse, such as New Zealand pastures, these sparrows revert to their ancestral weaver-finch habits, and build a freestanding, domed nest in dense bushes.

Native to Europe and Asia, the House Sparrow has been successfully introduced to all inhabited continents, and many remote oceanic islands, too. Throughout their introduced ranges, these highly variable and adaptable birds follow several rules first seen and described in native bird species: they are paler in plumage in drier habitats, larger in size further from the equator, and lay fewer eggs closer to the tropics.

Actual size

The egg of the House Sparrow is light white to greenish white or blueish white in background coloration, with variable gray or brown spotting, and is ⅞ x ⅝ in (22 x 16 mm) in size. This species seems unable to recognize its own eggs, as it does not practice selective removal of foreign eggs from its nests.

ORDER	Passeriformes
FAMILY	Passeridae
BREEDING RANGE	Temperate Europe, north, central, and Southeast Asia
BREEDING HABITAT	Rural areas, farmland, open forests and edges, parklands
NEST TYPE AND PLACEMENT	In a natural tree hole or old woodpecker cavity, also nestboxes, holes in walls, or cracks in buildings; a messy domed nest built of hay, grasses, stems, and lined heavily with feathers
CONSERVATION STATUS	Least concern
COLLECTION NO.	FMNH 10725

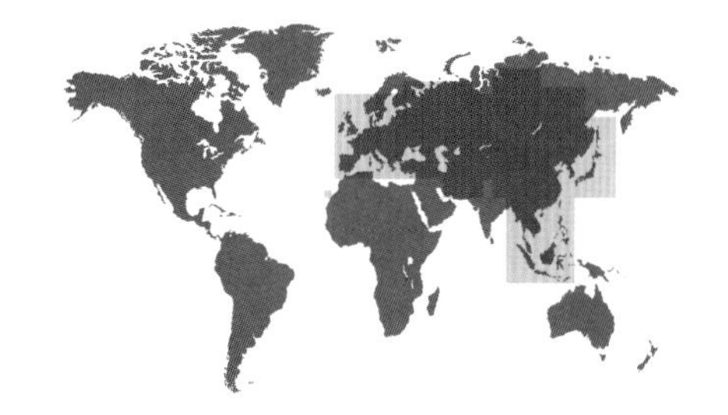

ADULT BIRD SIZE
5–5½ in (13–14 cm)

INCUBATION
10–14 days

CLUTCH SIZE
4–7 eggs

PASSER MONTANUS

EURASIAN TREE SPARROW

PASSERIFORMES

To assure mating success, the male Tree Sparrow often declares ownership of a suitable cavity in the winter by calling loudly and repeatedly to attract available females. Depending on nest-site availability and density, these birds may breed in loose colonies or solitarily. Colonial nesters tend to lay a large first clutch but smaller second clutches, whereas solitary nesters lay smaller first clutches but larger second clutches. An individual pair might start the season nesting colonially and end it breeding solitarily, shifting their nesting location to increase seasonal reproductive success. Instead of using a cavity, individuals sometimes will take up residence and lay eggs in the bulky twig-and-stick platform of large bird nests, including occupied nests of ospreys, magpies, storks, and herons.

Despite its name, this species is closely related to the House Sparrow, and not to the American Tree Sparrow. Accordingly, near St. Louis, in the United States, where a small introduced population has persisted for over 140 years, this species is referred to as the German Sparrow, to identify the species' introduction to the region by these European settlers.

Clutch

The egg of the Eurasian Tree Sparrow is white to pale gray or brown in background coloration, with heavy, darker maculation, and is ¾ x 9⁄16 in (19 x 14 mm) in dimensions.

Actual size

ORDER	Passeriformes
FAMILY	Passeridae
BREEDING RANGE	North Africa, southern Europe, and western and central Asia
BREEDING HABITAT	Barren, rocky hillsides, grassy fields with outcroppings; near farm houses and villages
NEST TYPE AND PLACEMENT	In crevices in cliffs, walls, or houses, also hollow tree trunks; messy cup or domed nest made of grasses, rootlets filaments, wool, and feathers
CONSERVATION STATUS	Least concern
COLLECTION NO.	FMNH 20622

ADULT BIRD SIZE
6–6½ in (15–17 cm)

INCUBATION
12–15 days

CLUTCH SIZE
4–6 eggs

PETRONIA PETRONIA

ROCK SPARROW

PASSERIFORMES

Clutch

The egg of the Rock Sparrow is white in background coloration, with heavy and cloudy brown markings all over the shell, and measures ¾ x ⅝ in (19 x 16 mm). In higher-elevation habitats, nest predators are rarer, and Rock Sparrows lay large eggs and fewer of them; nearly 90 percent of nests at those sites successfully produce at least one fledgling.

Drab in appearance, with a small yellow throat patch, the Rock Sparrow has become a role model to understand the reproductive complexities of seemingly unornamented species. The yellow throat patch is variable in size between individuals, and is present in both sexes. Males with larger patches usually win food fights, and a larger patch also indicates to females that he will engage in higher paternal provisioning rates at the nest. Females' feeding rates at the nest are not related to their throat patch size, but those with larger patches (even patches created experimentally) are more likely to be dominant in obtaining food.

Females also pay close attention to the quality of the males' song when making sexual decisions. The song of older males is higher pitched, but produced at a slower tempo; this seems to be attractive to females on nearby territories, whose nests end up populated with several eggs fertilized by the older neighbor. In response, males with unfaithful females sing louder while their mates are away from the territory.

Actual size

ORDER	Passeriformes
FAMILY	Passeridae
BREEDING RANGE	Montane regions from southern Europe through central Asia
BREEDING HABITAT	Barren, grassy and rocky mountain tops, alpine meadows
NEST TYPE AND PLACEMENT	Open cup built at the end of the burrow, made from grasses, stems, and rootlets; placed in crevices, under boulders, or in rodent burrows
CONSERVATION STATUS	Least concern
COLLECTION NO.	FMNH 20642

ADULT BIRD SIZE
6½–7½ in (17–19 cm)

INCUBATION
13–14 days

CLUTCH SIZE
3–4 eggs

MONTIFRINGILLA NIVALIS

WHITE-WINGED SNOWFINCH

PASSERIFORMES

These high-elevation, mountain-top birds have evolved behavioral and morphological features that enable them to succeed in this cold, snowy habitat. Compared to other alpine, open-cup-nesting passerine species, snowfinches start nesting relatively early in the spring, and many pairs fit in two complete nesting cycles. This may be possible because they nest in burrows and crevices that provide a well-protected and thermally insulated microhabitat. The eggs can be left here while the female feeds herself off the nest, without exposing them to prolonged periods of low temperature.

With human alteration of even high montane elevations, snowfinches have readily adapted to ski-slopes and vacation resorts, and remain there even throughout heavy winter snowfalls. At the same time, some breeding populations remain highly isolated in pristine mountains and are poorly studied, rarely monitored, or just recently discovered and documented.

Clutch

The egg of the White-winged Snowfinch is off- white to cream in coloration, immaculate, and 13⁄16 x 5⁄8 in (21 x 16 mm) in dimensions. The female alone builds the nest, incubates the eggs, and then both parents provision the nestlings.

Actual size

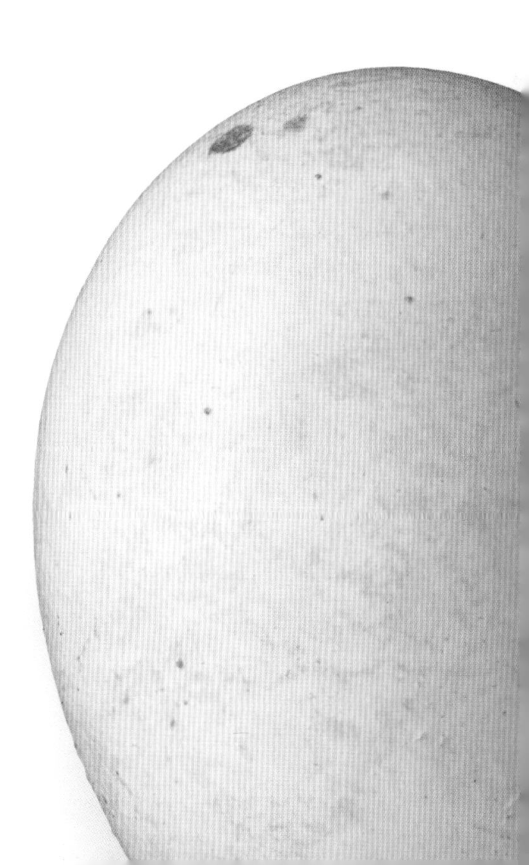

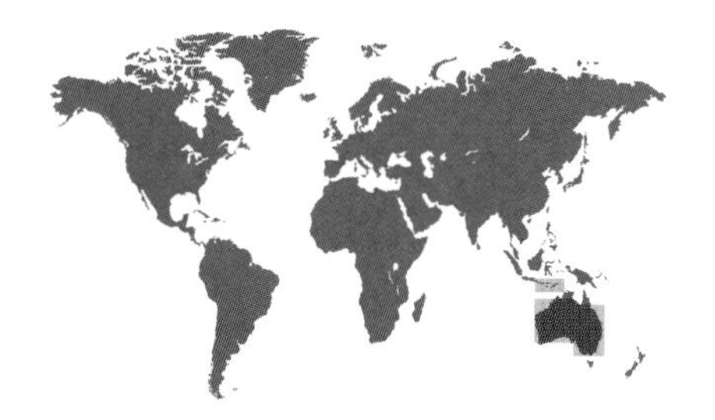

ORDER	Passeriformes
FAMILY	Estrildidae
BREEDING RANGE	Inland Australia, Indonesian archipelago
BREEDING HABITAT	Grassland, thickets, open forests, usually close to water; also artificial watering holes, farmland, and lawns
NEST TYPE AND PLACEMENT	Domed nest made of hay, stems, and fine twigs, lined with finer grasses, wool, and feathers
CONSERVATION STATUS	Least concern
COLLECTION NO.	WFVZ 159795

ADULT BIRD SIZE
4–4½ in (10–11 cm)

INCUBATION
14–16 days

CLUTCH SIZE
2–7 eggs

TAENIOPYGIA GUTTATA

ZEBRA FINCH

PASSERIFORMES

Clutch

Zebra Finches in the wild initiate breeding when encountering reliable access to water; in captivity, with unlimited access to water, these birds are ready to breed year-round. The socially monogamous pair bond, which lasts for a lifetime in captivity, is also present in the wild and represents a genetically faithful mating relationship; all eggs laid by a female are sired by the father with whom she is paired.

Zebra Finches are the avian equivalent of the white laboratory rat; due to the ease with which they can be bred in captivity, they are the subject of numerous genetic, morphological, developmental, neurophysiological, and behavioral studies. The full genome of the Zebra Finch has been sequenced, allowing researchers to investigate the genetic basis for developmental similarities between the ways male Zebra Finches learn to imitate their father's song and human children learn to babble and then speak their parents' language.

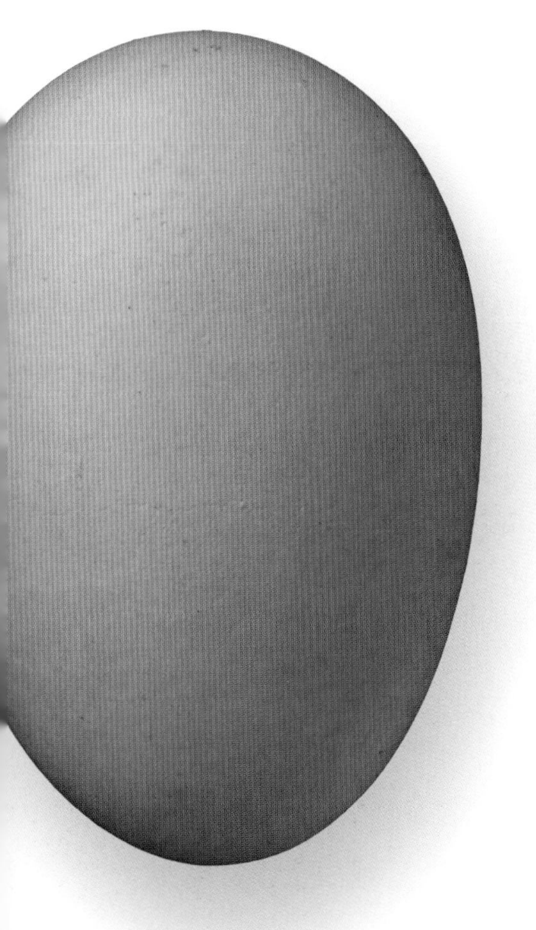

Actual size

The egg of the Zebra Finch is white or pale grayish blue in color, clear of markings, and is ⅝ x ⅜ in (16 x 10 mm) in size. Females readily deposit some of their eggs in the nests of other Zebra Finches, especially when their own nesting attempt is destroyed by weather or predators during the laying period.

ORDER	Passeriformes
FAMILY	Viduidae
BREEDING RANGE	Sub-Saharan Africa
BREEDING HABITAT	Dry, open wooded areas, thickets, grass fields
NEST TYPE AND PLACEMENT	Obligate brood parasite; lays eggs in the hay nests built in low shrubs by its common host, the Green-winged Pytilia
CONSERVATION STATUS	Least concern
COLLECTION NO.	WFVZ 60374

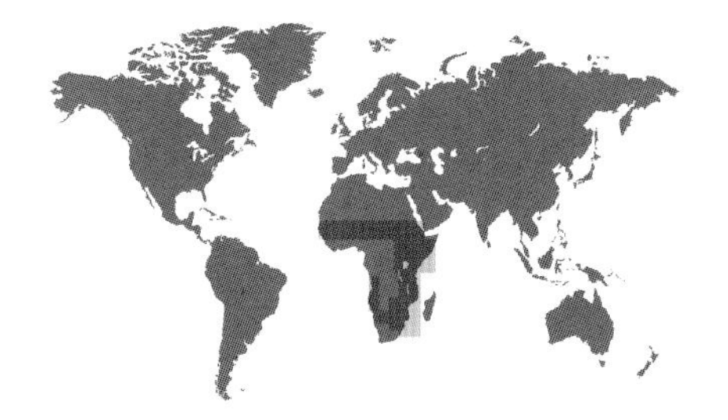

ADULT BIRD SIZE
5–6 in (13–15 cm)

INCUBATION
11–13 days

CLUTCH SIZE
1 egg per day

VIDUA PARADISAEA

EASTERN PARADISE WHYDAH

PASSERIFORMES

Whydahs are obligate brood parasites, laying eggs in estrildid finch nests throughout Africa. Having a different species as foster parents presents developmental conundrums for whydah young: how can they recognize their own species if they have never seen or heard anyone else like themselves before?

Clutch

Eastern Paradise Whydahs are songbirds, which learn to mimic tutor songs; other types of vocalizations, such as begging calls, are not learned from tutors. Thus, apparently songs learned from host males cannot serve as a reliable species-recognition signal for parasites. But whydahs have turned this situation to their advantage: young male whydahs sing the song of the host species but they also incorporate whydah begging calls into courtship vocalizations. In turn, the young females remember their foster father's song, but also recognize the whydah begging call notes, so they are only attracted to singing whydah males who grew up raised by the same host species as the parasitic female herself.

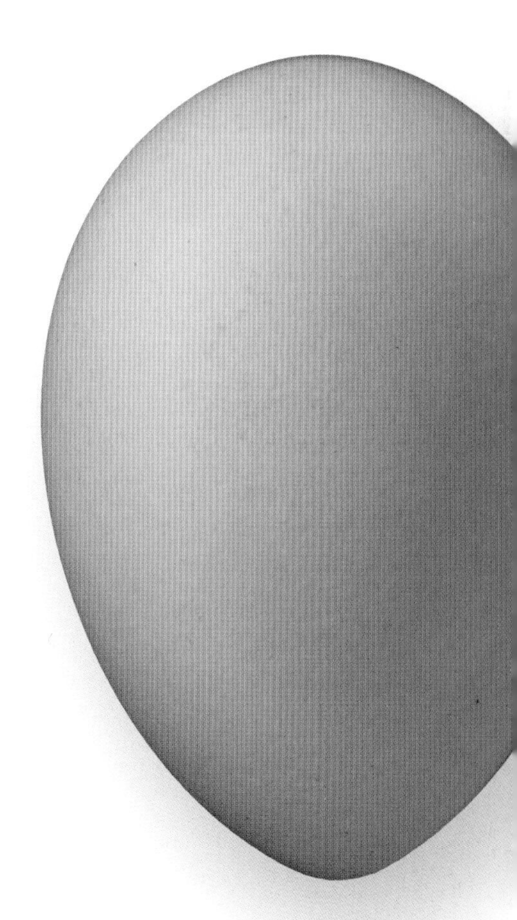

Actual size

The egg of the Eastern Paradise Whydah is white in background color, immaculate, and measures 11⁄16 x ½ in (18 x 13 mm) in size. The female typically lays her eggs in nests of the Green-winged Pytilia, but occasionally other hosts are used, leading to the formation of a new generation of whydahs mimicking and being attracted to the new host's songs and males who mimic those songs.

APPENDICES

GLOSSARY

Altricial The state of development of a chick that it needs to remain in the nest for prolonged periods as it is typically naked, blind, and depends on the parents fully for food. See Precocial.

Assortative mating Pairing and reproducing with an individual whose characteristics are similar ("likes attract"); the opposite of disassortative mating ("opposites attract").

Banding Placing a numbered metal band on the leg of a bird to identify and follow it as an individual throughout time. Placing one or more color bands on the same bird creates unique combinations that allow scientists to identify individual birds without having to recapture them.

Banding studies Scientific research using data on banded birds, often spanning multiple years with recaptures in multiple breeding seasons.

Benthic Living in the bottom layer of water and the sediment of a lake or ocean.

Brood patch A vascularized patch of featherless, bare skin that forms on the belly of many incubating birds during the breeding season, to allow efficient transfer of heat from adult to eggs.

Brood provisioning Dividing up fledgling chicks between the parents (and helpers) for more efficient care-giving after the young leave the nest.

Brooding An adult using its body and plumage to cover the chicks and keep them warm in or outside the nest.

Bycatch Unintentionally caught and killed organisms resulting from fishing, usually with nets or long-line hooks.

Camouflage Coloration and patterning that blends in well with the surroundings. See also Cryptic.

Color morph A distinct plumage coloration and pattern among several other colorations and patterns also occurring in the same population; a population may have several color morphs.

Commensal Coexisting with another organism without harm or benefit.

Consortship Escorting, following, and keeping near another individual, usually referring to behavior associated with a seasonal or long-term pair bond, but could also be seen in other mating or social groups.

Conspecific Of the same species. See Heterospecific.

Crepuscular Active at low light, including dusk and dawn.

Crop A muscular compartment just past the throat of a bird that helps to grind and break up hard foods, including seeds.

Cryptic Hidden against a background.

Cryptic species Two species that appear similar in morphology but possess other evolutionary (genetic, behavioral, etc.) differences that are the result of separate evolutionary paths.

Dabbling duck Collective common name for duck species that feed on or just below the surface of a pond, lake, river, or the sea; dipping their heads underwater, just their tails are visible.

Delayed plumage maturation The state of being reproductively ready but not yet having attained adult plumage. This is common in many bird groups, including gulls (Laridae) and manakins (Pipridae).

Dispersal The directional movement from one area into another. This can be "natal dispersal," which is movement by an individual away from its nest to its first breeding site, and "breeding" dispersal, between subsequent nesting sites.

Duetting Singing together, usually the male and female of a pair, with sounds overlapping or in close temporal proximity, often in a complex stereotyped format.

Embryogenesis The formation and development of a fertilized egg into an embryonic organism.

Endemic Occurring only within the confines of a specific locality, including islands, countries, continents, or other geographic regions.

Extirpation Being removed from an area, resulting in a local, regional, or total extinction.

Fecal sac Excrement produced by nestlings, enclosed in a gelatinous membrane, and easily carried in the beak and removed from the nest by parents.

Fledging The act of young departing from the nest, with or without powered flight.

Fledgling Chicks that have grown their initial flight feathers and left the nest, whether they are able to fly or not. Some fledglings still depend on parental care and feeding. See Nestling.

Fugitive species A species with good dispersal capabilities that allow it to occupy a variety of constantly changing habitats.

Genetic monogamy Faithfulness between the two members of a breeding pair, to the preclusion of extra-pair matings resulting in fertilizations. See Social monogamy.

Helpers at the nest Individuals other than the breeding pair who provide parental care including incubation, feeding, brooding, and/or protection of a nest site. Helpers in birds are often, but not always, young from a previous brood by the pair.

Heterospecific Of a different species. See Conspecific.

Hybridization The act of interbreeding between individuals from different species.

Immaculate Clear, having no spots.

Incubation The act of warming eggs to raise the internal temperature sufficiently high for embryonic development.

Kleptoparasite A bird that obtains food or prey by stealing it from another bird.

Lek Group of same-sex individuals displaying together to attract the opposite sex for mating.

Long-distance migrant A migrant bird engaging in a journey that takes it to another continent or across marine bodies of water. See Short-distance migrant.

Long-line fishing An oceanic fishing practice based on lines of baited hooks at regular intervals, lowered into great depths. Birds diving after the bait often get caught and drowned.

Maculation Spotting, speckling, blotching, or other patterning against a uniform background on an egg.

Melanin Pigment molecules responsible for many types of dark coloration in feathers and skin of birds and other organisms.

Mimetic Resembling (or mimicking) a model in appearance. Some brood parasitic bird species lay eggs that have evolved to mimic the appearance of the eggs of its host.

Natal A term referring to the locality where a bird hatched.

Nest failure Loss of the nest and its contents due to one or more factors, including predators, competition, or weather.

Nestling A chick that has hatched and is still in the nest.

Nidifugous Ready to leave the nest soon after hatching; the opposite of nidiculous.

Nocturnal Being active during nighttime.

Obligate Required, involuntary, or inflexible; generally under genetic control. A behavior that is present in all individuals of the same species.

Oscine Refers to the songbird lineage of Passerine birds, with a functional syrinx, and imitation-based song learning.

Pair bond The social attachment between a male and female for reproductive purposes.

Philopatry The tendency to return to the same place; for example natal philopatry refers to returning to the site of the nest where the bird was hatched.

Passerine Birds of the order Passeriformes, commonly known as perching birds. This order makes up roughly half of modern avian species.

Pelagic Of open water; the region of a lake or an ocean which is neither close to the shore, nor the bottom.

Phylogeny The tree of life, representing the evolutionary history and relationship of organisms as their lineages change through time.

Polyandry A breeding system with one female mating and producing young with multiple males.

Polygynandry A breeding system in which both polygyny and polyandry are present; multiple females may mate with one male and a female's eggs may be fertilized by multiple males.

Polygyny A breeding system with one male mating and producing young with multiple females.

Polymorphism Refers to a trait that has variation within a species.

Precocial The state of development of a chick such that it is ready to leave the nest soon after it hatches and dries; it may or may not be able to feed itself. See Altricial.

Raptor A term for a bird of prey.

Ratites A group of flightless birds native to the southern hemisphere; includes ostriches, emus, rheas, kiwis, and the extinct moas and elephantbirds.

Regenerating Regrowing lost or destroyed tissues; also can refer to vegetation patches.

Regurgitate To throw up food previously swallowed; in birds, this is a common method for parents to provide food to dependent young.

Sedentary Staying put in one place; for birds, this generally means the opposite of migratory.

Sexual dimorphism Differences in size, appearance, and/or behavior between females and males.

Short-distance migrant A seasonal traveller remaining within the confines of a continent. See Long-distance migrant.

Social monogamy A male and female pairing to share duties associated with caring for eggs and chicks; the young may or may not be sired by the male of the pair. See Genetic monogamy.

Sociality The need and ability to exist and interact in groups.

Stopover A location where migrant birds land and spend time en route to feed, rest, and refuel for the next leg of their aerial journey.

Suboscine Refers to the lineage of Passerine birds without a functional syrinx; includes tyrant flycatchers and antbirds.

Sympatric Occurring in the same, or overlapping, geographic ranges.

Synchrony Coordinated timing within an individual nest or between nests in a population; for example, a female may delay incubation of eggs laid on different days, so that they hatch on the same day.

Syrinx The vocal organ of birds, which is analogous to a mammal's larynx, but without vocal chords.

Tutor song A song heard and then learned by songbird chicks during their development, as opposed to a song that is genetically encoded. Usually it is the father's song.

Volant Able to fly, opposite of flightless.

RESOURCES & USEFUL INFORMATION

The collections, journals, books, and web sites listed below are a selection of the great number of resources available on the subject of birds and eggs.

COLLECTIONS

Many people encounter bird eggs unexpectedly: seeing a fallen shell on the ground from a nest high above in the tree or exposing a clutch of eggs while pruning the backyard bushes. If the nest is still active, it is best to walk away and let the owner(s) continue taking care of the eggs. For a more prolonged and less stressful exposure to eggs, a visit can be arranged by visiting museums, zoos, and natural history exhibitions that display specimens from their collections. Chances are that visitors will only see a small subset of what is in storage, and admittedly, colors fade over time even in dark and climate-controlled museum storage facilities, but viewing eggs in an exhibition avoids the concerns about interfering with successful incubation and hatching of a live egg and embryo. Most natural history museums hold and show bird eggs from their collections, including:

The Field Museum of Natural History (Chicago, USA), the Western Foundation of Vertebrate Zoology (near Los Angeles, USA), the Natural History Museum (Tring, Hertfordshire, UK), the Peabody Museum of Natural History (New Haven, USA), University Museum of Zoology (Cambridge, UK), the Museum of Comparative Zoology (Cambridge, USA), and many others. Each of these museums has its own web site, with information about current exhibits and visiting hours.

Backyard breeding

Having a personal collection of sustainably collected eggs at home is also an option, but always remember that most bird species—including their nest, eggs, and even fallen feathers—in nature are protected. Even more importantly, the reproductive biology of many birds is so poorly known that it is imperative to leave their nests alone and allow scientists to take a first look at them. Instead, consider keeping birds at home that will yield a selection of diverse eggs. For example, chickens, guineafowl, ducks, and geese in the backyard, and parakeets, finches, and canaries in an aviary, will often yield eggs that are best collected early and not allowed to hatch. Several chicken breeds lay eggs that are colored distinctively, from the familiar white and brown eggs of most egg-producing breeds to the blue and green eggs of the Araucana fowl. Also, whereas Bobwhite Quail and Painted Quail eggs are relatively uniformly white, and brown, respectively, Japanese Quail lay highly variable eggs that vary in background color and spotting pattern; a special mutation in Japanese Quail has also yielded a pure blue eggshell coloration.

SCIENTIFIC JOURNALS

To read about the latest discoveries in ornithology, behavioral biology, evolutionary history, and the chemical and structural basis of bird eggs, many primary scientific journals provide summaries or instant peeks into their latest content through their web sites; readers who would like to gain access to the full article, may do so through institutional libraries or personal subscriptions, or (the best-kept secret in academic publishing!) simply by sending a quick email to the author(s) of the article, requesting a personal PDF copy of the work. The following list of scientific journals is not complete, but will be helpful to get started:

General, biological coverage

Animal Behaviour; Behavioral Ecology; Behavioral Ecology & Sociobiology; Behaviour; Current Biology; Ecography; Ecology; Ethology; Ethology Ecology & Evolution; Evolution; Evolutionary Ecology Research; Functional Ecology; Journal of Animal Ecology; Journal of Evolutionary Biology; Journal of Experimental Biology; Nature; Nature Communications; Oecologia; PLoS ONE; Proceedings of the National Academy of Sciences of the USA; Proceedings of the Royal Society of London B; The Royal Society Journal Interface; Science; and the list goes on.

Specialist, ornithological journals

The Auk: Ornithological Advances; The Condor: Ornithological Applications; The Emu; The Ibis; Journal of Avian Biology; Journal of Field Ornithology; Journal of Ornithology; Notornis; and the *Wilson Journal of Ornithology.*

BOOKS

There are many thorough and locally helpful guide books written about bird nests and eggs, and these provide help with and clues for identifying an accidentally encountered clutch; other than a standard internet search engine, some of these resources include:

General reading

Birds' Eggs by Michael Walters (New York, Dorling Kindersley, Eyewitness Handbooks, 1994). For a worldwide survey of the diversity of avian eggs, with life-size illustrations, life history data, and an illustration of the diversity of eggs within many species, there is the classic work. Much has changed regarding the conservation status, classification and taxonomy, and the scientific knowledge of the species since this book was published, but its approach and content were definitely an inspiration for our book.

Eastern Birds' Nests and ***Western Birds' Nests*** by Hal H. Harrison are both available in hard copy (Boston, MA, Houghton Mifflin Harcourt, Peterson Field Guides, 2001 and 1998 respectively). Components of the titles are now also included in the searchable e-book formats of the Peterson Field Guide series.

Egg & Nest by Rosamond Purcell, Linnea S. Hall, René Corado, and Bernd Heinrich (Cambridge, MA, Belknap/Harvard University Press, 2008). For a visually enticing artist's impression, historian's account, conservationist's view, and scientist's tour of the Western Foundation of Vertebrate Zoology's bird nest and egg collection, this is the book to read.

Nests, Eggs, and Nestlings of North American Birds by Paul J. Baicich and Colin J. O. Harrison (Princeton, NJ, Princeton University Press, second edition 2005). This guide is to be consulted for truly detailed and informative text, data, and illustrations of not only eggs but also nests and nestlings.

Specialist reading

Two groups of birds do not incubate their own eggs directly. The first are the Megapodes, brush turkeys and their allies. They use biochemical, solar, and geothermal heat sources to hatch eggs buried in heaping piles of rotting vegetation or sand banks on beaches and volcanic slopes; these birds' unique natural history and conservation plight is fully covered in ***Mound-builders*** by Darryl Jones and Ann Goth, (Collingwood, VC, Australia, CSIRO Publishing Press, 2009).

The second group of birds that do not incubate the eggs are brood parasites; for an inspiring account of the unusual reproductive, behavioral, and cognitive traits of these species, the must-have book is ***Cuckoos, Cowbirds, and Other Cheats*** by N. B. Davies (London, A & C Black, Poyser Monographs, 2011).

To see a Cuckoo chick attempting to evict the eggs of its host parents, watch this short clip: **http://www.youtube.com/watch?v=yD18VFgM-Nw**

To learn about the structure, function, and development of avian eggs, there is much to be gained by examining the fossil record, including the reproductive and parental strategies of egg-laying dinosaurs. ***Eggs, Nests, and Baby Dinosaurs: A Look at Dinosaur Reproduction*** by Kenneth Carpenter (Bloomington, IN, Indiana University Press, 2000), makes the scientific literature on the recent developments and discoveries accessible to a non-specialist audience.

Architecture by Birds and Insects: A Natural Art by Peggy Macnamara (Chicago, IL, University of Chicago Press, 2008). Peggy Macnamara is the Field Museum's Artist-in-Residence, and her book features watercolors of bird and insect nests drawn from nature and from the Museum's collections.

Finally, for a glimpse into the life of a museum curator, involved in organizing, expanding, and studying eggs, both from living and extinct birds, there is a new personal account of real-life events and serendipitous anecdotes in ***The Owl that Fell from the Sky: Stories of a Museum Curator*** by Brian Gill (Wellington, NZ, Awa Press, 2012).

SOME USEFUL WEB SITES

The Field Museum of Natural History Bird Collection Database
www.fm1.fieldmuseum.org/birds/egg_index.php

The Western Foundation of Vertebrate Zoology Bird Collection
www.wfvz.org

The main resource for the Latin and common names of species included in this book is:

The Clements Checklist of Birds of the World by James F. Clements (Ithaca, NY, Cornell University Press, sixth edition, 2012 (online version 6.8)). Available at www.birds.cornell.edu/clementschecklist/

The Cornell Laboratory of Ornithology's Project Nest Watch site, where nests and eggs can be decoded using a geographic and a taxonomic guide: **http://nestwatch.org/learn/how-to-nestwatch/identifying-nests-and-eggs** and the subscription-only service of the Birds of North America series: **http://bna.birds.cornell.edu**

The CLASSIFICATION *of* BIRDS

Classifications of animals and plants are based on the best available data about their evolutionary relationships, and thus they often change as we learn more. The kinds of data on which to base bird classification are greater than what exists for most other organisms. Even so, we continue to learn more about relationships in birds all the time.

It has long been accepted that all living birds descend from a shared common ancestor. But the study of newly discovered dinosaur fossils has provided additional evidence to support the hypothesis that the ancestor of modern birds evolved from within the therapods, a specific group of dinosaurs. Studies suggest that a number of bird lineages may already have existed and survived the extinction of the dinosaurs 65 million years ago.

Major groupings of modern birds (orders and families), their composition, and their relationships have long been researched and debated. New fossils and genetic studies have greatly changed our understanding of the branching patterns among these lineages. The avian tree on the right is based on the largest data set of DNA data gathered to date (Hackett et al. 2008). DNA sequences totaling 32,000 base pairs from nuclear genes were compared for 169 species. The research was a collaborative effort of multiple lab groups led by Shannon Hackett at the Field Museum with support from the National Science Foundation's "Tree of Life" program. The different colored parts of the tree represent the best-supported lineages (clades) found in the study.

At the bottom of the figure is the oldest division in modern birds, which separates the Paleognathes (in purple: ostriches, rheas, cassowaries, emus, kiwis, tinamous) from the Neognathes (all other birds). In orange are the ducks and chicken-like birds; the branching of these two groups at the base of the tree has now been found in multiple studies. In brown is a set of birds including hummingbirds, swifts, and nightjars, which have long been put together although in this study hummingbirds and swifts may share a common ancestor with nightjars. The blue and yellow clades are groups that are totally or largely tied to water; the black clade is a mixture of some aquatic groups (rails and cranes) along with the decidedly non-aquatic cuckoos and bustards. Lastly, the green clade is land birds, the most diverse order of all, including the Passeriformes, or perching birds. The exact relationships of lineages in gray could not be determined in these analyses; there is still work to do.

A major result of such new thinking on relationships is insight into the evolution of all avian traits, including those of eggs. An example of a previously unrecognized sister relationship uncovered by recent DNA studies is that grebes and flamingos are each other's closest living relatives. Researchers also have noted that the eggs of these two groups share unique features as well, particularly a chalky layer of calcium phosphate that covers them. As with any aspect of avian biology, the shape, size, appearance and structure of eggs have evolved together with the lineages that lay them; this tree can help us build up our understanding of how eggs have changed through time.

LATIN & COMMON NAMES

Each species in this book has both a Latin name and a common name. The rules of scientific nomenclature dictate that each species must be described and named by a scientist, and the description published in a peer-reviewed journal. Latin or scientific names include a genus (e.g., *Turdus*) and a species epithet (e.g., *migratorius*); these are combined to form the unique name of the species (*Turdus migratorius*—the American Robin.) When a species is named, it is also placed in a genus with other species it is thought to be related to. For example, a great number of thrush species are grouped with the American Robin in the genus *Turdus*. A shared genus name indicates a consensus between scientists that this group of birds has a closer common ancestor than any of them has with birds outside the genus. As understanding of these relationships changes, the scientific names of the birds change.

Latin names provide a way for scientists around the world to share a common identifier for any species under discussion. International codes have been developed to help scientist adhere to rules about how to name new species and when necessary rename existing species as new research is done.

Common names have no scientific code governing them across countries, although each national may have a standardized list of species names. But because birds are familiar to so many people they have all been given common names. In different countries, a species may be known under different common names (e.g., Common Diver is called Common Loon in the U.S.). In this book, we have chosen to follow *The Clements Checklist of Birds of the World* as a standard for both our Scientific and Common names. This is a project that is regularly updated for new taxonomic findings under the direction of Tom Schulenberg at the Cornell Laboratory of Ornithology.

AUTHOR CITATIONS

The Clements Checklist of Birds of the World, James F. Clements (Ithaca, NY, Cornell University Press, sixth edition, 2012 (online version 6.8)). Available at www.birds.cornell.edu/clementschecklist/

Hackett, S.J., Kimball, R.T., Reddy, S., Bowie, R.C.K., Braun, E.L., Braun, M.J., Chojnowski, J.L., Cox, W.A., Han, K-L., Harshman, J., Huddleston, C.J., Marks, B.D., Miglia, K.J., Moore, W.S., Sheldon, F.H., Steadman, D.W., Witt, C.C., and Yuri, T. 2008. A phylogenomic study of birds reveals their evolutionary history. *Science.* 320:1763–8.

Passerines
Parrots
Falcons
Seriemas
Puffbirds, Jacamars
Barbets, Honeyguides, Woodpeckers
Kingfishers, Rollers, Bee-eaters, Motmots, Todies
Hornbills
Hoopoe, Woodhoopoes
Trogons
Cuckoo-Rollers
Owls
Mousebirds
New World Vultures
Hawks, Eagles, Secretary Bird
Plains-wanderer
Seedsnipes
Jacanas, Painted-snipes
Turnstones
Buttonquail
Gulls, Crab Plover
Thick-knees, Plovers, Oystercatchers, Sandpipers
Hammerkop
Shoebill
Pelicans
Herons
Ibises
Anhingas, Frigatebirds, Cormorants, Gannets
Storks
Penguins
Albatrosses, Shearwaters, Petrels
Loons
Turacos
Rails, Finfoots
Cranes, Limpkin, Trumpeters
Cuckoos
Bustards
Hummingbirds
Swifts
Owlet-Nightjar
Frogmouths
Nightjars
Oilbird
Potoos
Sunbittern, Kagu
Pigeons, Doves
Mesites
Tropicbirds
Hoatzin
Sandgrouse
Flamingos
Grebes
Pheasants, Quail, Guineafowl
Curassows, Chachalacas, Guans
Megapodes
Ducks, Geese, Other Waterfowl, Screamers
Kiwis
Cassowaries, Emus
Tinamous
Rheas
Ostriches
Neognathes
Paleognathes

INDEX *by* COMMON NAME

A
Acorn Woodpecker 366
African Stonechat 492
Alder Flycatcher 390
Amazonian Motmot 346
American Avocet 142
American Bittern 115
American Black Duck 84
American Coot 57
American Crow 436
American Dipper 473
American Flamingo 72
American Golden Plover 137
American Goldfinch 629
American Kestrel 278
American Redstart 543
American Robin 503
American Tree Sparrow 580
American White Pelican 128
American Wigeon 78
American Woodcock 163
Ancient Murrelet 192
Anhinga 133
Antarctic Giant Petrel 43
Apapane 635
Aplomado Falcon 281
Arctic Loon 53
Arctic Tern 180
Ash-throated Flycatcher 397
Asian Paradise Flycatcher 424
Atlantic Puffin 196

B
Bachman's Sparrow 579
Bachman's Warbler 528
Baird's Sparrow 585
Bald Eagle 258
Baltimore Oriole 619
Bank Swallow 449
Barn Owl 297
Barn Swallow 450
Barred Owl 304
Barrow's Goldeneye 92
Bar-tailed Godwit 155
Bay-breasted Warbler 549
Bearded Reedling 439
Bell's Sparrow 583
Bell's Vireo 415
Belted Kingfisher 349
Bendire's Thrasher 510
Bennett's Woodpecker 370
Berthelot's Pipit 518
Bewick's Wren 468
Black-and-white Warbler 531
Black-bellied Plover 136
Black-bellied Sandgrouse 289
Black-bellied Whistling-duck 94
Black-billed Magpie 432
Blackburnian Warbler 550
Blackcap 483
Black-capped Chickadee 453
Black-capped Vireo 416
Black-chinned Hummingbird 339
Black-collared Barbet 360
Black-crowned Night-heron 125
Black Curassow 218
Black-faced Grassquit 574
Black-footed Albatross 37
Black Francolin 225
Black Grouse 238
Black-headed Antthrush 385
Black Kite 256
Black-necked Grebe 70
Black-necked Stilt 141
Black Noddy 174
Black Phoebe 392
Blackpoll Warbler 553
Black Rail 58
Black Scimitarbill 357
Black Skimmer 183
Black-throated Accentor 516
Black-throated Blue Warbler 554
Black-throated Gray Warbler 561
Black-throated Green Warbler 565
Black Tinamou 206
Black Vulture 248
Black-winged Pratincole 166
Blue Bird-of-paradise 438
Blue-breasted Quail 228
Blue-cheeked Bee-eater 353
Blue-gray Gnatcatcher 472
Blue Grosbeak 603
Blue-headed Vireo 419
Blue Jay 429
Blue-winged Teal 82
Blue-winged Warbler 529
Boat-billed Heron 119
Bobolink 606
Bohemian Waxwing 520
Booted Eagle 275
Boreal Owl 307
Brant 86
Bridled Titmouse 455
Broad-billed Sandpiper 160
Broad-billed Tody 345
Broad-winged Hawk 270
Bronzed Cowbird 615
Bronze-winged Jacana 144
Brown-crested Flycatcher 399
Brown Eared-pheasant 231
Brown-headed Cowbird 616
Brown-headed Nuthatch 462
Brown Jay 426
Brown Noddy 173
Brown Pelican 129
Brown Thrasher 508
Budgerigar 291
Bulwer's Petrel 40
Burrowing Owl 303
Bushtit 460

C
Cabot's Tragopan 229
Cactus Wren 465
Calandra Lark 441
California Condor 250
California Quail 221
Canada Goose 85
Canada Warbler 566
Canyon Wren 466
Cape May Warbler 545
Cape Petrel 41
Carolina Chickadee 452
Carolina Wren 467
Caspian Tern 177
Cassin's Auklet 193
Cassin's Kingbird 403
Cassin's Sparrow 578
Cattle Egret 116
Cedar Waxwing 519
Cerulean Warbler 546
Cetti's Warbler 478
Chaffinch 621
Cheer Pheasant 232
Chestnut-backed Antbird 384
Chestnut-sided Warbler 552
Chimney Swift 334
Chipping Sparrow 581
Chuck-will's-widow 329
Cinereous Vulture 261
Cinnamon Teal 81
Citril Finch 632
Clapper Rail 63
Clark's Nutcracker 434
Cliff Swallow 451
Collared Flycatcher 490
Common Black-hawk 267
Common Bulbul 474
Common Cuckoo 320
Common Eider 102
Common Goldeneye 91
Common Grackle 613
Common Greenshank 150
Common Ground-dove 317
Common Kingfisher 347
Common Loon 52
Common Merganser 98
Common Murre 186
Common Nighthawk 327
Common Poorwill 328
Common Potoo 332
Common Raven 437
Common Sandpiper 147
Common Snipe 162
Common Starling 515
Common Swift 335
Common Tern 179
Common Tody-flycatcher 386
Common Yellowthroat 541
Connecticut Warbler 539
Cooper's Hawk 264
Coppery-bellied Puffleg 337
Costa's Hummingbird 340
Couch's Kingbird 402
Cream-colored Courser 165
Crested Caracara 277
Crested Lark 442
Crissal Thrasher 512
Curve-billed Thrasher 511

D
Dark-eyed Junco 596
Daurian Partridge 227
Demoiselle Crane 64
Desert Lark 440
Dickcissel 605
Dollarbird 355
Double-crested Cormorant 131
Downy Woodpecker 371
Dunlin 158
Dusky-capped Flycatcher 396

E
Eared Grebe 70
Eastern Bluebird 494
Eastern Kingbird 405
Eastern Meadowlark 609

Eastern Paradise Whydah 641
Eastern Phoebe 393
Eastern Screech Owl 298
Eastern Towhee 576
Eastern Whip-poor-will 331
Eastern Wood-pewee 389
Egyptian Vulture 259
Elegant Crested-tinamou 214
Elegant Tern 182
Elegant Trogon 343
Eleonora's Falcon 279
Elf Owl 302
Elliot's Pheasant 233
Emu 203
Eurasian Bullfinch 633
Eurasian Capercaillie 237
Eurasian Golden-oriole 423
Eurasian Griffon 260
Eurasian Hoopoe 356
Eurasian Jay 431
Eurasian Marsh-harrier 263
Eurasian Penduline Tit 458
Eurasian River Warbler 481
Eurasian Siskin 630
Eurasian Thick-knee 134
Eurasian Treecreeper 464
Eurasian Tree Sparrow 637
Eurasian Wren 470
Eurasian Wryneck 363
European Bee-eater 352
European Goldfinch 631
European Honey-buzzard 252
European Roller 354
European Storm-petrel 45

F

Ferruginous Pygmy-owl 301
Finsch's Wheatear 493
Firewood-gatherer 381
Florida Scrub-jay 430
Fork-tailed Storm-petrel 47
Forster's Tern 181
Fox Sparrow 588
Fulvous Chatterer 486
Fulvous Whistling-duck 95

G

Glossy Ibis 111
Golden-cheeked Warbler 564
Golden-crowned Kinglet 476
Golden Eagle 274
Golden-winged Warbler 530
Graceful Prinia 482
Grace's Warbler 560
Gray Catbird 505
Gray-cheeked Thrush 499
Gray-crowned Rosy-finch 622
Gray Heron 112
Gray-hooded Sierra-finch 572
Gray Jay 425
Gray-necked Wood-rail 55
Gray Parrot 294
Gray Vireo 417
Great Argus 235
Great Auk 189
Great Blue Heron 113
Great Bowerbird 409
Great Bustard 285
Great Cormorant 132
Great Crested Flycatcher 398
Great Egret 118
Great Elephantbird 205
Greater Honeyguide 362
Greater Painted-snipe 167
Greater Prairie-chicken 246
Greater Rhea 201
Greater Roadrunner 324
Greater Sage-grouse 241
Great Horned Owl 299
Great Kiskadee 400
Great Reed-warbler 480
Great Skua 184
Great-tailed Grackle 614
Great Tinamou 207
Green Heron 117
Green Jay 427
Green Kingfisher 351
Green Peafowl 236
Green-winged Teal 80
Guira Cuckoo 323
Gull-billed Tern 176
Gunnison Sage-grouse 242

H

Hairy Woodpecker 372
Harlequin Duck 96
Harris's Hawk 268
Harris's Sparrow 594
Hawfinch 634
Hazel Grouse 239
Helmeted Guineafowl 219
Henslow's Sparrow 586
Hermit Thrush 501
Hermit Warbler 563
Herring Gull 172
Hoatzin 326
Hooded Oriole 618
Hooded Warbler 542
Horned Grebe 69
Horned Lark 445
Horned Screamer 73
House Finch 626
House Sparrow 636
Hutton's Vireo 420
Hyacinth Macaw 295

I

Indigo Bunting 604
Ivory-billed Woodcreeper 383

J

Jackass Penguin 50

K

Kagu 287
Kentish Plover 138
Killdeer 140
King Penguin 49
King Rail 61
Kirtland's Warbler 544
Kittlitz's Murrelet 191

L

Lapland Longspur 522
Lark Bunting 584
Lark Sparrow 582
Least Bittern 123
Least Flycatcher 391
Least Grebe 71
Least Tern 175
Le Conte's Thrasher 513
Lesser Goldfinch 628
Lesser Scaup 89
Lesser Yellowlegs 152
Levaillant's Woodpecker 377
Lewis's Woodpecker 364
Light-mantled Albatross 39
Limpkin 67
Lincoln's Sparrow 590
Little Blue Heron 120
Little Bustard 286
Little Corella 290
Little Penguin 48
Little Shearwater 44
Little Tinamou 208
Loggerhead Shrike 413
Long-billed Curlew 154
Long-billed Thrasher 509
Long-eared Owl 305
Long-tailed Duck 97
Long-tailed Jaeger 185
Long-tailed Manakin 407
Long-tailed Sylph 336
Long-tailed Tit 459
Louisiana Waterthrush 526
Lucy's Warbler 536

M

Magnificent Frigatebird 127
Magnolia Warbler 548
Maleo 215
Mallard 83
Marsh Wren 471
Merlin 280
Miombo Scrub-robin 488
Montezuma Oropendola 620
Montezuma Quail 223
Mountain Bluebird 496
Mountain Chickadee 454
Mountain Quail 220
Mourning Dove 315
Mourning Warbler 540
Muscovy Duck 93

N

Nashville Warbler 537
Northern Bobwhite 222
Northern Cardinal 601
Northern Flicker 375
Northern Fulmar 42
Northern Gannet 130
Northern Goshawk 266
Northern House Wren 469
Northern Jacana 145
Northern Mockingbird 506
Northern Parula 547
Northern Pintail 77
Northern Rough-winged Swallow 446
Northern Saw-whet Owl 308
Northern Screamer 74
Northern Shoveler 79
Northern Shrike 412
Northern Waterthrush 527

O

Oilbird 333
Olive-sided Flycatcher 387
Orange-crowned Warbler 535
Orchard Oriole 617
Osprey 251
Ostrich 200
Ovenbird 524

P

Pacific Loon 54
Paint-billed Crake 59
Painted Redstart 568
Pale-billed Hornbill 309
Pallas's Sandgrouse 288
Pallas's Sea-eagle 257
Palm Tanager 571
Palm Warbler 555
Parakeet Auklet 194
Passenger Pigeon 316
Peregrine Falcon 282
Phainopepla 521
Pheasant-tailed Jacana 143
Pied Kingfisher 350
Pigeon Guillemot 190
Pileated Woodpecker 376
Pine Grosbeak 624
Pine Warbler 556
Pinyon Jay 428
Plain Chachalaca 217
Prairie Falcon 283
Prairie Warbler 559
Prothonotary Warbler 532
Purple Finch 625
Purple Martin 447

R
Razorbill 188
Red-bellied Woodpecker 367
Red-billed Chough 435
Red-billed Gull 171
Red-billed Leiothrix 485
Red-billed Tropicbird 126
Red-breasted Merganser 99
Red-breasted Nuthatch 463
Red-cockaded Woodpecker 373
Red-crested Pochard 100
Red Crossbill 627
Red-eyed Vireo 422
Redhead 90
Red-headed Woodpecker 365
Red-legged Kittiwake 169
Red-legged Seriema 284
Red-necked Nightjar 330
Red-shouldered Hawk 269
Red-tailed Hawk 272
Red-winged Blackbird 607
Red-winged Parrot 292
Red-winged Tinamou 211
Resplendent Quetzal 344
Rhinoceros Auklet 195
Ringed Kingfisher 348
Ring-necked Pheasant 234
Rock Bunting 598
Rock Partridge 224
Rock Pigeon 312
Rock Sandpiper 157
Rock Sparrow 638
Roseate Spoonbill 108
Rose-breasted Grosbeak 602
Rose-throated Becard 408
Rosy-faced Lovebird 293
Rosy Starling 514
Rough-legged Hawk 273
Ruby-crowned Kinglet 475
Ruby-throated Hummingbird 338
Ruddy Duck 101
Ruddy Quail-dove 318
Ruddy Shelduck 104
Ruddy Turnstone 156
Ruff 161
Ruffed Grouse 240
Rufous-collared Sparrow 592
Rufous-crowned Sparrow 577
Rufous Hornero 380
Rufous Hummingbird 341
Rufous-tailed Jacamar 359
Rufous-tailed Rock-thrush 491
Russet-throated Puffbird 358
Rusty Blackbird 612

S
Sabine's Gull 170
Sage Thrasher 507
Sandhill Crane 66
Say's Phoebe 394
Scarlet Ibis 109
Scarlet Tanager 599
Scissor-tailed Flycatcher 406
Seaside Sparrow 587
Secretary Bird 276
See-see Partridge 226
Semipalmated Plover 139
Sharp-shinned Hawk 265
Sharp-tailed Grouse 245
Shining Bronze-cuckoo 321
Short-eared Owl 306
Short-tailed Albatross 38
Short-toed Snake Eagle 262
Sky Lark 443
Small Ground-finch 575
Smoke-colored Pewee 388
Smooth-billed Ani 325
Snail Kite 255
Snow Bunting 523
Snow Goose 87
Snowy Egret 121
Snowy Owl 300
Solitary Sandpiper 148
Song Sparrow 589
Sora 60
Southern Brown Kiwi 204
Southern Cassowary 202
Southern Lapwing 135
Spectacled Eider 103
Spotted Flycatcher 487
Spotted Nothura 213
Spotted Redshank 149
Spruce Grouse 243
Squacco Heron 114
Steely-vented Hummingbird 342
Stilt Sandpiper 159
Streaked Scrub-Warbler 477
Sulphur-bellied Flycatcher 401
Summer Tanager 600
Superb Fairywren 410
Swainson's Hawk 271
Swainson's Thrush 500
Swainson's Warbler 533
Swallow-tailed Gull 168
Swallow-tailed Kite 253
Swamp Sparrow 591
Swinhoe's Pheasant 230

T
Tabon Scrubfowl 216
Tennessee Warbler 534
Terek Sandpiper 146
Thick-billed Murre 187
Thicket Tinamou 210
Thrush Nightingale 489
Townsend's Solitaire 497
Townsend's Warbler 562
Tree Swallow 448
Tricolored Blackbird 608
Tricolored Heron 122
Trumpeter Finch 623
Tufted Puffin 197
Tufted Titmouse 456
Tundra Swan 75
Turkey Vulture 249

U
Undulated Tinamou 209
Upland Sandpiper 153

V
Varied Thrush 504
Veery 498
Verdin 457
Vermilion Flycatcher 395
Violet-crested Turaco 319
Virginia Rail 62
Virginia's Warbler 538

W
Wandering Albatross 36
Warbling Vireo 421
Western Bluebird 495
Western Grebe 68
Western Kingbird 404
Western Meadowlark 610
White-bellied Nothura 212
White-breasted Nuthatch 461
White-collared Seedeater 573
White-crowned Pigeon 313
White-crowned Sparrow 595
White-eyed Vireo 414
White-fronted Goose 88
White-headed Vanga 411
White-headed Woodpecker 374
White Ibis 110
White-lined Tanager 570
White Stork 105
White-tailed Kite 254
White-throated Sparrow 593
White-throated Toucan 361
White Wagtail 517
White-winged Dove 314
White-winged Snowfinch 639
White-winged Tern 178
Whooping Crane 65
Wild Turkey 247
Willet 151
Williamson's Sapsucker 368
Willow Ptarmigan 244
Willow Warbler 479
Wilson's Phalarope 164
Wilson's Storm-petrel 46
Wilson's Warbler 567
Wood Duck 76
Wood Lark 444
Wood Stork 106
Wood Thrush 502
Worm-eating Warbler 525
Wren-like Rushbird 382
Wrentit 484

Y
Yellow-bellied Sapsucker 369
Yellow-billed Cuckoo 322
Yellow-billed Loon 51
Yellow-billed Magpie 433
Yellow-billed Stork 107
Yellow-breasted Chat 569
Yellow-crowned Night-heron 124
Yellow-crowned Parrot 296
Yellow-headed Blackbird 611
Yellowhammer 597
Yellow Rail 56
Yellow-rumped Warbler 557
Yellow-throated Vireo 418
Yellow-throated Warbler 558
Yellow Warbler 551

Z
Zebra Finch 640

INDEX *by* SCIENTIFIC NAME

A
Accipiter cooperii 264
Accipiter gentilis 266
Accipiter striatus 265
Acrocephalus arundinaceus 480
Actitis hypoleucos 147
Aechmophorus occidentalis 68
Aegithalos caudatus 459
Aegolius acadicus 308
Aegolius funereus 307
Aegypius monachus 261
Aepyornis maximus 205
Aethia psittacula 194
Agapornis roseicollis 293
Agelaius phoeniceus 607
Agelaius tricolor 608
Aglaiocercus kingi 336
Aimophila ruficeps 577
Aix sponsa 76
Alauda arvensis 443
Alca torda 188
Alcedo atthis 347
Alectoris graeca 224
Amazilia saucerrottei 342
Amazona ochrocephala 296
Ammodramus bairdii 585
Ammodramus henslowii 586
Ammodramus maritimus 587
Ammomanes deserti 440
Ammoperdix griseogularis 226
Amphispiza belli 583
Anas acuta 77
Anas americana 78
Anas clypeata 79
Anas crecca 80
Anas cyanoptera 81
Anas discors 82
Anas platyrhynchos 83
Anas rubripes 84
Anhima cornuta 73
Anhinga anhinga 133
Anodorhynchus hyacinthinus 295
Anous minutus 174
Anous stolidus 173
Anser albifrons 88
Anser caerulescens 87
Anthropoides virgo 64
Anthus berthelotii 518
Anumbius annumbi 381
Aphelocoma coerulescens 430
Aprosmictus erythropterus 292
Aptenodytes patagonicus 49
Apteryx australis 204
Apus apus 335
Aquila chrysaetos 274
Aramides cajaneus 55
Aramus guarauna 67
Archilochus alexandri 339
Archilochus colubris 338
Ardea alba 118
Ardea cinerea 112
Ardea herodias 113
Ardeola ralloides 114
Arenaria interpres 156
Argusianus argus 235
Artamella viridis 411
Asio flammeus 306
Asio otus 305
Athene cunicularia 303
Auriparus flaviceps 457
Aythya affinis 89
Aythya americana 90

B
Baeolophus bicolor 456
Baeolophus wollweberi 455
Bartramia longicauda 153
Bombycilla cedrorum 519
Bombycilla garrulus 520
Bonasa bonasia 239
Bonasa umbellus 240
Botaurus lentiginosus 115
Brachyramphus brevirostris 191
Branta bernicla 86
Branta canadensis 85
Bubo scandiacus 300
Bubo virginianus 299
Bubulcus ibis 116
Bucanetes githagineus 623
Bucephala clangula 91
Bucephala islandica 92
Bulweria bulwerii 40
Burhinus oedicnemus 134
Buteogallus anthracinus 267
Buteo jamaicensis 272
Buteo lagopus 273
Buteo lineatus 269
Buteo platypterus 270
Buteo swainsoni 271
Butorides virescens 117

C
Cacatua sanguinea 290
Cairina moschata 93
Calamospiza melanocorys 584
Calcarius lapponicus 522
Calidris alpina 158
Calidris himantopus 159
Calidris ptilocnemis 157
Callipepla californica 221
Calypte costae 340
Campethera bennettii 370
Campylorhynchus brunneicapillus 465
Caprimulgus carolinensis 329
Caprimulgus ruficollis 330
Caprimulgus vociferus 331
Caracara cheriway 277
Cardellina canadensis 566
Cardellina pusilla 567
Cardinalis cardinalis 601
Carduelis carduelis 631
Cariama cristata 284
Casuarius casuarius 202
Cathartes aura 249
Catharus fuscescens 498
Catharus guttatus 501
Catharus minimus 499
Catharus ustulatus 500
Catherpes mexicanus 466
Catreus wallichii 232
Centrocercus minimus 242
Centrocercus urophasianus 241
Cepphus columba 190
Cercotrichas barbata 488
Cerorhinca monocerata 195
Certhia familiaris 464
Ceryle rudis 350
Cettia cetti 478
Chaetura pelagica 334
Chalcites lucidus 321
Chamaea fasciata 484
Charadrius alexandrinus 138
Charadrius semipalmatus 139
Charadrius vociferus 140
Chauna chavaria 74
Chiroxiphia linearis 407
Chlamydera nuchalis 409
Chlidonias leucopterus 178
Chloroceryle americana 351
Chondestes grammacus 582
Chordeiles minor 327
Ciconia ciconia 105
Cinclus mexicanus 473
Circaetus gallicus 262
Circus aeruginosus 263
Cistothorus palustris 471
Clangula hyemalis 97
Coccothraustes coccothraustes 634
Coccyzus americanus 322
Cochlearius cochlearius 119
Colaptes auratus 375
Colinus virginianus 222
Columba livia 312
Columbina passerina 317
Contopus cooperi 387
Contopus fumigatus 388
Contopus virens 389
Coracias garrulus 354
Coragyps atratus 248
Corvus brachyrhynchos 436
Corvus corax 437
Coturnicops noveboracensis 56
Coturnix chinensis 228
Crax alector 218
Creagrus furcatus 168
Crossoptilon mantchuricum 231
Crotophaga ani 325
Crypturellus cinnamomeus 210
Crypturellus soui 208
Crypturellus undulatus 209
Cuculus canorus 320
Cursorius cursor 165
Cyanocitta cristata 429
Cyanocorax yncas 427
Cygnus columbianus 75
Cyrtonyx montezumae 223

D
Daption capense 41
Dendrocygna autumnalis 94
Dendrocygna bicolor 95
Diomedea exulans 36
Diomedea nigripes 37
Dolichonyx oryzivorus 606
Dromaius novaehollandiae 203
Dryocopus pileatus 376
Dumetella carolinensis 505

E
Ectopistes migratorius 316
Egretta caerulea 120
Egretta thula 121
Egretta tricolor 122
Elanoides forficatus 253
Elanus leucurus 254
Emberiza cia 598

Emberiza citrinella 597
Empidonax alnorum 390
Empidonax minimus 391
Eremophila alpestris 445
Eriocnemis cupreoventris 337
Erithacus luscinia 489
Eudocimus albus 110
Eudocimus ruber 109
Eudromia elegans 214
Eudyptula minor 48
Euphagus carolinus 612
Eurystomus orientalis 355

F
Falcipennis canadensis 243
Falco columbarius 280
Falco eleonorae 279
Falco femoralis 281
Falco mexicanus 283
Falco peregrinus 282
Falco sparverius 278
Ficedula albicollis 490
Formicarius nigricapillus 385
Francolinus francolinus 225
Fratercula arctica 196
Fregata magnificens 127
Fringilla coelebs 621
Fulica americana 57
Fulmarus glacialis 42
Furnarius rufus 380

G
Galbula ruficauda 359
Galerida cristata 442
Gallinago gallinago 162
Garrulus glandarius 431
Gavia adamsii 51
Gavia arctica 53
Gavia immer 52
Gavia pacifica 54
Gelochelidon nilotica 176
Geococcyx californianus 324
Geospiza fuliginosa 575
Geothlypis agilis 539
Geothlypis philadelphia 540
Geothlypis trichas 541
Geotrygon montana 318
Glareola nordmanni 166
Glaucidium brasilianum 301
Grus americana 65
Grus canadensis 66
Guira guira 323
Gymnogyps californianus 250
Gymnorhinus cyanocephalus 428
Gyps fulvus 260

H
Haemorhous mexicanus 626
Haemorhous purpureus 625
Haliaeetus leucocephalus 258
Haliaeetus leucoryphus 257
Helmitheros vermivorus 525
Hieraaetus pennatus 275
Himantopus mexicanus 141
Himatione sanguinea 635
Hirundo rustica 450
Histrionicus histrionicus 96
Hydrobates pelagicus 45
Hydrophasianus chirurgus 143
Hydroprogne caspia 177
Hylocichla mustelina 502
Hypnelus ruficollis 358

I
Icteria virens 569
Icterus cucullatus 618
Icterus galbula 619
Icterus spurius 617
Indicator indicator 362
Ixobrychus exilis 123
Ixoreus naevius 504

J
Jacana spinosa 145
Junco hyemalis 596
Jynx torquilla 363

K
Kuehneromyces mutabilis 178

L
Lagopus lagopus 244
Lanius excubitor 412
Lanius ludovicianus 413
Larus argentatus 172
Larus delawarensis 171
Laterallus jamaicensis 58
Leiothrix lutea 485
Leucosticte tephrocoti 622
Limicola falcinellus 160
Limnothlypis swainsonii 533
Limosa lapponica 155
Locustella fluviatilis 481
Lophura swinhoii 230
Loxia curvirostra 627
Lullula arborea 444
Lunda cirrhata 197
Lybius torquatus 360

M
Macrocephalon maleo 215
Macronectes giganteus 43
Malurus cyaneus 410
Megaceryle alcyon 349
Megaceryle torquata 348
Megapodius cumingii 216
Megascops asio 298
Melanerpes carolinus 367
Melanerpes erythrocephalus 365
Melanerpes formicivorus 366
Melanerpes lewis 364
Melanocorypha calandra 441
Meleagris gallopavo 247
Melopsittacus undulatus 291
Melospiza georgiana 591
Melospiza lincolnii 590
Melospiza melodia 589
Mergus merganser 98
Mergus serrator 99
Merops apiaster 352
Merops persicus 353
Metopidius indicus 144
Micrathene whitneyi 302
Micropalama himantopus 159
Milvus migrans 256
Mimus polyglottos 506
Mniotilta varia 531
Molothrus aeneus 615
Molothrus ater 616
Momotus momota 346
Monticola saxatilis 491
Montifringilla nivalis 639
Morus bassanus 130
Motacilla alba 517
Muscicapa striata 487
Myadestes townsendi 497
Mycteria americana 106
Mycteria ibis 107
Myiarchus cinerascens 397
Myiarchus crinitus 398
Myiarchus tuberculifer 396
Myiarchus tyrannulus 399
Myioborus pictus 568
Myiodynastes luteiventris 401
Myrmeciza exsul 384

N
Neocrex erythrops 59
Neophron percnopterus 259
Netta rufina 100
Nothura boraquira 212
Nothura maculosa 213
Nucifraga columbiana 434
Numenius americanus 154
Numida meleagris 219
Nyctanassa violacea 124
Nyctibius griseus 332
Nycticorax nycticorax 125

O
Oceanites oceanicus 46
Oceanodroma furcata 47
Oenanthe finschii 493
Opisthocomus hoazin 326
Oreortyx pictus 220
Oreoscoptes montanus 507
Oreothlypis celata 535
Oreothlypis luciae 536
Oreothlypis peregrina 534
Oreothlypis ruficapilla 537
Oreothlypis virginiae 538
Oriolus oriolus 423
Ortalis vetula 217
Otis tarda 285
Oxyura jamaicensis 101

P
Pachyramphus aglaiae 408
Pandion haliaetus 251
Panurus biarmicus 439
Parabuteo unicinctus 268
Paradisaea rudolphi 438
Parkesia motacilla 526
Parkesia noveboracensis 527
Passer domesticus 636
Passerella iliaca 588
Passerina caerulea 603
Passerina cyanea 604
Passer montanus 637
Pastor roseus 514
Patagioenas leucocephala 313
Pavo muticus 236
Pelecanus erythrorhynchos 128
Pelecanus occidentalis 129
Perdix dauurica 227
Perisoreus canadensis 425
Pernis apivorus 252
Petrochelidon pyrrhonota 451
Petronia petronia 638
Peucaea aestivalis 579
Peucaea cassinii 578
Phaethon aethereus 126
Phainopepla nitens 521
Phalacrocorax auritus 131
Phalacrocorax carbo 132
Phalaenoptilus nuttallii 328
Phalaropus tricolor 164
Pharomachrus mocinno 344
Phasianus colchicus 234
Pheucticus ludovicianus 602
Philomachus pugnax 161
Phleocryptes melanops 382
Phoebastria albatrus 38
Phoebetria palpebrata 39
Phoenicopterus ruber 72
Phrygilus gayi 572
Phylloscopus trochilus 479
Pica hudsonia 432
Pica nuttali 433
Picoides albolarvatus 374
Picoides borealis 373
Picoides pubescens 371
Picoides villosus 372
Picus vaillantii 377
Pinguinus impennis 189
Pinicola enucleator 624
Pipilo erythrophthalmus 576
Piranga olivacea 599
Piranga rubra 600

Pitangus sulphuratus 400
Platalea ajaja 108
Plectrophenax nivalis 523
Plegadis falcinellus 111
Pluvialis dominica 137
Pluvialis squatarola 136
Podiceps auritus 69
Podiceps nigricollis 70
Poecile atricapillus 453
Poecile carolinensis 452
Poecile gambeli 454
Polioptila caerulea 472
Porzana carolina 60
Prinia gracilis 482
Progne subis 447
Protonotaria citrea 532
Prunella atrogularis 516
Psaltriparus minimus 460
Psarocolius montezuma 620
Psilorhinus morio 426
Psittacus erithacus 294
Pterocles orientalis 289
Ptychoramphus aleuticus 193
Puffinus assimilis 44
Pycnonotus barbatus 474
Pyrocephalus rubinus 395
Pyrrhocorax pyrrhocorax 435
Pyrrhula pyrrhula 633

Q
Quiscalus mexicanus 614
Quiscalus quiscula 613

R
Rallus elegans 61
Rallus limicola 62
Rallus longirostris 63
Ramphastos tucanus 361
Recurvirostra americana 142
Regulus calendula 475
Regulus satrapa 476
Remiz pendulinus 458
Rhea americana 201
Rhinopomastus aterrimus 357
Rhynchotus rufescens 211
Rhynochetos jubatus 287
Riparia riparia 449
Rissa brevirostris 169
Rostratula benghalensis 167
Rostrhamus sociabilis 255
Rynchops niger 183

S
Sagittarius serpentarius 276
Saxicola torquatus 492
Sayornis nigricans 392
Sayornis phoebe 393
Sayornis saya 394
Scolopax minor 163
Scotocerca inquieta 477
Seiurus aurocapilla 524
Selasphorus rufus 341
Serinus citrinella 632
Setophaga americana 547
Setophaga caerulescens 554
Setophaga castanea 549
Setophaga cerulea 546
Setophaga chrysoparia 564
Setophaga citrina 542
Setophaga coronata 557
Setophaga discolor 559
Setophaga dominica 558
Setophaga fusca 550
Setophaga graciae 560
Setophaga kirtlandii 544
Setophaga magnolia 548
Setophaga nigrescens 561
Setophaga occidentalis 563
Setophaga palmarum 555
Setophaga pennsylvanica 552
Setophaga petechia 551
Setophaga pinus 556
Setophaga ruticilla 543
Setophaga striata 553
Setophaga tigrina 545
Setophaga townsendi 562
Setophaga virens 565
Sialia currucoides 496
Sialia mexicana 495
Sialia sialis 494
Sitta canadensis 463
Sitta carolinensis 461
Sitta pusilla 462
Somateria fischeri 103
Somateria mollissima 102
Somateria nigra 102
Spheniscus demersus 50
Sphyrapicus thyroideus 368
Sphyrapicus varius 369
Spinus psaltria 628
Spinus spinus 630
Spinus tristis 629
Spiza americana 605
Spizella arborea 580
Spizella passerina 581
Sporophila torqueola 573
Steatornis caripensis 333
Stelgidopteryx serripennis 446
Stercorarius longicaudus 185
Stercorarius skua 184
Sterna forsteri 181
Sterna hirundo 179
Sterna paradisaea 180
Sternula antillarum 175
Strix varia 304
Struthio camelus 200
Sturnella magna 609
Sturnella neglecta 610
Sturnus vulgaris 515
Sylvia atricapilla 483
Synthliboramphus antiquus 192
Syrmaticus ellioti 233
Syrrhaptes paradoxus 288

T
Tachybaptus dominicus 71
Tachycineta bicolor 448
Tachyphonus rufus 570
Tadorna ferruginea 104
Taeniopygia guttata 640
Tauraco porphyreolophus 319
Terpsiphone paradisi 424
Tetrao tetrix 238
Tetrao urogallus 237
Tetrax tetrax 286
Thalasseus elegans 182
Thraupis palmarum 571
Thryomanes bewickii 468
Thryothorus ludovicianus 467
Tiaris bicolor 574
Tinamus major 207
Tinamus osgoodi 206
Tockus pallidirostris 309
Todirostrum cinereum 386
Todus subulatus 345
Toxostoma bendirei 510
Toxostoma crissale 512
Toxostoma curvirostre 511
Toxostoma lecontei 513
Toxostoma longirostre 509
Toxostoma rufum 508
Tragopan caboti 229
Tringa erythropus 149
Tringa flavipes 152
Tringa nebularia 150
Tringa semipalmata 151
Tringa solitaria 148
Troglodytes aedon 469
Troglodytes troglodytes 470
Trogon elegans 343
Turdoides fulva 486
Turdus migratorius 503
Tympanuchus cupido 246
Tympanuchus phasianellus 245
Tyrannus couchii 402
Tyrannus forficatus 406
Tyrannus tyrannus 405
Tyrannus verticalis 404
Tyrannus vociferans 403
Tyto alba 297

U
Upupa epops 356
Uria aalge 186
Uria lomvia 187

V
Vanellus chilensis 135
Vermivora bachmanii 528
Vermivora chrysoptera 530
Vermivora cyanoptera 529
Vidua paradisaea 641
Vireo atricapillus 416
Vireo bellii 415
Vireo flavifrons 418
Vireo gilvus 421
Vireo griseus 414
Vireo huttoni 420
Vireo olivaceus 422
Vireo solitarius 419
Vireo vicinior 417

X
Xanthocephalus xanthocephalus 611
Xema sabini 170
Xenus cinereus 146
Xiphorhynchus flavigaster 383

Z
Zenaida asiatica 314
Zenaida macroura 315
Zonotrichia albicollis 593
Zonotrichia capensis 592
Zonotrichia leucophrys 595
Zonotrichia querula 594

ACKNOWLEDGMENTS

DR. MARK HAUBER

My entry into this project could not have been possible without an initial recommendation from Dustin Rubenstein at Columbia University, to link with the amazing scientific, editorial, and photography team at the Field Museum of Natural History in Chicago: John Bates, Barbara Becker, John Weinstein, and their associates, as well with as the helpful team at Ivy Press, Lewes, UK: Stephanie Evans, Caroline Earle, and Jason Hook. Researching each of the species entries in this book was made possible and pleasant by the dedicated work of John Oursler and Jessica Schwartz. Many colleagues, mentors, students, and friends helped with information, insights, and inspiration prior to and during the writing of this book, including Patricia Brennan, Nick Davies, Phill Cassey, Caren Cooper, Brian Gill, Tomas Grim, Daniel Hanley, Brani Igic, Rebecca Kilner, David Lahti, Arnon Lotem, David McDonald, Csaba Moskat, Marion Petrie, Rebecca Safran, Peter Samas, Rachael Shaw, Matthew Shawkey, Paul Sherman, Thomas Seeley, and Geoffrey Waterhouse. My mother, Zsuzsanna Berenyi, always made it clear to me as a child, growing up in Hungary, that watching White Storks and House Martins nest on buildings each spring was the right thing to do. It is an honor to have a (mostly) day-time job where writing a book about bird eggs is encouraged by the bosses and colleagues; this privilege has been provided by the Department of Psychology, Hunter College and the Graduate Center (City University of New York). My after-hours work schedule was kindly accepted and enthusiastically encouraged by my partner, John Oursler. Generous funding for our research on avian egg coloration was provided by the Human Frontier Science Program.

JOHN BATES & BARBARA BECKER

We would like to thank Jason Hook and everyone else at Ivy Press for initiating the Book of Eggs project; Mark Hauber for his marvelous knowledge of bird eggs and his ability to write about them so compellingly; and Field Museum photographer John Weinstein for brilliantly bringing out the beauty and uniqueness of each egg. Michael Hanson prepared the maps and Luke Campillo and Emma Solomon helped with databases and work in the collection. We are grateful to everyone associated with the Field Museum who supported the project throughout: Tom Gnoske, Shannon Hackett, Mary Hennen, Ben Marks, Doug Stotz, William Simpson, and Dave Willard; and particularly to Francie Muraski-Stotz, who first helped conceptualize the book. We also thank Linnea Hall and René Corado of the Western Foundation of Vertebrate Zoology in Camarillo, California for generously allowing us to access their wonderful collection and their expertise. John would like to thank Barbara for taking the organizational reins of the project and for her cheerful and substantial editorial skills. Barbara would like to thank John for his passion about birds and his critical insights on this journey; and the Field Museum for allowing access to a most magical hidden corner of their collection: the Egg Room.

PICTURE CREDITS

The publisher would like to thank the following individuals and organizations for their kind permission to reproduce the images in this book. Every effort has been made to acknowledge the pictures, however we apologize if there are any unintentional omissions.

CORBIS/GALLO IMAGES: 30; Mitsuaki Iwago/Minden Pictures: 6T; Mike Jones/Frank Lane Picture Agency: 23; JASON FROGGATT/Auckland Museum: 30; NATURE PICTURE LIBRARY/Mark Payne-Gill: 25; Nature Production: 26; Yuri Shibnev: 24; SCIENCE PHOTO LIBRARY/Sinclaire Stammers: 10; SHUTTERSTOCK/AdStock RF: 27; Antonio Abrignani: 73T, 107T, 167T, 287T, 346T, 380T, 616T; Stacey Ann Alberts: 21B; Denis Barbulat: 125T, 355T; Chaiwatphotos: 6B; Jiri Foltyn: 7BR; Trevor Kelly: 11; S.R. Maglione: 21T; Maria Rita Meli: 20; Hein Nouwens: 49T, 83T, 112T, 118T, 126T, 127T, 128T, 135T, 177T, 187T, 201T, 202T, 203T, 204T, 239T, 261T, 412T; PhotosByNancy: 22; Pictureguy: 18; Sergey Uryadnikov: 15T.

Unless credited above, all engravings courtesy of BIODIVERSITY HERITAGE LIBRARY www.biodiversitylibrary.org

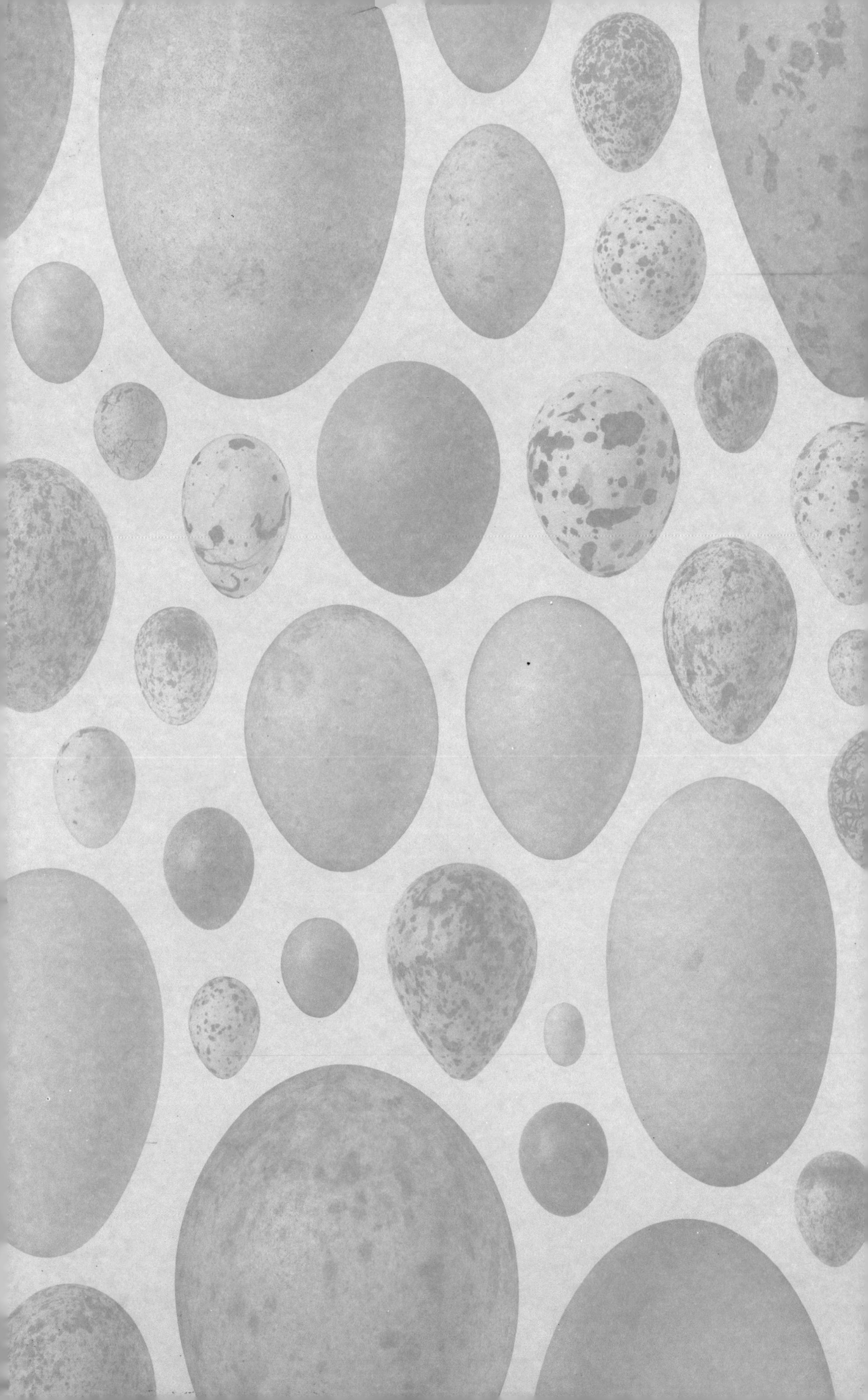